Advances in Fiber Optics Communications

The Artech House Telecommunication Library

Vinton G. Cerf, *Series Editor*

Digital Cellular Radio by George Calhoun

E-Mail by Stephen A. Caswell

Design and Prospects for the ISDN by G. DICENET

Introduction to Telecommunication Electronics by A.Michael Noll

Introduction to Telephones and Telephone Systems by A. Michael Noll

Integrated Services Digital Networks by Anthony M. Rutkowski

Signal Theory and Processing by Frederic de Coulon

Digital Signal Processing by Murat Kunt

Principles of Secure Communication Systems by Don J. Torrieri

Digital Image Signal Processing by Friedrich Wahl

Advances in Computer Systems Security: 3 volume set, Rein Turn, ed.

Techniques in Data Communications, by Ralph Glasgal

Measurement of Optical Fibers and Devices by G. Cancellieri and U. Ravaioli

Codes for Error Control and Synchronization by Djimitri Wiggert

Mathematical Methods of Information Transmission by K. Arbenz and J.C. Martin

Telecommunication Systems by Pierre-Girard Fontelliet

Optical Fiber Transmission Systems by Siegried Geckeler

Traffic Flow in Switching Systems by G. Hebuterne

Optimization of Digital Transmission Systems by K. Trondle and G. Soder

Quality Measures and the Design of Telecommunications Systems by John H. Fennick

Troposcatter Radio Links by G. Roda

Radiodetermination Satellite Services and Standards by Martin Rothblatt

Innovation in Internetworking, Craig Partridge, ed.

Principles of Signals and Systems: Deterministic Signals by B. Picinbono

Television Technology: Fundamentals and Future Prospects by A. Michael Noll

Manager's Guide to CENTREX by John R. Abrahams

Voice Teletraffic System Engineering by James R. Boucher

Advances in Fiber Optics Communications, Henry F. Taylor, ed.

Broadband LAN Technology by Gary Y. Kim

Advances in Fiber Optics Communications

Henry F. Taylor
Texas A&M University

Artech House

Library of Congress Cataloging-in-Publication Data

Advances in fiber optics communications
1. Optical communications. 2. Fiber optics.

I. Taylor, Henry F.
TK5103.59.A27 1988 621.38'0414 88-8193
ISBN 0-89006-328-1

ARTECH HOUSE, INC.
685 Canton Street
Norwood, MA 02062

International Standard Book Number: 0-89006-328-1
Library of Congress Catalog Card Number: 88-8193

10 9 8 7 6 5 4 3 2 1

Contents

Preface

Five years ago I edited an anthology for Artech House titled *Fiber Optics Communications.* That volume is now out of print and the publisher has requested a similar book reflecting the recent advances in the field. The earlier collection appeared to satisfy a real need in the research and development community, and we hope that this new volume will also find a receptive audience.

A collection of this type can be a useful reference for engineers, scientists, and technicians actively engaged in research and development in the field. It can also serve as an information source for people from other disciplines interested in learning more about the subject. Although many texts have been published on fiber optics, a book prepared by one or two authors usually cannot match a good review article in providing in-depth coverage of a particular subtopic. A collection of recent papers can also serve as a supplement for textbooks which can be quickly outdated in a fast moving field like fiber optics.

Selecting fifty or so papers on the subject of optical fiber communication from among the tens of thousands which have appeared in the literature was obviously not an easy task. Fortunately, a large number of quality review articles have been published during the past few years, and we were able to obtain excellent coverage of the subject by using recent papers. The breakdown according to year of publication is: one from 1984, two from 1985, 13 from 1986, 24 from 1987, and six from 1988. Thus, the emphasis is on the current technology and upon recent advances in the field. Review articles account for about seventy percent of the pages, with the balance devoted to shorter contributions describing new research results.

The three articles in Chapter 1 are general reviews of the field of fiber optics communication. Chapters 2 through 7 are concerned with the fabrication of components and their characteristics, and Chapters 8 and 9 deal with systems issues. The reader will thus find a wide range of perspectives, from semiconductor device physics and optical waveguide theory to the design of systems, and experience in installing and maintaining them. The technology is now relatively mature ("high on the learning curve") in most areas, leading to the belief that the majority of the selections will continue to be useful for years to come.

Finally, the editor would like to thank the authors and publishers of all the papers for giving permission for their work to be included in this collection.

Introduction

One of the most important technological developments during the 1980s has been the emergence of optical fiber communication as a major international industry. One indication of the extent of this development is the total length of installed fiber, which was estimated to be 3.2 million kilometers in the U.S. alone by the end of 1987 [1]. Over 90% of this fiber was placed in service during the time period of 1982–1987. Long-haul trunk installations have dominated, accounting for about 95% of the fiber in the U.S. [1]. The period of rapid expansion in the worldwide fiber optics industry has recently slowed, as most of the need for new long-haul telecommunication capacity has already been satisfied and in some areas a situation of overcapacity already exists. More of the focus within the industry is now on markets which require shorter cable lengths, with many observers feeling that local area networks and wideband subscriber loops represent the next major opportunities for dramatic growth.

Although telecommunication is the rationale for most of the current interest in fiber optics, this was not the case during the early days of the technology. The researchers who produced the first clad glass optical fibers in the early 1950s were not thinking of using them for communications; they wanted to make imaging bundles for endoscopy. During the 1950s and 1960s, methods for making fiber optic bundles were developed: incoherent bundles for illumination, and coherent bundles for image transmission. Automobile instrument panel illumination, decorative lighting, cathode ray tube faceplates, cockpit displays for aircraft, and medical endoscopes were among the commercial applications. Fiber optics was already a well established commercial technology when the famous paper by Kao and Hockham [2], suggesting the use of low–loss optical fibers for communication, appeared in 1966.

The first low-loss (20 dB/km) silica fiber was described in a publication [3] which appeared in October of 1970. (A segment of this fiber has been accorded a place alongside many other technological firsts in the Smithsonian Museum of History and Technology in Washington, DC.) The date of this publication is sometimes cited as the beginning of the era of fiber communication. Although this development did receive considerable attention in the research community at the time, it was far from inevitable that a major industry would evolve. The 20 dB/km loss figure was still too high for long-haul telecommunication systems. The fibers were fragile, and a way to protect them would have to be found. (Breakage factors in flexible bundles of that period were expressed in percent per foot!) There were no suitable light sources. Connectors with much closer mechanical tolerances than their electrical or microwave counterparts would be needed, and researchers did not know whether field termination and splicing of optical cables would ever be practical. Finally, there were serious doubts as to whether these components could ever be produced economically enough for the technology to play a major role in the marketplace.

Although the technological barriers appeared formidable, the economic potential was very significant. As a consequence, research and development activity expanded rapidly, and a number of important issues were resolved during the early 1970s. Success in material purification and the use of new core dopants led to further dramatic loss reductions in silica fibers. Development of graded-index multimode fibers provided a favorable combination of transmission bandwidth and source coupling efficiency. Researchers learned to protect the fibers with polymer coatings and to cable them without breakage. Diode lasers were operated continuously at room temperature for the first time and high-radiance *light-emitting diodes* (LEDs), capable of efficient coupling to multimode fibers, were produced. The arc fusion method for splicing fibers and a variety of connector designs were proved in practice.

During the middle and late 1970s, the rate of progress toward marketable products accelerated as the emphasis shifted from research to engineering. Fibers with losses approaching the Rayleigh limit of 2 dB/km at a wavelength of 0.8 μm, 0.3 dB/km at 1.3 μm, and 0.15 dB/km at 1.55 μm, were produced in the laboratory. Microbend loss problems were overcome through the use of improved fiber coatings and cabling techniques. Rugged cables and multifiber connectors were produced for field installation. Room temperature threshold currents for commercial gallium aluminum arsenide (GaAlAs) lasers operating in the 0.8 μm to 0.85 μm spectral region were reduced to the 20–30 mA range, and projected lifetimes in the 100,000 to 1,000,000 hour range were claimed for both lasers and LEDs. Light sources and improved photodetectors which operated near 1.3 μm were developed to take advantage of the low fiber loss and dispersion in this "longer wavelength region." Encouraging data on the reliability for longer wavelength lasers and LEDs fabricated in the indium gallium arsenide phosphide (InGaAsP) quartenary alloy system were also obtained. Several major field trials were undertaken during this period, including AT&T's Atlanta experiment (1976) and Chicago installation (1977), and Japan's Hi Ovis subscriber access project (1977).

Improvements in component performance, cost, and reliability by 1980 led to major commitments on the part of telephone companies. Fiber soon became the preferred transmission medium for long-haul trunks. Some early installations used 0.8 μm light sources and graded-index multimode fiber, but by 1983, designers of intercity links were thinking in terms of 1.3 μm, single-mode systems. The single-mode fiber, used in conjunction with a 1.3 μm laser, provides a bandwidth advantage which translates into increased repeater spacings for high-data-rate systems.

Data rates for installed fiber optic systems have recently moved into the gigabit per second range. Such systems use the spectrally pure distributed–feedback lasers to minimize fiber dispersion effects. Fibers designed for low dispersion at the 1.55 μm wavelength, which corresponds to minimum fiber loss, are now commonly used in long distance transmission. The use of wavelength multiplexing to further increase the fiber information capacity is becoming more widespread. Undersea fiber optic cables have linked Japanese islands for several years, and the first transoceanic fiber systems will soon be put into service.

In contrast to the recent expansion in long-haul systems, the potential of fiber optics in other areas is only beginning to be realized. After a slow start, fiber optic networks for computer systems and offices are becoming more prominent. Fiber optics is now used in numerous military systems. In the telephone system, the use of fiber optics for interconnecting central offices within a metropolitan area and for lower levels in the switching hierarchy is still increasing rapidly. Fiber links to the home have only been used in demonstration projects, since economics still favors conventional electrical wire and cable. However, many observers believe that national telephone systems will eventually be upgraded to handle video bandwidths by using fiber optics. These wideband subscriber loop systems would provide access on a switched basis to services such as picturephone, video entertainment, tronic components must decrease considerably before widespread installation of these broadband services will become economically feasible.

REFERENCES

1. S.B. Weinstein, "Telecommunications in the Coming Decades," *IEEE Spectrum,* November 1987, 62–67.
2. K.C. Kao and G.A. Hockham, *Proc. IEE* **113,** 1151 (1966).
3. F.P. Kapron, D.B. Keck, and R.D. Maurer, *Appl. Phys. Lett.* **17,** 423 (1970).

Chapter 1
Reviews

The emergence of fiber optics as a major industry has been paced by the ability of industry to produce high-quality, reliable components at a reasonable cost. The current technology in components and point-to-point and network systems is summarized in the first two papers, "**Advances in Lightwave Systems Research,**" by **T. Li,** and "**Fibre Optic Transmission Systems—Status and Trends in Japan,**" by **T. Kimura.** The technologies for wideband and high-data-rate applications are emphasized in "**The Application of Microwave Techniques in Lightwave Systems,**" by **H. Sobol.**

ADVANCES IN LIGHTWAVE SYSTEMS RESEARCH

AT&T TECHNICAL JOURNAL

Tingye Li

***Tingye Li** is head of the Lightwave Systems Research Department at AT&T Bell Laboratories in Holmdel, New Jersey. He is responsible for research on various lightwave technologies and systems for long-distance and local-area communications applications. Mr. Li joined the company in 1957 and has a B.Sc. from the University of Witwatersrand, and an M.S. and Ph.D. from Northwestern University, all in electrical engineering.*

Twenty years have passed since the first serious proposal was made to use a glass-fiber waveguide as a telecommunications transmission medium. Since then, vigorous research and development efforts have led to widespread applications of optical-fiber communication throughout the telecommunications industry. Today, lightwave systems research is exploring the vast potential bandwidth of low-loss single-mode fibers for long-haul and local-network applications. Laboratory systems with experimental single-frequency lasers and low-noise avalanche photodiodes have attained multi-gigabit-per-second transmission in single-mode fibers that are well over 100 km long. Coherent technology promises high receiver sensitivity and offers an efficient means for channel selection in a densely packed wavelength-multiplexed system. The first coherent-systems experiments have confirmed theoretical expectations but many challenges lie ahead. Other novel systems for local distribution and networking are also under study. This paper reviews recent research progress in these areas.

Perspective

Historically, telecommunications research has always been motivated to explore new technologies for expanding transmission capacity. Thus, the realization of the first laser in 1960 stirred intense interest in optical communications. The laser was envisaged as the requisite coherent source of an optical carrier on which information could be impressed at an unparalleled rate.

Research began on a broad front. But not until ten years later was a feasible, low-loss, transmission medium demonstrated. It was a doped high-silica optical fiber with a loss of 20 dB/km. Coincidentally, continuous-wave operation at room temperature was achieved in the AlGaAs (aluminum-gallium-arsenide) semiconductor laser. The attainment of these two basic elements of an optical communications system was soon followed by a profusion of research advances that, in less than a decade, made optical-fiber communications a practical reality.[1]

Deployed Fiber Systems. The first-generation fiber systems—introduced commercially in the late 1970s—operated at wavelengths of about 0.82 μm. They used AlGaAs injection lasers and multimode fibers of 3- to 4-dB/km transmission loss. Interoffice trunking was the main arena of application.

As these systems were being tested in the field just before their commercial introduction, two research advances were announced that affected the course of lightwave technology.

- The transmission loss of fiber was lowered dramatically to 0.5 dB/km in wavelength regions around 1.3 and 1.5 μm.
- Continuous-wave, room-temperature operation was achieved in an InGaAsP (indium-gallium-arsenide-phosphide) laser that could emit in the 1.0- to 1.7-μm wavelength range.

The wavelength region near 1.3 μm is especially significant. At this wavelength, the pulse spreading caused by material and waveguide effects (chromatic dispersion) in the fiber is minimal,[2,3] making it possible to transmit over long distances at exceedingly high data rates. Also, because chromatic dispersion is proportional to source spectral width, opening the 1.3-μm transmission window allowed light-emitting diodes (LEDs) to be used in systems with moderate speed and distance requirements. For some applications (such as data links), the simpler LEDs are favored over lasers because they afford superior reliability, lower cost, and much less sensitivity to temperature variations.

Hence, the second-generation of lightwave systems operate at a wavelength of 1.3 μm, using either LEDs or lasers as sources and multimode fibers as the transmission medium. For subscriber-loop applications, both LED- and laser-based systems operate at bit rates of up to 200 Mb/s. Laser-based systems, which can operate at higher speeds over longer distances, are deployed in metropolitan areas for interoffice trunking and in the long-haul plant.

Second-generation fiber systems have now been superseded by 1.3-μm, laser-based, single-mode systems that are designed for intercity and intercontinental service. These third-generation systems operate at several hundred megabits per second with repeater spans that exceed 40 km, and can be upgraded to speeds beyond 1 Gb/s (which is being planned). Many of these third-generation systems are now in commercial operation.

Fiber Systems Research. Current research in lightwave systems is exploring the vast potential bandwidth of low-loss, single-mode fibers. A minimum loss of 0.16 dB/km at $\lambda = 1.57$ μm (which is close to the theoretical limit of Rayleigh scattering in silica) has been demonstrated in the laboratory. Typical high-quality production fibers exhibit minimum losses close to 0.2 dB/km.[2,3]

For the 1.45- to 1.65-μm wavelength region of minimum loss, the potential bandwidth is 25,000 GHz. Imaginative and effective use of even a minute portion of this immense bandwidth can radically affect the telecommunications industry. The challenge to the research community is, therefore, intriguing and complex.

Meanwhile, researchers are trying to approach the limits that fiber loss and dispersion impose on transmission speed and distance. They have done systems experiments at data rates well beyond 1 Gb/s over single-mode fibers that exceed 100 km in length. Experiments that involved direct (intensity) modulation and detection required single-frequency lasers that have low frequency deviations (chirp) during modulation and high-sensitivity, broadband avalanche photodiodes (APDs). Such experiments have obvious implications for near-future long-haul systems.

Coherent lightwave transmission is being investigated for possible application in both long-haul trunking and local networks because:

- It offers higher receiver sensitivity.
- Simple optical amplifiers could be used for boosting the signal.
- It provides an efficient way to select a particular channel from many closely spaced wavelength-multiplexed channels.

Coherent transmission requires narrow-linewidth single-frequency lasers. Initial successes are encouraging, but many challenges lie ahead. Other novel systems for local distribution and networking are also under study.

This paper reviews recent progress in lightwave systems research, emphasizing high-speed transmission and coherent technology. It discusses the limits that single-mode fiber imposes, as well as the requirements of the relevant devices. Applications in the long-haul and local arenas are considered.

Transmission Bandwidth

The maximum rate of data transmission in a single-mode fiber is determined by its chromatic dispersion, which depends on the fiber geometry, refractive-index profile, and material properties.[2,3]

In a conventional single-mode fiber with a step refractive-index profile, the geometry and material effects combine to produce a dispersion minimum at a wavelength λ_0 slightly above 1.3 μm. Thus, the single-channel transmission bandwidth of a fiber that operates at the wavelength λ_0 can be very large, especially if the source has a narrow spectral width.

This is illustrated in Figure 1, which gives the maximum single-channel transmission bandwidth—or data rate—of a conventional single-mode fiber, 100 km long, as a function of wavelength for sources of various spectral widths, $\Delta\lambda$. However, a single-mode fiber with a refractive-index profile that is more complicated than a simple step can have maximum transmission bandwidths that are very different from those shown.[3]

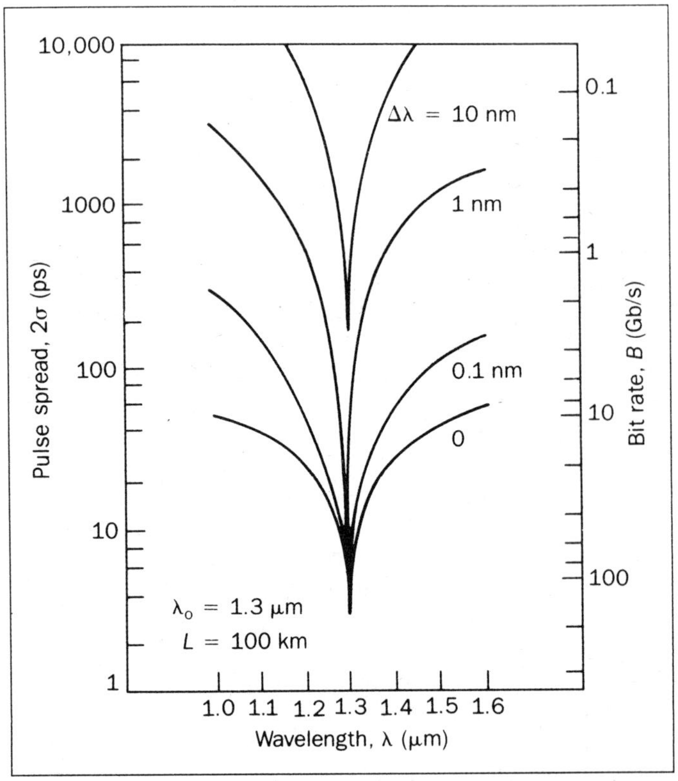

Figure 1. Theoretical maximum transmission bandwidth of a conventional single-mode fiber, 100 km long. The four curves represent sources of different spectral widths $\Delta\lambda$. The wavelength of minimum dispersion λ_o is 1.3 μm.

The calculations[4] that Figure 1 represents take into account higher-order waveguide and material effects, as well as performance degradation from intersymbol interference. The case labeled $\Delta\lambda = 0$ is akin to radio transmission in that the spectral width of the optical carrier is much narrower than the signal bandwidth, or bit rate, B. When this condition prevails, the maximum bit rate B_{max} for transmission over a distance L satisfies the relation $(B_{max})^3 L = constant$, for wavelengths near the dispersion minimum λ_0. Away from λ_0, the relation is

$(B_{max})^2 L = constant$.

When the spectral width of the optical carrier is much greater than the bandwidth of the signal, the maximum bit rate is related to the distance by $(B_{max})^2 L = constant$ near λ_0, and $B_{max} L = constant$ away from λ_0. Using these relations and Figure 1, one can scale maximum transmission distances for various data rates.

Figure 1 clearly shows that a source with wide spectral width would severely limit the transmission bandwidth of a conventional single-mode fiber operating at 1.55 μm where fiber attenuation is minimal. The chromatic dispersion there is about 18 ps/km·nm. Therefore, to maximize repeater spacing, one must use lasers with narrow spectral widths or shift the wavelength of minimum dispersion to 1.55 μm. The latter has been achieved but not without the penalty of some increased minimum loss.[2,3]

Single-Frequency Lasers

Conventional InGaAsP semiconductor lasers that operate at 1.3 or 1.5 μm tend to emit in several longitudinal modes (or lines) spread over a spectral width of several nanometers. These multifrequency lasers are not suitable as sources for high-speed, long-distance transmission (Figure 1).

Furthermore, the power in each mode tends to fluctuate, especially under direct modulation, even though the total power remains constant from pulse to pulse. This power fluctuation among modes is converted through chromatic dispersion of the fiber into pulse-shape variations at the receiver, a phenomenon known as *mode partition noise*. Partition noise can impair system performance even for moderate transmission distances.[5] Hence, sources for high-speed systems must operate in a single longitudinal mode.

Single-frequency operation can be obtained in a laser by using a coupled-cavity or distributed-feedback structure. The cleaved-coupled-cavity (C^3) laser, used in early long-distance transmission experiments at $\lambda =$ 1.5 μm, has two cavity sections that are coupled optically but isolated electrically.[6] Single-frequency operation is obtained by suitably adjusting the drive current through each section. The distributed-feedback (DFB) laser uses a grating to provide a distributed, but frequency-selective, feedback mechanism that confines emission to a single mode.[7] Single-frequency operation is more robust in DFB lasers.

Under direct modulation, the instantaneous frequency of the semiconductor laser will vary with drive current, resulting in pulse spreading at the receiver.[8] This phenomenon, known as *chirp*, arises because the refractive index of the laser depends on carrier density.[9] As the refractive index varies, so does the wavelength (or frequency) of emission.

If one uses a structure to confine the optical field more tightly, nonlinear processes in the laser—when high optical field densities are present—tend to damp the large amplitude and frequency fluctuations, and leave a smaller, steady frequency shift. (A steady frequency shift with negligible transition time will produce negligible pulse spreading.) A low-chirp InGaAsP DFB laser that was amenable to high-speed direct modulation was built[10] and used in long-distance transmission experiments at speeds up to 4 Gb/s.

APDs and High-Sensitivity Receivers

Simple, InGaAs PIN photodiodes have excellent frequency response and perform well in 1.5-μm receivers that operate at multigigabit-per-second data rates. (PIN stands for positive intrinsic negative; some authors may use p-i-n instead.) However, they lack the internal gain needed to enhance receiver sensitivity for high-speed transmission over long distances. Early attempts to make InGaAsP APDs were not successful because of the high dark current from tunneling and the degradation of frequency response from charge accumulation at heterojunction interfaces. Germanium APDs with low dark current and high gain have been developed and are now available commercially, but they contribute more excess noise than InGaAsP devices.

To solve the problems of tunneling dark current

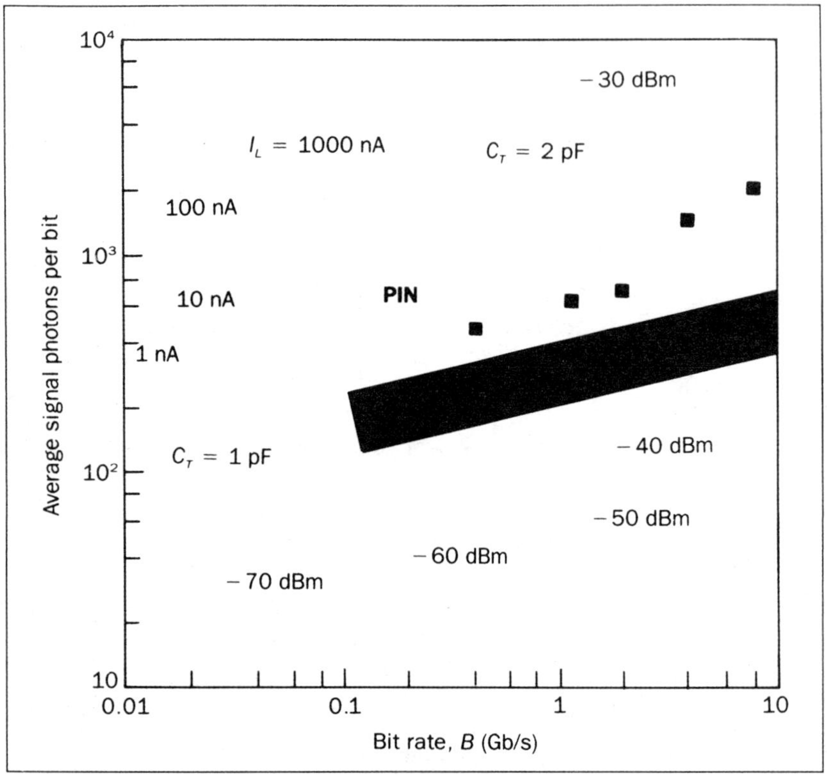

Figure 2. Receiver sensitivities of state-of-the-art optical repeaters, plotted as a function of bit rate. Sensitivities are expressed as the average number of signal photons (per bit) required to achieve an error probability of 10^{-9}. The blue curves are calculated for a PIN photodiode receiver with a GaAs FET preamplifier, while the red band gives the expected performance for an APD receiver. The blue dots and red squares represent the best known-experimental results for PIN and APD receivers, respectively.

and interface charge accumulation, a structure that consists of several layers of InGaAsP in varying compositions was tried and proved successful.[11] Its use of a narrow-bandgap InGaAs layer for light absorption and a wide-bandgap InP (indium-phosphide) layer for carrier multiplication overcame the problem of dark current. A grading layer of InGaAsP added at the heterojunction reduced charge accumulation. The resulting device—known as the separate absorption, grading, multiplication (SAGM) avalanche photodiode—has worked well in experimental high-speed systems. With its gain-bandwidth product as high as 60 GHz, the SAGM APD is the best performing photodetector for high-speed operation at 1.5 μm.

Wideband, low-noise preamplifiers are as crucial as high-performance photodetectors for attaining high sensitivity in optical receivers. (A useful definition of receiver sensitivity in a digital repeater is the minimum input power required to achieve an error probability of 10^{-9}.) The sensitivity of an optical receiver is limited by the quantum (or shot) noise associated with the primary photogenerated signal current in the detector. In practice, excess noise from the photodetector, thermal noise from the input circuit, device noise from the transistor preamplifier, and shot noise from leakage currents all contribute in varying degrees to prevent receiver performance from reaching the quantum-noise limit.

Figure 2 summarizes the performance of experimental optical receivers designed to operate in the spectral

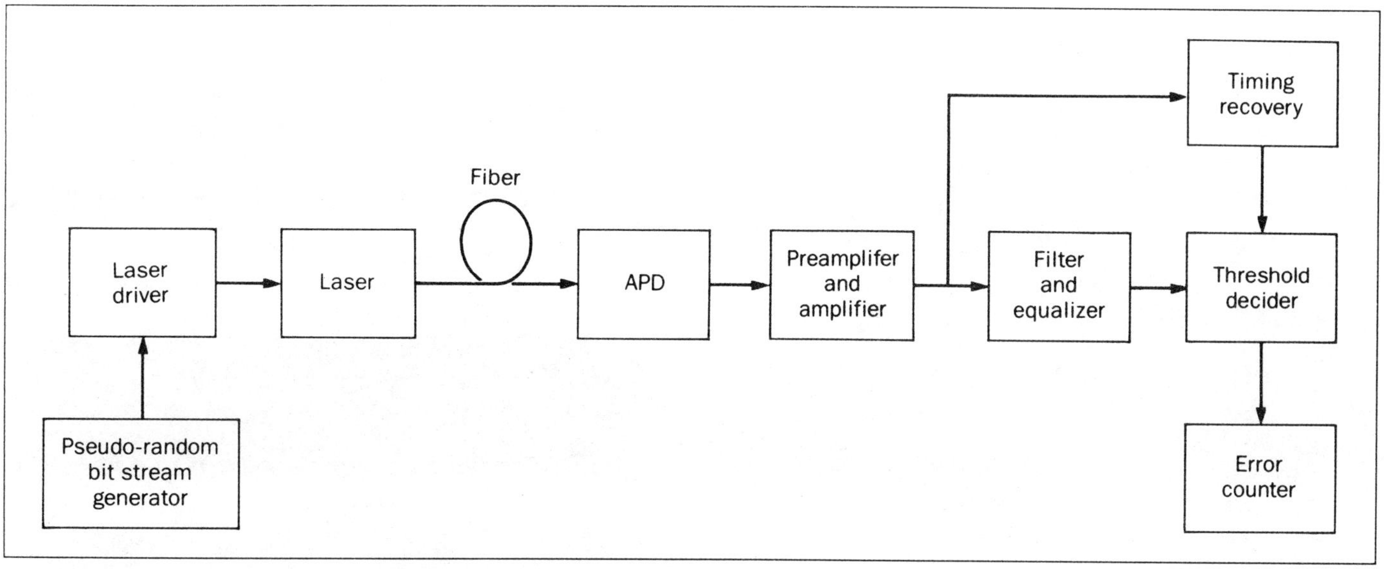

Figure 3. A typical lightwave system experiment.

region of 1.3 to 1.5 μm. The ordinate is the average (minimum) number of photons required to achieve an error probability of 10^{-9} in a time interval equal to the reciprocal of the bit rate, assuming an equal distribution of ones and zeros. This number is proportional to the average energy per bit. The abscissa is the bit rate.

The blue curves are the results of calculations based on parameters of currently available PIN photoreceivers that use GaAs (gallium-arsenide) field-effect transistors (FETs) as preamplifiers. The red shaded band represents the expected performance for APD receivers. The green diagonal grid lines convert the energy per bit into power in decibels referenced to 1 mW (dBm), assuming a 1.5-μm operating wavelength. The horizontal line at the bottom represents the ultimate sensitivity of ten photons per bit, as limited by the fundamental quantum noise. Noises that arise from thermal effects in circuits and devices cause degradations from this limit.

The straight blue lines whose slopes are proportional to the square root of the bit rate represent the theoretical performance for a receiver that consists of an ideal PIN photodiode followed by a GaAs FET preamplifier with a 50-mA/V transconductance, a total capacitance C_T of either 1 or 2 pF, and no leakage current. The blue curves that slope upward to the left represent performance when various leakage currents I_L are present. As can be seen, leakage currents are not important at high bit rates where the determining factor is the FET's channel conductance noise, which is proportional to the square of the total capacitance divided by the transconductance. The blue dots represent the best experimental values for PIN photoreceivers reported to date in the literature, while the red squares represent values for APD receivers.[12]

As Figure 2 shows, the best experimental APD receivers are about 5 dB more sensitive than the best experimental PIN receivers. This sensitivity difference will increase to about 7 dB for production receivers because ultra-high-performance PIN photoreceivers are more difficult to manufacture. Further improvements in device performance may soon enhance the best sensitivities of APD receivers to a level about 15 dB above the quantum-noise limit. Additional improvements will likely require device breakthroughs that will provide considerably less excess noise from the APD.

High-Speed Systems Experiments

Systems experiments are conducted to demonstrate the feasibility of transmission and explore the limits of technology regarding how individual components behave and affect the performance of an entire system. Figure 3 illustrates the basic elements of a system experiment.

To evaluate the behavior of each component in a system and assess the system's performance, one measures the bit error rate (BER) as a function of the received power for various operating conditions. Eye diagrams, which consist of superpositions of traces of the signal in a time slot (of one bit), are used as a diagnostic tool for analyzing component or subsystem behavior.

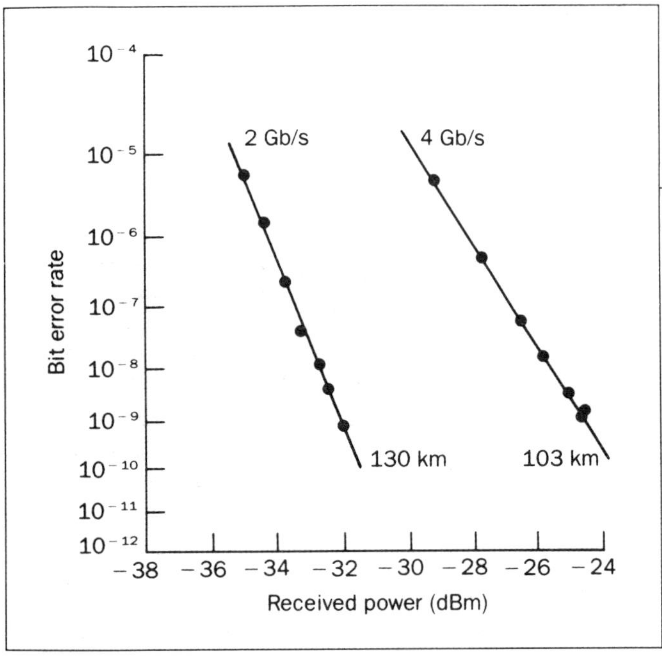

Figure 4. Measured error probability as a function of received power for two high-speed, long-distance system experiments.[13,14] For both, the lasers were biased above threshold to minimize chirp and, therefore, maximize transmission distance.

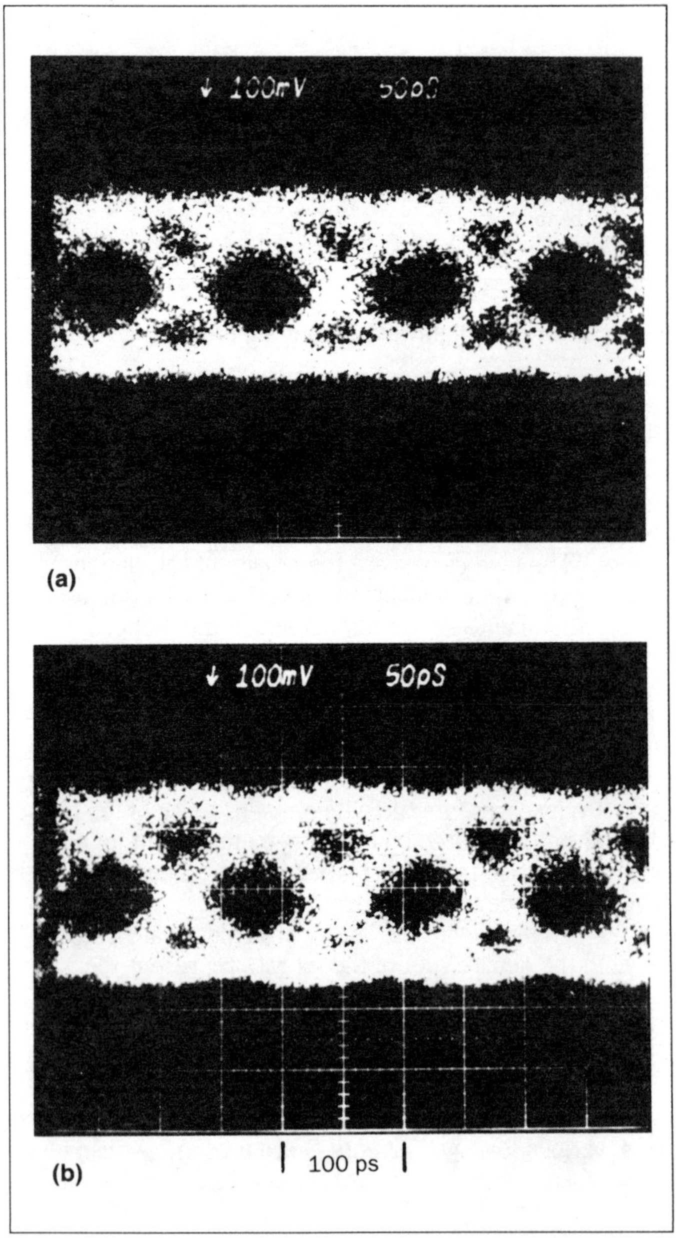

Figure 5. Eye diagrams of (a) transmitted and (b) received signals observed in the 8-Gb/s system experiment.[16] The *open* eye of the received signal after traversing 68 km showed that a bit-error rate of 10^{-9} was obtained.

Figure 4 shows measured BER data taken in two high-speed experiments that operated at 2 and 4 Gb/s and achieved transmission distances of 130 and 103 km, respectively.[13,14] The experimental low-chirp single-frequency DFB laser, high-speed SAGM APD receiver, and very-low-loss single-mode fiber with a core of nearly pure silica (0.21 dB/km) all contributed to the successes of the record-setting experiments.

To avoid laser chirp, one could let the laser operate continuously and gate the optical carrier with an external modulator. Indeed, further record-setting high-speed transmission experiments at 4 and 8 Gb/s attained transmission distances of 117 and 68 km, respectively.[15,16]

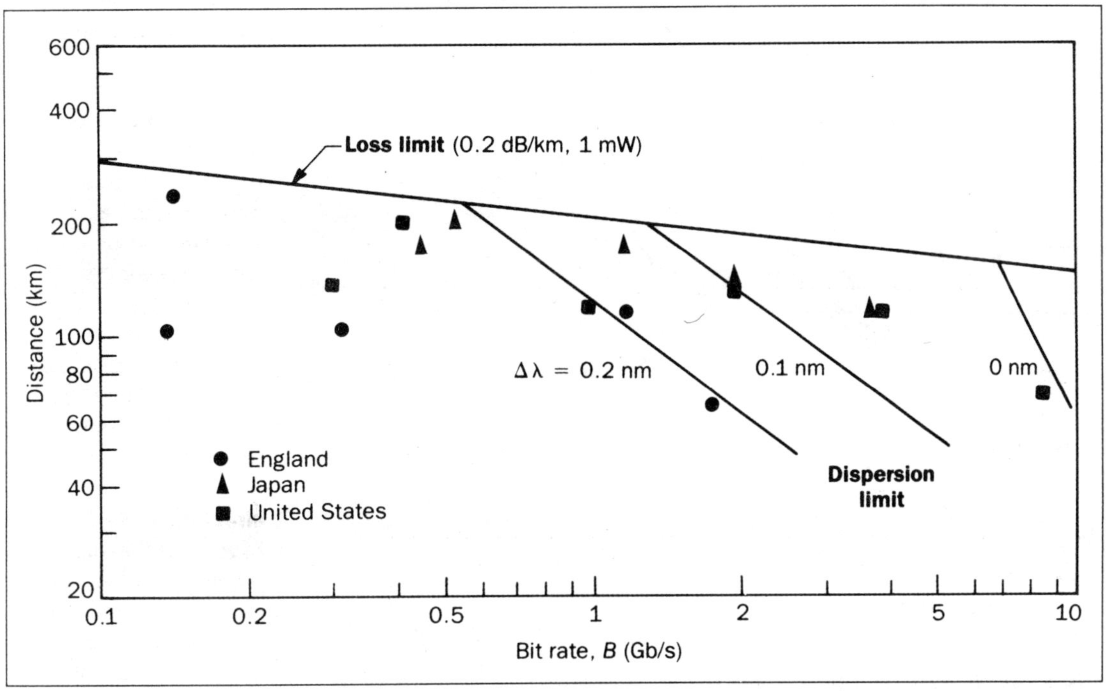

Figure 6. Results of long-distance transmission experiments conducted in laboratories in the United States, Japan, and England in the last few years. The solid lines represent theoretical limits imposed by fiber loss and dispersion, for a system that uses conventional single-mode fiber and state-of-the-art devices, and for various laser spectral widths ($\Delta\lambda$).

High-power C^3 lasers and titanium-diffused lithium-niobate waveguide modulators were used in both experiments. The 8-Gb/s experiment is especially significant because it was the first demonstration that the information bandwidth of the signal was a significant factor in limiting the transmission distance. (See the curve for $\Delta\lambda = 0$ in Figure 1.) Eye diagrams of the 8-Gb/s signal are shown in Figure 5.

High-bit-rate transmission experiments over long distances have been conducted in various laboratories in the last few years. Figure 6 plots the achieved transmission distances versus bit rate. The solid lines represent theoretical limits from fiber loss and dispersion, assuming a conventional single-mode fiber with 0.2-dB loss, 1-mW launched laser power, experimental APD receivers, and laser spectral widths of $\Delta\lambda = 0.2$, 0.1, and 0 nm. Interestingly, even though experimental conditions were diverse, the best observed results were close to the theoretical limits. Clearly, long-distance (>100 km) transmission at ultra-high (>10 Gb/s) data rates will require *dispersion-shifted* fibers, with λ_0 in the 1.5-μm region of minimum loss.

Of course, the transmission capacity of a lightwave transmission system can always be increased by using wavelength-division-multiplexing (WDM). The ability of a WDM system to grow gracefully with traffic demand makes it attractive where traffic growth is unpredictable. The disadvantage is that each multiplexed channel requires its own repeater, resulting in multiplied costs. However, the cost increase may be alleviated somewhat by possible large-scale integration of devices and circuits in a multiple-repeater package.

In a laboratory WDM experiment, light from ten carefully chosen DFB lasers were multiplexed onto a single-mode fiber to simulate a system with a total capacity

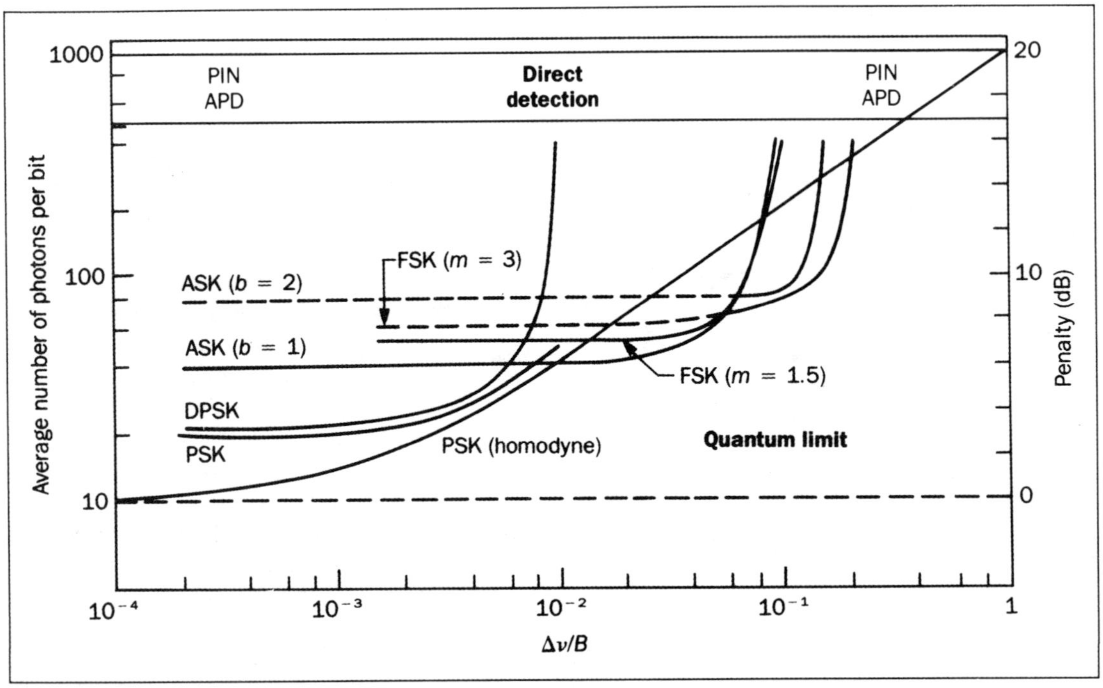

Figure 7. Receiver sensitivities for coherent modulation and detection as functions of the combined transmitter and local-oscillator laser linewidth ($\Delta\nu$) normalized to the bit rate *B*. Receiver sensitivity is expressed as the average number of photons (per bit) required to achieve an error probability of 10^{-9}, or as a power penalty relative to the quantum limit of direct detection. Except for the curve labeled PSK (homodyne), all others are calculated for heterodyne detection. The shaded band shows performance of experimental receivers that use direct detection.

of 20 Gb/s.[17] The grating-type wavelength multiplexer and demultiplexer had a channel spacing of 1.35 nm (17 GHz) centered around 1.545 μm. A transmission distance of 68.5 km was demonstrated.

Coherent Transmission

In coherent lightwave transmission, modulation is applied as amplitude, frequency, or phase variations in the optical carrier. The received signal and the output from a local-oscillator (LO) laser are mixed in the photodetector of a heterodyne receiver to generate an intermediate-frequency (IF) signal that typically falls in the microwave-frequency range. Ordinary microwave techniques can then be used to demodulate the IF signal.

If the LO power is large enough, quantum (or shot) noise in the detector will predominate over thermal and device noise in the preamplifier, thus affording high receiver sensitivity. The following motivates interest in coherent modulation and detection:

- The promise for increased receiver sensitivity
- The ability to select among densely packed WDM channels
- The potential for using wideband optical amplifiers to boost *en masse* a large group of WDM signals.

The theoretical sensitivity of a coherent-detection receiver can approach the *quantum limit* of direct detection of ten photons per bit. For example, the sensitivity limit of an ideal homodyne receiver (in which the intermediate frequency is zero) for phase-shift-keyed (PSK) signals is nine photons per bit. If heterodyne detection is used, the ideal sensitivities are 18 photons per bit for PSK

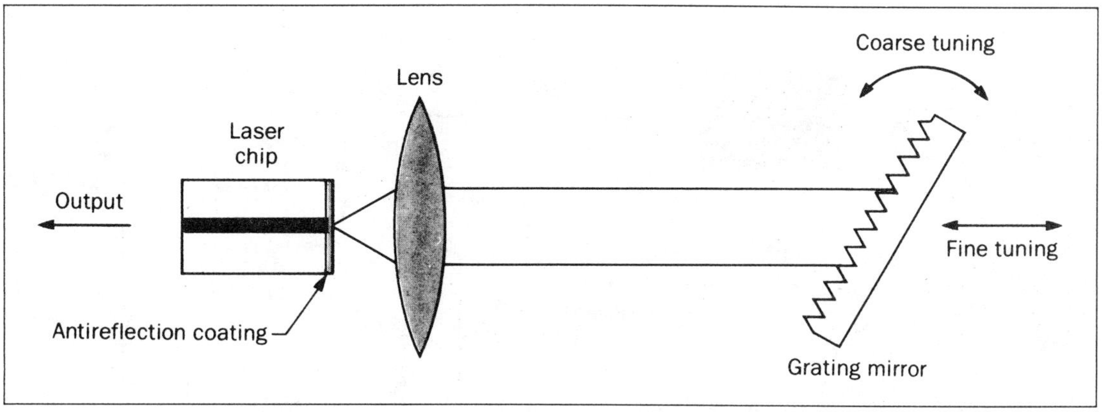

Figure 8. An extended cavity laser used in laboratory system experiments. One facet of the semiconductor laser chip is antireflection coated to extend its cavity into the long, passive region terminated by the grating mirror. Linewidths as narrow as a few kilohertz have been obtained in such lasers.

modulation and 36 photons per bit for both amplitude-shift-keyed (ASK) and frequency-shift-keyed (FSK) modulation.

Many problems must be solved before these ideal sensitivities can be approached in practice. Foremost among them is laser spectral purity. The linewidths of single-frequency semiconductor lasers are very large (compared to microwave sources, for example), typically from several megahertz to a few hundred megahertz. These large spectral widths are caused by phase noise that arises from refractive-index fluctuations (caused by carrier-concentration fluctuations), and from the high level of spontaneous emission in the laser cavity.[18] Phase noise can dramatically degrade receiver performance. The degradation can become so large that a 10^{-9} BER is impossible to attain, no matter how high the received power.

Figure 7 illustrates receiver performance for a variety of modulation and detection conditions. The ordinate is the average (minimum) number of photons per bit required to achieve 10^{-9} error probability. The abscissa is the combined linewidth of the transmitter and LO laser, normalized to the bit rate B. The quantum limit and experimental direct-detection sensitivities are shown as a dashed horizontal line and shaded region, respectively.

Except for the curve labeled PSK (homodyne), all other curves are calculated for heterodyne detection.[19-23] The ASK curves represent two values of b, which is the IF bandwidth normalized to the bit rate B. Those marked FSK are for two values of m, which is the frequency deviation normalized to the bit rate B.

It is possible to narrow the linewidth of a conventional or DFB laser by extending the length of its cavity. The linewidth reduction factor is approximately equal to the square of the ratio of the total cavity length to the active cavity length.[18]

Figure 8 is a schematic of an extended-cavity laser in which a grating mirror is used to achieve single-frequency operation. Coarse tuning of the wavelength of operation is obtained by tilting the grating, while fine tuning or frequency tracking is done by changing the cavity length or the laser drive current. Linewidths around 10 kHz are commonly observed in such extended-cavity lasers designed for laboratory experiments.[24] *Practical* narrow-linewidth *tunable* lasers that are suitable for field applications are not yet available.

Besides laser spectral purity, tunability, and stability, researchers also have been studying other important

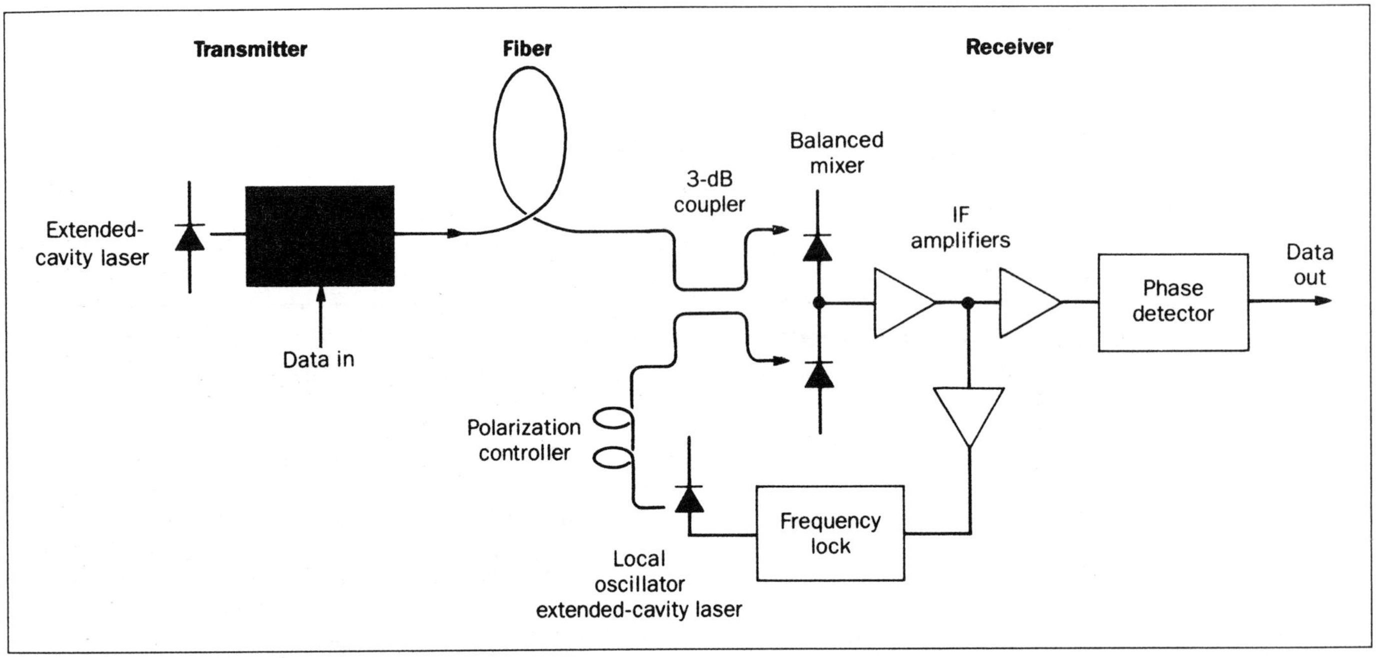

Figure 9. This coherent-transmission system experiment used differential-phase-shift-keyed modulation and heterodyne detection at bit rates of 400 Mb/s and 1 Gb/s.

issues specific to coherent transmission, including:

- Direct frequency modulation of injection lasers
- LO laser power and intensity noise
- Optical phase-locked loops for homodyne detection
- Nonlinear effects in the fiber that impair system performance
- Polarization properties of fibers and components.

Reflected power that is fed back into the laser cavity can cause linewidth variations and must be minimized using isolators.[25]

The issue of polarization is especially important because efficient coherent detection requires that the polarization state of the received signal be matched to that of the LO signal at the photodetector. Because the output polarization of conventional single-mode fibers tends to fluctuate randomly, one must find a way to maintain polarization in the fiber or compensate at the receiver for changes in the polarization. Experimental observations suggest that polarization variations are slow enough to make tracking and compensation possible.

The first coherent-lightwave-transmission experiments, which used *semiconductor* (rather than gas) lasers as transmitters and local oscillators, were reported in 1984.[26,27] The modulation formats were ASK and FSK at bit rates near 100 Mb/s. Later experiments at higher speeds attained receiver sensitivities that were closer to ideal.[28,29]

Figure 9 is a schematic of one of these experiments that operated at 400 Mb/s and 1 Gb/s with differential-phase-shift-keyed (DPSK) modulation.[28] With a balanced mixer to suppress LO intensity noise and use LO power efficiently, the system achieved a record receiver sensitivity of 45 photons per bit at a 400-Mb/s bit rate.

Operation at higher bit rates for coherent transmission is difficult because of the stringent requirements for flat response over a large bandwidth in the receiver front end and the following linear channel. Additionally, higher LO power is required to overcome thermal noise. Nevertheless, a transmission distance of 200 km at 1 Gb/s has been achieved.[28] The longest distance for high-speed coherent transmission reported to date is 290 km, in an experiment at 400 Mb/s using an optimal form of FSK, called continuous-phase (CP) FSK.[29]

Local Distribution and Networking

Until now, lightwave technology has not had a widespread effect in the local arena. Successful introduction of future integrated wideband services will depend on how various photonics technologies will be applied to solve the problems of local distribution and networking.

Therefore, much attention is focused on this area of research.

Coherent transmission is envisioned as a candidate for local distribution of many WDM channels. Channels can be selected by tuning the local oscillator. Recently, a ten-channel WDM experiment was done in a laboratory to show this principle.[30] A tunable heterodyne receiver was used to select a signal from among ten 0.83-μm lasers, each modulated at a 70-Mb/s data rate and separated by a 6-GHz channel spacing. System margins in this experiment permitted a maximum of 64 channels to be distributed.

A novel scheme for channel selection in a densely packed WDM system without resorting to coherent technology is the *phonon star*. It exploits backward Brillouin scattering in a single-mode fiber to provide amplification in a narrow bandwidth.[31] Because the required pump power is around 10 mW, tunable single-frequency semiconductor lasers can be used at the receiver for channel selection. The advantages of this scheme are:

- Polarization is not important.
- Lasers with extremely narrow linewidth are not necessary.
- Simple direct detection receivers with baseband electronics can be used.

A disadvantage is that a requisite length of single-mode fiber (<5 km) is needed to provide the Brillouin gain. In a recent experiment, data at 45 Mb/s were detected error-free in the presence of a signal in an adjacent channel as close as 140 MHz.[31]

A few channels (say about 50 channels of 50 Mb/s each) could be sent using microwave subcarrier multiplexing (SCM).[32,33] In this scheme, the lightwave carrier is intensity-modulated by signals that ride on microwave subcarriers. For example, several 50-Mb/s channels can individually modulate several microwave subcarriers at different microwave frequencies. The combined microwave signal can then intensity-modulate a laser, or be grouped in different frequency bands to modulate different lasers. The lasers need not operate in a single frequency, but their modulation bandwidths must be sufficiently large. A microwave demultiplexer that follows the wideband receiver separates the various subcarriers.

A recent experiment showed SCM with three 44-Mb/s signals multiplexed in a 4-GHz band and modulating a single laser.[32] Intermodulation distortion, which is present when two or more subcarriers modulate the laser, can be a limiting factor in the performance of such systems. However, acceptable transmission quality was obtained in the experiment.

Local networking requires intelligent switching to establish communications channels. A possible scenario for future local networks could be one where high-bit-rate time-division-multiplexed (TDM) lightwave signals carried by fibers are detected, demultiplexed, and then switched by electronic switching systems as lower-bit-rate channels of various services. For example, a 1-Gb/s lightwave channel with multiplexed voice, data, and video would serve each subscriber. Electronic switches that use high-speed silicon or gallium-arsenide integrated circuits would switch the various services that run at bit rates up to 100 or 200 Mb/s. Both circuit and packet switching techniques could be employed. This evolutionary approach can easily be made compatible with today's networks and systems that are under development.

On the other hand, the vast bandwidth of optical fibers challenges the imagination and ingenuity of research workers to innovate revolutionary systems that promise greatly enhanced technical capability and offer important economic advantages. Both WDM and TDM techniques are being explored for local switching and networking. Different network topologies and various protocols with distributed intelligence for accessing a network are being investigated. These system pursuits are exerting strong influences on device research. For example, *tunable* narrow-bandwidth optical filters and *tunable* narrow-linewidth lasers are being pursued for WDM applications. The work on local distribution and networking is just beginning. Which technologies will eventually prevail is difficult to predict.

Conclusions

Current vigorous research in lightwave technology and systems will provide many options for a variety of applications in the long-haul and local-networking arena. High-speed and wavelength-multiplexing techniques are being pushed to explore the vast transmission bandwidth of single-mode fibers.

Progress has been both rapid and abundant, and the fruits of research are being developed into commercial products. Photonics will surely become the dominant technology for information movement and management in the information-dominated society of the future.

References

1. T. Li, "Advances in Optical Fiber Communications: An Historical Perspective," *IEEE Journal on Selected Areas in Communications*, Vol. SAC-1, April 1983, pp. 356-372.
2. S. R. Nagel, J. B. MacChesney, and K. L. Walker, "Modified Chemical Vapor Disposition," *Optical Fiber Communications*, Vol. 1, Fiber Fabrication, edited by T. Li, Academic Press, Orlando, Florida, 1985, pp. 1-64.
3. D. Kalish and L. G. Cohen, "Single-Mode Fiber: From Research and Development to Manufacturing," *AT&T Technical Journal*, Vol. 65, No. 6, November/December 1986, pp. 19-32.
4. D. Marcuse and C. Lin, "Low Dispersion Single-Mode Fiber Transmission—The Question of Practical Versus Theoretical Maximum Transmission Bandwidth," *IEEE Journal on Quantum Electronics*, Vol. QE-17, June 1981, pp. 869-877.
5. R. A. Linke et al., "Mode Power Partition Events in Nearly Single-Frequency Lasers," *IEEE Journal of Lightwave Technology*, Vol. LT-3, June 1985, pp. 706-712.
6. W. T. Tsang, N. A. Olsson, and R. A. Logan, "High-Speed Direct Single-Frequency Modulation with Large Tuning Rate and Frequency Excursion in Cleaved-Coupled-Cavity Semiconductor Lasers," *Applied Physics Letters*, Vol. 42, April 1983, pp. 650-652.
7. H. W. Kogelnik and C. V. Shank, "Stimulated Emission in a Periodic Structure," *Applied Physics Letters*, Vol. 18, February 1971, pp. 152-154.
8. R. A. Linke, "Modulation Induced Transient Chirping in Single-Frequency Lasers," *IEEE Journal of Quantum Electronics*, Vol. QE-21, June 1985, pp. 593-597.
9. T. L. Koch and J. E. Bowers, "Nature of Wavelength Chirping in Directly Modulated Semiconductor Lasers," *Electronics Letters*, Vol. 20, December 1984, pp. 1038-1040.
10. T. L. Koch et al., "1.55-μm InGaAsP Distributed Feedback Vapor Phase Transported Buried Heterostructure Lasers," *Applied Physics Letters*, Vol. 47, July 1985, pp. 12-14.
11. J. C. Campbell et al., "High-Performance Avalanche Photodiode with Separate Absorption Grading and Multiplication Regions," *Electronics Letters*, Vol. 19, September 1983, pp. 818-819.
12. B. L. Kasper, "Sensitivity Limits on Direct Detection Receivers," *Technical Digest*, Conference on Optical Fiber Communication, OSA, Washington, D.C., 1986, paper WJ4.
13. B. L. Kasper et al., "A 130 km Transmission Experiment at 2 Gb/s Using Silica-Core Fiber and a Vapor Phase Transported DFB Laser," *Conference Proceedings* (Post Deadline Papers), 10th European Conference on Optical Communications, VDE-Verlag GmbH, Berlin, 1984, paper PD-6.
14. A. H. Gnauck et al., "4 Gb/s Transmission Over 103 km of Optical Fiber Using a Novel Electronic Multiplexer/Demultiplexer," *Technical Digest* (Post Deadline Papers), Conference on Optical Fiber Communication, OSA, Washington, D.C., 1985, paper PD-2.
15. S. K. Korotky et al., "4 Gb/s Transmission Experiment Over 117 km of Optical Fiber Using a Ti:$LiNbO_3$ External Modulator," *Technical Digest* (Post Deadline Papers), Conference on Optical Fiber Communication, OSA, Washington, D.C., 1985, paper PD-1.
16. A. H. Gnauck et al., "Information-Bandwidth-Limited Transmission at 8 Gb/s Over 68.3 km of Single-Mode Optical Fiber," *Technical Digest* (Post Deadline Papers), Conference on Optical Fiber Communication, OSA, Washington, D.C., 1986, paper PDP-9.
17. N. A. Olsson et al., "68.3 km Transmission with 1.37 Tbit·km/s Capacity Using Wavelength Division Multiplexing of Ten Single-Frequency Lasers at 1.5 μm," *Electronics Letters*, Vol. 21, January 1985, pp. 105-106.
18. C. H. Henry, "Theory of the Linewidth of Semiconductor Lasers," *IEEE Journal of Quantum Electronics*, Vol. QE-18, February 1982, pp. 259-264.
19. J. Salz, "Coherent Lightwave Communications," *AT&T Technical Journal*, Vol. 64, No. 10, December 1985, pp. 2153-2209.
20. B. Glance, "Performance of Homodyne Detection of Binary PSK Optical Signals," *IEEE Journal of Lightwave Technology*, Vol. LT-4, February 1986, pp. 228-235.
21. G. Nicholson, "Probability of Error for Optical Heterodyne DPSK System with Quantum Phase Noise," *Electronics Letters*, Vol. 20, November 1984, pp. 1005-1007.
22. G. Jacobsen and I. Garrett, "Error-Rate Floor in Optical ASK Heterodyne Systems Caused by Nonzero (Semiconductor) Laser Linewidth," *Electronics Letters*, Vol. 21, March 1985, pp. 268-270.
23. I. Garrett and G. Jacobsen, "Influence of (Semiconductor) Laser Linewidth on the Error-Rate Floor in Dual-Filter Optical FSK

Receivers," *Electronics Letters*, Vol. 21, March 1985, pp. 280-282.

24. R. Wyatt and W. J. Devlin, "10 kHz Linewidth 1.5 μm InGaAsP External Cavity Laser with 55 nm Tuning Range," *Electronics Letters*, Vol. 19, February 1983, pp. 110-112.
25. R. W. Tkach and A. R. Chraplyvy, "Regimes of Feedback Effects in 1.5-μm Distributed Feedback Lasers," *IEEE Journal on Lightwave Technology*, Vol. LT-4, November 1986, pp. 1655-1661.
26. M. Shikada et al., "100 Mbit/s ASK Heterodyne Detection Experiment Using 1.3 μm DFB-Laser Diodes," *Electronics Letters*, Vol. 20, February 1984, pp. 164-165.
27. R. Wyatt et al., "140 Mbit/s Optical FSK Fibre Heterodyne Experiment at 1.54 μm," *Electronics Letters*, Vol. 20, October 1984, pp. 912-913.
28. R. A. Linke et al., "Coherent Lightwave Transmission Over 150 km Fiber Lengths at 400 Mbit/s and 1 Gbit/s Data Rates Using Phase Modulation," *Electronics Letters*, Vol. 22, January 1986, pp. 30-31.
29. K. Iwashita et al., "Linewidth Requirement Evaluation and 290 km Transmission Experiment for Optical CPFSK Differential Detection," *Electronics Letters*, Vol. 22, July 1986, pp. 791-792.
30. E.-J. Bachus et al., "Ten-Channel Coherent Optical Fibre Transmission," *Electronics Letters*, Vol. 22, September 1986, pp. 1002-1003.
31. A. R. Chraplyvy and R. W. Tkach, "Narrowband Tunable Optical Filter for Channel Selection in Densely-Packed WDM Systems," *Electronics Letters*, Vol. 22, September 1986, pp. 1084-1085.
32. T. E. Darcie et al., "Lightwave System Using Microwave Subcarrier Multiplexing," *Electronics Letters*, Vol. 22, July 1986, pp. 774-775.
33. J. E. Bowers, "Optical Transmission Using PSK Modulated Subcarriers at Frequencies to 16 GHz," *Electronics Letters*, Vol. 22, October 1986, pp. 1119-1121.

Fiber Optic Transmission Systems—Status and Trends in Japan

TATSUYA KIMURA, SENIOR MEMBER, IEEE

Abstract —**Development of large-capacity optical fiber transmission systems in Japan is reviewed and futuristic optoelectronic technologies are discussed including optical devices and their integration, optical fibers, coherent transmission, and optical switching studies.**

I. Introduction

THE extensive and rapid introduction of fiber optic transmission systems into telecommunications networks over the last several years owes its flexible fiber cable and optical repeater adaptabilities to the conventional installation and maintenance environment. Furthermore, this accelerated introduction can be directly contributed to the high performance and reliability of the systems, which are supported by the progress of optoelectronics technology.

As of the beginning of 1985, optical fiber cables totaling nearly 7000 km in length have been installed for commercial telecommunications services in Japan. Data-rate and repeater spacing, which are the typical transmission system performance measures, have been, respectively, improved by adopting long wavelength carrier waves up to several Gbits/s and a spacing exceeding 100 km. These improvements have been confirmed by laboratory tests.

The performance of the installed commercial system, mainly based on a 1.3 μm carrier, is expected to be further enhanced by using the 1.5 μm wavelength band. This is because this wavelength band decreases fiber loss by approximately two-fifths. Further improvement is anticipated by employing a sophisticated modulation/demodulation/amplification scheme otherwise known as coherent optical fiber transmission technology.

Another area of focus with respect to optical fiber transmission technology is how to effectively distribute individual information to customers in a certain regional area. Although the subscriber system is mainly based on the same optoelectronic technology as that of the trunk system, it has different features which center on the requirements for low-cost fibers and equipment and on those for practical installation devices and procedures.

The Information Network System (INS) project [1], [2], which is promoted by Nippon Telegraph and Telephone Corporation (NTT) to provide an advanced, reliable, and economic digital telecommunications network supporting a variety of functions, will be incorporating optical fiber systems to meet its goal.

In this context, this paper reviews the recent Japanese system development activities, and then discusses several topics of importance in forecasting future fiber optic technologies.

Manuscript received February 25, 1985; revised March 10, 1986.

The author is with the NTT Musashino Electrical Communication Laboratory, Musashino-shi, Tokyo 180, Japan.

IEEE Log Number 8608815.

II. Broad-Band Transmission System Development

A. Terrestrial Trunk Transmission System

Large-capacity system development followed that of medium- and small-capacity digital fiber systems [3], [4]. The single-mode fiber capability of transmitting large-capacity information with a long wavelength carrier [5], [6] is effectively utilized in NTT's F-400M system [7], which transmits digital signals at the fifth hierarchical speed of 400 Mbits/s at the 1.3 μm wavelength. The system features an error rate of less than 10^{-8} and an r.m.s. jitter of less than 15° over a 2500 km hypothetical circuit length having intermediate repeaters located every 25 km. It employs cables containing a maximum of 24 single-mode fiber strands. Optical repeaters are usually accommodated in existing telephone station buildings. When a suitable building is not available, however, these repeaters are located in a hut and are powered locally. They employ InGaAsP/InP semiconductor lasers and Ge avalanche photodiodes for optical/electrical conversions. The use of these devices yields a transmitting and receiving power level difference of 26 dB, which is allotted for fiber cable losses and system margins. The line signal has an RZ waveform and a complementary pulse is inserted after every 10 signal pulses to secure timing signal retrieval.

Splicing, which has been one of the major technological difficulties in applying single-mode fibers, has been carefully managed by a combination of adjusting the fiber core axes, fusion splicing, and monitoring the transmitted light level. Relative to this, a study is currently being conducted on an automatic splicing machine for single-mode fibers [8]. With respect to the technology mentioned, a system was tested until March 1983 over the 76 km route between the Musashino and Atsugi Electrical Communication

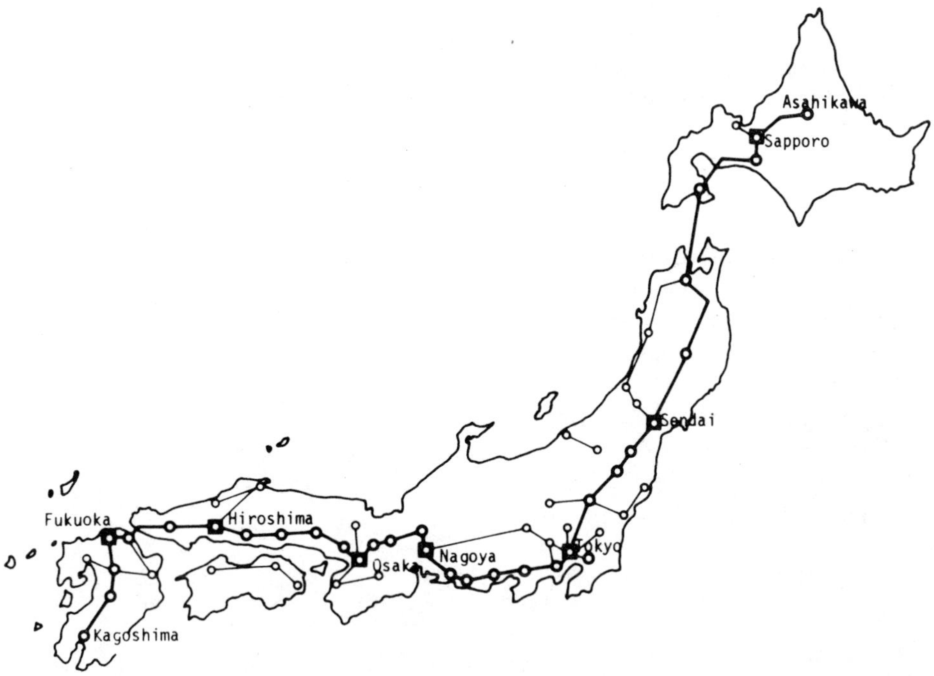

Fig. 1. Route of 400 Mbit/s systems introduced into Japanese public telecommunications networks.

TABLE I
MAIN FEATURES OF OPTICAL FIBER TRANSMISSION SYSTEMS

System	FS-1.5M	FS-6M	F-6M	F-32M	F-100M	F-400M
Line code	CMI	CMI	CMI	CMI	8B1C	10B1C
Information bit rate (Mbit/s)	1.544	6.312	6.312	32.064	97.728	397.200
Line bit rate (Mbit/s)	3.088	12.624	12.624	64.128	111.689	445.837
Capacity (ch)	24	96	96	480	1,440	5,760
Wavelength (μm)	1.3	1.2 1.3	1.3	1.3	1.3	1.3
Optical fiber	GI*	GI*	GI*	GI*	GI*	SM**
Repeater spacing (km)	25 35 45	20 30 40	15	10 15 25	10 15 25	25 30
Error rate (/rep)	10^{-9}	10^{-9}	10^{-10}	10^{-11}	10^{-11}	10^{-11}

* CCITT standard
** Core/cladding: 10μm/125μm

Laboratories. This test resulted in an extremely low splice loss of 0.11 dB on the average, with a standard deviation of 0.08 dB, proving the effectiveness of these splicing techniques for actual application. Following the conclusion of these tests, the system was put into full-scale operation which led the confirmation of its reliability for firm use.

NTT began constructing a nationwide trunk system traversing the length of Japan with the F-400M system in 1983. The installation, completed in March 1985, connects the city of Asahikawa in Hokkaido with the city of Kagoshima in Kyushu. As shown in Fig. 1, the system covers an overall length of approximately 3400 km. The system is expected to play a significant role as a trunk system in the forthcoming nationwide digital network [9].

The main features of fiber systems at various hierarchical data rates are summarized in Table I, together with those of the nonrepeatered undersea systems. The development target of the next system presently being planned, the F-1.6G, is to upgrade the transmission capacity four times while maintaining the same repeater spacing. The F-1.6G system is being field-tested at a 120-km route section between Tachikawa City and Kofu City in the above nationwide trunk network.

B. Submarine Systems

Optical fibers enable undersea systems to transmit a large capacity of digital information over a long span without increasing the cable diameter. Domestic undersea systems [10] have been developed at the 1.3 μm wavelength without undersea repeaters for each digital hierarchy. These employ graded-index (1.5, 6.3, and 32 Mbits/s) or single-mode (100 and 400 Mbits/s) fiber cables designed for laying at depths of up to 1500 meters. The circuit spans 33–48 km depending on the data rate. Several undersea systems have been recently installed between the islands of the Japanese archipelago.

Repeatered submarine systems are currently being developed by NTT [11] and KDD [12] at the respective data rates of 400 and 280 Mbits/s. Both systems were tested in 1984 off the coast of the Izu Peninsula. The FS-400M system [11] has been designed for domestic high-capacity use with 17 280 two-way voice channels. It features a 1000 km system length, 40 km repeater spacings, and a water depth of 8000 meters. Commercial use of the domestic system is scheduled to start in 1986. The OS-280M system

[8] is designed for transoceanic applications with the final goals being an 8000 km system length, a 50 km nominal repeater spacing and an 8000 meter water depth. The short-haul commercial test is for the system scheduled for 1986, with the long-haul commercial system planned to be laid in 1988.

The strength of fibers [13] and their splices [14], particularly important in repairing submarine cables, is enhanced by improving the fabrication and splicing processes. The MTBF of both systems mentioned is designed to be 10 years by providing a cold standby semiconductor laser in each system coupled with an optical coupler or a switch. One of the principal features of the systems being developed in Japan is that a germanium avalanche photodiode is used in the receiver instead of an InGaAs p-i-n photodiode, because the former device gives a significantly lower receiving level.

TABLE II
FIBER OPTIC LOCAL NETWORK SYSTEMS DEVELOPED IN JAPAN

System name	BRANCH 4780	SOLAR-net	FACOM 2883	SIGMA	MELNET R32	TOTAL-LAN	OPALnet-II
Company	NEC	NTT	Fujitsu	Hitachi	Mitsubishi	Toshiba	NTT
Topology	active tree	passive star	active ring	active ring	active ring	active ring	hybrid ring
Access	CSMA/CD	CSMA/CD	TDM	TDMA	TDMA	token passing	TDMA/token passing
Bit rate Mbit/s	10	32	33	31	32	100	100
Max span (Max loop length) km	1	1	9 (576)	2 (256)	2 (128)	4 (100)	1 (100)
Max number of nodes	1100	100	64	128	64	64	100
Source	LED	LED	LED	LED	LED	LD	LD
Detector	PIN-PD	APD	APD	PIN-PD	APD	APD	APD

Fiber: graded index, 50/125 μm
Wavelength: 0.85 μm

C. *Subscriber Systems and Local Area Networks*

Following the pioneering attempt of the Hi-OVIS project [15] by the Visual Information System Development Association in 1977, NTT fiber optic subscriber system study initially started from relatively simple but basic service applications. In 1980, a field test was conducted between Yokosuka Electrical Communication Laboratory and Yokosuka Telephone Repeater Station, separated by 7.8 km, to confirm basic subscriber system functions. The next trial was carried out in 1982 with 27 subscribers at Kichijoji, a suburb of Tokyo [16].

The subscriber system for business premises transmits high-speed data and video signals over 5 km using 4-wave wavelength-division multiplexing. The system confirmed a differential gain of less than 8 percent and a differential phase of less than 4 degrees for an NTSC video signal. These values are sufficient for teleconferencing service applications. The subscriber systems for local area applications aim at providing broad-band services such as CATV and the on-demand Video Response System. This system consists of two wavelength-division-multiplexed waves for video signals and one wave for control signals. In both systems, the analog video signals were transmitted by the direct intensity modulation of semiconductor lasers.

Design and installation of two principal subscriber cable types are among the test targets of the current field trial [9], [17], [18]. The first type involves two alternative cable types, which reflect the progress in cable advances. The first of these types involves subscriber cables, which from the beginning, incorporated up to 100 strands, constituting a 10-fiber unit. The second type, based on the 5-fiber tape unit, was introduced and is being tested as a high-density optical fiber cable having 200 fiber strands. With respect to the down-lead wires, we designed two special configurations. One fiber, intended for outdoor use, is surrounded by Kevler threads and has a sustaining steel wire. The other, intended for indoor use, is reinforced with four steel wires. A fiber switch box, located at the top of a pole, contains fiber connectors to allow for flexible future distribution. The switch box has been devised to neatly store the extra length required for splicing. Mechanical joints for cable splicing, multiple-fiber splicing, and a multiple-fiber connector are technologies expressly intended for the subscriber system installation [19].

In an attempt to construct an efficient and effective telecommunication network which fully meets the needs of the expected future information society, NTT is making every effort to actualize the so-called INS, or information network system. In this system, the principal feature is the effective blending of telecommunications and information processing technologies [1], [2]. Optical fiber transmission systems are expected to play a major role in this actualization by transmitting digitally encoded voice, data, facsimile, and still and moving pictures. A pilot system, called the INS Model System [20]–[22], was put into operation by connecting Kasumigaseki, a business district in central Tokyo, with Mitaka, a western suburb approximately 20 km away, employing the F-400M transmission system. The test started in September 1984 with 80 subscribers and was extended to 200 subscribers by March 1985. The test period will conclude in March 1987. The optical fiber technologies employed in the trunk and subscriber systems are applied in the Model System to assess the functionability and impact of advanced services.

Analog and pulsed-analog optical fiber systems have been developed for video transmission service [23]. With respect to analog systems, one of the unique applications of the fibers is the transmission of broad-band high-definition television signals. Here, 20 MHz bandwidth time-compression-multiplexed signals of the high-definition television are transmitted by an intensity-modulated AlGaAs laser. Results indicate that a signal-to-noise ratio of 40 dB is available at the received power level of about −32 dBm [24], [25]. The systems were tested in conjunction with the INS Model System, and are being used together with the F-400M system trunk to transmit high-quality television pictures between the Tsukuba Expo'85 site and several major Japanese cities.

Optical fiber local area networks are used for office and factory information processing. Since 1978, tests have been made on a ring-bus network operating at 100 Mbits/s [26], on a WDM optical accessing loop at 8 Mbits/s [27], and on the Research Information Processing Network [28]. Communications equipment suppliers have their own LAN systems as a menu shown in Table II [28a]. NTT has developed two kinds of systems using graded-index fibers at the short wavelength: the 100 Mbit/s loop-configuration system for large-scale applications and the 32 Mbit/s star-configuration system for medium-scale applications [29]. Furthermore, an optical cascade star configuration has been proposed and tested [30] to ascertain the tradeoff between system size and cable cost.

III. Futuristic Optoelectronics Technologies

A. *Optical Devices and Their Integration*

The next generation of broad-band or large-capacity systems will utilize carriers at the 1.5 μm wavelength region where a still lower fiber loss is attainable than that at 1.3 μm [31]. This will be profitable especially to large-capacity digital systems in expanding the repeater spacing. Narrow spectral-width semiconductor lasers having a controlled single longitudinal mode are important for avoiding the finite fiber dispersion influence at this wavelength. The alternative is to use an optical fiber with the zero-dispersion wavelength shifted to the 1.5 μm band. This can be accomplished through rearranging the fiber structure configuration [32]. High-quality distributed feedback (DFB) lasers are now available at this wavelength [33], [34].

Integration of optical and electronic devices on the same substrate will bring about superior high-speed performance by reducing stray capacitance and increasing packaging compactness. The integration of an AlGaAs laser, a quantum well laser, and a p-i-n photodiode with FET's has been reported [35]–[40].

B. *Optical Fibers*

Hydrogen diffused into fibers has been confirmed to increase optical loss in the long wavelength. The degradation takes place when the fiber contains phosphor impurities or when the fiber cable is corroded. Diffused hydrogen molecules bring about the reversible loss increase remarkable in the 1.1–1.25 μm wavelength, while the increased oxyhydrogen content causes irreversible loss in and above the 1.4 μm wavelength [41]. The fiber impurities, fiber drawing procedure, coating materials, and cable structure have to be optimized to prevent loss increase in the important 1.5 μm region [42].

Subscriber system development must effectively deal with a variety of economic problems, particularly those concerning low-cost fibers, cables, connectors, multiplexers, and optical semiconductor devices. Low-cost fiber fabrication technologies are being developed. The high-speed deposition of graded-index fiber preforms using the VAD process shows a deposition rate of 3–5 g/min and a higher than 70 percent yield. This is accomplished by revising the burner structure [43]–[46]. One example of a large obtained porous preform is shown in Fig. 2. Such preforms produce fibers which exhibit no optical loss degradation. High-speed fiber drawing up to 1200 meter/min has been quite recently reported through the application of a pressurized plastic coating [47].

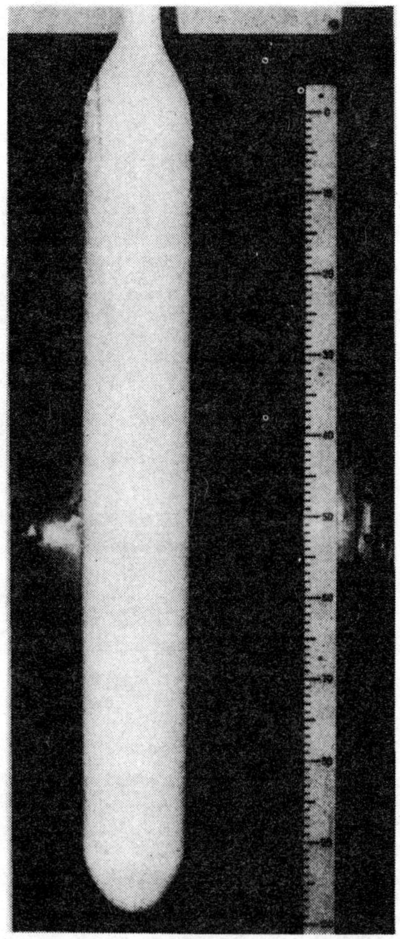

Fig. 2. Soot preform, 130 mm in diameter and 1 m in length, fabricated with the high-speed vapor phase axial deposition method.

Optical multiplexers are important devices for subscriber systems in which two-way and/or wavelength-multiplexed signals are transmitted [48]. Multiplexers made by deposited silica waveguides on a silicon substrate, as shown in Fig. 3, are being studied [49], [50].

One of the present technological targets is the application of single-mode fibers to broad-band subscriber systems to alleviate the multimode fiber modal noise problem. In this regard, fiber splicing and connection constitute the particular bottlenecks warranting solution. A fiber-splicing robot is being tested for more efficient installation of optical fiber cables. Although multimode fibers are still used in a test, average splice loss of 0.07 dB is obtained by using the prototype equipment shown in Fig. 4. The robot

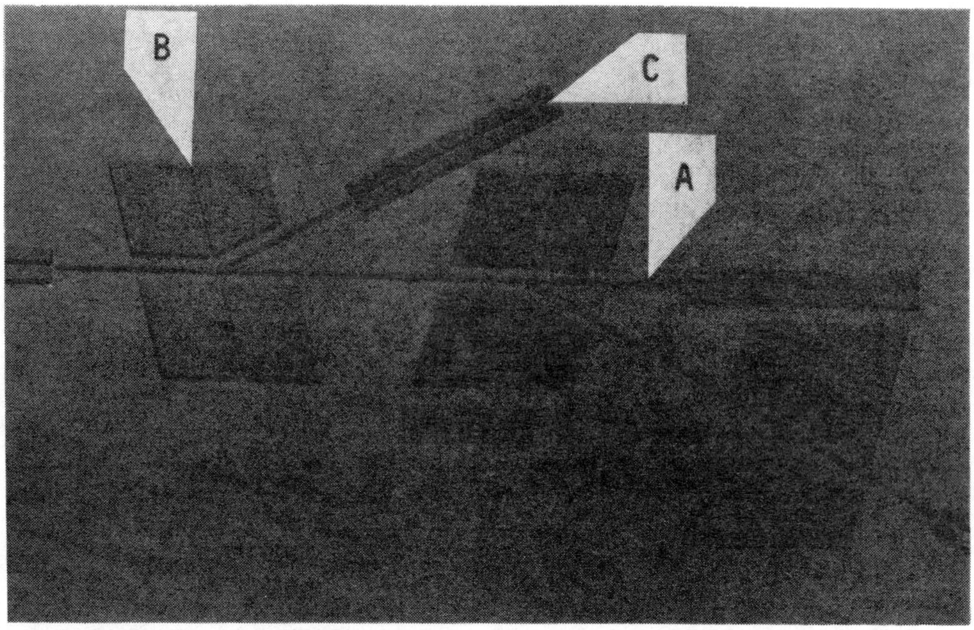

Fig. 3. Guided-wave optical multiplexer made by silica deposited on a silicon substrate. A: optical waveguide, B: groove to accommodate a multilayer optical filter, C: groove to accommodate a fiber.

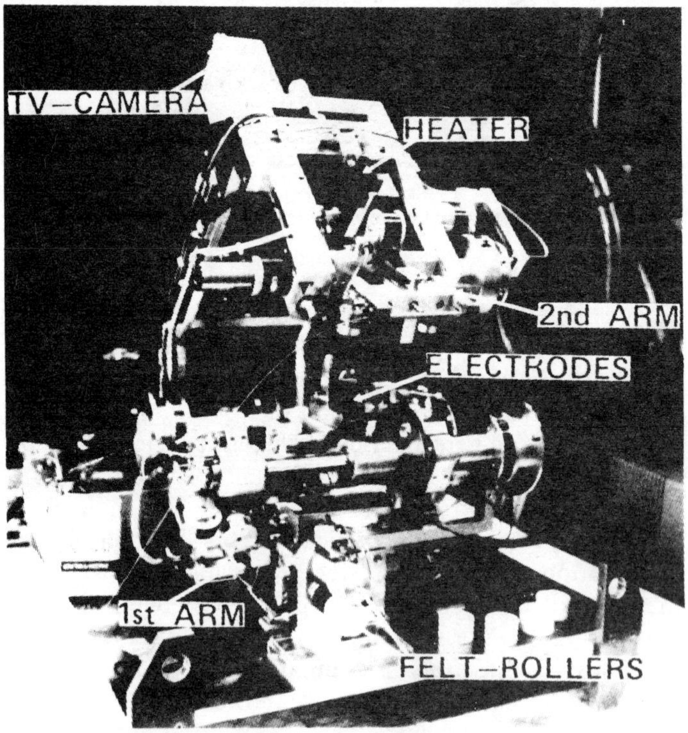

Fig. 4. A prototype of fiber-splice robot.

can strip, cut, align, arc-fuse, and reinforce a pair of fibers in 5.5 min [50a].

C. Coherent Optical Transmission Systems

The concepts of coherent optical transmission systems and optical amplification systems [51] have been tested in laboratory experiments [52], although there are practical arguments both for and against these systems. In the first case, homodyne or heterodyne detection of angular modulated coherent optical waves provides a low receiving level for broad-band signal transmission. Such systems, however, still need elaboration especially in terms of light source characteristics, such as linewidth reduction and frequency tuning/stabilization [53], [54]. By using presently available devices and fibers, the best laboratory test result seems to be 270-km single-mode fiber transmission. The experiment consists of 400 Mbit/s direct frequency-shift-keying and heterodyne detection in the 1.546 μm band. The laser linewidth was narrowed to 200 kHz by loading an external grating to an antireflection-coated facet. Receiver sensitivity was -49 dBm at an error rate of 10^{-9} [54a].

In the second case, several approaches to coherent optical transmission center on optical amplification with an interest toward transmission system applications [55]. One concerns antireflection coating affixed to the amplifier facets as being applicable to a traveling wave optical amplifier [56]. Another optical amplification scheme involves the stimulated Raman effect in optical fibers [57], [58]. Although its noise characteristics remain under study, the fiber amplifier has been shown to exhibit a high saturation-level output.

It has additionally been proven that optical frequency multiplexing involving narrow-channel spacing [59] will provide efficient utilization of the optical frequency spectrum when combined with the coherent transmission systems. Such coherent systems carrying digital or analog signals can be, in principle, applied to single-mode subscriber systems.

Furthermore, a semiconductor laser stabilized by negative feedback has been shown to exhibit a reduced line width from 200 MHz to 1 MHz as well as flat FM frequency characteristics [60]. Finally, it is interesting to note that the feedback scheme presents the possibility of

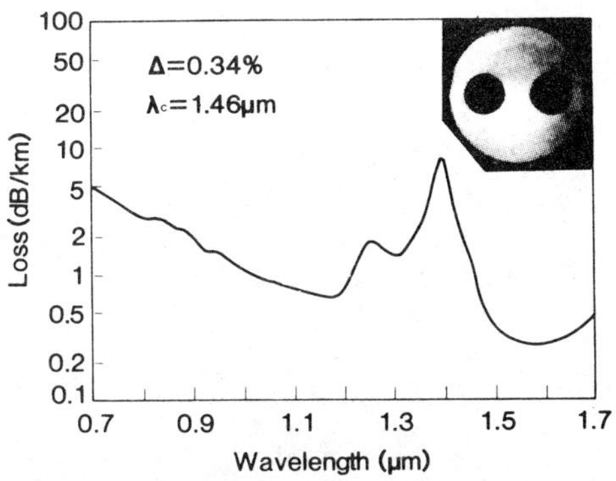

Fig. 5. Loss and cross section of a 2-km-long polarization maintaining fiber.

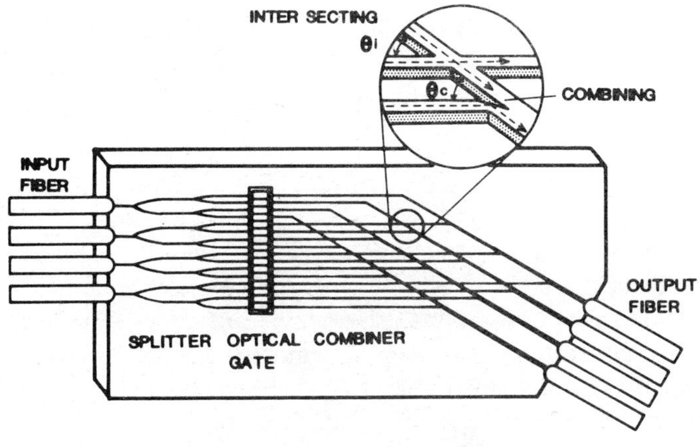

Fig. 6. A 4×4 optical switch configuration with silica waveguides.

reducing FM noise below the photon coherent state limit, that is, realizing a squeezed state [61]. Application of optoelectronic integration technology to such feedback noise reduction circuits will be useful for practical broadband devices.

Polarization maintaining fibers, in which stress is applied to the core, has been improved with respect to optical loss and polarization coupling. The fiber shown in Fig. 5 has 0.25 dB/km loss at 1.57 μm. Extrapolated crosstalk between two polarization states is expected to be -12 dB at 100-km length [61a].

D. Optical Switching Systems

Several attempts are currently being made to extend optical transmission technologies to switching system applications. One attempt involves directional couplers fabricated on a $LiNbO_3$ crystal used for spatially switching guided optical waves [62]. Another alternative scheme using photoelectronic conversion has been verified by using heterostructure switching photodiodes and fiber couplers [63]. In this scheme, a 3×3 matrix version switches the broadband signals of high-definition television within 10 ns. A semiconductor laser amplifier can also be used as an on–off switch [64] when its bias current is controlled.

In addition, the $LiNbO_3$ waveguide switch combined with a fiber delay line has been used for the time-division switching device of broad-band TV signals [65]. A bistable laser diode [66] having a saturable absorber region included in its cavity has also been used as a memory device for time-division switching applications [67].

Deposited silica waveguide, which was described in terms of guided-wave optical multiplexers, has also been applied to an optical switching circuit. A 4×4 optical switch configuration is shown in Fig. 6. An InGaAsP laser diode is hybrid-integrated in a groove traversing the optical waveguide for switching function as shown in Fig. 7 [68]. The optical signal is on–off switched by gain/loss in the laser active medium. Alternative switching device was proposed by using refractive index change induced by electric current in a semiconductor [69]. A guided-wave switching device, exemplified in Fig. 8, shows a 20.5-dB extinction ratio at a 91 mA drive without polarization direction dependence [70].

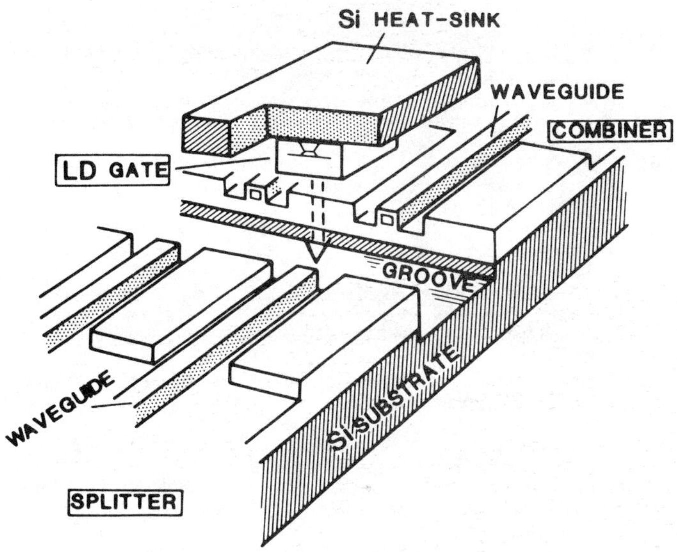

Fig. 7. Hybrid integration of InGaAsP laser diode switch on a silica waveguide circuit. The figure is a magnified view of the optical gate area in Fig. 6.

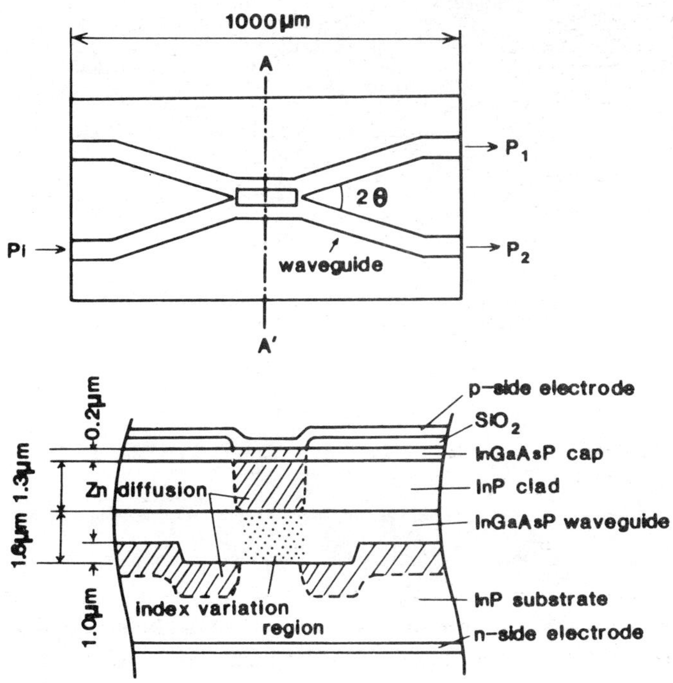

Fig. 8. Optical waveguide switch by current-induced refractive index change in a semiconductor. (a) Top view. (b) Cross-sectional view.

Although optoelectronic switching technology is still far from practical application, this switching may become useful in such special fields as broad-band video switches.

IV. Conclusion

Optical fiber transmission technologies were reviewed emphasizing their roles in the Japanese telecommunications networks. Additionally, several recent topics aimed at future generations of optical communication technologies were presented with respect to constituent optical devices and fibers as well as system configurations.

References

[1] G. Kaplan, "Japan's information network system," *IEEE Spectrum*, vol. 21, no. 5, pp. 50–52, May 1984.

[2] Y. Kitahara, "Information network system—Infrastructure for an advanced information society," *Japan Telecommun. Rev.*, vol. 4, no. 2, pp. 96–103, Apr. 1982.

[3] T. Horiguchi, T. Miki, and H. Ishio, "F-32M and F-100M fiber optic transmission systems," *Japan Telecommun. Rev.*, vol. 23, no. 2, pp. 125–134, Apr. 1981.

[4] K. Hashimoto, "F-6M optical fiber transmission system," *Japan Telecommun. Rev.*, vol. 26, no. 4, pp. 245–249, Oct. 1984.

[5] T. Kimura, "Single-mode systems and components for longer wavelengths," *IEEE Trans. Circuits Syst.*, vol. CAS-22, no. 12, pp. 987–1010, Dec. 1979.

[6] T. Kimura, "Single-mode digital transmission technology," *Proc. IEEE*, vol. 68, no. 10, pp. 1263–1268, Oct. 1980.

[7] Y. Hakamada, S. Tsuyuki, and K. Suto, "Design and performance of the F-400M repeatered line," *Rev. Electron. Commun. Labs. NTT*, vol. 31, no. 3. pp. 244–252, may 1983.

[8] T. Uenoya, "SM optical fiber splicing technique," *Japan Telecommun. Rev.*, vol. 26, no. 4, pp. 245–249, Oct. 1984.

[9] H. Ishihara, "Subscriber optical fiber cable technology—Developmental history and present status," *Japan Telecommun. Rev.*, vol. 26, no. 3, pp. 174–180, July 1984.

[10] Y. Kobayashi, K. Aida, and N. Tokura, "Non-repeatered submarine optical transmission terminal equipment," *Rev. Electron. Commun. Labs. NTT*, vol. 31, no. 6, pp. 844–852, Nov. 1983.

[11] H. Fukinuki, T. Ito, M. Aiki, and Y. Hayashi, "The FS-400M submarine system," *J. Lightwave Technol.*, vol. LT-2, no. 6, pp. 754–760, Dec. 1984.

[12] Y. Niiro, "The OS-280M optical-fiber submarine cable system," *J. Lightwave Technol.*, vol. LT-2, no. 6, pp. 767–772, Dec. 1984.

[13] S. Sakaguchi, "Drawing of high-strength long-length optical fibers for submarine cables," *J. Lightwave Technol.*, vol. LT-2, no. 6, pp. 808–815, Dec. 1984.

[14] Y. Miyajima, S. Furukawa, and M. Ohnishi, "High strength fiber splicing technique for submarine fiber cable," *Rev. Electron. Commun. Labs. NTT*, vol. 31, no. 5, pp. 808–814, Nov. 1983.

[15] M. Kawahata, "Hi-OVIS (Higashi Ikoma Optical Visual Information System) development project," presented at the *1977 Int. Conf. Integrated Opt. Opt. Fiber Commun.*, Tokyo, Japan, July 18–22, 1977.

[16] M. Koyama, H. Ishio, F. Nihei, and A. Sakamoto, "Optical subscriber loop system for business premises and local area applications—Field trial," *Rev. Electron. Commun. Labs. NTT*, vol. 32, no. 4, pp. 554–562, July 1984.

[17] M. Kawase, H. Shinohara, and F. Nihei, "Design and performance of optical subscriber cables," *Rev. Electron. Commun. Labs. NTT*, vol. 32, no. 4, pp. 626–635, July 1984.

[18] S. Kukita, K. Hogari, and Y. Katsuyama, "Design and performance of optical drop and indoor cables," *Rev. Electron. Commun. Labs. NTT*, vol. 32, no. 4, pp. 636–645, July 1984.

[19] N. Kashima, M. Tachikura, and M. Hirai, "Splicing and connecting techniques for optical subscriber cables," *Rev. Electron. Commun. Labs. NTT*, vol. 32, no. 4, pp. 676–683, July 1984.

[20] M. Kuwabara, "Japan tests its model communication system," *IEEE Spectrum*, vol. 21, no. 5, pp. 53–56, May 1984.

[21] K. Habara and H. Ikeda, "INS model system framework," *Rev. Electron. Commun. Labs. NTT*, vol. 32, no. 1, pp. 105–117, Jan. 1984.

[22] T. Murakami, "Inception of INS experience—Model systems sets in service," *Japan Telecommun. Rev.*, vol. 27, no. 1, pp. 2–16, Jan. 1985.

[23] M. Yamashita, "Optical fiber video transmission systems—FV-4M-A and FV-4M-P systems," *Japan Telecommun. Rev.*, vol. 25, no. 1, pp. 32–38, Jan. 1983.

[24] K. Sato and K. Asatani, "Fiber optic analog video transmission using semiconductor laser diodes," *Rev. Electron. Commun. Labs. NTT*, vol. 32, no. 6, pp. 1124–1133, Nov. 1984.

[25] T. Miki and J. Ishida, "Transmission technologies on HDTV," *J. Inst. Electron. Commun. Eng. Japan*, vol. 68, no. 6, pp. 635–642, June 1985.

[26] N. Okuda, T. Kunikyo, and T. Kaji, "Ring century bus—An experimental high speed channel for computer communications," in *4th Int. Council Comput. Commun.*, 1978, p. 161.

[27] M. Saruwatari, K. Nosu, T. Miki, and K. Asatani, "Optically accessing loop (OPAL) network using wavelength-division-multiplexing technology," in *Int. Conf. Commun.*, vol. 28.6, 1980.

[28] K. Yada, T. Ochiai, and M. Honda, "Optical fiber makes research information processing system," in *IEEE Comput. Soc. Conf. '81 Spring*, San Francisco, CA, 23–26 Feb. 1981, pp. 450–453.

[28a] I. Kobayashi and M. Koyama, "Recent status and trends of subscriber networks and LANs," presented at the Int. Conf. Integrated Optics Opt. Commun., Oct. 1–4, 1985, Venice, Italy.

[29] J. Minowa, Y. Hakamada, N. Tokura, and M. Kawase, "Development of fiber-optic local area network," *Electron. Comm. Labs. Tech. J.*, vol. 34, no. 5, pp. 851–860, May 1985.

[30] T. Tamura, M. Nakamura, S. Ohshima, T. Ito, and T. Ozeki, "Optical cascade star network—A new configuration for a passive distribution system with optical collision detection capability," *J. Lightwave Technol.*, vol. LT-2, no. 1, pp. 61–66, Feb. 1984.

[31] J. Yamada, S. Machida, T. Mukai, H. Tsuchiya, and T. Kimura, "Long-span single-mode fiber transmission characteristics in long wavelength region," *IEEE J. Quantum Electron.*, vol. QE-16, no. 8, pp. 874–884, Aug. 1980.

[32] N. Imoto, A. Kawana, S. Machida, and H. Tsuchiya, "Characteristics of dispersion free single-mode fiber in the 1.5 μm wavelength region," *IEEE J. Quantum Electron.*, vol. QE-16, no. 10, pp. 1052–1058, Oct. 1980.

[33] O. Mikami, "1.55 um GaInAsP/InP distributed feedback lasers," *Japan. J. Appl. Phys.*, vol. 20, no. 7, pp. L488–L490, July 1981.

[34] Y. Suematsu, "Advances in semiconductor lasers," *Phys. Today*, vol. 38, no. 5, pp. 32–39, May 1985.

[35] H. Matsueda, S. Sasaki, and M. Nakamura: "GaAs optoelectronic integrated light sources," *J. Lightwave Technol.*, vol. LT-1, no. 1, pp. 261–269, Mar. 1983.

[36] S. Yamakoshi, T. Sanada, O. Wada, T. Fujii, and T. Sakurai, "Low threshold AlGaAs/GaAs MQW laser integrated with FET's," in *Fourth Int. Conf. Integrated Opt. Opt. Fiber Commun.*, 27B3-1,Tokyo, Japan, June 1983, pp. 27–30.

[37] T. P. Tanaka, M. Hirao, and M. Nakamura, "Optoelectronic integrated light sources for high-speed optical communication systems," in *Eighth Conf. Opt. Fiber Commun.*, TuC2, San Diego, CA, Feb. 11–13, 1985.

[38] T. Horimatsu, T. Izawa, Y. Oikawa, H. Hamaguchi, and T. Nakagami, "Compact transmitter module with a monolithic laser/driver circuit," in *Eighth Conf. Opt. Fiber Commun.*, TuC3, San Diego, CA, Feb. 11–13, 1985.

[39] T. Nakagami, M. Sasaki, and T. Sakurai, "Optical receiver front-end modules using optoelectronic integrated circuits," in *Eighth Conf. Opt. Fiber Commun.*, TuC5, San Diego, CA, Feb. 11–13, 1985.

[40] T. Horimatsu *et al*, "400 Mbit/s transmission experiment using two monolithic optoelectronic chips," *Electron. Lett.*, vol. 21, no. 8, pp. 319–321, Apr. 11, 1985.

[41] K. Mochizuki, Y. Namihira, and H. Yamamoto, "Transmission loss increase in optical fibers due to hydrogen permeation," *Electron. Lett.*, vol. 19, no. 18, pp. 743–745, Sept. 1983.

[42] N. Uesugi *et al*, "Infrared optical loss increase for silica fiber in cable filled with water," *Electron. Lett.*, vol. 19, no. 19, pp. 762–764, Sept. 1983.

[43] H. Kawazoe, K. Orimo, and A. Iino, High-rate deposition VAD process using a double-flame burner," presented at the Eighth Conf. Opt. Fiber Commun., San Diego, CA, Feb. 11–13, 1985.

[44] M. Miyamoto, T. Ohashi, R. Yamauchi, O. Fukuda, and K. Inada, "High-rate deposition of VAD graded-index fibers," presented at the Eighth Conf. Opt. Fiber Commun., San Diego, CA, Feb. 11–13, 1985.

[45] T. Danzuka, H. Yokota, and I. Tsuchiya, "High-speed deposition of graded-index fiber by the multiflame vapor axial deposition method," presented at the Eighth Conf. Opt. Fiber Commun., San Diego, CA, Feb. 11–13, 1985.

[46] H. Suda, K. Chida, and S. Sudo, "High-rate fabrication technique for VAD fiber preforms," *Rev. Electron. Commun. Lab. NTT*, vol. 32, pp. 418–424, May 1984.

[47] S. Sakaguchi and T. Kimura, "A 1200-m/min speed drawing of optical fibers with pressurized coating," presented at the Eighth Conf. Opt. Fiber Commun., San Diego, CA, Feb. 11–13, 1985.

[48] H. Ishio, J. Minowa, and K. Nosu, "Review and status of wavelength-division multiplexing technology and its application", *J.*

Lightwave Technol., vo. LT-2, no. 4, pp. 448–463, Aug. 1984.
[49] T. Izawa, H. Mori, Y. Murakami, and N. Shimizu, "Deposited silica waveguides for integrated optical circuits," *Appl. Phys. Lett.*, vol. 38, no. 7, pp. 483–485, Apr. 1, 1981.
[50] M. Kawachi, Y. Yamada, M. Yasu, and M. Kobayashi, "Guided-wave optical wavelength-division multi/demultiplexer using high silica channel waveguides," *Electron. Lett.*, vol. 21, no. 8, pp. 314–315, Apr. 11, 1985.
[50a] T. Haibara, M. Tachikura, M. Matsumoto, and R. Arioka, "A fully automatic optical fiber arc-fusion splice machine," *Trans. IECE Japan*, vol. E68, no. 12, pp. 836–843, Dec. 1985.
[51] Y. Yamamoto and T. Kimura, "Coherent optical fiber transmission systems" *IEEE J. Quantum Electron.*, vol. QE-17, no. 6, pp. 919–935, June 1981.
[52] S. Saito, Y. Yamamoto, and T. Kimura, "Optical FSK heterodyne detection experiment using semiconductor laser transmitter and local oscillator," *IEEE J. Quantum Electron.*, vol. QE-17, no. 6, pp. 935–941, June 1981.
[53] Y. Yamamoto, S. Saito, and T. Mukai, "AM and FM quantum noise in semiconductor lasers—Part II: Comparison of theoretical and experimental results for AlGaAs lasers," *IEEE J. Quantum Electron.*, vol. QE-19, no. 1, pp. 47–58, Jan. 1983.
[54] S. Saito, O. Nilsson, and Y. Yamamoto, "Oscillation center frequency tuning, quantum FM noise, and direct frequency modulation in external grating loaded semiconductor laser," *IEEE J. Quantum Electron.*, vol. QE-18, no. 6, pp. 961–970, June 1982.
[54a] K. Iwashita, T. Iwai, T. Matsumoto, and G. Motosugi, "400 Mb/s optical FSK transmission experiment over 270 km of single-mode fiber," to be published.
[55] S. Kobayashi and T. Kimura, "Semiconductor optical amplifiers," *IEEE Spectrum*, vol. 21, no. 5, pp. 26–33, May 1985.
[56] T. Saitoh, T. Mukai, and O. Mikami, "Theoretical analysis and fabrication of antireflection coatings on laser diode," *J. Lightwave Technol.*, vol. LT-3, no. 2, pp. 288–293, Apr. 1985.
[57] Y. Aoki, S. Kishida, K. Washio, and K. Minemura, "Bit-error rate evaluation of optical signals amplified via stimulated Raman process in an optical fiber," presented at the Eighth Conf. Opt. Fiber Commun., San Diego, CA, Feb. 11–13, 1985.
[58] K. Nakamura, M. Kimura, S. Yoshida, T. Hidaka, and Y. Mitsuhashi, "Raman amplification of 1.50-um laser diode light in a low fiber loss region," *J. Lightwave Technol.*, vol. LT-2, no. 4, pp. 379–381, Aug. 1984.
[59] N. Imoto, N. Shimizu, and M. Ikeda, "A grating filter for embedded silica waveguides," in *Fourth Int. Conf. Integrated Opt. Optical Fiber Commun.*, 29A2-4, June 1983, Tokyo, Japan, pp. 27–30.
[60] S. Saito and Y. Yamamoto, "Negative feedback stabilized FM semiconductor laser for coherent optical communication systems," presented at the 10th European Conf. Optical Commun., Stuttgart, Germany, Sept. 3–6, 1984.
[61] Y. Yamamoto, O. Nilsson, and S. Saito, "Theory of a negative frequency feedback semiconductor laser," to be published.
[61a] T. Hosaka, Y. Sasaki, J. Noda, and M. Horiguchi, "Low-loss and low-crosstalk polarization-maintaining optical fiber," *Electron. Lett.*, vol. 21, no. 20, pp. 920–921, Sept. 1985.
[62] H. Haga, M. Izutsu, and T. Sueta, "An integrated 1×4 high-speed optical switch and its application to a time demultiplexer," *J. Lightwave Technol.*, vol. LT-3, no. 1, pp. 116–120, Feb. 1985.
[63] E. H. Hara, S. Machida, M. Ikeda, H. Kanbe, and T. Kimura, "A high-speed optoelectronic matrix switch using heterojunction switching photodiodes," *IEEE J. Quantum Electron.*, vol. QE-17, no. 8, pp. 1539–1546, Aug. 1981.
[64] M. Ikeda, "Laser diode switch," *Electron. Lett.*, vol. 17, no. 3, pp. 899–900, Nov. 1981.
[65] M. Kondo *et al.*, "High-speed optical time switch with integrated optical 1×4 switches and single-polarization fiber delay lines," in *Fourth Int. Conf. Integrated Optics Opt. Fiber Commun.*, Tokyo, Japan, June 27–30, 1983, pp. 29D3–7.
[66] H. Kawaguchi, "Bistable operation of semiconductor lasers by optical injection," *Electron. Lett.*, vol. 17, no. 20, pp. 741–742, Oct. 1, 1981.
[67] H. Goto *et al.* "An experiment on time-division digital switching using bistable laser diodes and optical switches," in *IEEE Global Telecommun. Conf.*, vol. 26-4, Atlanta, GA, Nov. 26–29, 1984.
[68] M. Kobayashi, A. Himeno, and H. Terui, "Guided-wave optical gate matrix switch," presented at the Int. Conf. Integrated Optics Opt. Commun., Oct. 1–4, 1985, Venice, Italy.
[69] O. Mikami and H. Nakagome, "Waveguide optical switch in InGaAsP/InP using free-carrier plasma dispersion," *Electron. Lett.*, vol. 20, no. 6, pp. 228–229, Mar. 1984.
[70] K. Ishida, H. Nakamura, H. Inoue, S. Tsuji, and H. Matsumura, "InGaAsP/InP multimode optical switches operated by a carrier induced refractive index change," presented at the Int. Conf. Integrated Optics Opt. Commun., Oct. 1–4, 1985, Venice, Italy.

Tatsuya Kimura (S'63–M'68–SM'78) was born in Osaka, Japan on February 20, 1940. He received the B.S. degree in electrical engineering, the M.S. degree in electronics engineering, and the Ph.D. degree in quantum electronics from the University of Tokyo, Japan, in 1962, 1964, and 1967, respectively.

Since he joined the Electrical Communication Laboratory, Nippon Telegraph and Telephone Corporation (NTT) in 1967, he has been engaged in research on optical transmission systems for public telecommunication including long-wavelength optical fiber transmission and coherent optical fiber transmission, and on optical equipment and devices, such as gas, solid-state and semiconductor lasers, light-emitting diodes, modulators, PM–AM converters, optical FDM filters, optical isolators, photodetectors, and optical fibers. He was engaged in development of a guided millimeter-wave transmission system during 1972–1973. He is presently Associate Director, Physical Science Division, Musashino Electrical Communication Laboratory.

Dr. Kimura is a member of the Institute of Electronics and Communication Engineers of Japan, the Japan Society of Applied Physics, and the American Physical Society. He is an Associate Editor of the JOURNAL OF LIGHTWAVE TECHNOLOGY. He received the Inada Memorial Award from the Institute of Electronics and Communication Engineers of Japan in 1963 for research on a parametric photodetector, and the NTT President Award in 1978 for long-wavelength optical fiber transmission system proposal and research.

The Application of Microwave Techniques in Lightwave Systems

HAROLD SOBOL, FELLOW, IEEE

(Invited Paper)

***Abstract*—Microwave techniques have a significant role in advancing the state-of-the-art of lightwave systems. Applications include the analysis and design of structures to support lightwave propagation, the development of devices and circuits to modulate, demodulate, and process microwaves bandwidth signals on lightwave carriers, and the development of circuits to process microwave bandwidth baseband signals. The merging of microwave and lightwave technologies will open new opportunities in system design.**

I. Introduction

THE VERY large bandwidth potential of coherent optical carriers has been recognized since the invention of the laser, more than two decades ago. However, only recently has this potential been realized. Microwave techniques and workers have played a significant role in advancing the state-of-the-art, and the future offers many opportunities for continuing contributions. This paper presents an overview of the use of microwave techniques and technology in lightwave system applications.

The following sections review the current status of wide bandwidth lightwave development, areas where microwave techniques have been employed, areas where fundamental differences in microwave and lightwave technologies exist due to the character of the devices employed and the quantum nature of light, and future applications where the merger of lightwave and microwave technologies may result in new system advantages.

II. Status of Wide Bandwidth Lightwave Development

Telecommunication transmission of broad bandwidth digital signals is currently the largest market for lightwave systems. This market also represents the first practical application for the transport of microwave bandwidth signals on lightwave carriers. The growth rate of the application of single-mode optical fibers in the transmission plant has been very large during the past two years. Nearly 1 500 000 km of optical fiber were produced in the U.S. in 1985 [1]. A very significant percentage of the fiber installed in the U.S. today is single mode and nearly all transmission is at 1.3-μm wavelength. The typical loss of the installed single-mode fiber is less than 0.5 dB/km at 1.3 μm.

TABLE I
REPRESENTATIVE COMMERCIAL HIGH BANDWIDTH LIGHTWAVE TRANSMISSION SYSTEMS
(1.3-μm single-mode fiber)

DATA RATE MB/S	EQUIVALENT 64 KB/S VOICE CIRCUITS	SPAN LENGTH KM	MANUFACTURER
139	2016	60	ROCKWELL
405	6048	45	FUJITSU
565	8064	45	ROCKWELL

TABLE II
REPRESENTATIVE RESULTS FOR GIGABIT TRANSMISSION SYSTEMS

DATA RATE GB/S	MICRONS	LASER TYPE	MODULATION	SPAN LENGTH KM	REFERENCE
0.140	1.55	DFB	DIRECT	220	15
0.420	1.55	DFB	DIRECT	203	16
4.00	1.53	C^3	EXT	117	17
4	1.54	DFB	DIRECT	103	18
8	1.53	C^3	EXT	68	9

Wide bandwidth transmission systems today utilize directly modulated InGaAsP lasers with spectral linewidths of about 4 nm, and InGaAs p-i-n or Ge APD photodiodes. Table I lists the performance of typical commercially available systems. Recently, new systems became available that have data rates of 810 Mbit/s and 1130 Mbit/s. Somewhat farther in time, after reliable high-performance single frequency lasers are produced, 1.55-μm systems with even higher data rates will be placed in service.

Laboratory and field trial systems using single frequency lasers (coupled cavity, external cavity, and distributed feedback) with direct and external modulation have been operated at multigigabit per second rates. The majority of the experimental systems operate at 1.55-μm wavelengths where the fiber loss is typically less than 0.2 dB/km. Table II lists the status of some representative experimental systems.

The above results emphasize the progress that has been made in the development of very wide bandwidth, direct, and external modulated CW lightwave sources. Both short wavelength GaAlAs and long wavelength InGaAsP lasers with direct modulation bandwidth in excess of 6 GHz are commercially available [3]. Experimental InGaAsP lasers with modulation bandwidth approaching millimeter wavelengths have been reported [4]. Photodiodes for use in wide bandwidth receivers have also been reported [5].

An experiment [2] to demonstrate fiber bandwidth capability was conducted by wavelength division multiplexing ten DFB 1.5-μm 2-Gbit/s direct-modulated lasers, each with a slightly different wavelength on a single fiber, to achieve a transmission rate of 20 Gbit/s over a 68.3-km span. Thus it appears that in the future, if the market requirements develop, systems with millimeter wave bandwidths may be employed.

III. Optical Waveguides

The analysis of the transmission of lightwaves in optical waveguides is a prime example of the application of microwave technique.

Optical fibers are the most popular lightwave waveguide, but other forms are utilized in devices, optical integrated circuits (OIC), and functional elements such as directional couplers, modulators, and switches. Silica based glass is used for fibers in the 1-μm band, $LiNbO_3$ is used for hybrid OIC's and functional elements, and GaAs and InP based structures are used for monolithic circuits. Guidance is achieved by field confinement due to index of refraction discontinuities. Propagation in the structures is analyzed by the use of Maxwell's equations and appropriate boundary conditions at discontinuities.

Optical waveguides and hollow metallic microwave guides have some common properties including a finite number of guided modes and the susceptibility to geometric distortions to cause mode conversion. In addition to the guided modes, the dielectric or optical waveguide also supports a continuum of unguided radiation modes that are also solutions to Maxwell's equations. The radiation modes contribute to loss; therefore, the geometric imperfections and bends that excite these modes should be avoided.

For the linear range of materials, the analysis of optical waveguides is similar to techniques for microwave structures, however, losses and effects of material nonlinearities must be treated considering quantum effects. Single-mode propagation is similar to the microwave case, but the number of modes in large size multimode fibers far exceeds the number of modes typically encountered in microwave structures.

The analysis of optical waveguides is well documented [6]–[8] and will be summarized. Various geometries used in optical waveguides are illustrated in Fig. 1.

Maxwell's equations have exact solutions for the dielectric cylinder, but the resulting expressions are extremely cumbersome for practical design use, even with infinitely thick cladding layers. The practical, weakly guiding case with very small differences between the core and cladding indices of refraction does result in easily applied design equations. The fields in the core are expressed in terms of Bessel functions, and the fields in the cladding are expressed in terms of modified Hankel functions. The arguments of the circular functions are expressed in terms of a normalized frequency V:

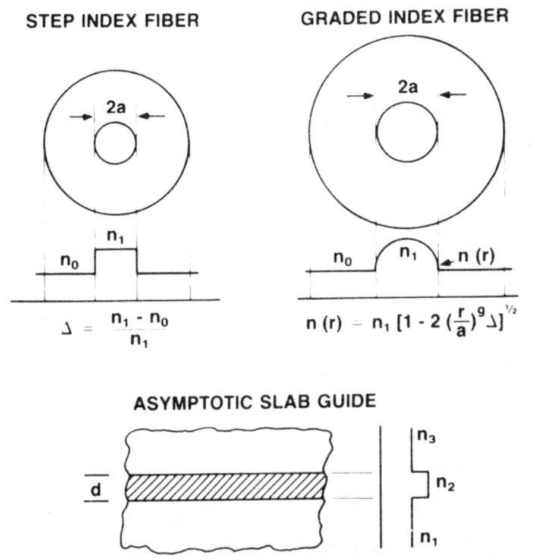

Fig. 1. Lightwave waveguides.

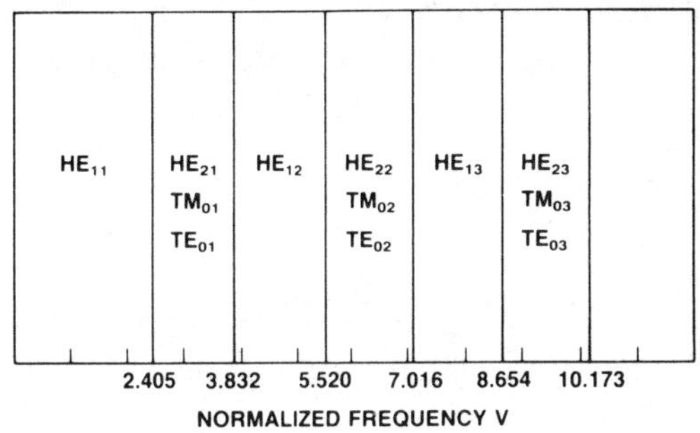

Fig. 2. Some lower-order mode cutoffs for optical fiber.

$$V = 2\pi a n_1 \sqrt{2\Delta}/\lambda \tag{1}$$

where the parameters are shown in Fig. 1.

The lowest order guided modes in a small diameter step index fiber are shown in Fig. 2. Maxwell's equations, in general, cannot be satisfied by independent TE and TM modes in the optical waveguide since all six wave components are required. The fundamental mode is an HE_{11} mode, and it is the only mode that exists for a range of normalized frequencies between $V = 0$ and $V = 2.405$, the first zero of a zeroth order Bessel function. For normalized frequencies greater than 2.405, multimode propagation occurs.

The single mode is actually two modes with orthogonal polarizations. The number of modes in a large size step index multimode fiber is given approximately by

$$N \simeq 1/2V^2. \tag{2}$$

The multimode graded index fiber shown in Fig. 1 is used to obtain higher bandwidth than the step index multimode fiber. The first higher order mode cutoff is given by

$$V_c \simeq 2.405\,(1 + 2/g)^{1/2} \tag{3}$$

and the number of modes is given by

$$N \simeq 1/4V^2 \qquad (g = 2). \tag{4}$$

Table III lists some typical characteristics for single mode and multimode step index fibers. Note that the mul-

TABLE III

	MULTIMODE	SINGLE MODE
λ	1.3μ	1.3μ
n_1	1.48	1.48
Δ	0.014	0.003
2a	50μ	7.5μ
V	30	2.07;< 2.405
N	≈ 450	1
		λc = 1.122μ

timode fiber has 450 modes, considerably more than any microwave metallic guide.

The normalized frequency for the asymptotic slab waveguide shown in Fig. 1 is

$$V = (n_1^2 - n_2^2)^{1/2} 2\pi d/\lambda. \tag{5}$$

The higher order mode cutoff wavelength, λc is given by $V_c = \Pi$ and the number of modes is given by

$$N \simeq 2V/\pi. \tag{6}$$

The losses in typical microwave structures are due to the metallic conductors, but the major losses in an optical waveguide result from scattering, absorption, and waveguide bending. The most important scattering effect is Rayleigh scattering which results from intrinsic interaction with the lattice of the material and varies as λ^{-4}. The primary absorption losses in silica glass fibers are due to the tails of the far infrared and ultraviolet absorption spectrums and are due to interactions with electronic and molecular transition bands. Other absorption losses arise from vibrational levels due to metallic and OH ions, and atomic defects. Macrobending of the fiber will cause radiation modes to be excited.

IV. Microwave Bandwidth Optical Devices

The laser operates in a manner fundamentally different than traditional sources of microwave energy, since it produces an optical signal by stimulated emission of radiation. The optical signal may be modulated using external modulators or in the case of semiconductor injection laser may be directly modulated by a signal in series with the laser bias supply. Since the radiative recombination time of carriers in the semiconductor laser during stimulated emission is nanoseconds or less, the laser has the inherent property of large direct modulated bandwidths.

A key factor in determining the modulation bandwidth of lasers is the presence of a relaxation frequency or resonance as shown in Fig. 3. The peaking in the response is due to interaction between the optical field in the laser and the injected electron density distribution which results in oscillations of both the light output and the electron density. The modulation response of the laser falls at a rapid rate at frequencies higher than the relaxation frequency; therefore, the resonance point represents the maximum modulating frequency. The relaxation frequency is accompanied by a large increase in laser noise

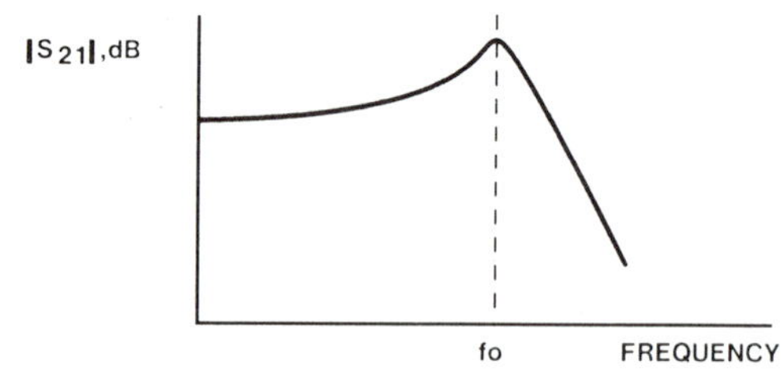

Fig. 3. Directly modulated laser response.

and should be avoided for satisfactory system performance.

The relaxation frequency f_0 varies as [9]

$$f_0 \propto \sqrt{\frac{A}{L}\frac{d_T P_b}{d_M}} \tag{7}$$

where

P_b	laser output power,
d_T and d_M	optical and mirror loss, respectively, in the laser,
A	differential gain constant, and
L	cavity length.

The modulation response [9] can be increased by using short cavities, biasing at high power, and increasing differential gain by increasing active region doping. The maximum power is limited by thermal effects and in GaAlAs diodes by failures at the crystal facets. Parasitic effects due to diode capacitance and lead inductance also limits the maximum modulation frequency. The parasitic capacitance may be reduced by using an inactive layer below the bonding pads [4]. Facet burnout in GaAlAs diodes is improved by using windows between the ends of the active regions of the laser and the facets [10]. InGaAsP [11] lasers have exhibited room temperature CW modulation bandwidths of 25 GHz and GaAlAs [10] lasers have shown bandwidths in excess of 10 GHz.

Photodiodes inherently have a high bandwidth capability. The major limitations are carrier transit time in the diode, capacitive loading, and parasitics. A trade-off exists between signal sensitivity and bandwidth since greater area yields higher sensitivity to impinging light beams but also results in higher capacitance. Optimum choices must be made, depending on performance required. GaAs photodiodes (0.9 μm) have been fabricated with 20-GHz bandwidths and InGaAs p-i-n diodes have been made with 36-GHz bandwidths [5].

Bias and electrical circuit design is extremely important in achievement of the bandwidth potential of both sources and detectors. The laser modulation drive circuit must provide maximum current to the low impedance of the laser without introducing resonances in the drive circuit passband. The resonance peaking in the laser modulation characteristics may be treated as a second-order low pass network. Wideband low noise amplifiers are required to process the current generated by the photo detector. GaAs FET's are usually used for this purpose. Obviously, good microwave circuit design is necessary for both devices.

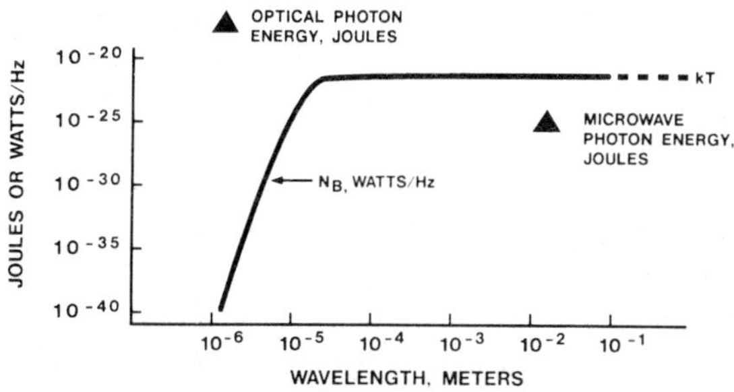

Fig. 4. Photon energy and thermal background noise power.

V. The Communication Channel

The lightwave channel differs from the microwave channel because of the quantum nature of light. Further, the noise generated in optical sources is different than the noise of microwave sources. An optical transmitter produces about 10^{15} photon/s, each with an energy of about 2×10^{-19} J. A microwave transmitter produces about 10^{24} photon/s, each with energy of about 4×10^{-24} J. The thermal background noise at a receiver is described by the distribution

$$N_B = hf\left[\exp\left(hf/kT\right) - 1\right]^{-1} \qquad (\mathrm{W/Hz}) \qquad (8)$$

and is shown in Fig. 4.

Since $hf << kT$ in the microwave receiver, the noise spectral density is given by the familiar kT or about 4×10^{-21} W/Hz. Thus, each microwave photon has considerably less energy than the background thermal noise. Since $hf >> kT$, in the lightwave receiver the energy of each optical photon is very much greater than the thermal noise. As a consequence, the thermal noise is a limiting factor in microwave channels, but it is not in lightwave channels. Another limit, the quantum noise limit, must be utilized in the lightwave channel. The quantum limit is derived from the statistics of photo detection and is described by

$$P_E = e^{-Ed/hf} \qquad (9)$$

where P_E is the probability of an error (not producing an electron–hole pair when a photon signal with energy E_d is present). For an error rate of 10^{-9}, E_d is equal to $21hf$, or 21 photons must be received for each electron–hole pair produced. The quantum limit corresponds to an average received optical power level of $10.5hfB$ W where B is the data rate. Thus, the limit at a 1-Gbit/s rate is -57 dBm. Direct detection receivers usually require receive signal levels 15–20 dB above the quantum limit. Coherent detection (heterodyne receivers) offer opportunities for detection sensitivities close to the quantum limit. The microwave and the optical channel are compared in Table IV. The "noise-figure" for the optical receiver is not defined in terms of temperature and represents degradation from the quantum noise limit.

Quantum noise, longitudinal mode-partition effects, abrupt jumps of longitudinal and tranverse lasing modes, mode-partition noise, and effects of reflections from the fiber are performance limiting factors in lightwave system, due to the laser source. These effects are usually small or nonexistent in microwave sources because of better mode control and absence of quantum effects. In addition, source stability and linewidth in microwave sources is considerably better than lasers due to the use of phase locked loops. Proper design of optical sources has minimized the noise effects but at this time not to the level of the microwave source. This is an area where microwave technology can make contributions.

TABLE IV

	MICROWAVE (THERMAL LIMITED)	LIGHTWAVE (QUANTUM NOISE LIMITED)
λ	5 cm	1μ
BANDWIDTH	30 MHz	1 GHz
NF	6 dB	"20 dB" (QUANTUM NOISE)
MINIMUM SIGNAL	48×10^{-11} FOR 30 dB S/N	2×10^{-7} WATTS FOR 10^{-10} BER
MODULATION	64 QAM	2 ASK
SYSTEM GAIN	100 dB	30 dB
TRANSMIT POWER	4.8 WATTS	0.2 mW
ENERGY/PHOTON	4×10^{-24} JOULES	2×10^{-19} JOULES
PHOTONS/SEC	1.2×10^{24}	10^{15}
CAPACITY PCM VOICE CIRCUITS	2,000	16,000

VI. High Speed Circuits

Just as in a microwave system, the bulk of the circuits in lightwave terminals and repeaters are not optical functions, but are baseband processing circuits. Typical high speed circuits required in wideband lightwave systems are wideband low-noise amplifiers, clock recovery, retiming, regeneration circuits, multiplexers, demultiplexers, and modulation drivers. Silicon logic and linear circuits, and GaAs FET low noise amplifiers are used in current systems. In the future as data rates are increased to 2 Gbit/s and above, GaAs logic circuits and MMICS will play an important role in digital and linear applications. Indeed 8 : 1 and 12 : 1 GaAs multiplexers and demultiplexers are available for 1-Gbit/s systems. Circuit boards that in the past have adhered to UHF design standards will be replaced by full microwave designs. Monolithic circuits that integrate photodetectors and wideband amplifiers will be applied in receivers.

VII. Optical Integrated Circuits

Extensive research has been conducted on OIC's over the past decade. These circuits employ various forms of the slab guide described in Section III. Both hybrid OIC's on $LiNbO_3$ substrates and monolithic circuits on GaAs substrates have been developed. Material doping is used to achieve index of refraction discontinuities for guidance. Work is currently underway on InP-based OIC's for longer wavelength applications. Directional couplers, hybrids, switches, modulators, integrated p-i-n FET's, and laser arrays for high power sources have been investigated. These above functions all resemble microwave counterparts. Continuing efforts in this area should benefit from contributions by microwave workers.

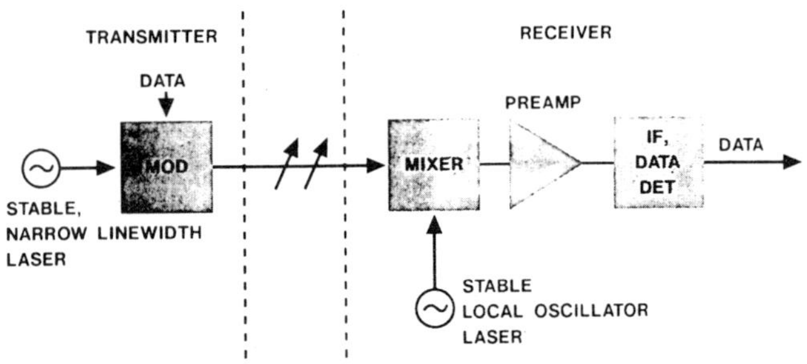

Fig. 5. Coherent detection lightwave system.

VIII. Coherent Detection

The systems dicussed in earlier parts of this paper utilize direct detection or direct conversion of optical photons to a baseband signal. As mentioned above, the resulting performance is 20 dB worse than the quantum limit. These systems are analogous to a radio utilizing an envelope detector. Many years ago, heterodyne radios were introduced that considerably improved sensitivity. Coherent detection is the lightwave heterodyne system and is illustrated in Fig. 5.

Considerable research activity is underway to develop coherent lightwave systems and much progress has been made, but many problems remain to be resolved before practical application takes place. At this point, 1-Gbit/s transmission in a coherent system has been reported over a 148-km link at 1.5-μm wavelength [13]. The signal sensitivity represents more than 7-dB improvement over direct detection and is about 11 dB short of the quantum limit.

The coherent system is structured analogously to a microwave radio with a front end mixer that heterodynes (or homodynes) the received signal to an IF signal where it is amplified and prepared for further processing. In addition, higher level modulation schemes such as PSK or FSK can be employed in place of the simple two-level ASK used in current lightwave systems. The result, reported above, utilized DPSK.

The problems [14] that must be solved in the future include the single-mode fiber polarization degeneracy, laser phase noise, and stability.

The single-mode fiber can propagate lightwaves in either of two polarizations. Conversion can occur with very small temperature changes or other effects in a random manner. Single polarization fibers (elliptical cross section) have not yet produced the low-loss performance of circular fibers. In order to avoid random polarization differences between the LO and the signal, means of tracking and correcting polarization problems must be used.

Noise caused by spontaneous emissions in the laser cavity considerably broadens the source linewidth (5–100 MHz) in today's DFB lasers. This results in performance degradation, particularly during PSK modulation. External cavities to narrow the bandwidth can result in substantial reduction. In the results reported above, the linewidth was narrowed to 5 kHz. However, this is still orders of magnitude worse than the linewidth of typical phase-locked loop stabilized microwave sources.

Laser frequency stability is also critically important to insure good tracking between the signal and the local oscillator. Both the noise and stability problems may be solved through the use of phase-locked loops at optical frequencies.

Experience gained from the design of digital and analog microwave radio should be directly applicable to the heterodyne optical system. The lightwave system using fiber as a transmission media will not suffer fading in the traditional sense of a microwave radio, but the concepts of multipath countermeasures developed for digital radio may well be applied to treating random polarization conversion in the single-mode fiber.

We have not addressed line of sight transmission of lightwave signals in space or in the atmosphere in this paper, but these are also areas where techniques from microwave systems can be applied.

IX. Additional Applications

The topics addressed to this point concerned the use of microwave techniques in lightwave systems. In this section, the application of lightwave techniques in microwave systems is considered. The discussion includes waveguide replacement, phased away radar master oscillator manifolds, and radio entrance links.

The transmission loss for microwaves transported on lightwave through optical fiber is considerably less than the loss of microwaves through a coaxial or waveguide transmission line as shown in Fig. 6. Thus an optical link is an interesting alternative to standard transmission lines. The weight of the fiber line will be less than alternate systems, and it is not subject to EMI considerations. However, one must also consider the cost of the links. In the all-microwave system, a direct interface is used between the microwave equipment and the transmission line. In the optical case, the microwave signal is used to modulate an optical source at the transmitter which is then interfaced to the fiber. Similarly an optical detector must be used at the receiver. Thus, the total insertion loss of the optical link including the electrooptic transducers must be taken into account. This loss may be in the 0–10 dB range depending on the system design and devices used. In cases where weight, size, etc., are dominant, cost may not be an issue, but it well may be in other cases. Further, power level considerations in short fiber runs may be important to avoid detector overload or burn out. Noise added by the optical units must also be taken into account before selecting an approach. The optical receiver may be simplified by radiating a GaAs FET directly with 0.9-μm light or an InP based FET with 1.3-μm light and utilizing the FET as both a photodetector and microwave amplifier.

An excellent opportunity for lightwave microwave transmission lines is in distribution of master oscillator signals to the elements of an active phased array radar as shown in Fig. 7. Considerable work is underway on this application with large potential being forecast [20]. The

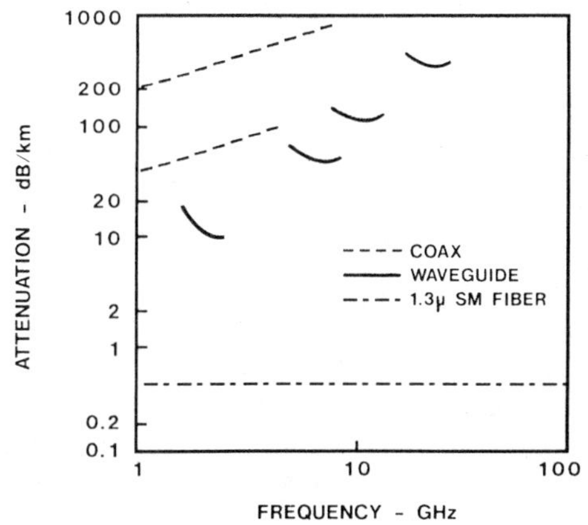

Fig. 6. Waveguide loss as a function of frequency.

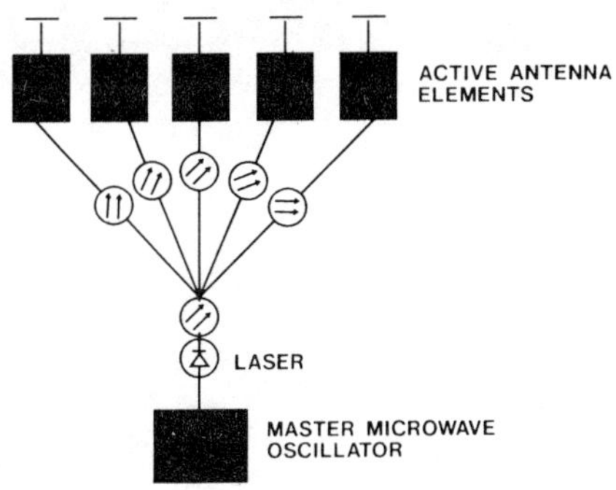

Fig. 7. Fiber manifold for phased-array radar.

branching circuitry for the manifold may be a good application for optical integrated circuits.

Another potential application for microwave modulated lightwaves is for microwave system entrance links. Lightwave entrance links may be operated by demodulating the microwave signal to baseband, or by down converting to IF for application to the lightwave entrance link. An alternate approach would be to place an LNA at the microwave antenna and amplifying the signal to the level required for modulating a laser without requiring any heterodyning or demodulation at the microwave terminal, but down conversion is required at the far end termination of the lightwave link. Once again, cost and added noise must be considered. The latter approach could also make use of direct detection in an FET amplifier at the receiver. The entrance links may be directly integrated into the microwave radio.

Future applications of lightwave devices in microwave systems and applications will involve the use of optical signal processing, but this area is beyond the scope of this paper.

X. Conclusion

The microwave and lightwave technologies are merging through the application of microwave techniques for high-speed lightwave transmission, and also for lightwave elements that will play a role in future microwave systems. This trend will continue for the forseeable future.

References

[1] H. Plasse, "Users perspective emphasized at 8th Newport Conference on Fiber Optics," *Laser Focus/Electrooptics*, pp. 62–63, Jan. 1986.

[2] N. A. Olsson, J. Hegarty *et al.*, "68.3 transmission with 1.37-TB/km capacity using wavelength division multiplexing of ten single frequency lasers at 1.5 μm," *Electron. Lett.*, vol. 21, no. 3, pp. 105–106, Jan. 1985.

[3] K. Y. Lau, N. Bar-Chaim, and I. Ury, "Wideband laser diodes spark new designs," *Microwaves and RF*, pp. 109–116, Nov. 1984.

[4] J. E. Bowers, B. R. Hemenway *et al.*, "26.5-GHz bandwidth InGaAsP lasers with tight optical confinement," *Electron. Lett.*, vol. 21, no. 23, pp. 1090–1091, Nov. 1985.

[5] J. E. Bowers, C. A. Burrus, and R. J. McCoy, "InGaAs p-i-n photodetectors with modulation response to millimeter wavelengths," *Electron. Lett.*, vol. 21, no. 18, pp. 812–814, Aug. 1985.

[6] S. E. Miller and A. G. Chynoweth, *Optical Fiber Communications*. New York: Academic, pp. 38–98, 1979.

[7] D. Gloge, "Weakly guiding fibers," *Appl. Opt.*, vol. 10, pp. 2252–2258, 1972.

[8] D. Marcuse, *Theory of Dielectric Optical Waveguides*. New York: Academic, 1974.

[9] C. B. Su, V. Lanzisera *et al.*, "15-GHz direct modulation bandwidth of vapor-phase regrown 1.3-μm InGaAsP buried-heterostructure lasers under CW operation at room temperature," *Electron. Lett.*, vol. 21, no. 13, pp. 577–579, June 1985.

[10] K. Y. Lau and A. Yariv, "Ultra-high speed semiconductor lasers," *IEEE J. Quantum Electron.*, vol. QE-21, no. 2, pp. 121–138, Feb. 1985.

[11] J. E. Bowers, "Millimeter-wave response of InGaAsP lasers," *Electron. Lett.*, vol. 21, no. 25/26, pp. 1195–1197, Dec. 1985.

[12] S. Y. Wang, D. M. Bloom, and D. M. Collins, "20-GHz bandwidth GaAs photodiode," *Appl. Phys. Lett.*, vol. 42, no. 2, pp. 190–192, Jan. 1983.

[13] M. Kachmar, "Coherent detection: Where light meets radio," *Microwaves and RF*, pp. 34–38, Feb. 1986.

[14] J. Salz, "Coherent lightwave communications," *AT&T Tech. J.*, vol. 64, no. 10, pp. 2153–2210, Dec. 1985.

[15] L. C. Blank, L. Bickers, and S. D. Walker, "Long span optical transmission experiments at 34 and 140 Mbit/s," *J. Lightwave Technol.*, vol. LT-3, no. 5, pp. 1017–1026, Oct. 1985.

[16] V. J. Mazurczyk *et al.*, "420-Mbit/s transmission through 203 km using very loss fiber and a DFB laser," in *Dig. ECOC 84* (Stuttgart, W. Germany), Sept. 1984.

[17] S. Korotky, G. Eisensteom *et al.*, "4-Gbit/s transmission experiment over 117 km of optical fiber using a Ti : $LiNbO_3$ external modulator," *J. Lightwave Technol.*, vol. LT-3, no. 5, pp. 1027–1030, Oct. 1985.

[18] A. H. Gnauck, B. L. Kasper *et al.*, "4-Gbit/s transmission over 103 km of optical fiber using a novel electronic multiplexer–demultiplexer," *J. Lightwave Technol.*, vol. LT-3, no. 5, pp. 1032–1034, Oct. 1985.

[19] A. H. Gnauck *et al.*, "Information bandwidth-limited transmission at 8 Gbit/s over 68.3 km of optical fiber," in *Dig. OFC 86 Post Deadline Papers* (Atlanta, GA), pap. PD9-1, Feb. 1986.

[20] B. Mitchell, "British Labs promote fiber for phased arrays," *Microwaves and RF*, pp. 39–42, Apr. 1986.

*

Harold Sobol (S'57–M'59–SM'69–F'73) received the B.S.E.E. degree from City College of New York and the M.S.E. and the Ph.D. degrees from the University of Michigan, Ann Arbor, MI.

He has worked on radar, missile guidance, superconducting devices, microwave tubes, plasmas, solid-state devices, microwave integrated circuits, and communication systems. He was with the University of Michigan's Willow Run and Electron Physics Laboratories from 1952–1959, IBM Watson Laboratory from 1960–1962, and

RCA from 1962–1973. He joined Collins Radio, Dallas, TX, in 1973 and was Director of Product Development of the Collins Transmission Systems Division of Rockwell International until 1985. He is currently Vice President of Engineering and Advanced Technology for Rockwell Telecommunications.

Dr. Sobol was first elected to the IEEE Microwave Theory and Techniques Society Ad Com in 1972 and has served as Chairman Technical Committees, Vice President, and was elected President for 1978. He served as General Chairman of 1974 Electronic Components Conference and of 1975 International Solid-State Circuits Conference. He has been a member of the Program Committees of the above and of the MTT Symposium since 1968. He was Vice-Program Chairman of 1973 IEEE Intercon. He was IEEE Microwave National Lecturer in 1970. He was the recipient of an IR-100 award in 1969 for his work on microwave integrated circuits and received the Dallas IEEE Section Award for Outstanding Engineer of 1975, an IEEE Centennial Medal in 1984, and the MTT-S Distinguished Service Award in 1986.

Chapter 2
Fibers, Cables, Connectors, and Splicing Techniques

The ability of industry to devise economical methods of producing large quantities of doped silica fiber with low optical loss, high bandwidth, and excellent mechanical properties has been a key factor in the emergence of the fiber optics industry. Of equal importance has been the development of practical methods for protecting the fibers in cables and for joining them with demountable connectors and permanent splices. A general overview of the topics covered in this section is given by **K. Inada** in "**Basic Components and Fiber Optic Passive Components: Status and Trends in Japan.**" The fabrication of fibers is reviewed in "**Optical Fiber Manufacturing Techniques,**" by **D.P. Jablonowski, U.C. Paek, and L.S. Watkins.** The use of carbon coatings to protect fibers from the environment is described by **K.E. Lu, G.S. Glaesemann, R.V. Vandewoestine, and G. Kar** in "**Recent Developments in Hermetically Coated Optical Fiber.**" Experience with installed cables is summarized in "**Reliability of Optical Fibers, Cables, and Splices,**" by **O.I. Szentesi.** The performance of a fiber cable in the undersea environment is described by **Y. Negishi, Y. Miyajima, O. Kawata, and N. Yoshizawa** in "**First Sea Trial of 1.5-μm Submarine Optical Fiber Cable.**" Permanent and demountable fiber interconnection methods are reviewed by **J.M. Anderson, D.R. Frey, and C.M. Miller** in "**Lightwave Splicing and Connector Technology.**"

Basic Components and Fiber Optic Passive Components: Status and Trends in Japan

KOICHI INADA, MEMBER, IEEE

Abstract—This paper reviews the status and trends of Japanese activities in fiber optic passive components. It covers the fiber manufacturing techniques (VAD method), several special fibers such as polarization-holding fibers, dispersion-shifted fibers, infrared fibers, metal coated fibers, and hydrogen-induced loss problems, as well as various kinds of fiber cables such as trunk communication cable, subscriber cable, submarine cable, and optical fiber composite overhead ground wire. The paper also discusses the problems of splicing, connecting, and coupling with respect to future trends and provides information on progress in Japan in the expanding field of communication technique.

I. Introduction

RESEARCH and development of fiber optic passive components were started in the early 1970's. At that time, various kinds of optical fibers attracted the attention of investigations, such as multicompound-glass fibers, silicone-clad silica-core fibers, liquid-core fibers, plastic fibers, as well as silica fibers. In 1974, a low-loss silicone-clad silica-core fiber was developed [1]. The loss was as low as 2.4 dB/km at 0.85 μm. At present, however, because of the low bandwidth, the usage of the silicone-clad silicone-core fiber is only for short-distance computer systems.

In 1976, an silica-based fiber with a minimum loss of 0.47 dB/km at 1.2 μm was manufactured by using the MCVD method [2]. Some reasons for the achievement of the low-loss fibers were the improvement of fiber structure design to have a very thick cladding and the adoption of low-modulus silicone resin as the fiber coating to reduce the microbending loss. After that, silicone-coated fibers became the standard structure of silica-based optical fibers.

In 1977, Ibaraki ECL of NTT invented the VAD method, which made a porous preform in the axial direction with vapor phase materials [3]. The transmission loss reported at the IOOC'77 conference in Tokyo was 5 dB/km. However, much attention was paid from the viewpoint of mass production and economy because of the possibility of continuous fabrication of high-silica-fiber preforms with a high deposition rate. The VAD soot preform has a high OH content because the soot is formed in an oxy-hydrogen flame. Several dehydration techniques have been developed, and in 1980, an ultimately low-OH content fiber was achieved with 0.04 dB/km of OH absorption loss at 1.39 μm, and the residual OH content was estimated less than 1 ppb [4].

Actual large demand for fibers started in 1982. The total installed length of fibers in Japan in the last two or three years would be more than 200 000 or 300 000 km. Most of these fibers seem to be manufactured by the VAD technique. Recently the main interest in fiber manufacturing technology is moving somewhat from the low-loss fiber to low-cost fibers, and also to some special fibers, for example, high-strength fibers, dispersion-shifted fibers, polarization-holding fibers, silica-core fibers, metal-coated fibers, and plastic fibers.

Most Japanese cable companies started the research and development of fiber cabling technique by using their own manufactured fibers. Therefore, they modified the fiber structures to make the best cable structure. The typical cable structures they were interested in were the tight structures instead of loose structures. At present, many types of optical cables have been put to practical use with a tight structure, except in a few applications. For example, trunk cable, junction cable, subscriber cable, submarine cable, optical fiber composite ground wire, television camera cable, power wire composite fiber cable, cab-tire cable, elevator cable, ship cable, LAN cable, and computer cable are all made with a tight structure.

In the first stage of the development of fiber splicing technique, two different methods were extensively studied. One was arc fusion with a prefusion process [5]. The other was the V-groove splicing method. Both methods were tested in the first field trial of fiber cable by NTT in 1978 [6]. Although the average splicing losses for both were 0.2 dB in about 2100 total splicing points, the arc fusion was better in reliability and splicing time. As a result, no V-groove splicing has been used in the field since 1979.

In 1981 a computer-controlled automatic arc fusion splicer was developed and used in NTT's second field trial [7]. After that, the machine has been further improved to have a small body and light weight.

Now, the main items in the development of splicing techniques are automatic low-loss splicing techniques for single-mode fibers, high-strength splicing techniques, and cost reduction of splicing machines.

Manuscript received February 17, 1985; revised March 3, 1986.
The author is with Fujikura Ltd., Chiba-ken 285, Japan.
IEEE Log Number 8608450.

In the first stage of the connector development, two different types of the connectors were studied. One was the core adjustment method with double-cylindrical ferrules. The other was to use a ferrule with a precise dimension. At present, quite a high-quality ferrule is commercially available. Therefore, there is no need to use the adjustable-type connector, even for single-mode fibers.

This paper outlines the status and trend of several important technologies on fibers, cables, splicers, connectors, and couplers.

II. VAD Technique

One of the advantages of the VAD technique is high-rate deposition of a large-sized preform. The maximum deposition rate of 6 g/min was achieved in single-mode fibers and 4–5 g/min in multimode fibers [8], [9]. Fig. 1 shows the schematic diagram of soot preform manufacturing by the VAD method. After making the soot, it is dehydrated and consolidated. Fig. 2 shows the making of the high deposition rate preform with multiburners by the VAD technique. The so-called depressed-cladding single-mode fibers can also be made by the VAD method [10], [11]. However, the single-mode fibers are substantially more insensitive to microbendings than multimode fibers. Therefore, there is little necessity for the depressed-cladding fibers in Japan so far as the cable structures are identical for multimode fibers and single-mode fibers.

In the near future, multimode fibers for the subscriber loop will have a large market. The most important key to applying the fibers for the subscriber loop will be the cost reduction of the system. Therefore, the higher deposition rate and the larger preform manufacturing technique must be established.

III. Miscellaneous Fibers

Extensive development of manufacturing technique of polarization-holding fibers was started in 1980 [12]. At present, the fiber is strongly needed for fiber sensor applications such as the fiber gyroscope, acoustic fiber sensor, electric field sensor, or heterodyne transmission system, although the scale of the demand for the fibers is still not large. The types of polarization-holding fibers can be divided into two categories; one is an elliptical cladding type [13], and the other is a stress-induced type [14], as shown in Fig. 3. About −32 dB of the polarization-mode crosstalk in a 1 km length was achieved with a transmission loss of 0.35 dB/km at 1.54 μm [15]. This crosstalk, however, is not sufficiently low for the coherent transmission system because the repeater spacing should be quite long to make the merit of the system. Twisted-type fibers have also been developed for the heterodyne system by twisting the fibers in the nylon extrusion process [16].

The dispersion-shifted fibers will have the real advantages for a long-distance transmission system because of the low loss characteristic. The first report on this

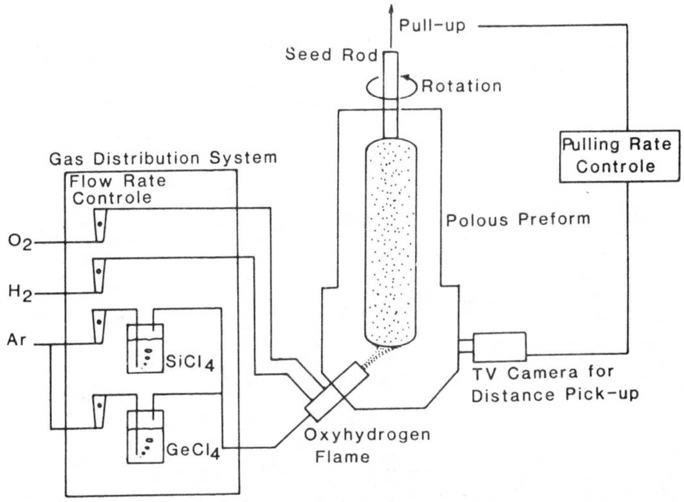

Fig. 1. Schematic diagram of soot preform making by the VAD method.

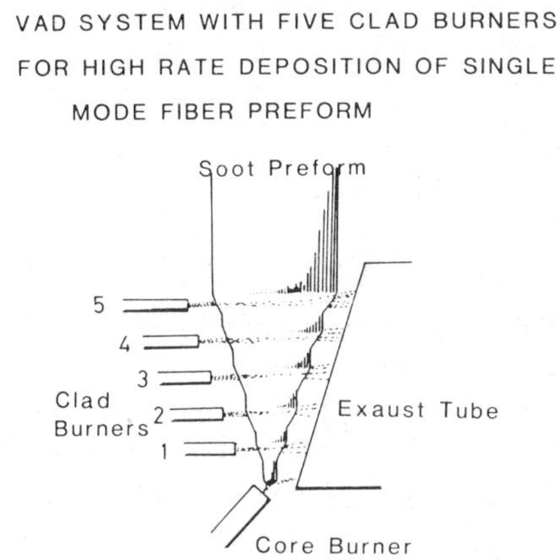

Fig. 2. High-speed deposition with multiburners by the VAD technique.

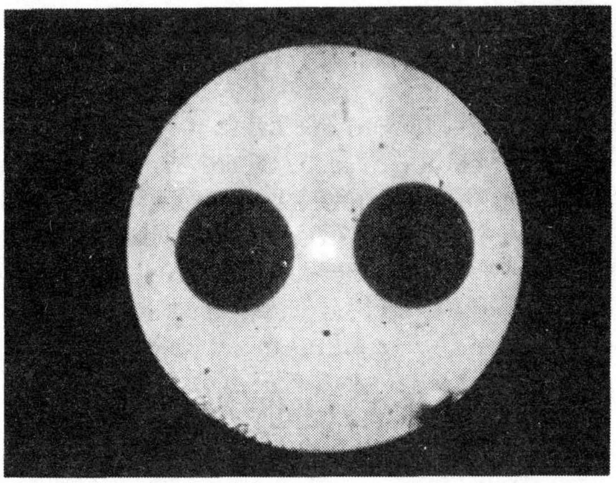

Fig. 3. Stress-induced "Panda-type" polarization-holding single-mode fiber.

fiber was done in 1982, although the loss was about 0.32 dB/km at 1.55 μm [17]. Recently, the lower-loss fibers have been made with a triangular profile by the VAD method.

So far as the silica-based fibers are used, the minimum loss is about 0.2 dB/km at 1.55 μm. In order to overcome this loss limit, several kinds of infrared fibers with new materials are now under development. The minimum loss of 6.3 dB/km was achieved at 2.13 μm for the fibers made by fluorozirconate glass with built-in casting method [18]. It is difficult to estimate when the fibers will really be used for long-distance communication. For the moment, the manufactured length of the fibers is not so long, the strength is not so high, and the material cost is quite expensive because the purification is not easy. Some breakthroughs in the techniques are definitely required.

Metal-coated fibers have been developed for the application of the fibers in a very-high-temperature environment such as 200 or 350°C [19]. The fibers are coated with aluminum with a dipping method.

IV. Hydrogen-Induced Loss

The first report on the hydrogen-induced loss was made in 1983 on the basis of the experimental research [20], [21]. The observed phenomena was that the loss of fibers in cable was increased in the long-wavelength region. The direct cause was found to be the hydrogen generated by the electrolysis of two different kinds of metals or from the silicone resin of the primary coating. The hydrogen reacts with phosphorus doped in the core, and the loss in the long-wavelength region is increased by the P–OH. Therefore, the easiest and the most effective method of avoiding the hydrogen effect has proved to be the elimination of phosphorus in fibers, silicone resin in coating, and metals in cable. At present, in Japan, the amount of phosphorus has been sufficiently reduced so as not to increase the loss for the next 20 years in ordinary applications.

V. Standard and Subscriber Cable

In 1977 the first field trial for optical cables was carried out by NTT in the center of Tokyo where fiber cable with 1000 km fiber length was installed. One of the purposes of the experiment was to find out the best cable structure [6]. Three types of cables were installed. As a result, it has been confirmed that the fiber structure and cable structure in Fig. 4 are most suitable from the viewpoint of long-time stability [22]. That is, the fiber has a silicone buffer coating and a nylon coating, and six fibers are stranded around a central member. The stability of this cable structure was confirmed by the subsequently conducted second field trial for VAD fibers and by the third field trial for single-mode fibers. After that, it was decided that this tightly stranded cable structure was to be one of the standard structures in Japan. At present, this type of cable has been installed through Japanese archipelagos for trunk and junction cables.

A field trial of the optical fiber subscriber network system was started at Mitaka, Musashino, and the Metropolitan area of Tokyo and still continues. The systems are intended to compose the information network system for ordinary subscribers.

In the future, the maximum number of fibers in a cable will exceed several thousand. On the other hand, the cost of the cable must be low enough for most subscribers to have an interest in the system. For the purpose of the reduction of cost and for easier splicing, a five-fiber-tape-type cable has been adopted for numbers of fibers exceeding 200 [23]. The cable structure is shown in Fig. 5 [24]. A distribution cable with a small number of fibers is also required in a subscriber system. In this case, the number of fibers in a unit is designed to be five or ten. The actual application of subscriber cable has recently been started. Therefore, the installed length is not yet large. In the near future, however, the demand will be quite increased if the cost of cable can be reduced. In order to achieve it, the cable structure might be changed.

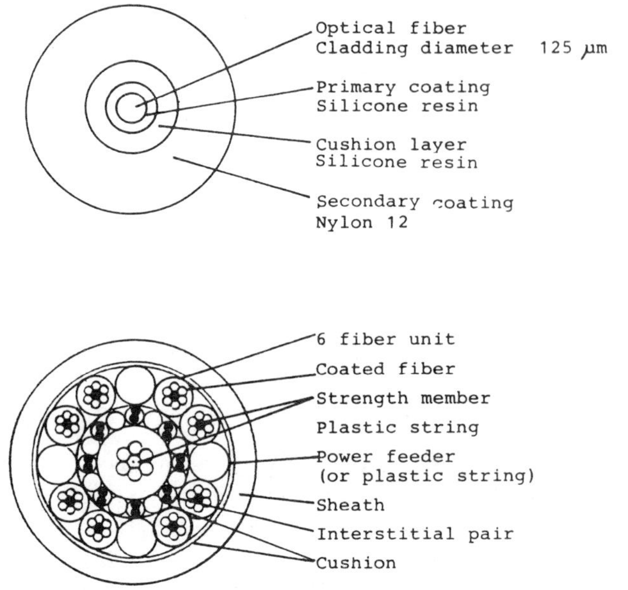

Fig. 4. Standard fiber and cable structures.

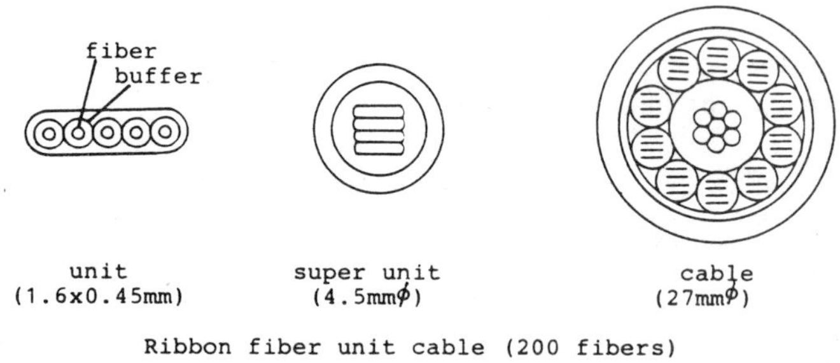

Fig. 5. Subscriber cable with 200 fibers.

VI. Submarine Cable

The maximum sea depth exceeds 8000 m in some areas around Japan. The submarine cable is attacked by the high water pressure and high tension in installation and restoration.

A submarine cable requires quite high-strength fibers as well as very low transmission loss in order to increase the reliability and reduce the number of repeaters. The primary key to making high-strength fibers is to use high-quality drawing furnace [25]. With this technique, an average survival length of 7 km has been achieved with a 2 percent

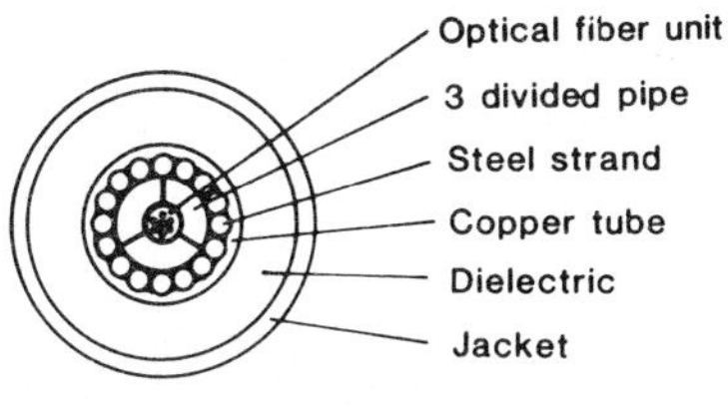

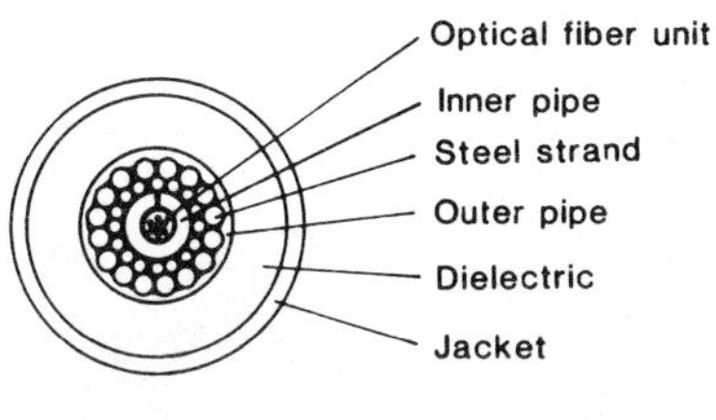

Fig. 6. Submarine cable structure.

proof test [26]. The length of a cable is expected to be about 55 km. Therefore, the fibers must be spliced with sufficiently high strength. The required proof test level for the splicing is more than 3 percent. It can be achieved by using an arc fusion splicing technique in a clean environment so that dusts may not attack the fiber surface [27].

The submarine cable is covered with a pressure-resistant copper pipe, tensile-strength steel wires, and a polyethylene sheath. Fig. 6 shows the typical submarine structures [28], [29].

In a shallow sea area, the cable is usually armored with steel wires to prevent the cable from attacks of troll or anchors of ships.

VII. Optical Fiber Composite Overhead Ground Wire (OPT-GW)

Long-distance power transmission lines have been widely installed in Japan. The line normally has an overhead ground wire on the top of the tower in order to protect the power line from lightning. The idea to using the OPT-GW is to build up a telecommunication system utilizing these already existing power lines. Recently, many Japanese power companies started the installation of OPT-GW. The environmental condition, however, is severer than that of any other land cable. For example, the OPT-GW is, in an electrical accident, instantly heated up to more than 350° C, and the cable is elongated by 0.6 percent due to thermal expansion. In winter it suffers from heavy snow and strong wind. The strength of the fibers must be sufficiently high. The fibers must be cabled with heat-resistant materials in order to endure these severe conditions [30]. Fig. 7 shows an example of an OPT-GW which will withstand these severe conditions. The first installation in Japan was in 1982 by Kansai Electric Power Company, and the system is operating well without any trouble [31].

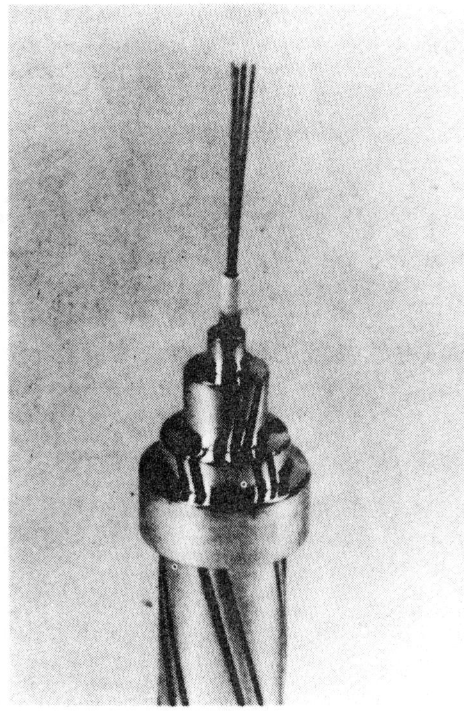

Fig. 7. Conventionally used OPT-GW.

VIII. Splicing

Multimode fiber splicing is much easier than that of single-mode fibers. Therefore, the multimode fiber splicing machine is simpler and less expensive than that of single-mode fibers. Moreover, at present the amount of installation of multimode fibers is much less than that of single-mode fibers. Accordingly, splicing technology of multimode fibers is now reviewed from the economical viewpoint, that is, short-time fiber cable end preparation, easy splicing with a light-weight small body machine, and short-time cable sheath jointing.

Precise core axis alignment is required for single-mode fiber splicing if the fibers have some eccentricities or outer diameter variations. In other words, it may be better if the machine can splice with a low loss, even for the fibers with a large core eccentricity or large outer-diameter variation. Fig. 8 shows the commonly used single-mode fiber splicing system in Japan [32]. The core axes are automatically aligned with a computer by monitoring the transmitted power so that the transmitted power is maximized. However, this system needs a light source in a telephone office, and also needs a power monitor in the far end.

To avoid this complication, the local power monitoring method has been developed. It can align the cores by bending the fiber sharply near the machine and detecting the leaked power from the cladding. A splicing loss of 0.09 dB on the average was obtained in the field [33].

If the core positions are directly observed, the fibers are aligned without using a light source and a detector. One of the methods is to observe the germanium fluorescence in the core under the UV ray [34]. The other is to observe the core's shadow under ordinary white light [35]. Fig. 9 shows

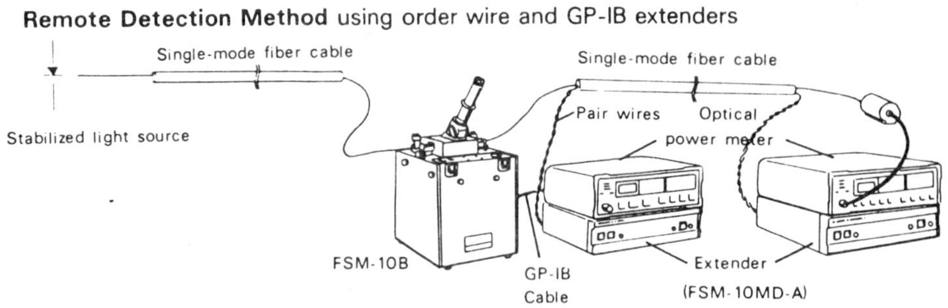

Fig. 8. Conventionally used single-mode fiber splicing machine.

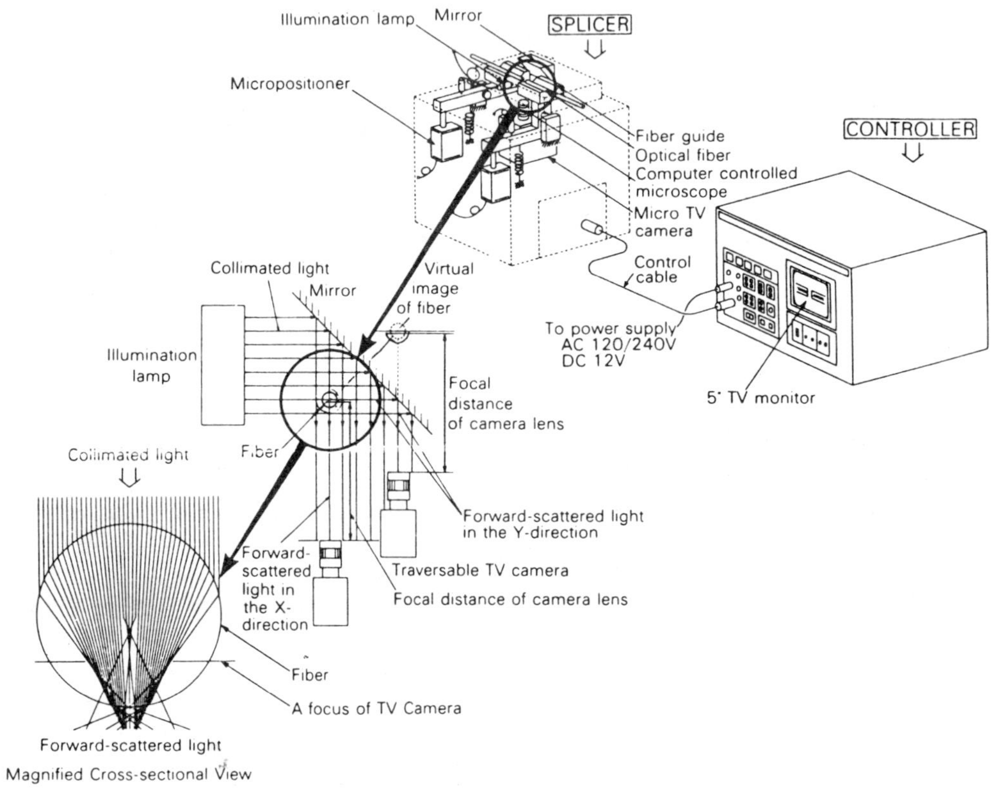

Fig. 9. Direct core observation type single-mode fiber splicing machine.

the machine designed by improving the latter method. It finds the centers of cores automatically with an installed video analyzer and automatically aligns the cores, splices the fibers, and estimates the splicing loss by the installed computer. The splicing loss is about 0.05 dB on the average. The machine can splice almost any kind of fibers, such as matched-cladding fibers, depressed-cladding fibers, pure-silica-core fibers, and dispersion-shifted fibers [36]. The mechanical strength of the spliced point is higher than that of the fiber spliced with an ordinal arc fusion machine because the machine can apply a long arc fusion time by applying an intentional core axis shift to show a minimum loss and maximum strength.

After the splicing, the spliced portion is protected with a reinforcement member, a hot melt inner tube, and a heat-shrinkable outer tube. Fig. 10 shows the standard protection method for the splicing point.

For the splicing of five-fiber-tape subscriber cable, the arc fusion technique is also applied [37]. The average splicing loss is 0.55 dB. This loss is almost similar to that of a single-fiber splicing. Recently, a ten-fiber-tape splicing machine has also been developed.

For future application, a fully automated splicing robot has been developed, although it is on only a laboratory scale. Almost all the processes, such as removing the nylon jacket, cutting the fiber ends, fiber alignment, arc fusion splicing, and reinforcement, can be done automatically. The splicing loss is 0.07 dB on the average [38].

IX. Connector

The FC-type connector is most commonly used in Japan. The ferrule is made of a ceramic capillary and a stainless pipe with very small tolerances [39]. Fig. 11 shows the structure and dimension of the connector. A connector made of plastic has been also developed in order to reduce the cost. The connector is compatible with the FC-type connector, and its polishing work is more speedy [40].

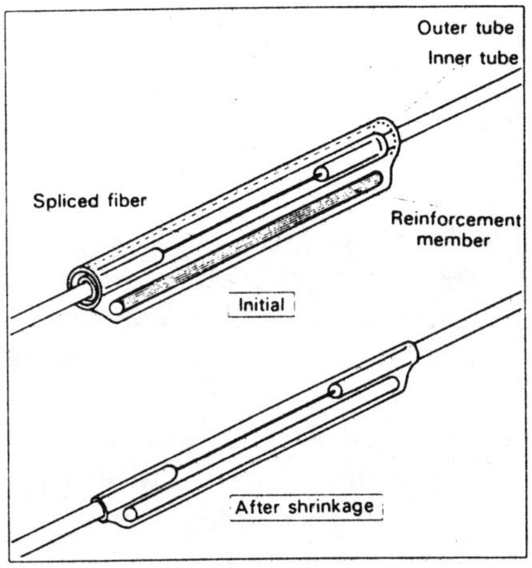

Fig. 10. Standard protection method for the splicing point.

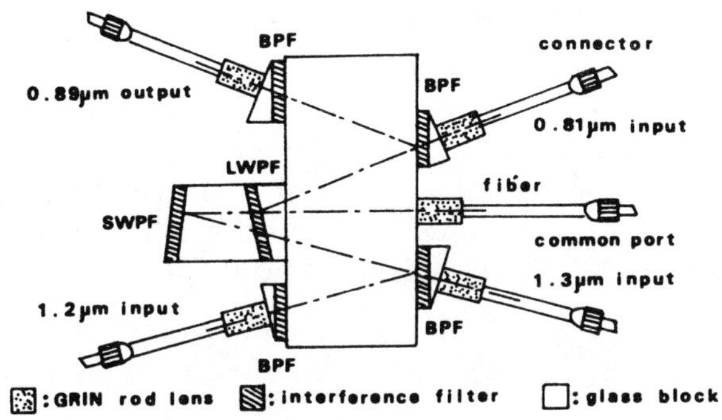

Fig. 13. Wavelength multi/demultiplexer.

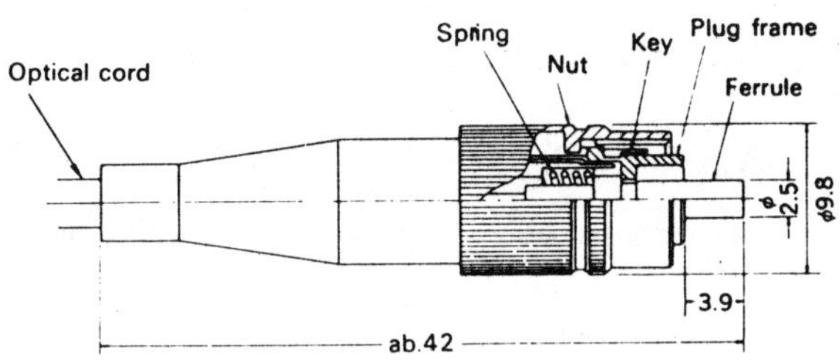

Fig. 11. Standard FC connector for multimode and single-mode fiber.

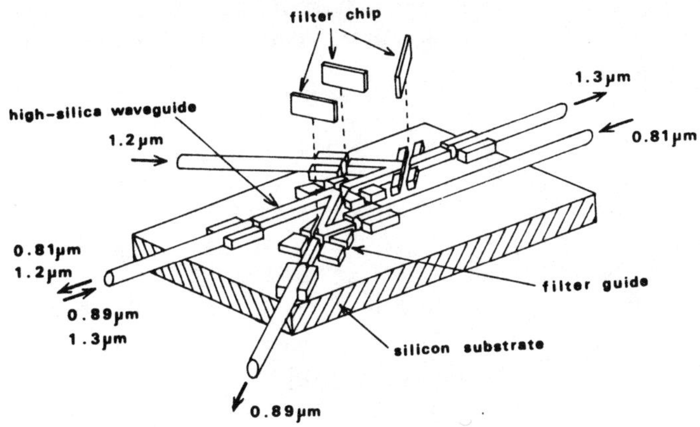

Fig. 14. Integrated circuit type multi/demultiplexer.

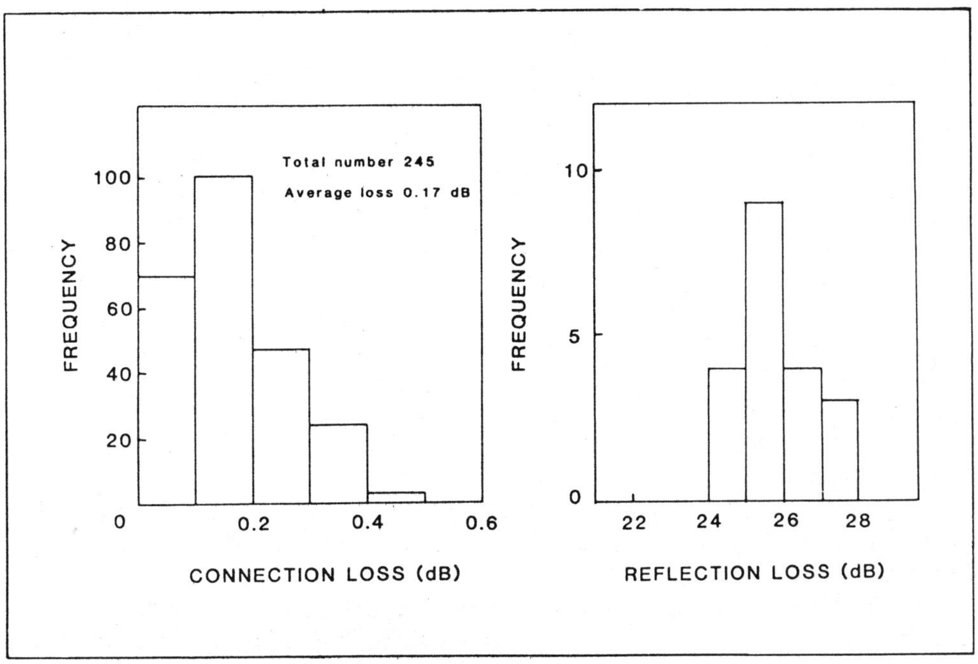

Fig. 12. Histogram of typical connection loss and reflection loss of the standard multimode fibers.

Recently, the end surface of the FC-type connector has been lapped hemispherically so that the end surface of the fiber can touch directly the other end surface of the mated fiber. The connection loss and reflection quality are improved by removing the Fresnel reflection. The connector is called the FC-type connector [41]. Histograms of typical connection loss and reflection loss for the standard multimode fibers are shown in Fig. 12.

Five-fiber-tape cable for a subscriber system requires the multifiber connector [42]. The connector is made by injecting the plastic resin into the precisely manufactured molding machine. The average connection loss is 0.14 dB with matching oil.

Recently, the ferrule-type connector for a ten-fiber tape has been developed for the future applications [43]. This is also made by molding the resin.

There are more than ten types of connectors actually used for silica-based fibers. Now standardization is strongly requested.

X. Couplers

Several kinds of fiber optic components such as directional couplers, star couplers, splitters, wavelength multi/demultiplexers, switches, isolators, circulators, and modulators are required for optical communication systems. Among these components, wavelength multi/demultiplexers, star couplers and switches are now widely used. For example, multi/demultiplexers are used for subscriber systems to apply a subscriber fiber for multichannel transmission. Fig. 13 shows the wavelength multiplexer [44]. Star couplers are mainly used for local area networks to connect many terminals, such as 32 or 100 [45].

Changeover of optical paths for the mechanical switch is carried out by moving the prism with electromagnetic force [46]. The rise time of the switch is on the order of milliseconds.

In these components it has been found that the degradation of reliability is mainly caused by the adhesive resin for fixing the optical elements. The reliability has been improved by soldering or welding these elements [47]. In order to apply these components widely to communication systems, improvements in productivity and reliability and cost reduction are strongly required.

Recently, integration of several functions on a monolithic substrate has been extensively studied by employing a semiconductor integrated circuit manufacturing technique. This will improve the reliability and productivity and simplify the assembling process. The hybrid integrated circuit for demultiplexing is shown in Fig. 14 [48]. It can fix the optical fibers and filters precisely to the optical guides and guiding grooves on a silicone substrate. It is expected to simplify the assembling process. A monolithic optical switch has been developed by using the optical effect of PLZT ceramics [49]. The rise time of the switch is about 1 ns.

There are many problems left in the intergation of fiber optic components. However, the real merit of integration will be brought by the combination with the OEIC technique.

XI. Summary

Fundamental components for optical communications have been mostly developed. At present, improvement of the manufacturing techniques for cost reduction is one of the most important items so far as commercial telecommunication is concerned. Hereafter, as many fundamental transmission lines such as pair copper wires, coaxial cables, and millimeter waveguides pursued their development courses, the fundamental optical components will pursue simplicity in structure and various kinds of other components will pursue complexity in function. Besides these, research and development on infrared fibers should be steadily continued for future application.

References

[1] S. Tanaka *et al.*, "Silicone-clad fused-silica-core fiber," *Electron. Lett.*, vol 11, no. 7, Apr. 3, 1975.

[2] M. Horiguchi and H. Osanai, "Spectral losses of low-OH-content optical fibers," *Electron. Lett.*, vol. 12, pp. 310–312, 1976.

[3] T. Izawa, S. Kobayashi, S. Sudo, and F. Hanawa, "Continuous fabrication of high silica fiber preform," presented at the 1977 IOOC, Tokyo, Japan.

[4] T. Moriyama, O. Fukuda, K. Sanada, K. Inada, T. Edahiro, and K. Chida, "Ultimately low-OH content VAD optical fibers," *Electron. Lett.*, vol. 16, pp. 689–699, 1980.

[5] M. Hirai and N. Uchida, "Melt splice of multimode optical fiber with an electric arc," *Electron. Lett.*, vol. 13, 1977.

[6] S. Shimada and K. Masuno, "Short-haul optical fiber transmission system field trial," *Japan Telecommun. Rev. NTT*, vol. 21, no. 2, Apr. 1979.

[7] T. Horiguchi, T. Miki, and H. Ishio, "F-32M and F-100M fiber optic transmission systems," *Japan Telecommun. Rev. NTT*, vol. 21, no. 2, Apr. 1981.

[8] T. Shioda *et al.*, "High speed production of VAD single mode fibers," presented at the IOOC'83, Tokyo, Japan, 1983, paper 27A3-2.

[9] H. Suda *et al.*, "Deposition mechanism of fine glass particles in the VAD process," presented at the 8th ECOC, Cannes, France, Sept. 1982, paper A2-4.

[10] M. Miyamoto, K. Akiyama, T. Shioda, K. Sanada, and O. Fukuda, "Fabrication and transmission characteristics of VAD fluorine doped single mode fibers," in *Proc. ECOC* 1983, paper II-A, p. 9.

[11] T. Miya, M. Nakahara, F. Hanawa, and Y. Ohmori, "Newly developed fluorine doping techniques for VAD process," presented at the 10th ECOC, Stuttgart, West Germany, 1984, paper 14A2.

[12] H. Matsumura, T. Katsuyama, and T. Sugnuma, "Fundamental study of single polarization fibers," in *Proc. 6th ECOC*, 1980, pp. 49–52.

[13] T. Katsuyama *et al.*, "Low-loss single polarization fibers," *Electron. Lett.*, vol. 17, no. 13, pp. 473–474, 1981.

[14] Y. Sasaki, K. Okamoto, T. Hosaka, and N. Shibata, "Polarization maintaining and absorption-reducing fibers," presented at the OFC'82, Phoenix, AZ, 1982.

[15] Y. Sasaki, T. Hosaka, and J. Noda, "Low-loss polarization-maintaining optical fiber with low-crosstalk," *Electron. Lett.*, vol. 21, no. 4, pp. 156–157, 1985.

[16] S. Machida, J. Sakai, and T. Kimura, "Polarization conservation in single mode fibers," *Electron. Lett.*, vol. 17, no. 14, pp. 494–495, 1981.

[17] T. Moriyama, M. Miyamoto, M. Akiyama, *et al.*, "Fabrication of low-loss and long-length VAD fibers for 1.55 μm wavelength region," in *Proc. 8th ECOC*, Cannes, France, Sept. 1982, pp. 31–35.

[18] T. Miyshita and T. Manabe, "Progress in fluoride glass fiber research and development in Japan," presented at the Second Int. Symp. Halide Glasses, 1983.

[19] S. Tanaka, T. Shiota, R. Yamauchi, and K. Inada, "Loss increase characteristics of various kinds of hermetically coated fiber at high temperature," presented at the 10th ECOC, Stuttgart, West Germany, 1984.

[20] N. Uesugi, T. Kuwabara, Y. Koyamada, Y. Ishida, and N. Uchida, "Optical loss increase of phosphor-doped silica fiber at high temperature in the long wavelength region," *Appl. Phys. Lett.*, vol. 43, no. 4, pp. 327–328, Aug. 1983.

[21] K. Mochizuki, Y. Namihira, and H. Yamamoto, "Transmission loss increase in optical fibers due to hydrogen permeation," *Electron. Lett.* vol. 19, no. 18, pp. 743–745, Sept. 1983.

[22] T. Horiguchi, M. Nakahara, T. Tanifuji, K. Omoto, and M. Tokuda, in *Nat. Conf. Rec. Commun., IECE Japan*, 1980, p. 335.

[23] Y. Katsuyama, S. Hatano, K. Hogari, T. Matsumoto, and T. Kokubun, "Single mode optical fiber ribbon cable," *Electron. Lett.*, vol. 21, no. 4, pp. 134–135, 1985.

[24] M. Kawase, M. Tachikura, and F. Nihei, "Mass fusion splices for high density optical fiber units," in *Proc. 8th ECOC*, Cannes, France, Sept. 1982, pp. 348–353.

[25] S. Sakaguchi, M. Nakahara, and Y. Tajima, "Drawing of high-strength long-length optical fibers," presented at the IOOC'83, Tokyo, Japan, 1983, paper 29A4-5.

[26] S. Araki, A. Wada, R. Yamauchi, and O. Fukuda, "Fabrication of high strength long-length fibers," presented at the OFC'84, New Orleans, LA, 1984, paper WF3.

[27] M. Oshawa, T. Yamada, and O. Watanabe, "High tensile strength arc-fusion splice for submarine optical fiber cable and its reliability assurance," in *Nat. Conv. Rec.*, 1984, paper S5-17.

[28] Y. Negishi, K. Ishihara, Y. Murakami, and N. Yoshizawa, "Design of deep sea submarine optical fiber cable," *J. Lightwave Technol.*, vol. LT-2, pp. 839–845, Dec. 1984.

[29] Y. Niiro, "The OS-280 optical-fiber submarine cable system," *J. Lightware Technol.*, vol. LT-2, pp. 767–772, Dec. 1984.

[30] Y. Honma, A. Okazato, K. Abe, T. Hayakawa, T. Kobayashi, and H. Ode, "Composite overhead ground wire with optical fibers," *Fujikura Tech. Rev.* no. 14, pp. 30–41, Jan. 1985.

[31] H. Nabeshima, K. Tsujimoto, T. Kobayashi, O. Watanabe, K. Inada, and A. Okazato, "Development and application of optical fiber composite overhead ground wire (OPT-GW)," in *Proc. 31st IWCS*, Nov. 1982, pp. 45–50.

[32] Y. Toda, M. Ogai, O. Watanabe, and S. Seikai, "Single mode fiber splicing system" *Int. Conf. Commun., 1981*, June 1981, paper 27.7.1-6.

[33] K. Kato and M. Miyauchi "preformance of a arc-fusion splicing apparatus using local power monitoring," in *IECE Nat. Conv. Rec. IECE Japan*, 1985, p. 2229.

[34] K. Tachikura, H. Yamamoto, and M. Nunokawa, "Novel core alignment method for low loss splicing of single mode fibers utilizing UV excited fluorescene of Ge-doped silica core," *Electron. Lett.*, vol. 18, no. 16, Aug. 1982.

[35] O. Kawata, K. Koshino, and K. Ishihara, "Low-loss single mode fiber splicing technique using core direct monitoring," *Electron. Lett.*, vol. 19, Nov. 24, 1983.

[36] T. Onodera, "Splicing characteristics of SM fibers by the direct core detection splicer," in *Nat. Conv. Rec., IECE Japan*, Mar. 1985, p. 2237.

[37] M. Tachikura and N. Kashima, "Fusion mass-splices for optical fibers using high frequency discharge," *J. Lightwave Technol.*, vol. LT-2, pp. 25–31, 1984.

[38] R. Arioka, T. Haebara, and M. Tachikura, "Fully-automatic optical fiber splice machine," in *Proc. IWCS*, 1984, pp. 50–57.

[39] K. Nawata, "Multimode and single-mode fiber connectors technology," *IEEE J. Quantum Electron.*, vol. QE-16, June 1980.

[40] S. Nagasawa *et al.*, "Reliability of sphere insert transfer-molding connector," in *Nat. Conv. Rec., IECE Japan*, Mar. 1984, p. 2234.

[41] N. Suzuki *et al.*, "Convex polished optical fiber connector," in *Nat. Conv. Rec., IECE Japan*, Mar. 1985, p. 2606.

[42] T. Satake *et al.*, "Basic characteristics of 5-fiber plastic connector," in *Nat. Conv. Rec., IECE Japan*, Mar. 1984, p. 2237.

[43] ——, "Design of 10-optical-fiber connector for subscriber cable splicing," in *Nat. Conv. Rec., IECE Japan*, Mar. 1985, p. 2254.

[44] R. Watanabe, K. Sano, and J. Minowa, "Design and preformance of multi/demultiplexer for subscriber loop system," presented at the 4th IOOC, Tokyo, Japan, 1983, paper 30C1-2.

[45] Y. Fujii, N. Suzuki, and J. Minowa, "A 100 input/output-port star coupler composed of low-loss slab-waveguide," presented at the 4th IOOC, Tokyo, Japan, 1983, paper 29C2-4.

[46] Y. Fujii, Y. Nagata, J. Minowa, T. Miki, T. Aoyama, K. Doi, and S. Nonaka, "Optical switching network for optical fiber communication," presented at the 5th ECOC, paper 13.3-1.

[47] K. Asanuma, H. Isono, T. Yokota, K. Terai, and H. Furuta, "4-channel optical fiber by laser welding," in *Nat. Conv. Rec. IECE Japan*, Mar. 1985, pp. 2620.

[48] M. Kawachi, Y. Yamada, M. Yasu, and M. Kobayashi, "High-silica guided-wave optical multi/demultiplexer," in *Nat. Con. Rec., IECE Japan*, Mar. 1985, pp. 1006.

[49] K. Wasa, H. Adachi, T. Kwaguchi, K. Ohji, and K. Setsune, "Optical tire switches using PLZT thin film waveguides on sapphire," presented at the 4th IOOC, Tokyo, Japan, 1983, paper 29C4-1.

Koichi Inada (M'73) was born in Tochigi, Japan, on February 24, 1941. He received the B.S. degree in electrical engineering from Yokohama National University, Japan in 1963 and the Ph.D. degree from the Tokyo Institute of Technology, Tokyo, Japan in 1976.

In 1963, he joined Fujikura Ltd., Tokyo, Japan, where he is currently General Manager of the Optical Fiber, Cable and Transmission System Development Division. He has been engaged in research and development of high-frequency coaxial cables, super conducting cables, millimeter waveguides, and optical fibers.

Dr. Inada is a member of the Institute of Electronics and Communication Engineers of Japan.

OPTICAL FIBER MANUFACTURING TECHNIQUES

Donald P. Jablonowski, Un Chul Paek, and Laurence S. Watkins

AT&T TECHNICAL JOURNAL

Donald P. Jablonowski *is with AT&T Network Systems in Atlanta.* ***Un Chul Paek*** *and* ***Laurence S. Watkins*** *are with the AT&T Engineering Research Center, Princeton, New Jersey. Mr. Jablonowski is department chief, Preform and Draw Process Development. In his department, fabrication techniques for advanced optical fibers are perfected. He received a B.S., M.S., and Ph.D. in electrical engineering from Carnegie-Mellon University. He joined AT&T in 1973. Mr. Paek is a consulting member of the technical staff and a Fellow of AT&T Bell Laboratories. He has specialized in the areas of fiber drawing and coating, fiber strength, and light propagation in fibers. He received a B.S. degree from Korea Merchant Marine Academy and M.S. and Ph.D. degrees in mechanical engineer-*
(continued on page 44)

The cost of fiber has been steadily decreasing because of the experience learned in production and developments in fiber processing. In this paper we will review the processes for making optical fibers, dwelling mostly on the fiber drawing process. Our results show that the fiber draw rate has been increased to more than 10 m/s without any degradation in fiber performance and specifications. The technology is not fully mature and it is likely that even higher rates should be achievable.

Introduction

The cost of fiber has been steadily decreasing since large-scale production was started in 1979. This has been due to both the experience learned in production and a series of major developments in fiber processing and the associated equipment. There is no doubt that this trend will continue for some time, resulting in continued reduction in the cost of fiber. Coupled with this trend has been the continuing demand to maintain and improve the performance specifications of the fiber.

In optical fiber technology there have been significant developments to increase the production rate. The shape and size of preforms depend to some extent on the fabrication technique; current diameters can range between 1 and 2 cm. Further efforts to scale up the preform process are likely to result in diameters in excess of 2 cm.

Since most of the glass in the preform is cladding, current developments are focusing on techniques to produce cladding glass as fast as possible. Parallel developments are required in fiber drawing technologies which will require larger furnaces to accommodate the larger diameter preforms and higher draw speeds to process massive amounts of glass.

In this paper we will review the processes to make optical fibers, dwelling on the fiber drawing process. The modified chemical vapor deposition (MCVD) process to make preforms is briefly described below; a more extensive article dealing with this process may be found in Reference 1.

Figure 1. Schematic of the modified chemical vapor deposition process.

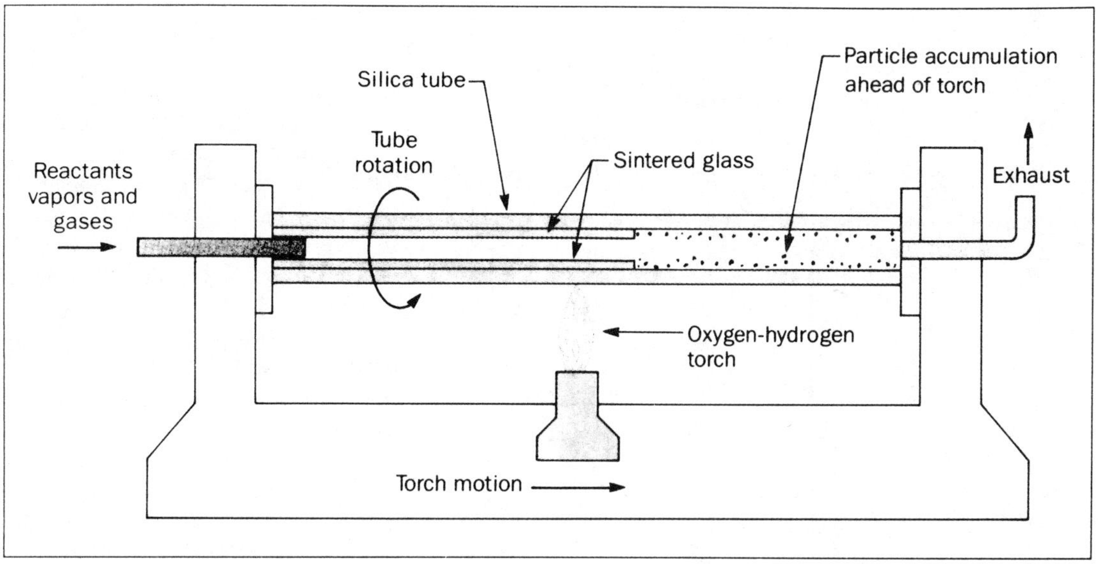

The MCVD Process

The MCVD process (Figure 1) was invented at AT&T Bell Laboratories in the early 1970s.[1-3] High-purity dry vapor of $SiCl_4$ with various combinations of vapors of $GeCl_4$ and fluorinated hydrocarbons are passed through a rotating silica substrate tube along with pure oxygen. An external oxyhydrogen torch heats the tube and the contents react to form silica glass particles doped with the desired combinations of germanium and fluorine. Downstream from the reaction the particles are driven to the tube walls by thermophoretic forces, where they deposit. The moving torch that first caused the reaction to occur then passes over this deposit and sinters it into a thin layer of doped glass. A number of layers are deposited by repeated passes of the torch. The relative proportion of the vapors is adjusted at each pass to provide the desired composition profile and so give the fiber its index profile.

The next step involves heating the quartz tube to a high temperature by slowly moving the same oxyhydrogen torch along the length of the tube. At a high enough temperature, the glass approaches its softening point, and surface tension causes the tube with the deposited glass layers to collapse uniformly into a solid rod called a preform.

The AT&T manufacturing facility was started up in Atlanta in 1979, and by the end of 1986, over 1.7 Mkm of fiber in lightguide cable had been shipped. The MCVD process is exceptionally well-suited for producing high-performance single-mode fibers of various designs. The layer structure and the ability to incorporate a variety of dopants at various concentrations makes it convenient to fabricate designs of any complexity. In addition, the manufacturing equipment is readily adjusted through software modifications to produce any design. Finally, the process is inherently reproducible and leads to tight distributions of fiber properties.[1]

Subsequent to the MCVD process, the preform is inserted into a high-temperature furnace and a fiber is drawn. The standard outside diameter is 125 μm. An in-line operation adds a 60-μm-thick polymer coating to the

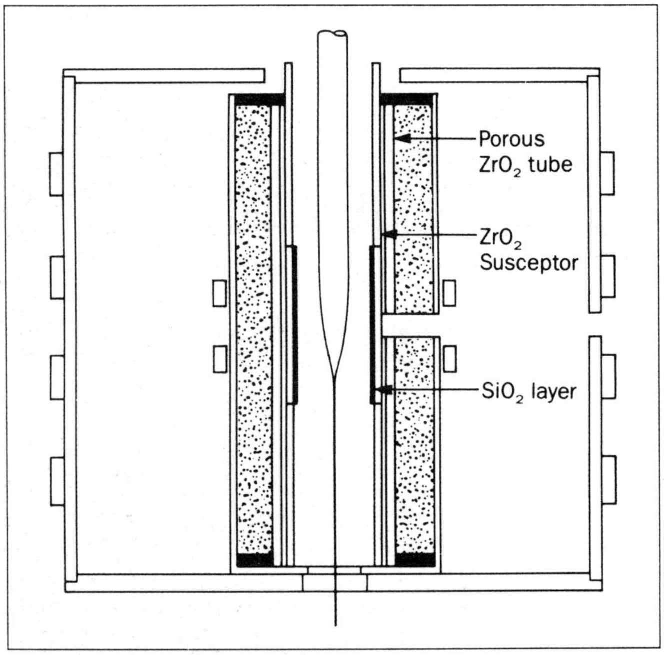

Figure 2. Configuration of the new zirconia induction furnace. The middle piece of the inner tube (susceptor) that is loosely sleeved with a low-density zirconia tube is covered with a thin glassy film.

fiber in order to protect the glass surface from damage. The physical properties of the coating are selected to enhance subsequent cabling and splicing operations and to mitigate microbending deformations that could lead to optical loss. The entire fiber length is tested for strength (proof tested), typically at 50 ksi. The proof test guarantees a minimum level of mechanical integrity for subsequent cabling, installation, and service operations. The local strength in a short length (a few centimeters) is much higher than the proof test level, typically 800 ksi, and this serves to make the fiber easy to handle, for example, in splicing operations. A detailed discussion of the drawing process is presented in the next section.

The Fiber Drawing Process

The fiber drawing process has seen significant developments in recent years. Not only the drawing speed but also furnace size is increasing as larger diameter preforms are being made. At the same time, diligence is required to design furnaces to ensure that they are not sources of particle contamination which will degrade fiber strength.

Coating speeds are already in the 5-m/s range in production and up to 20-m/s speeds have been published in research work.[4] This also requires advances in furnace design to ensure the correct draw tension, fiber cooling techniques, and application developments to properly apply multilayer protective coatings. Finally, it requires dimensional measurement and control systems to produce fiber and coating geometries within very tight specifications.

Zirconia Draw Furnace. Ultra-pure zirconia, fully stabilized with yttria, has desirable properties for use as a high-temperature furnace.[5] In operation, however, the zirconia susceptor tube develops cracks that can introduce zirconia particles into the heat zone via two routes: (1) particles that break off the cracked susceptor and (2) particles from the insulating grog that migrate through the cracks. A new furnace (Figure 2) was designed to prevent these particles from depositing on the preform surface.[6]

During operation of a furnace, the cracks are usually initiated in the midsection of the susceptor tube where the temperature is the highest. These cracks gradually increase in size during the life of the furnace. Finally, the tube reaches such an aggravated condition that the radio-frequency field is no longer able to couple to the fractured tube and the furnace has to be rebuilt.

Double jacket. To prevent zirconia particles migrating through the cracks, the susceptor tube is loosely jacketed with a low-density zirconia tube. This outer tube, because of its low density, permits the RF field to be transmitted through and be absorbed by the dense inner susceptor tube. The absorbed energy will be converted to radiant energy that heats the preform placed inside the tube. Furthermore, the outer tube is allowed to freely expand to a certain extent and maintain a crack-free condition. Therefore the zirconia insulating grain surrounding the outer tube will be completely isolated from the inner tube and is prevented from migrating into the inside of the furnace.

Silica layer. The second feature of the furnace is a silica glass film which is deposited on the inner surface of the susceptor tube. To do this the middle section of the susceptor tube is mounted in the head stock of a glass lathe. A silica ring torch is placed in front of the tube. The torch is connected to a chemical delivery system to generate silica soot particles which deposit to form a uniform layer on the inside of the tube as it rotates. The thickness of the layer is between 0.5 and 1 mm (Figure 3). The susceptor is replaced in the furnace and this layer will be sintered into a glassy film as it is raised to its operating temperature.

After 3 months' operation, a sample piece was taken from the middle ring to examine the condition of this silica film with a scanning electron microscope (SEM). Figure 4a is the SEM picture of the sample showing a silica glass film about 50 μm thick resulting from the deposited soot layer.

Figure 4b is a silica map of Figure 4a showing the silica particles throughout the cross-section of the sample. The white dots represent the silica distribution, and the number of dots at a given area correspond to the silica concentration at that point. This figure clearly shows that the silica particles diffused into the zirconia substrate during the sintering process and possibly lodged in the grain boundaries of the zirconia crystals or filled in the minute voids existing between the crystals. Particularly, it is important to note that the cracks in the tube were filled with silica glass and stayed intact without noticeable progression of deterioration.

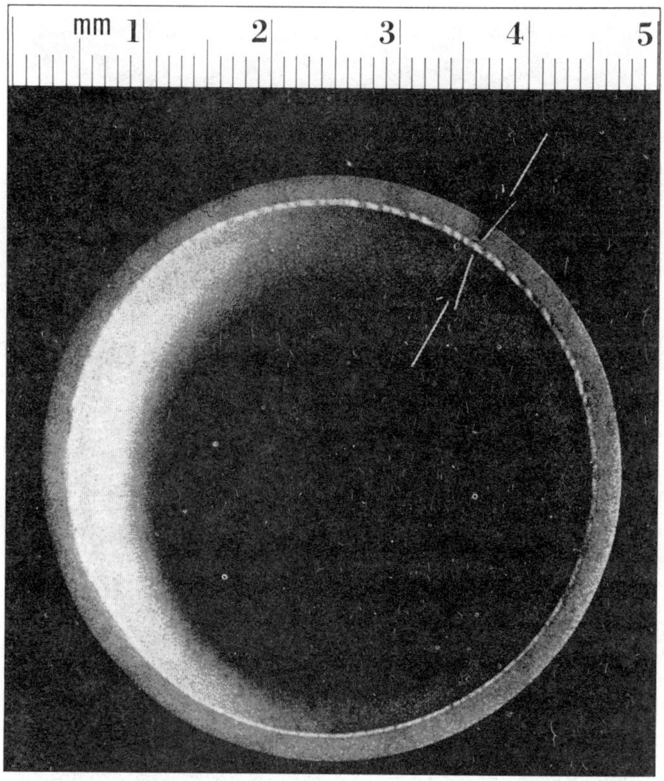

Figure 3. Cross-sectional view of deposited soot layer inside the zirconia ring. The thickness of the layer is approximately 0.7 mm.

High-Speed Drawing. During drawing the fiber is continuously cooled by the surrounding air as it moves away from the furnace, but if the draw speed is increased this raises the temperature at the coating cup. Above a certain point the coating material adjacent to the fiber will be excessively heated and finally reach a temperature at which the meniscus formation no longer occurs. This meniscus collapse inhibits wetting of the coating material, resulting in incomplete coating.

To avoid excessive heating at high draw speed,

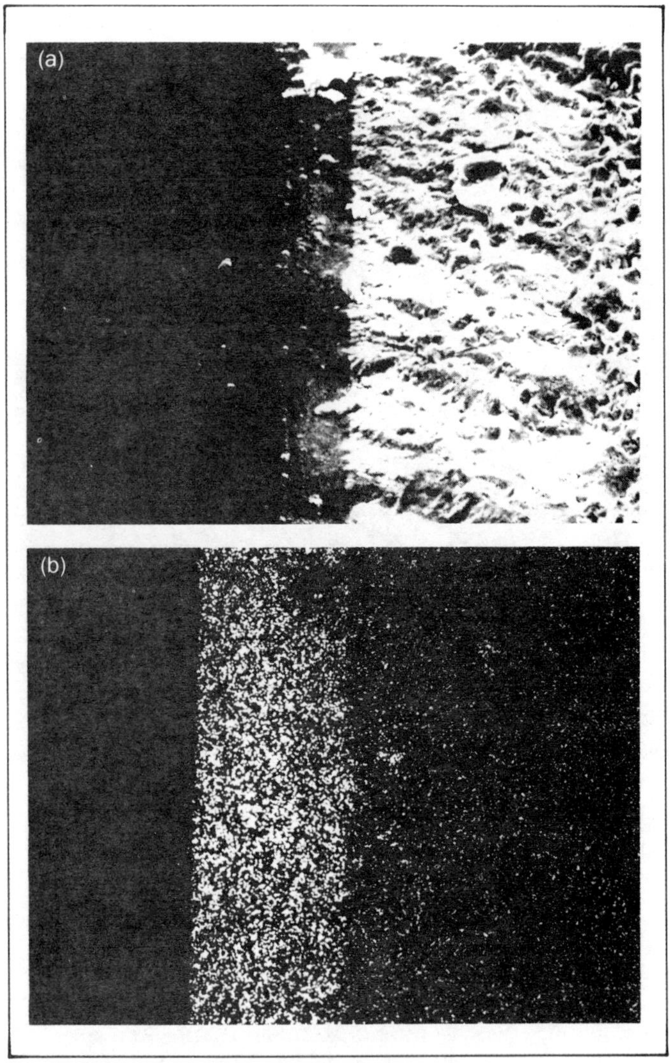

Figure 4. (a) SEM micrograph showing the sintered silica glass film on the zirconia substrate. (b) Silica map showing silica particle density distribution across the section.

two preventive measures can be taken: forced cooling[7] of the fiber by means of a gas and/or increasing the distance between the furnace and the coating applicator.[8] The relation between the distance H, measured from the bottom of the furnace to the coating cup, and draw speed V is given by

$$H = \frac{\rho C_p d\theta}{4h} V \tag{1}$$

where ρ is density of fiber material, C_p its specific heat, d fiber diameter, and h the heat transfer coefficient. θ in the above expression is a dimensionless temperature which is defined by

$$\theta = \ln\left[\frac{T_s - T_o}{T_c - T_o}\right] \tag{2}$$

where T_s is the lower softening point of silica, T_o ambient temperature, and T_c the fiber temperature at the coating cup.

Even in the case when a chiller is used for cooling the fiber, the above expression can predict the fiber temperature from the chiller on down. When the temperatures at the entrance and the exit of the chiller are T_1, T_2 respectively for a length L, the following expression can be obtained:

$$\frac{T_2 - T_o}{T_1 - T_o} = \exp\left[-\frac{4htL}{\rho C p d V}\right] \tag{3}$$

where h_t = heat transfer coefficient within the chiller.

High-Speed Coating. The purpose of applying a coating is to protect the surface from external damage and preserve the original strength. The theoretical strength of silica fiber is approximately 7 GN/m² (1 × 10⁶ lb/ft²) in air.

However, one of its shortcomings is that it can be easily damaged on its surface in contact with foreign particles or matter. Thus proper application of a thin coating on the fiber surface is needed.

Coating materials to be used for fibers can be categorized into two groups: organic material, for example ultraviolet-curable material,[9-12] thermally curable resin,[10,11] and hot melt coating,[12,13] and inorganic material such as metallic coating[14-15] and silicon nitride.[16] To assure good performance of a coated fiber, the physical properties of the candidate materials must meet stringent requirements for the physical properties of the coating.[17] In practice, ultraviolet or thermally curable materials are commonly used.

There are two types of coating application techniques that are commonly practiced in fiber manufacture. The first one is open-cup type application in which no external pressure is applied to the system. Therefore the only force existing in the applicator is the drag force from the fiber passing through the viscous material. In this case the coating thickness t can be predicted by

$$t = \sqrt{\frac{(R^2 - a^2)}{2\ln(R/a)}} - a \tag{4}$$

where a is the fiber radius and R is the applicator radius. Ths relation shows the important fact that the coating thickness is independent of coating speed when an unpressurized applicator is used, and is determined only by the die size and fiber diameter.

In high-speed coating, minimum shear stresses should be induced in the coating applicator. Otherwise, high shear flow may inhibit wetting if the coating material on the fiber surface or the non-Newtonian flow effect causes beading of the coating. Also, above a certain speed with an open cup, the fiber starts to slip because adhesion of the coating material at the fiber interface cannot sustain the drag force. To prevent this slippage, a mechanism that will provide a reduction of relative velocity between the fiber and the fluid in the cup must be introduced. The pressurized applicator[18] effectively eliminates this problem with the acceleration of fluid motion by the external pressure.

Pressure coating application is now the most prevalent technique in high-speed coating. The coating thickness is found to follow a linear relationship with a dimensionless quantity S, for a given fiber diameter and die size:

$$t = A \cdot S + B \tag{5}$$

In the above expression, A and B are constants depending on a and R. The dimensionless quantity S is defined by

$$S = \frac{\Delta p R^2}{8\eta V L} \tag{6}$$

where Δp is the pressure differential applied to the applicator, η the material viscosity, and L the applicator land length.

Therefore, Equation (5) shows the functional dependence of the various process parameters as well as system parameters. For instance, the thickness increases linearly with increasing pressure but depends inversely on the draw speed V and the viscosity η. A more explicit description of the thickness variation and draw speed can be found in Figure 5.[19] The viscosity of polymer material responds sensitively to temperature T, and decreases exponentially with increasing temperature.[20] Thus, in pressurized coating the heating effect of coating material in a coating cup is equivalent to an increase in pressure because the dimensionless pressure S is increased by decreasing the viscosity, therefore resulting in an increase in the coating thickness.

Strength of High-Speed-Coated Fiber. One of the noticeable differences in high-speed coating is the draw

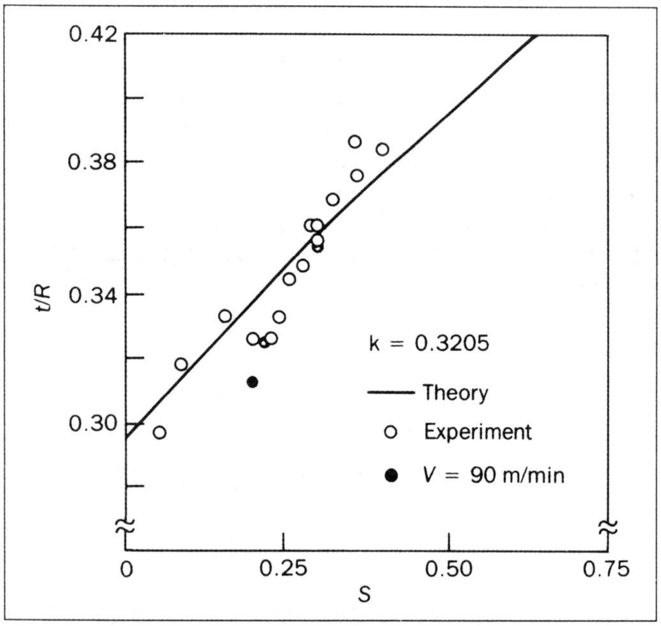

Figure 5. Relation between normalized coating thickness and dimensionless quantity S for a pressurized coating applicator; *k* = *a*/*R* (from Reference 17).

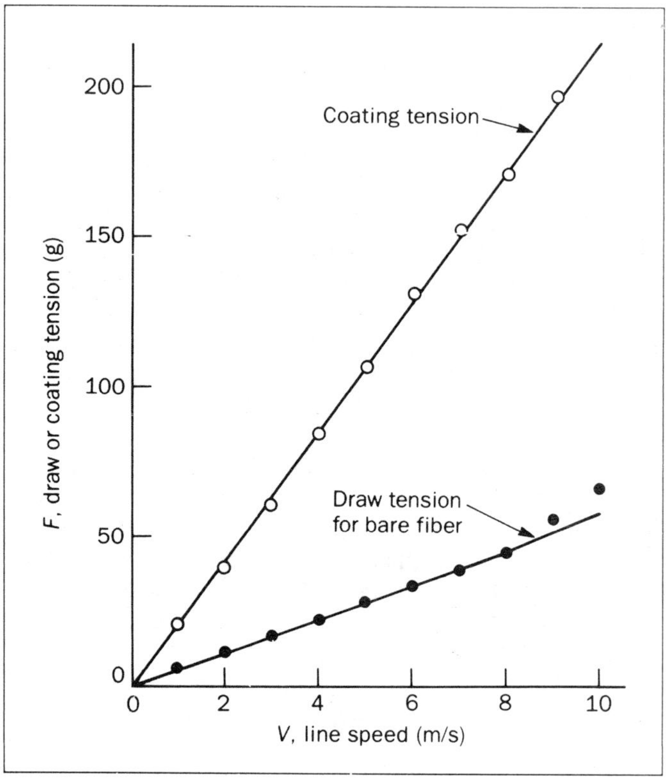

Figure 6. Relationships of draw and coating tension to coating speed. Both increase linearly with the line speed at constant temperature.

tension, which is maintained at a much higher level than for low speed. The important fact that has to be remembered is that the draw tension increases linearly with draw speed and will thus imbed increased residual stresses in the composite structured drawn fiber.[21] Such stresses in the fiber core will not only affect the strength characteristics but also influence the optical properties. Therefore it was important to study the effect of this high tension on the fiber transmission loss as well as on the strength.

The draw tension depends on several parameters, i.e., preform diameter, fiber diameter, furnace temperature, and draw speed.[22] We measured the draw tension for different line speeds at a fixed furnace temperature and found that the tensions with and without applying the coating are both linearly dependent on draw speed, as shown in Figure 6. This figure also shows that in the case of coating application, tension reaches about 200 g at the speed of 10 m/s, while the tension for bare fiber drawing at the same speed is about 50 g.

Another problem to note here is bubble inclusion in the coating. During coating the fiber drags air into the fluid, and these air bubbles accumulate in the coating mate-

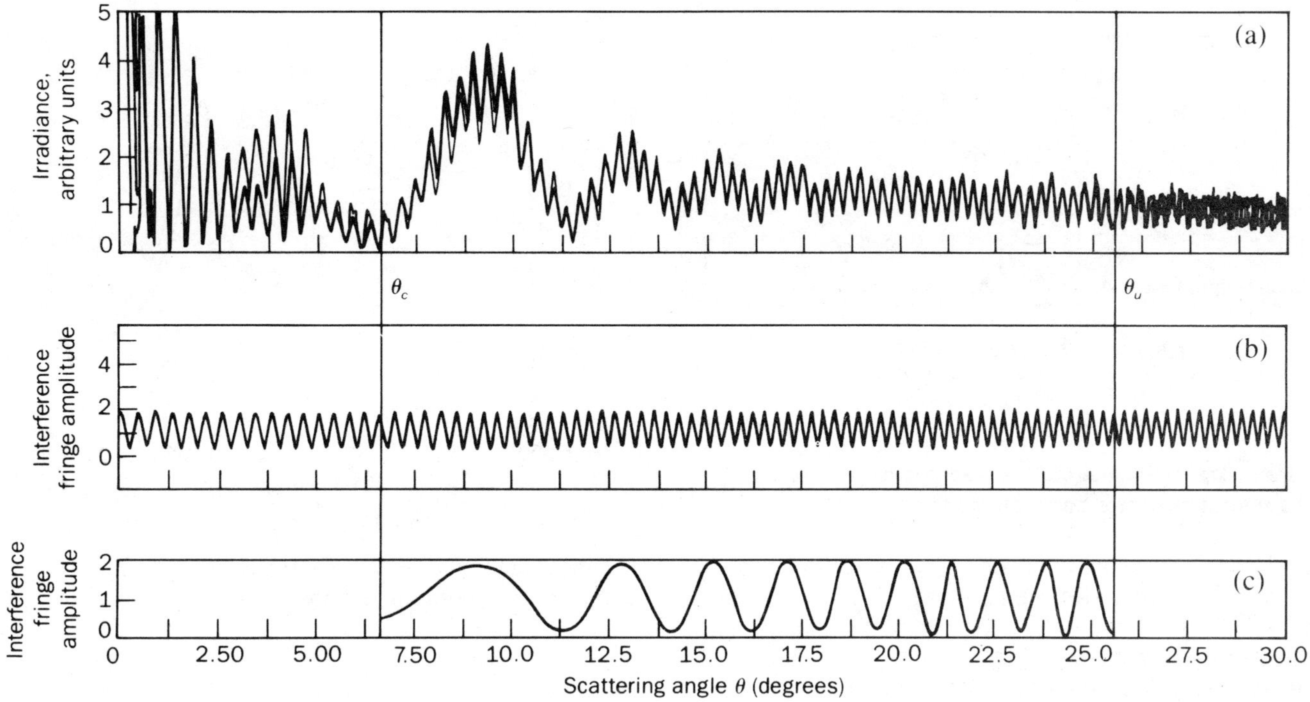

Figure 7. Composite graph of experimental and theoretical forward-scattering patterns for a step-index fiber. (Data for angles between 30° and 80° have been omitted since there is little change in structure in this range.) (a) Comparison of wave theory to experimentally measured patterns (heavy line is experimental data.). (b) Fringe position and (c) fringe modulation calculated from geometric ray tracing. The number of fringes is proportional to the diameter.

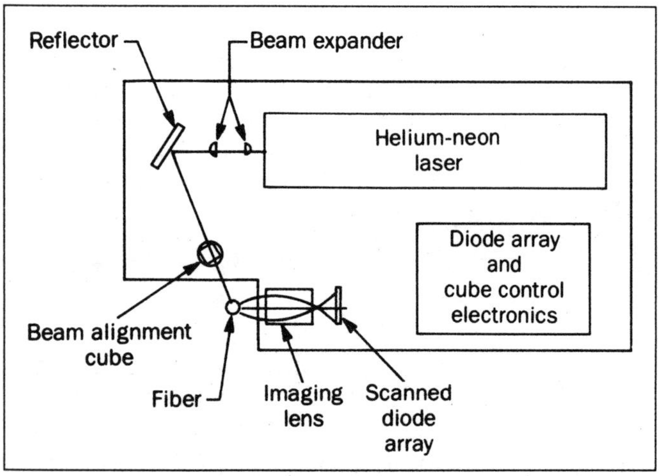

Figure 8. Optical layout of fiber measurement apparatus. Cube rotates to center beam on the fiber.

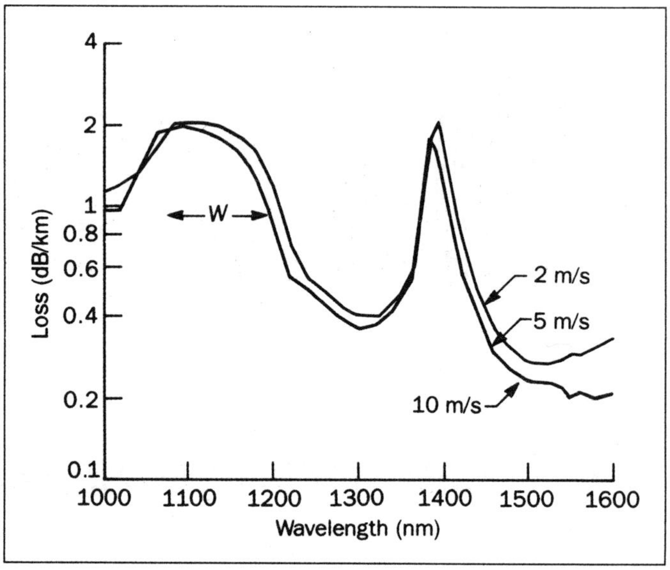

Figure 9. Transmission losses of single-mode fibers at λ = 1.32 μm for three different coating speeds.

rial. These bubbles are small enough to pass through the die, and after curing, remain permanently in the coating layer. Obviously, this will lead to a problem in maintaining a good coating concentricity and ultimately result in poor fiber strength. At high speed, bubble entrainment increases significantly, and so the applicator must be designed in such a way that it prevents this.

Control of Fiber Drawing. The diameter of the fiber in a steady-state drawing process is governed by mass conservation. The time to attain steady state, however, can be significant, and sometimes steady state is never achieved because of local fluctuations. In addition the preform diameter may not be constant. As a consequence, there is a requirement for a diameter measurement and control system if fibers with tight dimensional tolerances are to be made.

There are a number of reasons for minimizing diameter fluctuations. In most splicing techniques,[23] the outside surface of the fiber is used as a reference when the core regions of the fiber are aligned. For single-mode fibers, the dispersion and cutoff characteristics are very sensitive to the core diameter, as well as the refractive index profile.[24,25] Variations in the fiber diameter cause the same proportional variations in the core diameter.

Between the furnace and the coating cup it is very important that the fiber not be touched or abraded, since this causes minute flaws which act as a source for crack propagation under load.[26] For similar reasons, and to minimize microbending loss[27] in the cable, it is important that the fiber be well centered in the coatings. The fact that these two measurement and control operations must be done without fiber contact led to the use of laser light scattering.

Application of light scattering. There have been a number of techniques developed for measuring fiber diameter by shadow techniques.[28] The most accurate method for control, however, is the forward light scattering

Table I. Proof-Test Results

Sample no.	No. of breaks at 0.7 GN/m^2	No. of breaks at 1.4 GN/m^2	Drawn fiber length, km
1	0	0	7.0
2	0	1	7.0
3	0	0	6.7
4	0	1	6.5
5	0	1	7.0
6	0	3	7.0
7	0	0	8.0
	0	6	49.2

NOTE: V = 10 m/s.

approach.[29-32] When a collimated single-mode laser beam is directed at a fiber perpendicular to its axis, light is scattered in the plane perpendicular to the fiber axis. The light flux scattered as a function of the angle measured with respect to the forward direction of the incident beam is characteristic of the size of the fiber and the refractive index. If it is a clad fiber or has a transparent coating on it, the pattern will be modified by these structures.

Analysis and measurements of various patterns have been reported in a number of papers.[29-32] Figure 7 shows the intensity plot of the scattering pattern for a simple step index structure without plastic coating. The pattern basically consists of interference fringes of varying intensity in the angular range of approximately $\pm$ 5 to 90°. Changes in fiber diameter result in both movement of the fringes and a change in their angular frequency. The theoretical pattern in Figure 7 was obtained from calculated solutions of Maxwell's equations which predict all the characteristics of the pattern. Some of the dominant features, however, can be described using geometric ray tracing procedures and show that the number of fringes is directly proportional to the fiber diameter.[33]

Fiber diameter measurement. Figure 8 shows the optical layout for a system to measure and control fiber diameter. The beam from a He-Ne laser is expanded by cylindrical lenses and directed toward the fiber through an alignment cube. The light is collected by a low f/number lens and displayed on a 1024-diode array. The optical design displays the same precise angle range on the diode array even though the fiber can move around in the window.

The diode array provides a voltage-time display of the scattering pattern, and this is analyzed electronically to provide a fringe count which is proportional to fiber diameter. The system has 0.2-μm accuracy and can tolerate movement of the fiber within a 1-cm window. The diameter is measured at a 500-Hz rate and used through a suitable control algorithm to adjust the fiber draw motor velocity.[34]

Fiber coating centering. The same basic principles and geometric ray trace were used to analyze scattering from plastic-coated fibers.[35,36] The differences are that the coating can have either a smaller or larger refractive index than the core glass. The measurement is also different since it is concentricity rather than diameter which is of primary interest.

The effect of an eccentric fiber in the coating is to cause an eccentricity in the pattern. An automatic control system[36] measures these optical patterns, determining the eccentricity, and then adjusts the position of the coating applicator to center it to the fiber.

Conclusions

The application of this technology has resulted in significant increases in fiber draw speed while maintaining the strength and optical properties of the fiber. Table I presents data on a strength experiment on a synthetic glass rod drawn at 10 m/s, showing six breaks for 49 km of fiber tested at 1.4 GN/m^2. More recent data indicate even better strength results.[37]

Figure 9 shows transmission loss data for fiber drawn and coated at various speeds. The data indicate a slight reduction in loss at high speeds.

Our results show that the fiber draw rate has been increased to more than 10 m/s without any degradation in fiber performance and specifications and our analysis indicates even higher rates should be achievable.

Acknowledgment

The authors wish to express their gratitude to R. J. Klaiber for his guidance and encouragement and for his help with preparing this manuscript.

References

1. W. M. Flegal et al., "Making Single-Mode Preforms by the MCVD Process," *AT&T Technical Journal*, Vol. 65, Issue 1, January-February 1986, pp. 56-61.
2. J. B. MacChesney, P. B. O'Connor, and H. M. Presby, "A New Technique for the Preparation of Low Loss and Graded Index Optical Fibers," *Proceedings of IEEE*, Vol. 62, 1974, pp. 1278-1279.
3. T. Li, ed., *Optical Fiber Communications*, Vol. 1, Academic Press, New York, 1985.
4. R. B. Runk, "A Zirconia Induction Furnace for Drawing Precision Silica Waveguides," Topical Meeting on Optical Fiber Communication, Williamsburg, Virginia, 1977, paper TuB5-1.
5. S. Sakaguchi and T. Kimura, "A 1200 m/min Speed Drawing of Optical Fiber with Pressure Coating," Conference on Optical Fiber Communication, San Diego, February 11-13, 1985, paper MG2.
6. U. C. Paek and C. M. Schroeder, "Silica Coated Dual-Tube ZrO_2 Induction Furnace for High Strength Optical Fibre Production," *Electronics Letters*, Vol. 22, 1986, p. 72.
7. C. M. Darcangelo, M. R. Montieth, and Y. M. Ni, "Fiber Cooling Apparatus," U.S. Patent 4514205.
8. U. C. Paek and C. M. Schroeder, "High Speed Coating of Optical Fibers with UV Curable Materials at a Rate of Greater than 5 m/sec," *Applied Optics*, Vol. 20, 1985, p. 135.
9. U. C. Paek and C. M. Schroeder, "Calculation of Photopolymerization Energy Required for Optical Fiber Coating," *Applied Optics*, Vol. 20, 1981, p. 1230.
10. N. Inagaki and S. Yamakawa, "High Speed Coating and Coating Materials for Optical Fibers," Conference on Optical Fiber Communication, New Orleans, January 23-25, 1984, paper WF1.
11. S. Araki et al., "Fabrication of High Strength Long Length Fibers," Conference on Optical Fiber Communication, New Orleans, January 23-25, 1984, paper WF3.
12. L. L. Blyler et al., "A New Dual Coating System for Optical Fibers," Eighth European Conference on Optical Fiber Communication, Cannes, France, September 21-24, 1982, paper AV11.
13. T. J. Miller et al., "Silicone and Ethylene-Vinyl-Acetate Coated Laser Drawn Silica Fibers with Tensile Strengths Greater than 3.5 GN/m^2 (500 ksi) in Greater than 3 km Lengths," *Electronics Letters*, Vol. 14, 1978, p. 603.
14. D. A. Pinnow, J. A. Wysocki, and G. D. Robertson, "Hermetically Sealed High Strength Fiber Optical Waveguides," International Conference on Integrated Optics and Optical Fiber Communication (IOOC'77), Tokyo, 1977, paper B9-5.
15. S. Tanaka et al., "Low Loss Hermetically Coated Optical Fibers," Conference on Optical Fiber Communication, New Orleans, January 23-25, 1984, paper WF7.
16. R. Hiskes, "Improved Fatigue Resistance of High Strength Optical Fibers," Topical Meeting on Optical Fiber Communication, Washington, D.C., 1979, paper WF6.
17. L. L. Blyler, Jr., et al., "Fiber Drawing and Coating," Topical Meeting on Optical Fiber Communication, New Orleans, February 28-March 2, 1983, paper TuF1.
18. C. R. Taylor, "Fiber with Multiple Coatings," U.S. Patent 4480898.
19. A. Panoliaskos, W. L. H. Hallett, and I. Garis, "Prediction of Optical Fiber Coating Thickness," *Applied Optics*, Vol. 24, 1985, p. 2309.
20. K. Chida et al., "High Speed Coating of Optical Fibers with Thermally Curable Silicone Resin Using a Pressurized Die," *Electronics Letters*, Vol. 18, 1982, p. 713.
21. U. C. Paek and C. R. Kurkjian, "Calculation of Cooling Rate and Induced Stresses in Drawing of Optical Fibers," *Journal of the American Ceramic Society*, Vol. 58, 1975, p. 330.
22. U. C. Paek and R. B. Runk, "Physical Behavior of the Neck Down Region during Furnace Drawing of Silica Fibers," *Journal of Applied Physics*, Vol. 49, 1978, p. 4417.
23. J. F. Dalgleish, "Splices, Connectors and Power Couplers for Field and Office Use," *Proceedings of IEEE*, Vol. 68, 1980, p. 1226.
24. U. C. Paek, G. E. Peterson, and A. Carnevale, "Parametric Effects to the Bandwidth of a Single Mode Fiber with Experimental Verification," *Applied Optics*, Vol. 21, 1982, p. 704.
25. L. G. Cohen et al., "Transmission Studies of a Long Single Mode Fiber—Measurements and Considerations for Bandwidth Optimization," *Bell System Technical Journal*, Vol. 60, 1981, p. 1713.
26. L. L. Blyler, Jr., and F. V. DiMarcello, "Fiber Drawing, Coating

and Jacketing," *Proceedings of IEEE*, Vol. 68, 1980, pp. 1194-1198.
27. W. B. Gardner, "Microbending Loss in Optical Fibers," *Bell System Technical Journal*, Vol. 54, 1975, p. 457.
28. Product literature for Milmaster electronic micrometer, The Electron Machine Corp., Umatilla, Florida; Model 1160A Lasermike™ optical micrometer, Systems Research Labs, Inc., Dayton, Ohio; Anritsu M417 SLB diameter measuring system, Anritsu Electric Co. Ltd., Tokyo.
29. M. Kerker and E. Matijevic, "Scattering of Electromagnetic Waves from Concentric Infinite Cylinders," *Journal Optical Society of America*, Vol. 51, 1961, p. 506.
30. H. C. van der Hulst, *Light Scattering from Small Particles*, Wiley, New York, 1957.
31. J. L. Lundberg, "Light Scattering from Image Fibers at Normal Incidence," *Journal of Colloid and Interface Science*, Vol. 29, 1969, p. 565.
32. L. S. Watkins, "Scattering from Side-Illuminated Clad Glass Fibers for Determination of Fiber Parameters," *Journal Optical Society of America*, Vol. 65, 1974, p. 767.
33. D. H. Smithgall, L. S. Watkins, and R. E. Frazee, Jr., "High Speed Non-Contact Fiber Diameter Measurement Using Forward Light Scattering," *Applied Optics*, Vol. 16, 1977, p. 2395.
34. D. H. Smithgall, "Application of Optimization Theory to the Control of Optical Fiber Drawing Process," *Bell System Technical Journal*, Vol. 58, No. 6, July-August 1979, p. 1425.
35. B. R. Eichenbaum, "The Centering of Optical Fiber Coatings by Monitoring Forward Scattering Patterns," *Bell System Technical Journal*, Vol. 59, No. 3, March 1980, p. 313.
36. D. H. Smithgall and R. E. Frazee, Jr., "High Speed Measurement and Control of Coating Concentricity," *Bell System Technical Journal*, Vol. 60, No. 9, November 1981, p. 2065.

Biographies (continued)

ing from the University of California, Berkeley. He joined AT&T in 1969. Mr. Watkins is supervisor of the Fiber and Integrated Optics Group. He joined AT&T in 1966, and has specialized in the development of optical measurement techniques. He received the B.S. in electronic engineering and the Ph.D. in quantum electronics, both from Southampton University, England.

(Manuscript received November 12, 1986)

Recent Developments in Hermetically Coated Optical Fiber

K. E. LU, G. S. GLAESEMANN, R. V. VANDEWOESTINE, AND G. KAR

***Abstract*—This paper deals with recent progress made at Corning Glass Works on hermetically coated optical fiber. This fiber provides improved fatigue resistance, high usable strength, excellent resistance to hydrogen permeation, 25–90°C water, and corrosive chemicals such as hydrofluoric acid and boiling sodium hydroxide. These attributes make the hermetically coated fiber an excellent candidate for use in undersea cables, and other special applications where a harsh environment prevails.**

I. Introduction

IN RECENT YEARS, there has been an increased need for using optical fibers in harsh environments where the fiber must withstand high stress levels and must resist fatigue over long periods of time. Some examples include undersea applications where the cable could be exposed to very high stress for a period of several days during its development and/or repair. It is well known that the strength of optical fibers decreases with time in the presence of moisture and tensile stresses of sufficient severity [1]–[4]. This degradation can be increased with elevated temperature. By preventing the moisture from reaching the fiber surface, it is possible to allow higher in-service stresses and achieve longer lifetime.

Several researchers [5]–[10] have attempted coating fibers with metallic coatings such as aluminum, lead, and tin to serve as moisture barriers. Processing complexities, slower application speeds, work hardening of metals resulting in localized stress concentrations, and excessive microbending losses due to shrinkage of the metallic coatings are some of the reasons why these types of coatings have not yet found wide acceptance. On the other hand, inorganic and ceramic hermetic coatings investigated at several laboratories have been found to improve fiber fatigue resistance, as well as reduce its susceptibility to attenuation increases from hydrogen effects [11]–[14].

In this paper, recent progress made at Corning Glass Works on a hermetically coated optical fiber is discussed.

II. Hermetic Coating Process

The hermetic coating consists of a thin (<500-Å) layer of amorphous carbon coating which is deposited directly on the surface of the freshly-drawn optical fiber through a chemical vapor deposition process. The reactor is situated below the draw furnace into which the reactant is introduced and the by-products are removed through an exhaust port. A schematic of the apparatus is shown in Fig. 1. The fiber is subsequently overcoated with a protective polymeric coating by conventional methods. In this study, both UV acrylate and thermally cured silicone coatings were used. The rationale was to cover a wide spectrum of use temperature. A particle-free environment was maintained to ensure uniform high strength throughout the entire length of fiber manufactured.

Fibers were made in ~10-km lengths and prooftested at 2.5-percent strain level. In the subsequent sections, the strength and fatigue behavior of the hermetic coated optical fiber, as well as its resistance to hydrogen permeability and a variety of chemicals, will be discussed. The optical properties of the hermetically coated fiber were found to be unchanged from those of the conventional single and multimode fibers and, therefore, will not be discussed in this paper.

III. Strength and Fatigue

It is generally believed that the fatigue failure of optical fibers is caused by stress-enhanced crack growth of preexisting flaws to a critical dimension at which time failure occurs rapidly. The power law crack velocity relationship is most widely used to describe fatigue crack growth in optical fibers and is given by [15]

$$V = AK_I^n \tag{1}$$

where V is the crack velocity, K_I is the stress intensity factor which is related to the magnitude of the crack tip stress field and A and n are crack growth parameters. The parameter n, better known as the fatigue susceptibility parameter, provides a measure of materials resistance to fatigue, e.g., a low n value (10) indicates little fatigue resistance and a high n value (100) substantial fatigue resistance.

Equation (1) can be integrated for a given glass and temperature to give fracture strength (σ_f) as a function of stressing rate ($\dot{\sigma}$)

$$\sigma_f^{n+1} = (n + 1)\, BS_i^{n-2}\dot{\sigma} \tag{2}$$

where S_i is the inert strength and $B = 2K_c/AY^2(n - 2)$. Y is a dimensionless constant related to the crack loading geometry and crack shape, and K_c is the critical stress intensity factor. A least-square regression of (2), with σ_f as the dependent variable and $\dot{\sigma}$ as the independent vari-

Manuscript received May 26, 1987; revised August 18, 1987.
The authors are with Corning Glass Works, Corning, NY 14831.
IEEE Log Number 8718344.

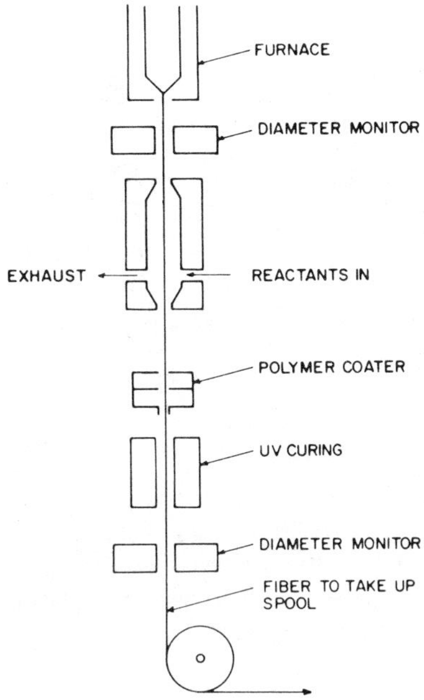

Fig. 1. A schematic showing the hermetic coating process.

able, gives the crack growth parameters n and B, knowing the inert strength.

A. Inert Strength

The fracture strength of the hermetically coated fiber was first measured in an inert environment in which no subcritical crack growth is expected to occur. Test specimens were placed in a vacuum and evacuated at 10^{-5} torr for 200 h to remove any moisture in the coating and then stored in a vacuum of 10^{-2} torr immediately prior to testing. Individual specimens were removed from the vacuum chamber and loaded to fracture in an Instron Universal Tester[1] in less than 40 s to minimize time for moisture diffusion through the polymeric coating. All vacuum chambers were backfilled with either argon or dry nitrogen before removing the specimen. For comparison, similar inert strength measurements were carried out on standard optical fibers.

Table I is a summary of the inert strength results which were fitted to a two parameter Weibull distribution function [16],

$$\ln \ln (1/1 - F) = m \ln (S_i/S_0) \qquad (3)$$

where F is the cumulative failure probability and m and S_0 are the Weibull slope and scale parameter, respectively. Also included in Table I are inert strength results from a silicon-oxynitride coated fiber [2]. Whereas the strength of Corning's hermetically coated fiber is higher than that of the silicon-oxynitride coated fiber, it is substantially less than that of the standard fiber indicating most likely a tensile residual stress on the fiber surface due to the hermetic coating.

B. Dynamic Fatigue In Water

The fatigue resistance of the hermetically coated fiber was measured in water for a range of temperatures (25, 55, and 90°C) using the dynamic fatigue technique [1] where fracture strength is determined as a function of strain rate. The testing technique involved threading the optical fiber through a 25-cm glass tube containing distilled water with a serum bottle stopper at the bottom [2]. The ends of the optical fiber were gripped top and bottom to two-inch diameter capstans fitted to an Instron Universal Tester. The water was circulated through the tube at a given temperature using a constant temperature circulator bath and the flow rate was regulated so as not to disturb the fiber during strength testing. The fibers were soaked for a minimum of 10 min prior to testing to allow the water to reach the fiber surface and bring the surface temperature into equilibrium with the surrounding water. The fracture strength of the hermetically coated fiber was then measured over four strain rates ranging from 0.015 percent/min to 15 percent/min, corresponding to crosshead speeds of 0.1–100 mm/min. A minimum of ten specimens were strength tested at each strain rate for the 25 and 55°C temperatures and 20 specimens per strain rate for the worst case condition of 90°C.

TABLE I
INERT STRENGTH DATA

FIBER	WEIBULL PARAMETERS m	S_o	STRENGTH (kpsi) HIGH	MEDIAN	LOW
Hermetically Coated:					
- Corning Hermetic Fiber	107.2	565	593	553	535
- Silicon-Oxynitride (Reference 2)	41.1	480	507	473	447
Standard	62.8	978	831	815	798

The results of the dynamic fatigue tests in water as a function of temperature for Corning's hermetically coated fiber are summarized in Fig. 2 along with similar data for standard fiber from [2]. The lines in Fig. 2 represent a best fit of all the data at a given temperature and the crack growth parameter n from (2) is given in Table II. The fact that n in the worst case (90°C) is 80 indicates that hermetically coated fiber is far less susceptible to fatigue than standard fibers which have n values near 20. In addition, the high n values for the hermetically coated fiber are accompanied by relatively little change in strength as a function of temperature for a given stressing rate. The relative insensitivity of the strength to temperature is indicative of a high activation energy for the crack growth process [2]. However, it should be noted that n decreased from 180 at 25°C to 80 at 90°C, suggesting a more activated crack growth process at elevated temperatures. The high resistance to fatigue of the hermetically coated fiber is attributed to the effectiveness of the coating in limiting the availability of moisture at the glass surface.

C. Dynamic Fatigue Using 10-m Gauge Length

Strength and fatigue tests on short gauge length specimens provide useful information on the relatively small flaws in the overall flaw distribution on the fiber since the

[1]Instron Corporation, Canton, MA.

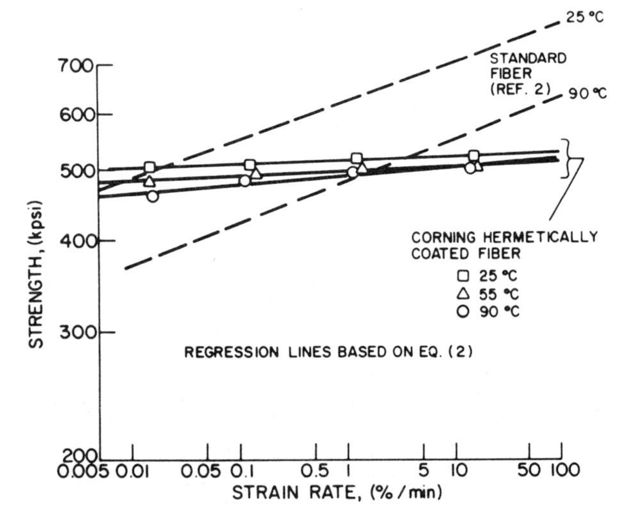

Fig. 2. Dynamic fatigue of hermetically coated optical fiber in water.

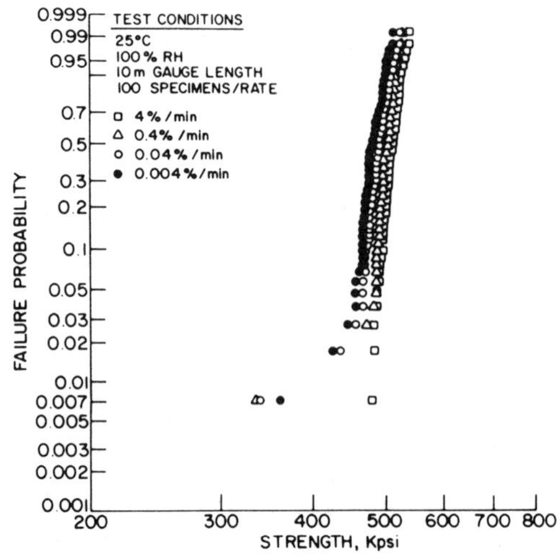

Fig. 3. Weibull strength distribution plots for hermetically coated fibers in 10-m gauge lengths.

TABLE II
SUMMARY OF FATIGUE SUSCEPTIBILITY PARAMETER N FOR HERMETICALLY COATED FIBER IN WATER

T (°C)	SPECIMENS/RATE	n
25	10	179 (240-143)*
55	10	142 (185-115)
90	20	79 (91-70)

* Values in parentheses represent the range in n for ±1 standard deviation of the slope of the dynamic curves in Figure 2.

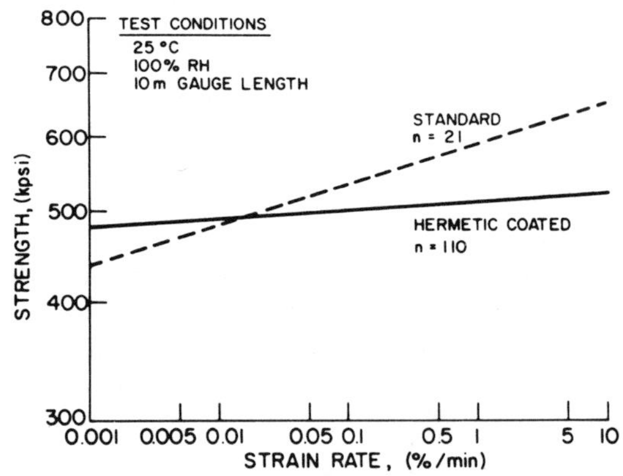

Fig. 4. Dynamic fatigue of hermetically coated and standard fibers in 10-m gauge lengths.

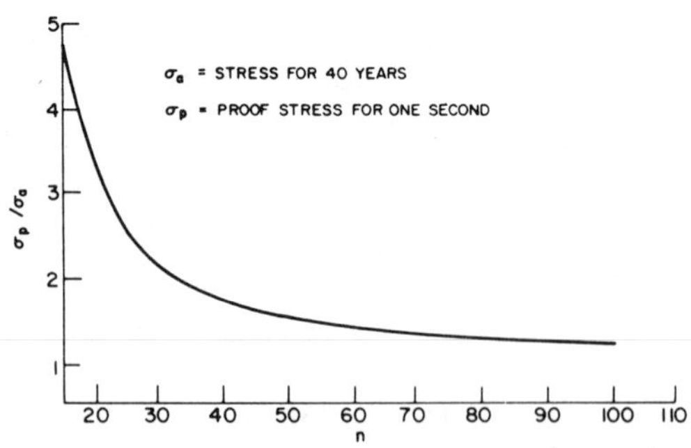

Fig. 5. Allowable applied stress after prooftesting as a function of fatigue suseptibility factor [17].

probability of encountering a large flaw is small. For long length applications, it is necessary to examine the strength and fatigue behavior of long gauge length specimens where flaw sizes are more representative of in-service lengths. Therefore, the dynamic fatigue behavior of the hermetically coated fiber was examined using 10-m gauge length specimens tested on a rotating capstan fiber tester (RCFT) at 25°C and 100-percent relative humidity. Strain rates ranged from 0.004 percent to 4 percent per minute with a minimum of 100 specimens per rate. In addition, 10-m dynamic fatigue experiments were carried out on a standard fiber using similar sample sizes per strain rate for comparison.

The strength distributions of the dynamic fatigue results for the 10-m hermetically coated fiber specimens are summarized in the Weibull graph in Fig. 3. Using all the data in Fig. 3, a dynamic fatigue plot was constructed for the hermetically coated fiber, and is shown in Fig. 4 along with similar data for a standard fiber. Again, the lines in the dynamic fatigue plots represent a best fit of all the data. From (2) the fatigue susceptibility parameter n for the hermetically coated fiber is 110 compared to a value of 21 for the standard fiber.

The contrasting fatigue performance of the hermetically coated fiber and standard fibers can be appreciated by considering allowable in-service and handling stresses as a function of fatigue behavior. Product designers usually account for in-service strength degradation by choosing an allowable stress based on a fatigue factor, which is the ratio of proof stress to allowable in-service stress for a given lifetime. For standard fibers a fatigue factor of 3 or an allowable stress equal to 30 percent of the proof stress is generally used as a rule of thumb. Gulati [17] has determined that the fatigue factor depends on the material's resistance to fatigue as shown in Fig. 5. As can be seen from Fig. 5, an n of 20, representative of a standard fiber and a lifetime of 40 years, gives the fatigue factor of 3. For n values of 100, a fatigue factor of 1.25 or an allowable stress, nearly 80 percent of the proof stress, is more appropriate for the same 40 year lifetime. This example illustrates the advantage of hermetically coated fiber for applications where long-term reliability under stress is required.

IV. Environmental Studies

A. Hydrogen Resistance

Single-mode hermetically coated fiber was tested in 100 percent H_2 at 1-atm pressure and room temperature for a period of 200 days. No increase was seen at the operating wavelengths of 1300 and 1550 nm. H_2 penetration of the core was measured by the growth of the interstitial peak at 1240-nm wavelength as shown in Fig. 6. The data suggests that the hermetically coated fiber can significantly retard H_2 permeation into the fiber under these conditions.

Currently, tests are in progress to determine the kinetics of hydrogen diffusion through the hermetic coating. Preliminary data suggests that at elevated temperatures there is some attenuation increase in presence of hydrogen, confirming an earlier report by Hiskes *et al.* [12]. Tests done on several selected fibers show that the activation energy of the rate of hydrogen incorporation is higher for the hermetically coated fibers (18 $\pm$ 3 kcal/mole) compared to that of the standard optical fibers (10.5 kcal/mole). A report will be published at a future date upon completion of this detailed study.

B. Chemical Durability

The chemical durability of the hermetically coated fiber was determined by exposing it to chemicals which are corrosive to glass, such as hydrofluoric acid (49-percent concentration at 23°C) and hot sodium hydroxide (5-percent concentration at 95°C) solutions. For these tests, 1-m lengths of fiber were wrapped on a 0.25-in diameter Teflon mandrel, placed in the solutions, and the time to failure was recorded. The stress on the coated fibers was calculated to be approximately 250 kpsi. While the nonhermetically coated fibers failed in a few hours, the hermetically coated fibers are yet to fail after eight months in HF solution and ten days in hot sodium hydroxide solution, respectively.

V. Summary

In summary, the hermetically coated optical fiber developed at Corning Glass Works exhibits significant improvement in fatigue resistance over nonhermetic fibers, excellent resistance to hydrogen permeation, at room temperature, to 25°C to 90°C water, and corrosive chemicals. This fiber can be made in long lengths, prooftested at 250 kpsi, and its long-term useable strength is expected to be approximately 80 percent of the prooftest level.

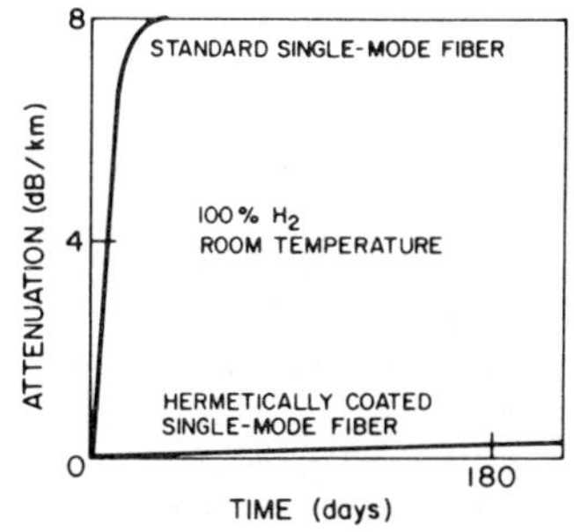

Fig. 6. Attenuation change at 1240 nm as a function of time (linear scale).

These attributes make this hermetically coated fiber an excellent candidate for use in undersea cables, fiber-optic guided missiles, and other special applications where a harsh environment is prevalent.

Acknowledgment

The authors wish to acknowledge the help of M. Donovan, D. Howe, Dr. J. Helfinstine, K. Giavara, and Dr. D. Powers for their measurements and analytical support.

References

[1] J. E. Ritter, Jr., J. M. Sullivan, Jr., and K. Jakus, "Application of fracture-mechanics theory to fatigue failure of optical glass fibers," *J. Appl. Phys.*, vol. 49, no. 9, pp. 4779–4782, 1978.

[2] J. E. Ritter, Jr., G. S. Glaesemann, and K. Jakus "Effect of temperature on the strength and fatigue behavior of optical fibers," *J. Mat. Sci.*, vol. 19, pp. 4087–4092, 1984.

[3] H. S. Chandan and D. Kalish, "Temperature dependence of static fatigue of optical fibers coated with a UV-Curable polyurethane acrylate," *J. Amer. Ceram. Soc.*, vol. 65, no. 3, pp. 171–173, 1982.

[4] S. Sakaguchi and T. Kimura, "Influence of temperature and humidity of dynamic fatigue of optical fibers," *J. Amer. Ceram. Soc.*, vol. 64, no. 5, pp. 259–262, 1981.

[5] J. Wysocki, "Reduction in static fatigue of silica fibers by hermetic jacketing," *App. Phys. Letts.*, vol. 34, no. 1, p. 17, 1979.

[6] P. G. Hale, J. B. Almedia, and C. J. R. Sheppard, "On-line metal coating of silica fibers using magnetron sputter," *Phys. Fiber Optics*, vol. 2, p. 115, 1981.

[7] M. L. Stein, S. Aisenberg, and J. M. Stevens, "Ion plasma deposition of hermetic coatings for optical fibers," *Phys. Fiber Optics*, vol. 2, p. 124, 1981.

[8] J. A. Wysocki, G. R. Blair, and G. D. Robertson, "Long-term strength of metal coated fibers in harsh environments," *Phys. Fiber Optics*, vol. 2, p. 134, 1981.

[9] J. C. Williams *et al.*, "Method of an arrangement for coating optical fibers with metallic materials," U.S. Patent 4 485 122, Nov. 1984.

[10] S. Tanaka, T. Siota, R. Yamauchi, and K. Inada, "Low loss hermetically coated optical fibers," in *Proc. OFC*, 1984, p. 104.

[11] K. J. Beales, D. M. Cooper, W. J. Duncan, and J. D. Rush, "Practical barrier to hydrogen diffusion into optical fibers," presented at OFC, 1984, post deadline pap.

[12] R. Hiskes, "High performance hermetic optical fibers," presented at OFC, 1984, post deadline pap.

[13] D. R. Biswas and S. Raychaudhuri, "Hermetic coating for optical fiber and method of forming the same," Eur. Patent 0 166649 A2, Dec. 1985.

[14] P. C. Schultz and R. S. Chaudhuri, "Hermetic coating on optical fibers," *Proc. SPIE*, 1986.

[15] A. G. Evans, "Slow crack growth in brittle materials under dynamic loading conditions," *Int. J. Fract.*, vol. 10, no. 2, pp. 251–259, 1974.

[16] W. Weibull, "A statistical distribution function of wide applicability," *J. Appl. Mech.*, vol. 18, no. 3, pp. 293–297, 1951.

[17] S. Gulati, "Crack kinetics during static and dynamic loading," *J. Non-Crys. Solids*, vol. 38, pp. 475–480, 1980.

K. E. Lu, photograph and biography not available at time of publication.

*

G. S. Glaesemann, photograph and biography not available at time of publication.

*

Robert V. VanDewoestine received the B.S., M.S., and Ph.D. degrees in mechanical engineering from Purdue University, West Lafayette, IN, in 1963, 1966, and 1969, respectively.

He joined the research and development laboratory at Corning Glass Works in 1969 where he has worked on a variety of glass and ceramic development projects. He is currently Manager of Optical Waveguide Development.

Dr. VanDewoestine holds 12 patents in a number of subject areas.

*

G. Kar, photograph and biography not available at time of publication.

Reliability of Optical Fibers, Cables, and Splices

OTTO I. SZENTESI, MEMBER, IEEE

Abstract—**This paper discusses fiber characteristics and cable design considerations that affect cable reliability. Installed cable and splice reliability data are presented, and it is concluded that intrinsic cable and splice reliability is very high. Cable failures are dominated by uncontrollable factors, and the highest failure mode is "backhoe blackout." Finally, restoration issues are discussed.**

I. Introduction

EVER since the success of the early field trials [1]–[3], fiber optic transmission has gained wide acceptance in such diverse applications as telephony, instrumentation, computer, CATV, and the military. As a matter of fact, in telephony trunking, and more recently in new feeder applications, it has become the transmission system of choice. The availability of optimized design for the specific application, correct installation practices, and long-term reliability and maintainability have been the prerequisites for broad use of fiber optic transmission systems.

In general, installed fiber optic cable systems have an enviable record for reliability, although failures have been reported. Intrinsic fiber reliability and the effects of cabling have been studied extensively [4]–[9]. The limitations are well understood, and the design principles for any given set of environmental and installation conditions are known [10], [11].

As far as the user is concerned, the reliability of the entire cable system is of importance. In this context, the cable system is defined as the fiber optic transmission medium between the fiber distribution frames or between transmitters and receivers. Cable system reliability is influenced by the reliability of the system components, namely, the input/output connectors/pigtails, splices, and the cables themselves, whose reliability in turn is limited by that of the fibers.

In this paper, fiber characteristics affecting cable reliability are presented. This is followed by cable design considerations. Finally, installed cable and splice reliability and restoration are discussed.

II. Fiber Characteristics

There are three key characteristics of optical fibers which may be affected by environmental conditions: strength, attenuation, and resistance to losses caused by microbending. The reliability of fibers for a given application is ensured by selection of the appropriate fiber design, coating, control of the manufacturing process, and extensive environmental testing prior to qualification.

A. Strength and Fatigue

Contrary to common perception, glass is an inherently strong material. Optical fibers elongate nearly elastically under tensile loading until brittle failure occurs. The strength of optical fibers is mainly determined by randomly distributed surface defects [12], most often mechanically or chemically induced cracks or flaws [13].

The fracture probability depends on the fiber stress, fiber length, and the loading time, and is usually represented as a Weibull distribution. The parameters of this distribution are obtained by experimentation, i.e., by loading a statistically significant amount of randomly selected samples. Because there is not a single value of strength, but rather a probability of strength, fibers must be tested to guarantee a minimum strength or a corresponding minimum lifetime. In fiber manufacturing, all fibers are proof-tested for a short test time so that flaws larger than a given size are detected by failure at the position of the flaw. For 125 μm O.D. fibers proof-tested to 0.35 GN/m^2, the critical flaw size is about 1.6 μm.

In addition to the initial strength of the fiber, fatigue must be considered. It is generally accepted that under normal conditions, stress-enhanced interaction of the glass with moisture allows the flaw to grow, thus reducing the stress at which fracture can occur. Long life is achieved by minimizing both stress and moisture.

To illustrate the sensitivity of static fatigue to initial strain capability, one model [14] predicts that a fiber with an initial strain capability of 0.33 percent, when exposed to 0.2 percent strain at 2 percent relative humidity, will break in a few hours, whereas a fiber with 0.47 percent initial strain capability under the same conditions will survive 100 years.

B. Attenuation

Recently, a good deal of interest has concentrated on the effects of hydrogen on the optical attenuation of fiber [15]–[21]. Initial research was driven by submarine applications because undersea installations may expose the fiber to hydrogen gas, but the results are also applicable to terrestrial systems. It has been found that in hydrogen atmospheres, the fiber's attenuation increases at the longer wavelengths due to the diffusion of the hydrogen into the glass core. At least three components have been identified

Manuscript received May 1, 1986.
The author is with the Siecor Corporation, Hickory, NC 28603.
IEEE Log Number 8610778.

[22]. The effect of interstitial molecular hydrogen (H_2) is reversible and is insensitive to glass composition. Reactions forming short wavelength edge effects and hydroxyl (OH) effects are permanent. The latter effect is also composition dependent. Permanent effects are associated with defects in glass structure caused by composition and processing conditions.

Multimode fibers doped with high percentages of phosphorus (7 percent by weight) are particularly susceptible. By reducing the P_2O_5 concentration, loss increases at 1300 nm can be suppressed to a sufficient extent without any hindrance for practical usage [23]. For low phosphorus doped multimode fiber (less than 1 percent by weight), up to approximately 0.06 dB/km at 1300 nm can be expected after 30 years [24]. For single-mode fibers, slow long-term loss increments at 1310 nm are expected to remain less than 0.02 dB/km after 25 years at 20°C in one atmosphere of hydrogen [20].

C. *Microbending*

Microbending, sharp curvatures involving local axial displacements of a few micrometers, and spatial wavelengths of a few millimeters, can cause significant losses in optical fibers. The fiber coating, which is applied during manufacturing to protect the fiber surface and preserve the intrinsic high strength, plays a dominant role in minimization of microbending.

Contributing factors include improperly selected coating, improperly applied coating, variation of coating modulus or dimension with temperature or humidity, coating asymmetry, inclusions, or outside distortions transferred to coating. Some of the desirable properties of fiber coatings are good abrasion resistance, ease of stripping, chemical resistance, ease of subsequent processing during cabling, and environmental stability.

The environmental tests performed on optical waveguides primarily examine the effectiveness of the coating materials. If the coating is not performing its function, degradation in the strength or microbend resistance of the fiber will be quite evident. Fiber treatments include acidic, basic and water soaks, temperature/humidity cycling, fungus testing, extended dry heat and abrasion, and flammability testing [25].

The objective of fiber optic cabling is to package the fibers without degrading their initial transmission characteristics in such a way that these characteristics remain stable throughout the design life of the cable under specified environmental and mechanical loading conditions.

Unlike copper cables where electrical isolation is an important design parameter, the only factors affecting the design of fiber optic cables are mechanical and environmental. All forces, whether axial or radial, acting on an optical fiber, and of course any bending moment, will cause the transmission characteristics to deviate. To prevent this, the cable design must either substantially isolate the fibers from forces acting externally on the cables or at least cushion the fibers so that these forces are not converted into serious deformations.

There are three major sources of fiber stress that must be analyzed as part of any optical cable design. Tensile stress occurs when the fiber is axially strained. This can occur during the cable-making process and at any time when a tensile load is applied to the cable. Torsional stress occurs if the fiber is twisted. Special stranding techniques (planetary, neutralizing, or SZ) are frequently employed to ensure that the cable-making process does not contribute a torsional component. Bending or flexural stress occurs when the fiber is subjected to a bend radius. Such stress is often unavoidable in cables where stranding is used, but must be recognized and included in the analysis together with cable bending.

As a matter of practicality to meet requirements of 40 year life and 100 percent relative humidity, the short-term stress (e.g., typical processing and installation times) should be limited to 67 percent and long-term, residual stress to 20 percent of the proof-test level. Lifetime is further improved by reducing the relative humidity level in the cable.

The temperature performance of the cable is a function of the coefficients of thermal expansions and Young's moduli for the component materials. Because glass has a low coefficient of thermal expansion and high modulus, at low temperatures the fiber may be subjected to microbending and at high temperatures to stress.

One of the important considerations in the selection of fiber coating and cabling materials is the aging characteristics. Many of the qualification tests are aimed at predicting long-term stability of the critical properties. For example, for cables designed for high residual long-term tension, the "creep" of the tensile members must be taken into account.

In order to ensure the integrity of fiber optic cables, industry-wide verification tests have been accepted and others are in preparation [26]. These include

1) Outer Jacket Tensile Strength and Elongation Test (FOTP-89)
2) Low and High Temperature Bend Test (FOTP-37)
3) Impact Resistance (FOTP-25)
4) Compressive Strength (FOTP-41)
5) Tensile Strength of Cable (FOTP-33)
6) Cable Twist (FOTP-85)
7) Cable Cyclic Flexing Test (FOTP-104)
8) Resistance to Water Penetration (FOTP-82)
9) Cable Filling/Flooding Compound Drip Test (FOTP-81)
10) Temperature Cycling (FOTP-72; in preparation).

IV. Reliability

It must be pointed out at the outset that published reliability data of installed, traffic-carrying fiber optic cable systems are limited [10], [27], [28]. The data discussed in this section are dominated by North American experience and are based on inputs received from a variety of sources including field reports, industry sources, and end-user reports.

A. *Installed Cable Reliability*

In Table I, previously reported [27] data on 38 Siecor cable systems are summarized. The 312 sheath · km consist of approximately 54 percent underground duct, 26 percent direct buried, and 20 percent aerial plant. The study covered the period from January 1977 to July 1982. For purposes of the study, failure was defined as performance outside the specified limits and failures were classified into three categories: extrinsic, avoidable extrinsic, and intrinsic.

Extrinsic failures include uncontrollable breaks such as those due to digups, breaking poles, collapsing ducts, etc. All types of cable, fiber or copper, are subject to such failures. Avoidable extrinsic failures are those due to improper installation. Intrinsic failures are fiber failures that occurred while the cable was handled, installed, and operated within specifications.

The largest number of incidents (10 out of a total of 14) were in the "uncontrollable extrinsic" category. The breakdown is as follows: 1) 5 cuts, backhoe, direct buried; 2) 1, rodent, direct buried; 3) 3, shotgun, aerial, and 4) 1, duct, severed, underground. Not unexpectedly, the underground duct environment seems the least susceptible to uncontrollable failures. Avoidable failures are usually due to placement errors and can be reduced by training and following established practices.

The one and only intrinsic failure, a factory splice, occurred in 1978. Factory splices are arbitrarily classified as fiber failures instead of field splice failures. The issue of factory splices will be discussed later.

More recent data from MCI are summarized in Table II. No intrinsic fiber/cable failures were reported. The 18 incidents are all extrinsic and are dominated by cable cuts, as before. The failure rate is one incident per year per 146 sheath · km of cable. This number is on the same order of magnitude as the one calculated from the number of hits/year/sheath · km for long-distance coaxial transmission systems. A 30 percent improvement can be achieved by improving parallel construction practices.

In addition to the typical incidents described in [27] and in Table II, reported failures include those due to a tunnel fire, burst stream pipes, ducts cut by piles being driven through them, and clam dredger cutting underwater cable. In other words, all the familiar outside plant problems.

TABLE I
FIBER OPTIC CABLE RELIABILITY—EARLY DATA (JANUARY 1977–JULY 1982)

Number of Systems	38
Sheath-km	312
Sheath-km Hours	4, 731, 263
Fiber-km Hours	39, 925, 656
Number of Incidents	
Uncontrollable extrinsic	10
Avoidable extrinsic	3
Intrinsic	1

TABLE II
FIBER OPTIC CABLE RELIABILITY—MCI (DECEMBER 1984–JANUARY 1986)

Sheath-km	5300
Sheath-km Hours (Est.)	23,000,000

TRAFFIC AFFECTING INCIDENTS

Description	Number	Average Restoration Time (Minutes)
Parallel Construction	6	518
Contractor Activity	1	450
Backhoe, Auger, Digging	3	567
Gunshot	2	260
Hurricane	3	489
Flood Damage	1	10,080
Ice Crush (Bridge Crossing)	1	840
Vandalism	1	53
TOTAL	18	

B. *Discussion of Specific Fiber Cable Failure Modes and Mechanisms*

Fiber Strength: The easiest way to minimize cable failures due to fiber strength is to follow correct installation practices. When placing underground cable, tension monitoring is essential so that the short-term maximum rating of the cable is not exceeded. Furthermore, the residual tension in the cable after installation must decrease to a level below the long-term tensile rating.

The author is aware of two unconfirmed spontaneous fiber failures after installation involving two different cables made by two different manufacturers. It is not clear if the recommended cable installation or long-term tension specifications have been exceeded. Relative to other failures, these are considered insignificant.

Hydrogen Effects: The implications of the effects of hydrogen on optical fibers will depend on the concentration of hydrogen which can build up inside cables [18], [29]. This, in turn, depends on the materials used in cable manufacturing and the cable construction. In one trial system [23], fiber loss increases were observed after two years. The loss increase for the multimode, high P_2O_5 concentration fibers amounted to 0.15 dB/km at 1300 nm, whereas the value at 1500 nm was 1.5 dB/km. There are two other field-related increases of multimode fiber loss where the most probable cause was hydrogen. In the case of a terrestrial system, the attenuation at 1270 nm increased by 0.7 dB/km over a period of two years. In another instance, the loss of a 3.3 km length of submarine cable increased about 1.8 dB/km at 1300 nm. Adjacent land sections, which contained fiber of the same design, were unaffected.

For today's low-phosphorus, conventional single-mode fibers, hydrogen pressures in well-designed cables are predicted to be sufficiently low such that background effects will not cause any deterioration in the long-term performance of systems operating at 1.3 or 1.55 μm. Loss increases at 1.3 and 1.55 μm are predicted to be less than 0.01 dB/km after 20 years [21]. This number is lower than the prediction referenced earlier [20], primarily because the hydrogen pressure in cables is expected to be lower than one atmosphere.

Lightning: Historically, metallic shielded telecommunications cables buried in earth have been damaged by transient energy associated with lightning discharge. The typical damage is pinholes due to arcing [30] in the cable shield, but on occasion, additional crushing damage occurs. The classical explanation of this phenomenon has been termed the "steamhammer" effect [31]. Recent work suggests that mechanical crushing of armored cables is more likely the result of magnetic fields generated by lightning passing over or in near proximity of cables [32]. Results showed that cables with shields or armor with circumferential conductivity are subject to severe damage, while those without suffer little or no damage. Thus, bare metallic tapes or solid tubes used for shield or armor functions may be very poor performers in the natural environment, while shields and armor made from longitudinally applied plastic coated metal tapes having dielectric coatings will be virtually undamaged.

To the author's knowledge, there has not been any reported fiber optic cable damage due to lightning. This should not come as a surprise since most fiber optic cables are filled and pinhole damage is not likely to impair the transmission properties of fibers. Furthermore, all dielectric cables and those with metallic central members are not susceptible to magnetic crush. Finally, most armored fiber optic cables are also magnetic crush resistant [32].

Rodents: Many cable parameters influence the effects of rodent attack. The major considerations in cable design and placement for best rodent protection are cable size, placing depth, and sheath material. Tests have shown that for cables in excess of 53 mm in diameter, gopher attack is rarely observed [33]. On the other hand, cables below 20 mm in diameter (most fiber optic cables fall into this category) are subject to the most severe attack because certain rodents are able to bite directly down on the cable.

Another important consideration in reducing the risk of rodent attack is cable placement. Studies have shown that the probability of rodents digging to 1.2 m is only 15 percent. The depth to which rodents penetrate the soil is largely a function of the location of the frostline which rarely reaches 1.2 m in the continental United States.

For rodent protection, stainless steel tape armoring 125 μm thick or greater appears to be necessary to provide protection against rodent attacks [33], [34]. A metal sheath material that has consistently resisted gopher and squirrel penetration in the laboratory and in service (in the absence of corrosion damage) is 150 μm thick steel [35].

Some users in areas of dense rodent population and high lightning incidence have opted for extruded inner-duct cable because of its larger outside diameter and its all-dielectric construction. Inner-duct cable consists of an all-dielectric fiber optic cable with an extruded high-density polyethylene conduit surrounding it. The typical inner diameter is 25 mm and the outer diameter is 35 mm.

The following cable failures due to rodent damage have been reported: buried all-dielectric cable (one failure), dielectric cable in underground duct (one failure), inner-duct cable (one failure), buried cable in corrugated inner-duct (two failures). There are no failures of steel armored cables reported.

Based on published data and field experience, corrugated plastic duct is not an effective deterrent. It would also seem that 35 mm outside diameter duct is not 100 percent effective. Appropriate steel armoring is effective. Considering that certain types of metallic armor are resistant to lightning, and that no lightning damage has been reported, it appears that the preferred choice is armored cable, with longitudinally applied plastic coated steel tape, in areas of dense rodent population, even in areas of high lightning incidence.

Shotgun Damage: Aerial installed telecommunication cables are subject to shotgun blasts resulting in damage by shotgun pellets. The rural areas are more likely to sustain cable damage due to recreational hunting. For aerial fiber optic cables, this is the most frequent failure mechanism.

In order to investigate the resistance of various sheaths, a few cables were tested [36]. The samples consisted of standard duct/aerial, single-armor, and double-armor cables. The three designs were also tested with an inner-duct cover. Twelve gauge modified choke and 20 gauge full choke shotguns were used at distances of 5, 10, and 20 m.

At a distance of 5 m, the dual-armor/inner-duct cable passed, but was greatly deformed by the shot. A distance of 10 m produced results showing that the cables passed in the samples with inner duct or double armor. The cables passed in each sample at 20 m. The results are not surprising, and there is no doubt that with additional layers of armor, cables with increased shotgun resistance can be manufactured. The additional costs may be justifiable along short sections of an aerial route.

Factory Splices: During cable manufacturing, it is sometimes convenient and economical to include factory splices in the cables. Because of size and reliability considerations, these splices are invariably made by fusion. The strength of fusion splices has been studied extensively. Early studies had shown that the primary source of the low strength of fusion splices was the damage done to the fiber by the fiber stripping process. By avoiding extensive mechanical damage, strengths of approximately 2.8 GPa could be obtained with several heat sources [37].

TABLE III
SIECOR FACTORY FUSION SPLICE SUMMARY

TOTAL NUMBER OF SPLICES	1,027
TOTAL SPLICE-HOURS	11,363,352
TOTAL NUMBER OF REPORTED FAILURES	0

Such strength is several times greater than the proof-test level of fibers employed in terrestrial systems.

Factory splicing techniques have improved to the point that they do not constitute a failure mode. The results of one study are shown in Table III. During the period of the study from 1979 to August 1984, there were 1027 factory fusion splices (11 363 352 splice · hours) and no reported failures [38].

C. Field Splice Reliability

The two splicing techniques that dominate field splicing are the mechanical mass splice which uses etched silicon linear arrays [39] and the electric-arc fusion splice [40], [41].

For over 228 000 fiber splices using linear arrays, no instances of problems resulting in the permanent loss of a single fiber path due to installation breaks or high loss in splicing have been reported [9].

The long-term reliability of a fusion splice is strongly influenced by the method of splice protection [42], [43]. The easiest way of eliminating the consequences of reduced strength, if it exists, is to use a splice protection method that does not subject the splice to stress. Inappropriate procedures and splice protection techniques have been identified in the early field trials.

Aging tests of Siecor fusion splices revealed no measurable change in performance [44]. In one study [27], which has been currently updated, there were a total of two failures out of 2889 field fusion splices representing over 135 million splice hours. Both failures were avoidable, occurred in the closure, but not at the actual fusion splice. It is estimated that there are in excess of 500 000 Siecor fusion splices, and to date there have been no reported outages of operating systems due to intrinsic failures in the outside plant environment.

In addition to the two dominant splicing techniques already mentioned, individual mechanical splices have been employed. Of these, the GTE elastomeric splice is the most widely used. There have been some reported problems with splices using early procedures. Recently, the procedure was modified by adding a UV-cured bonding agent to stabilize the splice. Field reliability data on using the latest procedure have not been reported.

V. RESTORATION

To maintain systems in service, it is clear that troubleshooting and repair procedures have to be established. A study of working fiber systems conducted in 1982 [28] indicated that fiber cable-related troubles constituted less than 10 percent of the total number of troubles, yet the majority of the total plant clearing time (88 percent) was spent on repairing fibers. The average plant clearing time per fiber cable incident was 2145 min. These long clearing times were attributed to the relative newness of the technology in 1982, and it was expected that the experience factor should improve these figures considerably.

A. Troubleshooting

Troubleshooting is an obvious prerequisite to restoration. Once it is determined that the trouble is in the cable, the problem fibers are accessed with an optical time domain reflectometer. Cable faults, increased attenuation, and failed splices or connectors are readily identifiable.

B. Splice Repair

Based on the data in Section IV, it appears that splice reliability is more than adequate to meet service needs. Failures in all instances were limited to one or two fiber splices and were repairable by reentering the closures. As long as the system is protected or there are a few spare fibers, service is not affected.

C. Repair of Partially Failed Cables

Similarly, cable failures which impair the transmission performance of a small fraction of the fibers in a cable can be handled without service interruption as long as the damaged area is accessible. This is normally the situation in aerial or direct buried plants and in manholes. It is relatively straightforward to open the cable sheath, repair the damaged fibers, and repair the sheath without disturbing the good fibers. In one instance, two 6-fiber buffer tubes in a 46-fiber cable were damaged by gunshot, and the cable was repaired without disturbing the undamaged fibers [45].

D. Emergency Restoration

Since most fiber cable failures are uncontrollable, emergency restoration procedures are necessary. The essential requirements of emergency repair are: 1) rapid temporary restoration, and 2) the ability to transition to permanent repair without disturbing any of the fibers other than the one being worked on. Procedures and emergency restoration kits have been developed to meet these requirements [46].

Analysis of case histories of actual cable restoration emphasizes the importance of advance planning. In one case, service was restored (four fibers operating) on a cable just 4 h after it had been completely severed by a backhoe. Splicing on the remaining four fibers in the cable was completed 1 h later. The cable was repaired using the fusion splicing method. In another case of a severed six-fiber cable, restoration efforts took over 24 h. The major difficulty encountered proved to be the location of the necessary hardware.

Table II shows the average restoration times for single-mode fiber systems. Most of the cables in these systems contained either 22 or 44 fibers. Neglecting the incident due to flood damage, the average restoration time was 479 min for the remaining 17 incidents. This represents a four times improvement over the 1982 results mentioned earlier.

E. Offset Breaks

In any restoration, it is of great concern where the fiber breaks are relative to the actual cable break. In any freely suspended fiber optic cable subjected to tensile forces acting axially, fibers will break at randomly distributed points over the cable length when pulling forces exceed specified values. In case of a cable break, there is no correlation between the location of the cable break and the locations of the fiber breaks.

In most installations the cable breaks occur very differently from above. The cables are not freely suspended, are restricted to move either by friction or lashing, and cable breaks are not usually caused by axial forces. Even so, certain cable designs are more susceptible to offset breaks than others [47]. The most susceptible designs are cables with high tensile rating, with fibers located in the center of the cable. With some pressurized, nonstranded cables, offset breaks hundreds of meters away have occurred. On the other hand, stranded filled loose tube cables are the least susceptible.

Several different types of tests were conducted in the laboratory on stranded filled loose tube cables. The intent was to simulate conditions of "real" failures. The largest offset break ever observed in Siecor cables was 154 cm from the cable break. In none of the field failures were offset breaks reported.

VI. Conclusions

Installed fiber systems have excellent reliability. Most of the failures to date were in the "uncontrollable extrinsic category" and the majority of these were due to complete cable cuts. It is expected that existing data on the frequency of copper cable cuts in a given service area are applicable to fiber cable cuts. Splice reliability has also been found to be excellent, provided that the appropriate techniques and equipment were used.

Emergency restoration procedures to deal with the dominant failure mode of cables have been established to assure a rapid return of service. Based on data to date, it is essential that the user have an established plan for restoration. Training and periodic retraining of maintenance personnel at six month intervals is strongly recommended.

Acknowledgment

The author is indebted to a large number of people at numerous operating telephone companies and colleagues at Siecor for discussions on the topics covered in this paper. In particular, the contributions of G. Arnos (MCI), R. DeWitt (Contel), D. Hanna (GTE Service Corporation), H. Hohne (New York Tel), and E. Loytty, T. McMahon, P. Bark, and D. Lawrence (Siecor) are gratefully acknowledged.

References

[1] J. S. Cook and O. I. Szentesi, "North American field trials and early applications in telephony," *IEEE J. Select. Areas Commun.*, vol. SAC-1, pp. 393–397, Apr. 1983.

[2] A. Moncalvo and F. Tosco, "European field trials and early applications in telephony," *IEEE J. Select. Areas Commun.*, vol. SAC-1, pp. 398–403, Apr. 1983.

[3] H. Ishio, "Japanese field trials and applications in telephony," *IEEE J. Select. Areas Commun.*, vol. SAC-1, pp. 404–412, Apr. 1983.

[4] J. E. Midwinter, *Optical Fibers for Transmission.* New York: Wiley, 1979, ch. 12.

[5] T. Li, Ed., *Optical Fiber Communications, vol. 1, Fiber Fabrication.* New York: Academic, 1985.

[6] H. Liertz and U. Oestreich, "Continuous screening of optical waveguides for quality control," *Siemens Res. Develop. Rep.*, vol. 6, pp. 277–279, 1977.

[7] P. R. Bark, U. Oestreich, and G. Zeidler, "Stress-strain behavior of optical fiber cables," in *Proc. 28th Int. Wire Cable Symp.*, Cherry Hill, NJ, Nov. 13–15, 1979, pp. 385–389.

[8] M. I. Schwartz, P. F. Gagen, and M. R. Santana, "Fiber cable design and characterization," *Proc. IEEE*, vol. 68, pp. 1214–1219, Oct. 1980.

[9] C. M. Miller, "Optical fiber cables and splices," *IEEE J. Select. Areas Commun.*, vol. SAC-1, pp. 533–540, Apr. 1983.

[10] G. H. Zeidler, P. R. Bark, D. O. Lawrence, and O. I. Szentesi, "Reliability of fiber optic cable systems," in *Proc. Int. Conf. Commun.*, Seattle, WA, June 8–12, 1980, pp. 10.3.1–10.3.4.

[11] P. R. Bark and D. O. Lawrence, "Design and performance of filled fiber optic cables," in *Proc. GLOBECOM*, Miami, FL, Nov. 29–Dec. 2, 1982.

[12] R. D. Mauer, "Strength of fiber optical waveguides," *Appl. Phys. Lett.*, vol. 27, pp. 220–221, Aug. 15, 1975.

[13] F. V. DiMarcello, C. R. Kurkjian, and J. C. Williams, "Fiber drawing and strength properties," in *Optical Fiber Communications*, T. Li, Ed. Orlando, FL: Academic, 1985, pp. 179–248.

[14] D. Kalish, B. K. Tariyal, and H. C. Chandan, "Effect of moisture on the strength of fiber," presented at the 27th Int. Wire and Cable Symp., Nov. 1978.

[15] N. Uesugi, T. Kuwabara, Y. Koyamada, Y. Ishida, and N. Uchida, "Optical loss increase of phosphor-doped silica fiber at high temperature in the long wavelength region," *Appl. Phys. Lett.*, vol. 43, pp. 327–328, Aug. 15, 1983.

[16] K. Mochizuki, Y. Namihira, and H. Yamamoto, " Transmission loss increase in optical fibers due to hydrogen permiation," *Electron. Lett.*, vol. 19, pp. 743–745, Sept. 1, 1983.

[17] N. Uesugi, Y. Ishida, Y. Mitsunaga, Y. Murakami, Y. Negishi, and N. Uchida, "Infra-red optical loss increase for silica fiber in cable filled with water," *Electron. Lett.*, vol. 19, pp. 762–764, Sept. 15, 1983.

[18] K. J. Beales, D. M. Cooper, and J. D. Rush, "Increased attenuation in optical fibers caused by diffusion of molecular hydrogen at room temperature," *Electron. Lett.*, vol. 19, pp. 917–919, Oct. 27, 1983.

[19] Y. Namihira, K. Mochizuki, M. Kuwazuru, and Y. Iwamoto, "Effects of hydrogen diffusion on optical fiber loss increase," *Electron. Lett.*, vol. 19, pp. 1034–1035, Nov. 24, 1983.

[20] N. J. Pitt and A. Marshall, "Long-term loss stability of single-mode optical fibers exposed to hydrogen," *Electron. Lett.*, vol. 20, pp. 512–514, June 7, 1984.

[21] A. Tomita and P. J. Lemaire, "Hydrogen-induced loss increases in germanium-doped single-mode optical fibers: Long-term predictions," *Electron. Lett.*, vol. 21, pp. 71–72, Jan. 17, 1985.

[22] A. J. Morrow, "Minitutorial: Intorduction to optical fibers," presented at Opt. Fiber Commun. Conf., Atlanta, GA, Feb. 24–26, 1986.

[23] T. Tanifuji, M. Matsumoto, M. Tokuda, and M. Miyauchi, "Wavelength-dependent optical loss increase in graded-index optical fiber transmission lines," *Electron. Lett.*, vol. 20, pp. 13–14, Jan. 5, 1984.

[24] S. Hornung, S. A. Cassidy, and M. H. Reeve, "The distribution of H_2 gas along an inland optical fiber cable," in *Tech. Dig., Symp. Opt. Fiber Measurements*, Boulder, CO, Oct. 2–3, 1984, pp. 85–88.

[25] R. C. Nevins and D. H. Taylor, "Testing for a long service life," *Telephony*, vol. 208, pp. 116–124, Feb. 11, 1985.
[26] Electronic Industries Association, *Reference Guide for Fiber Optic Test Procedures*, Component Bull. 9-E, Nov. 1985, EIA Eng. Dep., 2001 Eye Street, N.W., Washington, DC 20006.
[27] O. I. Szentesi, "Field experience with fiber-optic cable installation, splicing, reliability, and maintenance," *IEEE J. Select. Areas Commun.*, vol. SAC-1, pp. 541–546, Apr. 1983.
[28] Y. Singh and H. Yip, "Fiber optics: Exploitation of strategic benefits in planning telco network evolution," in *Proc. GLOBECOM*, San Diego, CA, Nov. 28–Dec. 1, 1983, pp. 126–130.
[29] Y. Mitsunaga, T. Kuwabara, T. Abe, and Y. Ishida, "Molecular hydrogen behavior for loss increase of silica fiber in cable filled with water," *Electron. Lett.*, vol. 20, pp. 76–78, Jan. 19, 1984.
[30] H. D. Campbell and D. O. Lawrence, "Lightning pinholes in cable sheaths," in *Proc. 27th Int. Wire Cable Symp.*, Cherry Hill, NJ, Nov. 14–16, 1978, pp. 273–280.
[31] W. C. L. Weinraub, D. D. Davis, and M. D. Kinard, "A rodent and lightning protective sheath for fiber optic cables," in *Proc. Int. Wire Cable Symp.*, Cherry Hill, NJ, 1983, pp. 243–249.
[32] P. U. Bakhru, K. E. Bow, D. Fischer, and E. Schrom, "Crushing metallic shielded telecommunications cables with dynamic magnetic fields," in *Proc. Int. Wire Cable Symp.*, Cherry Hill, NJ, 1985, pp. 173–186.
[33] R. A. Connolly and N. J. Cogelia, "The gopher and buried cable," *Bell Lab. Rec.*, vol. 48, pp. 99–103, Apr. 1970.
[34] W. E. Howard, "Tests of pocket gophers gnawing electric cables," *J. Widelife Management*, vol. 17, pp. 296–300, July 1953.
[35] N. J. Cogelia, G. K. LaVoie, and J. F. Glahn, "Rodent biting pressure and chewing action and their effects on wire and cable sheath," in *Proc. 25th Int. Wire Cable Symp.*, Cherry Hill, NJ, Nov. 16–18, 1976, pp. 117–124.
[36] D. Keller, D. Moser, and J. Thompson, "Shotgun resistant evaluation," Siecor Internal Rep., Mar. 12, 1986.
[37] J. T. Krause, C. R. Kurkjian, and U. C. Paek, "Strength of fusion splices for fiber lightguides," *Electron. Lett.*, vol. 17, pp. 232–233, Mar. 19, 1981.
[38] T. McMahon, "Factory fusion splice overview," Siecor Internal Rep., Aug. 14, 1984.
[39] C. Miller, "Fiber optical array splicing with etched silicon chips," *Bell Syst. Tech. J.*, vol. 57, pp. 75–90, Jan. 1978.
[40] D. L. Bisbee, "Splicing silica fibers with an electric arc," *Appl. Opt.*, vol. 15, pp. 796–798, Mar. 1976.
[41] Y. Kohanzadeh, "Hot splices of optical waveguide fibers," *Appl. Opt.*, vol. 15, pp. 793–795, Mar. 1976.
[42] T. Arai, O. Watanabe, K. Inada, S. Seikai, and M. Hirai, "Protection and reliability of optical fiber arc fusion splicing," in *Proc. 28th Int. Wire Cable Symp.*, Cherry Hill, NJ, Nov. 13–15, 1979, pp. 344–351.
[43] S. Stueflotten, "Protection of optical fiber arc fusion splices," *J. Opt. Commun.*, vol. 3, pp. 19–25, 1982.
[44] H. Liertz, "Fusion splicing machine for optical fibers," *Electron. Lett.*, vol. 16, pp. 426–427, May 22, 1980.
[45] "Repair of partially damaged single mode mini-bundle cable in an aerial installation," Siecor Recommended Procedure, SRP-000-010.
[46] O. I. Szentesi, "Reliability and restoration of fiber optic cables," in *Proc. GLOBECOM*, Miami, FL, Nov. 29–Dec. 2, 1982.
[47] F. Krahn, W. Meininghaus, D. Ritich, K. Serapins, and J. Gladenbeck, "Installation and field measurement equipment for optical communication cables," in *Proc. 28th Int. Wire Cable Symp.*, Cherry Hill, NJ, Nov. 13–15, 1979, pp. 357–363.

Otto I. Szentesi (S'64–M'70) received the B.Sc. degree in electrical engineering from the University of Alberta, Edmonton, Canada, in 1966 and the Ph.D. degree from University College, London, England, in 1970.

From 1970 to 1971 he was a Research Associate in the Department of Physics at the University of Calgary, Alta., Canada. In 1971 he joined Bell-Northern Research, Ottawa, Ont., Canada where his work included acousto- and electrooptics, high-power light-emitting diode and laser characterization and processing, fiber optic system design and optimization, and interactive audio/visual/graphic terminal development. Since 1978 he has been with Siecor Corporation, Hickory, NC, where he was responsible for fiber optic cable installations, training, project management, and the development and manufacturing of splicing equipment, interconnect hardware, and field test sets. Currently, he is Vice President and General Manager of the Optical Cable Accessories and Test Equipment Division.

Dr. Szentesi is a member of the IEEE Optical Communications Committe and the Association of Professional Engineers of Ontario, Canada.

First Sea Trial of 1.5-μm Submarine Optical Fiber Cable

YUKIYASU NEGISHI, YOSHIAKI MIYAJIMA, OSAMU KAWATA, AND NOBUYUKI YOSHIZAWA

***Abstract*—This paper describes the first sea trial of submarine optical fiber cable employing dispersion shifted fibers. It includes fiber structural design for dispersion shifted fiber, cable manufacturing results, fiber splicing, cable laying results, and a high accuracy measurement technique for optical loss and chromatic dispersion. Based on these results, the feasibility of long-span and high-capacity 1.5-μm submarine optical fiber cable is demonstrated.**

I. Introduction

SUBMARINE optical fiber cables are being used in a number of countries to realize long-length high-bit-rate optical fiber transmission systems utilizing the 1.3-μm wavelength region [1]–[5]. For these systems, optimum fiber parameter design [6], low-loss optical fiber and cable manufacturing techniques [7], low-loss splicing technique [8], and system and repeater circuit design [9] have been examined.

With the development of transoceanic systems, the eighth transatlantic telephone cable (TAT-8), and the third transpacific cable (TPC-3/HAW-4) in 1988 [10], work on submarine optical fiber cables for the 1.3-μm wavelength is nearly complete. Presently, research has begun on submarine optical fiber cable for the 1.5-μm wavelength region to realize the minimum loss point of silica fiber and to increase repeater spanning to almost twice that of 1.3-μm systems. Advantages of system economy and higher reliability may lead to the introduction of 1.5-μm submarine optical fiber cable systems as the next generation submarine cable systems [11].

To realize a long-haul high-bit-rate 1.5-μm undersea lightwave transmission system, our solution is to develop both a single longitudinal mode laser diode (LD) and dispersion shifted fiber to negate the effect of large chromatic dispersion in conventional fiber at 1.5 μm. A few reports dealing with distributed feedback LD (DFB-LD) and dispersion-shifted fiber have been presented to date [12], [13]. Based on these successes, submarine optical fiber system in the 1.5-μm region seems feasible.

In developing submarine optical fiber cable for 1.5-μm operation, we must carefully consider the increased sensitivity to bending loss during cabling and laying operations. In addition, we must establish a high accuracy measurement technique of optical loss and chromatic dispersion for the purpose of investigating long-length and high-bit-rate transmission potential. This paper describes design and manufacturing results of submarine optical fiber cable using dispersion shifted fibers and, also, the results of the first sea trial of such a cable. Further, a high accuracy measurement technique of optical loss and chromatic dispersion is described.

Manuscript received March 11, 1987; revised May 14, 1987.
The authors are with NTT Electrical Communications Laboratories, Tokai, Ibaraki-ken, 319-11, Japan.
IEEE Log Number 8717616.

II. Fiber Parameter Design

To realize a submarine optical fiber cable for using in the 1.5-μm wavelength region, two types of fibers can be considered feasible: conventional fiber with 1.3-μm zero dispersion wavelength or dispersion shifted fiber with zero dispersion in the 1.5-μm wavelength region. Conventional fiber for 1.5-μm use has already been developed [14]. Further, an ultralow-loss of 0.154 dB/km is reported with pure silica core single-mode fiber [15]. When applying conventional fiber to a 1.5-μm submarine cable systems, it is necessary to optimize fiber parameters (for example, the LP_{01} mode cutoff wavelength, mode field diameter) from that in 1.3-μm use. This is done to suppress increased bending loss caused by cabling and/or fiber storage in a jointing box.

On the other hand, it is difficult to manufacture low-loss dispersion-shifted fiber because of the high refractive index difference and complicated fiber profile required for canceling material dispersion with waveguide dispersion. However, low-loss dispersion shifted fiber with less than 0.20 dB/km at a 1.5-μm wavelength has been reported [16]. However, fiber parameter design considering bending loss characteristics and fiber splicing loss characteristics is still open.

A determination method for appropriate fiber parameters has been reported. In this method, mode field diameter (MFD) and cutoff wavelength (λ_{ce}) are determined to make the overall transmission-line loss in one repeater span minimum [6]. Taking into account intrinsic fiber loss, splicing loss, and bending loss, total loss in one repeater span α_T is given by

$$\alpha_T = (\alpha_R + \alpha_i + \alpha_{b1})L + (L/l - 1)\alpha_s + 4\alpha_{b2} \quad (1)$$

where L is a repeater span length, l is a fiber piece length between splices, and L = 120 km and l = 15 km were used. Symbols in (1) are summarized in Table I. We assumed two joint boxes in an initial 120-km-long cable and

TABLE I
OPTICAL LOSS FACTORS AND TYPICAL VALUES

Loss factor		Functional formula	Typical value: Conventional fiber	Typical value: Dispersion shifted fiber
	○ Rayleigh scattering	$\alpha_R = (A \cdot \Delta + B)/\lambda^4$	0.17 dB/km	0.19 dB/km
	○ Radiation	αi	0.04 dB/km	0.02 dB/km
Splice	○ Core axis offset and field mismatch	$\alpha_S = -10 \log [(2W_1W_2/(W_1^2+W_2^2))^2 \times \exp(-2d^2/(W_1^2+W_2^2))]$	~0.05 dB	~0.05 dB
Bending	○ Cabling	α_{b_1} : Uniform bending	0.02 dB/km	0.01 dB/km
	○ Storage in J.B.	α_{b_2} : 45mm diameter bending	< 0.01 dB	< 0.01 dB

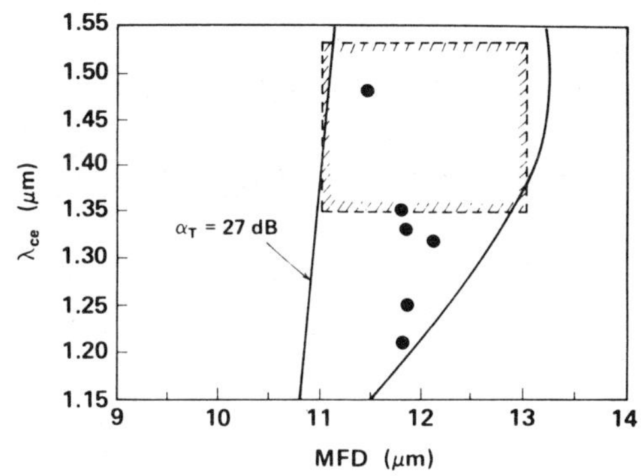

Fig. 1. Appropriate fiber parameters (for 1.3-μm dispersion-free fiber).

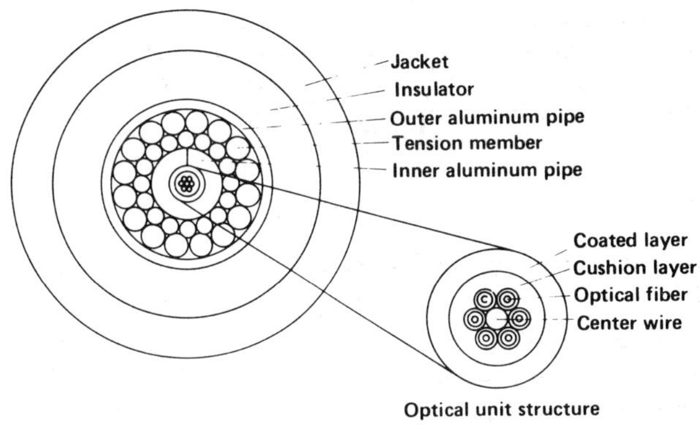

Fig. 3. Cable structure.

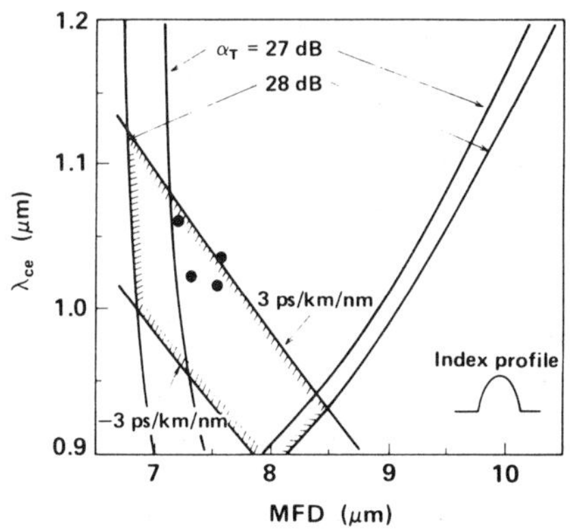

Fig. 2. Appropriate fiber parameters (for dispersion shifted fiber).

an additional two joint boxes after repairing it. Therefore, four $\alpha_{b2}s$ are considered in total in (1).

Appropriate fiber parameters of 1.3-μm zero dispersion fiber and 1.5-μm dispersion shifted fiber are shown in Figs. 1 and 2, respectively. Fig. 1 is based on [14]. Appropriate fiber parameters for dispersion shifted fiber considering the dispersion limit [17] is shown by the hatched area in Fig. 2. The fiber parameters in manufactured cable are plotted in Figs. 1 and 2, respectively.

III. Manufactured Cable Characteristics

A. Cable Structure

The manufactured cable structure is shown in Fig. 3. The cable consists of an optical unit with single-mode fibers, inner aluminum pipe, steel tension members, an outer aluminum pipe, dielectric layer, and jacket. Two cables were used for sea trial, one containing six pure-silica fibers with 1.3-μm zero dispersion wavelength, the other containing dispersion shifted fiber with zero dispersion in the 1.5-μm wavelength region. The cable lengths are 9 and 7 km, respectively. Of the 7-km length, 2.5 km was armored.

B. Loss and Dispersion Characteristics

The spectral loss of cabled dispersion shifted fiber and loss change during the manufacturing process are shown in Fig. 4. AL(1) and AL(2) in the inset of Fig. 4 correspond to the inner and outer aluminum pipe manufacturing processes, respectively. Both types of fibers were cabled without loss increase during the manufacturing process.

Dispersion characteristics for dispersion shifted fibers are shown in Fig. 5, along with the fiber profiles used. Profile (a) is a simple step-like profile with a fluorine cladding area, (b) consists of α-like profile core, and (c) is a gaussian profile with silica-based cladding. The dispersion slope is about 0.06 ps/km/nm^2.

Bending loss characteristics for dispersion-shifted fiber and 1.3-μm zero-dispersion pure-silica fiber are summarized in Fig. 6. Theoretical bending loss for conventional fiber is also shown in this figure [17]. Dispersion shifted fiber with step-like high refractive index difference exhibits less bending loss than conventional fiber. Zero dispersion 1.3-μm pure silica core fiber can decrease bending loss by adopting high cutoff wavelength. Therefore, both types of fiber have better bend performance than conventional 1.3-μm fiber.

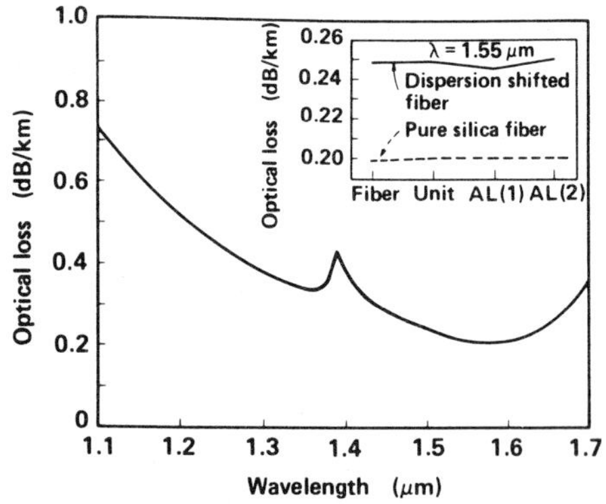

Fig. 4. Fiber loss spectrum and loss change in cable manufacturing process.

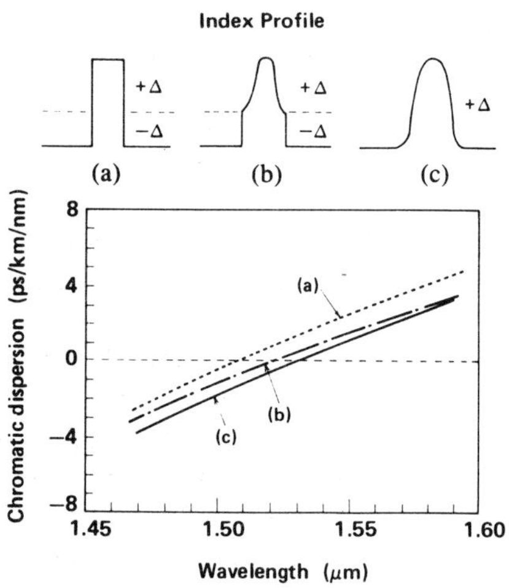

Fig. 5. (a)-(c) Dispersion characteristics.

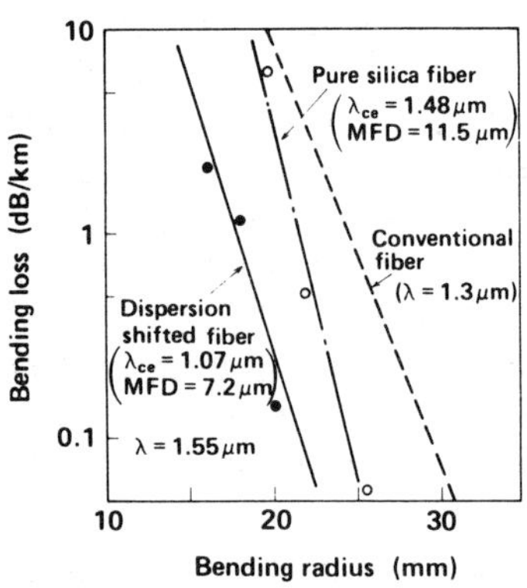

Fig. 6. Bending loss characteristics.

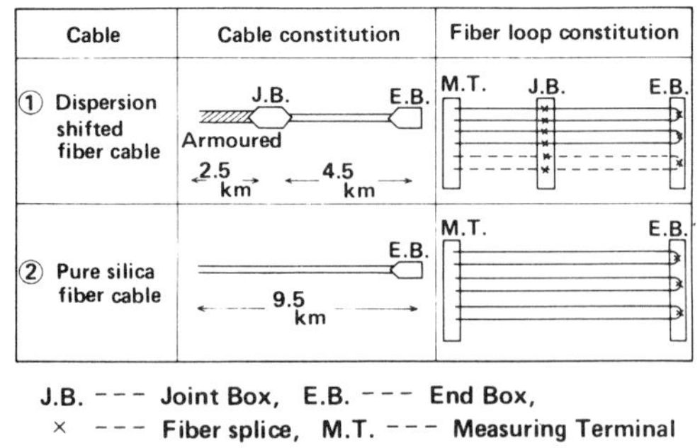

Fig. 7. Cable configuration for sea trial.

IV. Sea Trial Results

A. Outline of Sea Trial

A sea trial was carried out in the Pacific, near the Amami islands in June 1986 [18]. The cable configuration for the sea trial is shown in Fig. 7. Cable 1 consisted of dispersion shifted fibers, and 2.5-km armored cable was jointed to 4.5-km nonarmored cable by a joint box (JB). Cable 2 consisted of 1.3-μm zero dispersion pure silica fibers, and was 9.5-km nonarmored cable. Main subjects in this sea trial are 1) cable laying and recovery experiment and 2) the transmission experiment in 120-km fiber loop with a submerged repeater.

B. Cable Laying and Recovery Experiment

Results of the cable laying and recovery experiment are shown in Fig. 8, where (a) is optical loss change during cable laying and recovery process, (b) is cable tension, (c) is cable laying and recovery velocity. Maximum loss increase in laying and recovery process is 0.007 dB/km at 1.55 μm, which almost coincides with loss increase caused by temperature change of cable in water. No spectral loss change between 1.0 and 1.8 μm was observed before and after laying. Also, no change was observed in chromatic dispersion (2.33 ps/km/nm at 1.55-μm wavelength).

Maximum tension applied to the cable was 4.5 tons, which corresponds to 3 wh, where w is a cable weight in unit length and h is a water depth. Cable laying speed was usually kept to 2 knots (3.7 km/s), but was reduced to 1 knot when a joint box and repeater were laid out, as shown in Fig. 8(c). From these results, loss increase caused by laying can be estimated from the temperature change between cable tank and sea bed. Loss increase in fiber caused by temperature change is 0.0003 dB/km/°C, so that the laying effect is estimated as 0.01 dB/km considering a 30-°C temperature change. Loss increase in pure silica fiber cable was about 0.006 dB/km in maximum.

C. Constitution of 120-km Fiber Loop

For conducting the transmission experiment, a 120-km-long fiber loop was constituted by splicing an optical unit and optical fibers to the cable. The 120-km fiber loop contained twelve splices. A splice loss histogram for dispersion shifted fiber is shown in Fig. 9, where an average splice loss of 0.121 dB can be seen. The average value of mode field diameters in fibers is 7.34 μm at a 1.55-μm wavelength. Maximum splice loss was obtained for fiber with a minimum mode field diameter.

Splice loss was investigated for various fibers with different mode field diameters. The relationship between mode field diameters and splice loss is described in Fig. 10. The solid line is a calculated value assuming core axis misalignment of 1 μm and mode field mismatch of 0.5-μm. Splice loss decreases with large mode field diameter fiber. In order to attain low-loss splice less than 0.1 dB, the mode field diameter must be larger than 7.5 μm.

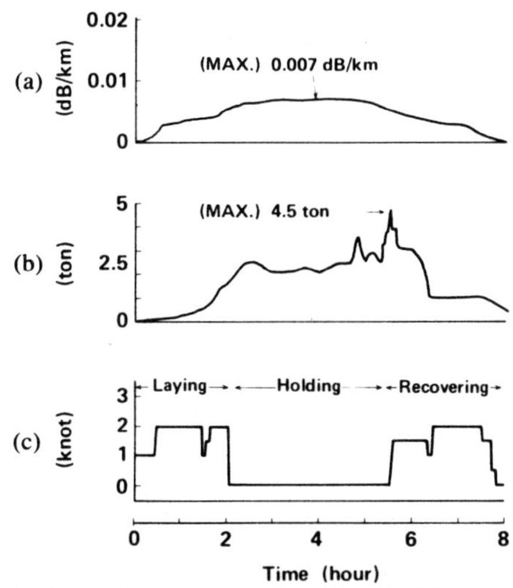

Fig. 8. Cable laying and recovery results: (a) optical loss change, (b) cable tension, (c) laying and recovery velocity.

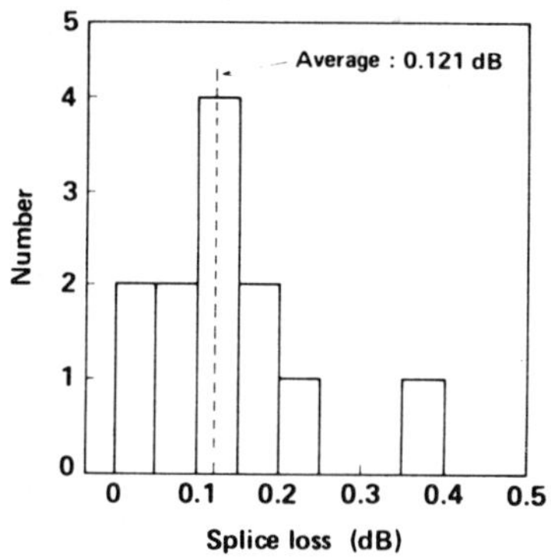

Fig. 9. Splice loss histogram.

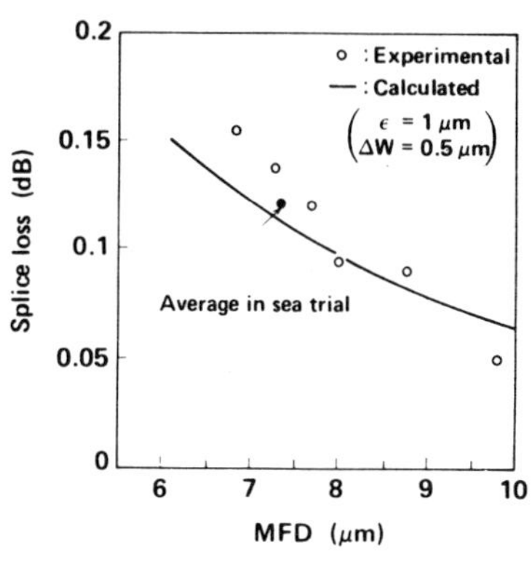

Fig. 10. The relationship between MFD and splice loss.

Average fiber loss including splicing loss is 0.25 dB/km at the 1.55-μm wavelength for 120-km-long dispersion shifted fiber, and is 0.20 dB/km for 120-km-long pure silica fiber.

V. Optical Loss and Chromatic Dispersion Measurement Technique

A. Optical Loss Measurement

Optical loss at the 1.5-μm wavelength region is much lower than at the 1.3-μm wavelength region. To utilize low loss characteristics to the maximum, loss change in the cable laying process should be evaluated with a higher degree of accuracy than that required for the 1.3-μm system.

During loss change measurement in the laying and recovery process, the optical light source is accommodated in the thermal control chamber to stabilize optical output power. As a thermally conductive medium, air and nonactivated liquid were investigated. Measured results of optical power change for both methods are shown in Fig. 11. Because of the high thermal conductivity of liquid material, optical power stability less than 0.003 dB is confirmed by using nonactivated liquid as the conductive medium. This level of accuracy is sufficient for evaluating the 0.01 dB/km laying effect.

B. Chromatic Dispersion Measurement Technique

For realizing long-length and high-bit-rate optical transmission in the 1.5-μm wavelength region, it is necessary to evaluate chromatic dispersion for long-length optical fiber accurately. In the following section, the chromatic dispersion measurement result in a 120-km-long fiber loop is described [19].

The experimental setup consisted of four sinusoidally modulated laser diodes (LD's), an InGaAs-APD, and a phase detector. The LD center wavelengths were $\lambda_1 = 1496$, $\lambda_2 = 1520$, $\lambda_3 = 1560$, and $\lambda_4 = 1590$ nm. When an LD is sinusoidally modulated, the modulated light signal travels along the fiber with a phase change $\phi = 2\pi f L \tau(\lambda)$, where f is modulation frequency, $\tau(\lambda)$ is the wavelength dependent time delay for fiber unit length, and L is fiber length. By detecting phase differences among LD's and using the least-mean-square fitting approximation for the Sellmeier equation $\tau(\lambda) = a\lambda^2 + b + c\lambda^{-2}$, the wavelength-dependent relative time delay is obtained. Then, chromatic dispersion is evaluated by differentiating wavelength-dependent time delay $\tau(\lambda)$.

The measured result of chromatic dispersion for 123-km fiber is shown in Fig. 12. A dispersion-free wavelength is found to be 1.51-μm, and the dispersion at 1.55 μm is 2.33 ps/km/nm. Standard deviation of dispersion for ten times measurements at 1.55 μm is 0.01 ps/km/nm.

The chromatic dispersion for long length fiber was investigated by comparing a total dispersion value with a sum of dispersion values of constituent fibers. Measured length dependence of the dispersion at 1.55-μm wavelength is indicated in Fig. 13 by open circles. Measurements were conducted by cutting the fiber in order at spliced points from the output end of fiber. Solid lines represent the plots of the calculated values using dispersions in constituent fibers. Calculated values are obtained by summing dispersion values of individual fibers considering the weight of fiber length. Filled circles indicate the difference between experimental and calculated values.

Dispersion difference $\sigma_{ex} - \sigma_{cal}$ is almost independent of fiber length. This is because the modulation frequency

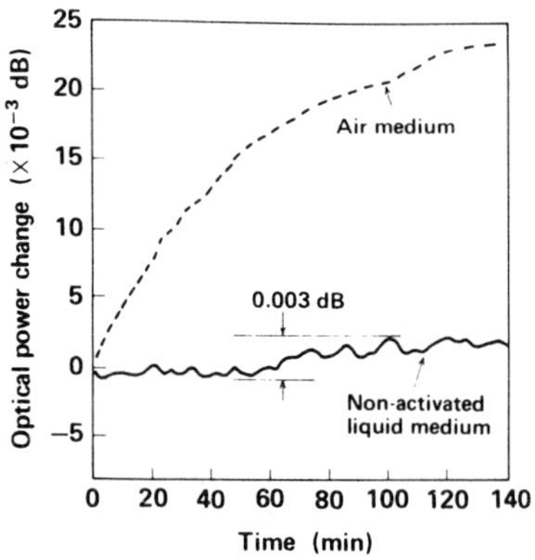

Fig. 11. Measured result of optical power change.

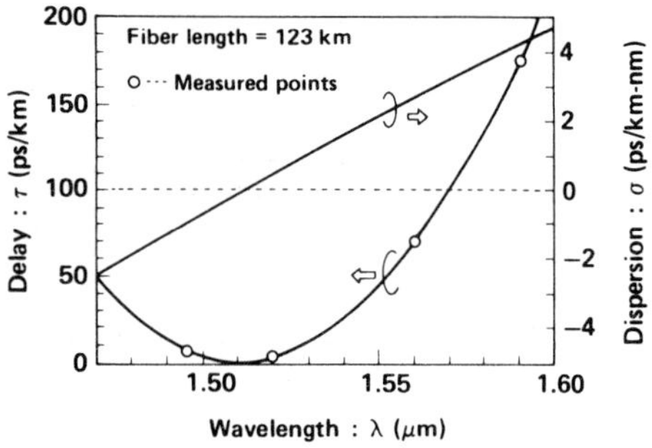

Fig. 12. Measured result of chromatic dispersion for 123-km fiber.

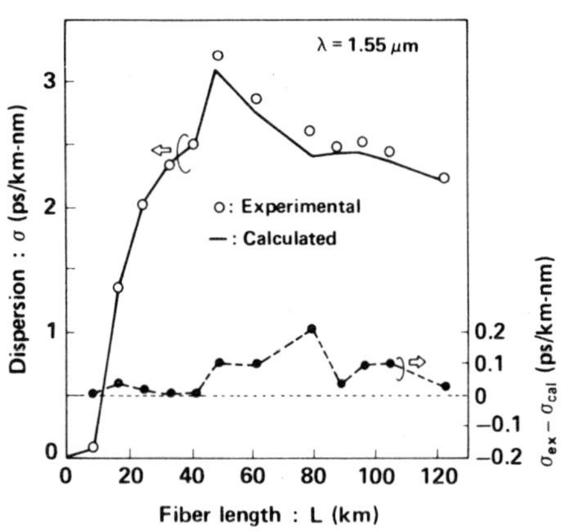

Fig. 13. Length dependence of chromatic dispersion.

f is chosen as fL = constant in this measuring equipment. The difference becomes somewhat larger in fibers 6–8, but the maximum difference is only 0.2 ps/km/nm. This implies that the total chromatic dispersion of a long-length spliced fiber is predictable from dispersion values of the constituent fibers.

VI. Conclusion

Results of the first sea trial with 1.5-μm submarine optical fiber cable assembling dispersion shifted fiber and 1.3-μm dispersion-free pure silica core fiber was reported. The conclusions can be summarized as follows.

1) Appropriate fiber parameters for dispersion shifted fiber which provides minimum optical loss in a repeater span is clarified.

2) From the sea trial result, loss change during the cable laying and recovery process is less than the value estimated from temperature change. The feasibility of the use of dispersion shifted fiber and 1.3-μm dispersion-free pure silica fiber is confirmed.

3) The optical loss change is measured with measurement error less than 0.003 dB by applying a new thermal control method for optical light source. Chromatic dispersion for 120-km-long fiber is measured and the additive rule of chromatic dispersion for such long dispersion shifted fiber in the 1.5-μm-wavelength region was confirmed.

Acknowledgment

The authors wish to express their thanks to N. Uchida for useful discussions and to N. Kojima for his continuous encouragement.

References

[1] Y. Negishi, K. Ishihara, Y. Murakami, and N. Yoshizawa, "Design of deep-sea submarine optical fiber cable," *IEEE J. Selected Areas Commun.*, vol. SAC-2, no. 6, pp. 879–885, 1984.

[2] Y. Niiro, "The OS-280M optical-fiber submarine cable system," *IEEE J. Selected Areas Commun.*, vol. SAC-2, no. 6, pp. 807–812, 1984.

[3] A. Adl, T. M. Chien, and T. C. Chu, "Design and testing of the SL cable," *IEEE J. Selected Areas Commun.*, vol. SAC-2, no. 6, pp. 864–872, 1984.

[4] P. Worthington, "Cable design for optical submarine systems," *IEEE J. Selected Areas Commun.*, vol. SAC-2, no. 6, pp. 873–878, 1984.

[5] J. P. Trezeguet, P. Ollion, P. Franco, and J. Thiennot, "The submarine optical cable of the submarcom S 280 system," *IEEE J. Selected Areas Commun.*, vol. SAC-2, no. 6, pp. 886–895, 1984.

[6] K. Kitayama, Y. Kato, M. Ohashi, Y. Ishida, and N. Uchida, "Design considerations for the structural optimization of a single-mode fiber," *J. Lightwave Technol.*, vol. LT-1, no. 2, pp. 363–369, 1983.

[7] K. Ishihara, O. Kawata, and N. Yoshizawa, "Design and characteristics of submarine optical fiber cable," in *Rev. Electrical Commun. Lab.*, vol. 33, no. 6, pp. 911–916, 1985.

[8] O. Kawata, K. Hoshino, Y. Miyajima, M. Ohnishi, and K. Ishihara, "A splicing and inspection technique for single-mode fibers using direct core monitoring," *J. Lightwave Technol.*, vol. LT-2, no. 2, pp. 185–191, 1984.

[9] H. Fukinuki, T. Ito, M. Aiki, and Y. Hayashi, "The FS-400M submarine system," *IEEE J. Selected Areas Commun.*, vol. SAC-2, no. 6, pp. 794–800, 1984.

[10] P. K. Runge and P. R. Trischitta, "The SL undersea lightwave system," *IEEE J. Selected Areas Commun.*, vol. SAC-2, no. 6, pp. 784–793, 1984.

[11] R. E. Wagner, "Future 1.55-μm undersea lightwave systems," *IEEE J. Selected Areas Commun.*, vol. SAC-2, no. 6, pp. 1047–1055, 1984.

[12] Y. Itaya, K. Wakita, G. Motosugi, and T. Ikegami, "Phase control by coating in 1.5-μm distributed feedback lasers," *IEEE J. Quantum Electron.*, vol. QE-21, no. 6, pp. 527–533, 1985.

[13] N. Kuwaki, M. Ohashi, C. Tanaka, and N. Uesugi, "Characteristics of convex-index fiber with index center core," in *Proc. OEC'86* (Tokyo, Japan), July 29–31, 1986, pp. 234–235.

[14] K. Kitayama, N. Uesugi, M. Ohashi, S. Seikai, and K. Ishihara, "Design and performance of ultra-low-loss single-mode fiber cable in 1.5-μm wavelength region," *J. Lightwave Technol.*, vol. LT-3, no. 3, pp. 579–585, 1985.

[15] H. Yokota, H. Kanamori, Y. Ishiguro, G. Tanaka, S. Tanaka, H. Takada, M. Watanabe, S. Suzuki, K. Yano, M. Hoshikawa, and H. Shimba, "Ultra-low-loss pure-silica-core single-mode fiber and transmission experiment," in *Proc. OFC'86* (Atlanta, GA), February 24–26, 1986.

[16] T. D. Croft, J. E. Ritter, and A. Bhagavatula, "Low-loss dispersion-shifted single-mode fiber manufactured by the OVD process," in *Proc. OFC'85* (San Diego, CA), Feb. 11–13, 1985, pap. WD2.

[17] M. Ohashi, N. Kuwaki, C. Tanaka, N. Uesugi, and Y. Negishi,

"Characteristics of dispersion-shifted dual shape core fibers," to be published (In Japanese).
[18] S. Tsutsumi, H. Okamura, Y. Miyajima, and M. Ohnishi, "1.5-μm submarine transmission system sea trial," *Electron. Lett.*, vol. 22, no. 24, pp. 1297–1299, 1986.
[19] Y. Miyajima, M. Ohnishi, and Y. Negishi, "Chromatic dispersion measurement over a 120-km dispersion-shifted single-mode fiber in the 1.5-μm wavelength region," *Electron. Lett.*, vol. 22, no. 22, pp. 1185–1186.

*

Yukiyasu Negishi was born on March 9, 1941. He received the B.S. degree in electronics engineering from Tohoku University, Sendai, Japan.

He started his career with the Ibaraki Electrical Communications Laboratory, NTT, Ibaraki, Japan, in April 1964. Since then, he has mainly engaged in development research on communication cables. Since 1978 he has studied submarine optical-fiber cables. He is now Head of the Trunk Transmission Line Section. Currently, his major responsibilities include developmental research on optical cables for land trunk transmission systems and submarine transmission systems.

Mr. Negishi is a member of the Institute of Electronics, Information, and Communication Engineers of Japan and the Japan Society of Applied Physics.

*

Yoshiaki Miyajima was born on March 2, 1953. He received the B.S., M.S., and Ph.D. degrees in electrical engineering from Kyushu University, Fukuoka, Japan, in 1976, 1978, and 1986, respectively.

He joined the Ibaraki Electrical Communications Laboratories, NTT, in 1978. He has been engaged in research and development on submarine optical-fiber cable, especially on the mechanical reliability of optical fibers for submarine cable. He is presently engaged in developmental research on 1.5-μm submarine optical-fiber cable. He is now Research Engineer in the Trunk Transmission Line Section, Transmission Line Division, NTT Ibaraki ECL.

Dr. Miyajima is a member of the Institute of Electronics, Information, and Communication Engineers of Japan and of the Institute of Electrical Engineers of Japan.

*

Osamu Kawata was born in Osaka, Japan, on January 29, 1952. He received the B.S. and M.S. degrees in electronics engineering from Tokyo Institute of Technology, Tokyo, Japan, in 1974 and 1976, respectively.

He joined the Ibaraki Electrical Communications Laboratory, NTT, Ibaraki, Japan, where he has been engaged in research and development on submarine optical-fiber cables and their joints.

Mr. Kawata is a member of the Institute of Electronics, Information, and Communication Engineers of Japan.

*

Nobuyuki Yoshizawa was born in Tokyo, Japan, on March 6, 1954. He received the B.S., M.S., and Ph.D. Degrees in mechanical engineering from Waseda University, Tokyo, Japan, in 1972, 1974, and 1986, respectively.

He joined the Ibaraki Electrical Communications Laboratory, NTT, Ibaraki, Japan, where he has been engaged in the mechanical design of submarine optical-fiber cables and their optical units.

Dr. Yoshizawa is a member of the Institute of Electronics, Information, and Communication Engineers of Japan.

REPORT:

Jerry M. Anderson, Dean R. Frey, *and* ***Calvin M. Miller*** *are with AT&T Bell Laboratories in Atlanta. Mr. Anderson is a supervisor in the Cable Joining Department, where he is responsible for connector systems development. He joined AT&T in 1979. He holds a B.S. and M.S. from Auburn University and a Ph.D. from Stanford University, all in mechanical engineering. Mr. Frey is head of the Cable Joining Department. He joined AT&T in 1959. His degrees are in mechanical engineering: a B.S. from Penn State University and an M.S. from New York University. Mr. Miller is supervisor of the Exploratory Lightguide Components Group. He is responsible for passive lightguide component design. He joined AT&T in 1967 and holds a B.S. from North Carolina State University and M.S.E. from Akron University in electrical engineering.*

LIGHTWAVE SPLICING AND CONNECTOR TECHNOLOGY

Introduction

The functions of splices and connectors are to interconnect the transmission media and to connect the media to optical devices. Splices are for the most part intended to stay permanently connected, although some designs offer the opportunity of separation and rejoining. Connectors, on the other hand, are designed for frequent separation and rejoining. Connectors may possibly require cleaning when rejoined, whereas splices require replacement of matching gels or adhesives and possibly repolishing. The most operationally significant system parameter of these components is optical loss. The loss is dependent on alignment of the fibers, the end conditions of the fibers, and the fiber-core parameters (primarily core diameter and peak index difference in multimode fibers and mode-field diameter in single-mode fibers). These fiber-core parameters are not under the control of the splice or connector design and will not be covered further here.

In the few years since lightwave systems have been in use, the technology has shifted from the graded-index multimode (MM) fiber to single-mode (SM) fiber that operates at 1310 nm and/or 1550 nm in essentially all systems. Present exceptions are local area networks (LANs), computer interconnections, and possibly residential distribution, but these may be SM in a few years. The smaller core size of SM media makes splicing and connecting much more difficult. Most SM cores are about 8 μm in diameter compared to 50 or 62.5 μm for MM cores. To get a feel for how small this is, an ordinary sheet of notebook paper is about 50 μm thick. Therefore, six SM-fiber cores laid side by side are less than the thickness of a sheet of notebook paper.

The optical loss of a splice or a connector is directly related to the accuracy of the alignment, and for a given process, the cost increases with accuracy of alignment. The cores of SM fibers must be aligned to within 1 μm in a connector, with no Fresnel reflection loss, to obtain a loss of a few tenths of a dB. Misalignment can be up to 3 μm if a loss of 2 dB is allowed. The cost of the connectors tends to differ inversely by about the same ratio.

Optical-loss requirements are not the same for all applications. Low-loss components are needed for long-haul intercity trunks to get maximum spacing between regenerators.

In metro trunks, loops, LANs, and computer interconnections, low optical loss can be traded for components with reduced cost and/or shorter installation times. The choice between optical performance and cost can be optimized for any given application.

Development programs and AT&T's product line allow the customer to choose from a full range of high-quality products for splicing and connecting lightwave media optimized for various applications.

A review of the designs and performance of AT&T splices and connectors is presented.

Splicing

The two fundamental approaches to splicing optical fibers are fusion welding and mechanical joining. Both methods require three basic steps:

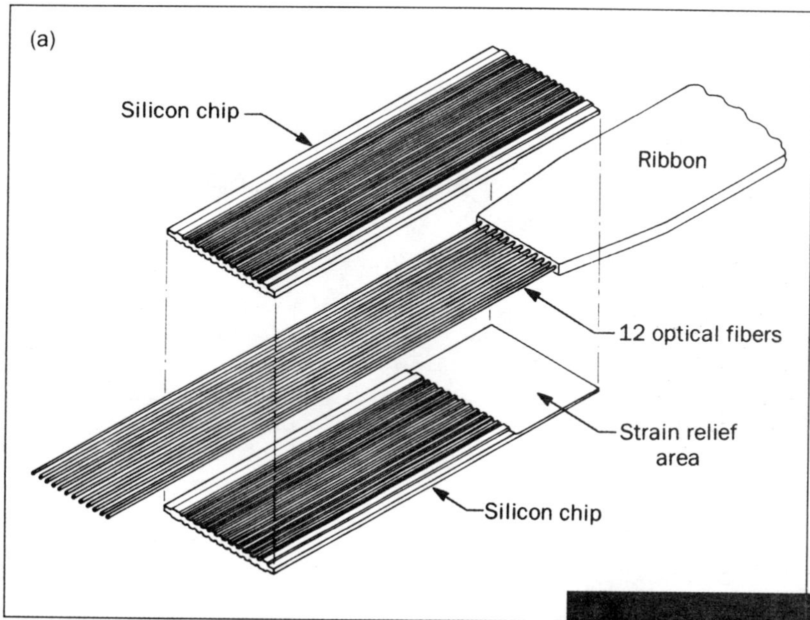

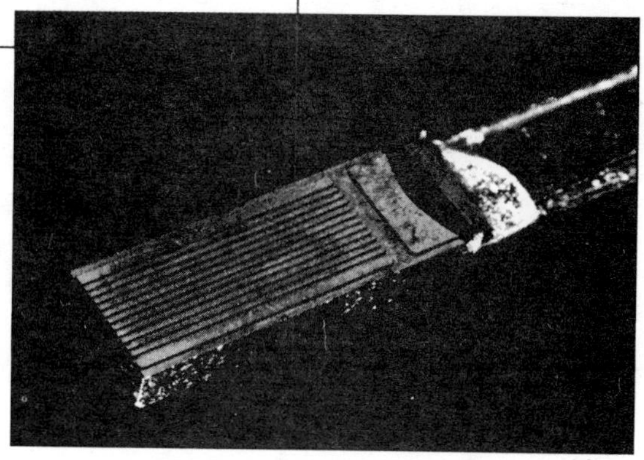

Figure 1. (a) Light ribbon array connector. (b) Silicon chip array splice schematic.

1. Fiber end preparation
2. Alignment of the fibers
3. Retention of the fibers in the aligned position.

Arc fusion is used by many for splicing optical fibers. In this method, the fiber ends are cleaved and aligned, and the ends are melted and fused to retain the alignment. In single-mode fusion, the fibers are aligned with x-y micropositioners that are driven with a feedback signal to peak the transmission of the splice. Despite automated alignment and fusion processes in the newer machines, the quality of the splices is still sensitive to small variations in the execution of the three basic steps and to unfavorable environmental conditions. Average field-splicing times vary greatly, from 15 to 40 minutes per splice, due mostly to environmental effects.

To achieve average splice losses of 0.1 to 0.25 dB with fusion in the field, resplicing is frequently required to eliminate high-loss splices. Fusion splices are physically the smallest and when properly made have extremely stable loss over a wide temperature range. However, for field splices of terrestrial cables, the size of a fiber splice is usually of little con-

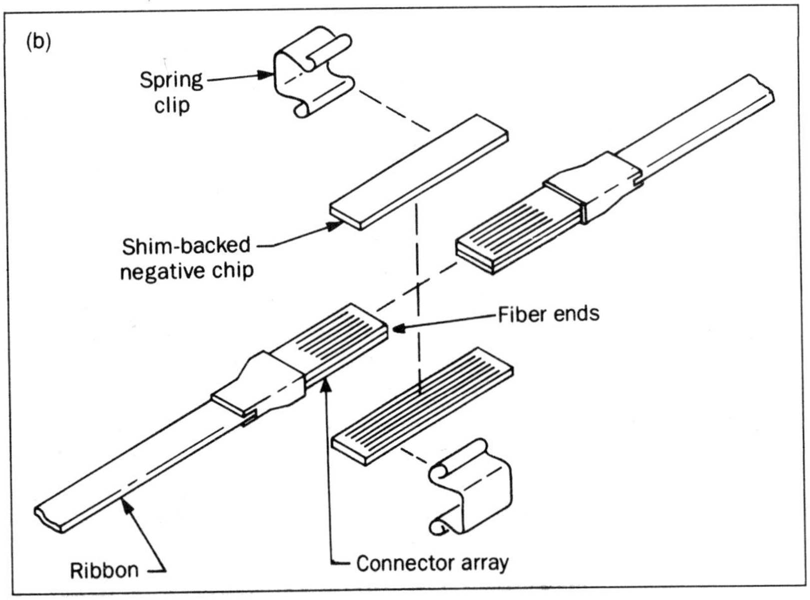

sequence because the size of a cable splice is dictated by the minimum bend radius of the fibers, not the size of the fiber splice.

To prevent fiber breakage due to static fatigue and excess loss at long wavelengths, the minimum bend radius for depressed-cladding fibers is $1^1/_2$ inches. For matched-cladding fiber designs, the minimum bend radius must be substantially larger to avoid excessive loss at long wavelengths (1550 nm).

In mechanical splices, the ground and polished ends are aligned and butted together with a matching gel or adhesive between the ends. Alignment is maintained mechanically. Some designs combine and simplify the three basic operations of end preparation, alignment, and retention to achieve increases in performance and productivity. Reflections from the fiber end faces can be made small enough that they do not interfere with the most sophisticated transmission systems by using proper index matching gels or adhesives. The remaining small reflections are an aid in analyzing faults along a fiber route with an optical time domain reflectometer.

During the evolution of lightwave technology, AT&T developed reliable, high production, craft friendly, and long-term stable mechanical splices that are matched to the application. To this end, the first lightguide cables were designed with flat linear arrays (ribbons) of 12 fibers.[1,2] These cables were very receptive to splicing en masse (12 fibers at a time) with an array splice[3] that could be partially assembled in the factory. The array splice has application for both MM and SM as discussed below. For individual fiber splicing, the fibers are placed in glass-ferrule supports that are easier to handle, grind, polish, and align than bare fibers. The ferrule supports, the matching gels, and the alignment-retention structures are chosen to be environmentally compatible with the fiber. This is essential for long-term stability over wide temperature and humidity changes.

Array Splices. In the array splice, 12 fibers of a ribbon are sandwiched in the grooves of two precisely etched silicon chips (Figure 1). The sandwich is held together with epoxy, and the end faces of the fibers and array are polished flat and perpendicular to the array. The arrays are placed on the cable in the factory. In the field, two mating arrays are butted and aligned with two negative chips that mate with the grooves on the outside of the arrays to form a splice as shown in Figure 1b. The alignment is maintained with mechanical springs. A matching gel placed between the arrays maintains low and stable loss over a wide range of temperature and humidity.

The array splice was introduced in 1979 for MM applications. Over half a million splices have been made in the field with an outstanding reliability record; no splice failures have been reported. The MM splice loss averages 0.15 dB with a standard deviation of 0.15 dB. Field productivity is higher and installed cost is lower than corresponding measures for individual fiber splicing. An array splice (12 fibers) can be joined and tested in the field in 5 to 10 minutes.

The high productivity and low splicing cost stimulated interest in the array splice for SM applications. This would have been a wishful thought in 1980, but better dimensional control of the silicon chips and fibers have made SM array splicing feasible.[4]

The loss distribution of splices made in the laboratory is close to that of splices made in the field.[5] Average loss is about 0.35 dB with a standard deviation of about 0.3 dB. The similarity of the two distributions indicates consistency between lab and field results.

Besides showing that the technology is transferable to the field, consistency between laboratory and field results indicates that the method is insensitive to craft skills and environmental conditions during splicing.

Rapid Ribbon Splicing. While factory-installed arrays yield the fastest field splice, fabrication of the entire splice in the field, which includes placing the fibers in the arrays, takes about 2 hours for the first splice and nearly 1 hour to complete each additional splice. This is still a very reasonable time and is as fast or faster than most individual splicing methods. However, 2 hours is a long time to restore the first pair of fibers of a damaged cable that is carrying service, even if six pairs of fibers are repaired at the same time. This shortcoming of the array splice for rapid restoration stimulated the development of the rapid ribbon splice, a field method for splicing a 12-fiber ribbon in 20 to 30 minutes.[6] Note, this is about the average time required for splicing one fiber with arc fusion.

In the rapid ribbon splice, all 24 fibers of a splice are ground and polished simultaneously. Then a portion of the ribbon is removed, and the bare fibers are placed into a grooved substrate with the ends butted as shown in Figure 2. The continuous grooves of the substrate align the fibers, and a matching gel is placed between the fiber ends to reduce the splice loss and reflections. A molded cover is placed over the fibers and secured with spring fasteners to retain the alignment. Only simple hand tools, a vacuum pump, and a fixture with a low-power microscope are required to make the splice. The rapid ribbon splicing operations

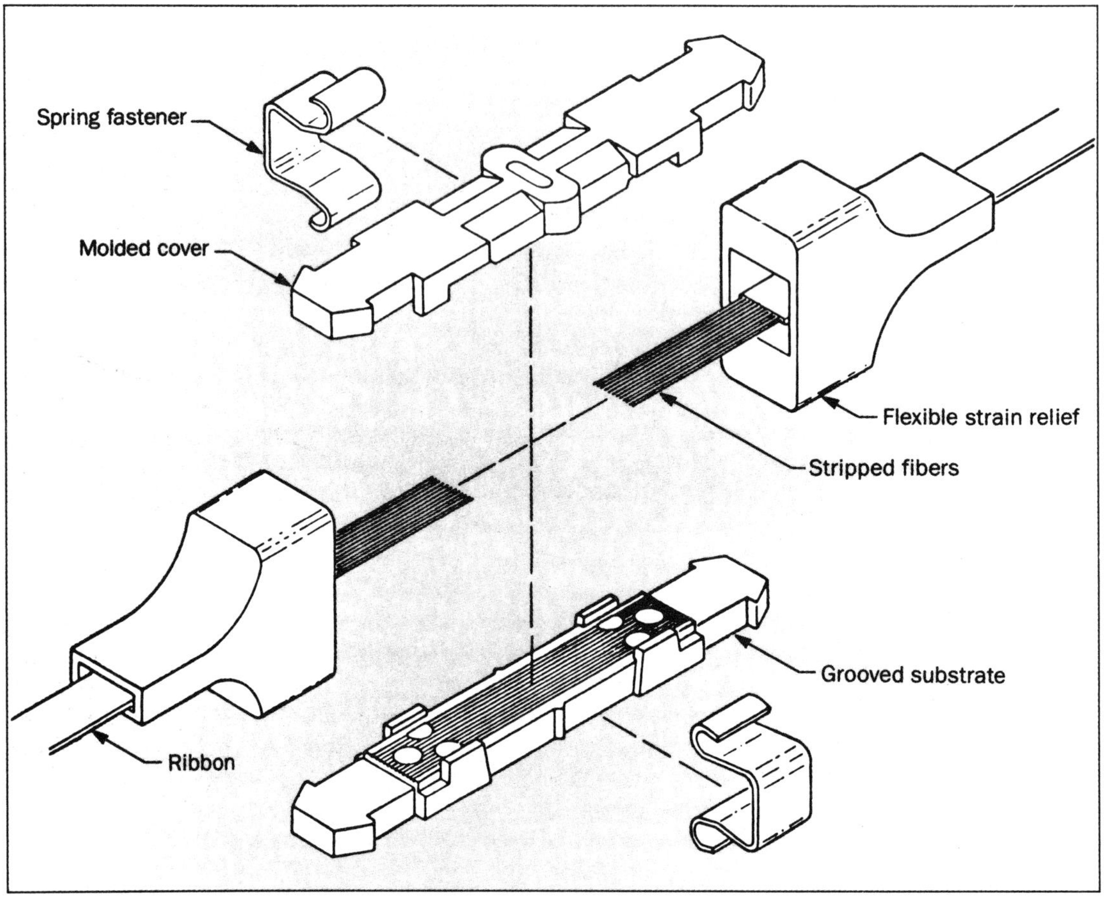

Figure 2. Rapid ribbon splice.

are easy and craft friendly.

The rapid ribbon splice was first developed for multimode fibers, but it has since been enhanced so that SM fibers can be spliced as well.[7] Like the array splice, the field loss data tracks the laboratory loss data very well.

Average splice loss is about 0.35 dB with a standard deviation of about 0.3 dB. The splices are very stable; added loss over a temperature cycle from –40°F to 170°F is less than 0.05 dB with no residual loss after cycling.

While the rapid ribbon splice design objective was to develop a quick temporary repair splice, the completed design surpasses the objectives and is a permanent splice that has application in new installations. It is particularly suited for LANs, building distribution systems (BDS), service drop cables, and other uses where factory-installed arrays are impractical. In these applications, it is difficult

Figure 3. Principle of operation of the rotary splice.

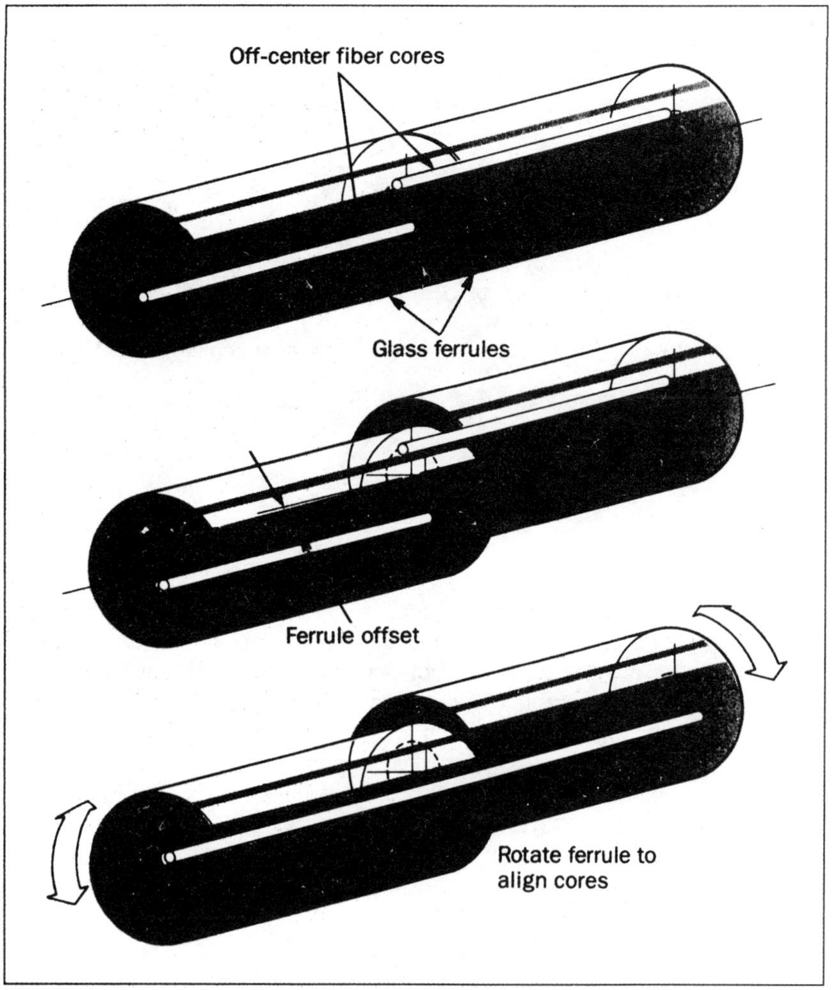

to determine the exact length of the cables, and the installations do not provide slack storage consistent with preterminated ends.

Individual-Fiber Splices. AT&T's individual-fiber splices use glass capillary tubes to provide robust, easy-to-handle terminations for individual fibers in much the same way as silicon chips do for multiple fibers.

Accuracies and tolerances of glass tubes combined with careful mechanical design allow MM and SM fibers to be spliced with average splice losses of 0.20 dB without using any test equipment. Active alignment allows tuning of rotary SM splices to less than 0.05 dB

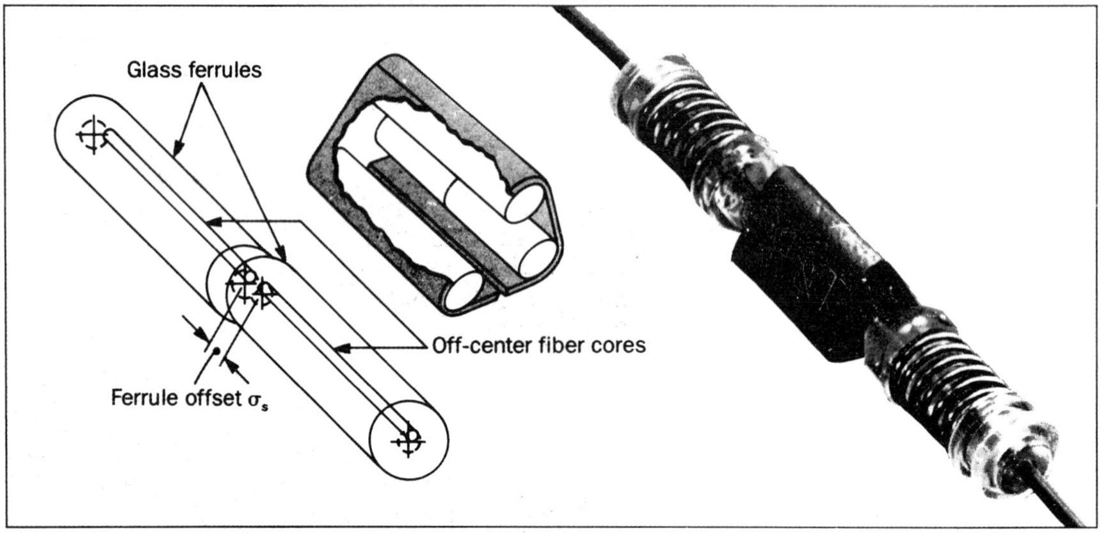

Figure 4. Alignment sleeve for rotary mechanical splice.

even when splicing is done in the field.

The MM mechanical splice was the first glass tube splice.[8] MM fibers are terminated in glass-tube assemblies using ultraviolet-curable cement, and the ends are ground and polished. Each ferrule is inserted in opposite ends of a cyclindrical alignment sleeve and index matched with the same gel used with silicon chip arrays. Splice losses average 0.2 dB with no tuning or can be tuned to below 0.1 dB by active alignment. Losses are acceptable for MM fibers but not for SM fibers.

To achieve SM splice losses of 0.05 dB requires core alignment to within 0.4 μm. The rotary mechanical splice[9] consistently achieves such alignment with no precision positioners of any kind.

The principle of operation for the rotary mechanical splice is shown in Figure 3. Each fiber is mounted slightly off center in a glass ferrule, and the two ferrules are offset relative to each other so that two positions of perfect alignment are available when the ferrules are rotated. The rotary mechanical splice is manually adjusted to one of these positions without sophisticated positioners.

Figure 4 shows the alignment sleeve for the rotary mechanical splice. This sleeve is designed to obtain extremely stable splice loss (0.02 dB variation) over a wide temperature range (–40°F to +170°F) with no residual loss. Three-point contact with the glass tubes that contain the fiber ensures a stable, balanced geometry. By using glass rods in the alignment sleeve, mechanical and thermal properties of the rods and tubes are closely matched to those of the fiber. The same matching gel used in arrays and other mechanical splices is used to couple the light optically from one fiber to the other without Fresnel reflections. This soft gel expands or contracts as needed to offset slight differences in the thermal or mchanical properties of assembly components.

Although array splicing and rapid rib-

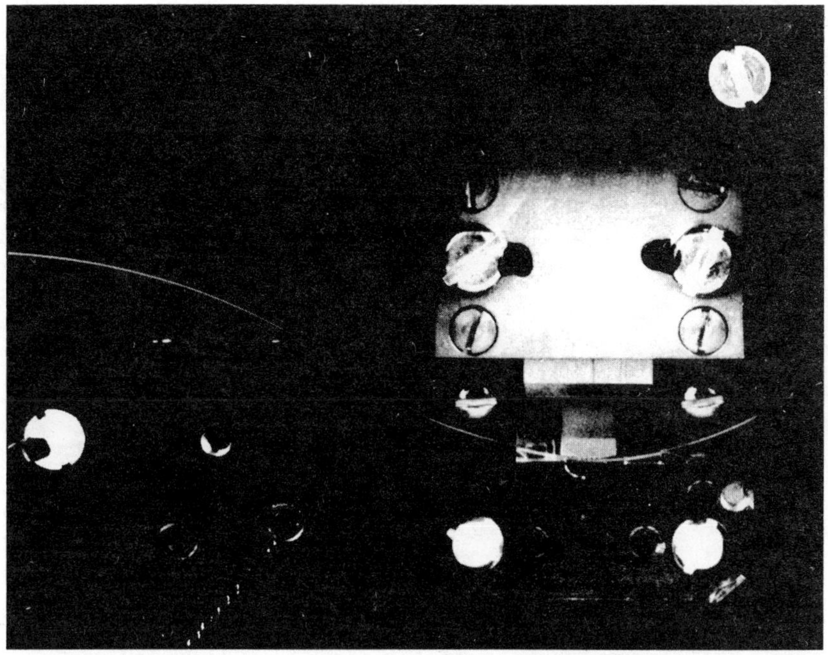

Figure 5. Microbend local-injection unit.

bon splicing are highly productive techniques for groups of 12 fibers, the rotary splice is the most quickly assembled high-performance individual-fiber mechanical splice available. Productivity is typically 10 minutes per splice including termination, tuning, testing, and splice storage. This splicing technique also achieves consistent performance; i.e., splice losses in the field with minimally trained craft almost exactly match laboratory results with highly trained operators. Systems can be better optimized because consistent splice performance is assured.

An enhanced, second-generation rotary splice—called the prealigned rotary (PAR) SM splice—achieves low loss (0.20 dB) without tuning. Splicing procedures and tolerances of the glass tubes have been optimized to achieve these results; however, the splice can be optionally tuned to values below 0.05 dB with active alignment. The three largest offset-producing effects in glass-tube SM splices are:

- Eccentricities in the bore of the glass tubes
- Oversize diameter of the bore
- Eccentricities in the core of the fiber.

The PAR splice almost totally eliminates the offsets that result from the first two of these mechanisms, leaving only the fiber core eccentricity. Stability is the same as in the standard rotary splice and productivity is improved by a factor of 2 (5 minutes total assembly time).

Local Injection Detection. Both fusion and mechanical splices often use active alignment to minimize offsets. Although far-end transmission monitoring has been used in the field, it is a cumbersome technique compared to the con-

Figure 6. Schematic of microbend region.

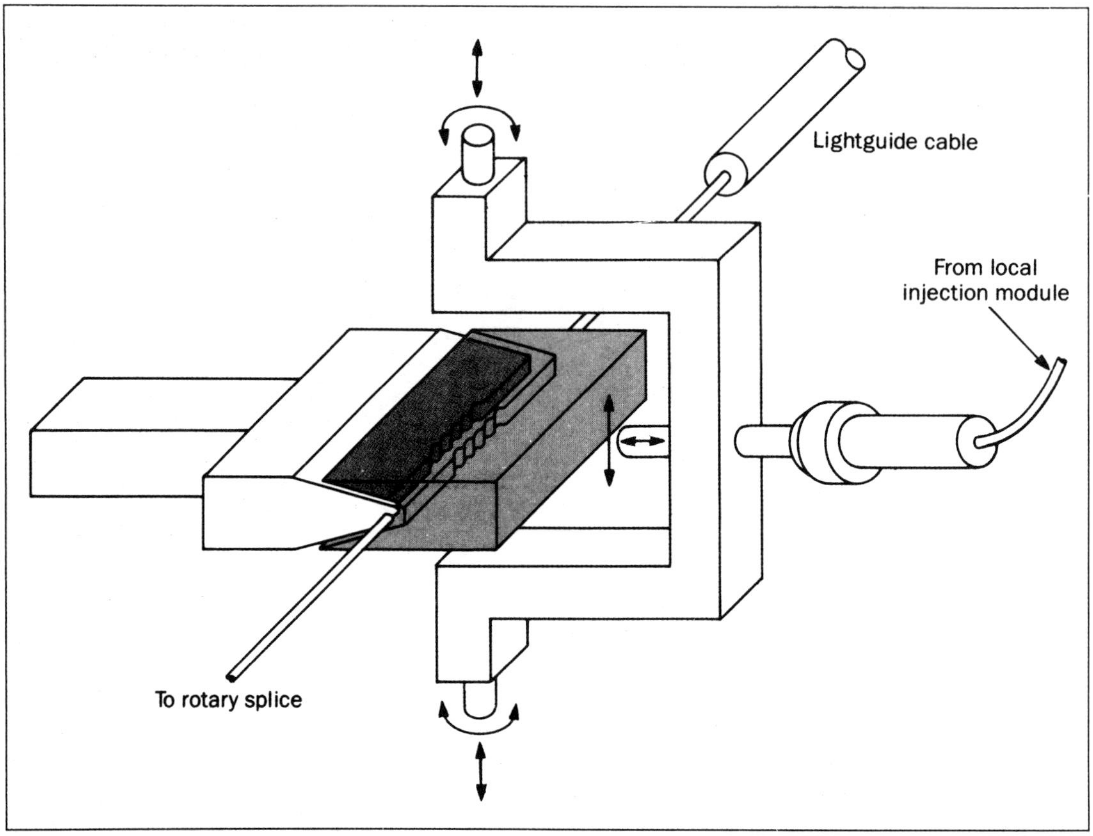

venience of local light detection. Local detection can be done by sensing the light transmitted through the splice or sensing the light lost at the splice. The sensitivity of local loss detection is 200 times greater than the sensitivity of local transmission. This increased sensitivity is used in the rotary splice to achieve extremely accurate alignment and accompanying low loss. Because local loss detection equipment can be easily calibrated, accurate splice-loss measurements can also be accomplished with this technique.

There are two fundamental processes used to inject or couple light into the core of a single-mode fiber: microbending and macrobending. These bends couple light from the fundamental LP_{01} mode to the LP_{11} leaky mode and vice-versa. Conventionally, light from a 0.8- to 0.9-μm LED source is focused on to a relatively small radius (1 to 5 mm) macrobend to inject typically –50 dBm or less light into the fiber. Although this technique will work satisfactorily with transmission detection, most fibers are not single-mode at 0.8 to 0.9 μm.

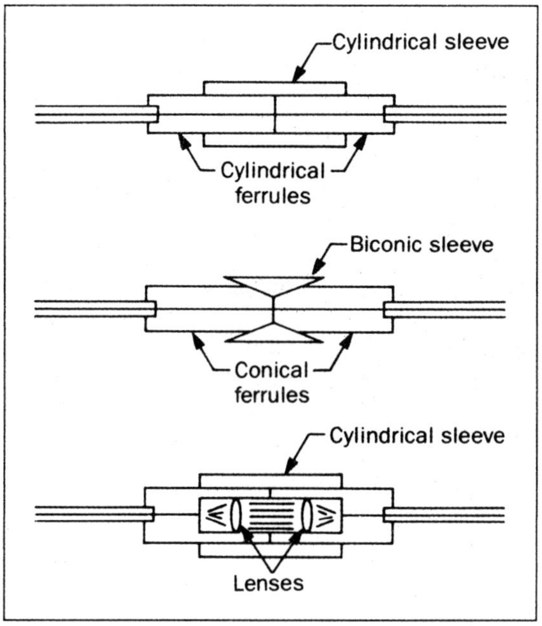

Figure 7. Three types of connectors. Top to bottom: cylindrical ferrule, conical ferrule, and expanded-beam lens.

Moreover, bending stresses are high—125 to 625 kpsi; a large portion of the –50 dBm is sometimes in the cladding; and fiber-to-fiber variation is usually high. These factors combine to yield poor splice alignment and give no indication of splice loss.

Local light injection using short-period, small-amplitude microbends[10] has several advantages:

1. The modulated 1.3-μm laser source excites the single-mode fiber properly.
2. Greater than –35 dBm can be injected almost totally in the core.
3. Bending stresses are less than 50 kpsi.
4. Fiber-to-fiber variation is low.
5. The technique is compatible with local loss-detection techniques.

A photograph of a microbend local-injection unit is shown in Figure 5, and a schematic of the microbend region itself appears in Figure 6.

Connectors

Optical fiber connectors are used to couple light from a source to fiber, from one fiber to another, and from fiber to a detector. The first and third are special applications and will not be discussed here.

A few connectors use bare fibers with scored and broken ends, but by far the most popular configuration for optical fiber connectors is the precision-aligned butt joint with ground and polished ends. Most butt connectors use secondary and tertiary alignment. This means the fibers are positioned and held in supporting ferrules as the ends are ground and polished. These ferrules, which are usually cylindrical or conical, are then aligned in sleeve couplings (Figure 7). Good alignment of the fiber cores requires precise diameter and concentricity control of the cylindrical ferrule and sleeve, and precise taper length and concentricity control of the conical parts.

Some connectors use an expanded-beam lens, as shown in Figure 7. This connector has relaxed concentricity requirements, but this is offset by more stringent angular requirements. Because the ends can be sealed with glass disks, the connector is less susceptible to damage during multiple reconnects, and its performance is less sensitive to debris. This has made the expanded-beam connector attractive for some applications in dirty environments. But to date, it has not been successful in capturing a sizable market and will not be discussed further here.

Connectors generally have greater loss than splices because they are not tuned, and matching gels are not ordinarily used. Further-

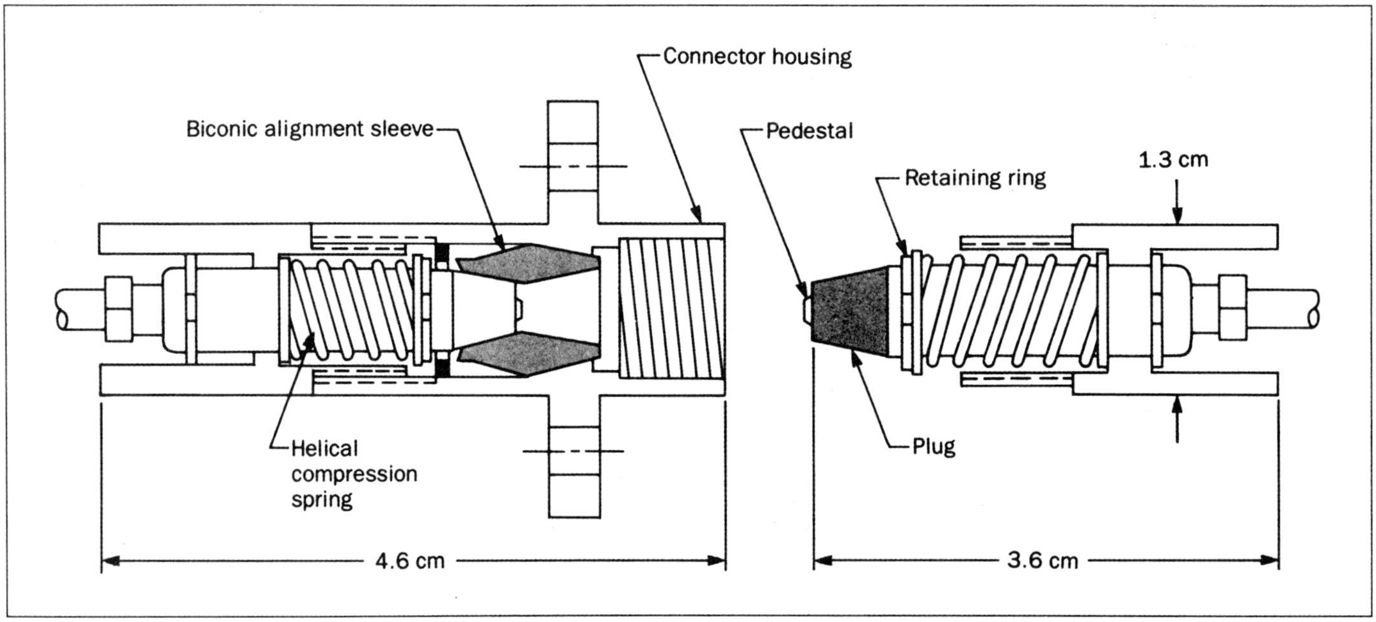

Figure 8. Biconic connector.

more, connector performance can suffer from wear and end-face damage. The fiber end faces must be flat, perpendicular to the alignment-surface axis, and in contact to get the lowest loss in a dry, butt-joint connection. With contacting end faces, the loss of high-performance MM and SM connectors is typically 0.3 to 0.5 dB. These figures will increase by about 0.3 dB from Fresnel reflections if there is an air gap between the fiber ends.

Other connector contributions to loss that may be discussed without specifying the shape of alignment surfaces are:

1. Eccentricity between bore and alignment-surface axes
2. Diametral clearance between the largest bore and smallest fiber
3. Angular misalignment between bore and alignment-surface axes.

Typical limits for high-performance MM connectors are 5 μm for items (1) and (2), and 1° for item (3). Approximate limits for similar-quality SM connectors are 1 μm for (1), 2 μm for (2), and 1/2° for (3).

Several million AT&T biconic connectors are in use today in both MM and SM systems, and several companies, in addition to AT&T, now make the biconic connector. Ruggedized multiple-fiber connectors have been designed for military applications that incorporate the biconic connector.

AT&T and many other manufacturers throughout the world also offer cylindrical connector designs. These are currently being introduced in simplex and duplex configurations for local area networks.

Biconic Connector. In simple description, the biconic connector for single-fiber lightguide

Figure 9. Setup for grinding biconic plugs.

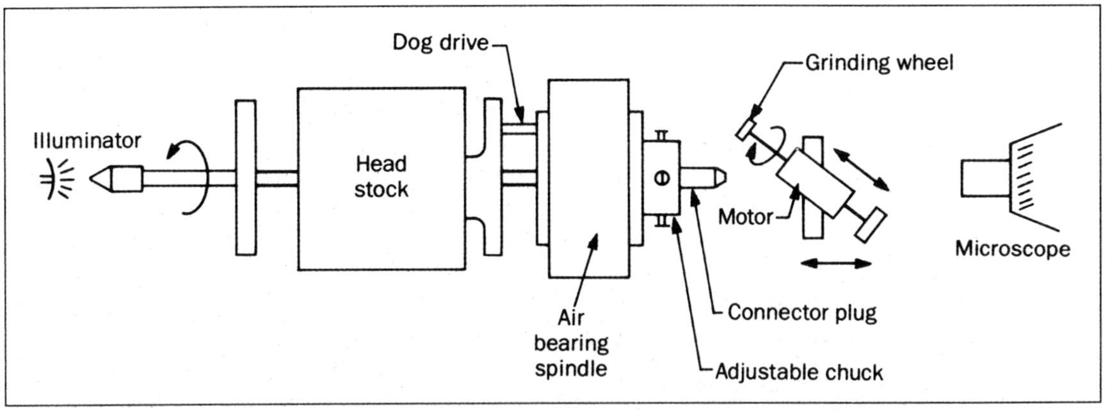

cable, illustrated in Figure 8, consists of a plug on each cable end, an alignment sleeve, and some peripheral hardware to contain and mount the sleeve and hold the assembly together under load. The plug ends have the shape of truncated cones, and the sleeve has two corresponding internal tapers. Ideally, the transfer-molded plugs, fiber cores, and both tapers in the alignment sleeve have a common axis.

The initial design of the connector was based on the observation that transverse offset of fiber ends contributes about 20 times more loss in a lightguide connection than longitudinal end separation of the same amount. A taper was chosen for the aligning surface so that lot-to-lot isotropic shrinkage variations in nominally identical molding materials would not manifest themselves in transverse offset but in the less critical end separation.[11] A taper also provides for relatively easy backplane insertion and low abrasion.

Anticipated variations in end separation were accommodated in the 1977 Atlanta MM Fiber System Experiment through elastic contact of index-matching cushions that adhered to each plug tip and provided a protective covering for protruding fiber ends. The connector patent[12] credits this "principle of flexible interfacing" with attaining the low light loss typical of aligned fiber ends in "close approach" without tedious adjustment or stringent end-separation tolerances. The alignment sleeves used in the Atlanta Experiment were machined from brass. After the Atlanta Experiment, a lapping and polishing procedure was adopted, which produced fiber ends that were flush with the body of the silica-filled epoxy plug and permitted precise adjustment of the taper length. From the plug material, Bell Laboratories developed a transfer-molded biconic sleeve whose dimensions could be held more closely than those of machined sleeves. Eccentricities were less, and tighter control on the length parameter allowed nominal separation between fiber ends to be reduced to 25 μm. Reported mean loss and standard deviations, obtained without the index-matching cushions, suggested that some connections must have had contacting plug ends.[13]

In 1980, Bell Laboratories reported mean connector losses of just over 0.20 dB with standard deviations less than 0.10 dB by purposely selecting plug and sleeve lengths to ensure contact of plug ends in the assembled connector. Impressive figures like these drove the evolution of specifications of the two crucial dimensions—plug taper length and sleeve ball separation—in a direction that corresponds to contacting plugs. Given the reality that reasonable manufacturing tolerances were essential, it was geometrically obvious that the biconic design could not achieve seated alignment and contacting plug ends in all connections without some compliance in the connector components. An analysis showed that this compliance was available in the elastic deformation of the meshing conical alignment surfaces as the plugs are seated by the controlled compression of the helical spring. The plugs were found to approach one another by about 25 μm during seating if the cone angles differed slightly to dispose alignment to begin at the plug ends and promote the necessary deformation by intensifying the contact stresses between the plug and sleeve.

The MM biconic connector set a new industry standard for precision molding,[14] and AT&T's initial strategy for connecting SM fibers was to use the MM biconic production for suitable ultraprecise connector components. However, it became apparent in the early 1980s that this scheme would not meet the rapidly increasing demand for SM connectors, and production plans to grind MM biconic plugs to meet SM concentricity requirements were implemented.

In this operation, a slightly oversized biconic connector is held in the rotating jaws of a grinder as shown in Figure 9, and microscopic observation of the illuminated bore dictates the plane and angular adjustments necessary to make the axis of rotation coincident with the bore axis. A grinding wheel then trims the conical surface, leaving the taper length just long enough to permit polishing when the fiber is subsequently installed.

Essentially the same procedure can be followed with even more precise results when the SM fiber is already in place and the core is illuminated.[15]

This micromachining process is well-suited to automation, and current production yields are well over 90 percent. The balance of current production plugs still meet required MM tolerances. Factory automation with such high yields, virtually no rejected product, and stabilized quality have steadily driven the cost of manufacturing downward while performance levels have risen. AT&T's SM customers can expect connector loss to be typically less than half a decibel with a maximum of 1 dB.

Environmental performance of biconic connectors in MM applications has always been good, but the stability of a SM connection during a thermal cycle is the sternest test. In 10 temperature cycles between –20°C and 60°C over a 5-day period, the SM biconic attenuation typically varies less than 0.1 dB and never more than 0.3 dB. This variation is cyclic with no residual accumulation.

The biconic connector has also recently found an MM military application. The tactical fiber optic cable assembly (TFOCA) connector shown in Figure 10a with its lanyard-attached dust cover was developed by AT&T during 1985 under contract to the U.S. Army Communications Electronic Command. It is a hermaphroditic duplex embodiment of the biconic connector ruggedized to meet stringent military requirements. The development included design of the sleeve retainer visible in

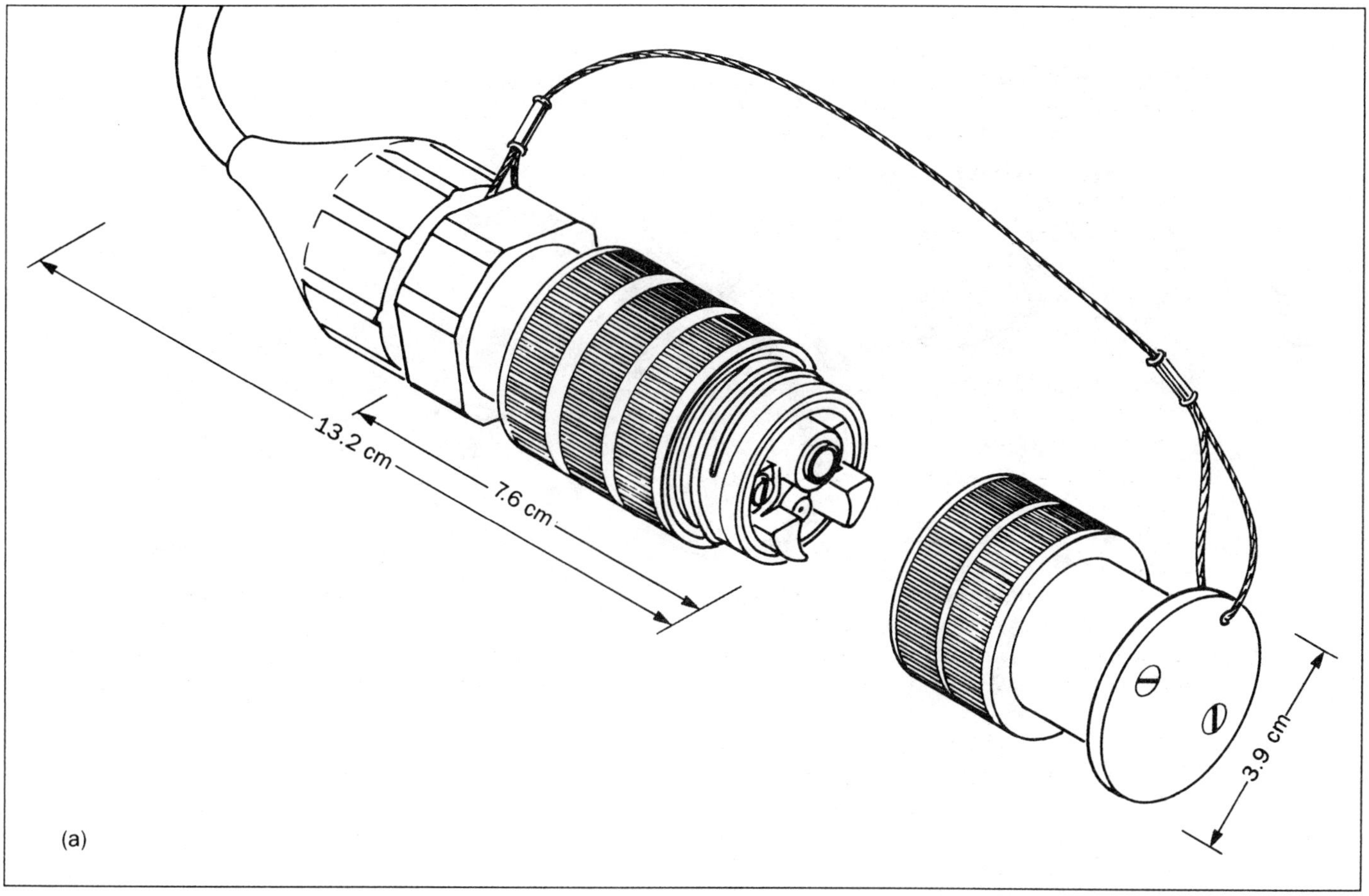

Figure 10. Tactical fiber optic cable assembly. (a) Connector with dust cover. (b) Exploded view.

Figure 10b at the forward end of an exploded view of the connector-insert assembly. It is removable to allow cleaning of the biconic alignment surfaces, provides the hermaphroditic geometry, and furnishes impact protection for an otherwise exposed biconic plug. The TFOCA connector is completely sealed against casual water entry when it is mated or when the hermaphroditic dust cover is in place. The confirming test requirement was a 24-hour immersion in 2 meters of water.

The development contract specified a battery of more or less typical connector tests, but two of these are novel enough to mention here. The first of these was the mud test. In this test, the unprotected connector was used to scoop a generous quantity of mud following an initial optical-loss baseline measurement. The connector was then cleaned using only tap water and a cloth. In many repeated demon-

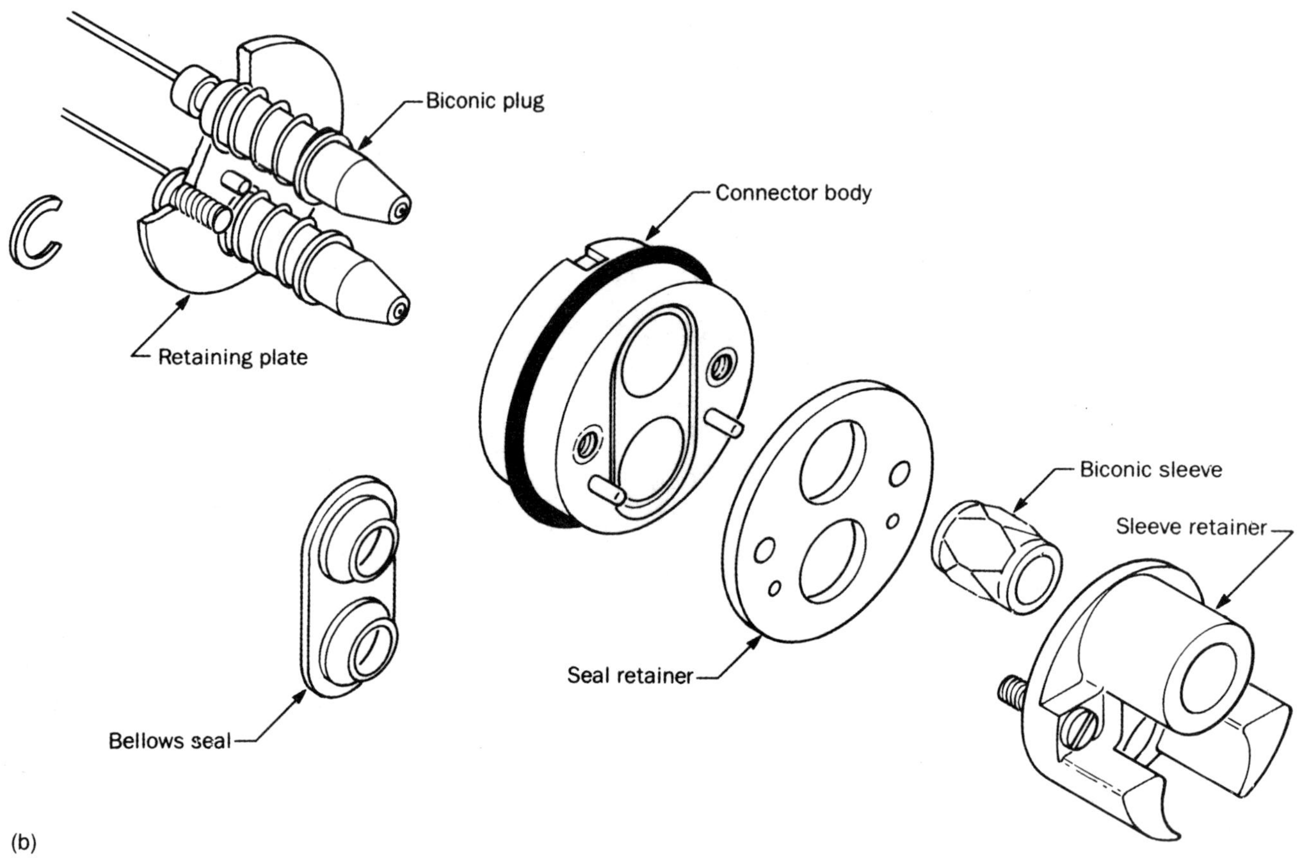

(b)

strations, the maximum observed change in attenuation was 0.05 dB. The second test specifies that the connector be able to sustain a 400-pound cable tension with no mechanical or optical degradation. This was done without epoxy, crimping, or the use of any special tools by capturing the cable's aramid-yarn strength members between locking steel tapers that tighten under tension. The design exceeded expectations; the mean tensile failure load was 999 pounds, well above specification.

ST® Connector. A market is opening up for short-haul, connector-intensive applications calling for a small, low-cost multimode connector. Such applications include fiber distribution of wideband services within large buildings and development of fiber-based local area networks. In spite of the biconic connector's previously discussed attributes and track record, it is perceived among these nontradi-

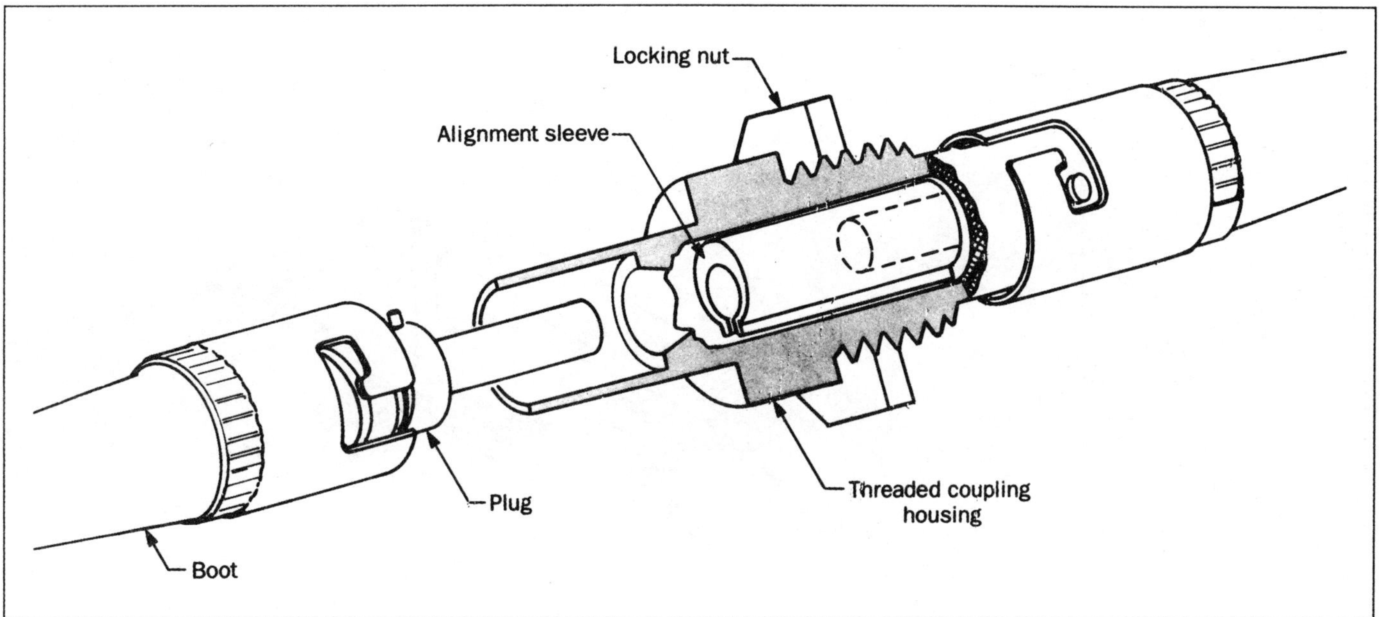

Figure 11. ST® series connector.

tional potential customers to be larger and more expensive than their applications allow. The ST connector designed by Bell Laboratories and shown in Figure 11 is AT&T's entry in this new market. It is smaller and less expensive than the biconic connector without sacrificing the high-performance characteristics of its predecessor.

The connector is directly compatible with AT&T's 50- and 200-Mb/s optical data links and, like the biconic, can be applied in the factory or terminated in the field. It consists of two identical robust plugs with cylindrical ceramic tips that are aligned by a uniquely configured split sleeve in four available housings. The sleeve is a patented[16] molded part with a cross section that is tailored to expand uniformly and exert a constant pressure when the plug tips are inserted. The design provides accurate alignment even with slightly mismatched tip diameters. A twist-lock bayonet latch joins the plug and coupling, which are keyed to produce highly repeatable readings when reconnected. The plug tips in a connection make contact under a spring-controlled axial thrust.

The fiber is secured in the plug with a specially developed heat-cured epoxy, and a combination of adhesive bonding and mechanical crimping is used to attach the plug to single-fiber cable. An elastic boot limits the bend radius of the cable at the connector entrance. A special tool kit allows an operator to assemble the connector in the field. Up to 12 connectors can be cured at once in the oven that is part of the tool kit.

Figure 12. ST duplex connector.

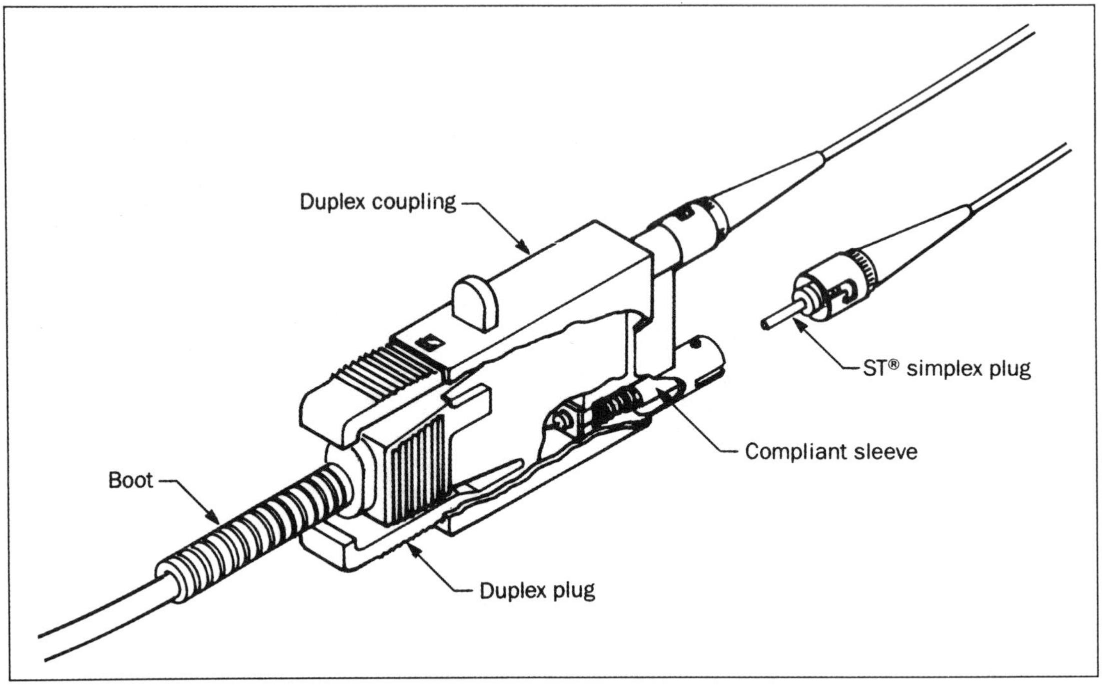

Average assembly time is about 6 minutes per connector for multiple installations; a single connector can be installed in about 15 minutes.

Loss measurements were made with 62.5-μm core cables arranged in concatenations made of typical production cables and couplings. Concatenations of 10 jumpers of similar fiber were built incrementally to determine connection losses. From 260 random interconnections of 35 jumpers and 50 alignment sleeves, the mean loss and standard deviation were 0.22 and 0.15 dB, respectively.[17,18]

The two most active groups working on standards development for LANs are ANSI X3T9.5 FDDI (Fiber Distributed Data Interface) and IEEE 802.6 MAN (Metropolitan Area Network). Both have decided to specify a standard duplex receptacle, and any supplier who attempts to respond to either standard development must provide a duplex connector.

Figure 12 shows AT&T's ST connector in its duplex configuration with a duplex-to-simplex coupling. The duplex plug is polarized and keyed to mate only with matching receptacles or couplings; the keying scheme may be changed in the field. The plug snaps into the receptacle or coupling and is released by depressing the clothespin latches. A spring-loaded bumper extends to protect the plug tips as the plug is removed. The ST duplex connec-

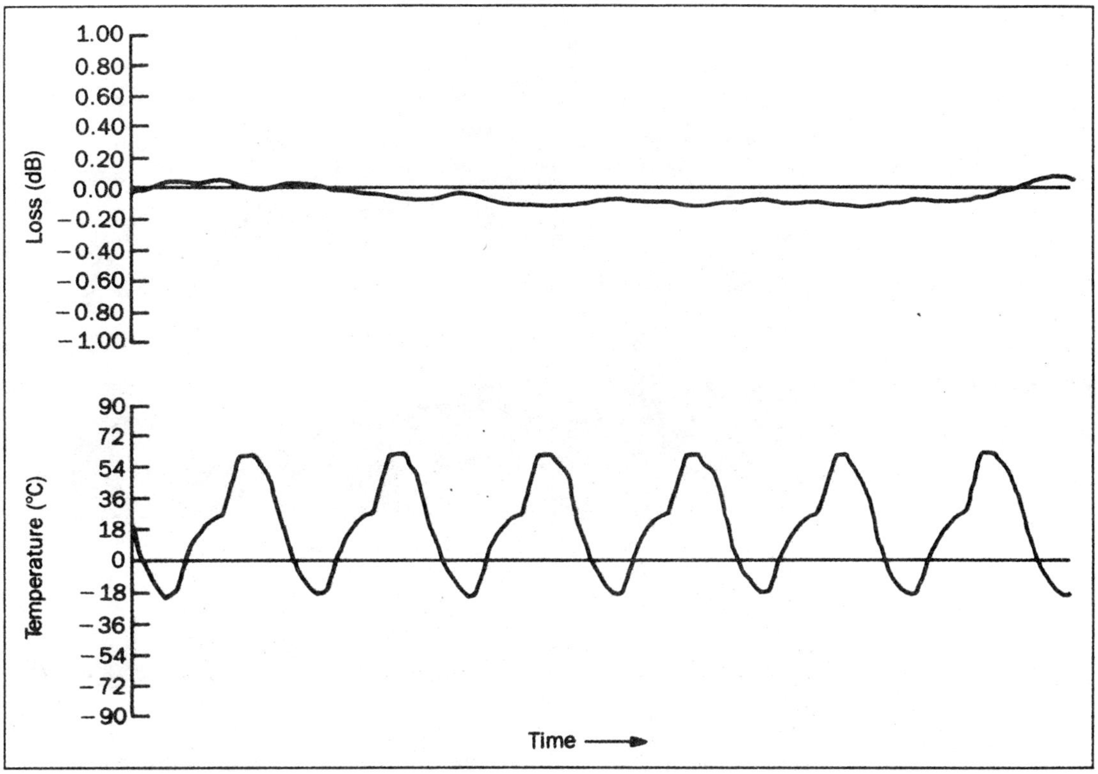

Figure 13. Typical thermal cycle test data for the ST duplex connector.

tor uses an adaptation of the TFOCA military connector's cable-retention scheme; i.e., locking tapers capture the cable's aramid yarn. The optical performance of the connectors in duplex configuration approaches that of a single connector. A random concatenation of cables that yield 200 loss measurements has a mean loss of 0.28 dB and a standard deviation of 0.17 dB. The environmental performance of the ST duplex connector is equally impressive. Figure 13 shows typical loss variation in parallel with corresponding temperature variations ranging from –20°C to 60°C.

A multifiber array connector (MAC) has been developed by AT&T Bell Laboratories[19] for use with the Fastech® packaging system (Figure 14), as well as with other types of packaging systems. This connector allows designers to interconnect from 2 to 18 optical fibers through a backplane. The MAC is compact, with a maximum envelope of 0.5 by 0.6 by 2.26 inches, and requires a minimum circuit-pack spacing of 0.75 inch. It can be used to interconnect 50/125 μm or 62.5/125 μm multimode fiber. When 50/125 μm multimode fiber is used, an average loss of 0.5 dB occurs, with 0.87 μm wavelength and short launch and receive conditions. A single-mode version of

Figure 14. Multifiber array connector for use with the Fastech® packaging system.

the connector is under development with a projected average loss of 1 dB.

The MAC hardware consists of protective housings for the arrays (refer to Figures 1a and 1b) and a bend-radius limiter for ribbon jumpers. This hardware has alignment and mating features that automatically align and optically connect the arrays when a circuit pack is plugged into the backplane. The housings are constructed of flame-retardant molded plastic, and are designed for complete snap-fit assembly. This allows fanout or ribbon-jumper arrays to be removed and replaced without removing the housing from the backplane or the circuit pack. Fiber mismatch is prevented by keying the housing.

References

1. M. J. Saunders and W. L. Parham, "Adhesive Sandwich Optical Fiber Ribbons," *The Bell System Technical Journal*, Vol. 56, No. 6, July-August 1977, pp. 1013-1014.
2. M. J. Buckler, M. R. Santana, and M. J. Saunders, "Lightguide Cable Manufacture and Performance," *The Bell System Technical Journal*, Vol. 57, No. 6, July-August 1978, pp. 1745-1757.
3. C. M. Miller, "Fiber Optic Array Splicing with Etched Silicon Chips," *The Bell System Technical Journal*, Vol. 57, No. 1, January 1978, pp. 75-90.
4. C. H. Gartside, III, and J. L. Baden, "Single-Mode Ribbon Cable and Array Splicing," Conference on Optical Fiber Communication, San Diego, February 13, 1985, *Technical Digest*, Paper No. WJ2, p.106.
5. J. L. Baden, J. A. Aberson, and M. J. Swiderski, "Mass Splicing of Single-Mode Fibers," Conference on Optical Fiber Communication, San Diego, February 25, 1986, *Technical Digest*, Paper No. TUL1.
6. N. E. Hardwick and S. T. Davies, "Rapid Ribbon Splice for Multimode Fiber Splicing," Conference on Optical Fiber Communication, San Diego, February 12, 1985, *Technical Digest*, Paper No. TUQ27.
7. N. E. Hardwick and S. T. Davies, "Single-Mode Rapid Ribbon Splice," *Proceedings*, 35th International Wire and Cable Symposium, Reno, Nevada, November 1986.
8. J. A. Aberson and K. M. Yasinski, "Multi-mode Mechanical Splices," *Proceedings*, 10th European Conference on Optical Communication, Stuttgart, Germany, September 1984, p.182.
9. C. M. Miller and G. F. DeVeau, "Simple High-Performance Mechanical Splice for Single-mode Fibers," Conference on Optical Fiber Communication, San Diego, February 11, 1985, *Technical Digest*, Paper No. MI2, 1985, p. 26.
10. J. A. Aberson and I. A. White, "Low-Stress High-Efficiency Local-Power Injection Into Single-mode Fibers," *Proceedings*, 35th International Wire and Cable Symposium, Reno, Nevada, November 1986.
11. P. K. Runge and S. S. Cheng, "Demountable Single-Fiber Optic Connectors and Their Measurement on Location," *The Bell System Tech-*

nical Journal, Vol. 57, No. 6, July-August 1978.
12. United States Patent 4,107,242; August 15, 1978.
13. P. Kaiser, W. C. Young, N. K. Cheung, and L. Curtis, "Loss Characterization of Biconic Single-Fiber Connectors," *Technical Digest*, Symposium on Optical Fiber Measurements, Boulder, Colorado, 1980.
14. W. C. Young et al., "A Transfer-Molded Biconic Connector with Insertion Losses Below 0.3 dB without Index Match," Sixth European Conference on Optical Communication, York, England, 1980, Conference Publication Number 190.
15. W. C. Young, L. Curtis, and P. Kaiser, "Biconic Single-Mode Connectors with Insertion Losses Below 0.3 dB," Topical Meeting on Optical Fiber Communication, Phoenix, Arizona, April 13-15, 1982.
16. United States Patent 4,541,685; September 17, 1985.
17. A. Wallace Carlisle, "AT&T's New Lightwave Connector Cuts Installation Time and Cost," *AT&T Bell Laboratories Record*, March 1985, p. 15.
18. A. Wallace Carlisle, "Small Size, High Performance Lightguide Connector for LAN's," Optical Fiber Conference, San Diego, February 1985, *Technical Digest*,
19. W. J. Parzygnat, "A Multifiber Circuit-Pack to Back-Plane Optical Connector," Conference on Optical Fiber Communication, San Diego, February, 1986, *Technical Digest*,

Chapter 3

Light Sources and Transmitters

Advances in semiconductor light sources for optical fiber communication during the past fifteen years or so have been as impressive as progress in the fibers themselves. High-radiance light-emitting dioides and low-threshhold laser diodes for both 0.8 μm and 1.3 μm spectral regions have been available for the past five to ten years. More recently, distributed-feedback lasers for 1.3 and 1.55 μm have become commercial products. High reliability and economical manufacturing methods have been established for each of these types of emitters. A great deal of effort has also been devoted to the development of transmitters that contain driving circuitry as well as the light source. Such transmitters are available for use with a variety of standard electronic interfaces for virtually every digital and analog transmission application.

Two general reviews which cover photodetectors as well as the light sources are provided by **D.H. Newman and S. Ritchie** in **"Sources and Detectors for Optical Fiber Communication Application: The First 20 Years,"** and by **H. Haupt and O. Hildebrand** in **"Lasers and Photodetectors in Europe."** The use of a light-emitting diode with single-mode fiber for high-speed transmission is described by **G.K. Chang, H.P. LeBlanc, and P.W. Shumate** in **"Novel High Speed LED Transmitter for Single Mode Fibre and Wideband Loop Transmission Systems."** Reliability, always an important issue with semiconductor lasers, is discussed by **M. Hirao, K. Mizuishi, and M. Nakamura** in **"High-Reliability Semiconductor Lasers for Optical Communications."** The next two papers describe recent results in semiconductor laser technology: **"Room-Temperature cw Operation of MBE-Grown GaInAs/AlInAs MQW Lasers in 1.5 μm Range,"** by **Y. Matsushima, U. Utaka, K. Sakai, and O. Takeuchi,** and **"High-Power, Wide-Bandwidth 1.55 μm-Wavelength GaInAs/InP Distributed Feedback Laser,"** by **K. Kihara, K. Kamite, H. Sudo, T. Tanahashi, T. Kusunoki, S. Isozumi, H. Ishikawa, and H. Imai.**

Sources and detectors for optical fibre communications applications: the first 20 years

D.H. Newman and S. Ritchie

Indexing terms: *Optical fibres, Optoelectronics*

Abstract: The evolution of semiconductor optical sources and photodetectors for optical communication systems is described. From rudimentary GaAs *p-n* junction emitters and silicon photodiodes in the mid-1960s, the paper charts progress towards the much more sophisticated multilayer laser structures and III–V avalanche photodetectors being made in the mid-1980s. Technological breakthroughs are identified, as well as some of the major problems that have had to be solved, such as well controlled epitaxial growth and device reliability.

1 Introduction

The paper reviews the development of semiconductor sources and detectors for optical fibre communications. Their evolution has been long and tortuous, spanning over half a working lifetime. From the very beginning, the potential for semiconductor lasers in the telecommunications field was recognised because they were rugged, compact and relatively easy to modulate directly by way of the input current drive. However the difficulties in turning this potential into reality were vastly underestimated, and this has not been helped by the pattern of fibre development. Fibre technology has had the advantage of continuous evolution over the last two decades with one basic material (silica). However the demands made on sources and detectors have changed considerably from single-mode and multimode systems operating at 850 nm, to multimode followed by monomode systems operating at 1300 and 1500 nm. Thus the semiconductor technologist has had to cope with a severe perturbation on the way, having to move from GaAs-based and Si technologies to the much newer InP-based technologies.

This paper is not meant to be an exhaustive review of the evolution of devices for optical communications systems but rather represents the authors personal views on factors that have been important in influencing progress.

2 Sources

2.1 Early devices

The genesis of activity relating to sources for optical communications began as far back as the 1950s in materials development work on GaAs. This is a semiconducting compound which had been recognised as having potential as a superior transistor material, because of its wider bandgap and higher electron mobility than silicon or germanium. Studies of zinc diffusion paved the way for the fabrication of *p-n* junctions which, on recognition of the potential for efficient photon emission at a forward-biassed *p-n* junction in direct gap semiconductors [1, 2, 3] (like GaAs), resulted in the demonstration of LED [4, 5] and laser [6, 7] action in 1962. These first lasers were a far cry from the communications lasers we know today — they operated pulsed (typically 10 μs pulses) at very low duty cycles and then only at temperatures below 100 K. However, optimism prevailed among the small group of workers in the field that somehow it would be possible to reduce operating currents by several orders of magnitude to enable continuous operation to be achieved at room temperature, and the identification of the possibilities for optical fibre transmission by Kao and Hockham [8] in 1966 encouraged the continuing drive for better performance. It is interesting that the predictions being made for optical fibres revolved around the use of monomode fibre in spite of the obvious difficulties of fabrication. A result was that the potential of LEDs was temporarily overlooked because of the very small coupling efficiency that would result — no one at that time was seriously predicting low loss fibre of the type in existence today.

2.2 Heterojunctions

Probably the most significant step in optical source and detector technology occurred in the late 1960s. Following on from Kroemer's suggestions [9] in 1963 that additional semiconductor layers with wider bandgaps and lower refractive index could be included in laser structures to provide better carrier and optical confinement, Alferov [10] and Panish [11] both showed how it was possible to grow lattice-matched heterojunctions of $Al_xGa_{1-x}As$ on GaAs that exhibited true heterojunction properties — that is, the interfaces were not controlled by defects resulting from poor morphology and lattice mismatch. GaAs had had its first lucky break with nature providing a situation where, at 900°C, GaAs and AlAs had an identical lattice constant, implying that true epitaxial layer growth was possible close to this temperature. The use of the liquid-phase epitaxy technique (which still survives today as the production workhorse of the semiconductor communications laser industry) enabled such epitaxial growth to be performed. Liquid-phase epitaxy involves the placing of a supersaturated melt (containing, for example, Ga, Al, As and dopants in the correct ratios) over a cooling substrate in a hydrogen atmosphere at around 850°C, so that material from the melt is epitaxially deposited to form a layer of the required composition. The first use of this heteroepitaxy was in the fabrication of single heterostructure (SH) lasers [12], where the reduced laser threshold currents achieved in these nGaAs-pGaAs-pAlGaAs junction devices indicated that the GaAs/AlGaAs heterojunction was having the beneficial effect predicted. It was soon realised, however, that to achieve the best laser structures, multiple epitaxial layers of different aluminium and dopant compositions in the form of a double heterostructure were required, and the science (or was it art!) of multiwell liquid-phase epitaxy was born. With this technique, a

Paper 4628J (E13) received 23rd January 1986

The authors are with British Telecom Research Laboratories, Martlesham Heath, Ipswich IP5 7RE, United Kingdom number of melts of different composition are successively passed over the cooling substrate in a hydrogen atmosphere (Fig. 1) to build up a sequence of layers, some of which need to be of the order of, or less than, 1 μm thick.

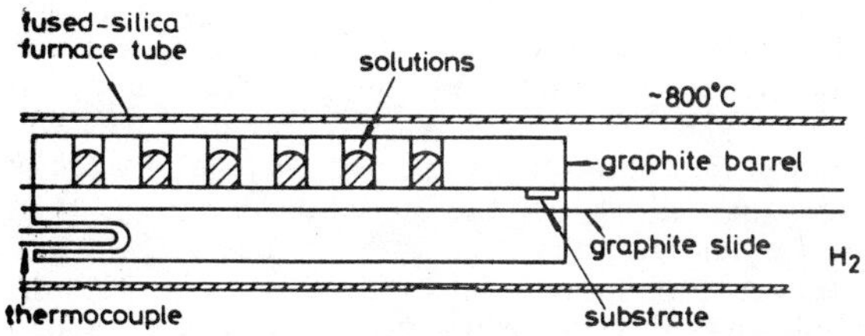

Fig. 1 *Multiwell liquid-phase epitaxy for growing double heterostructure lasers*

A large measure of technical skill was required of those who were able to control the process in those early days. Although the process has been brought under better control by design improvements and the use of microprocessors, LPE still demands a degree of operator skill as well as technical knowhow.

The reasons for the importance of a viable heterostructure technology was that it was then possible to achieve, simultaneously, both carrier and optical confinement in laser structures. Fig. 2*a* shows how radiative emission is achieved by electron and hole injection across a heavily doped forward-biased *p-n* (homo) junction, with emission occurring over a volume determined by minority carrier diffusion lengths (several microns in GaAs), and Fig. 2*b* indicates how the imposition of heterojunctions of suitable wider bandgap can be used to confine carriers to a much restricted volume in the double heterostructure (DH). An additional property of the wider bandgap confinement layers in the DH laser is that their refractive indices are less than that of the narrow bandgap recombination (or active) layer at the emission wavelength. Hence

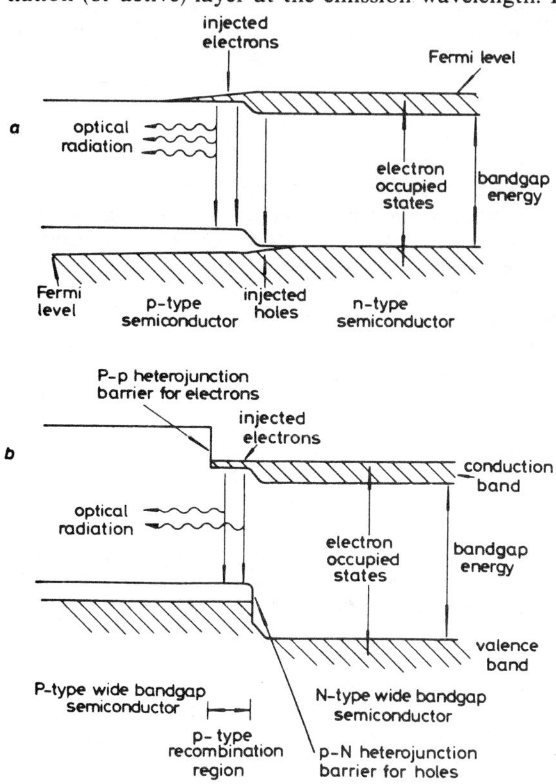

Fig. 2 *Band diagram*
a Injection across a *p-n* homojunction
b Injection and confinement in a double heterostructure configuration

the structure forms a strong optical slab waveguide that effectively guides the emitted radiation in the *p-n* junction plane, even with the active layer reduced to around 0.2 μm, when 30% Al confining layers are employed. The provision of cleaved end facets provides 30% semireflecting mirrors that give optical feedback and the basis for laser action. However, in order to achieve continuous (as opposed to pulsed) operation at room temperature, a stripe geometry configuration is adopted following the earlier example of Dyment and D'Asaro [13] with homojunction lasers. The most straightforward examples are the oxide-isolated or proton-isolated versions (Fig. 3), which

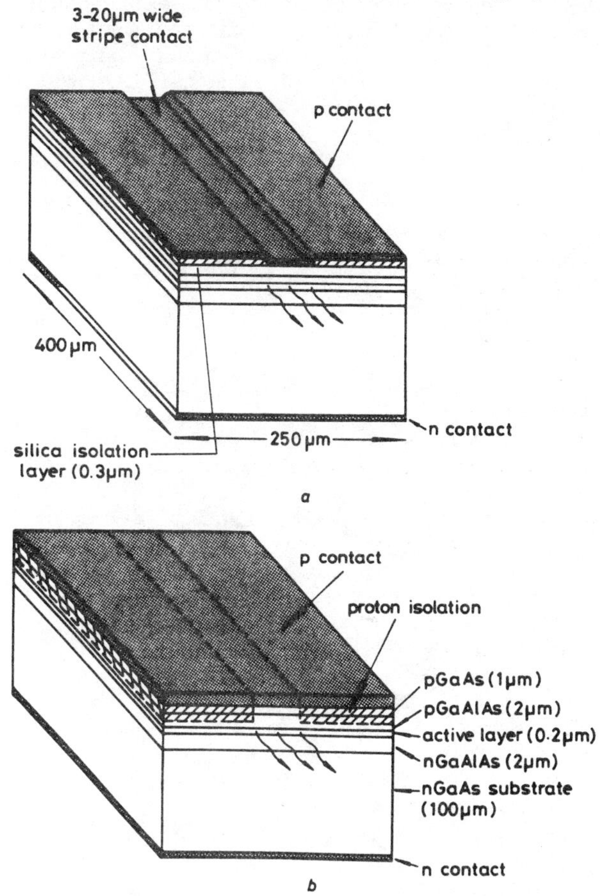

Fig. 3 *Simple GaAs stripe geometry double heterostructure laser structures*

confine current flow to a narrow stripe (approximately 3–20 μm wide), providing more efficient coupling into optical fibre and giving better thermal performance.

2.3 GaAs source reliability

It was at this point in the development of the GaAs laser that it began to demonstrate some unfortunate features. It proved to be a very unreliable component, with workers observing device failure in a matter of minutes or hours. Earlier work on GaAs LEDs [14] and single-heterojunction (SH) lasers [15] had indicated that reliability was a potential problem. However, the early observations on DH lasers [16, 17] provided the impetus for analytical studies of degraded lasers to be undertaken. Experience showed that the rate of degradation (measured by the increase in laser threshold with time) was adversely affected by stress [18], and so it became common to bond the laser chip to its heat sink by the very ductile indium solder. However, although this minimised one degradation mechanism, it raised another much later on. Indium solder and gold, both evaporated directly on the laser chip and plated on the heat sink, could form InAu intermetallics and associated voids even at temperatures below 100°C in the presence of current flow. This causes thermal resistance to deteriorate, making devices prone to overheating [19–21]. This episode is certainly the result of workers concentrating so much on the possible internal laser chip degradation mechanisms that little attention was given, initially, to the metallurgy of the external bond.

However, there is no doubt that the study of the internal laser chip degradation processes yielded an intriguing insight into the interactions of dislocations and point defects in the presence of electron-hole recombination. Early analytical studies showed that the rapid degradation (observed as an increase in current threshold) was associated with the growth of dark-line defects (DLDs) which propagated during operation along ⟨100⟩ crystallographic directions (Fig. 4) for devices fabricated on (100) orientated

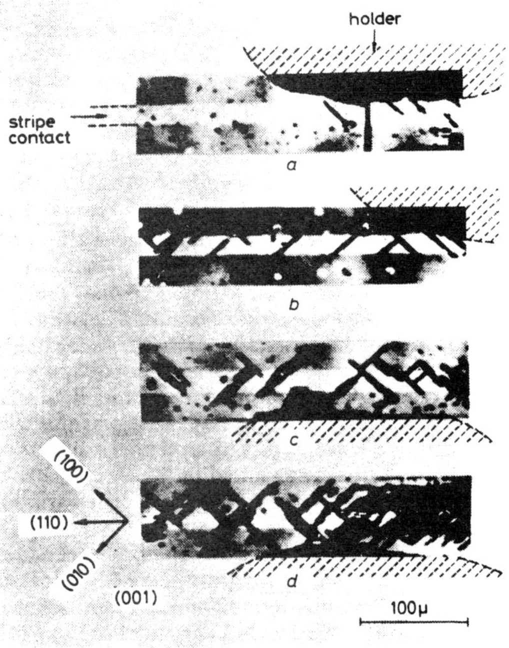

Fig. 4 *Growth of dark line defects in stripe lasers* [22] (*Japanese Journal of Applied Physics*)

material [17, 22]. The dark line is a region of high nonradiative recombination so that carriers injected into that region do not contribute to the radiative output. In addition, the regions are optically absorptive, so absorb radiation emitted elsewhere in the laser active layer, which further inhibits laser action. It transpired, following transmission electron microscopy (TEM) investigations [23, 24], that dislocations present in the original GaAs substrate, which threaded up through the grown epitaxial layers, were the initiation point for the growth of DLDs by a process involving nonradiative electron-hole recombination. This provided (in the case of GaAs) sufficient energy to cause the motion of defects and the formation of dislocation dipoles by a climb process [25]. The solution to this particular problem was, in the end, straightforward — elimination of threading dislocations through the use of better quality substrates and more careful attention to the details of epitaxial growth.

Other reliability problems in GaAs lasers were also identified, including catastrophic mirror damage which limited the peak power outputs to the order to 1–5 × 10^5 W/cm^2 at the output facet (i.e. typically a fews 10 s of mW for stripe geometry lasers) due to local heating by optical absorption close to the facet and resultant burn-off [26, 27].

Having eliminated DLD degradation, limited output powers to stop facet burn-off and modified metallisation and soldering procedures, lasers still degraded, albeit much more slowly, with operating lives being predicted in excess of 10^5 hours [28–30] (over 10 years). It became clear that native defect migration could still occur in the absence of threading dislocations in the presence of strain fields induced by, for example, oxide stripe windows [31] and that noncatastrophic laser mirror deterioration [32] was also possible. The use of facet coatings (e.g. Al_2O_3) to alleviate the latter problem, and further attention to the finer points of materials growth to produce more uniform and higher quality layers, eventually resulted in the arrival of products fit for the rigors of first generation (850 nm) optical fibre systems in the very early 1980s. Of course, the effort expended on lasers has spun off into GaAs high-radiance LEDs, so that reliable LEDs also existed to compete with lasers for use in the graded-index fibre systems being installed, especially in the lower bit rate applications.

It is pertinent to observe that the semiconductor laser gained a reputation as a very unreliable component as a result of its early performance, from which it is only just recovering.

2.4 Gallium arsenide laser designs

Of course the 10 year long reliability investigation saga does not tell by any means the whole story of the GaAs emitter. However, the availability of a reliable emitter was a prerequisite for use in fibre systems. The same 10 year period was full of studies of the physics of lasers and a proliferation of designs. These were aimed at achieving the lowest threshold currents in combination with stable optical outputs, both temporally and, importantly, in terms of coupling into 50 μm core fibre.

The early stripe lasers showed several anomalies in their characteristics, including 'kinks' in their relative output against injection current characteristics sometimes associated with multitransverse mode operation, which could give unstable coupling characteristics into fibres [33]. In addition, interactions associated with injected carriers, their diffusion in stripe structures and optical fields, manifested themselves in variable temporal performance, including the presence of oscillatory outputs and slow laser turn-off [34]. These effects limited the modulation speeds attainable. Hence structures were devised to overcome some or all of the above problems, which, in general, introduced further complications into epitaxial layer growth and device processing. However, some unique features of LPE were used to assist in the growth of these more complex structures, making it possible for certain of the structures to be grown controllably. Techniques of growth into shallow (say 1–2 μm deep) grooves cut into substrates (Fig. 5*a*) or over ridges (Fig. 5*b*) were perfected, which enabled controlled variations in the thickness of some of the overgrown layers to be engineered in such a way that optical guiding and/or current confinement effects in the resulting stripe laser structures were incorporated. The technique of two-stage epitaxial growth was developed, which enabled buried heterostructure (BH) lasers (Fig. 5*c*) to be constructed so that reverse *p-n* junctions on either side of the active stripe blocked current flow whilst compositional changes provided a lateral waveguide. Such lasers enable threshold currents of less than 10 mA to be achieved. There is no doubt that the inherent properties of LPE have played an essential part in allowing variants of

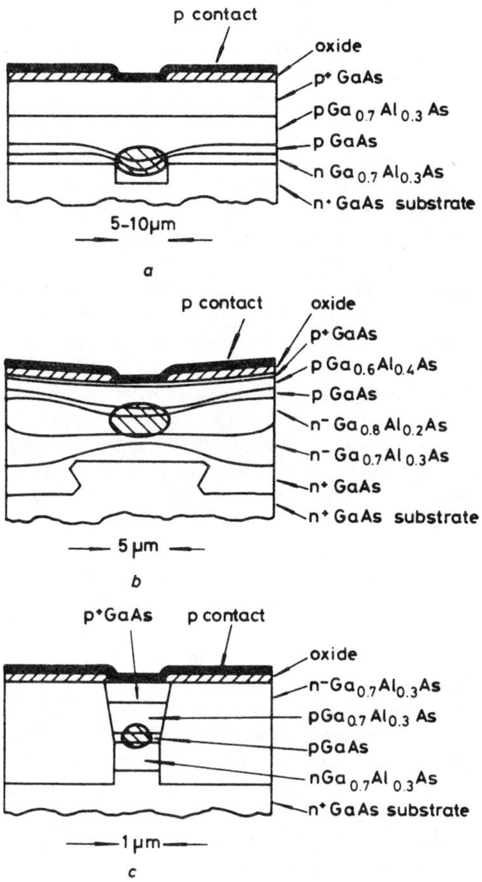

Fig. 5 *Advanced GaAs double heterostructure laser structures*
a Channel substrate
b Constricted double heterostructure large optical cavity (CDH-LOC) [35]
c Buried heterostructure

the above types of lasers to be fabricated with yields sufficient for commercial production.

Other growth techniques have emerged over the period in the form of molecular beam epitaxy (MBE) and metallorganic vapour phase epitaxy (MOVPE) [36] which begins to offer the promise of more uniform, larger area growth and control over the deposition of very thin layers. That is not to say that some of the most successful device structures grown by LPE are going to be easily remodelled using these techniques. However, the emergence of these techniques has led to consideration of the potential of quantum well laser structures to provide even lower threshold devices [37]. This possibility is brought about by the quantisation of valence and conduction band levels as layers are reduced in width to the order of 200 Å or less. This results in a stepwise density of states, and calculations indicate that the optical gain is significantly enhanced.

The work on the development of better controlled and reliable laser structures would have been no use in itself without the ability to couple effectively and stably the laser output into the required 50 μm core graded-index fibre. This potential hurdle was recognised by only a small number of workers initially, who began to evaluate options for mounting fibre tails, lasers, feedback photodiode and even control electronics within packages [38] that could be handled in the same manner as silicon chip DIL packages, as far as the assembly of circuit boards and equipment subassemblies were concerned (Fig. 6). The success of these developments has played a large part in

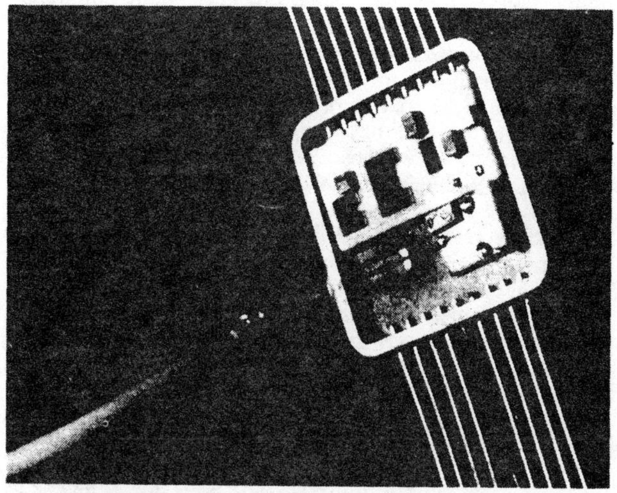

Fig. 6 *Hybrid integrated GaAs laser transmitter package*

making semiconductor lasers and LEDs fully acceptable as components to the communications system design engineer.

2.5 Long wavelength sources

At the same time that GaAs DH laser and LED technology were maturing, significant progress was being made in fibre technology. It was becoming obvious that the ultimate potential for silica fibres in telecommunications lay in systems operating in the 1300–1550 nm wavelength region [39, 40], because both loss and dispersion were much lower at these wavelengths. A dramatic change in materials technology was called for, as GaAs was not suitable for use at these wavelengths. A potential solution to this need had already been suggested in the early 1970s by Antypas [41], who had shown that independent variation of lattice constant and bandgap was possible in a range of quaternary solid solutions, in particular, that a range of InGaAsP alloys could be grown lattice-matched to InP. In the late 1970s first results on 1200–1300 nm surface-emitting DH high-radiance LEDs [42] and 1200 nm DH lasers [43] were reported. Surprisingly (at least to those having worked on lasers in the earlier days of GaAs) lives in excess of 1000 hours of room temperature operation were reported for these early InGaAsP devices. This was viewed as very promising since the threading dislocation content of the active region was presumed to be high due to the 10^4–10^5 cm^{-2} dislocation density in the InP substrates available at the time for overgrowth of laser structures. In addition, facet burn-off and erosion problems in lasers were also showing less of a tendency to occur than in the case of GaAs.

Partly as a result of the apparent absence of such intrinsic internal degradation mechanisms and partly because of all the experience built up on the physics of lasers and on growth and processing techniques for GaAs, the whole development cycle for these second-generation optical communications lasers and LEDs has been significantly compressed, so that, less than 4 years after decisions to install 850 nm laser systems in British Telecom's intercity network, 1300 nm laser systems were already being installed both in the UK and elsewhere, and decisions taken by international telecommunications operators to install undersea systems on both short-distance and trans-oceanic routes.

2.6 Problems for InGaAsP sources

The enthusiasm for 1300 nm fibre systems generated by fibre technologists and system engineers does not mean that LEDs and (particularly) lasers based on InP substrates have not had problems associated with them that have needed significant attention. This has been particularly so, as advances in fibre technology and the real system gains possible have returned the focus of attention to single-mode fibre transmission and, for the future, the possibilities for coherent fibre transmission techniques [44].

A prime area of concern has been that related to the enhanced temperature dependence of laser threshold current in comparison with GaAs devices — another topic that has provided the incentive to develop the theories of the associated physics. Intense study of the problem has resulted in a number of suggestions to explain the phenomenon relating to intrinsic physical properties of the InGaAsP system, such as Auger recombination, intervalence band absorption and carrier leakage effects over heterojunctions [45], but the relative importance of the various mechanisms is still not clearly defined. This insoluble enhanced-temperature dependence problem helped to push forward studies evaluating other potential quaternary systems (particularly AlGaAsSb lattice-matched to GaSb) for 1300–1500 nm operation [46, 47]. However, no alternatives have shown particular promise so far. In the continuing absence of any solution, it has been necessary to pay significantly more attention to thermal dissipation, both in terms of achieving low operating currents and efficient heat sinking arrangements. The need to minimise or eliminate thermal resistance degradation of the solder bond (an effect discussed earlier that could be tolerated to some extent in GaAs lasers and LEDs) has also attained a critical significance.

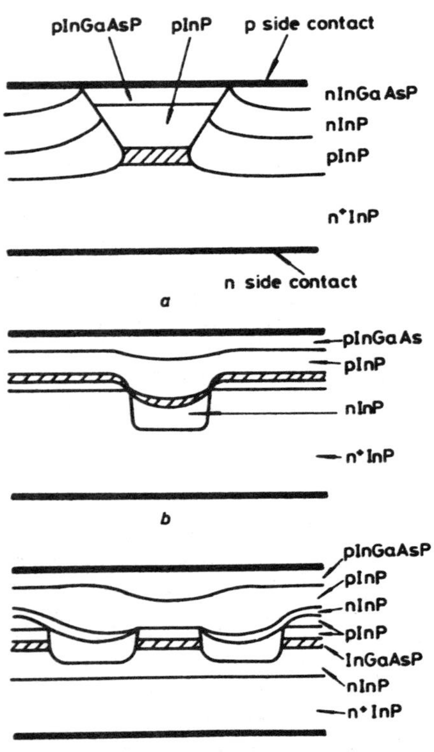

Fig. 7 *Buried heterostructure variations for InGaAsP laser structures*

a Conventional buried heterostructure
b Buried cresent
c Double channel planar

This need to minimise thermal problems, together with the requirements on efficient launching into single-mode fibre, has focused more attention on the buried heterostructure family of devices than has been the case with GaAs DH lasers, and a range of variations on the buried heterostructure (BH) theme have been promulgated (Fig. 7) primarily using LPE techniques. Significant problems have had to be overcome, for example the need for attainment of the precise melt composition is much more severe for the quaternary In-Ga-As-P system than in the case of the ternary Ga-Al-As, as the lattice constant of the quaternary varies significantly with compositional changes, unlike the ternary case. However, the problems associated with needing to eliminate the presence of oxygen to a high degree, because of its reactivity with aluminium, is avoided, but those associated with control of the very volatile phosphorus emerge. In spite of such potential hazards, the LPE technique has continued to be successful, and, in spite of the considerable progress made with VPE, MOVPE and MBE techniques for 1000–2000 nm emitting materials [36], LPE still holds sway as *the* production technique for long wavelength sources, which says something for the resilience of the original method and its flexibility.

Another problem, that has had to be readdressed with single-mode fibres coming into use with InGaAsP lasers, is laser-to-fibre coupling. For most applications efficient coupling is required, and this is a function of the laser design (dictating the angular distribution of the emission) and the coupling optics. The need for automated techniques for fibre alignment, and the requirement for package designs giving stability of coupling over a temperature range and with time, has consumed considerable effort. It is certainly going to need further advances in this area to achieve laser packaging costs that give scope for significant price reductions in the future.

2.7 Narrow linewidth devices

The developing growth techniques mentioned earlier are opening up new possibilities for the future in the form of integrated optics and integrated optoelectronics. They are also providing the basis for the next generation of laser emitters — that is those associated with coherent transmission or wavelength division multiplexed (WDM) fibre transmission links — specifically the development of distributed feedback (DFB) lasers. Such lasers were first demonstrated by Nakamura *et al.* [48] in 1975 in GaAs, but further developments were put in abeyance until the drive to single-mode long wavelength systems indicated the requirement for narrow linewidth sources to overcome the need to minimise fibre dispersion, as well as the potential for WDM in the broad low-loss fibre window beyond 1000 nm and the possibilities for coherent detection, given lasers with sufficiently narrow linewidths [44]. There are possible solutions to these requirements using a combination of a conventional DH laser chip and some form of external cavity. These could certainly become viable [49], but for a range of applications, the laser chip with inbuilt wavelength (frequency) control in the form of the DFB laser potentially offers the neatest solution, as it can essentially just replace the present generation of laser chips within the same package.

The requirements for fabrication of a DFB laser have demanded the development of an additional range of technologies based on spin-offs from developments in gas lasers and in electron-beam lithography. The principle of the DFB laser is the introduction of a longitudinal diffraction grating through the incorporation of a periodic refractive index variation within the conventional DH laser active region. The period of the grating is of the same order as an integral multiple of half the wavelength of the radiation within the cavity (Fig. 8). This implies a grating

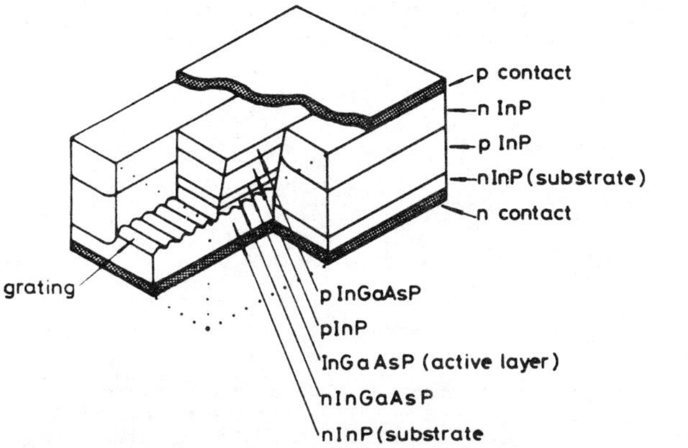

Fig. 8 *Buried heterostructure-distributed feedback (BH-DFB) laser structure*

period of about 0.23 μm (fundamental) or 0.46 μm (2nd order grating) for emission at 1500 nm. Under specific conditions, controlled single spectral mode output is achieved, with the additional significant advantage that the lasing wavelength is relatively insensitive to temperature in comparison with conventional cleaved facet (Fabry-Perot) devices.

The original GaAs DFB lasers, and many of the examples reported more recently in InP-based materials, were fabricated by the laser holographic exposure technique, where a large area of material covered with resist can be exposed to produce a uniform grating. This grating is then subsequently overgrown with additional epitaxial layers. The directly written electron beam exposure technique [50] is a serial process and, at the dimensions used, demands precise attention to the details of electron scattering in InP etc. However, it does enable grating patterns to be exposed only in areas required and can provide for different grating periods to be written on different areas of a wafer. This capability gives the potential for fabrication of an integrated array of emitters for use in WDM systems of the future. Thus this technique offers the potential for the cost-effective realisation of controlled frequency separation multiwavelength sources in the near future and could prove to be, in association with the increased sophistication in growth available through MOVPE, one of the most significant breakthroughs of the 1980s in the semiconductor laser story.

2.8 LEDs

Only scant mention of LEDs has been made, but significant developments aimed at communications applications have occurred in this area. The first step was the development of the Burrus diode [51] (so-called after its inventor) now called the surface emitting LED (SLED). Even in its early GaAs homojunction form, with its capability to locate a fibre within an etched well to achieve useful optical coupling, LEDs gained a firm place in optical communications. The later development of the DH edge-emitting LED (ELED) and DH SLED for 1300 nm operation (Fig. 9) has widened the potential for LEDs. However, to date, such devices have always been overshadowed by their laser counterparts for one reason or another. In general, the laser is capable of modulation at higher bit rates and can launch more power into fibres, especially for single-mode fibre systems, but LEDs do have the advantage of being cheap and easy to drive. Interactive CATV systems utilising fibre optics for primary links are

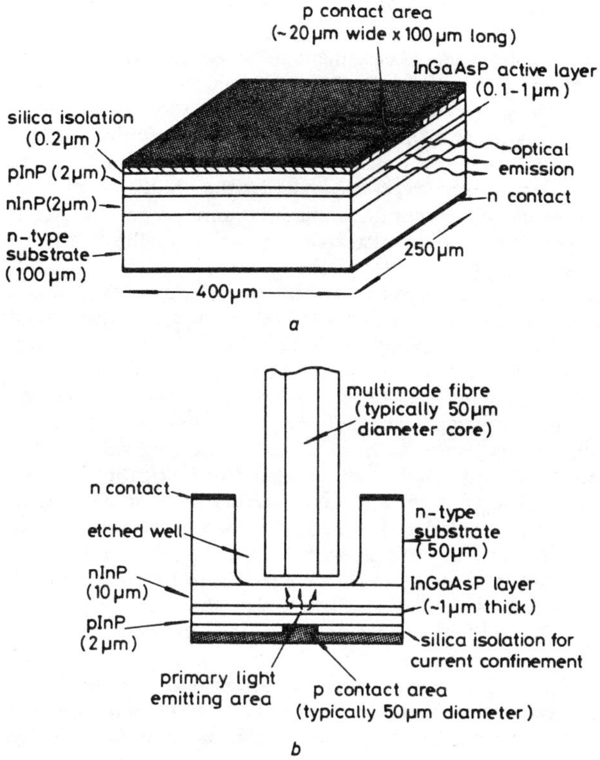

Fig. 9 *High-radiance LED structures*
a Surface emitting LED (SLED)
b Edge emitting LED (ELED)

being developed and decisions made to begin to install fibre optics in telecommunications local-loop environments to connect remote concentrators and business communities. The opportunity exists for eventual extension of fibres to subscribers for broadband services and also, as the current copper connections wear out, if such fibre systems can be engineered cost-effectively. Thus, with the reduced distance requirements of the local network, there is a possibility that LEDs, with their inherently simpler drive requirements, can, in conjunction with single mode fibre, be part of the needed cost-effective realisation. Transmission over 20 km at bit rates of at least 140 Mbit/s has already been shown to be feasible [52], which indicates that achievement of the performance required for such applications is no longer a technical problem.

3 Photodetectors

3.1 Optical receivers

Throughout the whole development period of optoelectronic components for fibre systems, the receiver end of an optical link seems to have been a poor relation of the transmitter — presumably because lasers are viewed with more glamour than silicon APDs and III–V PIN diodes. It has often been a source of slight bewilderment to those involved in developing photodetectors and receivers that the systems designer demands the last fraction of a dB sensitivity out of the receiver, but doesn't appear too worried about the several dB lost between the laser and the fibre. Despite these problems, the photodetector and receiver designers (one should not separate the two, as the performance of the receiver depends critically on the photodetector and vice versa) have risen to the challenge and made significant improvements over the years which have contributed towards optical systems becoming so cost effective.

The photodetector element must be considered in the context of the whole receiver. A very important feature of a receiver for most optical links is that it should be able to detect as low an optical signal as possible, without introducing too many errors. This basically means that the noise in the receiver, contributed by the photodetector and the electronic circuitry, should be minimised. In digital systems, noise gives rise to bit errors, and the signal-to-noise ratio should be sufficiently large to give an acceptable error rate. An error rate of 1 in 10^9 bits is often taken as giving satisfactory performance, and, in a conventional PCM system with equiprobable 1 s and 0 s, this corresponds to a signal-to-noise power ratio of about 20 dB.

Thus, in the choice of a photodetector for a receiver, noise performance is all important. However, there are other parameters that need to be considered, namely speed, quantum efficiency (QE), ease of fabrication and reliability. Also, it is important to realise that a detector cannot be treated in isolation; the noise of the front end of the receiver is equally important. A good low-noise photodetector is worthless without well designed low-noise circuitry. In some cases, notably the high-impedance PIN-FET receiver to be discussed later, the noise contribution associated with the PIN photodiode can be made insignificant compared with that associated with the FET.

Optical/electrical transducers have been demonstrated using a variety of physical effects. In the decades prior to 1966 the most common detectors were photomultipliers (using photoemission) and photoconductors. Neither of these are really suitable for optical fibre-based systems. Photomultipliers are large, need high voltages, have a limited gain-bandwidth product and have poor quantum efficiency at the wavelengths of interest (>850 nm). Photoconductors have, until recently, been slow and noisy. The speed problem has largely been associated with traps, and it is the relatively low resistance that determines the rather high thermal (Johnson) noise. A particular attraction of these two classes of device is the high internal gain that can be achieved. One incident photon creates one carrier in the device, and then an internal multiplication mechanism provides gain within the device. This gain occurs with very little, if any, excess noise, and so provides an enhanced signal-to-noise ratio.

The silicon *p-n* junction photovoltaic device has many of the features lacking in photomultipliers and photoconductors, notably high speed, low noise and size compatibility with optical fibres. Unfortunately it has unity gain. However, in the few years before 1966, work had started on the silicon avalanche photodiode (APD) [53, 54], a solid-state analogue of the photomultiplier tube based on the simple *p-n* photovoltaic device. Although this work had not been specifically aimed at optical communication. It was immediately clear that the silicon APD was the ideal device for this new application.

3.2 Silicon avalanche photodiodes

The silicon APD has a structure similar to that of the simple photovoltaic *p-n* junction diode. Carriers are generated by the absorbed photons, are separated by an electric field and give rise to a photocurrent in the external circuit. However, in the APD the device is operated very close to the breakdown point. Under these conditions, the photogenerated carriers may impact ionise producing more carriers and thus giving gain. This gain allows receivers to be made that have up to 15 dB better sensitivity than those with unity gain *p-n* diodes. As the loss of silica fibre at 800–950 nm is about 2 dB/km, this corresponds to an extra 7 km of fibre length — and makes all the difference between an economic and uneconomic optical system.

In the simplest possible device, a *p-n* junction (Fig. 10*a*),

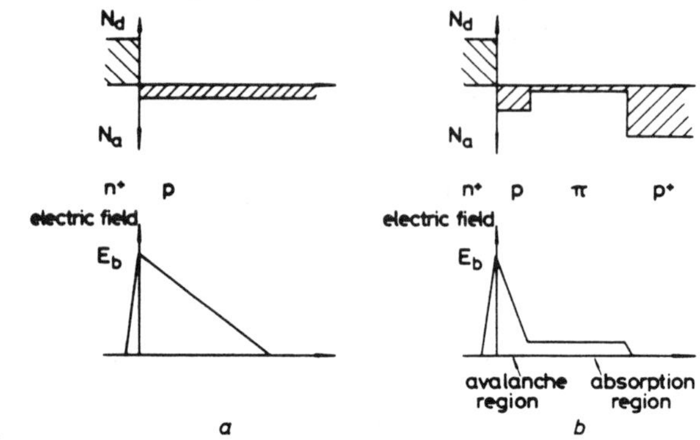

Fig. 10 *Doping profile and electric field distribution at breakdown*
a Simple *pn* structure
b Reach-through avalanche photodiode structure
E_b is the breakdown field and $\int E\,dx$ is the breakdown voltage

the electric field is held at the breakdown field E_b (which is dependent on doping, but is typically 5×10^5 V/cm) when the carriers can gain enough energy to multiply by impact ionisation. The space-charge region thickness should be large enough so that all the incident photons are absorbed in it. If it is too thin there will be a quantum efficiency penalty and carriers absorbed in undepleted regions may slowly diffuse into the high field region giving an unwanted tail at the end of the optical pulse. Unfortunately, the absorption coefficient of silicon is relatively low at wavelengths between 800 and 900 nm (1000 cm^{-1} at 800 nm, and 500 cm^{-1} at 900 nm). Thus the absorbing region often needs to be over 50 μm thick to give a good QE and reasonable speed. This means that not only are very low doped structures required, but that operating voltages can be extremely high, up to 1000 V. These problems led to the reach-through design, derived from the IMPATT diode structure, and this has now become an industry standard for silicon APDs.

A reach-through APD has a doping profile as shown in Fig. 10*b*. The absorbing region is effectively separated from the avalanche region. A field of sufficient magnitude to keep the carriers at their saturated drift velocity is required in the absorption region, whereas the multiplication region has a higher doping level. As the voltage is the integral of the electric field over distance, the breakdown voltage is much lower than that for the simple *p-n* junction case. The reach-through structure has another benefit in that the dependence of gain on applied voltage is much less than for the abrupt *p-n* junction, allowing the circuitry needed for gain control to be much simpler. A more detailed description of APD operation is given in References 55 and 56.

The signal-to-noise ratio (S/N) of a receiver incorporating an APD may be approximated by:

$$\frac{S}{N} = \frac{P^2 M^2 (e\eta/h\nu)^2}{2e[(Pe\eta/h\nu) + I_{dm}]M^2 F(M)B + 2eI_{du}B + \langle i_n \rangle^2} \tag{1}$$

where P is the received optical power, e the electronic charge, η the device quantum efficiency, $h\nu$ the photon energy, and B the bandwidth of the receiver, I_{dm} is the bulk dark current, i.e. that which is multiplied, and I_{du} is the unmultiplied dark current, e.g. edge and surface-induced currents. $\langle i_n \rangle^2$ is the noise current of the amplifier circuit. M is the gain of the device and $F(M)$ is an excess noise factor associated with the multiplication process, which has to be introduced because impact ionisation is a random statistical process, thus causing noise.

The noise term in eqn. 1 is made up of the amplifier noise and a term corresponding to the quantum noise of the input optical signal and the APD dark current. In practice, for silicon devices the dark current is usually very small ($\ll 1$ nA) and can be ignored. It can be seen that, for a specified $\langle i_n \rangle$ and S/N, there is an optimum value of M to minimise P, and this minimum value depends on the value of $F(M)$.

McIntyre [57] has shown that this excess noise term depends on M and the ratio of electron (α) and hole (β) ionisation rates k ($= \alpha/\beta$). k needs to be either $\ll$ or $\gg 1$ for lowest noise: i.e. the electron and hole ionisation ratios need to be very different. k is dependent on electric field, becoming nearer unity as the field increases. The value of $F(M)$ also depends on the electric field profile and the distribution of absorbed photons. For lowest noise it is necessary to have the carrier with the highest ionisation coefficient (electrons in the case of silicon) initiate the ionisation process. In fact, a figure of merit for an APD structure is k_{eff}, a value of k weighted by these other geometric factors. For typical Si APD structures [58] k_{eff} is 25 giving an excess noise factor of 5 at a gain of 100. This is extremely good in comparison with other materials, as will become clear later.

So it is clear that the design of an APD to have a low $F(M)$ consistent with high speed and high QE is quite complicated. The implementation of commercially acceptable devices is made even more difficult, because of the necessity of very uniform doping densities (laterally) and guard rings to prevent premature breakdown, and the need for high quality dielectric passivation layers and channel stop rings to prevent surface leakage. Fig. 11 is a

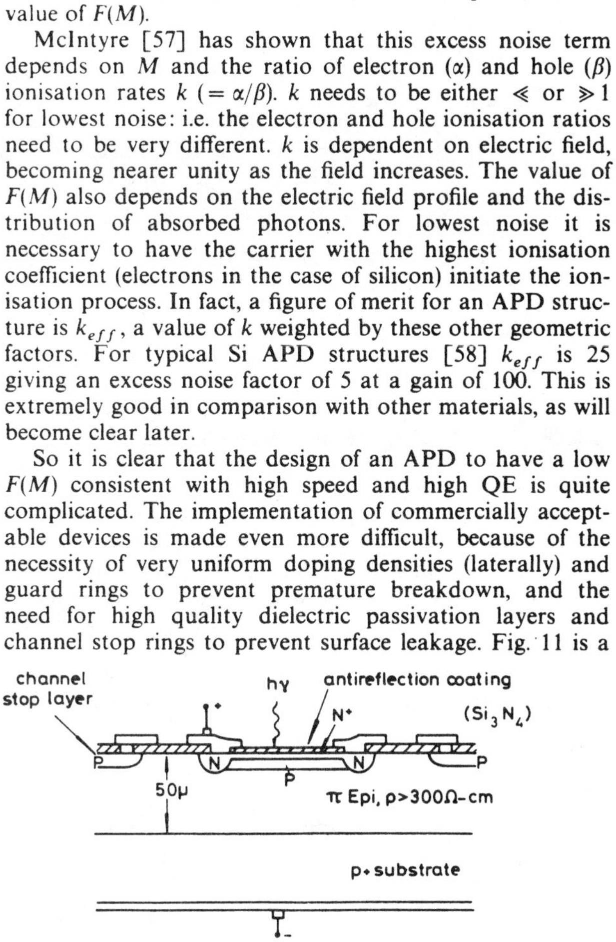

Fig. 11 *State-of-the-art reach-through silicon avalanche photodiode structure [58] (Reprinted with permission from the Bell System Technical Journal, Copyright © 1978 AT&T)*

schematic diagram of a state-of-the-art Si APD structure, similar to those in use in optical systems throughout the world [58]. In this device high resistivity silicon is grown on a p^+ substrate, and the guard ring and channel stop regions formed by diffusions. The p-type avalanche region is formed by a boron implantation and subsequent drive in. The temperature and time of the n^+ layer diffusion determines the exact thickness of the avalanche region, and the precise gain-voltage characteristics. As well as these steps to form the basic structure, other special procedures are employed to obtain the highest performance. For example a deep phosphorous diffusion into the p^+ substrate getters defects that can give high leakages, and an anneal in HCl reduces the mobility of the ions in the surface oxide.

It will be noticed that this structure is not optimised for lowest noise, because avalanche multiplication is not initiated solely by electrons. In fact, the noise penalty over the single carrier injection case is estimated to <1 dB. The fabrication of devices with true single carrier injection into the avalanche region involves either thinning the p^+ substrate and illuminating the substrate, or growing multiple epitaxial layers on an n^+ substrate. The process described above is much more amenable to mass production, and this is an important factor for a cost effective and reproducible process. Indeed, Si APDs with lower noise and higher performance than the one described here have been demonstrated, but at the expense of more complicated processing.

So it is clear that in the trend towards better and better APD performance, the APD structure has become quite complex. Although a schematic diagram suggests that it should be no more complex to make than a conventional silicon IC, the thickness and doping of the multiplication layer is critical. Some cynics have even suggested that the fabrication of commercially available Si APDs often involves a trial and error process of successive heat treatments until the avalanche layer thickness is correct. But there will always be workers on III–Vs who will question the inherent superiority of silicon technology.

3.3 Alternatives to the silicon APD for 800–950 nm systems

It appears that the silicon APD is an almost ideal photodetector. The ability to achieve high gains with little excess noise, combined with a reasonably well developed technology, gives a very attractive product. Its one drawback is the trade off between QE and speed, because of its low absorption coefficient. However, for most applications around 850 nm very high speed is not required, and with rise times of a few nsec reasonably easy to achieve, Si APDs may be used in systems operating at several 100 Mbit/s.

Alternative devices have been seriously considered in the past, notably APDs made of GaAs and Ge [59]. Although Ge has a much higher absorption coefficient in the 850 nm region than Si, its major weaknesses are that p-n junctions have dark currents, typically 3 to 4 orders of magnitude larger than those in silicon (in the μA range), the dark current and gain are very sensitive to temperature, and the k value is between 0.5 and 1. Although Ge has come into its own at the longer wavelengths where Si cannot be used, it is not competitive with Si at the shorter wavelengths on noise grounds alone.

GaAs can have dark currents as low as those in silicon, but, as with Ge, its k value is much nearer unity, resulting in a noisier device. In addition, its absorption coefficient changes rapidly just in the wavelength region where the GaAs emitters are operating. Although the benefits of higher speed have been demonstrated — 35 ps rise time [60] — a GaAs APD cannot compete on sensitivity with a Si device at bit rates of several 100 Mbit/s.

At this stage it may be worth mentioning an increasing trend in optical communications for inexpensive, low performance systems at wavelengths around 850 nm. A particularly attractive component for these applications is an integrated detector preamplifier [61]. This component has a unity gain silicon photodiode and its preamplifier integrated on the same Si substrate. Although not giving the highest performance, there are likely to be considerable cost benefits if this device can be mass produced. Other components that have been used for optical communication applications are the Si phototransistor and photo-Darlington [62]. These devices can exhibit high gains, but are much slower than photodiodes and APDs (several μs), have quite complex structures and their quantum efficiencies are low unless the wavelength is very short. However, because of low voltage operation and the potential for monolithic integration, these components might also be candidates for the lower performance systems.

So, to summarise; silicon APDs have no rivals for the high performance optical systems using GaAs-based emitters. These devices have probably presented less difficulties in their development than any other optoelectronic device for optical fibre systems. In part, this has been due to an established technology, but an extremely important factor in the success of the Si APD is its inherent low noise properties, i.e. very low dark currents and a high k value. It seems extremely fortuitous that the material with the highest k value measured to date should also offer an established processing technology and sensitivity to an important wavelength region for optical communications.

3.4 Detector considerations for longer wavelength systems

Silicon is effectively transparent at wavelengths greater than 1100 nm, so as the potential for the longer wavelength optical systems (1300 and 1550 nm) emerged, it was clear that a new device based on a different semiconductor was needed. The obvious choice was germanium, which has a direct bandgap suitable for use out to 1500 nm. Although the performance of a Ge APD at long wavelengths is poor compared with a Si APD at the shorter wavelengths, overall systems benefits can still be obtained because of the much lower fibre loss and dispersion. The Ge APD noise is limited by high dark currents, and it became clear that *p-n* photodiodes with lower dark currents and very low capacitances, combined with a very high performance FET amplifier stage, could give as good a performance, if not better, than the Ge APD. These *p-n* photodiodes were made with III–V materials, notably InGaAs on GaAs or InP, and were known as PIN diodes. The 'insulating' part of the PIN is a little misleading — although an insulating or very low doped region is highly desirable, the quest towards high purity material in these new III–Vs has been slow, and even now doping levels of about 10^{15} cm^{-3} are about the best that can be obtained routinely.

Initially it was not clear whether a Ge APD or III–V PIN photodiode would be best for the highest sensitivity long-wavelength systems. Although germanium had a considerable start because of its previous use for electronic devices, the effort on III–V materials for lasers promised good PIN diodes, and even III–V APDs, which were considered the best option for the future, but which presented quite considerable technological challenges.

The debate on the relative merits of Ge APDs, III–V APDs, PIN-FET and PIN-bipolar receivers seems neverending, and could fill many many pages of this journal. In general, the Japanese workers have favoured Ge APDs, Europeans the PIN-FET, and workers in the US started with Ge APDs before moving on to PIN-FETs. The USA and Japanese workers have also put considerable effort on the III–V APD, although at present these are not commercially available, unlike the other two options. Rather than give a detailed discussion of the relative merits of the different approaches, Table 1 highlights some of the important features of receivers based on PIN FETs, Ge APDs, and III–V APDs. Perhaps the most important point to mention is that the ultimate in receiver performance should not be taken as the sole criterion for assessing a photodetector. Other factors, such as dynamic range of the receiver, its temperature stability, cost and reliability, all need to be taken into account.

3.5 Germanium APDs

Germanium APDs were being made as early as 1965, when Melchior and Lynch [63] showed that the concepts already demonstrated in Si could be extended to Ge. It was clear at this early stage that the high dark currents in germanium APDs were a problem, and also that the excess noise factors were much higher than those in silicon. The dark currents were associated with edge and surface effects resulting from difficulties in passivating germanium, and

Table 1: Relative merits of receivers incorporating III–V PIN diodes, Ge APDs, and III–V APDs

	III–V PIN diode	Ge APD	III–V APD
Device structure	simple	difficult especially for 1.55 μm operation	very difficult
Fabrication	new technology but relatively simple	established technology	still in research stage, yields poor
Receiver sensitivity	very good at low bit rates progressively worse above 565 Mbit/s, trade-off with dynamic range	better at high bit rates, especially if device cooled	potentially gives best sensitivity, but by only a few dB. Still considerably inferior to Si APD
Associated circuitry	low voltages, high performance FETs needed, great care with circuit layout to minimise stray capacitances	high voltage operation simpler circuitry, but more of it, extra circuitry to stabilise gain, and compensate for temperature variations	
Applications	chosen for submarine systems because of simplicity, will be preferred for coherent systems	preferred for high bit rate land systems at present; problems with saturation in coherent systems	still at research stage

also were a fundamental consequence of the small bandgap. There was little further work on Ge APDs until a decade later, when detectors for the longer wavelength systems were needed.

The type of Ge APD being used at the outset of research into long wavelength systems is shown in Fig. 12*a*

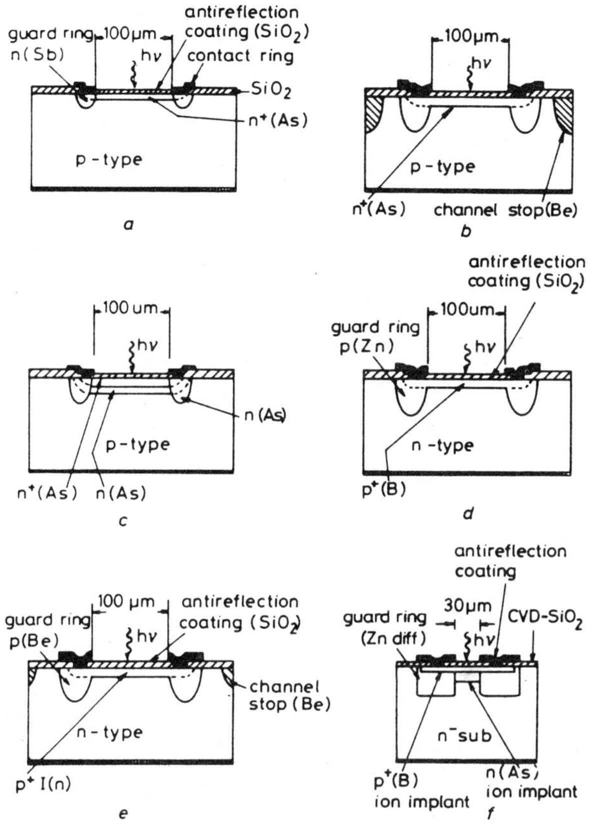

Fig. 12 *Evolution of germanium APDs*

a Basic n^+p structure (from reference 64, IEEE Journal of Quantum Electronics, Copyright © 1978 IEEE)
b Improved n^+p structure (from reference 65, Fujitsu Sci. & Tech. Journal)
c n^+np structure reference 66, IEEE Journal of Quantum Electronics Copyright, © 1980 IEEE)
d p^+n structure (from reference 67, Applied Physics Letters)
e all-implanted p^+n structure (from reference 68, Applied Physics Letters)
f p^+nn^- structure (Hi-Lo) (from reference 70)

[64]. This is an n^+p structure, with guard rings to prevent premature breakdown. A silica antireflection coating was used, and the junction diameter was 100 μm. Dark currents near breakdown were between 100 and 300 nA and were very sensitive to temperature. For these Ge APDs, unlike for silicon devices, the dark current had significant components of both bulk (multiplied) and surface (unmultiplied) current. It is the multiplied component (typically 100 nA in the device in Fig. 12*a*) which needs to be low ($\simeq 1$ nA) before dark current ceases to dominate the noise. The excess noise factor $F(M)$ was approximately equal to M, implying a $k_{eff} \simeq 1$, but it was appreciated that the n^+p structure was not optimum for low noise, as electrons rather than holes (which have the higher ionisation coefficient in Ge) initiated the multiplication. Noise became worse at the longer wavelengths as more absorption took place in the p Ge, producing even more electron initiated multiplication. Perhaps the only beneficial feature of these Ge APDs over their Si counterparts is that, because of germanium's relatively high absorption coefficient at 1300 nm, breakdown voltages are quite low, typically 25 V in the device discussed above.

Since 1978, there has been a gradual evolution and improvement in the Ge APD structure, mainly in Japanese laboratories. The structure in Fig. 12*b* [65] is aimed at reducing dark currents and improving stability by using a channel stop layer and a special double dielectric passivation. An n^+np structure in Fig. 12*c* [66] was employed to improve noise properties. In this, most of the absorption occurs in the $\simeq 2$ μm thick n layer, thus allowing multiplication to be initiated by holes. The excess noise factor at $M = 10$ was 6, a considerable improvement over the value of 10 for n^+p structures. However, the fabrication of the 2 μm deep n layer by diffusion was difficult, and the speed performance was limited by the low diffusion coefficient of carriers in the n region.

Perhaps the most significant breakthrough in the development of Ge APDs was the development of the p^+n structure shown in Fig. 12*d* [66, 67]. In this, the carriers are absorbed in the n substrate, and thus multiplication is initiated by holes, giving a similar improvement in noise performance to that of the n^+np structure. Although the problem of layer thickness control did not occur in this structure, the high temperatures and long diffusion times needed for the p diffusants gave irreproducible results and high dark currents. Dark currents could be reduced to levels of about 150–250 nA (100 μm diameter area) by using an all-implanted technology (Fig. 12*e*) [68]. As dark currents were near theoretical limits, it was clear that one approach was to use smaller area devices. A similar structure to that shown in Fig. 12*e*, but with a sensitive area of only 30 μm [69], has given a multiplied dark current of only 5 nA at room temperature, together with the improved noise performance of $F(M) \simeq 7$ for $M = 10$ expected for the p^+n structure. An obvious disadvantage with this small area device is that the fibre tail has to be monomode, and thus coupling losses may be greater.

The speed of these p^+n devices at 1550 nm is poor, because the absorption length of Ge at this wavelength is about 10 μm, and so most of the absorption takes place outside the depletion region. This suggests the use of a reach-through structure as used for Si APDs. Fig. 12*f* shows a p^+nn^- structure [70] known as a Hi-Lo structure which combines high cutoff frequency (700 MHz), low multiplied dark current (33 nA) with good excess noise performance $F(M) = 6.5$ at $M = 10$. As expected, breakdown voltage is higher, -85 V. Unmultiplied leakage currents are quite high (1 μA), but good receiver sensitivities have been obtained with this structure [71].

These Hi-Lo devices seem to be amongst the highest performance Ge APDs reported at present covering the 1550 nm wavelength range, and it is estimated that the dark current levels give only 1.5 dB receiver penalty (compared with zero dark current) at 25°C and 3 dB penalty at 45°C. So it seems that the inherent noise properties of the impact ionisation process is now determining the detector performance. Although these Ge APDs are very competitive with PIN-FETs and the III–V APDs that are now emerging, it is difficult to see much greater improvements because of these fundamental limitations. Although Ge APDs are likely to have an important role in optical systems in the next few years, it seems most likely that they will eventually be superseded by the III–V APD.

3.6 III–V PIN diodes

It was in 1978 that some detailed theoretical analysis of receiver performance by Smith, Hooper and Garrett [72] indicated that a simple unity gain PIN diode could be made to give as good a performance as a Ge APD, if it had sufficiently low dark current and capacitance, and was used in conjunction with a very high performance circuit. Fig. 13 shows the front end of a PIN-FET high impedance

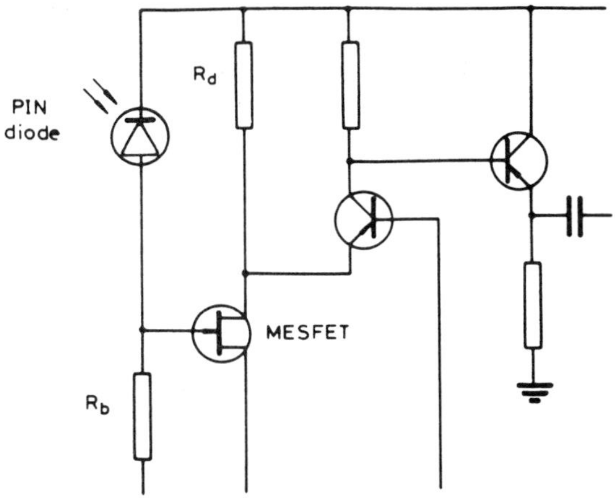

Fig. 13 *PIN-FET receiver front end*

receiver circuit. The bias resistor R_b is made sufficiently large that its thermal noise contribution is insignificant. Then the main noise source, assuming that the PIN dark current and FET gate leakage are low, is that associated with the FET channel. With R_b so high, the input of the circuit is capacitive, and the effective noise due to the FET channel has to be divided by the capacitive reactance when referenced to the input. Thus the noise figure of merit for this circuit is proportional to g_m/C^2, where g_m is the transconductance of the FET and C is the input capacitance of the circuit, consisting of FET gate capacitance, PIN photodiode capacitance and stray capacitances.

Calculations indicated that the values of g_m and C that would be needed for the PIN-FET to better a Ge APD at low bit rates were quite reasonable. Although the hypothesis that a unity gain PIN diode could give as good if not better performance than a Ge APD seemed surprising at the time, the validity of the concept was demonstrated using a Si PIN diode [73] before being applied to the longer wavelength systems.

High performance FETs were available because low noise GaAs MESFETs designed for microwave applications could be used, however PIN diodes with sufficiently low dark currents and capacitance needed to be developed. Ge PINs could be ruled out because of their high dark currents, and the first III–V device demonstrated in the PIN-FET configuration was a novel departure from the standard lattice-matched approach of most III–V optoelectronic devices. Using vapour phase epitaxy, a gradually increasing proportion of indium was added to the gas phase, grading the composition, and thus the lattice constant and bandgap, until the material was sensitive to the appropriate wavelength [74]. This approach had been developed for military applications at 1060 nm, but it was surprisingly successful at 1300 nm, giving low dark currents and capacitances. Some of the first long wavelength systems installed for the public network in the UK used these devices. However, this device has now been superseded by the lattice-matched $In_{0.53}Ga_{0.47}As/InP$ device, which can be used out to longer wavelengths. The graded structure was only really suitable out to 1300 nm, as dark currents became unacceptable at large mismatch. The lattice-matched InGaAs structure on the other hand detects out to 1670 nm.

InGaAs/InP PIN diodes commonly used at present are shown in Figs. 14*a* and *b*. The layer structures have been made using all the conventional growth techniques for III–V semiconductors — LPE, VPE, MOVPE and MBE — although LPE is generally favoured, because of the rela-

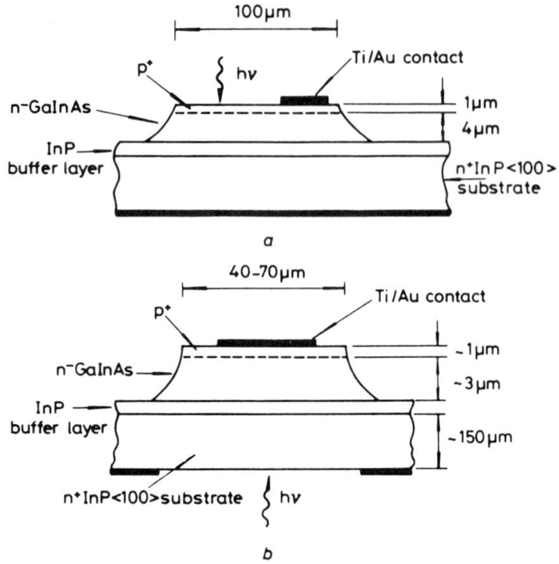

Fig. 14 *Mesa InGaAs/InP PIN photodiodes*
a Top-entry structure
b Substrate-entry structure

tive ease in obtaining the low doping levels needed ($\simeq 10^{15}$ cm^{-3}) to obtain sufficiently low capacitance (<0.2 pF). However, obtaining low impurity concentrations in the new III–V material is not easy, and with LPE it has been found necessary to use long (several days) baking procedures to purify the source material. More recently, MOVPE has produced high quality devices [75], and this technique seems much more appropriate for large scale production of these devices.

The top-entry device is the simplest structure, with the light being introduced through the top p^+ layer. The disadvantage of this structure is that there is a quantum efficiency penalty, with some of the light being absorbed in the undepleted *p* region. This can be alleviated by the use of a transparent *p* InP layer instead of the *p* InGaAs layer, but this presents problems for the LPE growth technique, due to meltback of the InGaAs layer by InP during growth. Another disadvantage of the top-entry structure is that there is a limit to how small the device can be made, with both light access and metallic contact needed on the top. To enable much smaller devices, and thus even lower capacitances, the substrate-entry device is used. Light enters through the transparent InP substrate and the device area can be made as small as is practical for bonding. The disadvantage of the substrate-entry device is that unconventional mounting techniques are needed to allow optical access to the substrate, but various methods have been demonstrated [76, 77].

Using the substrate-entry configuration with small area devices, capacitances may be made extremely small (<0.1 pF), quantum efficiencies near 100% and dark currents <1 nA. This means that with presently available FETs, the performance of the receiver is not limited by the PIN diode parameters, and is determined primarily by the subsequent electronics. Thus the PIN diode seems to be very well developed for this application. In principle, even lower capacitances will be possible as material purity improves, but there will be little performance benefit whilst FET and stray capacitances dominate receiver sensitivity.

However, an inherent problem with the mesa device is that it has an exposed *p-n* junction. Although the reliability of such devices can be fully acceptable, there will always be a preference for devices with a protected junction. Planar devices are now increasingly being used (Fig. 15), and some impressive lifetest results have been reported [78, 79].

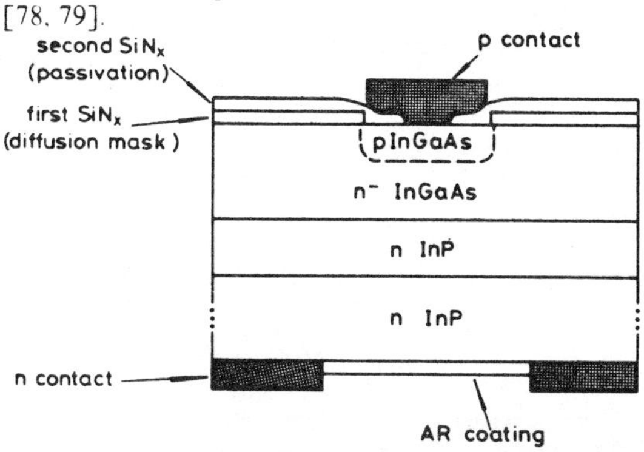

Fig. 15 *Planar InGaAs/InP PIN photodiode (from reference 79) (Reprinted with permission from the AT&T Technical Journal, Copyright © 1985 AT&T)*

InGaAs/InP PIN diodes are widely used in optical systems installed at the moment, and these devices are planned to be incorporated in the TAT8 transatlantic link [79]. Although it seems likely that III–V APDs will soon be commercially available, and that receivers incorporating these will be more sensitive than those with PIN diodes, PIN diodes are still likely to have a promising future. As mentioned earlier, receiver sensitivity is not the only factor to consider: the fact that PIN diodes will be relatively inexpensive, need only low voltage operation, and are relatively insensitive to temperature variations will make them very attractive for many applications.

3.7 III–V avalanche photodiodes

Conceptually, the III-V APD is a natural development from the III–V PIN diode, and there has been much work on this device over the past eight years to try and demonstrate its usefulness for optical systems. Such have been the difficulties that it is only now that these devices are emerging from the research laboratory and being demonstrated in optical systems. The number of problems that have been encountered and the extent of their difficulty seem out of all proportion to the relatively small sensitivity improvement that has been demonstrated to date by receivers incorporating III–V APDs rather than III–V PINs or Ge APDs. However, it is certainly true that the best receivers reported to date in the medium to high bit rate range (565 Mbit/s and above) have used a III–V APD. Although the receiver sensitivity benefits are only a few dB at present, if it is possible to reduce the inherent noise of the device — and there are several ideas being investigated — this device is liable to be preferred to the Ge APD for many applications.

Naively, one might expect APD operation by reverse biassing a simple InGaAs PIN diode. Unfortunately, because of a combination of the small bandgap and the low electron effective mass, it is found that III–V compounds with the smaller bandgaps needed for long wavelength response show Zener breakdown (band-to-band tunnelling) rather than the avalanche breakdown that is required to give gain [80]. Avalanche breakdown is possible by using very low doped material — this reduces the electric field needed for breakdown by impact ionisation to a level at which tunnelling currents are small. However, the doping needed for InGaAs to give avalanche breakdown is about 5×10^{14} cm^{-3} [81]. Not only is this level difficult to achieve, but the voltages needed to obtain breakdown become very high.

As with the Si reach-through APD earlier, the solution was to separate the absorption and multiplication regions [82, 83]—the so-called SAM (separate absorption and multiplication) APD. Fig. 16 shows the layer composition,

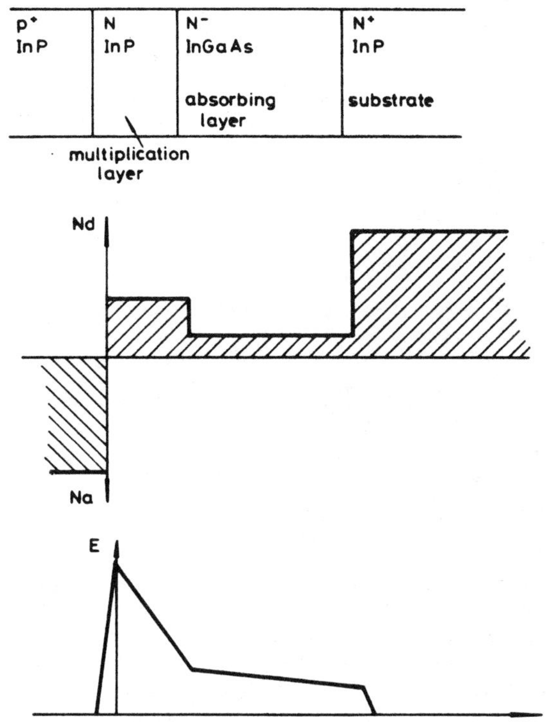

Fig. 16 *Layer composition, doping profile and electric field profile of SAM (separate absorption and multiplication) APD structure*

doping profile and electric field profile of a basic SAM APD structure. The breakdown occurs at the InP *p-n* junction where tunnelling is much less. The electric field in the InGaAs is sufficiently high to separate the photogenerated carriers, but not high enough to cause tunnelling. In principle this device looks relatively easy to make, but the control over the doping and thickness of the *n* InP layer is critical: if it is too thin the field in the InGaAs will be too high and there will be excessive leakages, if it is too thick, the electric field will not extend to the absorbing region, and there will be QE and speed penalties.

So, in principle, it is possible to obtain low dark currents in III–V APDs, and this is a good example of the use of heteroepitaxy to allow flexibility in device design — an approach not applicable to germanium. At present, the epitaxial growth techniques can only just provide the control over layer thickness, doping and layer uniformity that is required for these devices, and the development of a high yield process for the mass production of these devices presents considerable problems. However, obtaining low dark currents was only the first problem to be solved. Soon after the development of the SAM APD it became clear that there was a speed limitation. Initially this was identified as being associated with holes being trapped in the valence band 'notch' at the InGaAs/InP heterointerface, and being thermally emitted with a long time constant [84] (Fig. 17*a*). This problem can be alleviated by incorporating a layer of InGaAsP between the two layers (Fig. 17*b*) to 'smooth out' the notch, and this gives much improved speed performance. However, gain-bandwidth products were still only between 10 and 20 GHz, not quite high enough for the high bit-rate systems (up to 4 Gbit/s)

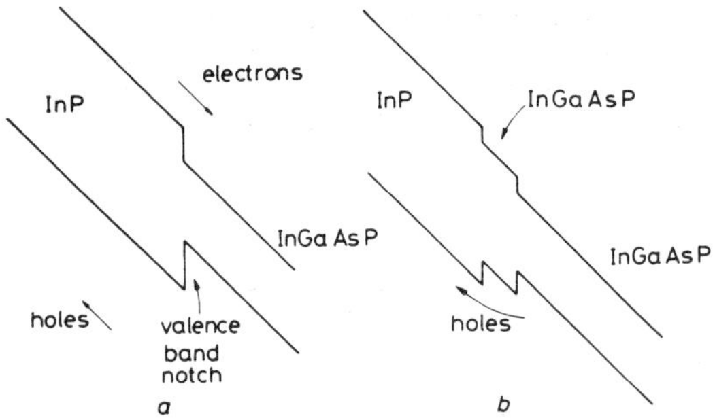

Fig. 17 *Band diagram*
a InP/InGaAs heterojunction showing 'notch' in which holes may be trapped
b InP/InGaAs heterojunction with InGaAsP layer to reduce effect of the notch

being investigated. Another factor limiting speed is the time for the avalanche process to build up [85], which is a function of the avalanche region width and the *k* value of the material. Further optimisation was needed by decreasing the avalanche region width (increasing the doping and reducing the width of the *n* InP region) and gain bandwidth products of 60 GHz have recently been reported [86].

The story is complicated even further when one considers the multiplication noise of these devices. As with germanium — and, it appears, practically every other material except silicon — the *k* value for these III–V materials is near unity. There is a variety of data on the values of *k* for III–V materials, and there is even disagreement as to whether *k* is greater than or less than unity for some materials. However, there seems general agreement that *k* for InP is between 1 and 0.2, a far cry from over 25 for silicon. Thus it is desirable to decrease the breakdown field as much as possible to reduce k_{eff}. This can be done by using lower doping and thicker avalanche regions, but this can compromise speed. Thus, at present, the design of a III–V APD is a delicate balance between optimising dark current, speed and noise. Results given in the literature do not often give all the relevant parameters, and the best way to judge an APD at present is to see how it performs in a receiver. It is extremely difficult to predict theoretically the ultimate in sensitivity from an 'optimised' APD because of all the conflicting requirements.

After this background to the evolution of III–V APDs, some of the most recent device structures and their performance can be described. Fig. 18*a* shows a SAM APD grown on a p^+ substrate [86]. This has the particular advantage that the structure can be grown by LPE, the best developed of all the growth techniques with respect to dopant control, without the problems of meltback of the InGaAs by InP during growth. The device is optimised for high speed, with two rather than one, intermediate InGaAsP layers to effectively 'grade' the bandgap discontinuity. In addition, the avalanche build-up time is reduced by using a thin InP multiplication region. A gain-bandwidth product of 60 GHz has been reported for this device. The structure in Fig. 18*b* is grown by VPE, and has the particularly attractive feature of being a planar device, i.e. it has no exposed *p-n* junction [87]. Implanted guard rings prevent premature breakdown and the multiplied dark current at unity gain is 10 nA, nearly low enough to have insignificant effect on receiver sensitivity. The excess noise factor at $M = 10$ is 5–6, which is slightly better than a Ge device, and the gain-bandwidth product is 35 GHz. This seems to be about the best allround performance of a III–V APD reported to date. The more complex device shown in Fig. 18*c* is designed to improve the guard-ring

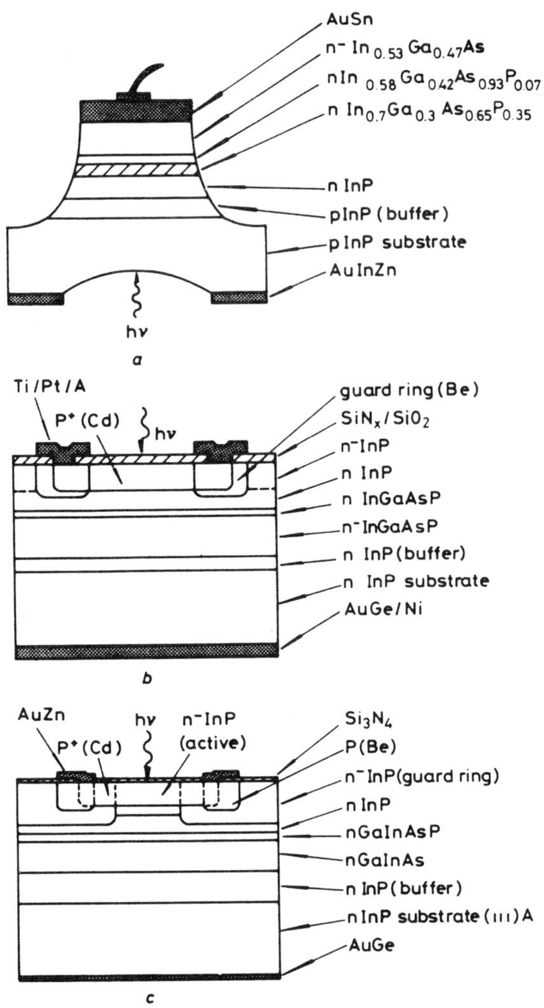

Fig. 18 *Recent SAM APD structures*
a LPE-grown with p^+ substrate (from reference 86, Copyright © Istituto Internazionale Delle Comunicazioni, 1985)
b VPE-grown planar structure (from reference 87)
c LPE-grown buried-structure APD (from reference 88, Applied Physics Letters)

structure and improve gain uniformity [88]. The basic SAM structure with an intermediate layer is used, but the final low-doped InP layer is grown over an etched mesa. This results in the edge of the space-charge region, where electric fields are the highest, being in low-doped material. Together with a graded junction in this region formed by the Be implantation, the guard-ring region has a much higher breakdown voltage than the main avalanche region. This device is grown by LPE, by controlling the amount of meltback, and seems more difficult to make than the other two structures. Although dark currents are not significantly better than the other structures yet, this is an example of the type of complex technology that will be needed in the future to demonstrate the highest performance devices.

So it is a very exciting time for the development of III–V APDs. Research devices are just showing better performance than their III–V PIN diode and Ge APD competitors, and it remains to be seen how much performance benefit will be realised, on what time scale, and perhaps of most importance, how easy it will be to make sufficiently high quality devices on a commercial basis.

3.8 Photodetectors for the future

Although this article has been aimed at tracing the history of photodetectors over the first 20 years of optical communications, it seems very relevant to mention briefly some of the research activities that have been in progress over the past few years that are liable to have a significant impact on detector performance in the future.

Dark currents in III–V APDs can, in principle, be controlled to acceptably low levels, assuming that sufficiently high quality epitaxial material can be grown with the required control over layer thickness and doping levels. So the real problem to be attacked is the multiplication noise, which is a fundamental material property. Whereas in germanium this seems an insurmountable obstacle, the use of heteroepitaxy with III–V materials offers several novel possibilities. Capasso *et al.* [89] has suggested a variety of structures to improve the value of k_{eff} by enhancing the impact ionisation rate of one carrier. Fig. 19 gives an

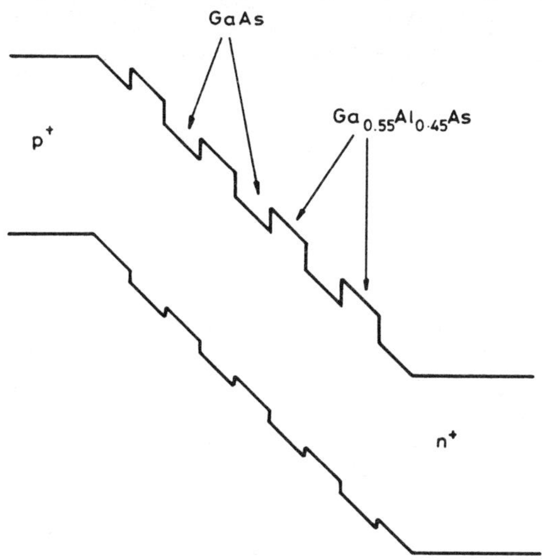

Fig. 19 *Multilayer APD to enhance ionisation rate of electrons*

example of one such structure, which consists of a GaAs/GaAlAs multilayer structure with each layer thickness about 0.05 μm. The principle of operation is that a carrier in the higher bandgap material will effectively gain energy with respect to the band edge as it crosses into the smaller bandgap material. Thus it will have a much higher probability of ionisation. The ionisation rate of holes on the other hand is relatively unaffected by the multilayer structure, because the band edge discontinuity in the valence band is much less than that in the conduction band. This effect has been demonstrated in the GaAs/GaAlAs material system [89], with $k = 8$ for the multilayer structure comparing with $k \simeq 1$ for bulk GaAs material. This result and other related concepts for improving k values, together with theoretical studies [90], are very exciting and could eventually lead to III–V APDs having as good performance as their Si counterparts. As yet, these effects have not been demonstrated in a material sensitive to the longer wavelengths, primarily because the materials growth technology is not yet developed sufficiently, rather than for any more fundamental reasons.

There are also suggestions that other, less well developed, materials could have k values much greater or less than 1, for example gallium antimonide [91] and cadmium mercury telluride [92]. It remains to be seen whether new materials or novel structures using established materials will give the best results.

Alternatives to APDs and PINs for the longer wavelength systems have also been studied in great detail. Phototransistors [93] and photoconductors [94] made in III–V materials are more attractive than their silicon counterparts were, because of the improvements that can be made using heterojunction structures. However, they have not given the sensitivity that APDs and PINs have, so are not widely used. However phototransistors and photoconductors can be much more suitable for monolithic integration with other components, and in the future may find an important role for applications where sensitivity is not of primary importance.

4 Final comments

This paper does not contain an exhaustive survey of source and detector work over the last 20 years, and it has certainly not attempted to mention all the device developments that are currently underway. There are a number of excellent reviews available [33, 34, 49, 56, 95, 96, 97] which should be consulted for more detailed analysis of particular devices. What has been emphasised in this paper however, is that the evolution of optical sources and detectors over the first 20 years of optical communications has involved a major effort to match device parameters to changing fibre specifications. As soon as the development of GaAs sources and Si APDs had reached the stage of moving into commercial production, everything changed as the longer wavelength operation began to be favoured. A change in fibre properties, i.e. reducing the fibre loss at 1300 nm an order of magnitude, resulted in a complete change in the device requirements. Those of us working on semiconductor optoelectronic devices often wished that the boot was on the other foot, and that the fibre workers could be persuaded to modify their fibre properties to suit our devices. But this was not to be. Fibre evolution seems reasonably stable at the moment, but further device development is still needed to maximise the potential of existing fibre (higher power and narrower linewidth lasers, more sensitive detectors etc.). Despite this, there is the possibility that the fibre workers will demonstrate yet more perversity by developing fibre optimised at wavelengths between 2 and 5 μm — and then a new generation of devices will have to be developed.

Of course there is one very important aspect that has hardly been covered in this survey: the problems of making commercially acceptable devices in large numbers. All the references describe research work, and it is often difficult to read between the lines to establish the likely yield. It is generally accepted that some of the exciting devices discussed in the literature are 'one offs' and are unlikely ever to be repeated again. However, devices are now being made in large numbers for commercial optical fibre systems, and the problems involved in this are very different from those encountered in the research laboratory.

A few years ago it was difficult to acquire high quality lasers and detectors. Most of the research and development effort was by companies whose main aim was the provision of telecommunications equipment, and it was not in their interest to give others access to their prime products. However, in the past few years there has been the appearance of many small companies (mainly in the USA) setting up primarily to make these devices. At the moment prices are still high, demand exceeds supply and long waiting lists and frustrated customers far exceed the number of bankrupt optoelectronic companies. Soon however, market conditions will result in more competition, lower prices and better availability. Or will they? The market for these devices will be huge, and the methods of producing them are primitive at the moment (compared with Si technology). Really large scale production seems a long way off. The major effort in the next few years must be for device manufacturers to develop the basic III–V device technology to the stage where dramatic cost reductions are possible. It will only be then that the providers of telecommunication systems will find it cost effective to implement fully all the exciting possibilities offered by optics, such as optical local loops and ultra-high capacity links into the home.

On the research front, the technology developed for discrete lasers and detectors is now being extended to an increasing range of optoelectronic components, particularly those involving monolithic integration of several optoelectronic and electronic devices. The expanding family of components incorporating optical devices will allow extra flexibility in system design and more cost-effective implementation. The first twenty years of optoelectronics has only been the start of a revolution in applications for optics, but the technology developed during this period will form a strong basis for future developments.

5 Acknowledgments

The authors extend their gratitude to their many colleagues at British Telecom Research Laboratories for useful discussions. In addition, interaction with colleagues in UK industrial research laboratories has been invaluable in gaining the overall perspective revealed in this paper. Particular mention should be made of the foresight of the early workers in the UK at the Services Electronics Research Laboratories, Baldock, which started the UK activity in semiconductor optoelectronic devices. Acknowledgment is also made to the Directors of Technology Applications and Research of British Telecom for permission to publish this paper.

6 References

1 BERNARD, M.G., and DURAFFOURG, G.: 'Laser conditions in semiconductors', *Phys. Status Soidi*, 1961, **1**, pp. 699–703
2 BASOV, N.G., KROKHIN, O.N., and POPOV, Y.N.: 'Production of negative-temperature states in p-n junction of degenerate semiconductors', *Soviet Phys.*, JEPT, 1961, **13**, pp. 1320–1321
3 PANKOVE, J.I.: 'Tunneling — assisted photon emission in Gallium Arsenide pn junctions', *Phys. Rev. Lett.*, 1962, **9**, pp. 283–285
4 NASLEDOV, D.N., ROGACHEV, A.A., RYVKIN, S.M., and TSARENKOV, B.V.: 'Recombination radiation of gallium-arsenide', *Soviet Phys. Solid State*, 1962, **4**, pp. 782–784
5 KEYES, R.J., and QUIST, R.M.: 'Recombination radiation emitted by gallium arsenide', *Proc. I.R.E.*, 1962, **50**, pp. 1822–1823
6 HALL, R.N., FENNER, G.E., KINGSLEY, J.D., SOLTYS, T.J., and CARLSON, R.O.: 'Coherent light emission from GaAs junctions', *Phys. Rev. Lett.*, 1962, **9**, pp. 366–368
7 NATHAN, M.I., DUMKE, W.P., BURNS, G., DILL, F.H., and LASHER, G.J.: 'Stimulated emission of radiation from GaAs p-n junctions', *Appl., Phys. Lett.*, 1962, **1**, pp. 62–64
8 KAO, K.C., and HOCKHAM, G.A.: 'Dielectric fibre surface waveguides for optical frequencies', *Proc. IEE*, 1966, **113**, (7), pp. 1151–1158
9 KROEMER, H.: 'A proposed class of heterojunction injection lasers', *Proc. IEEE*, 1963, **51**, pp. 1782–1783
10 ALFEROV, ZH.I., ANDREYEV, V.N., KOROL'KOV, I.V., PORTNOY, E.L., and TRET'YAKOW, D.N.: 'Injection properties of nAl_xGa_{1-x}As-pGaAs heterojunctions', *FTP*, 1968, **2**, pp. 1016–1017
11 PANISH, M.B., HAYASHI, I., and SUMSKI, S.: 'A technique for the preparation of low threshold room temperature GaAs', *IEEE J. Quantum Electron.*, 1969, **QE-5**, pp. 210–212
12 KRESSEL, H., and NELSON, H.: 'Close confinement gallium arsenide *pn* junction lasers with reduced optical loss at room temperature', *RCA Rev.*, 1969, **30**, pp. 106–113
13 DYMENT, J.C., and D'ASARO, L.A.: 'Continuous operation of GaAs junction lasers on diamond heat sinks at 200 K', *Appl. Phys. Lett.*, 1967, **11**, pp. 292–294
14 BIARD, J.R., PITTMAN, G.E., and LEEZER, J.F.: 'Degradation of quantum efficiency in gallium arsenide light emitters'. Gallium Arsenide: 1966 Symposium Proceedings, Reading, UK, 1966, pp. 113–117
15 KRESSEL, H., and BYER, N.E.: 'Physical basis of non-catastrophic degradation in GaAs injection lasers', *Proc. IEEE*, 1969, **57**, pp. 25–33
16 NEWMAN, D.H., and RITCHIE, S.: 'Gradual degradation of GaAs double-heterostructure lasers', *IEEE J. Quantum Electron.*, 1973, **QE-9**, pp. 300–305
17 DE LOACH, B.C., HAKKI, B.W., HARTMAN, R.L., and D'ASARO, L.A.: 'Degradation of C.W. GaAs double-heterojunction lasers at 300 K', *Proc. IEEE*, 1973, **61**, (7), pp. 1042–1044
18 HARTMAN, R.L., and HARTMAN, A.R.: 'Strain-induced degradation of GaAs injection lasers', *Appl. Phys. Lett.*, 1973, **23**, pp. 147–149
19 RITCHIE, S., GODFREY, R.F., WAKEFIELD, B., and NEWMAN, D.H.: 'The temperature dependence of degradation mechanisms in long-lived (GaAl) As DH lasers', *J. Appl. Phys.*, 1978, **49**, pp. 3127–3132
20 PLUMB, R.G., GOODWIN, A.R., and BAULCOMB, R.S.: 'Thermal impedance aging characteristics of CW stripe lasers', *IEE J, Solid State and Electron Devices*, 1979, **3**, pp. 206–209
21 FUJIWARA, F., IMAI, H., FUJITWARA, T., HORI, F., and TAKUSAGAWA, M.: 'Analysis of deterioration in In solder for GaAlAs DH lasers', *Appl. Phys. Lett.*, 1979, **35**, pp. 861–863
22 YONEZU, H., UENO, M., KAMEJIMA, T., and SAKUMA, J.: 'Lasing characteristics in degraded GaAs-GaAlAs double heterostructure laser', *Japan. J. Appl. Phys.*, 1974, **13**, pp. 835–842
23 PETROFF, P., and HARTMAN, R.L.: 'Defect structure introduced during operation of heterojunction GaAs lasers', *Appl. Phys. Lett.*, 1973, **23**, (8), pp. 469–471
24 HUTCHINSON, P.W., DOBSON, P.S., O'HARA, S., and NEWMAN, D.H.: 'Defect structure of degraded heterojunction GaAlAs-GaAs lasers', *ibid.*, 1975, **26**, pp. 250–252
25 O'HARA, S., HUTCHINSON, P.W., and DOBSON, P.S.: 'The origin of dislocation climb during laser operation', *ibid.*, 1977, **30**, pp. 368–371
26 IMAI, H., MORIMOTO, M., SUDO, H., FUJIWARA, T., and TAKUSAGAWA, M.: 'Catastrophic degradation of GaAlAs DH laser diodes', *ibid.*, 1978, **33**, pp. 1011–1013
27 HENRY, C.H., PETROFF, P.M., LOGAN, A.R., and MERRITT, F.R.: 'Catastrophic damage of Al_xGa_{1-x}As double-heterostructure laser material', *J. Appl. Phys.*, 1979, **50**, pp. 3721–3732
28 FURUKAWA, Y., KOBAYASHI, T., WAKITA, K., KAWAKAMI, T., IWANE, G., HORIKOSHI, Y., and SEKI, Y.: 'Accelerated lifetest of AlGaAs-GaAs DH lasers', *Japan. J. Appl. Phys.*, 1977, **16**, pp. 1495–1496
29 HARTMAN, R.L., SCHUMAKER, N.E., and DIXON, R.W.: 'Continuously operated (Al, Ga) As double-heterostructure lasers with 70°C lifetimes as long as two years', *Appl. Phys. Lett.*, 1977, **31**, pp. 756–759
30 KRESSEL, H., ETTENBERG, M., and LADANY, I.: 'Accelerated step-temperature aging of Al_xGa_{1-x}As heterojunction laser diodes', *ibid.*, 1978, **32**, pp. 305–308
31 WAKEFIELD, B.: 'Strain enhanced luminescence degradation in GaAs-GaAlAs double-heterostructure lasers revealed by photoluminescence', *J. Appl. Phys.*, 1979, **50**, pp. 7914–7916
32 YONEZU, H., ENDO, F., KAMEJIMA, T., TONKAI, T., YUASA, Y., and FURUSE, T.: 'Mirror degradation in AlGaAs double-heterostructure lasers', *ibid.*, 1979, **50**, pp. 5150–5156
33 For a review, see GOODFELLOW, R.C., and DAVIS, R.: 'Optical source devices', *in* HOWES, M.J., and MORGAN, D.V. (Eds.): 'Optical fibre communications' (Wiley, 1980), Chap. 2, p. 27
34 For a review, see CASEY, H.C., and PANISH, M.B.: 'Heterostructure lasers, Part B' (Academic Press, 1978), Chap. 7
35 BOTEZ, D.: 'High-power single-mode semiconductor diode lasers', *Proc. IEDM*, 1981, p. 184
36 For a review, see 'Semiconductors and Semimetals, Volume 22, Lightwave Communications Technology (Ed. TSANG, W.T.), Part A, Material Growth Technologies (Academic Press, 1985), Chaps. 2, 3 and 5
37 HOLONYAK, N., KOLBAS, R.M., DUPUIS, R.D., and DAPKUS, P.D.: 'Quantum-well heterostructure lasers', *IEEE J. Quantum Electron.*, 1980, **QE-16**, p. 170
38 CAMERON, K.H., MATTHEWS, M.R., PRESTON, K.R., and SMITH, D.W.: 'Transmitter modules for optical fibre transmission systems', SILA 1979, Cardiff, Wales
39 PAYNE, D.N., and GAMBLING, V.A.: 'Zero material dispersion in optical fibres', *Electron. Lett.*, 1975, **11**, pp. 176–178
40 HORIGUCHI, M., and OSANAI, H.: 'Spectral losses of low-OH-content optical fibres', *ibid.*, 1976, **14**, pp. 310–312

41 ANTYPAS, G.A., and MOON, R.L.: 'Growth and characterization of InP-InGaAsP lattice-matched heterojunctions', *J. Electrochem. Soc.*, 1973, **120**, pp. 154–157

42 DENTAI, A.G., LEE, T.P., BURRUS, C.A., and BUEHLER, F.: 'Small-area, high radiance CW InGaAsP LEDs emitting at 1.2–1.3 μm', *Electron. Lett.*, 1977, **13**, pp. 484–485

43 SHEN, C.C., HSIEH, J.J., and LIND, T.A.: '1500-h continuous CW operation of double-heterostructure GaInAsP-InP lasers', *Appl. Phys. Lett.*, 1977, **30**, pp. 353–354

44 HODGKINSON, T.G., SMITH, D.W., WYATT, R., and MALYON, D.J.: 'Coherent optical fibre transmission systems', *Br. Telecom Technol. J*, 1985, **3**, pp. 5–18

45 CASEY, H.C.: 'Temperature dependence of the threshold current density in InP-$Ga_{0.28}In_{0.72}As_{0.6}P_{0.4}$ ($\lambda = 1.3$ μm) double heterostructure lasers', *J. Appl. Phys.*, 1984, **56**, p. 1959

46 AARIK, A., DOLGINOV, L.M., DRAKIN, A.E., DRUZHININA, L.V., EILISEEV, P.G., LYUK, P.A., SVERDIOV, B.N., SKRIPKIN, V.A., and FRIEDENTKHAL, F.: 'Properties of AlGaAsSb-GaSb heterojunction injection lasers in the 1.4–1.8 μm wavelength range', *Sov. J. Quantum Electron.*, 1980, **10**, (1), pp. 50–53

47 LAW, H.D., HARRIS, J.S., WONG, K.C., and TOMESETTA, R.: 'GaAlAsSb/GaSb alloys: material preparation and applications to optoelectronic devices'. Institute Physics Conference, 1979, pp. 420–428

48 NAKAMURA, M., AIKI, K., UMEDA, J., and YARIV, A.: 'CW operation of distributed-feedback GaAs-GaAlAs diode lasers at temperatures up to 300 K', *Appl. Phys. Lett.*, 1975, **27**, pp. 403–405

49 For a review, see 'Semiconductors and Semimetals, Volume 22, Lightwave Communications Technology (Ed. TSANG, W.T.), Part B, Semiconductor injection lasers, 1 (Academic Press, 1985), Chaps. 1 and 5

50 WESTBROOK, L.D., NELSON, A.W., and DIX, C.: 'High-quality InP surface corrugations for 1.55 μm InGaAsP DFB lasers fabricated using electron-beam lithography', *Electron. Lett.*, 1982, **18**, pp. 863–865

51 BURRUS, C.A., and DAWSON, R.W.: 'Small-area high-current density GaAs electroluminescent diodes and a method of operation for improved degradation characteristics', *Appl. Phys. Lett.*, 1970, **17**, pp. 97–99

52 SHUMATE, P.W., GIMLETT, J.L., STERN, M., ROMEISER, M.B., and CHEUNG, N.K.: 'Transmission of 140 Mbit/s signals over single-mode fibre using surface and edge emitting 1.3 μm LEDs', *Electron. Lett.*, 1985, **21**, pp. 522–525

53 JOHNSON, K.M.: 'High speed photodiode enhancement at avalanche breakdown voltage', *IEEE Trans.*, 1965, **ED-12**, pp. 55–63

54 ANDERSON, L.K., McMULLIN, P.G., D'ASARO, L.A., and GOETZBERGER, A.: 'Microwave photodiodes exhibiting microplasma-free carrier multiplication', *Appl. Phys. Lett.*, 1965, **6**, pp. 62–64

55 WEBB, P.P., McINTYRE, R.J., and CONRADI, J.: 'Properties of avalanche photodiodes', *RCA Review*, 1974, **35**, pp. 234–278

56 SZE, S.M.: 'Physics of semiconductor devices'. 2nd edn. (Wiley Interscience, 1981), p. 766

57 McINTYRE, R.J.: 'Multiplication noise in uniform avalanche diodes', *IEEE Trans.*, 1966, **ED-13**, (1), pp. 164–168

58 MELCHIOR, H., HARTMAN, A.R., SCHINKE, D.P., and SEIDEL, T.E.: 'Planar epitaxial silicon avalanche photodiode', *Bell Syst. Tech. J.*, 1978, **57**, (6), pp. 1791–1807

59 GRIERSON, J.R., and O'HARA, S.: 'A theoretical assessment of low and high bandwidth gallium arsenide, germanium and silicon avalanche photodiodes for optical communication systems', *Solid-State Electron.*, 1975, **18**, pp. 1003–1011

60 LAW, H.D., NAKANO, K., and TOMASETTA, L.R.: 'State-of-the-art performance of GaAlAs/GaAs avalanche photodiodes', *Appl. Phys. Lett.*, 1979, **35**, (2), pp. 180–182

61 HARTMAN, D.H., GRACE, M.K., and RYAN, C.R.: 'A monolithic silicon photodetector/amplifier IC for fibre and integrated optics application', *IEEE J Lightwave Technol.*, 1985, **LT-3**, (4), pp. 729–738

62 SZE, S.M.: 'Physics of semiconductor devices', 2nd edn. (Wiley Interscience, 1981), p. 784

63 MELCHIOR, H., and LYNCH, W.T.: 'Signal and noise response of high speed germanium avalanche photodiodes', *IEEE Trans.*, 1966, **ED-13**, (12), pp. 829–838

64 ANDO, H., KANBE, H., KIMURA, T., YAMAOKA, T., and KANEDA, T.: 'Characteristics of germanium avalanche photodiodes in the wavelength region of 1–1.6 μm', *IEEE J Quantum Electron.*, 1978, **QE-14**, (11), pp. 804–809

65 MIKAWA, T., KAGAWA, S., and KANEDA, T.: 'Germanium avalanche photodiodes for optical communication systems', *Fujitsu Scientific & Technical J.*, 1980, pp. 95–117

66 MIKAMI, O., ANDO, H., KANBE, H., MIKAWA, T., KANEDA, T., and TOYAMA, Y.: 'Improved germanium avalanche photodiodes', *IEEE J Quantum Electron.*, 1980, **QE-16**, (9), pp. 1002–1007

67 KANEDA, T., FUKUDA, H., MIKAWA, T., BANBA, Y., and TOYAMA, Y.: 'Shallow-junction p^+-n germanium avalanche photodiodes (APDS)', *Appl. Phys. Lett.*, 1979, **34**, (13), pp. 866–868

68 KAGAWA, S., KANEDA, T., MIKAWA, T., BANBA, Y., TOYAMA, Y., and MIKAMI, O.: 'Fully ion-implanted p^+-n germanium avalanche photodiodes', *ibid.*, 1981, **38**, (6), pp. 429–431

69 MIKAWA, T., KANEDA, T., NISHIMOTO, H., MOTEGI, M., and OKUSHIMA, H.: 'Small-active-area germanium avalanche photodiode for single-mode fibre at 1.3 μm wavelength', *Electron. Lett.*, 1983, **19**, (12), pp. 452–453

70 NIWA, M., TASHIRO, Y., MINEMURA, K., and IWASAKI, H.: 'High-sensitivity Hi-Lo germanium avalanche photodiode for 1.5 μm wavelength optical communication', *ibid.*, 1984, **20**, (13), pp. 552–553

71 NIWA, M., ABE, Y., TASHIRO, Y., MINEMURA, K., and IWASAKI, H.: 'Receiver sensitivity temperature dependence of small detectable area Hi-Lo germanium avalanche photodiode', *ibid.*, 1984, **20**, (24), pp. 993–994

72 SMITH, D.R., HOOPER, R.C., and GARRETT, I.: 'Receivers for optical communications: A comparison of avalanche photodiodes with PIN-FET hybrids', *Optical & Quantum Electron.*, 1978, **10**, pp. 293–300

73 SMITH, D.R., and HOOPER, R.C.: 'High performance digital optical receivers with PIN photodiodes'. Proc Int Symp on Circuits and Systems, Tokyo, 1979, pp. 511–514

74 AHMAD, K., and MABBITT, A.W.: 'Gallium indium arsenide photodiodes', *Solid State Electron.*, 1979, **22**, pp. 327–333

75 NELSON, A.W., WONG, S., RITCHIE, S., and SARGOOD, S.K.: 'GaInAs PIN photodiodes grown by atmospheric-pressure MOVPE', *Electron. Lett.*, 1985, **21**, (19), pp. 838–840

76 MacDONALD, B.M., and SAUNDERS, A.G.: 'Coupling of an optical fibre to a substrate-entry photodiode', *ibid.*, 1981, **17**, (22), pp. 832–833

77 SUSSMANN, R.S., ASH, R.M., MOSELEY, A.J., and GOODFELLOW, R.C.: 'Ultra-low capacitance flip-chip-bonded GaInAs PIN photodetector for long-wavelength high-data-rate fibre-optic systems', *ibid.*, 1985, **21**, (14), pp. 593–595

78 ISHIHARA, H., MAKITA, K., SUGIMOTO, Y., TORIKAI, T., and TAGUCHI, K.: 'High-temperature aging tests on planar structure InGaAs/InP PIN photodiodes with Ti/Pt and Ti/Au contact', *ibid.*, 1984, **20**, (16), pp. 654–656

79 SAUL, R.H., CHEN, F.S., and SHUMATE, P.W.: 'Reliability of InGaAs photodiodes for SL applications', *AT & T Tech. J.*, 1985, **64**, (3), pp. 861–882

80 FORREST, S.R., DiDOMENICO, M., SMITH, R.G., and STOCKER, H.J.: 'Evidence for tunneling in reverse-biased III–V photodetector diodes', *Appl. Phys. Lett.*, 1980, **36**, (7), pp. 580–582

81 TAKANASHI, Y., KAWASHIMA, M., and HORIKOSHI, Y.: 'Required donor concentration of epitaxial layers for efficient InGaAsP avalanche photodiodes', *Japan. J. Appl. Phys.*, 1980, **19**, (4), pp. 693–701

82 NISHIDA, K., TAGUCHI, K., and MATSUMOTO, Y.: 'InGaAsP heterostructure photodiodes with high avalanche gain', *Appl. Phys. Lett.*, 1979, **35**, (3), pp. 251–253

83 CAMPBELL, J., DENTAI, A.G., HOLDEN, W.S., and KASPER, B.L.: 'High-performance avalanche photodiode with separate absorption 'grading' and multiplication regions', *Electron. Lett.*, 1983, **19**, (20), pp. 818–820

84 FORREST, S.R., KIM, O.K., and SMITH, R.G.: 'Optical response time of $In_{0.53}Ga_{0.47}As$/InP avalanche photodiodes', *Appl. Phys. Lett.*, 1982, **41**, (1), pp. 95–98

85 EMMONS, R.B.: 'Avalanche-photodiode frequency response', *J. Appl. Phys.*, 1967, **38**, (9), pp. 3705–3714

86 CAMPBELL, J.C., HOLDEN, W.S., FERGUSON, J.F., DENTAI, A.G., and JHEE, Y.: 'A high-speed InP/InGaAsP/InGaAs avalanche photodiode exhibiting a gain-bandwidth product of 60 GHz'. Post Deadline paper at IOOC-ECOC 1985, Venice.

87 SUGIMOTO, Y., *et al.*: 'High-speed planar-structure InP/InGaAsP/InGaAs avalanche photodiode grown by VPE', *Electron. Lett.*, 1984, **20**, (16), pp. 653–654

88 KOBAYASHI, M., YAMAZAKI, S., and KANEDA, T.: 'Planar InP/GaInAsP/GaInAs buried-structure avalanche photodiode', *Appl. Phys. Lett.*, 1984, **45**, (7), pp. 759–761

89 CAPASSO, F., TSANG, W.-T., and WILLIAMS, G.F.: 'Staircase solid-state photomultipliers and avalanche photodiodes with enhanced ionisation rates ratio', *IEEE Trans.*, 1983, **ED-30**, (4), pp. 381–389

90 RIDLEY, B.K.: 'Factors affecting impact ionisation in multilayer avalanche photodiodes', *IEE Proc.*, 1985, pp. 177–183

91 HILDEBRAND, O., KUEBART, W., BENZ, K.W., and PILKUHN, M.H.: '$Ga_{1-x}Al_xSb$ avalanche photodiodes: Resonant impact ionisation with very high ratio of ionisation coefficients', *IEEE J Quantum Electron.*, 1981, **QE-17**, (2), pp. 284–288

92 ALABEDRA, R., ORSAL, B., LECOY, G., PICHARD, G.,

MESLAGE, J., and FRAGNON, P.: 'An $Hg_{0.3}Cd_{0.7}Te$ avalanche photodiode for optical-fibre transmission systems at $\lambda = 1.3$ μm', *IEEE Trans.*, 1985, **ED-32,** (7), pp. 1302–1306
93 BRAIN, M.C., and SMITH, D.R.: 'Phototransistors, APD-FET and PINFET optical receivers for 1–1.6 μm wavelength', *ibid.*, 1983, **ED-30,** (4), pp. 390–395
94 FORREST, S.R.: 'The sensitivity of photoconductor receivers for long-wavelength optical communications', *IEEE J. Lightwave Technol.*, 1985, **LT-3,** (2), pp. 347–360
95 BUTLER, J.K. (Ed.): 'Semiconductor injection lasers' (IEEE Press, New York, 1980)
96 KRESSEL, H. (Ed.): 'Semiconductor devices for optical communications', 2nd edn., *Topics Appl. Phys.*, Vol. 39 (Springer-Verlag, 1982), Chap. 2
97 KRESSEL, H., and BUTLER, J.K.: 'Semiconductor lasers and heterojunction LEDs' (Academic Press, 1977)

Lasers and Photodetectors in Europe

HERBERT HAUPT AND OLAF HILDEBRAND

Abstract—This paper reviews the European activities on fiber optic semiconductor devices for optical communication. Progress in the development of lasers, LED's, and photodetectors and also first attempts of monolithic integration are reported. Emphasis is put on materials and devices for the wavelength range of 1200–1600 nm and for high-speed transmission. Results of recent work are highlighted. So, for example, extremely high reliability and high optical output power of LED's and Fabry–Perot lasers have been achieved. MOCVD growth techniques are increasingly applied for device fabrication; for lasers, highly reproducible characteristics have been demonstrated. Development of dynamic single-mode lasers (DSM) is under progress with low threshold currents and good optical characteristics reported. First electrically tunable single-mode lasers and DSM lasers with parallel coupled cavities have been realized. Long wavelength p-i-n photodetectors have reached maturity. InGaAsP and CdHgTe avalanche photodiodes show excellent properties for high-speed long-haul transmission. First steps of monolithic integration of several optical and electrical functions have been made and specifically for p-i-n–FET combinations good results have been achieved.

The plans of European PTT authorities to implement fiber optics in trunk lines and subscriber loops will further push the activities on optoelectronic device technology.

I. Introduction

TWENTY years ago, two laboratories in Europe proposed independently to use optical fibers for broadband information transmission [1], [2]. Shortly after that date, reasonably low-loss fibers [3] and the first suitable double-heterostructure semiconductor lasers [4] had been realized. Appropriate silicon detectors for the fiber-loss window around 850 nm were already available. So semiconductor work concentrated on lasers and the related GaAlAs/GaAs material system. This situation changed around the middle of the 1970's, when fiber losses in the long-wavelength range close to 1300 and 1550 nm approached the Rayleigh scattering limit and zero-dispersion of single-mode fibers around 1300 nm became obvious. Thus, new material systems were required for lasers as well as for photodiodes.

In this paper, we will mainly concentrate on devices for the wavelength range of 1200–1600 nm and the respective technologies. We try to give an overview of the main streams and some highlights of the work in Europe on sources, detectors, and integrated optoelectronics. In order to limit the extent of this paper, we have restricted ourselves to actually realized devices suitable for application in commercial optical transmission systems at present or in the near future.

The paper is not intended to give a complete overview on all activities and laboratories engaged in optoelectronic semiconductor devices. The numerous basic research activities on special device or material properties are not covered, as well as packaging-related activities or details of system tests. GaAlAs/GaAs devices are only mentioned in cases where the more advanced technology of this material is used for modeling of new structures or where a current interest exists for short-range optical fiber communication.

II. Light Sources

In this section, we attempt to review the large area of European activities on light-emitting diodes and lasers for optical communication systems. Light sources suitable for short-range data transmission, like GaAlAs/GaAs lasers or InGaAsP/InP 1300 nm and 1550 nm LED's, are included in this review as well as 1300 nm and 1550 nm Fabry–Perot laser and dynamic single-mode laser devices for long-haul high-bit-rate systems. Finally, operating quantum well devices will be addressed.

A. Light-Emitting Diodes

The continous progress in development of low-loss large bandwidth optical fibers during the last years, together with the large number of short-range links required for optical communication systems in Europe, suggests the application of LED's operating close to the minimum fiber dispersion wavelength region. For such short-range systems where the large optical output power of laser diodes is not required, an LED is an attractive low-cost highly reliable alternative. The advantages are lower temperature sensitivity and simple driving and control circuitry. No expensive parts like a Peltier cooler and monitor photodetector are required for a transmitter module.

Therefore, LED-based systems are under development or even in production for short-haul telecommunication applications including CATV and local area networks. The LED structures are either surface or edge emitters.

1) Surface Emitters: Different types of GaAlAs/GaAs and InGaAsP/InP LED's have been developed by Plessey (United Kingdom) [5] with special emphasis on optimization of the coupling efficiency by microlenses directly mounted onto the light-emitting surface of the devices.

Manuscript received March 19, 1985; revised March 3, 1986.
The authors are with SEL Research Centre, D-7 Stuttgart 40, West Germany.
IEEE Log Number 8608807.

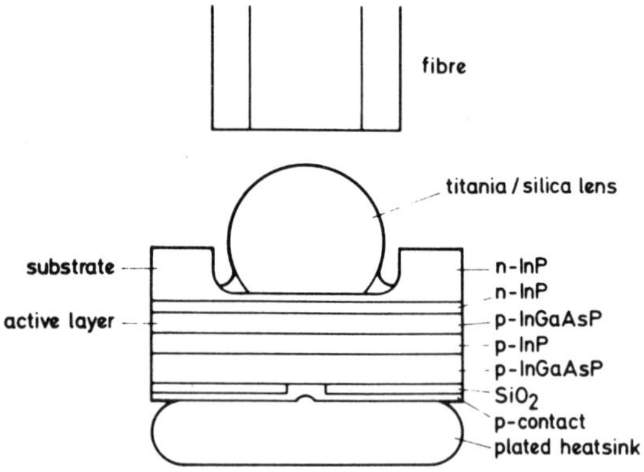

Fig. 1. Surface-emitting LED structure with microlens [5].

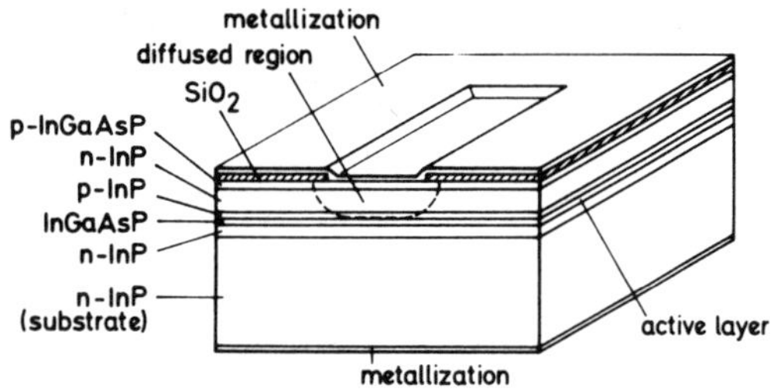

Fig. 2. Edge-emitting LED structure with diffusion-isolated active stripe [7].

The surface emitting 1300 nm LED (SLED) is shown in Fig. 1. As usual, current is confined to flow through a small central area with 10–50 μm diameter, defined either by dielectric isolation (as shown in Fig. 1) or by proton implantation. Power levels up to 250 μW (at 150 mA drive current) have been coupled into a graded-index fiber for the GaAlAs/GaAs device and 70 μW (at 100 mA) for the 1300 nm InGaAsP/InP device. The power of the 1300 nm device is limited due to saturation effects typical for InGaAsP/InP surface emitters, but it is sufficient, e.g., for 34 Mbit/s applications with more than 10 km fiber links.

At TU Munich, a surface emitting high-speed 1300 nm LED has been developed with a 3 dB modulation cutoff frequency above 900 MHz. The high-frequency modulation behavior of this device is enhanced by means of a thin strongly Mg-doped active layer, resulting in a short effective carrier lifetime. Using this LED, a transmission experiment at 560 Mbits/s over 4.4 km was performed, demonstrating the usefulness of LED sources even for higher bit rates [6].

2) Edge Emitters: The principal power saturation effect of surface emitters can be overcome by use of edge emitting structures. Plessey has developed two types of ELED's, the first (as shown in Fig. 2) with an undoped active layer and diffusion-isolated active stripe, and a second one with a p-doped active layer with proton implant isolation [7]. In both cases, an absorbing unpumped region prevents the device from laser action.

Whereas the first device has the higher optical output performance of 150–300 μW coupled into graded-index fiber (at 150 mA drive current), the second device shows improved temperature and switching characteristics due to the p-doped active layer. Both LED's have been designed for 140 Mbit/s multimode system applications.

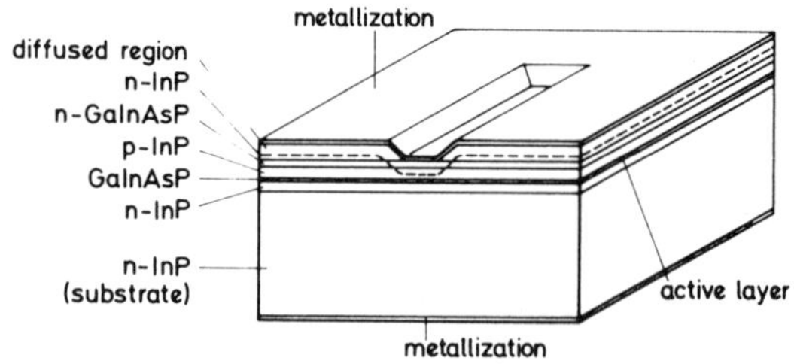

Fig. 3. V-groove edge-emitting LED structure [8].

At AEG in Germany, considerable effort has been made for development of 1300 nm V-groove LED's [8]. The device structure of this edge emitting LED is shown in Fig. 3. Essentially, this is a gain-guided structure where laser operation is suppressed by an absorbing region at the rear side. Typically, an optical power of 70 μW is coupled into a standard graded-index fiber at a drive current of 100 mA. These diodes have been shown suitable for bit rates up to 280 Mbits/s.

LED's are also attractive light sources for short-range *single-mode* fiber systems where, at first sight, only very small coupled light power is achievable. Nevertheless, data transmission over 10 km at 140 Mbits/s has been demonstrated by AEG in the laboratory [9].

Using an index-guided edge emitting LED, an optical power as high as 250 μW could recently be coupled into a *single-mode* fiber at 100 mA drive current. With this power level, a transmission length over 80 km (without margin) at 140 Mbits/s is expected [10]. However, for high power levels, these diodes become unstable and tend to laser oscillation.

Due to the broad emission spectrum of LED's, emission wavelength should be close to the dispersion minimum of the fiber. Using a dispersion shifted monomode fiber, a transmission experiment over 107 km at 16 Mbits/s has been demonstrated by Plessey with a 1550 nm LED light source [11]. Though a common DCPBH laser was used and heated to suppress laser action, this experiment demonstrates the use of LED's even for long-haul low-cost systems.

B. Fabry–Perot Laser Diodes

1) GaAlAs/GaAs Laser Diodes: GaAlAs/GaAs lasers are still attractive although the primary interest with respect to optical fiber communication has shifted to InGaAsP/InP laser diodes. Technology of GaAlAs/GaAs lasers is highly advanced. They offer special features such as a small temperature coefficient of threshold current and emission wavelength which is well matched to the sensitivity of inexpensive silicon photodetectors. Furthermore, these lasers are now available from industrial large-scale production. Therefore, they are still of current interest not only for short-range optical communication, but additionally for applications which require their relatively short emission wavelength, e.g., optical information processing. Further applications arise from the development of high-power laser arrays [8], e.g., for laser-beam printers, optical triggering of thyristors, free-space optical communication, direct pumping of YAG lasers, CD players, etc.

Since, however, the most attractive applications of GaAlAs/GaAs lasers deviate from broad-band optical communication, we will only briefly report recent activities on individual laser devices.

Due to extensive studies of degradation mechanisms at Siemens in Germany, LPE grown GaAlAs/GaAs gain-guided oxide stripe lasers have reached an extremely high reliability: degradation-free operation has been observed for more than 30 000 hours at an operation temperature as high as 100°C with 5 mW output power per facet [13]. From further degradation data, the authors conclude that degradation phenomena previously observed for GaAlAs/GaAs lasers are primarily processing induced and are not inherent to the material or laser structure and are also not principally pertinent to laser action.

One step to reduce processing induced effects may be the application of the very homogeneous metal organic vapor phase epitaxy (MOVPE) instead of LPE. Recently, MOVPE-grown oxide stripe GaAlAs/GaAs lasers have been processed with approximately half the standard deviation of threshold current and efficiency as compared to LPE-grown devices. Degradation rates as low as 1×10^{-5} h^{-1} at 100°C and 5 mW optical output per facet have been measured over an operation period exceeding 16000 h [14].

An index-guided GaAlAs/GaAs laser type with reduced threshold current has also been developed at Munich University [15]. This so-called metal-clad ridge-waveguide (MCRW) laser will be described below, for the InGaAsP/InP material system.

2) InGaAsP/InP Laser Diodes: There is no definite decision which of the two optical fiber windows, 1300 nm or 1550 nm, is more attractive for optical communication systems in Europe. According to the various demands for fiber links ranging from local area networks to long-haul undersea transmission systems, a variety of laser structures has been developed. Whereas the considerable dispersion at 1550 nm necessitates the use of single-frequency lasers for long-haul high bit-rate systems, the minimum dispersion at 1300 nm allows the use of either multimode lasers with a narrow spectral envelope, or low-grade single-frequency lasers.

Therefore, depending on the special system demands, the overall power budget and cost optimization may lead to different optimum solutions.

The most advanced InGaAsP/InP 1300 nm and 1550 nm laser structures developed and investigated throughout Europe are as follows.

a) V-Groove Laser Diodes: The structure of this gain-guided laser type developed by AEG is essentially the same as for V-groove LED's (Fig. 3), except that the absorbing region is omitted. As typical for gain-guided lasers, the threshold current of V-groove lasers is relatively large with lowest values of 80 mA. Kink-free output power up to 35 mW and CW operation up to 70°C have been obtained. Due to the broad spectral envelope and low sensitivity on light reflections from the fiber, this type of laser is suitable for graded-index fiber systems [16].

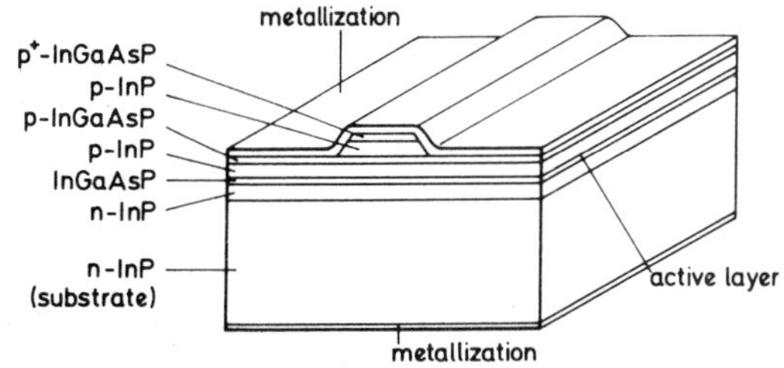

Fig. 4. Metal-clad ridge-waveguide (MCRW) laser structure [18].

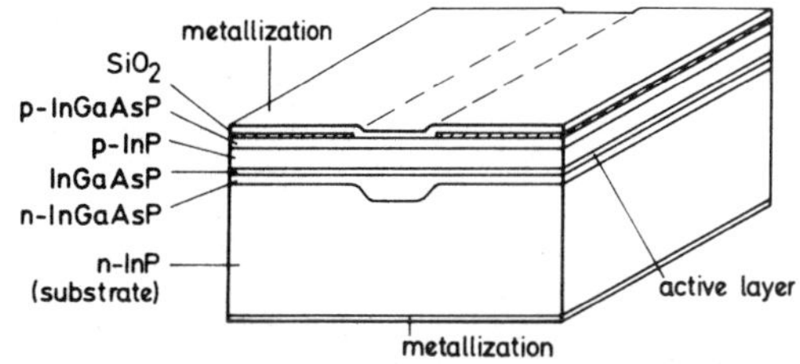

Fig. 5. Inverted rib waveguide (IRW) laser structure [19].

b) Metal-Clad Ridge-Waveguide (MCRW) Laser Diodes: Based on the development of GaAlAs/GaAs MCRW lasers [15], Siemens has demonstrated the successful extension of the index-guiding mechanism [17] and device technology to the quaternary InGaAsP/InP material system [18] (Fig. 4). The lasers have been prepared by LPE for 1300 nm emission wavelength. CW threshold currents as low as 28 mA and efficiencies above 0.25 mW/mA per facet have been achieved with stable operation in the fundamental lateral mode and with linear light-current characteristics up to more than 10 mW. The maximum CW operation temperature was 75°C.

If CW operated, these lasers show essentially a single-longitudinal mode spectrum which, however, ceases at high-speed modulation. The lasers show a totally flat response within a 3 dB bandwidth of larger than 4 GHz.

Due to the comparatively simple fabrication technique and their high-frequency modulation characteristics these MCRW lasers are attractive candidates for high-bit-rate 1300 nm single-mode fiber transmission systems.

c) Inverted Rib Waveguide (IRW) Laser Diodes: The inverted rib waveguide structure investigated by STL/STC (UK) was extensively discussed in [19]. The structure shown in Fig. 5 again is relatively simple to fabricate. A real dielectric waveguide is incorporated by etching a channel into the substrate prior to a single-stage liquid-phase epitaxial growth. The laser has a typical room temperature threshold of 80 mA and operates in single-longitudinal mode to above 10 mW per facet. The IRW laser can be directly modulated at more than 1.2 Gbits/s.

This laser type was developed for 1300 m long-haul use, particularly for undersea applications where the repeater units are hardly accessible and a system life of 25 years is specified. Such systems require extremely reliable laser diodes.

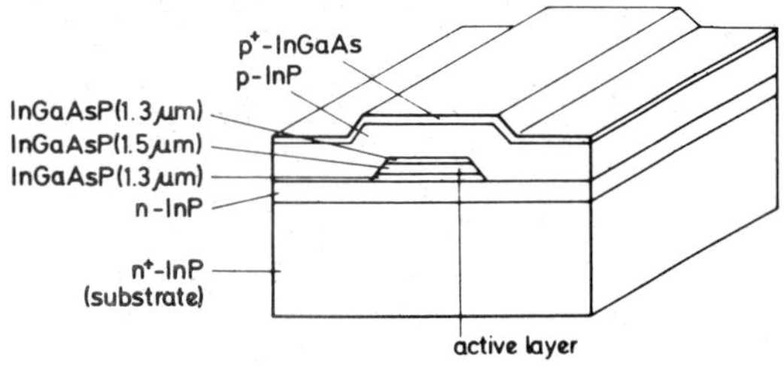

Fig. 6. Buried ridge structure (BRS) [21],[22].

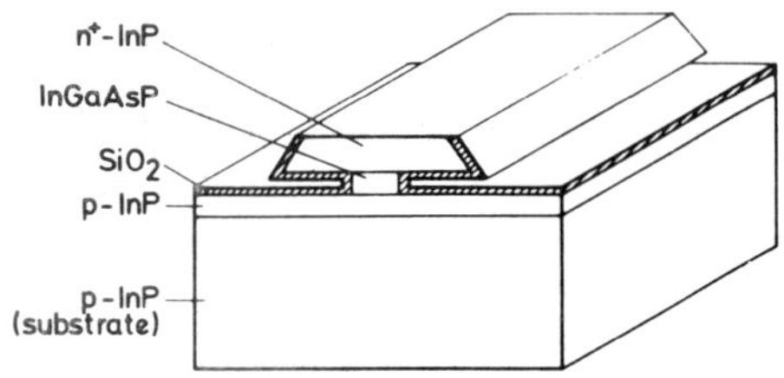

Fig. 7. Mushroom laser structure [23].

Life tests on IRW lasers have demonstrated very high reliability with a medium change in threshold current of 7.0 percent in 25 years at 50°C [20]. These results indicate extremely slow intrinsic degradation of the InGaAsP/InP laser material.

d) Buried Ridge Structure (BRS) Laser Diodes: InGaAsP/InP buried ridge structure (BRS) lasers have been developed by Thomson-CSF for 1300 nm [21] and for 1550 nm [22] emission wavelength (Fig. 6). These lasers were fabricated of material completely grown by low-pressure metalorganic chemical vapour deposition (LP-MOCVD). An excellent reproducibility and homogeneity with high processing yield has been obtained, demonstrating superiority of LP-MOCVD as compared to conventional liquid phase epitaxy. Recently, Thomson reported fabrication of 1700 laser chips from 12 wafers with an extremely low tolerance of threshold current of $I_{th} = 22$ mA ± 1 mA and similar homogeneity of efficiency. Also, first lifetime data are very promising [21].

The BRS laser structure is fabricated by a two-step epitaxial process with a photolithographic mesa-etching process in between. Localization of the current injection is achieved by deep proton implantation. Therefore, this structure can be fabricated relatively simply as compared to other index-guided buried heterostructures, and does not rely on special features of epitaxial processes like many other buried structures. It seems that the BRS laser structure is very attractive for low-cost mass production.

e) Mushroom Laser Diodes: The so-called mushroom structure as shown in Fig. 7 represents a real index-guided type laser where lateral index steps with a small width of the active region are achieved in conjunction with a relatively large contact area on top of the laser structure. As for the mainly index-guided MCRW structure and in contrast to buried laser structures, no further epitaxial processes are required after lithographic definition of lateral current and light confinement.

Since very promising electrical and optical data have been reported for this laser structure by FTZ, Germany

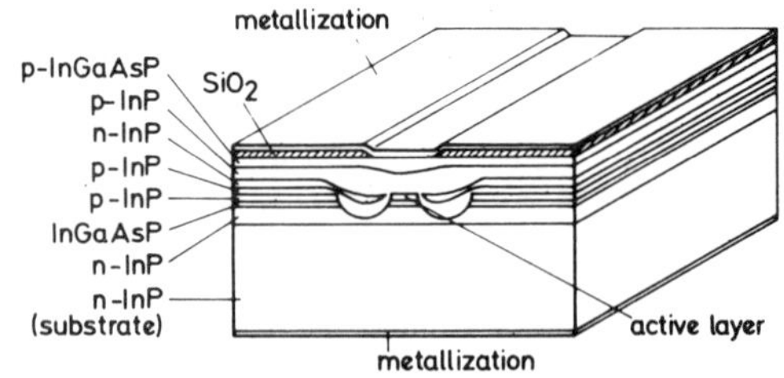

Fig. 8. Double-channel planar buried heterostructure (DCPBH) laser [26]–[29].

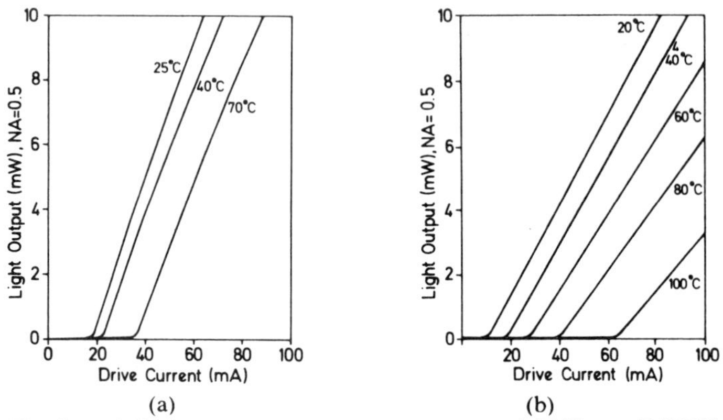

Fig. 9. (a) Light output versus current curves of 1300 nm DCPBH laser.[1] (b) Light output versus current curves of 1580 nm DCPBH laser [29].

[23] (low threshold currents, high output power, and high efficiency), many European laboratories have studied this laser type. It appears however, that special care has to be taken for a proper passivation of the originally laterally naked active region, either by deposition of dielectric materials [23] or by additional mass transport processes [24]. Nevertheless, the achievable lifetime of this laser structure is still an open question.

f) Double Channel Planar Buried Heterostructure (DCPBH) Lasers: The DCPBH laser as first proposed by Mito *et al.* [25] is one of the most promising laser diode structures for optical fiber systems (Fig. 8). It offers low threshold current, high efficiency, and high maximum operating temperature. Consequently, this laser structure, as well as modified structures, are extensively studied in numerous European laboratories; see e.g., [26]–[30].

DCPBH lasers are investigated for both optical fiber windows, e.g., for 1300 nm at SEL and Philips [26],[27], and for 1550 nm at STL and BTRL [28],[29]. For 1300 nm laser diodes, threshold currents as low as 8 mA, efficiencies of more than 0.3 mW/mA [27],[1] and an optical output power above 30 mW[1] have been achieved.

For 1550 nm laser diodes, the best values yet achieved were reported recently [29]: maximum CW output power of 50 mW per facet with 0.24 mW/mA efficiency at 20°C and CW operation up to 125°C. For lasers with slightly inferior high-power performance, threshold currents as low as 7.8 mA were reported. Typical light output versus current curves are shown in Fig. 9 for 1300 nm[1] and 1520 nm DCPBH lasers.

[1]Unpublished results of SEL-Research Centre obtained by: B. Gohla, O. Hildebrand, K. Lösch. G. Luz, G. Schemmel, and K. Wünstel.

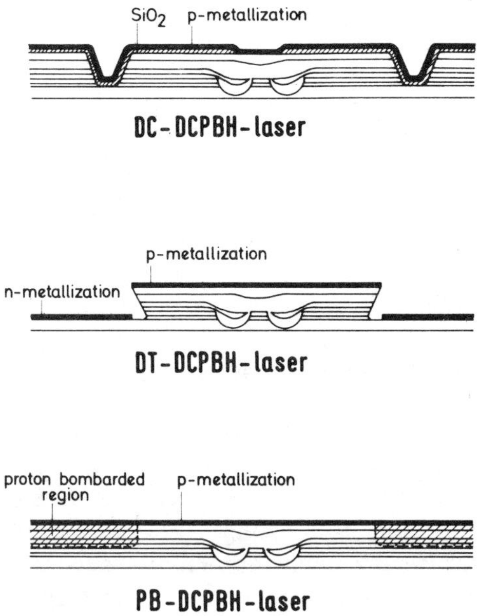

Fig. 10. Different approaches for improved high-speed modulation behavior of DCPBH lasers.

For a four-channel wavelength-division-multiplexing system, DCPBH lasers with emission wavelengths of $\lambda =$ 1200, 1240, 1280, and 1320 nm, respectively, have been fabricated by SEL. The laser structure used is the oxide-isolated DCPBH allowing for a system transmission rate of 4×565 Mbits/s.[1]

One of the major development goals of DCPBH lasers is to improve the high-frequency modulation behavior. The common DCPBH structure as reported in [25] typically shows a -3 dB cutoff frequency on the order of a few hundred MHz, i.e., the response is too slow for system applications beyond 565 Mbits/s. This slow response is caused by the dc current blocking p-n-p-n layers which give a large parallel capacitance.

Different approaches have been made to reduce the parasitic chip capacitance; the most effective solution is to really isolate the parasitic chip area from the active area either by proton isolation of the structure outside the active region (proton bombarded DCPBH = PB-DCPBH) [30], or by etching a pair of channels outside the active region (double channel DCPBH = DC-DCPBH) [30][2] as shown in Fig. 10.

The same effect of removing parasitic chip capacitance is achieved with the dovetail DCPBH (DT-DCPBH) laser also developed by SEL as shown in Fig. 10. It should be noted that this structure offers the possibility of monolithic integration, since both n- and p-contact are located on top of the chip, and as they are evaporated in one step. Therefore, this structure can also be grown on semiinsulating substrates.

In Fig. 11, the high-frequency response of a DT-DCPBH laser with a large mesa width of 70 μm is shown at a bit

[2]Unpublished results of SEL-Research Centre obtained by: B. Gohla, O. Hildebrand, K. Lösch, G. Luz, A. Mozer, G. Schemmel, M. Schilling, and K. Wünstel.

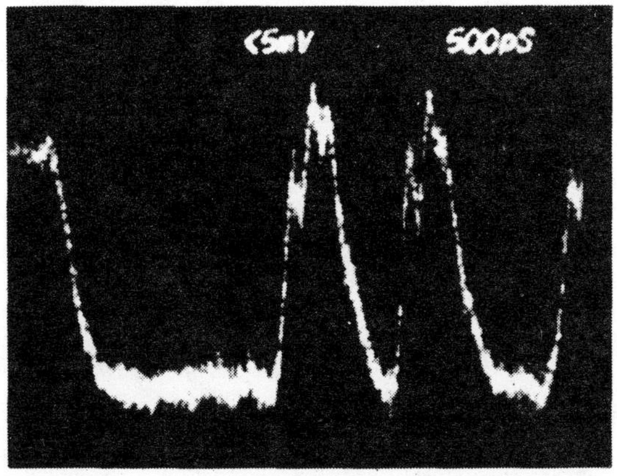

Fig. 11. High-frequency response of a DT-DCPBH laser with 70 μm mesa width. Modulation rate 2 Gbits/s, horizontal scale: 500 ps per unit.

rate of 2 Gbits/s. Further reduction of mesa width will allow even higher bit rates.

C. Dynamic Single-Mode Laser Diodes

The limits for long-haul high bit-rate data transmission are either imposed by fiber absorption or fiber dispersion. Again, depending on system requirements, wavelength tolerances, and fiber data, transmission either in the 1300 nm or in the 1550 nm fiber window can be the better solution even if dynamic single-mode (DSM) lasers are used. Therefore, European activities on DSM lasers are not only concentrated on usually dispersion-limited 1550 nm applications, but also on 1300 nm high-power DSM lasers overcoming absorption limitations.

It is interesting to note that most results published in Europe on DFB DSM lasers are based on ridge waveguide structures as originally developed by Kaminow *et al.* [31], a structure which in principle is capable of high-frequency modulation. Several activities are under way for buried heterostructure DFB lasers, mainly DCPBH lasers. No activities are known on lasers with distributed feedback bragg reflectors (DBR) lasers. Alternative solutions for dynamic single-mode lasers make either use of external cavities [32]—although not reported in the following—or of a novel approach, the laterally coupled waveguide lasers mentioned below.

1) Ridge Waveguide DFB Lasers: CW operation of 1550 nm ridge-waveguide DFB lasers was reported for the first time by British Telecom Research Laboratories (BTRL) [33]. These devices also represent the first CW operation of DFB lasers fabricated by hybrid LPE/MOCVD epitaxial technique as well as with gratings produced by electron-beam lithography.

The structure shown in Fig. 12 was formed by a two-stage epitaxy process using LPE for the growth of the separate confinement heterostructure and atmospheric pressure MOCVD for the overgrowth of the chemically etched grating, thus maintaining the original V-shaped second-order grating profile. The wafers were then etched to form ridge-waveguide DFB lasers for lateral current and optical beam confinement.

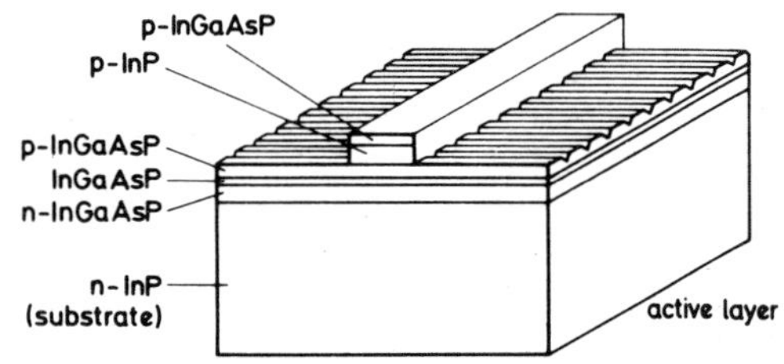

Fig. 12. Ridge waveguide DFB laser [33]–[36].

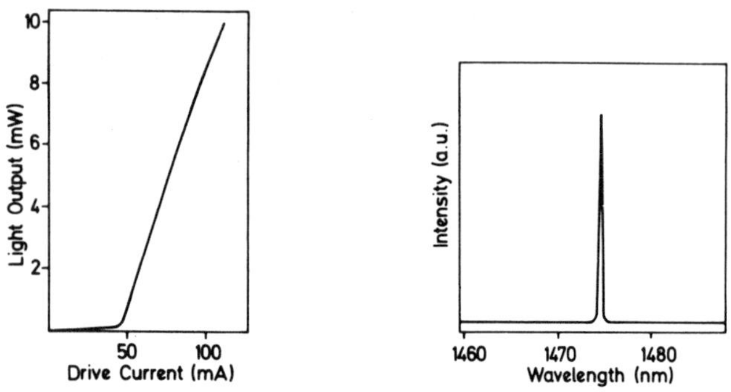

Fig. 13. Light output versus current curve and emission spectrum of a 1500 nm ridge-waveguide DFB laser [33].

Typical CW threshold currents of these devices are in the range of 38–60 mA with an external quantum efficiency of typically 15–27 percent [34], and with a maximum output power well in excess of 10 mW [33] (Fig. 13).

Under high-speed modulation at 2 Gbits/s, a side-mode suppression better than 38 dB was observed. The maximum chirp-induced wavelength shift was about 0.15 nm. These data indicate the suitability of ridge-waveguide DFB lasers for data transmission in the Gbit/s range.

The first CW operation of high-quality 1550 nm DFB lasers fabricated on material *completely* grown by LP-MOCVD was reported by Thomson-CSF and CNET [35]. They also use a two-step epitaxial process; the second-order grating in this case was made by holographic lithography. With a ridge waveguide DFB laser structure they achieved threshold currents of 60 mA with an optical output power of 6 mW per facet. Stable single-longitudinal-mode emission was observed under high-speed modulation at 1 GHz.

Fabrication of low-threshold 1300 nm ridge waveguide DFB lasers was recently reported by Laboratoires De Marcoussis-CGE, France [36]. Again, a two-step epitaxial process was used with second-order holographic grating formation in between. In this case, however, liquid phase epitaxy was used for both epitaxial steps, with no grating profile degradation observed. Lowest CW threshold currents of 31 mA have been achieved with stable DFB operation up to 5 mW output power per facet and good dynamical properties for direct modulation up to 2 GHz.

2) DFB-DCPBH Lasers: Recently, room temperature CW operation of 1300 nm DFB-DCPBH lasers was achieved at SEL [37]. Second-order DFB gratings were fabricated by holographic lithography on InP substrates. Except for the special adaption of liquid phase epitaxial processes, the further processing is the same as for common DCPBH lasers. In order to suppress the Fabry–Perot

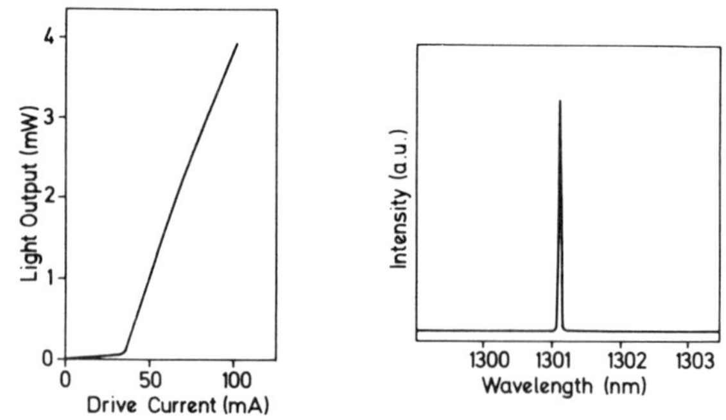

Fig. 14. Light output versus current curve and emission spectrum of a 1300 nm DFB-DCPBH laser.

modes, AR mirror coating was applied. These DFB-DCPBH lasers operate in a single longitudinal DFB mode at output powers up to 5 mW. Threshold currents are as low as 33 mA (Fig. 14). Though these lasers are not yet optimized for high-frequency modulation, the DFB mode is stable even at 3 GHz modulation frequency where the side-mode suppression is better than 14 dB. The chirp-induced linewidth broadening amounts to 0.14 nm which is well within the demands for 1300 nm systems.

Strong activities on 1300 nm and 1550 nm DFB-DCPBH lasers are under way throughout Europe. So publication of further results is expected for the near future.

3) Laterally Coupled Ridge Waveguide (LCRW) Laser Diodes: This laser structure, although not yet realized in the InGaAsP/InP material system, appears to be an interesting alternative to dynamic single-mode laser designs which use frequency-selective structures arranged along the optical axis, e.g., a DFB grating.

The LCRW laser structure developed by Siemens [38] consists of three MCRW lasers with 15 μm center spacing. They are placed on top of a GaAlAs-DH structure grown by MOCVD. The central laser has two cleaved facets, whereas the two side lasers have shorter resonator cavities with one of their facets formed by ion-beam etching.

If the three lasers are driven together with specific ratios of injection currents, they synchronize each other and oscillate in a single phase-locked supermode with optical output power in excess of 20 mW and a side-mode suppression of 24 dB; even at large signal modulation of the central laser at 1 GHz the linewidth still has its static value of 0.022 nm with a side-mode suppression of 20 dB.

According to these data, LCRW lasers, if realized with InGaAsP/InP, should be well suited for the 1550 nm wavelength region where chirp is a serious problem.

D. Quantum Well Laser Diodes

Although operating GaAlAs/GaAs quantum well lasers has been realized in Europe, the development of these lasers for the 1300 nm or 1550 nm wavelength range is still in an early stage.

As an example of GaAlAs/GaAs devices, Thomson-CSF has made GRIN-SCH structures with threshold current densities as low as 121Acm^{-2} and high T_o values [39]. With simple gain-guided stripe geometry laser diodes, the room temperature threshold current was only approximately half that of conventional double heterostructure lasers [40].

First quantum well structures in the InGaAs(P)/InP system have been grown and studied recently in several European laboratories (e.g., at Thomson-CSF [41] or at Stuttgarat University [42]).

III. Photodetectors

In Europe only three types of photodetectors are used today in commercial optical transmission systems operating at 1000–1600 nm:

1) InGaAs(P)/InP p-i-n photodiodes,
2) Germanium avalanche photodiodes (APD's),
3) CdHgTe p-i-n photodiodes and APD's.

While the first group of devices is extensively studied in many fiber optic laboratories, Ge-APD's appear to play only an intermediate role due to their high dark currents and unfavorable noise properties, and only little work has been done on these devices in Europe [43]. The last group of devices plays an important role in France.

In addition to these devices as used in operating systems, InGaAs(P)/InP APD's and monolithically integrated p-i-n FET's are extensively studied and are expected to be the high-performance detectors of the future.

A. InGaAs(P)/InP p-i-n Photodiodes

The p-i-n photodiodes based on quaternary InGaAsP or ternary InGaAs compound systems lattice matched to InP substrates have now reached maturity in terms of device characteristics and system performance [44]–[48]. They will therefore be described in more detail and thereafter some alternatives and future trends will be considered.

Fig. 15 shows the relatively simple but fully performing structure of such a p-i-n diode as e.g., developed by SEL [45]. It consists of a several μm thick $In_{0.53}Ga_{0.47}As$ layer grown by liquid-phase epitaxy on a (100) oriented InP substrate with a n^--InP buffer layer in between. The p-n junction is formed by Zn diffusion into the n-InGaAs top layer. The mesa is chemically etched down to the buffer layer and passivated by polyimide.

The n^- InGaAs layer represents the absorption zone. In order to achieve high quantum efficiency, this layer has to be more than 2 μm thick. For such a thick layer, the net doping level N_D-N_A has to be kept lower than 3×10^{15} cm^{-3} since full depletion is required already at the low operation voltage of -5 V for high-speed operation and high sensitivity of the diode [49].

Illumination through the substrate combines the advantages of lower recombination loss at the heterointerface as compared to surface recombination losses, largest possible contact area, and a p-n junction area as small as required to collect the light from the fiber. With a 75 μm mesa diameter, the junction capacitance is below 0.25 pF and the leakage current is less than 1 nA (Fig. 15). Together with the low series resistance, this allows for more

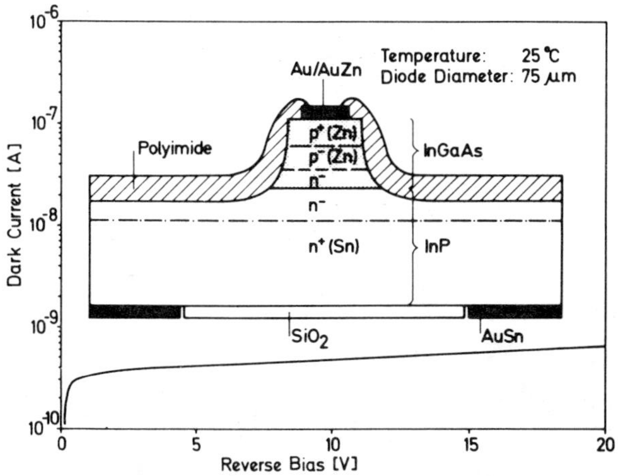

Fig. 15. Dark current characteristic of a InGaAs/InP mesa photodiode [44],[45]. Insert: Structure of the diode.

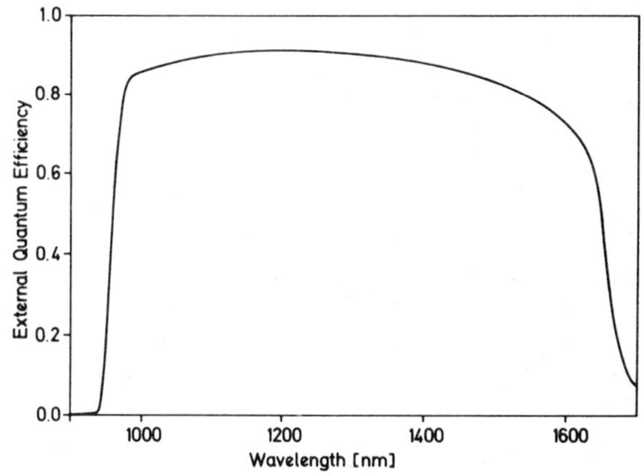

Fig. 16. External quantum efficiency of a InGaAs/InP photodiode [45].

than 6 GHz modulation and extremely high sensitivity. External quantum efficiencies above 80 percent in the full interesting wavelength range of 1000 nm–1550 nm and of more than 90 percent at the main transmission wavelength are achieved with a simple $\lambda/4$-SiO_2 antireflection layer (Fig. 16). Multilayer AR coating would still improve these figures. The short wavelength cutoff is given by band-to-band absorption in the InP substrate. Thus, the substrate acts as a filter in window multiplex systems.

These p-i-n diodes of the described type have been incorporated into receiver modules with a low-noise thin-film load resistor, GaAs-FET, and transimpedance preamplifier [44]. At a 168 Mbit/s line rate, a bit-error rate (BER) of 10^{-9} has been achieved reproducibly for a minimum light power of -42 to -44 dBm in the fiber pigtail. This comes within half a dB close to the theoretical limit for the given input capacitance and FET leakage current of the preamplifier.

With a high input impedance FET/bipolar cascode preamplifier and a 40 μm diameter mesa-diode designed for an incoming single-mode fiber, 10^{-9} BER was achieved at -36.5 dBm in a 1.2 Gbit/s system operating at 1528 nm (BTRL) [48]. The capacitance of the small area mesa-diode was only 0.07 pF when fully depleted.

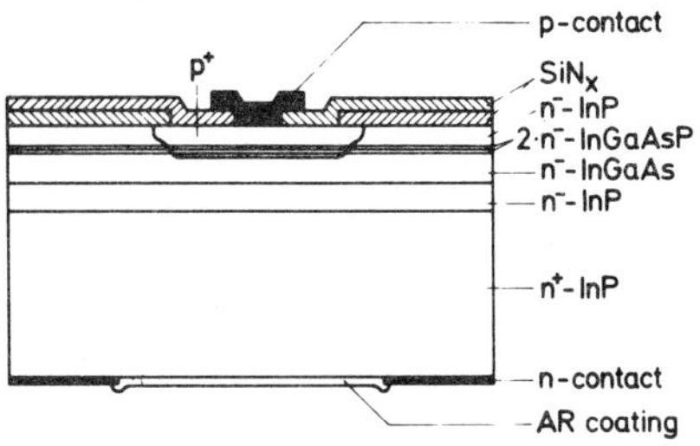

Fig. 17. Structure of a small-area planar InGaAs/InP photodiode [54].

B. Structural and Technological Alternatives

1) Layer Growth by MBE and MOCVD: Epitaxial buffer and absorption layers with low background doping level (N_D-$N_A = 3\times10^{15}$ cm^{-3}) have been grown by molecular beam epitaxy at CSELT, Italy [50] and by MOCVD in France [51], [52] and at BTRL [53]. Mesa-type devices fabricated from these layers show low dark currents indicating that these growth techniques are well suited for device fabrication. The quantum efficiency is not yet fully satisfactory. This might be due to the fact that the devices are illuminated from the top side [50]–[52]. In this configuration losses arise from surface recombinations and diffusion length reduction in the highly p-doped layer. Higher quantum efficiency of about 70 percent was actually observed with an additional InP cover layer [51].

Comparing the three epitaxy processes, LPE at present gives routinely the best results but, with respect to low-cost production MOCVD, has the higher potential. The MBE process is too expensive for diode fabrication. Further improvements of p-i-n diode characteristics that might be achievable by MBE would not have a benefit due to the limitation of receiver performance by GaAs-FET and circuit properties.

2) Planar Devices: Planar p-i-n diodes with illumination through the substrate have been grown at Siemens by LPE (Fig. 17) [54]. Extremely low dark currents down to 40 pA at −10 V could be achieved by application of an InP cap layer (with two thin quaternary layers in between in order to prevent melt-back during epitaxy). Passivation of the InP cap layer is easier as compared to InGaAs reducing the dark current flowing along the SiN_x/InP interface where the p-n junction merges the surface. The p-n junction was formed by Zn diffusion through a SiN_x mask. Since the diffusion coefficient of Zn in InP is distinctly larger than in InGaAs, pronounced lateral underdiffusion occurs; on the other hand control of the diffusion depth in InGaAs is relatively uncritical.

The planar diodes also have a very high quantum efficiency (above 90 percent) and show a pulse response down to 50 ps FWHM.

3) InGaAs Photodiodes for 1.0–2.4 μm: Mesa-type diodes have been fabricated at Thomson-CSF [55] with a long wavelength cutoff extended to beyond 1.7 μm, i.e., beyond the cutoff value for $In_{0.53}Ga_{0.47}As$ lattice-matched to InP. This was possible using mismatched material of $In_{0.8}Ga_{0.2}As$ composition. Due to a lattice mismatch of 2 percent, maximum quantum efficiency was only 21 percent at 2 μm layer thickness and leakage current is extremely high. This should be considerably improvable by growth of InAsP grading layers on both sides of the active device region. Photodetectors with extended infrared response could be valuable in context with expected low-loss fluoride glass fibers.

C. InGaAs(P)/InP Avalanche Photodiodes

These devices are being developed in several European laboratories, e.g., at SEL, Siemens, CGE, and AEG [45],[56]–[58] but are not yet stable enough for operating transmission systems. So far, mesa-type structures with polyimide passivation are used mostly with illumination through the substrate in order to achieve lowest leakage current and capacitance.

Multiplication is done in InP separated from the InGaAs(P) absorption zone in order to minimize dark current multiplication [59]. Due to an effective ionization coefficient ratio $\alpha/\beta = k \lesssim 0.4$, the depletion layer of the InP multiplication zone has to be n-type neighboring the n-absorption zone. The layer sequence is then given by the choice of substrate doping type. In case of p-type material the p-n junction is formed by Zn outdiffusion into the n-InP layer [56]. If the structures are grown on n-type substrates, the multiplication zone is on top of the absorption zone [45],[57] and the p-n junction can be formed by Cd diffusion or implantation from the top (Fig. 18). It is still an open question which type of doping is easier to control. Mesa-structures on n-type substrates need somewhat more care to prevent edge breakdown. On the other hand, they are the first step to the finally aspired planar structure with a guard ring.

In the above described SAM-APD's, the valence band barrier at the hetero-interface hinders the flow of holes into the multiplication region leading to slow photocurrent response [60] even at high reverse bias voltages. For quaternary devices optimized for 1300 nm this effect is less pronounced due to the smaller band-gap difference. Nevertheless, for this device as for the ternary ones, one or two intermediate quaternary layers have been incorporated resulting in a bandwidth far above 1 GHz.

Dark currents as low as 10 nA at 0.9 V_B (V_B = breakdown voltage) were achieved at SEL,[3] with smooth photocurrent multiplication characteristics at dC as well as 1 GHz modulation (Fig. 19).

The related spectral dependence of external quantum efficiency is shown in Fig. 20.

The dc avalanche multiplication up to 10 000 could be measured [56], but for low-noise operation, a gain of 10 is close to optimum. In system tests 6–7 dB improvement in receiver sensitivity as compared to p-i-n FET devices have been demonstrated. Achieved gain bandwidth products are 15–30 GHz.

[3]Unpublished results obtained at SEL Research Centre by: G. Busch, W. Kuebart, A. Schaupp, J. Scherb, U. Spalthoff, and P. Speier.

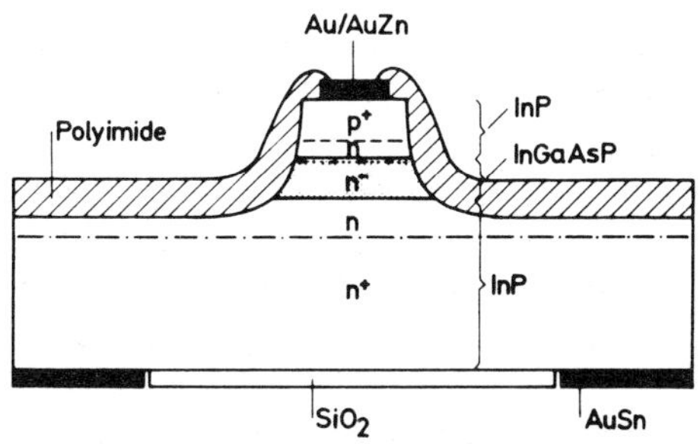

Fig. 18. Structure of a InGaAsP/InP mesa-avalanche-photodiode [45], [58].

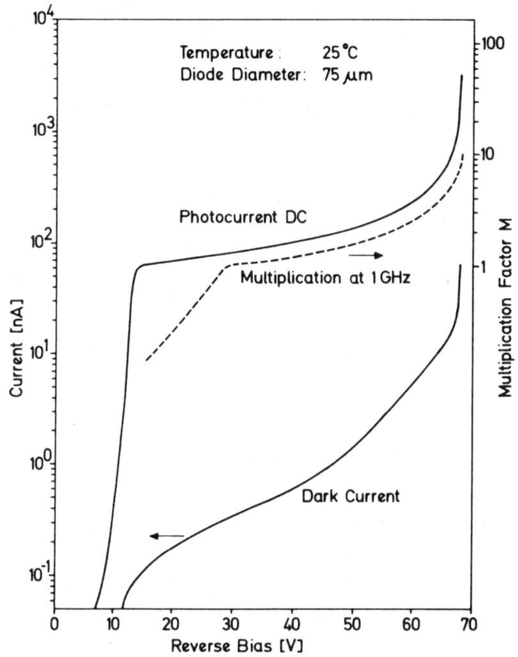

Fig. 19. Dark current and photocurrent of an APD as shown in Fig. 18.

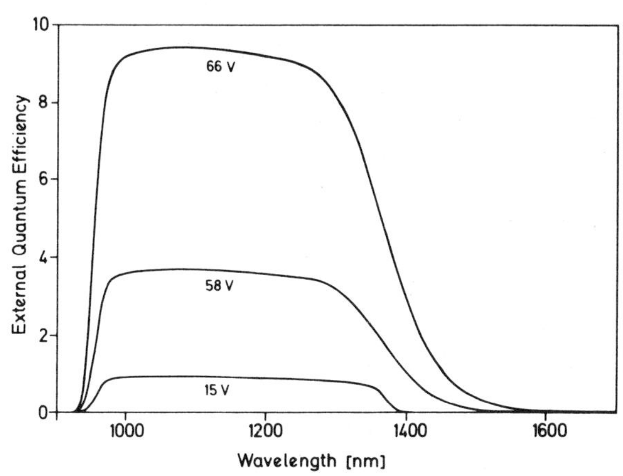

Fig. 20. External quantum efficiencies of an InGaAsP/InP APD at different reverse bias voltages [45].

D. CdHgTe Avalanche Photodiodes

CdHgTe photodiodes and APD's have been realized and operated in field systems by SAT in France [61],[62].

$Cd_xHg_{1-x}Te$ ternary alloys form a continuous family of semiconductors whose band-gap energy variation with x enables optical detection from 800 nm to far infrared. At a

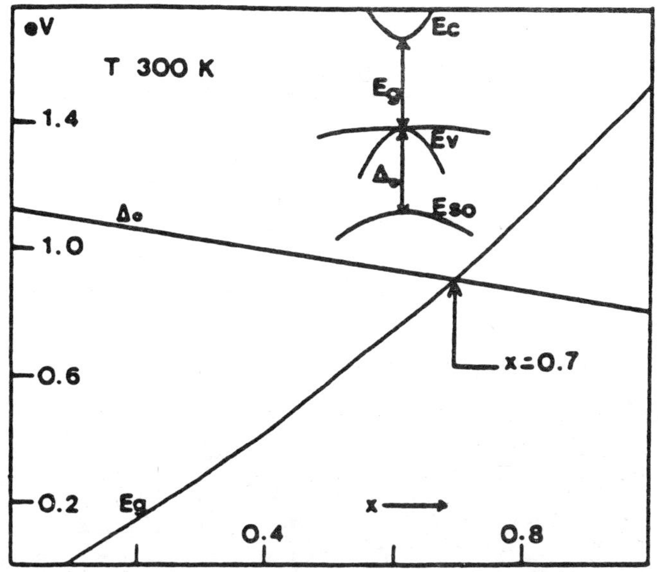

Fig. 21. Band-gap energy E_g and spin-orbit splitting Δ_0 of valence band as a function of x in $Cd_xHg_{1-x}Te$ [61].

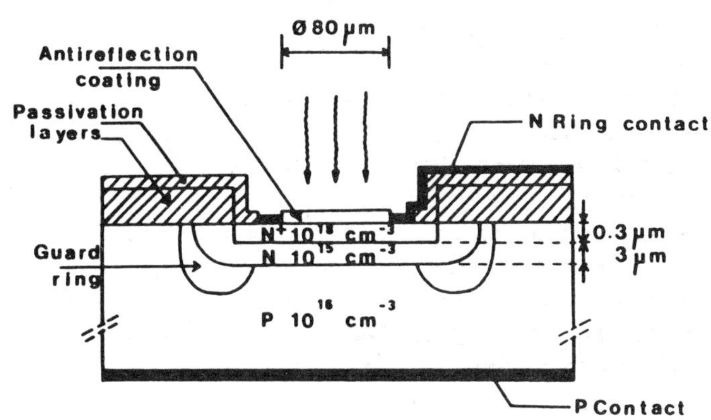

Fig. 22. $Cd_{0.7}Hg_{0.3}Te$ avalanche photodiode structure [61].

composition in the vicinity of $x = 0.7$, suitable for 1300 nm wavelength detection, the spin-orbit splitting of the valence band is close to the band-gap energy (Fig. 21). Resonant impact ionization processes in such a material yield a high ionization coefficient ratio which is estimated to be $\beta/\alpha \approx 10$ similar to the observations in GaAlAsSb at Stuttgart University [63].

CdHgTe photodiodes are realized by a conventional planar process on a bulk crystal material, with Al ion implantation of n^+ type region followed by low-temperature anneal under Hg vapor pressure. The same structure is used for p-i-n and avalanche photodiodes (Fig. 22). A large depleted region is pinned at 0.3 μm from the surface, even under high polarization, by the n^+ layer and edge breakdown due to the diffusion radius is eliminated by the diffused guard ring.

The photodiode sensitivity measured at low-bias voltage reaches 0.7 A/W as soon as the space charge region reaches the n^+ layer. The low-frequency multiplication of the photocurrent in the avalanche mode has been found to be homogeneous up to a multiplication factor of 30. Due to the high value of β/α, the excess noise factor is $F \lesssim 0.5$.

The high-frequency response does not exhibit a significant decrease up to 500 MHz.

$Hg_{1-x}Cd_xTe$ APD's with $x = 0.6$ for the detection of 1550 nm have also been realized [64].

Technological studies are in progress to improve HgCdTe detector's performance such as bandwidth and long-term reliability and to stabilize operation of 1550 nm APD's.

E. Photoconductive Detectors

First structures with 1–3 μm thick InGaAs epitaxial layers lattice matched to semiinsulating InP:Fe have been realized at RWTH Aachen [65]. The light has to be focussed onto the 10 μm-long and 30 μm-wide channel between the electrodes. The device works at a low electrical field as a conventional photoconductor. At higher fields, the field in the channel becomes nonuniform and the device behaves more similar to an FET. Also, the gain increases because of the higher velocity of the mobile charge carriers. The achievable gain depends on the carrier lifetime which, in turn, determines the pulse width. Thus, the gain can be optimized only to the debit of bandwidth and vice versa. Pulse widths of 50–100 ps have been obtained for an InGaAs detector.

Because of the similarity to FET's, these devices promise easy monolithic integration to an FET receiver front end.

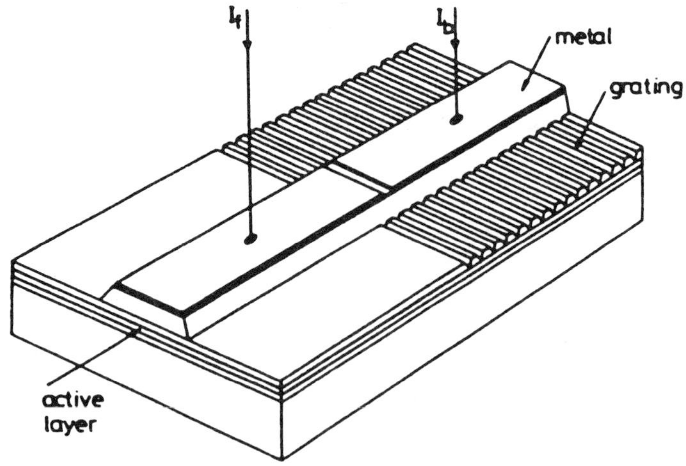

Fig. 23. Ridge-waveguide hybrid DFB/DBR laser structure [66].

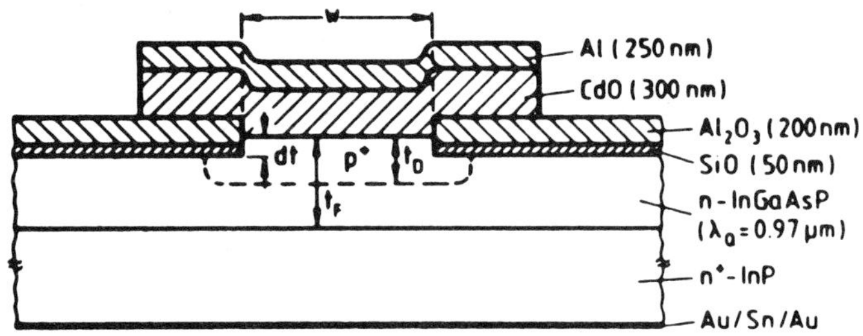

Fig. 24. Cross-section of the phase modulator device [67].

IV. First Steps to Monolithic Integration

Work towards integration of laser and detector diodes with further optical or with electrical functions is under way in Europe. The work concentrates not only on the design of discrete devices appropriate for monolithic integration, but also on realization of new required functions. Furthermore, actual monolithic integration has been started.

A. Integration of Optical Elements

BRTL [66] has realized a 1500 nm ridge-waveguide DFB laser with a separately contacted uncorrugated active ridge-waveguide regions (Fig. 23). This monolithic hybrid DFB/DBR laser shows a single-longitudinal-mode emission which can be tuned over a 5 nm wavelength range. This is obtained by adjusting the ratio of currents injected into the corrugated and uncorrugated regions, respectively.

HHI in Germany reported on a high-efficiency InGaAsP/InP phase modulator [67] (Fig. 24). The waveguide was formed by a InGaAsP ($\lambda_g = 970$ nm) epitaxial layer with a shallow Cd-diffused p^+ region. This was contacted first by a semitransparent CdO conductor stripe before the final Al electrode was applied. The CdO contact layer essentially reduces the optical losses as compared to a directly evaporated metal contact on top of the waveguide. With the device, a phase shift of 8 deg/V mm was achieved.

Siemens [68] developed a monolithic InGaAs photodiode array with an integrated slab waveguide (Fig. 25) which was grown by LPE on the bottom side of the

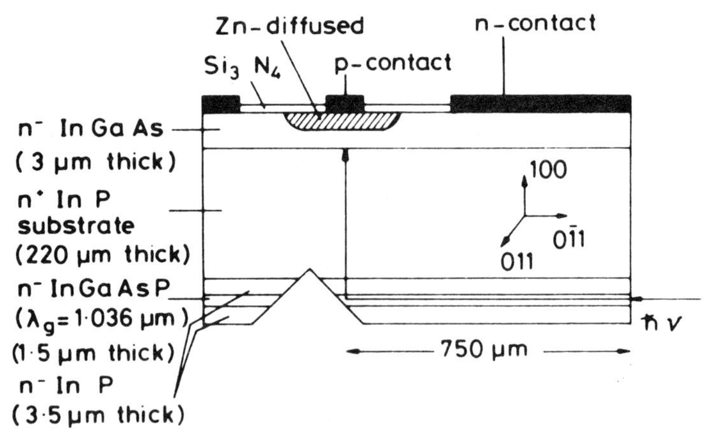

Fig. 25. Schematic cross-section through one of the photodiodes of the linear array, with indicated light path [68].

substrate, opposite to six planar p-i-n diodes with 500 μm spacing on top of the wafer. 1150 nm light of a HeNe laser was coupled into the waveguide from the $(0\bar{1}1)$ cleaved surface and reflected at the etched V-groove to the diodes. Absorption losses in the double-heterostructure waveguide are 2.5 dB/cm. The internal quantum efficiency of the photodiode including mirror loss was estimated to 87 percent the external efficiency of the device is 39 percent. Crosstalk between two neighboring diodes is better than -40 dB.

B. Monolithic Integration of Lasers and Detectors with Electronic Elements

Two rules have to be followed with the design of optical and electrical elements suitable for integration: 1) the layer structures should be compatible as far as possible; 2) for easy electrical separation the elements are normally grown on semiinsulating (si) substrates with both p- and n-contacts on top of the chip, e.g., shown above for the DT-DCPBH laser of SEL. Technical University of Munich reported recently a top-contacted MCRW GaAlAs/GaAs laser on an si-substrate with data comparable to a conventional MCRW laser [69].

For the electronic elements, specifically transistors, the semiconductor material system GaInAsP as used for long-wavelength lasers and photodetectors is also well suited. Due to the excellent electron transport properties especially for InGaAs field-effect transistors (FET's) and more recently heterobipolar transistors (HBT's) have attracted great attention for IC's and also for optoelectronic integration.

At present, HBT's appear more likely to be integrated with laser structures because of their ability to drive large capacitive loads and due to a certain similarity of its structure to that of a laser. HBT's are being developed at several places in Europe; see, for example, [70] and [71]. No results on successful integration with a laser diode have been published yet.

The integration of p-i-n FET detectors is somewhat more advanced. A major problem of monolithic integration is to maintain an essentially planar surface during wafer processing. This is required for high-resolution lithography. In order to achieve this goal, attempts have been undertaken to bury the relatively thick absorption zone into the semiinsulating InP substrate. At STL [72], long channels were etched along a crystallographic axis into the substrate and refilled by selective epitaxy. Overgrowth at the channel edges could be suppressed resulting in a planar surface. Unfortunately, the channels restrict the flexibility for further integration. Therefore, at SEL [73] cavities corresponding to the diode size were etched and refilled. Despite the different growth conditions on the different crystalline planes, flat surfaces were obtained but with poor yield only. It is felt that selective epitaxy is still far from being a suitable manufacturing process.

Therefore, a different concept has been followed by SEL.[4] On top of the semiinsulating InP substrate, a n^--InGaAs layer is grown which acts as absorption zone of the planar p-i-n diode and as a buffer layer of the FET. The $n-$InGaAs FET channel layer is less than 1 μm thick. In this way, a nearly planar surface is obtained without the difficult regrowth on structured substrates. The monolithic p-i-n FET combination was realized with an FET transconductance as high as 150 mS/mm for a rather large gate length of 4.5 μm. The dark current of the planar p-i-n diode is less than 20 nA.

V. Conclusion

State of the art and major trends of research and development on light sources and detectors in Europe have been reviewed. Also, first approaches towards monolithic integration of optical, electronic, and waveguiding devices have been described. These activities within the last few years reflect the strategies of the European PTT administrations to use fiber optics in trunk lines and also in subscriber networks. The need for components will push the research and development efforts further up in the near future, specifically in the areas of basic III-V compound materials technologies, of highly performing devices, of ultrahigh-speed lasers and detectors, and of monolithic integration of different optoelectronic functions.

[4] Unpublished results obtained at SEL Research Centre by F. J. Tegude, H. Eisele, and M. Schilling at SEL-RC.

References

[1] M. Börner, "Multi-level transmission system for pulsecode modulated informations," DBP, Patent 1254513, 1966.

[2] C. K. Kao and G. A. Hockham, "Dielectric-fibre surface waveguides for optical frequencies," *Proc. IEE*, vol. 113, pp. 1151–1158, 1966.

[3] F. P. Kapron, D. B. Keck, and R.D. Maurer, "Radiation losses in glass optical waveguides," *Phys. Lett.*, vol. 17, pp, 423–425, 1970.

[4] I. Hayashi, M.B. Panish, and P. W. Foy, "A low threshold room temperature injection laser," *IEEE J. Quantum Electron.*, vol. QE-5, pp. 211–212, 1969.

[5] R. C. Goodfellow, A. C. Carter, I. Griffith, and R. R. Bradley, "GaInAsP/InP fast, high-radiance 1.05-1.3 μm wavelength LED's with efficient lens coupling to small numerical aperture silica optical fiber," *IEEE Trans. Electron. Devices*, vol. ED-26, pp. 1215–1220, 1979.

[6] H. Grothe, W. Harth, and W. Proebster, "InGaAsP/InP surface emitting LED for 560 Mbit/s transmission over 4.4 km," in *Proc. 4th IOOC*, Tokyo, Japan, pp. 2–3, 1983 (post deadline paper).

[7] A. C. Carter, J. Ure, M. A Harding, and R. C. Goodfellow, "Reliable high power edge emitting LEDs for 1.3 um 140 Mb/s multimode communication systems," in *Proc. 4th IOOC*, Tokyo, Japan, pp. 210–211, 1983.

[8] G. Arnold, F.-J. Berlec, H. Gottsmann, and W. Pfister, "Emission and coupling properties of 1.3 μm V-groove edge-emitting LED's," to be published.

[9] G. Arnold, and O. Krumpholz, "Optical transmission with single-mode fibres and edge-emitting diodes," *Electron. Lett.*, vol. 21, pp. 390–392, 1985.

[10] G. Arnold, H. Gottsmann, O. Krumpholz, E. Schlosser, and E.-A. Schurr, "1.3 μm edge-emitting diodes launching 250 μW into a single-mode fibre at 100 mA," *Electron. Lett.*, vol. 21, pp. 993–994, 1985.

[11] R. Plastow *et al.*, "Transmission over 107 km of dispersion-shifted fibre at 16 Mbit/s using a 1.55 μm edge-emitting source," *Electron. Lett.*, vol. 21, pp. 369–370, 1985.

[12] F. Kappeler, H. Westermeier, R. Gessner, M. Druminski, and K.-H. Zschauer, "High cw power array of optically coupled (Ga,Al) As oxide stripe lasers with dc-to-light conversion efficiencies of up to 36%," presented at the 9th IEEE Int. Semiconductor Laser Conf., Rio de Janeiro, Brazil, Aug. 1984, paper G-3.

[13] K.-H. Zschauer, K. Mettler, and H. D. Wolf, "Reliability of GaAlAs lasers," to be published.

[14] M. Druminski *et al.*, "MOVPE (Al,Ga)As/GaAs 870 nm oxide stripe lasers with highly uniform laser characteristics," to be published.

[15] M. C. Amann and B. Stegmüller, "Low-threshold-current injection laser with built-in passive waveguiding," in *Topical Meeting on Integrated and Guided-Wave Optics*, Incline Village, pp. MC3-1–MC3-4, 1980.

[16] P. Marschall, W. Pfister, and E. Schlosser, "V-groove-laser for the 1.3 μm wavelength region," in *Proc. 8th ECOC*, Cannes, France, 1982, pp. 179–182.

[17] M.-C. Amann, "Calculation of metal-clad ridge-waveguide (MCRW) laser modes by mode coupling technique," to be published.

[18] M.-C. Amann and B. Stegmüller, "Low-threshold InGaAsP-InP metal-clad ridge-waveguide (MCRW) lasers for 1.3 μm wavelength," to be published.

[19] S. E. H. Turley *et al.*, "Properties of inverted rib-waveguide lasers operating at 1.3 um wavelength," *Electron. Lett.*, vol. 17, pp. 868–870, 1981.

[20] A. Rosiewiecz, B. R. Buttler, and R. E. P. Hinton, "Temperature-dependent lifetests of IRW lasers operating at 1.3 μm," *IEE Proc.*, vol. 132, pp. 97–100, 1985.

[21] M. Razeghi, R. Blondeau, K. Kazmierski, M. Krakowski, and J. P. Duchemin, "Very low threshold buried ridge structure lasers emitting at 1.3 μm grown by low pressure metal-organic chemical

vapour deposition," *Appl. Phys. Lett.*, vol. 46, pp. 131–133, 1985.
[22] R. Bondeau *et al.*, "cw operation of GaInAsP buried ridge structure laser at 1.5 μm grown by LP-MOCVD," *Electron. Lett.*, vol. 20, pp. 850–851, 1984.
[23] H. Burkhard and E. Kuphal, "Three- and four-layer LPE-InGaAs(P) mushroom stripe lasers for $\lambda = 1.30$, 1.55 and 1.65 μm with low threshold currents," presented at the 9th IEEE Int. Semiconductor Laser Conf., Rio de Janeiro, Brazil, Aug. 1984, paper D-5.
[24] D. Renner, P. D. Greene, A. Collar, D. Moule, and G. H. B. Thompson, "Design, fabrication, and performance of buried heterostructure lasers based on the mass transport effect," presented at the 9th IEEE Int. Semiconductor Laser Conf., Rio de Janeiro, Brazil, Aug. 1984, paper D-2.
[25] I. Mito *et al.*, "InGaAsP double-channel planar-buried-heterostructure laser diode (DC-PBHLD) with effective current confinement," *J. Lightwave Technol.*, vol. LT-1, pp. 195–202, 1983.
[26] B. Gohla *et al.*, "1.3 μm double channel planar buried heterostructure laser," in *Proc. 10th ECOC*, Stuttgart, Germany 1984, pp. 60–62.
[27] P. I. Kuindersma, A. Valster, and W. Baks, "1.3 μm buried heterojunction laser diodes under high electrical stress," in *Proc. 10th ECOC*. Stuttgart, Germany, 1984, pp. 64–65.
[28] A. W. Nelson *et al.*, "High performance DC-PBH lasers at 1.52 μm by a hybrid MOVPE/LPE process," *Electron. Lett.*, vol. 21, pp. 493–494, 1985.
[29] D. Renner, A. J. Collar, P. D. Greene, and G. D. Henshall, "High power operation of DCPBH lasers emitting at 1.52 μm wavelength," *Electron. Lett.*, vol. 21, pp. 1006–1007, 1985.
[30] A. Valster, L. J. Meuleman, and P. I. Kuindersma, "Improved high-frequency response of InGaAsP double-channel buried-hetero-structure lasers," *Electron. Lett.*, vol. 22, pp. 16–18, 1986.
[31] I. P. Kaminow *et al.*, "Low threshold ridge-waveguide lasers at 1.55 μm," *Electron. Lett.* vol. 19, pp. 877–879, 1983.
[32] R. Plastow and A. C. Carter, "Single-longitudinal-mode operation of a 1.55 μm laser with a novel self-aligned external cavity," presented at the 9th IEEE Int. Semiconductor Laser Conf., Rio de Janeiro, Brazil, Aug. 1984, paper E-6.
[33] L. D. Westbrook, A. W. Nelson, P. J. Fiddyment, and J. S. Evans, "Continous-wave operation of 1.5 μm distributed feedback ridge-waveguide lasers," *Electron. Lett.*, vol. 20, pp. 225–226, 1984.
[34] L. D. Westbrook, I. D. Henning, A. W. Nelson, and P. J. Fiddyment, "Spectral properties of strongly coupled 1.5 μm DFB laser diodes," *IEEE J. Quantum Electron.*, QE-21, pp. 512–518, 1985.
[35] M. Razeghi *et al.*, "Low threshold distributed feedback lasers fabricated on material grown completely by LP-MOCVD," *IEEE J. Quantum Electron.*, QE-21, pp. 507–510, 1985.
[36] C. Artigue, Y. Louis, P. Poingt, D. Sigogue, and J. Benoit, "Low-threshold second-order DFB laser emitting at 1.3 μm for high bit-rate applications," in *Proc. IOOC-ECOC 1985*, Venice, Italy, 1985, pp. 33–36.
[37] A. Mozer *et al.*, "Realisierung und Eigenschaften von 1.3 μm Dauerstrich-DFB-lasern," *Verhandl. DPG (VI)*, vol. 21, HL 300, 1986.
[38] F. Kappeler, "Improved dynamic single-frequency low-chirp operation of a laterally coupled ridge waveguide (LCRW) laser," in *Proc. IOOC-ECOC 1985*, Venice, Italy, 1985, pp. 771–774.
[39] S. D. Hersee, B. de Cremoux, and J. P. Duchemin, "Some characteristics of the GaAs/GaAlAs graded-index separate-confinement heterostructure quantum well laser structure," *Appl. Phys. Lett.*, vol. 44, pp. 476–478, 1984.
[40] M. Krakowski *et al.*, "Low threshold current GaAs/GaAlAs GRIN-SCH stripe lasers grown by OMVPE," *Electron. Lett.*, vol. 19, pp. 1082–1084, 1983.
[41] M. Razeghi, J. Nagle, and C. Weisbuch, "Optical studies of GaInAs/InP quantum wells," in *Inst. Phys. Conf.*, 1985, ser. no. 74, chap. 5, pp. 379–384.
[42] F. Scholz *et al.*, "GaInAs-InP-multi quantum well structures grown by metalorganic gas phase epitaxy with adducts," to be published.
[43] G. Ebbinghaus, M. Metzger, and T. Scherg, "Small-area ion-implanted p^+n Germanium avalanche photodiodes for a wavelength of 1.3 μm," *Siemens Forsch.- und Entwickl-Ber.*, vol. 14, pp. 284–288, 1985.
[44] K. Hess, R. Dorn, K. Lösch, G. Schemmel, and H. Haupt, "Performance characteristics and lifetest data of GaInAs/InP and GaInAsP/InP photodiodes," in *Proc. 8th ECOC 1982*, Cannes, France, 1982, pp. 300–303.
[45] B. Gohla *et al.*, "Components for a high speed optical communication system with large repeater spacing at 1200 nm wavelength," BMFT (German Ministry of Research and Technology), Final Rep., Ref. no. TK 0179, 1985.
[46] K. Böhm, R. Deufel, E. Schlosser, and A.-E. Schurr, "Pulse response of back-illuminated GaInAs pin-photodiodes," in *Proc. 9th ECOC 1983,* Geneva, Switzerland, pp. 163–166.
[47] G. R. Antell and R. F. Murison, "InGaAs/InP mesa photodetector passivated with silicon dioxide," *Electron. Lett.*, vol. 20, pp. 919–921, 1984.
[48] M. C. Brain, P. P. Smyth, D. R. Smith, B. R. White, and P. J. Chidgey, "Pin-FET hybrid optical receivers for 1.2 Gbit/s transmission systems operating at 1.3 and 1.55 μm wavelength," *Electron. Lett.*, vol. 20, pp. 894–896, 1984.
[49] R. C. Eden, "Heterojunction III-V alloy photodetectors for high-sensitivity 1.06 μm optical receivers," *Proc. IEEE*, vol. 63, pp. 32–37, 1975.
[50] P. Cinguino, F. Genova, C. Rigo, and A. Stano, "Low dark current InGaAs pin photodiodes grown by molecular beam epitaxy," *Electron. Lett.*, vol. 21, pp. 139–140, 1985.
[51] P. Poulain, M Razeghi, K. Kazmierski, R. Blondeau, and P. Philippe, "InGaAs photodiodes prepared by low-pressure MOCVD," *Electron. Lett.*, vol. 21, pp. 441–442, 1985.
[52] J. P. Andre *et al.*, "Low dark current and low capacitance InGaAs pin photodiode grown by MOVPE," in *Proc. IOOC-ECOC 1985*, Venice, Italy, pp. 541–544.
[53] A. W. Nelson, S. Wong, S. Ritchie, and S. K. Sargood, "GaInAs pin-photodiodes grown by atmospheric-pressure MOVPE," *Electron. Lett.*, vol. 21, pp. 838–840, 1985.
[54] R. Trommer *et al.*, "Planar small-area InGaAs/InP photodiodes for wavelengths of 1.3 μm and 1.55 μm," *Siemens Forsch.- und Entwickl-Ber.*, vol. 14, pp. 280–283, 1985.
[55] P. Poulain, J. P. Moy, and P. Radisson, "InGaAs photodiodes for the 1.0 to 2.4 μm spectral region prepared by low pressure MOCVD," in *Inst. Phys. Conf.*, 1984, ser. no. 4, chap. 6, pp. 421–425.
[56] R. Trommer, "Design and fabrication of InGaAs/InP avalanche photodiodes for the 1 to 1.6 μm wavelength region," *Frequenz*, vol. 38, pp. 212–216, 1984.
[57] M. Boulou, R. Vergnaud, G. Soulage, and J. Benoit, "InP/InGaAsP/InGaAs avalanche photodiode with front illumination for optical fiber transmission at Gbit rates in the 1-1.6 μm wavelength region," in *Proc. IOOC-ECOC 1985*, Venice, Italy, 1985, pp. 549–552.
[58] O. Krumpholz and H.-P. Vollmer, "Komponenten für langwellige Übertragung," *NTZ*, vol. 35, pp. 692–696, 1982.
[59] H. Kanbe, N. Susa, H. Nakagome, and H. Ando, "InGaAs avalanche photodiode with InP pn-junction," *Electron. Lett.*, vol. 16, pp. 163–165, 1980.
[60] S. R. Forrest, O. K. Kim, and G. R. Smith, "Optical response time of InGaAs/InP avalanche photodiodes," *Appl. Phys. Lett.*, vol. 41, pp. 95–98, 1982.
[61] G. Pichard, J. Meslage, T. Nguyen Duy, and F. Raymond, "1.3 μm CdHgTe avalanche photodiodes for fibre-optic applications," in *Proc. 9th ECOC 1983*, Geneva, Switzerland, 1983, pp. 479–482.
[62] T. Nguyen Duy, J. Meslage, and G. Pichard, "CMT—The material for fiber optic communication devices," *J. Crystal Growth*, vol. 72, pp. 490–495, 1985.
[63] O. Hildebrand, W. Kuebart, K. W. Benz, and M. H. Pilkuhn, "GaAlSb avalanche photodiodes: Resonant impact ionization with very high ratio of ionization coefficients," *IEEE J. Quantum Electron.*, vol. QE-17, ppl 284–288, 1981.
[64] R. Alabedra *et al.*, "Avalanche photodiodes $Hg_{1-x}Cd_xTe$ with $x = 0.6$ for the detection at $\lambda = 1.55$ μm," in *Opto 1985*, Paris, France, 1985, pp. 84–88.
[65] H. Beneking, "On the response behaviour of fast photoconductive optical planar and coaxial semiconductor detectors," *IEEE Trans. Electron. Devices*, vol. ED-29, pp. 1431–1441, 1982.
[66] L. D. Westbrook, A. W. Nelson, P. J. Fiddyment, and J. V. Collins, "Monolithic 1.5 μm hybrid DFB/DBR lasers with 5 nm tuning range," *Electron. Lett.*, vol. 20, pp. 957–959, 1984.
[67] C. Bornholdt *et al.*, "High-efficiency phase modulator in InGaAsP/InP," in *Proc. ECIO 1985*, Berlin, West Germany, 1985, pp. 121–125.
[68] R. Trommer, "Monolithic InGaAs photodiode array illuminated through an integrated waveguide," *Electron. Lett.*, vol. 21, pp. 382–383, 1985.
[69] G. Müller and M. Honsberg, "Fabrication and characteristics of MCRW GaAs/GaAlAs lasers on semi-insulating substrate," *J. Opt. Commun.*, vol. 6, pp. 42–43, 1985.
[70] L. M. Su, N. Grote, R. Kaumanns, and H. Schroeter, "NpnN double-heterojunction bipolar transistor on InGaAsP/InP," *Appl. Phys. Lett.*, vol. 47, pp. 28–30, 1985.
[71] H. Dämbkes, U. König, and B. Schwaderer, "InGaAs/InP hetero-bipolar transistors for integration on semi-insulating InP substrate," *Electron. Lett.*, vol. 20, pp. 955–957, 1984.
[72] P. J. G. Dawe, D. A. H. Spear, and G. H. B. Thompson, "Embedded GaInAs pin-diode and InP-based JFET for optoelectronic integration," in *Proc. IOOC-ECOC 1985*, Venice, Italy, 1985, pp. 307–310.
[73] M. Schilling, G. Schemmel, and F. J. Tegude, "Selective LPE growth of InGaAs on semi-insulating InP," presented at the 27th Electron. Materials Conf., Boulder, CO, June 1985.

Herbert Haupt was born on April 27, 1938 in Schwenningen a.N., Federal Republic of Germany. He received the Dipl. degree in 1966 and the Ph.D. degree in 1972, both from Technical University of Berlin.

His work at the University dealt with luminescence phenomena and with crystal growth of II-VI compound semiconductors.

In 1974, he joined SEL Research Centre at Stuttgart as a project manager for lasers, detectors and optical fibers. Since 1979, he has been leading the Materials and Components Division with activities on basic materials and technology research for fiber optic components and for displays, printers, sensors, and manufacturing techniques.

Dr. Haupt is chairing the Optical Communication Committee of NTG (Nachrichtentechnische Gesellschaft im VDE) in Germany and he is a member of ECOC and IOOC management. In 1984, he was chairman of the 10th ECOC (European Conference on Optical Communication) at Stuttgart, Germany.

Olaf Hildebrand was born on May 4, 1948 in Hamburg, Federal Republic of Germany. He received the Dipl. degree in 1973 and the Ph.D. degree in 1977, both from the University of Stuttgart. His thesis dealt with high excitation phenomena in III-V compound semiconductors and basic research in the field of semiconductor lasers.

Since 1978, his main research field has been physics and technology of avalanche photodetectors based on III-V compound semiconductors. In May 1983, he joined SEL Research Centre Stuttgart, Germany, where he is now Division Manager for optoelectronic components in the field of optical communication.

Dr. Hildebrand is a member of the German Physical Society (DPG).

NOVEL HIGH-SPEED LED TRANSMITTER FOR SINGLE-MODE FIBRE AND WIDEBAND LOOP TRANSMISSION SYSTEMS

Indexing terms: Optical transmission, Light-emitting diodes

A new high-speed LED transmitter using novel GaAs MESFET bias and drive circuits has been developed. It can modulate LEDs beyond 560 Mbit/s with up to 200 mA drive current. At 560 Mbit/s, extinction ratios of 15 dB were attained with devices that otherwise operated up to 250 Mbit/s. The transmitter was designed for single-mode loop applications.

There has been strong interest recently in the use of light-emitting-diodes (LEDs) as light sources for single-mode fibre loop applications.[1–5] Because of low cost and high reliability, LEDs offer substantial advantages over diode lasers. It is important to investigate the LED operation at high bit rates to provide future broadband integrated services to customer premises. Edge-emitting LEDs are preferred for single-mode fibre coupling over surface-emitting LEDs because of higher coupling efficiency, but inherently they also have slower speed.[6–7] Techniques such as large bias current are often employed to increase carrier concentration in the recombination region and enhance the frequency performance of the LEDs.[6] However, DC bias results in extinction-ratio penalties and higher device operation temperatures. We report here a novel active bias and driver circuit that enables us to modulate commercial edge-emitting LEDs at 560 Mbit/s with a high AC drive current (about 200 mA) and achieve excellent power extinction ratio (better than 15 dB).

Fig. 1 shows the schematic diagram of the LED driver circuit. It consists of a level shifter, a GaAs MESFET current switch, a high-speed equaliser and an active bias control circuit. The level shifter, which down-shifts the ECL input level (−0·8 V to −1·8 V) more negatively to the switching transistor logic level (−5·2 V to −6·2 V), consists of series-connected GaAs Schottky diodes and a resistor-transistor

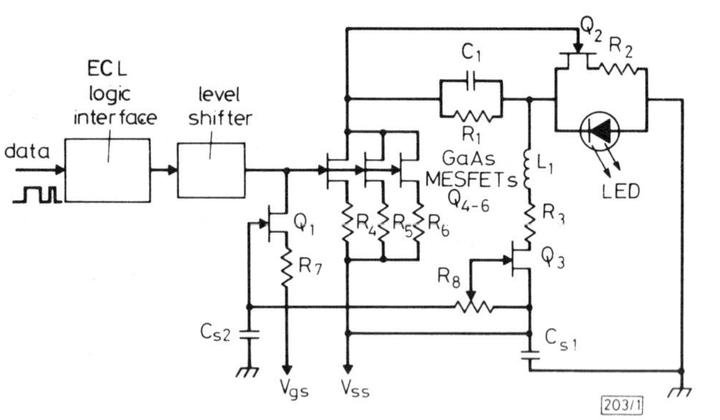

Fig. 1 *Schematic diagram of high-speed active-bias LED driver*

current matching network. Q_1 serves as a variable-voltage load to maintain nearly constant current through the diodes during 'on' and 'off' states so that inductance-related voltage transients and switching delay can be minimised. The current switch is made of three parallel MESFET transistors with low pinch-off voltages (−1 V) and 70 mA I_{dss}. Ballast resistors R_4, R_5 and R_6 are used to balance drain currents through negative feedback of the gate-to-source voltages. The high-speed compensation network formed by R_1, C_1, R_2 and Q_2 can enhance the high-frequency response and provide pulse-shaping for the LED driver. The active bias circuit is formed by a current source, with L_1, R_3, R_8 and Q_3 as the circuit elements, and the gate control circuit of Q_2 connected to drains of Q_4–Q_6. When the LED is 'off', Q_2 conducts most of the bias from the current source because of the low on-resistance at small V_{ds}. Q_2 can be turned off quickly when the LED is to be turned 'on'. The inductor L_1 will force the bias current to transfer from Q_2 to the LED, resulting in a high-current pedestal biased state. The combined LED driver output is shown in Fig. 2. Similarly, when the LED is to be turned 'off', Q_4–Q_6 are switching 'off'. This allows gate potential of Q_2 to discharge rapidly through the R_1, C_1 path to the

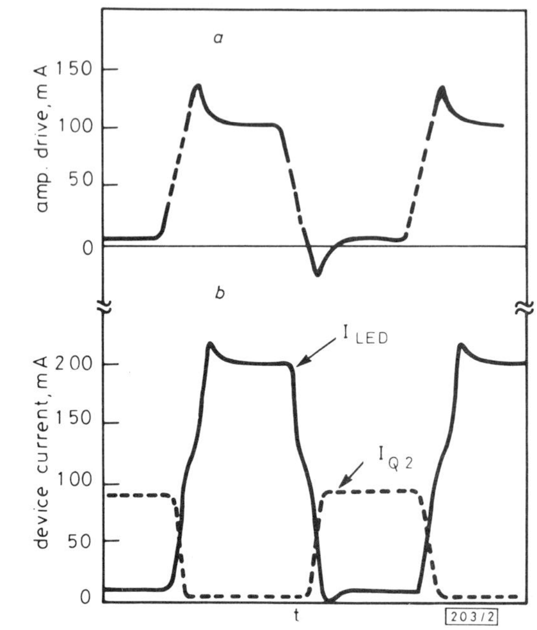

Fig. 2 *LED drive current from (a) current switch Q_4–Q_6 and (b) active-bias switch Q_2 and total combined current*

source potential, thus restoring the low-impedance path through Q_2. The LED then discharges rapidly through Q_2 and returns to the 'off' state. The turn-on and turn-off transient sequences described above are naturally gauged by the difference in gate capacitances of Q_4–Q_6 and Q_2, because we used the same transistor for both functions. Reduced bias in the LED quiescent state results in large extinction ratios and lower junction temperatures. The transmitter driver circuits are fabricated on a G-10, double-sided, printed circuit board with 50 Ω microstrip transmission lines. GaAs MESFETs were used for all switching and control functions. The LED module is mounted on the circuit board and attached to an aluminum heat sink. A complete transmitter is shown in Fig. 3.

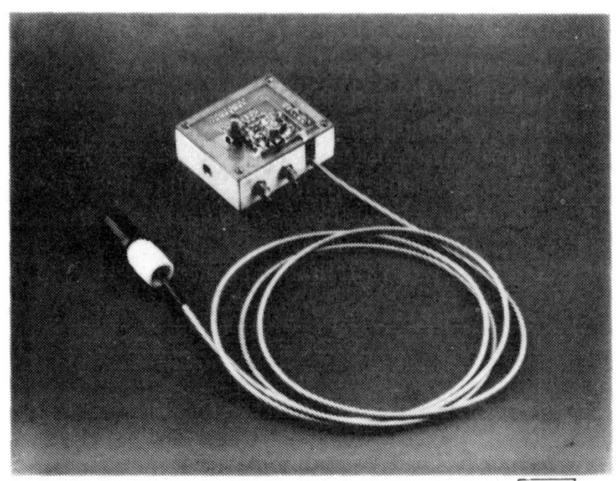

Fig. 3 *Prototype LED transmitter*

Results of tests using a 1·3 μm, Lasertron QL1300 edge-emitting LED are shown in Fig. 4. The LED has a 2·1 ns rise

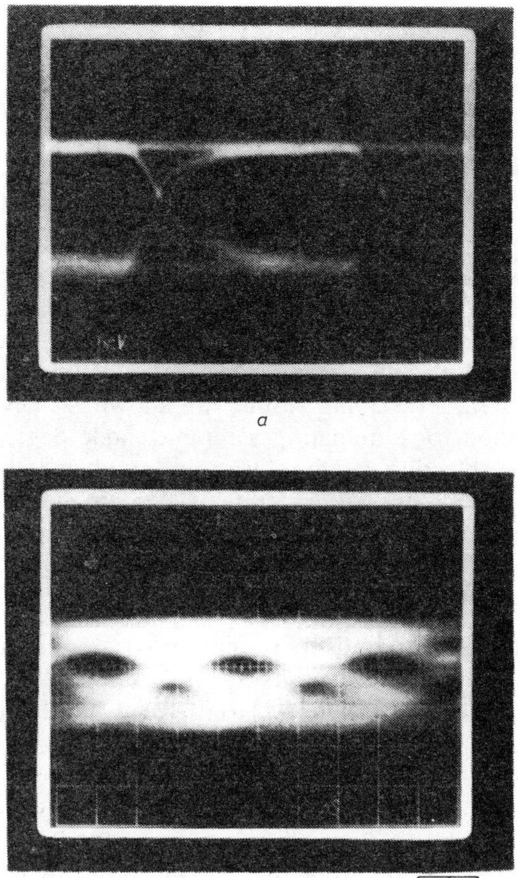

a

b

20 3/4

Fig. 4 *Transmission eyes of QL1300 LED with bias tee driver modulated at (a) 90 Mbit/s and (b) 560 Mbit/s*

and 4 ns fall time when driven through a bias tee at 90 Mbit/s (Fig. 4*a*). No useful eye exists at 560 Mbit/s (Fig. 4*b*). However, Fig. 5 shows the improved eye at 560 Mbit/s with our new transmitter operated with 70 mA active bias current and 110 mA amplifier drive. At 560 Mbit/s, the 'on' state peak power launched into single-mode fibre pigtail is −22 dBm and the 'off' state power is −37 dBm, for an extinction ratio of 15 dB. If we replace Q_2 with a 200 Ω shunt resistor to duplicate the same bias and drive conditions, the extinction ratio becomes only 4 to 1 and 30% more power is dissipated in the diode junction. This would result in a significant sensitivity penalty at the receiver. The new driver circuit has also been tested with a high-speed, zinc-doped, surface-emitting LED.[7] Rise and fall times of 350 ps were observed, as shown in Fig. 6.

We have designed and tested a new LED transmitter circuit for single-mode-fibre subscriber-loop applications. It can modulate high-power LEDs at 560 Mbit/s while achieving excellent extinction ratios and low device operating temperatures. It can be simply fabricated and requires two power supplies. It also interfaces with standard ECL logic levels.

G.-K. CHANG
H. P. LEBLANC

Bell Communications Research
Red Bank, NJ 07701, USA

P. W. SHUMATE

Bell Communications Research
Morristown, NJ 07960, USA

12th October 1987

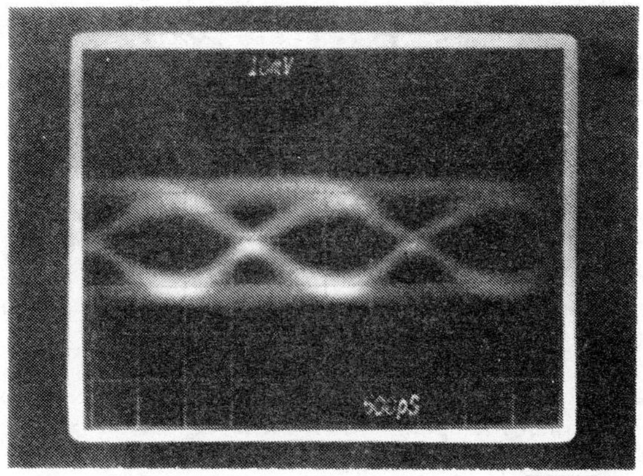

203/5

Fig. 5 *Transmission eye of Q11300 LED using the active-bias driver modulated at 560 Mbit/s, 500 ps/div*

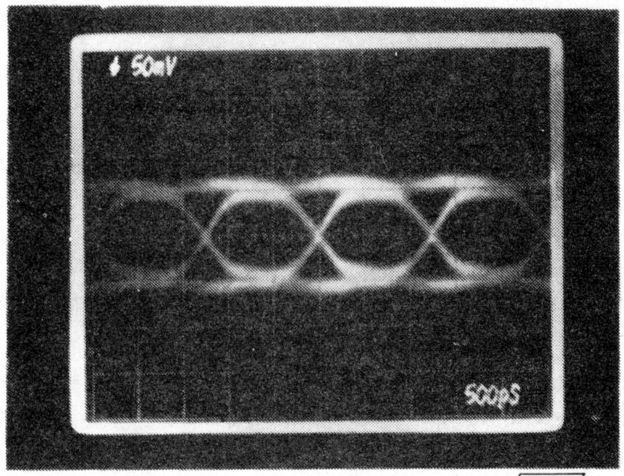

203/6

Fig. 6 *Rise- and fall-time measurements of new driver circuit using a fast surface-emitting LED, 500 ps/div*

References

1 SHUMATE, P. W., GIMLETT, J. L., STERN, M., ROMEISER, M. B., and CHEUNG, N. K.: 'Transmission of 140 Mbit/s signals over single-mode fibre using surface and edge-emitting 1·3 μm LEDs', *Electron. Lett.*, 1985, **21**, pp. 522–524

2 FUJIMOTO, N., *et al.*: '1·2 Gbit/s optical transmission experiment over 10 km of single-mode fiber using edge-emitting LED'. Late news paper of OFC '87, Reno, Nevada, Feb. 1987

3 ARNOLD, G., and KRUMPHOLZ, O.: 'Optical transmission with single-mode electron fibres and edge-emitting diodes', *Electron. Lett.*, 1985, **21**, pp. 390–392

4 GIMLETT, J. L., STERN, M., VODHANEL, R. S., CHEUNG, N. K., CHANG, G. K., LEBLANC, H. P., SHUMATE, P. W., and SUZUKI, A.: 'Transmission experiments at 560 Mbit/s and 140 Mbit/s using single-mode fibre and 1300 nm LEDs', *ibid.*, 1985, **21**, pp. 1198–1200

5 SAUL, R. H., KING, W. C., OLSSON, N. A., ZIPFEL, C. L., CHIN, B. H., CHIN, A. K., CAMLIBEL, I., and MINNECI, G.: '180 Mbit/s, 35 km transmission over single-mode fibre using 1·3 μm edge-emitting LEDs', *ibid.*, 1985, **21**, pp. 773–775

6 NAMIZAKI, H., *et al.*: 'Frequency response of GaAlAs light-emitting diodes', *IEEE Trans.*, 1984, **ED-21**, p. 688

7 SUZUKI, A., INOMATO, Y., HAYASHI, J., ISODO, Y., UJI, T., and NOMURA, H.: 'Gbit/s modulation of heavily Zn-doped surface-emitting InGaAsP/InP OH LED', *Electron. Lett.*, 1984, **20**, pp. 273–274

High-Reliability Semiconductor Lasers for Optical Communications

MOTOHISA HIRAO, KEN'ICHI MIZUISHI, AND MICHIHARU NAKAMURA

***Abstract*—InGaAsP/InP buried heterostructure (BH) lasers emitting at 1.3 μm have continued to operate stably for more than 3.3 × 10^4 h (3.8 years) at 50–60°C and at an output power of 5 mW/facet. A statistically estimated median lifetime exceeds 10^6 at 50°C. A relatively low activation energy of 0.32 eV is obtained for slow degradation. The saturable behavior of the aging characteristics is observed in many of the lasers. This mode is explained by the increased leakage current through the buried regions, and can be eliminated by electroluminescence (EL) mode aging at high temperature and current. Distributed feedback (DFB) lasers emitting at 1.55 μm are also subjected to accelerated aging at 60°C with a 3 mW/facet output after EL-mode aging. These DFB lasers demonstrate stable aging characteristics, for more than 2000 h of operating time being currently achieved.**

I. Introduction

THE reliability of long-wavelength InGaAsP/InP lasers emitting at 1.3 μm has been of great concern for practical applications to optical transmission systems. Long-term stable operation of these lasers was first reported by Yamamoto *et al.* [1] using narrow-stripe planar structure lasers. The results of this experiment suggested the possibility of much longer lifetimes than the previously developed AlGaAs/GaAs lasers. In addition, recent improvements in device structures, fabrication technology, and adequate device screening methods have made it possible to assure high reliability in 1.3 μm lasers.

Most of the 1.3 μm InGaAsP/InP lasers developed for practical use have narrow-stripe refractive-index guiding structures in order to operate continuously, even at high temperatures in a stable fundamental transverse mode. These structures include the buried heterostructure (BH) [2], the buried crescent (BC) [3], the double-channeled planar buried heterostructure (DC–PBH) [4], and the *V*-grooved substrate buried heterostructure (VSB) [5]. For example, the low threshold current (10–30 mA at 25°C) of the BH lasers permits them to lase continuously at ambient temperatures up to 110°C, and to operate stably over a long period at 70°C with an output power of 5 mW/facet [6], [7]. The long-term stable aging characteristics at such high temperatures have confirmed the high reliability of these narrow-stripe refractive-index guiding lasers [8]–[13]. However, erratic behavior in the aging characteristics, such as saturable mode degradation [8], has been found to exist in these long-lived lasers during the initial stage of aging. As a result, it is difficult to select "good" lasers only through conventional screening tests.

Manuscript received April 1, 1986.
The authors are with the Central Research Laboratory, Hitachi, Ltd., Kokubunji, Tokyo 185, Japan.
IEEE Log Number 8610776.

The purpose of this paper is to present reliability results for state-of-the-art 1.3 μm InGaAsP/InP BH lasers. The degradation modes and statistical lifetime estimation based on the long-term aging results are shown. A new screening method for eliminating saturable mode degradation and for selecting reliable lasers is introduced. In addition, the preliminary aging results of 1.55 μm distributed feedback (DFB) lasers are presented. These lasers have been developed for light sources in high-bit-rate and long-haul optical transmission systems.

II. Accelerated Aging Characteristics of 1.3 μm BH Lasers

The buried heterostructure (BH) laser has a well-known refractive-index guiding structure with a narrow active layer width (1–2 μm) for operating in the fundamental transverse mode. In this experiment, the laser chips were mounted junction-up on SiC submounts using Pb–Sn eutectice solder [14], [15]. Protective coating film was not applied to the mirrors of the laser. These BH lasers with threshold currents of 15–25 mA at room temperature can operate stably at an output power of 5 mW/facet at temperatures above 70°C.

Aging was carried out in dry nitrogen ambient for lasers prepared without preselection. The output power of individual lasers was kept constant at 5 mW/facet by automatically adjusting the driving currents. At first, the aging temperature was set at 50°C, and was later changed to 60°C after aging for 17 000–24 000 h. Typical long-term aging characteristics of the 1.3 μm BH lasers are shown in Fig. 1. The abrupt increases in driving current shown in the figure are due to such a temperature change. It should be noted that in a few lasers, relatively rapid increases in driving current appear during the first stage of aging, and thereafter the current increase rate reduces.

Such a saturable degradation behavior is found to depend on the aging temperature, as shown in Fig. 2. That is, the knee point to be noted in the figure varies with the aging temperature (also see Fig. 4). It takes about 1000 h of aging at 60°C to reach the knee point, while it requires more than 5000 h at 40°C. Mizuishi *et al.* [16] statistically analyzed the aging characteristics of BH lasers in order to estimate their maximum median lifetimes. The following equations for degradation acceleration were used [16]:

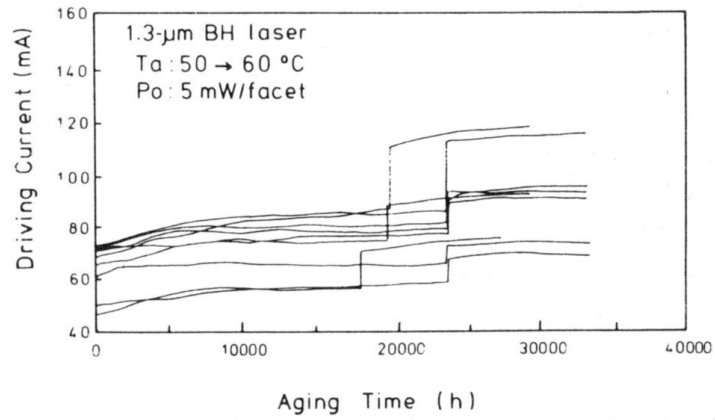

Fig. 1. Long-term aging characteristics of 1.3 μm BH lasers. Aging was performed initially at 50°C and then changed to 60°C, thus causing abrupt increase in driving current.

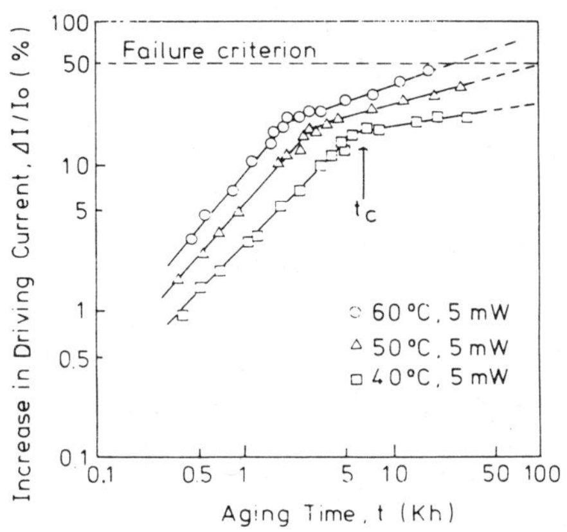

Fig. 2. Typical saturable mode degradation of BH lasers. The value t_c indicates a knee point in the degradation curve [16].

$$dI/I_0 = At^{m^1} \exp(-E_{a1}/kT_j) \quad (t < t_c) \tag{1}$$

$$dI/I_0 = At^{m^2} \exp(-E_{a2}/kT_j) \quad (t > t_c). \tag{2}$$

Here, dI/I_0 is the percentage increase in driving current at t hours, A_1 and A_2 are proportionality factors, m_1 and m_2 are operating time (t) powers, k is Boltzmann's constant, and T_j is the junction temperature. Variables E_{a1} and E_{a2} are the activation energies which indicate the degradation dependences on temperature before and after the t_c, respectively. From the aging results of more than 100 BH lasers aged at 40, 50, and 60°C for 4000–10 000 h, it is found that the most significant parameters are: $m_1 = 0.82$ (lowest), 1.20 (median), and 1.75 (highest), $m_2 = 0.17$, 0.38, and 0.60, $E_{a1} = 1.23$, 1.32, and 1.40 eV, and $E_{a2} = 0.30$, 0.32, and 0.35 eV.

Using the median values for these parameters, the median lifetimes of the BH lasers can be predicted to exceed 10^6 h at 50°C. This prediction is based on using the almost 50 percent increase in driving current as a failure criterion.

By utilizing the appropriate 1.3 μm DC–PBH or VSB lasers, Ikegami and Takahei [17] reported an activation energy of a few tenths of an electron volt, and estimated a mean time to failure of more than 10^5 h at 50°C with 5 mW/facet. For gain-guided 1.3 μm IRW (inverted rib waveguide) lasers, Rosiewicz *et al.* [18] also reported an activation energy of 0.25 eV for slow degradation. Thus, low activation energy values for slow degradation have often been observed in InGaAsP/InP lasers, irrespective of the type of laser structure being investigated.

III. New Screening Method for InGaAsP/InP Lasers

The saturable aging characteristics of InGaAsP/InP lasers make it difficult to select long-life lasers through conventional screening tests. A detailed examination of degraded lasers revealed that the rapid degradation is caused mainly by the increased carrier leakage through the buried regions surrounding the active region [8]. An analysis using an equivalent circuit model adequately explains the degradation behavior caused by such leakage current [8], [19], [20]. To eliminate the lasers with potential problems associated with carrier leakage and to stabilize the initial aging characteristics, electroluminescence (EL) mode aging is found to be of great use as a screening method for InGaAsP/InP lasers.

The results of typical EL-mode aging at 100°C and 200 mA for BH lasers are shown in Fig. 3. The results indicate a marked reduction in the degradation rate following the rapid increase in threshold current within about 20 h. This rapid increase is associated with a decrease in forward diode voltage at currents of 1–10 mA, suggesting the presence of carrier leakage through the buried regions surrounding the active region.

The relationship between the time duration for stabilization t_c and the reciprocal of the aging temperature $1/T_j$ for BH lasers is plotted in Fig. 4. The open circles show the median values of t_c for each laser-mode aging at ambient temperatures of 40, 50, and 60°C with a constant output of 5 mW/facet. The solid circle shows the median value of t_c for EL-mode aging at an ambient temperature of 100°C and at constant currents of 150, 200, and 250 mA. In EL-mode aging, t_c is defined as the knee point observed in the threshold current versus aging time plot. The solid line represents the best fit for the plots and can be expressed by the equation shown in Fig. 4. It should be noted that a stimulated emission was hardly observed in EL-mode aging used here because a junction temperature was high enough to prevent lasing operation. Therefore, current and temperature are considered to be main factors for degradation acceleration. The good linear relationship in the t_c and $1/T_j$ plot between the laser-mode and EL-mode aging suggests the presence of a single degradation mechanism activated by temperature and current rather than by optical power.

The long-term aging results performed at 60°C and 5 mW/facet after EL-mode aging are shown in Fig. 5. The lasers having driving currents of more than 120 mA to maintain a 5 mW/facet output at 60°C were rejected after EL-mode aging. This criterion was tentatively used by considering the thermal effect due to driving current on degradation acceleration. The lasers which passed EL-mode aging operate stably at a very low degradation rate for more than 5000 h without showing any signs of sat-

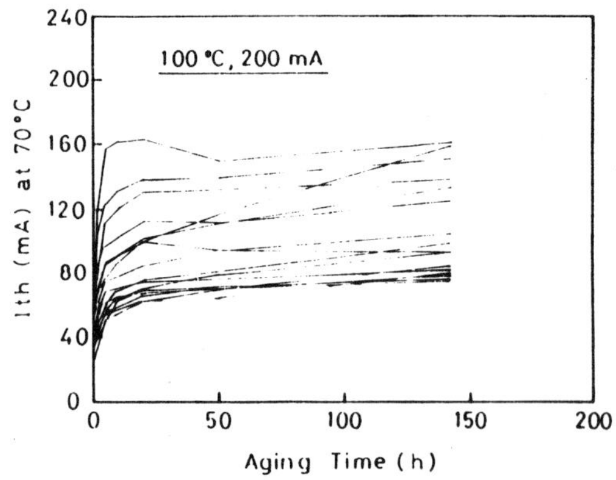

Fig. 3. Threshold current changes of 1.3 μm BH lasers during EL-mode aging. Threshold currents were periodically measured at 70°C [21].

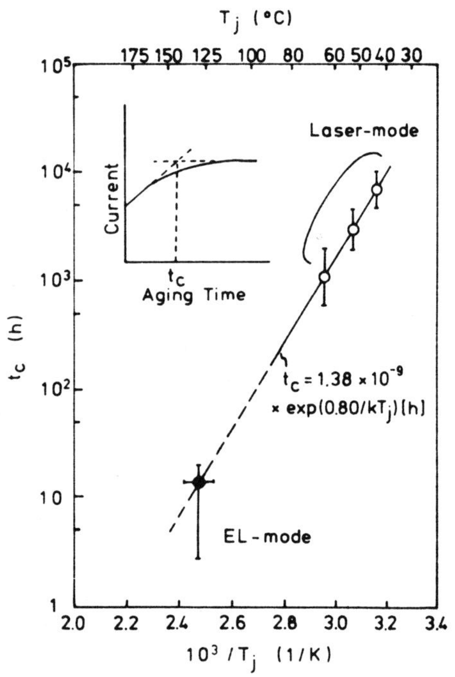

Fig. 4. The dependence of the knee point t_c on temperature in the degradation curves. Each plot indicates a median value for about 30 devices. Vertical and horizontal indicate standard deviation and scatter due to driving current variations of 150–250 mA.

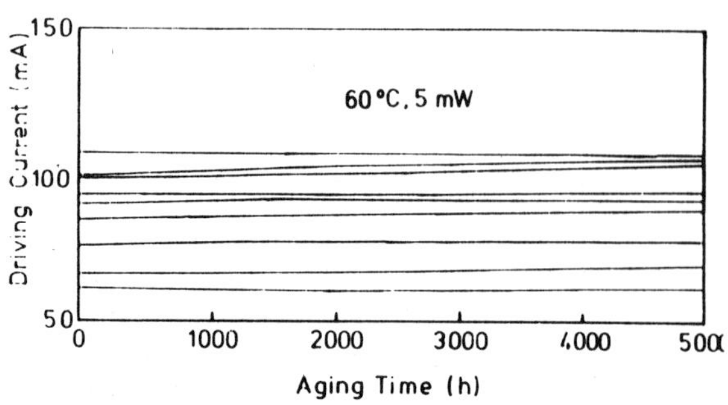

Fig. 5. Long-term aging characteristics of BH lasers after EL-mode aging. EL-mode aging was performed at 100°C at a constant 200 mA for 20 h [21].

urable mode degradation. The median degradation rate of these lasers is found to be at least seven times lower (e.g., 0.5 percent/kh versus 3.5 percent/kh [21]) than the median degradation rate of lasers not subjected to EL-mode aging. Thus, EL-mode aging at high temperature and current is an effective screening method for selecting stable and long-life lasers since it can eliminate saturable mode degradation and any erratic behavior during the initial stage of operation.

Ikegami *et al.* [17], [22] also performed EL-mode aging tests using DC–PBH and VSB lasers. Aging temperatures and currents used were 50–150°C and 100–250 mA. It was found that too much stress (>175 mA at 100°C or 150 mA at 150°C) during aging causes a reduced slope efficiency as well as increased leakage current through the buried regions. An activation energy in terms of increased threshold current at 70°C was estimated at 0.7 eV. While under moderate conditions (<250 mA at 50°C or 100 mA at 100°C), only observed were threshold current increases, which obeys an Arrhenius relationship with an activation energy of a few tenths of an electron volt.

Nash *et al.* [11] performed temperature–current overstress tests (250 mA, 100°C) for BH lasers to exclusively stress the current leakage path. They observed similar effects as ours previously mentioned, stabilizing the degradation rate within <20 h. Supposing that in overstress tests at 250 mA and 135–150°C the time duration f_{pr} stabilization (just as t_c in Fig. 4) obeys an Arrhenius relationship, an activation energy was estimated at 1.2 eV (larger than 0.8 eV shown in Fig. 4). On the other hand, an activation energy of 0.54 eV was obtained for lasing operation at 40–70°C using the same structure lasers [10].

As mentioned, the large difference between activation energies for degradation of InGaAsP/InP lasers has been observed. This is due to the presence of two different modes of degradation in addition to the difference between test conditions and laser structures being investigated. However, at least it can be said that relatively rapid degradation such as the saturable initially occurring mode of degradation is associated with mechanisms specified by activation energies larger than those for the long-term slow degradation (e.g., 1.23–1.40 eV versus 0.30–0.35 eV for BH lasers [16], 1.5 eV versus 0.54 eV for BH lasers [10], [11], 0.7 eV versus a few tenths of an electron volt for DC-PBH and VSB lasers [17], and 0.51 eV versus 0.25 eV for IRW lasers [18]). This implies that if high-temperature EL-mode aging is applied to InGaAsP/InP lasers with appropriate conditions, it becomes possible to select reliable lasers within a short period of aging.

IV. Degradation Modes of InGaAsP/InP Lasers

Degradation modes of InGaAsP/InP lasers are significantly different from those of AlGaAs/GaAs lasers. Specifically, facet damage and the accumulation of dark line defects in the active region are not serious causes of the rapid degradation in InGaAsP/InP lasers. Furthermore, facet oxidation is not a cause of long-term or saturable mode degradation [9], [23]. One of the main causes for the degradation in InGaAsP/InP BH-type lasers is found to be the carrier leakage through the buried regions surrounding the active region.

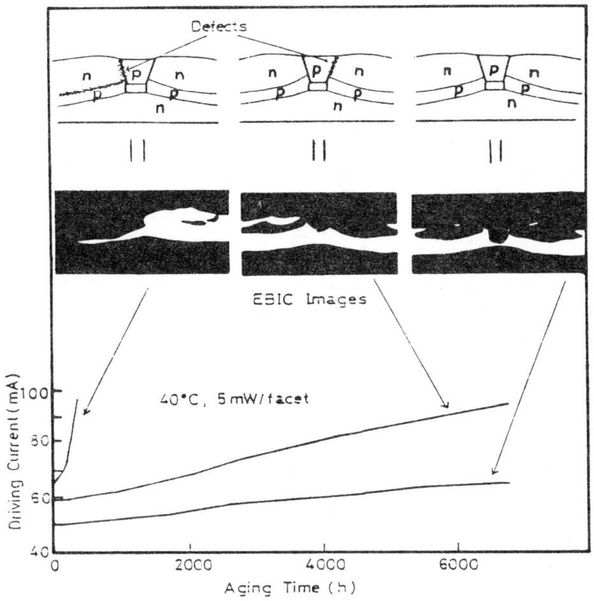

Fig. 6. EBIC images on BH laser mirror surfaces after aging. The dark portions, shown in the top drawings by X's, indicate the presence of imperfections at p–n junctions [9].

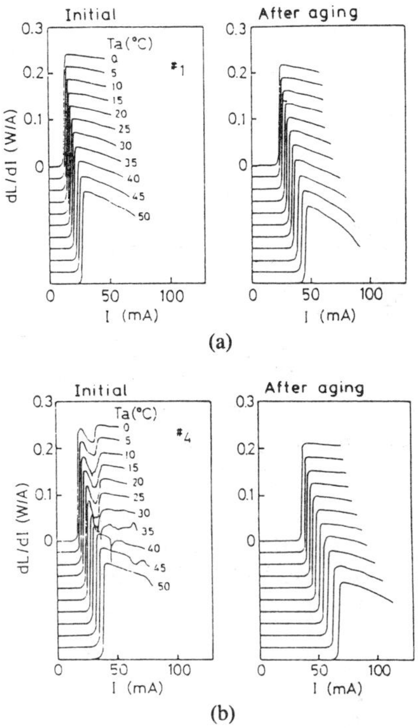

Fig. 7. *dL/dI–I* curves measured before and after aging at 50°C and 5 mW/facet for lasers with and without kinks. The initially kink-free laser remained as such despite a large increase in threshold current [9].

A conventional BH laser is fabricated by two-step crystal growth and has a reverse-biased barrier layer for preventing carrier leakage. The crystal defects and the misplacement at the interface between barrier and active layers can potentially cause carrier leakage. Such carrier leakage in the BH laser is furthered by aging, resulting in increased threshold current. The extent to which carrier leakage affects long-term reliability is not easy to accurately estimate. However, EL-mode aging is found to be a useful tool for eliminating potential failures associated with carrier leakage. That is, it is possible to select "good" lasers without significantly increasing the leakage current through EL-mode aging.

The electron-beam-induced current (EBIC) image contrasts on the mirror surfaces of BH lasers observed after aging are shown in Fig. 6. The bright region indicates the EBIC signals of the p–n junction, while the dark regions denoted by X's, show regions of incomplete p–n junctions due to crystal defects. The short-life laser produces an asymmetric EBIC image, indicating the presence of crystal defects incurred during the second stage of crystal growth. As previously mentioned, such shot-life lasers can be eliminated through EL-mode aging.

In a BH laser heavily degraded and showing an ohmic-like current–voltage curve during EL-mode aging, local imperfections along the active stripe are observed by SEM and EBIC images. That is, high-stress EL-mode aging also accelerates the degradation related to such latent imperfections. Consequently, the lasers undergoing EL-mode aging are expected to exhibit well-stabilized behavior during continous long-term operation.

In general, surface treatment of the crystal is an important factor for fabricating reliable InGaAsP/InP BH lasers. The forward voltage (V_f) of an InP p–n junction at currents of 1–10 mA is affected by surface treatment chemicals prior to the second crystal growth. The p–n junction grown continuously without any wet process does not exhibit any appreciable change in V_f during aging. Higuchi *et al.* [24] reported that the presence of a p–n junction at the grown interface caused initial degradation in the BC laser, and that the displacement of the junction position into the n-type cladding layer by thermal diffusion was effective in eliminating the initial rapid degradation.

Saturable mode degradation was also observed in gain-guided lasers [18]. This phenomenon was found to be strongly related to the method of assembly used; that is, its appearance probability depended on the mounting polarity and the kind of bonding solder [9]. In fact, it was observed that the reattachment of a laser chip from one heatsink to another is accompanied by slight saturable mode behavior during aging, even in lasers which successfully underwent EL-mode aging.

V. Changes in Laser Characteristics During Aging

The practical lifetime of a laser is determined by the changes in laser characteristics during aging. Among the laser characteristics measured during aging are light output–current (*L–I*) and first derivative (*dL/dI*)–*I* curves, far-field patterns (FFP), longitudinal mode spectra, and dynamic responses of light output. The measurements are performed at various ambient temperatures from 0 to 50°C sers are strongly related to the initial kinks. It should be noted that these kinks especially appear in lasers with oversized cavities (stripe width > 2 μm). The theoreti- by 5 or 10 degree steps.

Typical dL/dI–I characteristics of BH lasers before and after aging are shown in Fig. 7. Kinking behavior in the dL/dI–I curves is not found in the laser capable of lasing in the fundamental transverse mode (active layer width < 1.7 μm). This is despite a 43 percent increase in driving current after 2446 h of aging as shown in Fig. 7(a). On the other hand, initially appearing kinks disappear completely as a result of the 41 percent increase in driving current after 1409 h of aging, as shown in Fig. 7(b). In this case, the initial kinks were concealed below the subthreshold portion due to an increase in threshold current during aging. This was confirmed by measuring the laser cooled. Such a phenomenon was commonly observed in other lasers having initial kinks.

Abnormalities in longitudinal mode spectra (i.e., spectrum splitting and broadening, and peak spectrum shift to shorter wavelengths) are observed for lasers with initial kinks before aging. Self-pulsation and FFP deformation are also found. All of these abnormalities, together with the initial kinks, disappear during aging for the laser with the dL/dI–I curves shown in Fig. 7(b). Furthermore, the kinks tend to appear only in a limited current range at a given temperature and are not directly related to the optical output. In other words, an optical output level having kinks appears to be lower during aging as the threshold current increases.

As previously mentioned, all of the abnormalities observed in the aging characteristics of the InGaAsP/InP lacally calculated stripe width of BH lasers for kink-free operation in the fundamental transverse mode is less than 2 μm, which agrees well with the experimental results.

VI. Preliminary Aging Results of 1.5 μm DFB Lasers

Recently, distributed feedback (DFB) lasers have been extensively developed for light sources in high-bit-rate/long-haul optical transmission systems. This is because DFB lasers meet the requirements for narrow spectra and for eliminating mode partition noise of a conventional Fabry–Perot (FP) laser operating in the multilongitudinal mode. However, data on reliability of DFB lasers are still comparatively scarce. This section presents preliminary aging results for the 1.5 μm DFB lasers.

The tested DFB lasers have a buried heterostructure, as shown in Fig. 8. The corrugation for DFB-mode operation is formed holographically on an n-type InP substrate. The multilayer structure consists of an n-InGaAsP waveguide layer, undoped active layer, p-InGaAsP anti-meltback layer, p-InP cladding layer, and a p-InGaAsP cap layer. The wavelengths corresponding to the three quaternary layers (waveguide, active, and anti-meltback layers) are 1.27, 1.53, and 1.27 μm, respectively. The buried heterostructure is fabricated using the same technique as previously mentioned [2]. In order to eliminate the FP mode, the backside of the stripe is buried by InP during the second step regrowth process, resulting in a window structure.

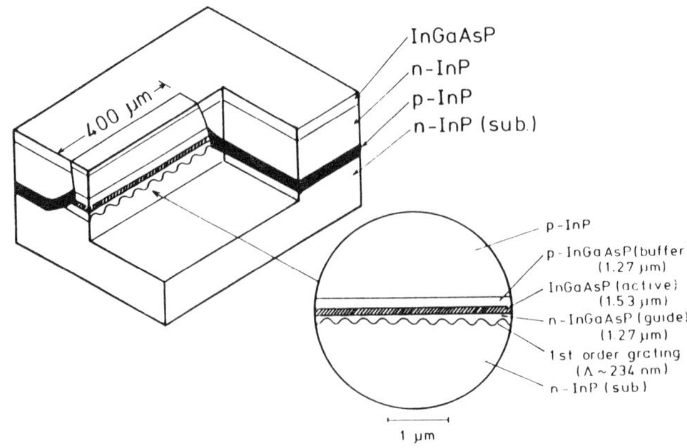

Fig. 8. Schematic structure of a 1.5 μm BH-DFB laser.

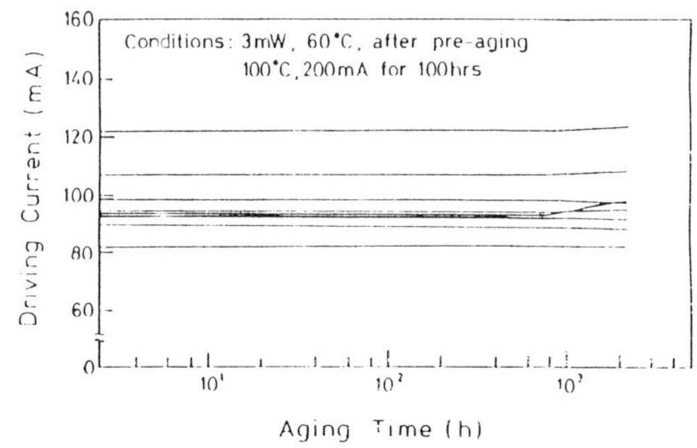

Fig. 9. Aging characteristics of 1.5 μm BH-DFB lasers after EL-mode aging. Aging was performed at 60°C and 3 mW/facet.

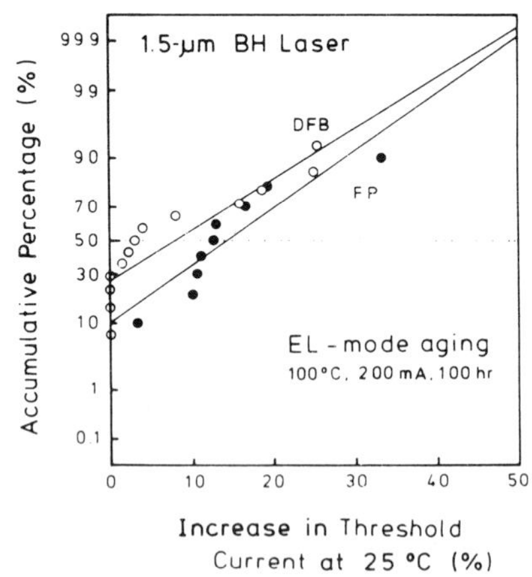

Fig. 10. Threshold changes of 1.5 μm BH-DFB lasers during the EL-mode screening test compared to those of 1.5 μm conventional BH lasers.

The lasers were first subjected to EL-mode aging at 100°C and 200 mA for 20 h. The subsequent aging was performed at 60°C with a constant power of 3 mW/facet. The aging results of the 1.5 μm BH–DFB lasers are shown in Fig. 9. As shown in the figure, stable aging characteristics can be obtained for up to 2000 without displaying any saturable mode degradation.

A comparison of EL-mode aging between conventional BH and BH–DFB lasers at 100°C and 200 mA is shown in Fig. 10. No significant difference in the threshold increase percentage is found between the lasers. Consequently, it is likely that undesirable effects on DFB laser

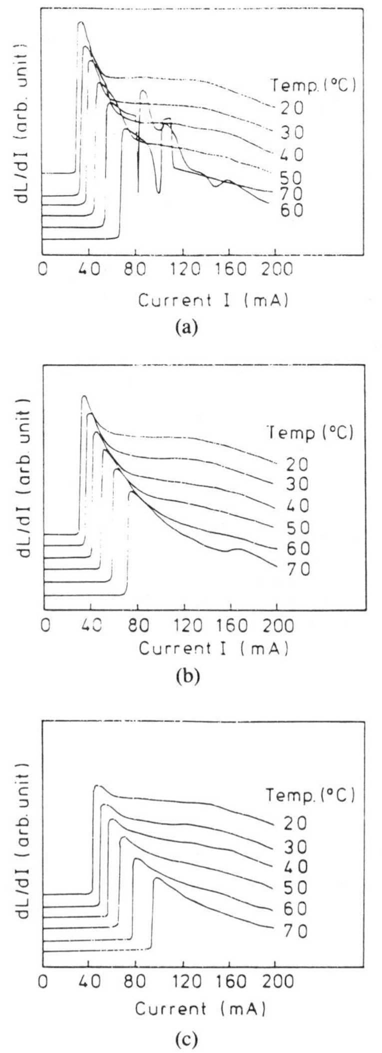

Fig. 11. *dL/dI–I* curves of a 1.5 μm BH–DFB laser before and after EL-mode aging. The sharp spikes in the *dL/dI–I* curve at 60 and 70°C indicate the change in longitudinal mode from a single to two modes.

reliability will not be produced due to the presence of the corrugation near the active layer. However, the corrugation is found to deteriorate during heat treatment in an LPE furnace just prior to crystal growth [20]. Moreover, when this corrugation is preserved under a GaAs cover crystal, a mass-transported quarternary layer in the concave of the corrugation can be identified using transmission electron microscopy [25]. Nevertheless, the aging results shown in Figs. 9 and 10 are encouraging for the reliability of DFB lasers.

Another factor in the reliability of DFB lasers is the longitudinal mode stability. It is known that longitudinal mode instabilities cause sharp spikes in the *dL/dI–I* curve. Hence, the influence of degradation on the longitudinal mode stability of DFB lasers has been investigated by measuring *dL/dI–I* curves at various temperatures. Consequently, EL-mode aging has been performed for the lasers at 100°C and 200 mA, and the *dL/dI–I* curves were periodically measured during the test.

The DFB laser did not show any spikes in the *dL/dI–I* curve below 50°C, as shown in Fig. 11(a). In fact, this laser initially operated in a single DFB mode below 50°C. On the other hand, at temperatures above 60°C, sharp spikes appeared in its *dL/dI–I* curves, and these spikes were found to exactly correspond to a mode change. This means that the mode change caused the laser to operate in two spectra modes (the wavelength difference between two modes is less than a few nanometers).

However, drastic changes in the *dL/dI–I* curves appeared after EL-mode aging for 1 h, during which the threshold current at 20°C increased by 7 percent. This is shown in Fig. 11(b). The sharp spike in the curve at 60°C up disappeared and a single-mode operation was maintained to 200 mA. In the *dL/dI–I* curve at 70°C, the current level corresponding to a mode change shifted from 90 to 155 mA, above which the laser continued to operate in two spectra modes.

After EL-mode aging for 3 h, the laser shows a complete single-mode operation in the temperature range from 20 to 70°C up to currents of 200 mA. This is shown in Fig. 11(c). The threshold current increased by 50 percent during aging, and this caused a shift in the occurrence of mode changes at higher current levels of more than 200 mA. Thus, the mode change mentioned above is somewhat different from the transverse mode instabilities related to kinks (Fig. 7). Similar results were also observed in ten other BH–DFB lasers.

On the other hand, DFB lasers which can completely operate in a single mode did not show any longitudinal mode changes, regardless of the more than 100 percent increase in threshold current during aging, except for the change in peak wavelength. This point illustrates the potential long lifetimes of 1.5 μm BH–DFB lasers, although further systematic reliability studies should be performed.

VII. Summary

The reliability of state-of-the-art 1.3 μm InGaAsP/InP BH lasers and 1.5 μm InGaAsP/InP BH–DFB lasers has been demonstrated. Many of the 1.3 μm BH lasers showed a saturable initially occurring mode of degradation, which is defined as relatively rapid degradation during the initial aging stage followed by slow degradation. A statistical analysis of the aging characteristics for about 100 lasers also showed that 1.3 μm BH lasers have a median lifetime of more than 10^6 h at 50°C. A relatively low activation energy of 0.32 eV was obtained for slow degradation.

The most significant degradation mode in these advanced 1.3 μm BH lasers was due to carrier leakage through the buried regions around the active layer. This leakage caused the saturable behavior in aging characteristics. Degradation due to mirror facet damage or dislocations in the active layer was not observed in the 1.3 μm BH lasers.

Electroluminescence (EL) mode aging at high temperature and current was quite useful for eliminating the lasers with potential defects which were responsible for the short lifetime. EL-mode aging under proper conditions stabilized saturable mode degradation during the initial aging stage.

EL-mode aging was also an effective screening method for BH–DFB lasers. The 1.5 μm BH–DFB lasers, which were subjected to EL-mode aging over a short period, stably operated for more than 2000 h at 60°C and 3 mW/facet. It was confirmed that 1.5 μm BH–DFB lasers have the potential of high reliability, although further study is desirable, especially on the mode stability related to aging-induced degradation.

Acknowledgment

The authors would like to thank Dr. Y. Koga and M. Sakamoto for their encouragement throughout this work. They are also grateful to S. Tsuji and A. Oishi for the fabrication of laser diodes and for technical discussions, and to Dr. H. Matsumura and M. Okai for developing the 1.5 μm BH–DFB lasers. The graphic design work provided by H. Muto is also appreciated.

References

[1] T. Yamamoto, K. Sakai, and S. Akiba, "10000-h continuous operation of InGaAsP/InP DH lasers at room temperature," *IEEE J. Quantum Electron.*, vol. QE-15, pp. 684–687, 1979.

[2] M. Hirao, S. Tsuji, A. Doi, and M. Nakamura, "Fabrication and characterization of narrow stripe InGaAsP/InP buried heterostructure lasers," *J. Appl. Phys.*, vol. 51, pp. 4539–4540, 1980.

[3] E. Oomura, T. Murotani, H. Higuchi, H. Namizaki, and W. Susaki, "Low threshold InGaAsP/InP buried crescent laser with double current confinement structure," *IEEE J. Quantum Electron.*, vol. QE-17, pp. 646–650, 1981.

[4] I. Mito, M. Kitamura, K. Kobayashi, S. Murata, M. Seki, Y. Odagiri, H. Nishimoto, M. Yamaguchi, and K. Kobayashi, "InGaAsP double channel planar buried heterostructure laser diode with effective current confinement," *J. Lightwave Technol.*, vol. LT-1, pp. 195–202, Mar. 1983.

[5] H. Ishikawa, H. Imai, T. Tanahashi, K. Hori, and K. Takahei, "V-grooved substrate buried heterostructure InGaAsP/InP laser emitting at 1.3-μm wavelength," *IEEE J. Quantum Electron.*, vol. QE-18, pp. 1704–1711, Oct. 1982.

[6] K. Mizuishi, M. Hirao, S. Tsuji, H. Sato, and M. Nakamura, "Accelerated aging characteristics of InGaAsP/InP buried heterostructure lasers emitting at 1.3 μm," *Japan. J. Appl. Phys.*, vol. 19, pp. L429–L432, 1980.

[7] M. Hirao, S. Tsuji, K. Mizuishi, A. Doi, and M. Nakamura, "Long wavelength InGaAsP/InP lasers for optical fiber communication systems," *J. Opt. Commun.*, vol. 1, pp. 10–14, 1980.

[8] K. Mizuishi, M. Hirao, S. Tsuji, H. Sato, and M. Nakamura, "Reliability of InGaAsP/InP buried heterostructure lasers," in *Proc. 13th Conf. Solid State Devices*, Tokyo, Japan, 1981; *Japan. J. Appl. Phys.*, vol. 21, suppl. 21-1, pp. 359–364, 1982.

[9] K. Mizuishi, M. Sawai, S. Todoroki, S. Tsuji, M. Hirao, and M. Nakamura, "Reliability of InGaAsP/InP buried heterostructure 1.3-μm lasers," *IEEE J. Quantum Electron.*, vol. QE-19, pp. 1294–1301, 1983.

[10] B. W. Hakki, P. E. Fraley, and T. F. Eltringham, "1.3-μm laser reliability determination for submarine cable systems," *AT&T Tech. J.*, vol. 64, pp. 771–807, 1985.

[11] F. R. Nash, W. J. Sundburg, R. L. Hartman, J. R. Pawlik, D. A. Ackerman, N. K. Dutta, and R. W. Dixon, "Implementation of the proposed reliability assurance strategy for an InGaAsP/InP planar mesa, buried heterostructure laser operating at 1.3 μm for use in a submarine cable," *AT&T Tech. J.*, vol. 64, pp. 809–860, 1985.

[12] H. Imai, H. Ishikawa, and K. Hori, "Aging tests of low threshold current InGaAsP/InP V-grooved substrate buried heterostructure lasers emitting at 1.3-μm wavelength," *Appl. Phys. Lett.*, vol. 41, pp. 583–584, 1982.

[13] O. Fujita, Y. Nakano, and G. Iwane, "Reliability of semiconductor lasers for undersea optical transmission systems," *IEEE Trans. Electron Devices*, vol. ED-32, pp. 2603–2608, 1985.

[14] S. Tsuji, K. Mizuishi, Y. Nakayama, M. Shimaoka, and M. Hirao, "InGaAsP/InP laser diodes mounted on semi-insulating SiC ceramics," in *Tech. Dig., 14th Int. Conf. Solid State Devices*, Tokyo, Japan, 1982, paper B-2-4.

[15] K. Mizuishi, "Some aspects of bonding-solder deterioration observed in long-lived semiconductor lasers," *J. Appl. Phys.*, vol. 55, pp. 289–295, Jan. 1984.

[16] K. Mizuishi, S. Todoroki, M. Sawai, M. Hirao, and M. Nakamura, "Statistical lifetime prediction of 1.3-μm InGaAsP/InP buried-heterostructure (BH) lasers," in *Tech. Dig., 4th Int. Conf. Integrated Opt. and Optical Fiber Commun. (IOOC)*, Tokyo, Japan, 1983, paper 29B-2-4.

[17] T. Ikegami and K. Takahei, "Reliability of light sources," in *Tech. Dig., 4th Int. Conf. Integrated Opt. and Optical Fiber Commun. (IOOC)*, Tokyo, Japan, 1983, paper 29B-2-3.

[18] A. Rosiewicz, B. R. Butler, and R. E. P. Hinton, "Temperature-dependent lifetests of IRW lasers operating at 1.3 μm," *Proc. IEEE*, vol. 132, pt. J, pp. 97–100, Feb. 1985.

[19] E. Oomura, H. Higuchi, R. Hirano, Y. Sakakibara, H. Namizaki, and W. Susaki, "Degradation mechanism in 1.3-μm InGaAsP/InP buried crescent laser diode at a high temperature," *Electron. Lett.*, vol. 19, pp. 407–408, 1983.

[20] P. D. Wright, W. B. Joyce, and C. C. Craft, "Electrical derivative characteristics of InGaAsP buried heterostructure lasers," *J. Appl. Phys.*, vol. 53, pp. 1364–1372, 1982.

[21] K. Mizuishi, M. Hirao, S. Thuji, T. Kajimura, and M. Nakamura, "Improved screening method for long-life InGaAsP/InP buried heterostructure (BH) lasers," in *Tech. Dig., 9th IEEE Int. Semiconductor Laser Conf.*, Rio de Janeiro, Brazil, 1984, paper N-2.

[22] T. Ikegami, K. Takahei, M. Fukuda, and K. Kuroiwa, "Stress tests on 1.3-μm buried heterostructure laser diode," *Electron. Lett.*, vol. 19, no. 8, pp. 282–283, 1983.

[23] M. Fukuda, "Facet oxidation of InGaAsP/InP and InGaAs/InP lasers," *IEEE J. Quantum Electron.*, vol. QE-19, pp. 1692–1698, 1983.

[24] H. Higuchi, E. Oomura, R. Hirano, Y. Sakakibara, H. Namizaki, W. Susaki, and K. Fujiwara, "Stable cw operation of improved InGaAsP/InP BC lasers at high temperatures," in *Tech. Dig., 9th ECOC*, Geneva, Switzerland, 1983, pp. 39–42.

[25] H. Nakamura, S. Tsuji, A. Ohishi, M. Hirao, H. Kakibayashi, and H. Matsumura, "TEM and IMA observation of thermal deformation of an InP surface corrugation," in *Tech. Dig., Int. Symp. Gallium Arsenide and Related Compounds*, Karuizawa, Japan, 1986.

Motohisa Hirao was born in Ibaragi, Japan, in 1938. He received the B.S. degree in metallurgy from the University of Tokyo, Tokyo, Japan.

He joined the Central Research Laboratory, Hitachi, Ltd., Tokyo, Japan in 1962. Since 1964 he has been engaged in research on III–V compound semiconductor crystal growth by vapor phase and liquid phase epitaxy. His current interests include the processing and the characterization of long wavelength optical devices such as lasers and photodetectors and integrated optical devices.

Mr. Hirao is a member of the Japan Society of Applied Physics.

Ken'ichi Mizuishi was born in Chiba, Japan, in 1948. He received the B.S. and Ph.D. degrees, respectively, in electrical engineering from Waseda University, Tokyo, Japan, in 1971, and in electronic engineering from the University of Tokyo, Tokyo, Japan, in 1984.

In 1971 he joined the Central Research Laboratory, Hitachi, Ltd., Tokyo, Japan. From 1971 to 1977 he was engaged in research on III–V compound semiconductor devices including GaAs DBD's, IMPATT diodes, and GaAs MESFET's. Since 1978 he has been engaged in reliability research of semiconductor lasers. His current interests include the R & D of assembly technologies for microelectronics and optoelectronic devices, e.g., LSI cooling modules, LD modules, hybrid optical IC's, etc.

Dr. Mizuishi is a member of the Japan Society of Applied Physics and the Institute of Electronics and Communication Engineers of Japan.

Michiharu Nakamura was born in Osaka, Japan, in 1942. He received the B.S., M.S., and Ph.D. degrees in physics from the University of Tokyo, Tokyo, Japan, in 1965, 1967, and 1975, respectively.

Since 1967 he has been engaged in research on III–V compound semiconductor devices at the Central Research Laboratory, Hitachi, Ltd., Tokyo, Japan. His current interests include semiconductor lasers, integrated optics, and high-speed GaAs LSI's. He is now the Department Manager of the Central Research Laboratory, Hitachi, Ltd.

Dr. Nakamura is a member of the Physical Society of Japan, the Japan Society of Applied Physics, and the Institute of Electronics and Communication Engineers of Japan.

ROOM-TEMPERATURE CW OPERATION OF MBE-GROWN GaInAs/AlInAs MQW LASERS IN 1·5 μm RANGE

Indexing terms: Semiconductor lasers, Quantum optics

We report the first successful room-temperature CW operation of GaInAs/AlInAs multiquantum-well (MQW) injection lasers fabricated by molecular beam epitaxy (MBE). The emitting wavelength and threshold current were 1·57 μm and 530 mA, respectively, for an SiN-defined stripe laser with a 12 μm-wide, 600 μm-long cavity.

Introduction: Recently, semiconductor lasers emitting in the 1·5 μm wavelength range are of great interest and importance for low-loss optical fibre communication systems. Most efforts so far have been concentrated on lasers based on the GaInAsP/InP material systems fabricated by liquid-phase epitaxy (LPE). The multiquantum-well (MQW) structure of $Ga_{0·47}In_{0·53}As/Al_{0·48}In_{0·52}As$ (well/barrier combination) is another promising candidate instead of GaInAsP alloy for the active layer of the 1·5 μm laser. From this point of view, the crystal growth and optical properties of the GaInAs/AlInAs MQW structure have been intensively investigated.[1,2] Few previous studies, however, have been reported on the GaInAs/AlInAs MQW laser,[3,4] where room-temperature pulsed operation has been demonstrated. In this letter room-temperature, continuous-wave (CW) operation of GaInAs/AlInAs MQW injection lasers emitting at 1·57 μm wavelength is reported for the first time.

Fabrication: A schematic diagram of a GaInAs/AlInAs MQW laser is illustrated in Fig. 1. The laser wafers were grown by conventional metal-source molecular beam epitaxy (MBE) on Sn-doped (100) InP substrates. The first layer consisted of a 0·8 μm-thick, n-AlInAs buffer layer doped with Si to $n = 1 \times 10^{18}\,cm^{-3}$. The nondoped MQW active layer was composed of ten GaInAs wells (thickness $L_z = 80$ Å) and nine AlInAs barriers (thickness $L_b = 20$ Å). Next, a 2·5 μm-thick, p-AlInAs cladding layer was grown, doped with Be to $p = 2 \times 10^{18}\,cm^{-3}$. Finally, a 0·15 μm-thick, p-GaInAs contact layer doped with Be to $p = 2 \times 10^{18}\,cm^{-3}$ was grown successively. The epitaxial growth was performed under an arsenic pressure of about 2×10^{-5} torr. The growth rate and growth temperature of the active layer were 0·8 μm/h and about 560°C, respectively. It was seen from a double-crystal X-ray rocking curve that the lattice matching between the epitaxial layer and the substrate was excellent, with $\Delta a/a \lesssim 5 \times 10^{-4}$. Details of the growth condition will be presented elsewhere.

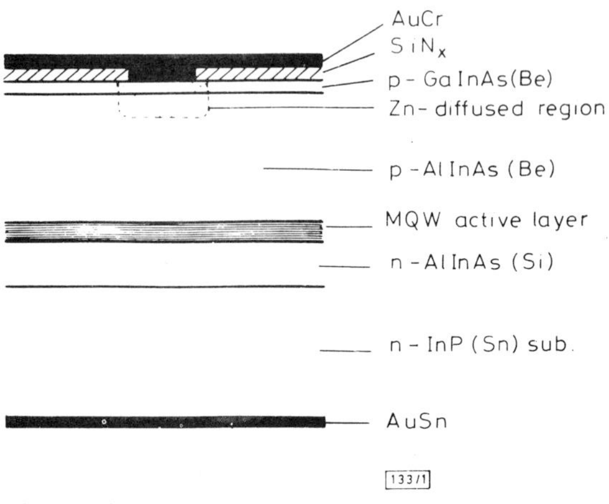

Fig. 1 *Schematic diagram of GaInAs/AlInAs MQW injection laser lattice-matched to InP substrate*

For the fabrication of current injection lasers, selective diffusion of Zn was performed at a temperature of 550°C, using a plasma-CVD SiN film as a mask. After the diffusion, ohmic contacts were formed by Au/Cr and Au/Sn films evaporated on to the p-contact layer and the n-InP substrate, respectively. The stripe width and cavity length of the laser reported here were 12 μm and 600 μm, respectively. A laser diode was mounted in p-side-up configuration on a diamond heat sink using Sn solder. These laser diodes exhibited excellent current/voltage characteristics. The reverse-bias leakage current was less than 2 mA at −5 V, and the total series resistance was about 3 Ω under forward bias.

Characteristics: CW operation of the MQW laser was successfully confirmed up to the heat-sink temperature of 31°C. Fig. 2 shows the CW light output power per facet against injection current at different heat-sink temperatures. The output power was detected by a calibrated Ge photodiode. This is the first demonstration of room-temperature CW operation of GaInAs/AlInAs MQW lasers. At a temperature of

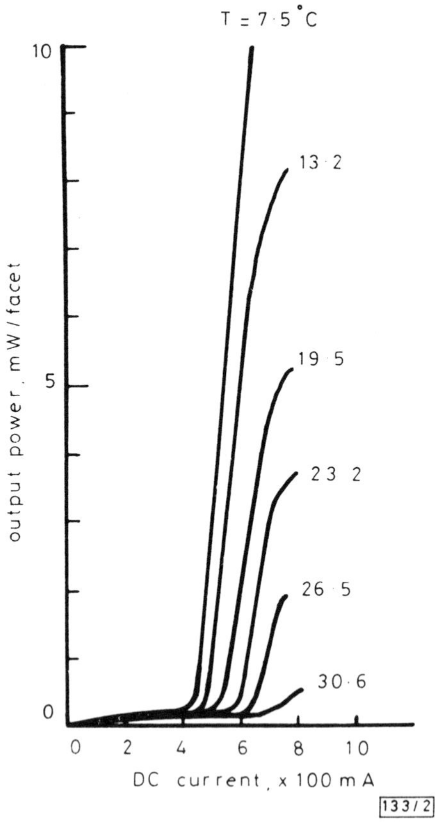

Fig. 2 *Light-output-power/injected-current characteristics of GaInAs/AlInAs MQW laser under CW operation at various heat-sink temperatures*

19·5°C, the threshold current was 530 mA, corresponding to a nominal current density of 7·4 kA/cm². This value is much lower than that of the InGaAs/InGaAlAs/InAlAs/InP SCH-MQW laser reported by Kawamura *et al.*[4] On the other hand, our obtained value is higher than that of 2·4 kA/cm² shown by Temkin *et al.*[3] However, it should be noted that they measured the threshold current density of broad-area-type MQW lasers under pulsed operation. Judging from the well known fact in oxide-stripe lasers that the injection current spreads from the contact to the active layer by approximately 2–3,[5] our obtained value is considered to be overestimated. In reality, the pulsed threshold current density was evaluated to be 2·2–3·3 kA/cm² for the effective stripe width of 20–30 μm.

The CW threshold current was approximately proportional to exp (T/T_0) with $T_0 = 62$ K, in the measured temperature range. T_0, determined from CW operation, has never been reported for 1·5 μm MQW lasers. In comparison, the T_0 of recently reported GaInAs/InP MQW lasers grown by chemical beam epitaxy (CBE) was 65–80 K under pulsed operation.[6] The external differential quantum efficiency was 4·4%/facet at a temperature of 19·5°C. An output power of more than 10 mW was achieved at a temperature of 7°C.

Fig. 3 shows the lasing spectrum of an MQW laser under CW operation at a heat-sink temperature of 18·7°C. This

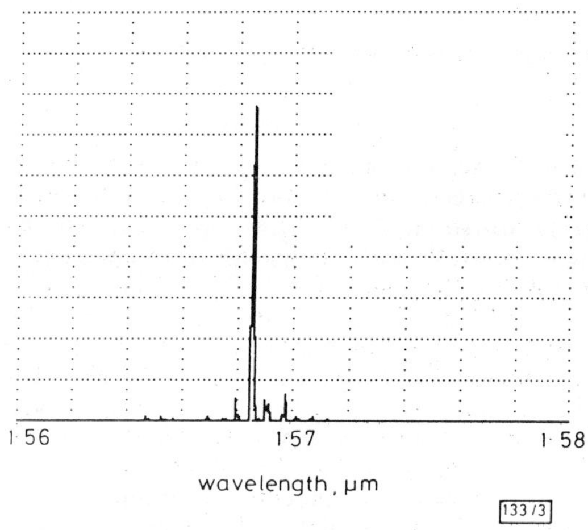

Fig. 3 *Lasing spectrum at room temperature (18·7°C) of GaInAs/AlInAs MQW laser under CW operation*

$I = 1{\cdot}15 I_{th}$

measurement was carried out using an optical spectrum analyser. A dominant peak appeared at 1·568 μm. The rather single-longitudinal-mode dominated spectrum was considered to indicate that the step-like density of states in the MQW active layer gives a narrow gain spectrum.

Conclusions: We have demonstrated the first room-temperature CW operation of an MBE-grown GaInAs/AlInAs MQW laser emitting at 1·57 μm. For the laser with an SiN stripe of 12 μm width and 600 μm cavity length, the threshold current and external differential quantum efficiency were 530 mA and 4·4%/facet, respectively, at the heat-sink temperature of 19·5°C. An output power of more than 10 mW was obtained at 7°C. The temperature-dependent coefficient of threshold T_0 was measured to be 62 K near room temperature. Further reduction of the threshold current density and detailed characterisation is necessary to develop these MQW lasers.

Acknowledgment: We are grateful to Dr. T. Muratani, Dr. T. Yamamoto, Dr. Y. Kushiro and Dr. S. Akiba of KDD for discussions and encouragement.

Y. MATSUSHIMA
K. UTAKA
K. SAKAI
O. TAKEUCHI*

KDD Meguro R & D Laboratories
2-1-23 Nakameguro
Meguro-ku, Tokyo 153, Japan

* On leave from Waseda University, Shinjuku-ku, Tokyo, Japan

28th September 1987

References

1 GOLDSTEIN, L., JEAN-LOUIS, A. M., MARZIN, J. Y., and ALLOVON, M.: 'MBE growth and characterization of $Ga_{0.4}In_{0.53}As/Al_{0.48}In_{0.52}As$ multiquantum well'. Proc. 11th symp. on GaAs and related compounds, Biarritz, 1984, pp. 133–137

2 WELCH, D. F., WICKS, G. W., and EASTMAN, L. F.: 'Luminescence line shape broadening mechanisms in GaInAs/AlInAs quantum wells', *Appl. Phys. Lett.*, 1985, **46**, pp. 991–993

3 TEMKIN, H., ALAVI, K., WAGNER, W. R., PEARSALL, T. P., and CHO, A. Y.: '1·5–1·6 μm $Ga_{0.47}In_{0.53}As/Al_{0.48}In_{0.52}As$ multiquantum well lasers grown by molecular beam epitaxy', *ibid.*, 1983, **42**, pp. 845–847

4 KAWAMURA, Y., ASAHI, H., and WAKITA, K.: 'InGaAs/InGaAlAs/InAlAs/InP SCH-MQW laser diodes grown by molecular-beam epitaxy', *Electron. Lett.*, 1984, **20**, pp. 459–460

5 THOMPSON, G. H. B.: 'Physics of semiconductor laser devices' (Wiley Interscience, Chichester, 1980), pp. 346–371

6 TSANG, W. T.: '$Ga_{0.47}In_{0.53}As$/InP double-heterostructure and multiquantum well lasers grown by chemical beam epitaxy', *IEEE J. Quantum Electron.*, 1987, **QE-23**, pp. 936–942

HIGH-POWER, WIDE-BANDWIDTH, 1·55 μm-WAVELENGTH GaInAsP/InP DISTRIBUTED FEEDBACK LASER

Indexing term: Semiconductor lasers

A high output power of 52 mW and wide bandwidth of 9·8 GHz have been achieved in a 1·55 μm-wavelength distributed feedback laser with a flat-surface buried heterostructure by optimising the coupling coefficient and reducing the parasitic capacitance.

Introduction: Distributed feedback (DFB) lasers have been developed as a light source for high-bit-rate and long-haul optical fibre transmission systems. For a DFB laser to be used at a wavelength of 1·55 μm where the transmission loss of an optical fibre is lowest, high output power and wide bandwidth are required to facilitate long-haul and high-bit-rate optical fibre transmission. In the 1·55 μm wavelength region the largest output power so far of 45 mW was reported for a BH-DFB laser with an antireflection/high-reflection (AR/HR) facet configuration,[1] and a bandwidth of 7·1 GHz was achieved in the HRO-DFB laser.[2]

We recently reported 1·3 μm-wavelength, flat-surface buried heterostructure (FBH) DFB lasers, and demonstrated high performance such as low threshold current, high efficiency, high yield for single-longitudinal-mode (SLM) operation and a very wide bandwidth of 13 GHz.[3–5] Here we have applied the structure to 1·55 μm-wavelength DFB lasers with a low-reflection/as-cleaved facet configuration and achieved a high output power of 52 mW and a wide bandwidth of 9·8 GHz. The SLM output power of 52 mW is the largest so far reported in 1·55 μm DFB lasers.

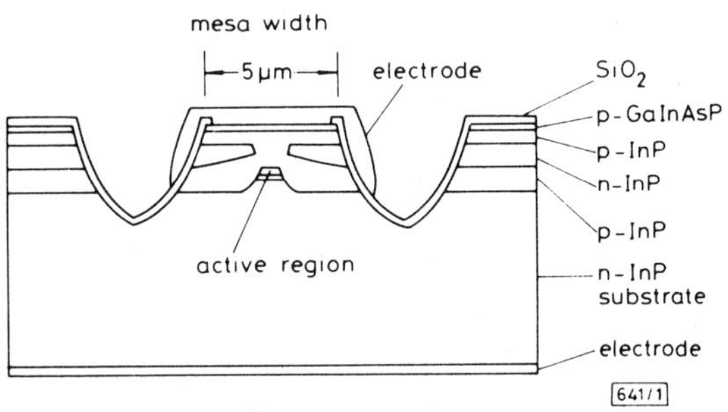

Fig. 1 *Schematic cross-section of a 1·55 μm-wavelength FBH-DFB laser*

Experiments and results: Fig. 1 shows a schematic cross-section of the 1·55 μm-wavelength FBH-DFB laser with a mesa structure. The mesa width is 5 μm. The mesa stripe is effective in reducing the capacitance associated with the *pn* junction in the current-blocking layers, and to reduce the pad capacitance we design a small-area electrode. The laser length and width are both 300 μm.

The front facet is coated with low-reflective Al_2O_3 film with a reflectivity of about 5%, and the rear facet is as-cleaved. The width of the active region is about 1 μm. A first-order corrugation is formed on an *n*-type InP substrate. The period and depth are about 200 nm and 35 nm, respectively. The depth gives the coupling coefficient of about 30 cm^{-1}; therefore $\kappa L \simeq 1$. A high yield for high-power SLM operation is expected for $\kappa L = 1$, because the light intensity distribution for $\kappa L = 1$ is very flat along the laser axis, which results in the reduction of the effect of spatial hole burning.[6]

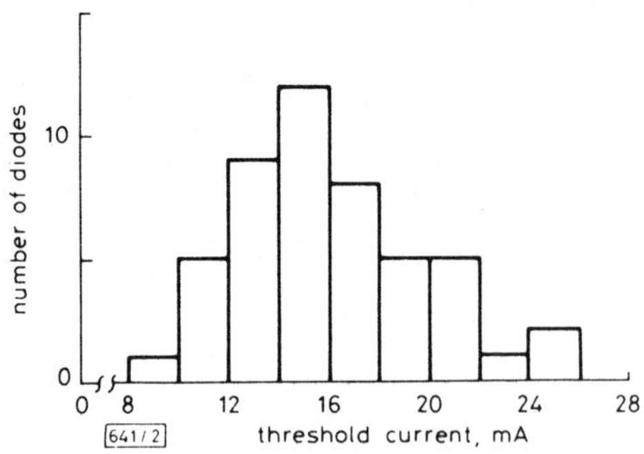

Fig. 2 *Histograms of CW threshold current of DFB lasers for SLM operation at 16 mW*

$N = 48$

127 chips were cleaved out of ten wafers at random, and the CW output-power/current characteristics and spectral characteristics were measured at 25°C. SLM operation without mode hopping up to 8 mW was obtained for 63 samples (50%) and up to 16 mW for 48 samples (38%). Figs. 2 and 3 show histograms of the CW threshold current and the external differential quantum efficiency at 8 mW, respectively, for 48 samples which showed SLM operation up to 16 mW. The lowest threshold current was 9 mA and 80% of laser diodes showed a threshold current below 20 mA. The highest external differential quantum efficiency was 29%/facet, and 77% of laser diodes showed an external differential quantum efficiency of greater than 22%/facet. Examples of the output-power/current and spectral characteristics are shown in Fig. 4. The threshold current and external differential quantum efficiency are 16 mA and 29%/facet, respectively, and SLM operation is achieved up to 52 mW without mode hopping.

The parasitic capacitance measured at 1 GHz was reduced to 8 pF owing to the 5 μm mesa structure and the small-area electrode. The series resistance was about 4 Ω. The DFB mode wavelength was tuned at the gain peak. The small-signal frequency response at 25°C is shown in Fig. 5. Bias currents are set at $2I_{th}$, $3I_{th}$ and $4I_{th}$. The −3 dB modulation bandwidth above 9·8 GHz was achieved at a bias current of $4I_{th}$. In the modulation measurement we used a low-capacitance PIN diode. The small-signal bandwidth did not increase compared with the 1·3 μm-wavelength FBH-DFB laser, even when the bias current of the laser was increased above $4I_{th}$. The frequency bandwidth of this laser is greater than 9·8 GHz.

Discussion: The external differential quantum efficiency of 1·55 μm-wavelength lasers is generally smaller than that of 1·3 μm-wavelength lasers.[7] In the case of the FBH-DFB laser, the external differential quantum efficiency is lower for 1·55 μm-wavelength lasers and is about 24%. We have achieved a high output power of 52 mW, maintaining SLM operation without an AR/HR facet configuration. The high output power is due to the efficient current confinement provided by the FBH structure. The FBH structure suppresses a thyristor phenomenon occurring at the *pnpn* InP current blocking layers, because the thickness of the current blocking layers is typically 1–1·5 μm. Owing to the thickness of current blocking layers, efficient current blocking is maintained up to a high injection current level. Maintenance of SLM operation up to 52 mW is due to the small κL of 1, which reduces the effect of spatial hole burning along the laser axis.

Conclusion: High-power, high SLM operation yield and wide bandwidth have been achieved in a 1·55 μm-wavelength DFB laser by application of the FBH structure developed in the 1·3 μm-wavelength range. A rather small κL of 1 enabled us to achieve high SLM operation yielding high power. The 5 μm

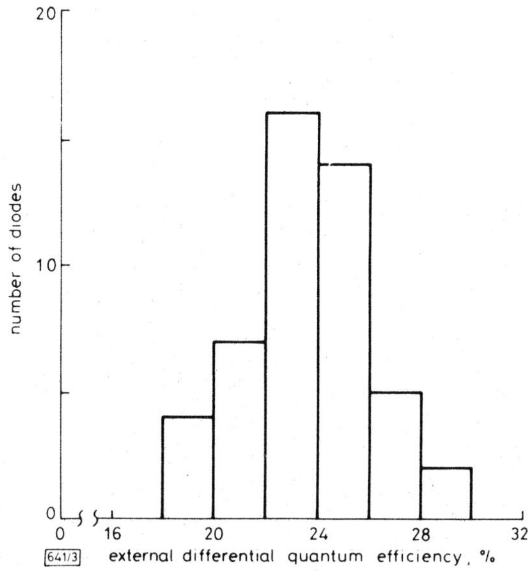

Fig. 3 *Histograms of external differential quantum efficiency at 8 mW of DFB lasers for SLM operation at 16 mW*

$N = 48$

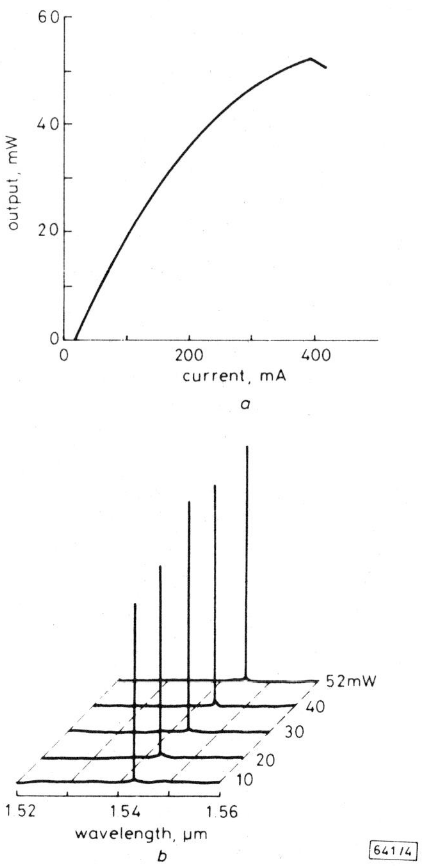

Fig. 4 *Examples of (a) high-power/current characteristics and (b) spectral characteristics*

25°C, CW

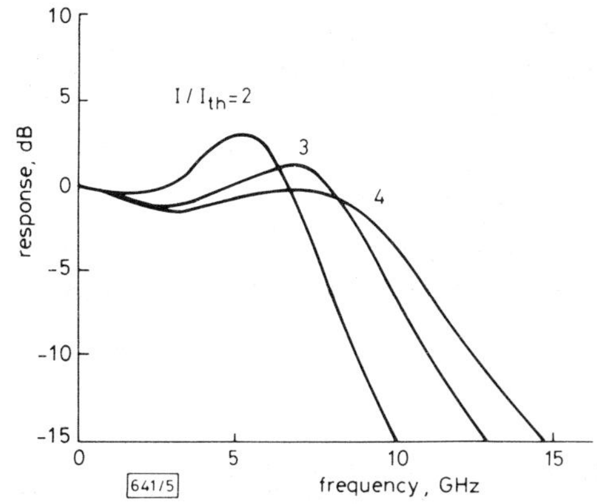

Fig. 5 *Small-signal frequency response of a 1·55 μm-wavelength DFB laser at 25°C*

mesa structure and small electrode realised a high-frequency response. High-power SLM operation is achieved up to 52 mW, owing to the efficient current blocking realised in the FBH structure and to the reduction of spatial hole burning.

Acknowledgment: We thank T. Misugi and T. Sakurai for their encouragement, and H. Soda for useful discussions. Thanks are also due to H. Sigemori and K. Sato for their technical support in this work.

K. KIHARA
K. KAMITE
H. SUDO
T. TANAHASHI
T. KUSUNOKI
S. ISOZUMI
H. ISHIKAWA
H. IMAI

Optical Semiconductor Device Laboratory
Fujitsu Laboratories Ltd.
10-1 Morinosato-Wakamiya
Atsugi, Kanagawa 243-01, Japan

29th June 1987

References

1 YOSHIDA, J., ITAYA, Y., NOGUCHI, Y., MATSUOKA, T., and NAKANO, Y.: 'Sufficiently side-mode-suppressed high-output-power 1·5 μm DFB lasers', *Electron. Lett.*, 1986, **22**, pp. 327–328

2 BOWERS, J. E., TSANG, W. T., KOCH, T. L., OLSSON, N. A., and LOGAN, R. A.: 'Microwave intensity and frequency modulation of heteroepitaxial-ridge-overgrown distributed feedback lasers', *Appl. Phys. Lett.*, 1985, **46**, pp. 233–235

3 ISHIKAWA, H., YAMAKOSHI, S., and ISOZUMI, S.: 'Distributed feedback laser emitting at 1·3 μm for high-bit-rate systems', *Fujitsu Sci. & Tech. J.*, 1986, **22**, pp. 451–460

4 ISHIKAWA, H., SODA, H., WAKAO, K., KIHARA, K., KAMITE, K., KOTAKI, Y., MATSUDA, M., SUDO, H., YAMAKOSHI, S., ISOZUMI, S., and IMAI, H.: 'Distributed feedback laser emitting at 1·3 μm for gigabit communication systems', *J. Lightwave Technol.*, 1987, **LT-5**, pp. 848–855

5 KAMITE, K., SUDO, H., YANO, M., ISHIKAWA, H., and IMAI, H.: 'Ultra high speed InGaAsP/InP DFB lasers emitting at 1·3 μm wavelength', *IEEE J. Quantum Electron.*, June 1987, **QE-23**, to appear

6 SODA, H., ISHIKAWA, H., and IMAI, H.: 'Design of DFB lasers for high-power single-mode operation', *Electron. Lett.*, 1986, **22**, pp. 1047–1049

7 KOBAYASHI, K., and MITO, I.: 'High light output-power single-longitudinal-mode semiconductor laser diodes', *J. Lightwave Technol.*, 1985, **LT-3**, pp. 1202–1210

Chapter 4
Photodetectors and Receivers

During the initial stages of development in optical communication, the photodetector was not perceived to be a major problem. The technology for making silicon (Si) photodiodes with high quantum efficiency and low noise was already well advanced, having been developed for applications other than fiber optics. Both PIN detectors and avalanche photodiodes (APDs) in Si performed quite well with 0.8 μm light sources. In the mid–1970s, when the potential importance of the longer wavelength region began to be appreciated, the performance of semiconductor photodetectors was relatively poor at wavelengths beyond 1 μm. The most suitable 1.3 μm detectors at that time were germanium (Ge) photodiodes, but the dark current (and hence the shot noise) in Ge photodiodes is much higher than in Si. The comparison is even less favorable for Ge APDs because the multiplication excess noise factor is also higher than for Si. The trend in recent years has been toward the use of photodiodes fabricated in III–V semiconductors for the longer wavelength region. Best performance has been achieved in InGaAs PIN photodiodes, and in InGaAsP APDs.

An overview of photodetectors and associated electronics is given in **"Optical Receivers for Lightwave Communication Systems,"** by **M. Brain and T.P. Lee.** The technology of photodetectors for fiber optic systems is also covered in the first two papers of Chapter 3. Recent developments in photodetectors and receivers for the long–wavelength region are described in **"Planar Embedded InP/GaInAs p-i-n Photodiode for Very High-Speed Operation,"** by **S. Miura, H. Kuwatsuka, T. Mikawa, and O. Wada,** and **"Multigigabit-per-Second Avalanche Photodiode Lightwave Receivers,"** by **B.L. Kasper and J.C. Campbell.**

Optical Receivers for Lightwave Communication Systems

MIKE BRAIN AND TIEN-PEI LEE

(Invited Paper)

Abstract—**As long-wavelength optical telecommunications systems are now being installed in countries throughout the world, this is an opportune time to review the successful developments in detectors and receivers for this application. We given prominence to receivers using p-i-n and avalanche photodiodes, describing in some detail the principles of their operation as well as the details of their development. However, we also consider bipolar phototransistors, photoconductive detectors, and the receiver requirements for coherent optical systems, concluding with a broad comparison of the various receiver designs.**

I. Introduction

IT IS NOW almost twenty years since optical fiber was proposed as a wider-bandwidth and potentially cheaper communication medium than coaxial cable and radio [1]. Today long-haul high-bit-rate optical telecommunications systems are in operation in a number of countries. For example, the U.K. trunk and junction telephone networks are equipped with 1.3-μm wavelength systems carrying traffic at bit rates between 8 and 565 Mbit/s; Japan has a 400 Mbit/s system extending over 2000 km; and while long-haul installations in the US also operate mainly at 400 Mbit/s, a comparable quantity of fiber is being installed in metropolitan areas for operation at bit rates of 90–565 Mbit/s (indeed, a further market is emerging in optical local area networks, providing high bandwidth at low cost within campus sites or areas within cities). Administrations elsewhere in the world also have systems in operation, with many more announcing their intentions to install optical fibers or move to the longer operating wavelength of 1.55 μm during the next few years. With regard to undersea systems [2], a submarine optical cable containing three fiber pairs to link the UK with Belgium has been announced, while the first transatlantic optical system (TAT8) is scheduled for 1988; the information rate for each fiber in these systems will be 280 Mbit/s [2]. This paper reviews the progress in photodetectors and receivers that has contributed to these successful applications of optical technology.

The first systems were developed for operation at a wavelength of 0.85 μm [3]: laser and LED sources in GaAs/GaAlAs were complemented by silicon photodetectors which proved to be well-suited to use in optical communications receivers. Because electrons and holes in silicon have very different ionization rates [4], avalanche photodiodes (APD's) of this material were able to provide large values of photocurrent gain (with good uniformity over the photosensitive area) for only a small increase in overall receiver noise [5], [6]. Since the optical absorption coefficient of silicon is low at 850 nm, a "reach-through" APD structure (resembling the IMPATT diode used in microwave systems) was developed to extend the depletion region to a width of typically 30–100 μm, to provide a high quantum efficiency and fast response [7]; and with very low leakage currents, such detectors provided excellent receiver performance at the bit rates then being considered [8].

The course of further developments was determined largely by the characteristics of silica fiber at 1.3-μm wavelength: pulse dispersion is at a minimum and fiber loss is as low as $\frac{1}{2}$ dB/km, falling further to $\frac{1}{4}$ dB/km at 1.55 μm [9]. Transmission over longer distances and at higher bit rates would therefore become possible by moving to a longer operating wavelength, but would require sources and detectors in a new materials system. Although germanium photodetectors could be used at 1.3 μm [10], they exhibit high leakage currents and have only a limited spectral response at 1.55-μm wavelength. In contrast, III-V quaternary compounds offered the prospect of a direct bandgap, wide enough to give a lower bulk-leakage current density, yet narrow enough to give a good quantum efficiency and fast response at 1.55-μm wavelength from a narrow depletion region [11]. This paper describes how the properties of these semiconductor materials have influenced the performance of photodetectors, and hence the design of optical receivers for operation at these wavelengths.

Of the III-V materials suitable for sources and detectors operating at 1.3 and 1.55 μm, the InGaAsP/InP system has received most attention with the ternary compound $In_{0.53}Ga_{0.47}As$ being chosen for photodetector applications. Although p-i-n photodiodes in this material have a lower leakage current than those in germanium, APD's in InGaAs exhibit a very high bulk leakage current density due to band-to-band tunneling at high electric field strength

Manuscript received July 1, 1985; revised August 2, 1985.
M. Brain is with the British Telecom Research Laboratories, Martlesham Heath, Ipswich IP5 7RE, U.K.
T.-P. Lee is with Bell Communications Research, Murray Hill, NJ 07974.

[12], leading to poor receiver performance. Much early attention was therefore devoted to the development of p-i-n photodiodes for use with low-noise GaAs MESFET preamplifiers, since such receivers would not only be simpler to operate, but could also offer better sensitivity than those using APD's [13]. These developments are described in Section II.

Section III describes how new structures have been devised for APD's in both germanium and InGaAs to overcome their earlier limitations. A reach-through structure for Ge APD's has extended their range of operation to beyond 1.55 μm [14], while the need for monomode fiber in high-bit-rate transmission systems has allowed the sensitive area to be reduced to 30 μm [15], giving a substantial reduction in leakage current. Tunneling in InGaAs APD's is avoided by placing the high-field region in InP [16], and consequent loss of frequency response through carrier accumulation at the step in the valence band has been minimized by careful design of the hetero-interface with the InGaAs absorption layer [17], [18]. Although such devices are extremely difficult to make and more complicated to use than p-i-n photodiodes, they do provide the best sensitivity in direct detection systems and have been demonstrated at bit rates up to 4 Gbit/s [19].

Bipolar phototransistors [20] and photoconductors [21] have also been considered for use in optical receivers, since they appear to offer a low-noise photocurrent gain mechanism and a structure which effectively integrates the photodetector with the first stage of the preamplifier. Theoretical analyses suggest [22], [23], however, that their performance would more closely resemble that of receivers based on p-i-n photodiodes than APD's, although they may offer advantages for certain applications. These are discussed in Section IV.

The above comments address direct-detection communication systems, in which the photodetector responds merely to the intensity of the incident optical signal. The attraction of exploiting the very low fiber loss at 1.55 μm has stimulated the development of laser sources with very narrow spectral linewidth [24] to minimize fiber dispersion and thus allow long-distance transmission at very high bit rates. This in turn has enabled the development of coherent optical transmission techniques which are discussed in Section V. The concluding section then reviews the state of the art and future prospects.

II. Receivers Using p-i-n Photodiodes

Present optical telecommunications systems have been developed mainly for long-distance transmission at relatively high bit rates (e.g., 34 Mbit/s and above). To allow large repeater separations, therefore, receivers have been designed to have a good sensitivity, that is, to require only a very low level of mean received optical power while still operating within the maximum specified bit error rate (BER). Receiver sensitivity is determined by the signal-to-noise ratio (SNR) at the input to the decision circuit in the regenerator [6]: since the noise is dominated by thermal noise sources in the preamplifier, the sensitivity can be improved either by applying gain within the detector, as in an APD, or by reducing the noise of the preamplifier, as in the p-i-n FET hybrid [13]. The best results are obtained by applying both approaches, especially where a large excess-noise factor is associated with the gain of an APD [25], but this section will concentrate on the design and performance of low-noise receivers using p-i-n photodiodes. For applications in junction networks, that is, at lower bit rates and over shorter distances, dynamic range assumes a greater importance; the conflicting requirements for good sensitivity and wide dynamic range in a single receiver design will be discussed.

A. Receiver Sensitivity

Personick [6] has shown that an integrating front-end design can improve the sensitivity of both APD- and p-i-n-based receivers, since this allows the spectral density of Johnson-noise sources at the input to be minimized. Simplifying assumptions have been applied to his analysis by Smith and Garrett [26], yielding the following expression for the sensitivity $\bar{P}$ of a receiver using a p-i-n photodiode, operating at a wavelength λ with a quantum efficiency η, for received pulses at bit-rate B:

$$\bar{P} = \frac{hcBQ}{2\lambda\eta}(QI_2 + 2\sqrt{Z}) \tag{1}$$

where

$$Z = \frac{I_2}{q^2B}\left[S_I + \frac{S_E}{R^2}\right] + \frac{(2\pi C)^2}{q^2} S_E B I_3. \tag{2}$$

In this expression, $\bar{P}$ is the mean received optical power level assuming equal numbers of ON and OFF pulses, h is Planck's constant, c is the velocity of light, q is the electronic charge, and Q is determined by the BER ($Q = 6$ for a BER of 10^{-9}).

The integrating front-end is formed by the photodiode bias resistance R and the input capacitance C. I_2 is the double-sided noise equivalent bandwidth of the receiver, normalized to the bit rate, and is used in conjunction with those noise sources for which the frequency response of the receiver is the same as for the received signal, principally the Johnson noise of the bias resistance R and the shot noise on the leakage and bias currents I_L at the input of the receiver. Together, these noise sources comprise the equivalent input mean-square noise current spectral density S_I

$$S_I = \frac{2kT}{R} + qI_L \tag{3}$$

where T is the absolute temperature and K is Boltzmann's constant.

Having integrated the received signal, the pulse waveform is recovered by a subsequent compensating stage in the receiver, as shown in Fig. 1. There are further noise sources within the preamplifier which are subject to this compensation, but were not subject to the integrating action of the receiver input: together they constitute the mean-square noise voltage spectral density S_E, to which the bandwidth integral I_3 applies. In the case of the p-i-n

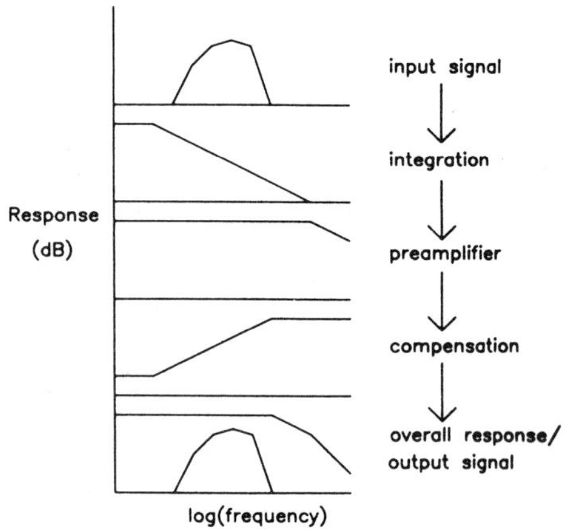

Fig. 1. Schematic illustration of the frequency response of each stage in an integrating receiver.

FET, S_E is dominated by the channel noise of the FET (characterized by the coefficient Γ), but also includes the Johnson noise of the drain resistance R_d and other noise sources associated with the preamplifier output buffer stage (all of which are referred back to the input of the receiver), as follows:

$$S_E = \frac{2kT\Gamma}{g_m} + \frac{2kT}{g_m^2 R_d} + \text{other terms.} \tag{4}$$

Since compensation to recover the original pulse waveform imposes a rising frequency characteristic on these noise sources, the latter become more significant at higher bit-rates and result in the sensitivity of p-i-n FET receivers varying at a rate of 4.5 dB/octave. We note also that S_E contributes to the term in I_2 at low frequencies $\omega RC << 1$, where the high-impedance front-end is non-integrating.

The second-stage amplifier also contributes noise that is subject to neither of the preceding integration and compensation stages, but may still be expressed in terms of I_2 and I_3 through being referred back to the input of the receiver. This will have little effect on the output SNR if the gain of the preceding FET stage is sufficiently high. This suggests a high value for the FET load resistance R_d, and a cascode configuration is therefore adopted for the preamplifier.

From the above expressions, it can be deduced that the p-i-n FET hybrid [27]–[30] offers a better sensitivity than other forms of p-i-n-photodiode receiver. A good sensitivity at higher bit rates requires a low input capacitance C: hybrid integration minimizes contributions from packages, and the sum of the capacitances of the p-i-n, FET, and the bias resistor R can be made less than 0.5 pF. For lower bit rates, the dark current I_L can be a dominant source of noise; hence the p-i-n and FET should be designed to have low leakage currents. The Johnson noise at the input can be made negligible by making R sufficiently large: this also tends to position the first pole of the frequency response below the range of frequencies being transmitted, greatly simplifying the compensation stage.

In comparison, a bipolar transistor is a cheaper component than a FET and offers higher values of transconductance g_m, but at the expense of a higher level of noise (arising from the shot noise on the base current) and a higher input capacitance [31]. The higher gain available using bipolars could, however, be exploited in a p-i-n-bipolar transimpedance preamplifier, whose input capacitance can be reduced by using a common-collector input stage [32]. Although the transimpedance requires a smaller value for R (and the additional Johnson noise would be significant), the main advantage of this non-integrating type of receiver would be a wider dynamic range, as will be discussed later.

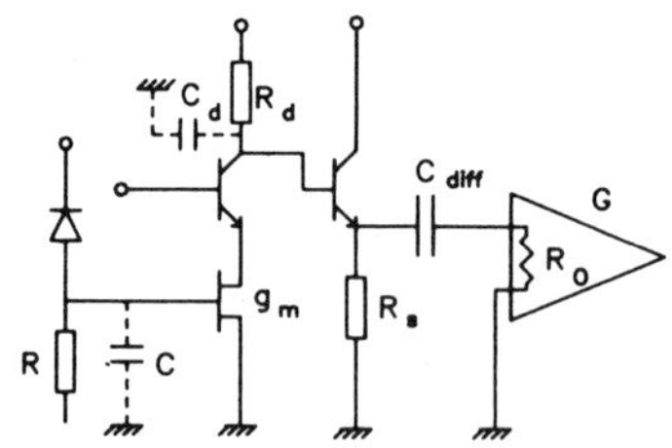

Fig. 2. Simplified circuit layout for high-impedance p-i-n FET receiver with FET/bipolar cascode preamplifier.

B. Frequency Response

To provide good sensitivity, the response of a digital optical receiver must be uniform for the range of frequencies over which the transmitted power is distributed. For bandwidths below 300 MHz, most of the parasitic impedances within the standard 14-pin DIL metal package have a negligible effect. The frequency response of the high-impedance p-i-n FET hybrid shown in Fig. 2, which comprises the integrating front-end, FET preamplifier, differentiator, and second-stage amplifier, can then be written as follows:

$$V_{\text{out}} = \frac{I_{\text{in}} R}{(1 + sCR)} \cdot \frac{g_m R_d}{(1 + sC_d R_d)} \cdot \frac{sC_{\text{diff}}}{(1 + sC_{\text{diff}} R_0)} \cdot G \tag{5}$$

where $s = j\omega = j2\pi f$. Here, V_{out} is the output voltage at frequency f from the second-stage amplifier, the latter having a voltage gain G and having an input impedance R_0 which together with the capacitance C_{diff} forms the differentiator. The photocurrent I_{in} for a received optical power P is given by

$$I_{\text{in}} = P \cdot \frac{\lambda\eta}{hc} q. \tag{6}$$

R_d should be large to minimize the effects of second-stage noise, but not so large as to limit the receiver bandwidth in combination with C_d, the total capacitance at the output of the FET stage. If we assume, therefore, that $sC_d R_d << 1$ and $sCR >> 1$ for all transmitted frequencies, the above expression for V_{out} reduces to the following:

$$V_{\text{out}} = I_{\text{in}} \frac{C_{\text{diff}}}{C} \cdot \frac{g_m R_d G}{(1 + sC_{\text{diff}} R_0)}. \tag{7}$$

Thus, the high-frequency roll-off is determined by the differentiator, and a uniform response at lower frequencies requires all the parameter values in the above equation to

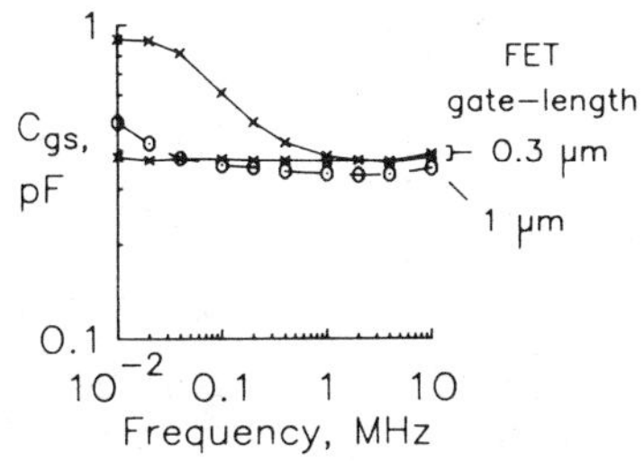

Fig. 3. Gate–source capacitance C_{gs} measured as a function of frequency for GaAs MESFET's, showing the effect of trapping centers at the junction.

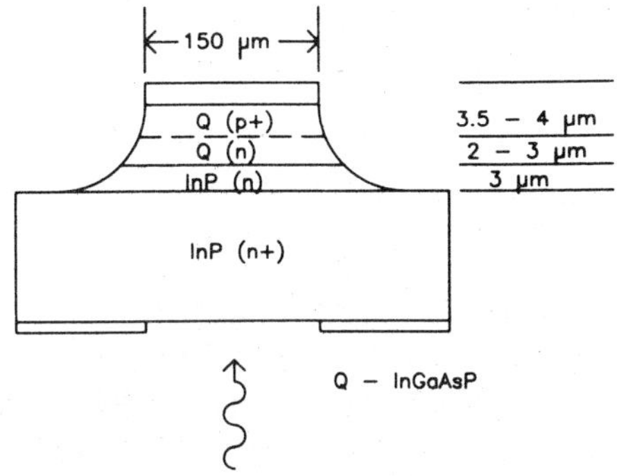

Fig. 4. Substrate-entry p-i-n photodiode [37].

be independent of frequency. Chip components satisfying this requirement are available for C_{diff} and R_d, and monolithic silicon amplifier chips provide uniform G and R_0. However, trapping centers at the gate–source junction of the FET appear electrically as a series RC combination in parallel with C_{gs}, causing C to increase at frequencies below a few megahertz. Fig. 3 shows capacitance/frequency measurements for a 1-μm dual-gate MESFET with little evidence of traps, and for submicrometer MESFET's from different batches with and without traps. Thus, FET's for use in integrating receivers must be free from such effects if loss of sensitivity due to pattern-dependent variations in the output pulse-height is to be avoided.

C. p-i-n FET Receiver Development

Having indicated how the device and circuit parameters influence the sensitivity and frequency response, this section describes the performance achieved in practice.

1) p-i-n Photodiodes: High-sensitivity optical receivers require photodiodes with a small junction area, typically a few tens of micrometers in diameter, and a low-doped depletion region, usually around 3-μm wide. The latter provides the required high quantum efficiency (QE) and low capacitance, without needlessly restricting the speed of response through excessive carrier transit time. Low doping permits full depletion at a low reverse bias voltage, and minimizes the tunneling leakage current. Since InGaAs junction devices are sensitive to moisture, the photodiode must be passivated, and/or the package hermetically sealed, to ensure long-term stability.

Although the bandgap and the wavelength of peak response of an InGaAsP detector could be optimized by appropriate choice of alloy composition [33], most long-wavelength photodetectors in this system have in fact been developed using $In_{0.47}Ga_{0.53}As$ [11], which is lattice-matched to InP and responds to wavelengths up to around 1.7 μm. Devices made by defining mesas on LPE-grown material have been reported with doping concentrations in the low 10^{15} cm^{-3} range [34]–[36]. The mesa provides both the sensitive area and a metallized area for the top contact, but as Fig. 4 shows, these functions can be performed by the same area if the junction is illuminated via an n-type substrate, thus minimizing the junction capacitance and giving values below 1 pF [37]. The responsivity is also high and relatively wavelength-independent since the carriers are no longer subject to surface recombination: quantum efficiencies in the range 0.5–0.7 without antireflection coatings are typical, with rise-times of around 100 ps or less. Such devices offer repeaterless transmission over 24 and 8 km at 45 and 274 Mbit/s, respectively, using a 1.3-μm wavelength LED source [38].

The leakage current in InGaAs is dominated by band-to-band tunneling at high reverse-bias voltages, and generation–recombination (GR) at lower voltages [39], [40]; most authors have reported best results for leakage current density in the range 2–5 $\times$ 10^{-5} A · cm^{-2} at 10 V reverse bias.

Although this is low enough for many purposes, further reduction has been demonstrated by forming the junction in a higher bandgap transparent material lattice-matched to the ternary absorbing region. The leakage current density in a top-entry photodiode was reduced to 1.5 $\times$ 10^{-6} A · cm^{-2} at 10 V reverse bias by locating the junction in an InGaAsP capping layer [41], and to 2.6 $\times$ 10^{-7} A · cm^{-2} at 1 V reverse bias in a substrate-entry device with the junction in InP [42]. Although the latter result required a largely undepleted absorbing layer, the high carrier mobility obtained by using p-type material still allowed operation at frequencies in excess of 2 GHz [43].

Thus, mesa devices on LPE material can be produced relatively simply with good results, but the questions of mass production and reliability must also be addressed. MBE has yielded devices with comparable doping concentrations but higher leakage-current density [44], and is an expensive process which is particularly advantageous for growing more complex structures such as superlattices. However, VPE kits can accommodate larger substrates, with possible improvements in the purity of the grown layers [45], [46]; and VPE has produced mesa devices which compare well with the LPE detectors described above, and have demonstrated a stable leakage current [47]–[49]. Junctions in InGaAs are known to be moisture-sensitive [50] (although stability is improved if the final etching process leaves the surface rich in As [51]), and while passivation of the exposed junction region using photoresist has shown short-term effectiveness [52], [53], the long-term stability of organic passivations is suspect, and planar structures are expected to offer better reliability by analogy with other materials systems [45], [54]. Unpassivated planar devices with p-contact metallizations in TiPt and TiAu have demonstrated stability [55], but attempts to passivate InGaAs using SiO_2 resulted in increased leakage current [45]. Mesa and planar methods have therefore been combined in a new structure in which the surface to be passivated is in InP (for which SiO_2 is suitable) rather than InGaAs, but the junction area is then larger than the sensitive area, resulting in a higher capacitance [56]. Recent work has demonstrated, however, that multilayer passivations using SiN_x can be effective [57].

MOVPE is also being developed for this application: devices grown at low pressure [58] already compare with those from LPE and VPE, and atmospheric-pressure growth systems are close to yielding good results [59].

To summarize, therefore, LPE and VPE have produced photodiodes of comparable performance in terms of low capacitance, low leakage, high QE and fast response. MOVPE may offer significant advantages for mass production [60]. To prevent impairment due to moisture, however, a hermetic package for the photodiode may be necessary.

2) Hybrid-Integrated p-i-n FET Receivers: p-i-n FET receivers have been assembled using individually packaged components to give sensitivities at 1.3 μm of -46.7 dBm at 44.7 Mbit/s and -36 dBm at 274 Mbit/s, but these values are strongly influenced by the relatively high input capacitance of 1.5–2.5 pF [36]. This could be reduced significantly by hybrid integration, as was demonstrated using a silicon p-i-n photodiode and a GaAs MESFET, hybrid integrated on a thick-film alumina substrate to give an input capacitance of 0.7 pF: the receiver sensitivity at 0.85-μm wavelength was -46 dBm at 160 MBd [61]. This section describes the design and performance of hybrid circuits for operation at up to 1.2 Gbit/s.

Longer wavelength operation was provided first at 1.24 μm [27], then at 1.3 and 1.5 μm [28], using mesa-geometry InGaAs p-i-n photodiodes prepared from VPE-grown material. A substrate-entry p-i-n/fiber subassembly was then developed to minimize junction capacitance. Comprised of a quartz cube with an integral fiber pigtail, it provided a simple means of actively aligning the InGaAs/InP photodiode with the fiber core to give very high coupling efficiency, while the lower dielectric constant of quartz reduced the stray capacitance [62]. With the total input capacitance reduced to less than 0.5 pF, p-i-n FET hybrid receivers were able to demonstrate better sensitivities at up to 565 Mbit/s than the Ge APD's then available [29], [63], with very little variation with temperature. These early p-i-n FET hybrids used a FET/bipolar cascode preamplifier which provided a -3-dB bandwidth of 350 MHz. Broadly comparable commercial products are now available in high-impedance and transimpedance form from a number of manufacturers.

For bandwidths greater than 350 MHz, the FET/bipolar cascode presents too large a value of drain capacitance C_d, and is replaced by a FET/FET cascode stage. Also, bond-wire inductances and metallization capacitances can have a considerable effect on the response of the receiver at these frequencies, and greater care is needed in the initial design of the hybrid circuit layout and in the grounding of the package to the PCB. For this reason, the p-i-n FET hybrid developed for 1.2 Gbit/s operation contained the differentiator and second-stage amplifier (in chip form) within the 14-pin DIL package [30]. Having developed the required layout, stable and predictable results have been obtained for receivers using commercial 1-μm-gate-length GaAs MESFET's, in which the second gate of the cascode is grounded within the package. The receiver sensitivity for a 10^{-9} BER is typically -33 to -34 dBm at 1.528 μm for NRZ pulses in a 2^{15}–1 PRBS at 1.2 Gbit/s, measured using a parallel-processing regenerator. The sensitivity of a similar receiver operating at 650 Mbit/s was -38 dBm at 1.3 μm, using a 2^7–1 PRBS. These results agree well with theoretical expectations.

A better sensitiviy has been obtained by using commercial submicrometer-geometry GaAs MESFET's in a high-impedance p-i-n FET hybrid. Compared with the 1-μm-gate-length MESFET's, these devices have a comparable gate–source capacitance C_{gs} (0.37 pF), but the higher transconductance of 45 mS improved the receiver sensitivity to -36.5 dBm at 1.2 Gbit/s and 1.528-μm wavelength, measured in the above-described manner.

Since C_{gs} dominates the input capacitance C, which in turn determines both the receiver noise level and the uniformity of the frequency response, we can now see that the performance of state-of-the-art p-i-n FET hybrids depends mainly on the FET. In Section II-B and Fig. 3, we noted the effect on C_{gs} of trapping centers at the gate–source junction. This gives rise to penalties in the receiver sensitivity measured at 1.2 Gbit/s with presently available commercial test equipment, because the latter uses a 2^{15}–1 PRBS which has a signal disparity of 251 and hence a significant spectral power content in the megahertz range, extending down to frequencies of a few tens of kilohertz. Similar penalties could clearly arise in practical systems, but could be avoided by using a bounded-disparity line transmission code: this would also improve the dynamic range of the receiver and ensure a strong spectral line for timing extraction purposes.

Thus, the high-impedance p-i-n FET hybrid presently offers a better sensitivity than other p-i-n photodiode-based receivers, but obtaining such performance reproducibly depends critically on the performance of commercially available submicrometer-geometry GaAs MESFET's. Equally, further improvement to the sensitivity of these receivers will depend on progress in FET development, both through novel structures (e.g., superlattices [64]) or new materials such as GaInAs or InP. Progress in FET and monolithically integrated p-i-n FET structures in these materials will be described next.

3) FET's and Monolithically-Integrated p-i-n FET's in InGaAs/InP: Because they provide a higher carrier mobility or a higher drift velocity, InGaAs or InP may yield FET's with a higher transconductance than GaAs. This in turn may lead to monolithic integration of the p-i-n photodiode with the FET, with the attendent advantages of lower cost, better reliability, and smaller physical size. The absence of bond–wire inductance and bond–pad capacitance may also be advantageous for transimpedance receivers operating at high frequencies.

The feasibility of monolithic p-i-n FET integration in InGaAs was first demonstrated in 1980 by Leheny *et al.*, who produced a 20-μm gate-length JFET with a transconductance of 1 mS [65]. Kasahara has more recently reported an InGaAs p-i-n photodiode integrated with an InP MISFET, giving up to 18 mS for a gate-length of 1.4 μm and a width of 400 μm; the sensitivity of this receiver at 100 Mbit/s was -34.5 dBm at 1.3-μm wavelength, for a QE of 0.51 and an input capacitance of 2.9 pF [66].

The main problems in producing good FET's in these new materials are in realizing the necessary high mobility in a thin channel layer, and in translating this into a high transconductance by means of a short gate-length. Schottky junctions in these materials have a low barrier potential, which is why JFET and MISFET structures have been pursued in preference to the MESFET structures used in GaAs. However, a value of 300-mS/mm gate-width has recently been reported for a self-aligned 1.5-μm-gate-length MISFET formed in MBE InGaAs, using PECVD silicon nitride as the gate insulator [67]. This falls to 150 mS/mm when a thin layer of e-beam-evaporated SiO_2 is used instead to form an enhanced Schottky-gate FET; but the device processing is much simpler, and there is still scope for increasing the transconductance [68]. A smaller value of 110 mS/mm has been reported for a 1-μm gate-length InP MESFET [69]. (For comparison, the transconductance of the commercial GaAs MESFET used in [30] was 150 mS/mm).

Thus, although the results reported for integrated p-i-n FET receivers are still some way behind those for the p-i-n FET hybrid, the results for FET's suggest that comparable performance may eventually be obtained more cheaply.

D. Dynamic Range

The dynamic range of a receiver is the range of input power levels over which the BER is acceptable. For some applications, the receiver is required to operate satisfactorily when the received signal is close to the launched power level, but such operation depends on the ability of each stage of the receiver to accommodate large variations in voltage or current. For an integrating receiver, it also depends on the characteristics of the received signal. This section therefore illustrates the influences on dynamic range, and demonstrates non-rigorously the conflict with the requirements for good sensitivity outlined above, using parameter values for the high-impedance p-i-n FET receiver developed for 1.2 Gbit/s [30].

The first limitation on dynamic range arises at the receiver input. If the received signal contains equal numbers of ON and OFF pulses, and the photocurrent I due to a peak optical pulse power P is as given by (6), then the mean dc voltage $RI/2$ across the photodiode bias resistance R must be less than the available dc offset voltage across R, typically 15 V. Thus for $R = 1\ \mathrm{M}\Omega$, we require $I < 30\ \mu$A. Now $\overline{P} = -36.5$ dBm corresponds to $I = 0.44\ \mu$A, which suggests a dynamic range of 18 dB; this is close to the value of 20 dB measured using a 24-bit test pattern. This could be improved by reducing R, provided the increase in Johnson noise and possible complication of the frequency compensation circuitry could be tolerated.

The next limitation to dynamic range applies to the signal voltage V_{op} at the output of the cascode stage. The voltage response of the integrating front-end to a current step function of amplitude I is given as a function of time t by

$$V = RI[1 - \exp(-t/RC)]. \tag{8}$$

If the bit period is T' and the mean input voltage is $RI/2$, then for an excursion in disparity n from the mean value, the input signal voltage V_{in} is given by

$$V_{\text{in}} = \frac{RI}{2}[1 - \exp(-nT'/RC)]. \tag{9}$$

If, for simplicity, we assume that the maximum positive and negative excursions from the mean value are equal to $N/2$, where N is the digital sum variation (DSV) for the PRBS, and that $NT' << RC$ (to allow compensation by simple differentiation), then expanding the above gives

$$V_{\text{op}} = g_m R_d V_{\text{in}} \simeq \frac{g_m R_d}{4C} \cdot NT'I \tag{10}$$

This must be less than the dc bias voltage V_{FET} at the cascode drain (for negative excursions from the mean), and across the drain resistance R_d due to the drain current I_d (for positive excursions). The latter gives

$$I < \frac{4CI_d}{g_m NT'} = I_{\max}. \tag{11}$$

For $I_d = 15$ mA and $N = 251$ (corresponding to a 2^{15}–1 PRBS), this gives $I_{\max} = 3\ \mu$A, implying a dynamic range of 8.3 dB relative to $I = 0.44\ \mu$A. This agrees reasonably well with the measured value of 9 dB.

This condition implies that I_d should be as large as possible. Remembering, however, that the negative excursions must be accommodated within V_{FET}, there is no point in having $R_d I_d > V_{\text{FET}}$, and R_d may need to be large to avoid sensitivity penalties due to second-stage noise.

We thus see the conflicting requirements for wide dynamic range and good sensitivity. Nine decibels is already a small value for dynamic range, and from (11) will be further diminished if C/g_m is reduced to improve sensitivity. Choosing R_d according to dynamic range requirements may also incur a sensitivity penalty. Since there is little scope for varying I_d to improve the dynamic range, the remaining options are as follows:

1) Limit N: The disparity cannot be allowed to be totally unbounded when using an integrating receiver, if only because of the response limitation imposed at low frequencies by dc-blocking capacitors. Moreover, efficient timing extraction suggests some limitation on the low-frequency content of the transmitted signal. Coding of the signal may therefore be needed to guarantee efficient regeneration, and could then be chosen to restrict N to a small value, giving a much bigger dynamic range.

2) Nonlinear R_d: Using a diode as the drain load would provide a low load to the signal while generating a large voltage drop from the dc bias current I_d. For large-disparity signals, however, this same nonlinearity may cause the equalized output pulse height to be pattern dependent, and would therefore need to be compensated elsewhere if penalties in sensitivity were to be avoided.

3) Input current shunt: For large values of input current I, the dynamic range could be extended by shunting a proportion of I away from the photodiode bias resistance R [70].

4) In-band equalization: Compensation using a simple capacitive differentiator requires $NT' << RC$; if the converse condition were applied, i.e., $NT' >> RC$ ($>> T'$), (10) would become

$$V_{op} \simeq g_m R_d RI/2 \tag{12}$$

and the required dynamic range could be provided by making R small. However, the compensation circuit would need to be designed according to the particular values of R and C: any discrepancy would lead to a sensitivity penalty from misequalization. The additional Johnson noise may also be significant.

5) Non-integrating receivers: The sensitivity of a high-impedance nonintegrating receiver (i.e., $2\pi RCB \sim 1$) would be appreciably worse than for an integrating design, due to the Johnson noise of R, but an intermediate level of performance could be provided by the more complex transimpedance design.

Thus, the receiver design most suitable for a given application will depend on the relative priorities of sensitivity, dynamic range, and signal disparity. The integrating p-i-n FET receiver offers the best sensitivity, and would provide a reasonable dynamic range for signals with a low disparity. If unbounded disparity or wide dynamic range are major design requirements, then modifications to the high-impedance design would necessitate careful frequency compensation, and a transimpedance receiver design may be preferable; but either option would entail penalties in receiver sensitivity.

III. Receivers Using Avalanche Photodiodes (APD's)

Photocurrent gain is generated in an APD by providing a region of high electric field strength at the p-n junction, in which photogenerated carriers can create further electron–hole (e–h) pairs through ionizing collisions with the lattice. Since the noise of a p-i-n photodiode receiver is due mainly to thermal noise sources in the preamplifier, substituting an APD can allow operation with appreciable avalanche gain before the multiplied shot noise on the photocurrent becomes significant. There is then an optimum gain at which the SNR is a maximum; thereafter, the excess noise (which reflects the probabilistic nature of the avalanche process) causes the SNR to deteriorate. Bulk leakage currents flowing through the high-field region also generate multiplied shot noise, and so tend to reduce the optimum gain and the maximum SNR for a given received power level.

Many of the characteristics of APD's are influenced by the ratio k of the carrier ionization rates. Silicon APD's with $k \simeq 0.02$ are characterized by low excess noise, a gradual increase of gain with bias voltage, and good uniformity of gain across the sensitive area. In contrast, holes and electrons in Ge and III-V materials have comparable ionization rates. This introduces a much higher level of feedback into the avalanche process: when the electric field strength is high enough for a photogenerated hole to create an e–h pair, the returning electron is also likely to undergo an ionizing collision, and so on. The gain in such diodes thus changes very rapidly with elecric field strength (hence also with bias voltage, and with doping concentration across the sensitive area), and the large uncertainty in the number of secondary e–h pairs created by each photogenerated e–h pair results in a high level of excess noise. Furthermore, gain in the latter case involves a large number of transits of the high-field region and therefore takes time to build up, whereas APD's with very different ionization rates will have a higher gain–bandwidth product since each ionizing carrier produces many other carriers during a single transit, giving a much lower mean carrier transit time.

Thus good performance is obtained from APD's in which one type of carrier ionizes much more strongly than the other, and where that carrier initiates the avalanche process. The first part of this section deals with the relative importance of the various noise sources in receivers using APD's. The second part then considers the level of preamplifier noise that can be tolerated at high frequencies where the gain–bandwidth limit is approached. In the final parts, the development and performance of practical devices in Ge and III-V materials are described.

A. APD Receiver Sensitivity

Using the same notation as for p-i-n-photodiode-based receivers in Section II, the sensitivity $\overline{P}$ of an APD-based receiver can be expressed as follows [13]:

$$\overline{P}_{APD} = \frac{hcBQ}{2\lambda\eta}\left[QI_2F_p + \frac{2}{M_p}\left[Z + \frac{I_2}{qB} I_{DM}M_D^2F_D\right]^{1/2}\right] \tag{13}$$

where Z retains the form of (2), but the components of the leakage term I_L in (3) are all independent of avalanche gain. F_p is the excess noise factor associated with the photocurrent gain M_p as

$$F_p = k_w M_p + (1 - k_w)(2 - 1/M_p). \tag{14}$$

This is obtained from McIntyre's derivation [5], where $k_w = k$ for an electron-initiated avalanche, $1/k$ for a hole-initiated avalanche, and intermediate values for mixed carrier injection. Similarly, the multiplied shot noise on the bulk leakage current I_{DM} is characterized by F_D, M_D, and k_{wd} [71]. If I_{DM} is sufficiently small, then increasing

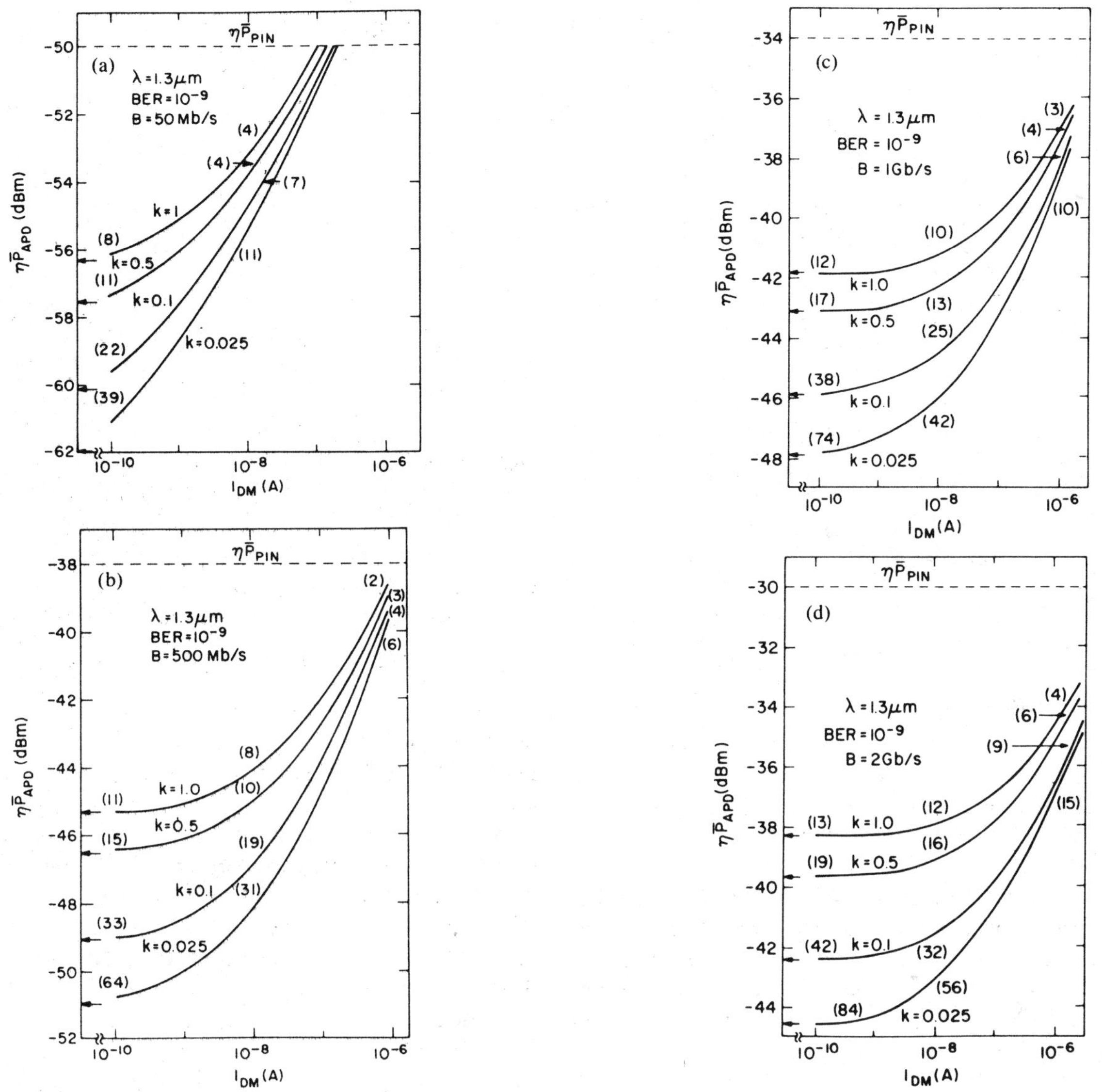

Fig. 5. Dependence of receiver sensitivity on the multiplied dark current and k value for several bit-rates [74]: (a) 50 Mbit/s; (b) 500 Mbit/s; (c) 1 Gbit/s; (d) 2 Gbit/s.

the photocurrent gain M_p reduces the effect of the thermal noise terms in (13) and improves $\bar{P}$; eventually the multiplied shot noise on the signal dominates, and a best-case value for $\bar{P}$ is obtained at an optimum value for M_p. APD's are therefore designed so that the avalanche is initiated by the more strongly ionizing type of carrier, since a low value for k_w tends to produce a high optimum gain and a good sensitivity.

For Ge APD's, $k_w \sim 0.5$–1, and the multiplied shot noise on the bulk leakage current I_{DM} may be significant [71], [72]. For III-V APD's $k_w \simeq 0.3$–0.5 [18], and I_{DM} may contain a large tunneling leakage component [73]. Both a high k_w and a high bulk leakage current tend to reduce the optimum gain and result in a worse sensitivity. In designing a receiver for a good sensitivity using these APD's, therefore, the following considerations must be made [25]:

1) Figure 5 [74] shows how the advantage in using an APD diminishes as the bulk leakage current I_{DM} increases, especially at lower bit rates. Band-to-band tunneling should therefore be minimized, and the sensitive area should be made small.

2) If I_{DM} is small but $k_w \sim 1$, the sensitivity $\bar{P}$ may still be improved by reducing Z as for p-i-n photodiode-based receivers: that is, through a low input capacitance, high first-stage gain, and reduced Johnson noise (although the very-high-impedance front-end design used for the PIN-FET would be unsuitable for use with APD's since the voltage drop across R would modulate the avalanche gain).

3) For receivers optimized as described above, the overall noise spectral density increases with bit rate, due to differentiated preamplifier noise in the case of integrating front-end designs, and due to increased Johnson noise from resistances taking smaller values to avoid RC time constant limitations. Fig. 6 shows that the sensitivity deteriorates more rapidly with increasing bit rate as k_w approaches unity, but in all cases the optimum gain is higher at higher bit rates. Contrary to this requirement,

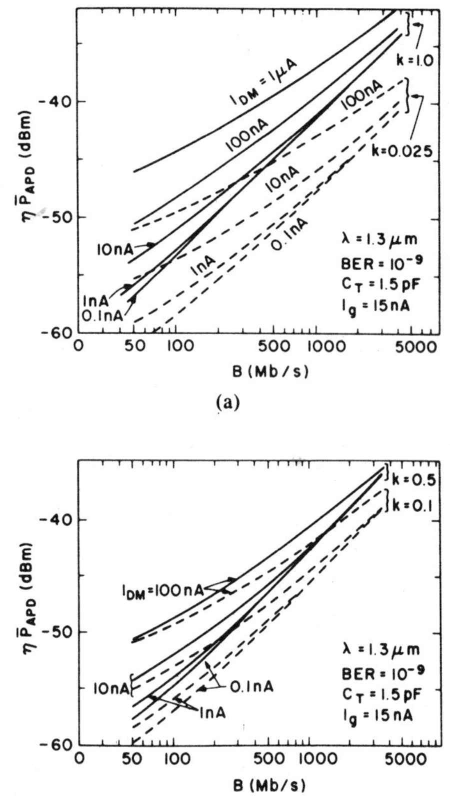

Fig. 6. Receiver sensitivity as a function of bit-rate for (a) $k = 0.025$ and 1; and (b) $k = 0.1$ and 0.5 [74].

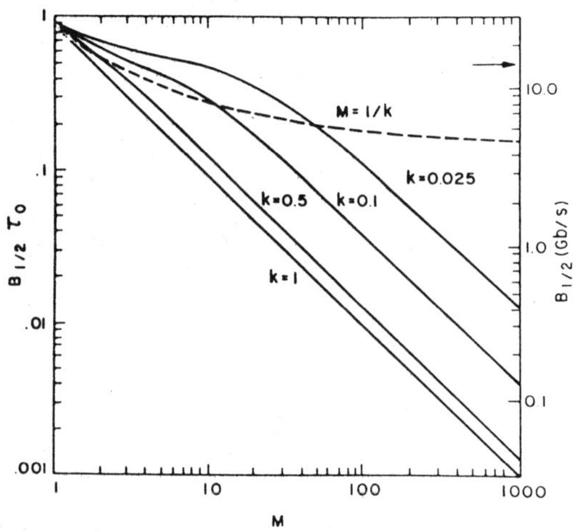

Fig. 7. Normalized 3-dB bit rate as a function of the dc gain of an avalanche photodiode. The right-hand coordinate is the 3-dB bit rate for $\tau_0 = 30$ ps [74], [77].

the high-frequency behavior of APD's is characterized by a gain-bandwidth product: eventually, providing the necessary bandwidth for higher bit rates requires the avalanche gain to be reduced. Sensitivity penalties will then be avoided only if the noise of the preamplifier is reduced: this is considered in the following section.

B. Gain-Bandwidth Limited Operation

Consider an APD in which $k_w < 1$; that is, the avalanche is initiated by the more strongly ionizing carrier. Initially as M increases from unity, each photogenerated carrier will in turn generate almost all of its secondary carriers within a single transit of the high-field region, and the response time τ_{tr} will be the sum of the transit times τ_T for the low-field region (the ternary region in a III-V APD) and τ_0 for the high-field region. However, the less strongly ionizing carrier will begin to contribute to the gain when $M > 1/Nk_w$, where $0.3 < N < 2$ [75], and the response time will lengthen according to the mean number MNk_w of transits of the high-field region (or feedback cycles) necessary for that value of gain. This process is thus characterized by a gain-bandwidth product: neglecting τ_T for the moment, the avalanche build-up time is $MNk_w\tau_0 = M\tau_1$, where τ_1 is the effective carrier transit time [76], and the ac gain at frequency ω is given by

$$M(\omega) = M_0/(1 + (M_0\tau_1\omega^2)^{1/2} \tag{15}$$

where M_0 is the dc photocurrent gain.

At the gain-bandwidth limit, therefore, the effective 3-dB bandwidth $(BW_{\text{eff}})_{1/2}$ of the APD and the bit rate $B_{1/2}$ are related to the gain M_0 as follows [76]:

$$(BW_{\text{eff}})_{1/2} = 0.5 I_2 B_{1/2} = 1/(2\pi M_0 \tau_1) \tag{16}$$

where I_2 is the double-sided noise-equivalent bandwidth used in Section II, and takes a value of 1.1277 for NRZ pulses. This relationship is illustrated in Fig. 7 for various k_w.

We have already seen that M_{opt}, influenced by the thermal noise of the preamplifier and by k_w, tends to increase with bit rate. By deriving an expression for M_{opt} and substituting it into (16), we can deduce the maximum level of preamplifier noise for which the optimum gain falls within the gain-bandwidth limit.

Neglecting I_{DM} for the moment, and putting $M_p = M$, $F_p = F$, (13) reduces to

$$\begin{aligned} \eta\bar{P}_{\text{APD}} &= \frac{hcBQ}{2\lambda}\left[QI_2F + \frac{2\sqrt{Z}}{M}\right] \\ &= \frac{hcBQ^2}{2\lambda} I_2 F + \frac{\eta\bar{P}_{\text{p-i-n}}}{M} \end{aligned} \tag{17}$$

where the preamplifier noise is characterized by the sensitivity $\eta\bar{P}_{\text{p-i-n}}$ that would result from a p-i-n photodiode feeding the same preamplifier. $\bar{P}_{\text{APD}}$ is then minimized with respect to M, to yield [77]

$$\begin{aligned} M_{\text{opt}} &= k_w^{-1/2}\left[\frac{2\eta\bar{P}_{\text{p-i-n}}}{Q^2(hc/\lambda)\, I_2 B} + k_w - 1\right]^{1/2} \\ &\simeq \left[\frac{2\eta\bar{P}_{\text{p-i-n}}}{k_w Q^2(hc/\lambda)\, I_2 B}\right]^{1/2} \end{aligned} \tag{18}$$

Putting $B = B_{1/2}$ in (18) and substituting for M_{opt} in (16), we obtain

$$\begin{aligned} (BW_{\text{eff}})_{1/2} &= 0.5 I_2 B_{1/2} = 1/(2\pi M_{\text{opt}}\tau_1) \\ &= \frac{1}{2\pi\tau_1}\left[\frac{k_w Q^2(hc/\lambda)\, I_2 B_{1/2}}{2\eta\bar{P}_{\text{p-i-n}}}\right]^{1/2} \end{aligned}$$

i.e.,

$$B_{1/2} = \frac{1}{(2\pi\tau_1)^2} \cdot \frac{2k_w Q^2(hc/\lambda)}{\eta \bar{P}_{\text{p-i-n}} I_2}. \qquad (19)$$

If this equation yields a value for $B_{1/2}$ which is less than the required bit rate for a given $\eta \bar{P}_{\text{p-i-n}}$, then sensitivity penalties will follow either from operating at $M < M_{\text{opt}}$ or from inter-symbol interference due to insufficient bandwidth at higher M. Since τ_1 is proportional to k_w, we thus require low values for k_w and $\eta \bar{P}_{\text{p-i-n}}$ to avoid such penalties, and it has been shown that $k_w = 0.5$ is expected to be the worst case [77].

At bit rates of several gigabits per second it should be possible to make the multiplied shot noise on I_{DM} negligible, as was assumed above; if this were not so, the value for M_{opt} would have to be deduced numerically. It can be seen, however, that such noise will reduce M_{opt} and thus increase the bandwidth capability of the APD. Thus, any given APD and preamplifier which together satisfy (19) will be able to provide sufficient bandwidth for the desired bit rate, regardless of I_{DM}.

C. InGaAs Avalanche Photodiodes

1) Homojunction Structure: In this section, we examine the different contributions to leakage current in InGaAs homojunction APD's, and calculate the maximum doping concentration at the p-n junction consistent with a low leakage current near breakdown. As Fig. 8 illustrates, the dark current can be described as the sum of three independent current sources: generation–recombination (G–R) via mid-band traps in the depletion region, tunneling of carriers across the bandgap, and a shunt path (surface leakage) across the p-n junction.

The G–R current, which is dominant at low voltage, is given by

$$I_{GR} = (qn_i AW/\tau_{\text{eff}})[1 - \exp(-qV/2kT)] \qquad (20)$$

where n_i is the intrinsic carrier concentration, τ_{eff} is the effective carrier lifetime, W is the depletion layer width, and A is the diode area.

Tunneling dominates the dark current at high voltages, and is given for direct-gap semiconductors as follows [39]:

$$I_{\text{tun}} = \gamma A \exp[-\theta m_0^{1/2} E_g^{3/2}/q\hbar E_m] \qquad (21)$$

where m_0 is the free electron mass, E_m is the maximum junction field, E_g is the energy gap, and $\hbar$ is Planck's constant divided by 2π. The parameter θ is given by $\theta = \alpha[m^*/m_0]^{1/2}$, where m^* is the electron effective mass, and α depends on the detailed shape of the tunneling barrier but is on the order of unity for band-to-band processes. The factor γ (not to be confused with Γ in (4)) depends on the initial and final state of the tunneling carrier; for band-to-band tunneling, $\gamma = (2m^*/E_g)^{1/2} q^3 E_m V/h^2$.

Finally, the shunt current $I_{sh} = VR_{sh}$ is important only at low temperatures. Here R_{sh} is the shunt resistance which is usually greater than 10^{10} Ω for typical diodes at room temperature [39].

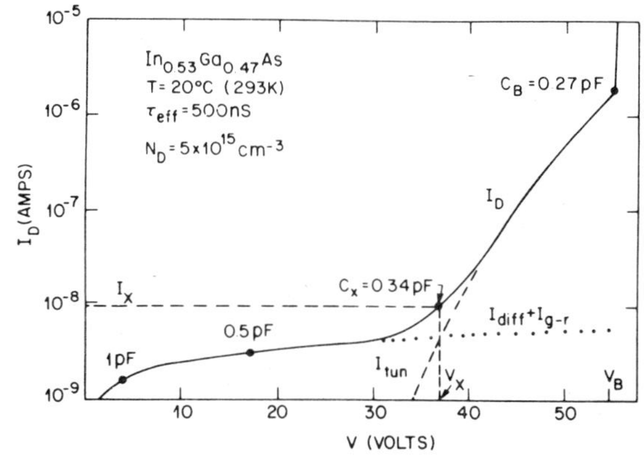

Fig. 8. Theoretical reverse-bias dark current characteristic for InGaAs showing the various current components [40].

The avalanche breakdown voltage depends on the bandgap energy and the background doping of the semiconductor, and is given for abrupt junctions by the following empirical formula [78]:

$$V_B = 60[E_g/1.1]^{3/2}\,[N_D/10^{16}]^{-3/4}\ \text{V} \qquad (22)$$

where E_g is in electronvolts and N_D in cm^{-3}. A theoretical reverse-bias dark current characteristic of an $In_{0.53}Ga_{0.47}As$ diode is shown in Fig. 8, where the various components of the dark current are indicated.

It is useful to define a cross-over voltage V_x at which $I_{\text{tun}} = I_{GR}$, as shown in Fig. 8. V_x is approximated as follows [40]:

$$V_x = 5.5 \times 10^{15} \frac{T\theta^2 E_g^2}{0.06 N_D \tau_{\text{eff}}}\ \text{V}. \qquad (23)$$

The importance of V_x is that when $V_B > V_x$, the avalanche gain is accompanied by a large tunneling current which also undergoes multiplication, resulting in a high level of shot noise. For low-noise APD receivers, therefore, we require $V_B < V_x$, which leads to the following restraint on doping density:

$$N_D < 1.4 \times 10^{18} \frac{\theta^2 E_g^2}{0.24\tau_{\text{eff}}}\ \text{cm}^{-3}. \qquad (24)$$

Figure 9 is a plot of V_x and V_B as functions of doping density N_D in $In_{0.53}Ga_{0.47}As$ for several values of τ_{eff}. Thus, for an InGaAs APD to have a low tunneling current prior to avalanche breakdown, the background doping density must be $N_D < 2 \times 10^{15}\ \text{cm}^{-3}$.

2) SAM and SAGM APD's: In order to reduce the tunneling current discussed in the previous section, a structure which contains separate absorption and multiplication regions (SAM) has been proposed [79]. The construction of this structure is illustrated schematically in Fig. 10. Here the incoming photons are absorbed by the InGaAs ternary layer, which is in a low-field region. The primary photogenerated carriers are swept into the high-field region of the p-n junction in the InP binary layers. Because of the large bandgap in InP, the tunneling current is reduced and the condition $V_B < V_x$ can be more easily satisfied, with doping density N_B in the binary compound ranging from 5×10^{15} to $5 \times 10^{16}\ \text{cm}^{-3}$.

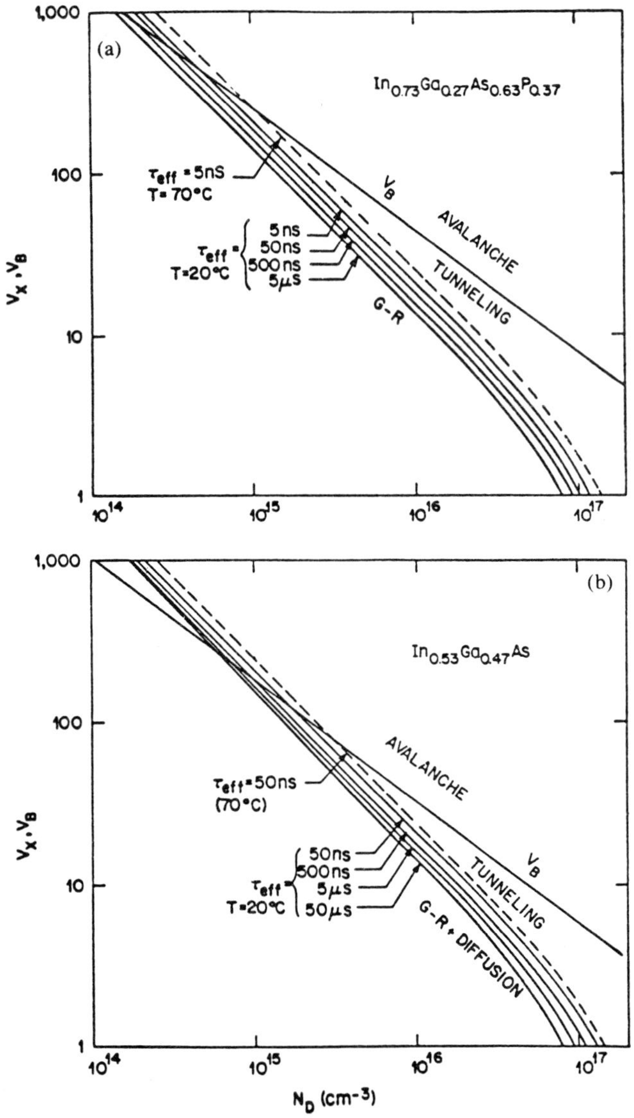

Fig. 9. Cross-over (V_x) and breakdown (V_B) voltages versus doping density for $In_{0.53}Ga_{0.47}As$ [40].

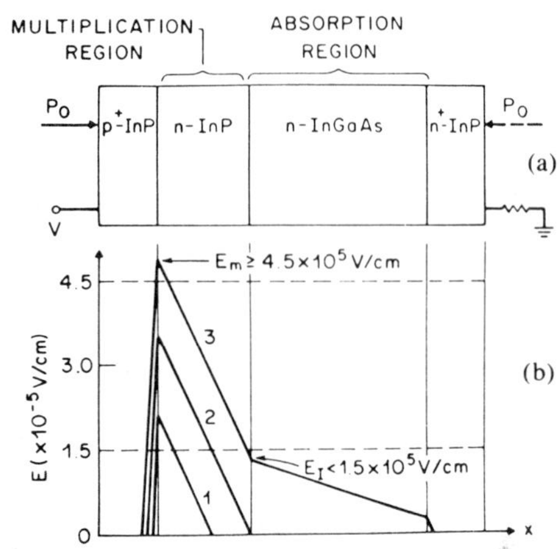

Fig. 10. Schematic of a separate absorption and multiplication avalanche photodiode (SAM APD).

However, there are two conditions which determine the thicknesses and doping densities of the binary and ternary layers. First, for negligible tunneling current, it is found that the interface field E_I between the ternary and binary layers must be less than 1.5×10^5 V/cm [80]. From Gauss' Law, E_I is given by

$$E_I = \frac{q}{\kappa_T} \sigma_T \tag{25}$$

where $\kappa_T = 1.04$ pF/cm is the permittivity of $In_{0.53}Ga_{0.47}As$ and σ_T is the total number of charges per unit area swept out in the ternary layer. Equation (25) leads to the first condition that $\sigma_T \lesssim 1 \times 10^{12}$ cm^{-2} at breakdown.

The second condition requires that the maximum field at the p-n junction must be $E_m = 4.5 \times 10^5$ V/cm at breakdown to ensure sufficient carrier multiplication. As in (25), E_m can be written as

$$E_m = q \left[\frac{\sigma_B}{\kappa_B} + \frac{\sigma_T}{\kappa_T} \right] \tag{26}$$

where σ_B is the total charge swept out in the n-type binary layer. Equation (26) yields the second condition that

$$\sigma_B + \sigma_T \geq 3 \times 10^{12} \text{ cm}^{-2}. \tag{27}$$

In obtaining (27), we have assumed $\kappa_B \simeq \kappa_T$. The sheet charge densities σ_B and σ_T can be related to the thickness and doping densities of binary and ternary layers, respectively, by the expressions

$$\sigma_B = \int_0^{X_I} N_B(x)\, dx \tag{28}$$

and

$$\sigma_T = \int_{X_I}^{W} N_T(x)\, dx \tag{29}$$

where W is the total thickness of the depletion region. Assuming uniformly doped layers with $N_B(x) = N_B$ and $N_T(x) = N_T$, the breakdown voltage can be written as [80]

$$V_B = \frac{q}{2\kappa} \left[\frac{1}{N_T} \left(\frac{\kappa E_m}{q} - N_B X_I \right)^2 + X_I \left(\frac{2\kappa E_m}{q} - N_B X_I \right) \right] \tag{30}$$

where $E_m = 4.5 \times 10^5$ V/cm is the maximum electric field at the onset of breakdown. The total depletion width at breakdown is

$$W_{BD} = \left[\frac{2\kappa V_{BD}}{q N_T} + X_I^2 \left(1 - \frac{N_B}{N_T} \right) \right]^{1/2}. \tag{31}$$

Figure 11 is a plot of the depth of the penetration W_T of the depletion region into the $In_{0.53}Ga_{0.47}As$ ternary layer at breakdown as a function of N_T for several values of $\sigma_B = N_B X_I$. The diagonal lines follow the two pre-conditions that $\sigma_B + \sigma_T \geq 3 \times 10^{12}$ cm^{-2} and $\sigma_T \leq 1 \times 10^{12}$ cm^{-2}. Thus, the line determines the maximum allowable doping density in the ternary layer for a given value of σ_B.

Assuming the absorption coefficient of $In_{0.53}Ga_{0.47}As$ is 10^4 cm^{-1}, a minimum ternary layer thickness of 2 μm is required for an internal quantum efficiency of 0.88. This

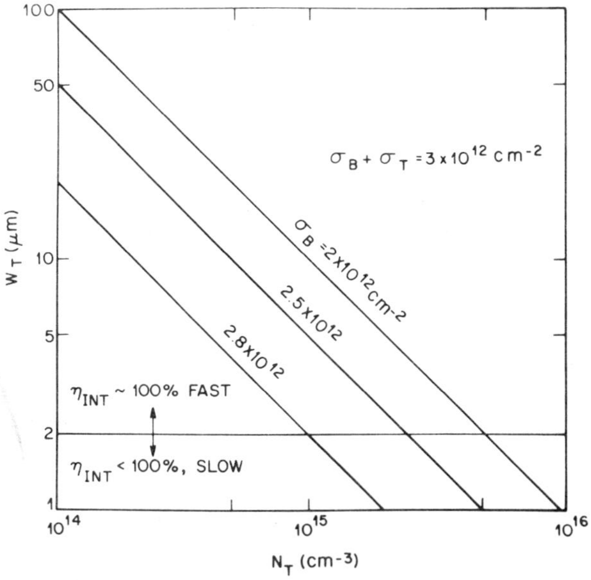

Fig. 11. Width of depleted $In_{0.53}Ga_{0.47}As$ at breakdown, as a function of ternary doping and binary fixed charge density. Diagonal lines satisfy the breakdown condition of $\sigma_B + \sigma_T = 3 \times 10^{12}$ cm^{-2}, where σ_T is the fixed charge swept out in InGaAs at breakdown [80].

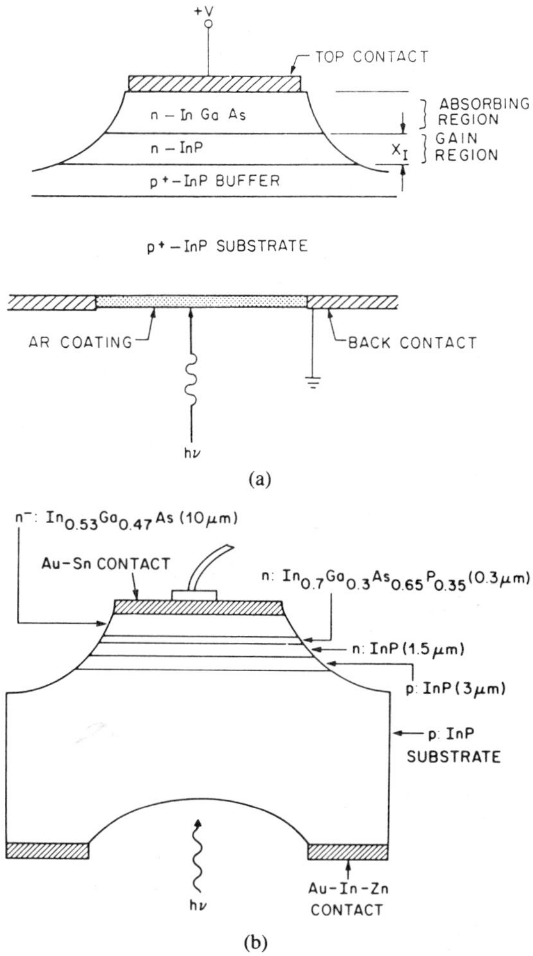

Fig. 12. (a) Schematic cross section of the mesa SAM APD [85]. (b) Schematic cross section of the mesa SAGM APD [91].

requirement is shown in Fig. 11 by the horizontal line. The intersection points of the horizontal and diagonal lines determine the maximum allowable values of N_T for $W_T = 2$ μm. If N_T were greater than the maximum allowable values, W_t would be less than 2 μm, resulting in a decrease in the internal quantum efficiency.

The speed of response of a SAM APD has been found to be limited by the charge pile-up at the large valence-band discontinuity at the InGaAs/InP heterointerface [81]. This charge pile-up also reduces the internal quantum efficiency at low bias voltages. As the bias voltage increases, the band tilts and the depth of the band discontinuity decreases, resulting in the increase of both internal quantum efficiency and speed. However, the remaining band discontinuity at voltages near breakdown ultimately limits the speed of response. A solution to this problem has been suggested [81] in the form of a thin graded band-gap layer of quaternary material incorporated by varying the composition from InP to $In_{0.53}Ga_{0.47}As$ in the heterointerface region. This separate absorption, grading, and multiplication (SAGM) structure reduces substantially the charge pile-up at the heterointerface, and thus increases the speed of response.

3) Performance of InGaAs APD's: Mesa-type InGaAs/InP SAM-APD's have been reported by several groups [16], [82]–[85]. The basic structure is shown in Fig. 12(a). Briefly, thin layers of p^+-InP, n-InP, and n-InGaAs were grown successively on p^+-InP substrate by liquid-phase epitaxy. The doping densities and layer thicknesses were approximately in agreement with the conditions discussed above. Typical thickness of the n-InGaAs absorption layer ranged from 2 to 4.3 μm with doping densities below 1×10^{16} cm^{-3}, and that of the n-InP layer varied from 1.5 to 2 μm with doping densities between 8×10^{15} and 2×10^{16} cm^{-3}. The preferred substrate was p^+-InP, because it allowed the use of n-type InGaAs for light absorption so that holes are injected into the high-field multiplication region of the InP p-n junction. This leads to a desired low excess noise since the ionization rate of holes in these materials is higher than for electrons ($\alpha/\beta < 1$). Experimentally determined effective k-values ($=\alpha/\beta$) ranged between 0.52 and 0.59 [86]–[90]. The dark current, the quantum efficiency, and the gain obtained in these early experimental devices are summarized in Table I.

These devices exhibited low dark currents and useful multiplication factors with excess-noise factors lower than those of Ge APD's. When excited by an optical pulse, however, the risetime and falltime have both fast and slow components. To increase the speed of response, a SAGM structure, as shown in Fig. 12(b), has been constructed [91]. The "grading" layer thickness was 0.3 μm, with doping density in the range 1–2 $\times 10^{16}$ cm^{-3}. This device had a dark current of 20 nA at a gain of 20. The 3-dB bandwidths of 1 GHz at a gain of 10 and 2 GHz at a gain of 5 imply a gain–bandwidth product of 10 GHz.

Recent effort has been devoted to the planar SAGM APD [79], [92]–[98] because of the potential advantages of a fully passivated surface region and suitability for manufacture. The challenge is to provide a stable dielectric–

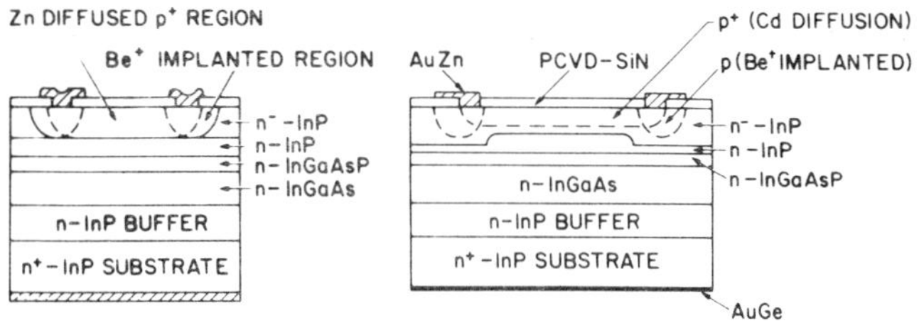

(a) (b)

Fig. 13. (a) Schematic cross section of the planar SAGM APD. (b) Schematic cross section of the planar buried structure SAGM APD.

TABLE I
MESA STRUCTURE InGaAs AVALANCHE PHOTODIODES

Diode structure	Dark Current Density (M=10) (A/cm^2)	Quantum Efficiency	Uniform Gain	Ref
SAM InGaAs/InP	5.2×10^{-4}	0.50	30 (45)	[83]
SAM InGaAs/InP	2.2×10^{-3}	0.52	45	[82]
SAM InGaAs/InP	7.8×10^{-4}	0.50	60	[16]
SAM InGaAsP/InP	3×10^{-6}	-	700	[84]
SAM InGaAs/InP	8×10^{-5}	0.60	20 (5500)	[85]
SAGM InGaAs/InGaAsP/InP	5×10^{-4}	0.70	42	[91]

semiconductor interface for high reliability. Silicon nitride film has been a suitable passivation for planar p-i-n devices [57].

Figure 13(a) is a cross section of a typical planar SAGM-APD structure. Here the substrate is n^+-InP to facilitate the formation of the p-n junction by p^+-diffusion. Most designs use a n^--InP top layer where a guard ring is formed either by Cd diffusion through a 500 Å thin SiO_2 layer [94] or by Be implantation [95]–[97]. The light-sensitive junction is formed in the underneath n-InP layer by Cd diffusion. The breakdown voltage of the guard ring is 10–15 V higher than that of the active region. The structure shown in Fig. 13(b) used a n-InP buried region to control the electric field strength so that high speed of response is obtained [99]. A gain–bandwidth product of 10 GHz, similar to that of the mesa structure, has been realized.

D. Germanium Avalanche Photodiodes

High-speed germanium APD's [100] have been made for some time and, recently, improvements have been made in reducing the dark current and decreasing the excess noise [15], [101]–[104].

Previously, an n^+-p structure has been used in Ge diodes for reasons of convenience in fabrication processes. When operating at 1.55-μm wavelength near the absorption edge, however, the absorption coefficient decreases and the proportion of electrons injected into the junction region increases [71]. For a diode with 0.4-μm junction depth, the fraction of electron injection will result in increased excess noise. If, on the other hand, the n^+ layer is made thick to increase hole injection, the quantum efficiency and the speed of response will deteriorate due to surface recombination and carrier diffusion toward the junction. Several improved Ge-APD structures [104] have been made to utilize either an n^+-n-p junction or a p^+-n junction. In the former case, the n^+ layer is made sufficiently thin (0.3–0.4 μm) that the light is absorbed almost entirely in the n-layer. The thick n-layer (21.5 μm) was formed by arsenic ion implantation at 150 keV with a 2×10^{13} cm^{-2} dose followed by a drive-in diffusion at 730°C for 1 h. A planar structure was adopted with an arsenic diffused guardring. For the latter structure, a thin (0.3 μm) p^+ layer was formed by boron ion implantation into an n-type wafer. The acceleration energy was 40 keV with a 1×10^{13} cm^{-2} dose, and a surface concentration of 1×10^{18} cm^{-3} was obtained. The guardring junction depth was 4 μm, made by zinc diffusion at 830°C for 3 h.

At 1.55 μm wavelength, it is necessary to have at least 10-μm depletion-layer width for high speed of response and simultaneously for high quantum efficiency. The depletion layer in the p^+-n structure was not wide enough to prevent carrier diffusion towards the junction region limiting the speed of response. A p^+-n-n^- (Hi-Lo) concentration profile was therefore used [15]: the n-layer was formed by As^+ implantation on a low impurity concentration n^- substrate. The latter results in a substantial increase in the dark-current density, but a low dark current and a low junction capacitance can be obtained by using a small-area (30-μm diameter) diode.

The performance characteristics of these Ge-APD's are summarized in Table III. The conventional n^+-p structure is also included for comparison. We note that a significant reduction of the effective k value has been achieved with these improved structures.

E. APD's in Other Material Systems

Photodiodes made in $Al_xGa_{1-x}Sb$, lattice matched to GaSb substrate material, have been reported for use in the 1.3–1.6-μm-wavelength region [105]–[108]. Although high

TABLE II
PLANAR SAGM AVALANCHE PHOTODIODES

Surface Layer	Junction Diffusion	Guard-Ring	Dark Current (A/cm^2)	Quantum Efficiency	Uniform Gain	Ref
n-InP	Cd	-	10^{-6} ($0.5V_b$)		3000	[79]
n-InP	Cd	-	1×10^{-4}	0.70	5.5	[92]
n-InP	Cd	Cd-diff	1×10^{-3}		20	[94]
n-InP	Cd	Be-implant	2×10^{-4}	0.50	110	[97]
n^--InP	Cd	Be-implant (Buried structure)				[98]

TABLE III
COMPARISON OF GERMANIUM AVALANCHE PHOTODIODE STRUCTURES

Diode Structure	Dark Current* μA	Breakdown Voltage V	Maximum Gain	Quantum Efficiency**	Effective k	Halfwidth ps
n^+-p	0.2	23 - 33	40	0.85	1.0	200
n^+-n-p	1	30 - 32	30	0.85	0.6 - 0.7	440
p^+-n	0.3	30 - 32	30	0.90	0.6 - 0.7	200
p^+-n-n^-	0.9 (30μm dia)	85		0.70	0.65	500 (M=10)

* at 25°C ** with AR coating, λ = 1.55 μm

quantum efficiency and moderate gains ($M \sim 50$) have been obtained, the leakage current was excessive. It was found [109] that the leakage current was dominated by tunneling at voltages above 10 V, and by surface rather than bulk leakage at lower voltages.

An interesting feature of this material system is that the hole ionization rate exhibits a resonant characteristic as a function of the alloy composition [110]. This is because the bandgap energy E_g and the spin–orbit splitting energy Δ become equal for $x = 0.65$, corresponding to the long-wavelength cut-off at 1.55 μm. An enhancement in β/α by a factor as large as 20 has been measured [110]. This low effective k value ($k_w = 0.05$) would lead to an extremely low-noise APD if the dark current could be substantially reduced in this material.

The resonance enhancement of the ionization-rate ratio has also been observed in the $Hg_{1-x}Cd_xTe$ material system. The condition $\Delta/E_g = 1$ occurs at $x = 0.73$ [111], which corresponds to $E_g = 0.97$ eV or $\lambda = 1.28\ \mu$. The effective k value of 0.11 was measured with a diode hav- a long-wavelength cut-off at 1.41 μm [111]. It is also possible for resonance enhancement to occur at longer wavelengths near 1.6 μm, using the quaternary $Hg_{1-x}Cd_xSe_yTe_{1-y}$ alloy system.

Another method of increasing α/β (or β/α) is by making a multi-quantum-well structure of alternating wide and narrow bandgap materials [112]–[115]. The enhancement of α/β arises from two factors. First, when the carriers enter the wide-bandgap material from the narrow-bandgap material, they possess an energy which is greater than the ionization energy in the narrow-gap material; therefore, when the conduction-band discontinuity exceeds the valence-band discontinuity, the ionization rate of electrons is enhanced. Second, holes are scattered and collected more effectively in a quantum-well structure than in bulk [116], leading to a small hole mean free path which significantly increases the α/β ratio [90].

Experimental devices [115] using alternative alloys of $Al_{0.45}Ga_{0.55}As$ (550 Å) and GaAs (450 Å) grown to a total thickness of 2.5 μm by molecular-beam epitaxy, have shown a value of $\alpha/\beta = 10$, compared to $\alpha/\beta \sim 1$ for bulk GaAs.

IV. PHOTOTRANSISTORS AND PHOTOCONDUCTIVE DETECTORS

A. *Heterojunction Bipolar Phototransistors (HPT's)*

HPT's offer enhanced current gain by virtue of a wide-bandgap emitter [117]; InGaAs/InP HPT's were therefore suggested as a low-noise alternative to APD's for long-wavelength optical receivers. However, early devices with electrically-floating bases showed high gain only at high optical power levels, and also suffered from a limited speed of response [118], [119]. By providing an external electrical bias to the base region, a gain-bandwidth product of 1.5 GHz was achieved and current gain could be traded against speed [120], but the extra metallization added to the input capacitance. A slightly higher value of 1.7 GHz was realized in a small-area (20-μm-diameter) mesa device with a floating base [121]: the collector–emitter capacitance was 0.1 pF, and the gain was 100 for power levels in the range 20–100 nW.

These early observations were consistent with a model by Milano *et al.* [122]. A high gain-bandwidth product requires low base-collector and base-emitter capacitances and a low Shockley base–emitter resistance r_e. Low optical signal levels do not provide a sufficient collector current for a low value of r_e; hence the need for a dc bias current, applied either electrically [120] or optically [123], [124]. Also, for fast response, the HPT should be designed for low current gain by broadening the base region.

Analysis of an HPT-based digital optical receiver suggests the following expression for the sensitivity $\overline{P}$ [125]:

$$\overline{P} = \frac{hcBQ}{\eta\lambda}\,[QI_2 + \sqrt{Z}]. \tag{32}$$

This expression is similar to (1) for p-i-n-photodiode-based receivers, although the shot noise on the signal photocurrent in an HPT is effectively doubled [126]; but clearly the current gain h_{fe} of an HPT is not analogous to the photocurrent gain M of an APD (13) and does not directly reduce the contribution of the preamplifier thermal noise term Z to the sensitivity $\overline{P}$. This is because h_{fe} also appears in the expression for the frequency response of the HPT, whose compensation is implicit in (32).

h_{fe} is nonetheless an important parameter in reducing Z, since it reduces the effect on the sensitivity of the Johnson noise of the collector load resistance. Also, since the term in I_2 is proportional to the collector current I_c while the I_3 term is inversely proportional to I_c, there is an optimum value of I_c which gives a minimum value for Z; a high current gain is then advantageous, but only in an integrating receiver. Taking the parameter values reported by Campbell *et al.*, however, (i.e., input capacitance of 0.4 pF and $h_{fe} \sim 100$), an integrating receiver would be expected to have a sensitivity at best comparable with that of the PINFET hybrid [125], [127], but with the need for an optical bias.

B. Photoconductive Detectors

High-purity $Ga_{0.53}In_{0.47}As$ has now been produced by both liquid-phase and vapor-phase epitaxy techniques. The high electron mobility ($\mu_e = 10^4$ cm^2/V · s) in n-type layers and the very high peak velocity ($v_p = 3 \times 10^7$ cm/s) make this material suitable for photoconductive detectors with large gain-bandwidth product. Recent papers [21], [128]–[130], indicate that the receiver performances using photoconductors made in InGaAs for the 1.3- and 1.55-μm wavelength region compare favorably with p-i-n detector receivers at bit rates up to 1 Gbit/s.

The advantages of photoconductive detectors are operation at only 2 V bias and ease of integration with FET's for the front end. Because of the small light-absorption area needed for high-speed operation, however, the light-coupling efficiency is inferior to that of p-i-n diodes.

A detailed calculation of photoconductor receiver sensitivity reported recently [23], [131] showed that the primary noise current comes from Johnson noise for bit rates up to 4 Gbit/s. At low bit rates, the receiver sensitivity varies as the square root of the channel resistance, as expected for the Johnson-noise-limited case. On the other hand, the improvement at high bit rates is degraded by the slow response of the photoconductive detector. The inherently slow response requires equalization in the receiver to minimize intersymbol interference, and this results in an increase in the total receiver noise and hence a loss in sensitivity.

Photoconductive detectors may be useful at wavelengths beyond 2 μm, where low-leakage p-n junctions are difficult to obtain in narrow-bandgap semiconductors.

V. Receivers for Coherent Optical Systems

The advantages of coherent detection are two-fold. First, the receiver sensitivity may approach the limit defined by the shot noise on the signal; and, second, the much greater wavelength selectivity allows many more channels to be carried at different wavelengths by the same fiber than by the conventional methods of wavelength multiplexing in direct-detection systems.

The first of these advantages relies on the incoming optical signal being amplified when it is mixed with a large local-oscillator (LO) signal, resulting in the total noise at the receiver being dominated by the shot noise on the received signal. The potential advantage is greater for higher bit rates as the margin between the launched power and the receiver sensitivity diminishes, and the need for narrow spectral linewidth in the laser relaxes. Heterodyne systems are potentially easier to implement than homodyne, but require a much larger receiver bandwidth to accommodate the signal spectrum centered on the intermediate frequency.

Laboratory demonstrations using PINFET receivers [132]–[137] have shown that the sensitivity of a coherent receiver can indeed approach within a few decibels of the shot-noise limit. Much of this success is due to the low noise of the receiver and freedom from saturation by the LO signal, but similar demonstrations at higher bit rates

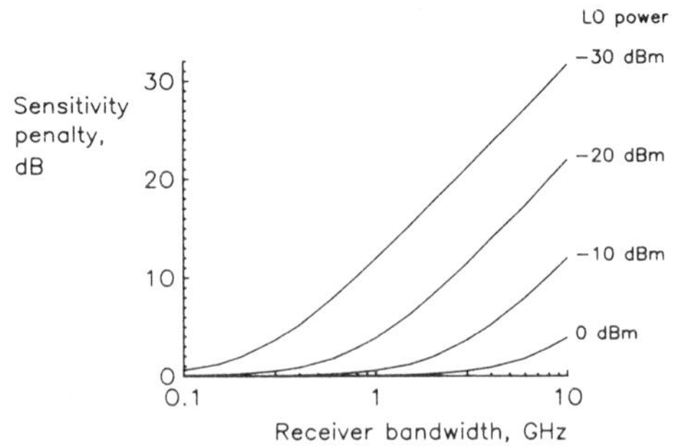

Fig. 14. Estimated deviation from shot-noise-limited receiver sensitivity due to limited LO power, as a function of receiver bandwidth, for an ideal p-i-n FET receiver with $C = 0.5$ pF, $g_m = 20$ mS, and a responsivity of 1 A/W.

will require a wider receiver bandwidth and more LO power. Figure 14 shows the expected deviation from the shot-noise-limited receiver sensitivity due to limited LO power as a function of receiver bandwidth, assuming an ideal integrating p-i-n FET receiver with responsivity of 1 A/W, input capacitance 0.5 pF, and FET transconductance of 20 mS. This is calculated simply from the deviation in the SNR from the shot-noise-limited case due to the thermal noise of the receiver, assuming the latter to be dominated by the I_3 term ((2) and (4)). Clearly, it may no longer be valid to ignore all other noise sources as the penalty becomes very large, but Fig. 14 serves to indicate the level of LO power needed at the detector in order to approach the shot-noise-limited sensitivity, and the rapidity with which the penalty increases when the LO power is in fact restricted.

The main requirements for a coherent receiver are therefore as follows. First, the detector must not be saturated by the required level of LO power. Second, heterodyne receivers require a greater bandwidth than homodyne or direct-detection receivers for the same transmitted bit rate. Third, in the event of the LO power at the detector being limited, the thermal noise of the receiver must be low to minimize the penalty in receiver sensitivity, and effective use must be made of the LO power that is available by using low-loss optical components and a balanced receiver.

VI. Comparison of Receiver Performance

Table IV summarizes the results of system experiments using the various types of receiver considered in this paper. The receiver sensitivity is defined as the average received optical power, $\overline{P}$, in decibels relative to 1 mW for achieving a bit error rate of 1 in 10^9.

Ge APD's were the first readily available detectors upon which long-wavelength receivers could be based, and much has been done to optimize their performance. At 1.3-μm wavelength, receiver sensitivities obtained using n^+-p-structured devices were initially poor and varied by several decibels between 20° and 60°C [29], due to the high level of multiplied shot noise on the bulk leakage current [72]; this has been reduced by cooling the detector [138] and, with the widespread adoption of monomode fiber, by reducing the junction area of the APD [139]. Hybrid-in-

TABLE IV
SUMMARY OF EXPERIMENTAL RECEIVER SENSITIVITIES
(Average received optical power $\bar{P}$, in dBm, for 10^{-9} BER.)

Bit-rate	InGaAs PIN		InGaAs SAM/ SAGM APD		Ge APD		InGaAs photoconductor	
	1.3 µm	1.55 µm	1.3 µm	1.55 µm	1.3 µm	1.55 µm	1.3 µm	1.55 µm
34 Mb/s	-52.5[i]				-46	-55.8[h]		
45	-49.9		-51.7[a]		-51.9[n]			
100					-40.5			
140	-46[i]				-45.2[h]	-49.3[h]		
274	-43		-45	-38.7[b]		-36[b]		
320	-43.5[i]							
420			-43[c]	-41.5[c]				
450			-42.6[d]		-39.5[d]	-39.1[l] -39.4[k] -40.5[m] -43.2[l]		
565					-33			
650	-35.5[i] -38							
1 Gb/s			-38[c]	-37.5[c]	-28		-34.4[f]	
1.2	-33.2[j]	-36.5[p]						
1.8			-31.3[d]		-30.1[d]			
2				-36.6[e]		-31[e]		-28.9[g]
4				-32.6				

References

a: [141]; b: [142]; c: [18]; d: [143]; e: [144]; f: [129]; g: [130]; h: [138]; i: [145]; j: [146]; k: [147]; l: [148]; m: [15]; n: [139]; p: [30]

tegration is also being applied to improve sensitivity [140], since the high k-value limits the ability of the APD to offset the thermal noise of the preamplifier [25]. For 1.55-μm wavelength, reach-through devices now offer a high-frequency capability unattainable with simpler structures, and have been demonstrated in receivers operating at up to 2 Gbit/s [14].

InGaAs APD's also suffered initially from high bulk leakage current [73], high excess noise [86]–[90], and poor frequency response [81]; but these difficulties have been overcome in the SAGM structure [18]. Of the detectors considered here, the InGaAs/InGaAsP/InP SAGM APD is at present the most difficult to make, but combines wide dynamic range with very good receiver sensitivity for 1–2 Gbit/s, and has been demonstrated at 4 Gbit/s [19]. Since the gain-bandwidth product of these devices is lower than for Ge APD's, the use of hybrid-integration has a much greater impact on receiver sensitivity in the gigabit per second range in reducing both the preamplifier noise and the optimum avalanche gain.

In comparison, the sensitivity of the photoconductor will always be inferior to that of the SAGM APD receivers; but its simplicity and potential for integration with FET amplifiers make it attractive for low-cost optical receivers.

Compared with detectors which incorporate gain, p-i-n photodiode receivers offer a number of attractive features. Integrating receivers have demonstrated good sensitivities at up to 1.2 Gbit/s in both the 1.3- and 1.55-μm wavelength windows, with negligible temperature dependence (except at the lowest bit rates) [29], [30]; and the structure and electrical bias requirements for the detector are relatively simple. Their main drawback is a limited dynamic range, except where a low-disparity line transmission code can be used. The freedom from saturation of the p-i-n photodiode and the low noise of the p-i-n FET hybrid have, however, proved beneficial for coherent optical systems, through which even better receiver sensitivity and greater information-carrying capacity may eventually be realized.

Let us now consider the ideal receiver: that is, one which offers good sensitivity and wide dynamic range at all wavelengths of interest; which uses cheap components and is simple to assemble and operate; which is reliable, readily available, and so on. Clearly none of the receivers considered here is ideal, but each offers the desired attributes in different proportion. Ge APD's already provide wide dynamic range and good sensitivity at 1.3-μm wavelength over a wide range of bit rates. InGaAs SAGM APD's offer better sensitivity, provided the growth technology can be developed to make them readily available. The attraction of the p-i-n FET hybrid lies in the relative simplicity of the structure and biasing of the p-i-n diode; it is less versatile in that sensitivity must be traded against dynamic range, but is a most promising receiver for coherent systems applications.

The performance of a receiver depends not only on the properties of the photodetector but also on the characteristics of the following amplifier. Low-noise front-ends for both APD and p-i-n-based receivers are provided by hybrid-integrating the detector with a GaAs FET preamplifier with low leakage current, low capacitance, (<0.5 pF), and high transconductance (>30 mS). FET's in InGaAs or InP may offer lower noise (benefiting p-i-n FET receivers more than APD FET's), and monolithic integration in these materials should reduce assembly costs.

Finally, progress in new optical fiber materials may call for a new generation of detectors for wavelengths beyond 2 μm, for which the photoconductor and/or coherent detection techniques may be the most promising.

ACKNOWLEDGMENT

M. Brain would like to thank colleagues at BTRL for their various contributions to this paper, notably D. R. Smith and R. C. Hooper on p-i-n FET hybrids, S. Ritchie and D. Wake on p-i-n photodiodes and integrated p-i-n FET's, and D. W. Smith on coherent receivers. He also thanks the Director of Research, British Telecom Research Laboratories, for permission to publish this paper. T. P. Lee is grateful to his former colleagues at AT&T Bell Laboratories, especially to S. R. Forrest, J. C. Campbell, C. A. Burrus, A. G. Dentai, B. L. Kasper, and M. A. Pollack for their contributions on APD's, and C. Y. Chen on photoconductive detectors.

REFERENCES

[1] K. C. Kao and G. A. Hockham, "Dielectric fiber surface waveguides for optical frequencies," *Proc. IEEE*, vol. 113, no. 7, pp. 1151–1158, July 1966.

[2] J. M. Horne and K. M. Langridge, "Submarine systems—Their optical future," *Brit. Telecommun. Eng.*, vol. 4, pp. 14–20, Apr. 1985; also *J. Lightwave Tech.*, vol. LT-2, Dec. 1984.

[3] J. E. Midwinter, "First generation trunk transmission systems: Capabilities and limitations," *IEEE J. Selected Areas Commun.*, vol. SAC-1, no. 3, pp. 381–386, Apr. 1983.

[4] C. A. Lee, R. A. Logan, R. L. Batdorf, J. J. Kleimack, and W. Wiegmann, "Ionization rates of holes and electrons in silicon," *Phys. Rev.*, vol. 134, p. A761, 1964.

[5] R. J. McIntyre, "Multiplication noise in uniform avalanche diodes," *IEEE Trans. Electron Devices*, vol. ED-13, no. 1, pp. 164–168, Jan. 1966.

[6] S. D. Personick, "Receiver designs for digital fiber optic communication systems," *Bell Syst. Tech. J.*, vol. 52, no. 6, pp. 843–886, July–Aug. 1973.

[7] P. P. Webb, R. J. McIntyre, and J. Conradi, "Properties of avalanche photodiodes," *RCA Rev.*, vol. 35, pp. 234–278, June 1974.

[8] R. W. Berry, D. J. Brace, and I. A. Ravenscroft, "Optical fiber system trials at 8 Mbits/s and 140 Mbits/s," *IEEE Trans. Commun.*, vol. COM-26, pp. 1020–1027, 1978.

[9] T. Miya, Y. Terunuma, T. Hosaka, and T. Miyashita, "Ultimate low-loss single-mode fiber at 1.55 μm," *Electron. Lett.*, vol. 15, pp. 106–108, 1979.

[10] J. Yamada and T. Kimura, "Characteristics of Gbit/s optical receiver sensitivity and long-span single-mode fiber transmission at 1.3 μm," *IEEE J. Quantum Electron.*, vol. QE-18, no. 4, pp. 718–727, Apr. 1982.

[11] K. J. Bachman and J. L. Shay, "An InGaAs detector for the 1.0–1.7 μm wavelength range," *Appl. Phys. Lett.*, vol. 32, no. 7, pp. 446–448, Apr. 1, 1978.

[12] S. R. Forrest, M. DiDominico Jr., R. G. Smith, and H. J. Stocker, "Evidence for tunneling in reverse-bias III-V photodetector diodes," *Appl. Phys. Lett.*, vol. 36, pp. 580–582, Apr. 1980.

[13] D. R. Smith, R. C. Hooper, and I. Garrett, "Receivers for optical communications: A comparison of avalanche photodiodes with PIN-FET hybrids," *Opt. Quantum Electron.*, vol. 10, pp. 293–300, 1978.

[14] J. Yamada, A. Kanawa, T. Miya, H. Nagai, and T. Kimura, "Gigabit/s optical receiver sensitivity and zero-dispersion single-mode fiber transmission at 1.55 μm," *IEEE Trans. Microwave Theory Tech.*, vol. MTT-30, no. 10, pp. 1525–1535, Oct. 1982.

[15] M. Niwa, Y. Tashiro, K. Minemura, and H. Iwasaki, "High-sensitivity Hi-Lo germanium avalanche photodiode for 1.5 μm-wavelength optical communication," *Electron. Lett.*, vol. 20, no. 13, pp. 552–553, June 21, 1984.

[16] N. Susa, H. Nakagome, H. Ando, and H. Kanbe, "Characteristics in InGaAs/InP avalanche photodiodes with separated absorption and multiplication regions," *IEEE J. Quantum Electron.*, vol. QE-17, no. 2, pp. 243–250, Feb. 1981.

[17] H. Ando, Y. Yamauchi, and N. Susa, "High-speed planar InP/InGaAs avalanche photodiode fabricated by vapour phase epitaxy," *Electron. Lett.*, vol. 19, no. 14, pp. 543–545, July 7, 1983.

[18] J. C. Campbell, A. G. Dentai, W. S. Holden, and B. L. Kasper, "High-performance avalanche photodiode with separate absorption 'grading' and multiplication regions," *Electron. Lett.*, vol. 19, no. 20, pp. 818–820, Sept. 29, 1983.

[19] S. K. Korotky, G. Eisenstein, A. H. Gnauck, B. L. Kasper, J. J. Veselka, R. C. Alferness, L. L. Buhl, C. A. Burrus, T. C. D. Huo, L. W. Stulz, K. Ciemiecki Nelson, L. G. Cohen, R. W. Dawson, and J. C. Campbell, "4 Gb/s transmission experiment over 117 km of optical fiber using a Ti : $LiNbO_3$ external modulator," Post-deadline paper PD1, OFC '85, San Diego, CA, Feb. 1985.

[20] J. C. Campbell and K. Ogawa, "Heterojunction phototransistors for long-wavelength optical receivers," *J. Appl. Phys.*, vol. 53, no. 2, pp. 1203–1208, Feb. 1982.

[21] C. Y. Chen, A. Y. Cho, N. A. Olsson and P. A. Garbinski, "Reduction of fall times in $Ga_{0.47}In_{0.53}As$ photoconductive receivers through back gating," *Appl. Phys. Lett.*, vol. 46, no. 3, pp. 296–298, Feb. 1, 1985.

[22] M. C. Brain and D. R. Smith, "Phototransistors in digital optical communication systems," *IEEE J. Quantum Electron.*, vol. QE-19, no. 6, pp. 204–213, June 1983.

[23] S. R. Forrest, "Photoconductor receiver sensitivity," *IEEE Electron Device Lett.*, vol. EDL-5, no. 12, pp. 536–539, Dec. 1985

[24] M. R. Matthews, K. H. Cameron, R. Wyatt, and W. J. Devlin, "Packaged frequency-stable tunable 20 kHz linewidth 1.5 μm InGaAsP external cavity laser," *Electron. Lett.*, vol. 21, no. 3, pp. 113–115, Jan. 31, 1985.

[25] M. C. Brain and D. R. Smith, "Phototransistors, APD-FET, and PINFET optical receivers for 1–1.6 μm wavelength," *IEEE Trans. Electron Devices*, vol. ED-30, no. 4, pp. 198–203, Apr. 1983.

[26] D. R. Smith and I. Garrett, "A simplified approach to digital optical receiver design," *Opt. Quantum Electron.*, vol. 10, pp. 211–221, 1978.

[27] D. R. Smith, R. C. Hooper, K. Ahmad, D. Jenkins, A. W. Mabbitt, and R. Nicklin, "p-i-n/F.E.T. hybrid optical receiver for longer-wavelength optical communication systems," *Electron. Lett.*, vol. 16, no. 2, pp. 69–71, Jan. 17, 1980.

[28] D. R. Smith, A. K. Chatterjee, M. A. Z. Rejman, D. Wake, and B. R. White, "p-i-n/F.E.T. hybrid optical receiver for 1.1–1.6 μm optical communication systems," *Electron. Lett.*, vol. 16, no. 19, pp. 750–751, Sept. 11, 1980.

[29] D. R. Smith, R. C. Hooper, P. P. Smyth, and D. Wake, "Experimental comparison of a germanium avalanche photodiode and InGaAs PINFET receiver for longer wavelength optical communication systems," *Electron. Lett.*, vol. 18, no. 11, pp. 453–454, May 27, 1982.

[30] M. C. Brain, P. P. Smyth, D. R. Smith, B. R. White, and P. J. Chidgey, "PINFET hybrid receivers for 1.2 Gbit/s transmission systems operating at 1.3 and 1.55 μm wavelength," *Electron. Lett.*, vol. 20, no. 21, pp. 894–895, Oct. 11, 1984.

[31] R. G. Smith and S. D. Personick, in *Semiconductor Devices for Optical Communications*, H. Kressel Ed. Berlin: Springer, 1980, pp. 89–160.

[32] M. J. N. Sibley, R. T. Unwin, and D. R. Smith, "The design of PIN-bipolar transimpedance preamplifiers for optical receivers," *Radio Electronic Eng.*, Feb. 1985.

[33] M. A. Washington, R. E. Nahory, M. A. Pollack, and E. D. Beebe, "High-efficiency $In_{1-x}Ga_xAs_yP_{1-y}$ photodetectors with selective wavelength response between 0.9 and 1.7 μm," *Appl. Phys. Lett.*, vol. 33, no. 10, pp. 854–856, Nov. 15, 1978.

[34] R. F. Leheny, R. E. Nahory, and M. A. Pollack, "$In_{0.53}Ga_{0.47}As$ p-i-n photodiodes for long-wavelength fiber-optic systems," *Electron. Lett.*, vol. 15, no. 22, pp. 713–715, Oct. 25, 1979.

[35] R. F. Leheny, R. E. Nahory, M. A. Pollack, E. D. Beebe, and J. C. De Winter, "Characterization of $In_{0.53}Ga_{0.47}As$ photodiodes exhibiting low dark current and low junction capacitance," *IEEE J. Quantum Electron.*, vol. QE-17, no. 2, pp. 227–231, Feb. 1981.

[36] T. P. Lee, C. A. Burrus, A. G. Dentai, and K. Ogawa, "Small area InGaAs/InP p-i-n photodiodes: fabrication, characteristics and performance of devices in 274 Mb/s and 45 Mb/s lightwave receivers at 1.31 μm wavelength," *Electron. Lett.*, vol. 16, no. 4, pp. 155–156, Feb. 14, 1980.

[37] C. A. Burrus, A. G. Dentai, and T. P. Lee, "InGaAsP p-i-n photodiodes with low dark current and small capacitance," *Electron. Lett.*, vol. 15, no. 20, pp. 655–657, Sept. 27, 1979.

[38] D. Gloge, A. Albanese, C. A. Burrus, E. L. Chinnock, J. A. Copeland, A. G. Dentai, T. P. Lee, T. Li, and K. Ogawa, "High-speed digital lightwave communication using LEDs and PIN photodiodes at 1.3 μm," *Bell Syst. Tech. J.*, vol. 59, no. 8, pp. 1365–1382, Oct. 1980.

[39] S. R. Forrest, R. F. Leheny, R. E. Nahory, and M. A. Pollack, "$In_{0.53}Ga_{0.47}As$ photodiodes with dark current limited by generation-recombination and tunneling," *Appl. Phys. Lett.*, vol. 37, no. 3, pp. 322–325, Aug. 1, 1980.

[40] S. R. Forrest, "Performance of $In_xGa_{1-x}As_yP_{1-y}$ photodiodes with dark current limited by diffusion, generation recombination, and tunneling," *IEEE J. Quantum Electron.*, vol. QE-17, no. 2, pp. 217–226, Feb. 1981.

[41] F. Capasso, R. A. Logan, A. Hutchinson, and D. D. Manchon, "InGaAsP/InGaAs heterojunction p-i-n detectors with low dark current and small capacitance for 1.3–1.6 μm fiber optic systems," *Electron. Lett.*, vol. 16, no. 23, pp. 893–895, Nov. 6, 1980.

[42] T. P. Pearsall, M. Piskorski, A. Brochet, and J. Chevrier, "A $Ga_{0.47}In_{0.53}As$/InP heterophotodiode with reduced dark current," *IEEE J. Quantum Electron.*, vol. QE-17, no. 2, pp. 255–259, Feb. 1981.

[43] T. P. Pearsall, R. A. Logan, and C. G. Bethea, "GaInAs/InP large bandwidth (>2 GHz) pin detectors," *Electron. Lett.*, vol. 19, no. 16, pp. 611–612, Aug. 4, 1983.

[44] T. P. Lee, C. A. Burrus, A. Y. Cho, K. Y. Cheng, and D. D. Manchon, Jr., "Zn-diffused back-illuminated p-i-n photodiodes in InGaAs/InP grown by molecular beam epitaxy," *Appl. Phys. Lett.*, vol. 37, no. 8, pp. 730–731, Oct. 15, 1980.

[45] N. Susa, Y. Yamauchi, H. Ando, and H. Kanbe, "Planar type vapor-

phase epitaxial $In_{0.53}GA_{0.47}As$ photodiode," *IEEE Electron Device Lett.*, vol. EDL-1, no. 4, pp. 55–57, Apr. 1980.

[46] S. B. Hyder, R. R. Saxena, S. H. Chiao, and R. Yeats, "Vapor-phase epitaxial growth of InGaAs lattice-matched to (100) InP for photodiode application," *Appl. Phys. Lett.*, vol. 35, no. 10, pp. 787–789, Nov. 15, 1979.

[47] G. H. Olsen, "Low-leakage, high-efficiency, reliable VPE InGaAs 1.0–1.7 μm photodiodes," *IEEE Electron Device Lett.*, vol. EDL-2, no. 9, pp. 217–219, Sept. 1981.

[48] D. G. Jenkins and A. W. Mabbitt, "The reliability of GaInAs photodiodes and GaAs FETs for use in PIN-FET fiber optic receivers," presented at the IEEE Specialist Conf. Light Emitting Diodes Photodetectors, Ottawa-Hull, Canada, Sept. 1982.

[49] M. Ettenberg, G. H. Olsen, I. Ladany, P. Webb, N. J. DiGiuseppe, T. J. Zamerowski, and J. Appert, "On the reliability of 1.3 μm lasers and 1.0–1.7 μm detectors grown by vapor phase epitaxy," presented at the 9th IEEE Int. Semicond. Laser Conf., Rio de Janeiro, Brazil, Aug. 1984.

[50] L. S. Lichtmann, P. A. Kohl, and R. H. Burton, "Degradation of silicon nitride-protected planar InGaAs pin photodiodes by an electrochemical etching reaction," presented at the IEEE Specialist Conf. Light Emitting Diodes Photodetectors, Ottawa-Hull, Canada, September 1982.

[51] H. J. Stocker and D. E. Aspnes, "Surface chemical reactions on $In_{0.53}Ga_{0.47}As$," *Appl. Phys. Lett.*, vol. 42, no. 1, pp. 85–87, Jan. 1, 1983.

[52] V. Diadiuk, C. A. Armiento, S. H. Groves, C. E. Hurwitz, "Surface passivation techniques for InP and InGaAsP p-n junction structures," *IEEE Electron Device Lett.*, vol. EDL-1, no. 9, pp. 177–178, Sept. 1980.

[53] H. Nickel and E. Kuphal, "Surface-passivated low dark current InGaAs pin photodiodes," *J. Opt. Commun.*, vol. 4, no. 2, pp. 63–67, 1983.

[54] S. R. Forrest, I. Camlibel, O. K. Kim, H. J. Stocker, and J. R. Zuber, "Low dark-current, high-efficiency planar $In_{0.53}Ga_{0.47}As$/InP p-i-n photodiodes," *IEEE Electron Device Lett.*, vol. EDL-2, no. 11, pp. 283–285, Nov. 1981.

[55] H. Ishihara, K. Makita, Y. Sugimoto, T. Torikai, and K. Taguchi, "High-temperature aging tests on planar structure InGaAs/InP pin photodiodes with Ti/Pt and Ti/Au contact," *Electron. Lett.*, vol. 20, no. 16, pp. 654–656, Aug. 2, 1984.

[56] G. R. Antell and R. F. Murison, "InGaAs/InP mesa photodetector passivated with silicon dioxide," *Electron. Lett.*, vol. 20, no. 22, pp. 19–920, Oct. 25, 1984.

[57] R. H. Saul, F. S. Chen and P. W. Shumate, "Reliability of InGaAs photodiodes for SL applications," *AT&T Tech. J.*, vol. 64, pp. 861–882, Mar. 1985.

[58] P. Poulain, M. Razeghi, P. Hirtz, K. Kamierski, and B. de Cremoux, "Zn diffused InGaAs/InP PIN-DH photodiodes grown by low-pressure-MOCVD," presented at the 9th IEEE Int. Semicond. Laser Conf., Rio de Janeiro, Brazil, Aug. 1984.

[59] R. H. Moss and E. A. D. White, "Current status of MOVPE for optoelectronic application," *Br. Telecomm. Technol. J.*, vol. 2, no. 4, pp. 74–83, Sept. 1984.

[60] R. H. Moss and S. Ritchie, "The development of gallium indium arsenide for photodetectors," *Br. Telecomm. Technol. J.*, vol. 1, no. 1, July, 1983.

[61] D. R. Smith, R. C. Hooper, and R. P. Webb, "High performance digital optical receivers with PIN photodiodes," in *Proc. Int. Symp. Circuits Syst.* (Tokyo), pp. 511–514, 1979.

[62] B. M. MacDonald and A. G. Saunders, "Coupling of an optical fibre to a substrate-entry photodiode," *Electron. Lett.*, vol. 17, no. 22, pp. 832–833, Oct. 29, 1981.

[63] M. C. Brain, "Comparison of available detectors for digital optical fiber systems for the 1.2–1.55 μm wavelength range," *IEEE J. Quantum Electron.*, vol. QE-18, no. 2, pp. 219–224, Feb. 1982.

[64] D. Arnold, R. Fischer, J. Klem, F. Ponse, and H. Morkoc, "Reduction of backgating in GaAs/AlGaAs MESFETs by optimization of active-layer/buffer-layer interface," *Electron. Lett.*, vol. 20, no. 9, pp. 376–378, Apr. 26, 1984.

[65] R. F. Leheny, R. E. Nahory, M. A. Pollack, A. A. Ballman, E. D. Beebe, J. C. DeWinter, and R. J. Martin, "Integrated $In_{0.53}Ga_{0.47}As$ p-i-n F.E.T. photoreceiver," *Electron. Lett.*, vol. 16, no. 10, pp. 353–355, May 8, 1980.

[66] K. Kasahara, J. Hayashi, K. Makita, K. Taguchi, A. Suzuki, H. Nomura, and S. Matsushita, "Monolithically integrated $In_{0.53}Ga_{0.47}As$-PIN/InP-MISFET photoreceiver," *Electron. Lett.*, vol. 20, no. 8, pp. 314–315, Apr. 12, 1984.

[67] C. L. Cheng, A. S. H. Liao, T. Y. Chang, R. F. Leheny, L. A. Coldren, and B. Lalevic, "Submicrometer self-aligned recessed gate InGaAs MISFET exhibiting very high transconductance," *IEEE Electron Device Lett.*, vol. EDL-5, pp. 169–171, May 1984.

[68] C. L. Cheng, A. S. H. Liao, T. Y. Chang, E. A. Caridi, L. A. Coldren, and B. Lalevic, "Silicon oxide enhanced Schottky gate $In_{0.53}Ga_{0.47}As$ FET's with a self-aligned recessed gate structure," *IEEE Electron Device Lett.*, vol. EDL-5, no. 12, pp. 511–514, Dec. 1984.

[69] C. L. Cheng, L. A. Coldren, B. I. Miller, A. S. H. Liao, R. F. Leheny, and B. Lalevic, "A new self-aligned recessed-gate InP MESFET," *IEEE Trans. Electron Devices*, vol. ED-31, no. 6, pp. 840–841, June 1984.

[70] B. Owen, "PIN-GaAs FET optical receiver with a wide dynamic range," *Electron. Lett.*, vol. 18, no. 14, pp. 626–627, July 8, 1982.

[71] M. C. Brain, "Characterisation and estimated performance of commercial n^+-p Ge APDs for long-wavelength optical receivers," *Opt. Quantum Electron.*, vol. 13, no. 5, pp. 353–367, 1981.

[72] ——, "Noise and responsivity of commercial p^+-n Ge avalanche photodiodes," *Electron. Lett.*, vol. 19, no. 20, pp. 813–815, Sept. 29, 1983.

[73] S. R. Forrest, M. DiDomenico, Jr., R. G. Smith, and H. J. Stocker, "Evidence for tunneling in reverse-biased III-V photodetector diodes," *Appl. Phys. Lett.*, vol. 36, pp. 580–582, 1980.

[74] S. R. Forrest, "Sensitivity of avalanche photodetector receivers for high bit rate long wavelength optical communication systems," in *Semiconductors and Semi-Metals*, vol. 22, part D, Willardson and Beer, Eds. New York: Academic, New York, 1985, ch. 4.

[75] R. B. Emmons, "Avalanche photodiode frequency response," *J. Appl. Phys.*, vol. 38, p. 3705, 1967.

[76] T. Kaneda and H. Takanashi, "Avalanche build-up time of the germanium avalanche photodiode," in *Japan. J. Appl. Phys.*, vol. 12, p. 1091, 1973.

[77] S. R. Forrest, "Gain-bandwidth-limited response in long-wavelength avalanche photodiodes," *J. Lightwave Tech.*, vol. LT-2, pp. 34–39, 1984.

[78] S. M. Sze, *Physics of Semiconductor Devices.* New York: Wiley, 1969.

[79] K. Nishida, K. Taguchi, and Y. Matsumoto, "InGaAsP heterostructure avalanche photodiodes with high avalanche gain," *Appl. Phys. Lett.*, vol. 35, pp. 251–252, 1979.

[80] S. R. Forrest, R. G. Smith, and O. K. Kim, "Performance of $In_{0.53}Ga_{0.47}As$/InP photodiodes," *IEEE J. Quantum Electron.*, vol. QE-18, pp. 2040–2048, 1982.

[81] S. R. Forrest, O. K. Kim, and R. G. Smith, "Optical response time of InGaAs/InP avalanche photodiodes," *Appl. Phys. Lett.*, vol. 41, pp. 95–98, 1982.

[82] H. Kanbe, N. Susa, H. Nakagome, and H. Ando, "InGaAs avalanche photodiode with InP p-n junction," *Electron. Lett.*, vol. 16, pp. 163–164, 1980.

[83] N. Susa, H. Nakagome, O. Mikami, H. Ando, and H. Kanbe, "New InGaAs/InP avalanche photodiode structure for the 1–1.6 μm wavelength region," *IEEE J. Quantum Electron.*, vol. QE-16, pp. 864–869, 1980.

[84] V. Diadiuk, S. H. Groves, C. E. Hurwitz, and G. W. Iseler, "Low dark-current, high-gain, GaInAs/InP avalanche photodetectors," *IEEE J. Quantum Electron.*, vol. QE-17, pp. 260–263, 1981.

[85] O. K. Kim, S. R. Forrest, W. A. Bonnor, and R. G. Smith, "A high-gain $In_{0.53}Ga_{0.47}As$/InP avalanche photodiode with no tunneling leakage current," *Appl. Phys. Lett.*, vol. 39, pp. 402–404, 1981.

[86] T. Shirai, F. Osaka, S. Yamasaki, T. Kaneda, and N. Susa, "Multiplication noise of InP avalanche photodiodes," *Appl. Phys. Lett.*, vol. 39, p. 168, 1981.

[87] T. Shirai, S. Yamasaki, F. Osaka, K. Nakajima, and T. Kaneda, "Multiplication noise in planar InP/InGaAsP heterostructure avalanche photodiodes," *Appl. Phys. Lett.*, vol. 40, p. 532, 1981.

[88] L. W. Cook, G. E. Bulman, and G. E. Stillman, "Electron and hole impact ionisation coefficients in InP determined by photomultiplication," *Appl. Phys. Lett.*, vol. 40, p. 589, 1982.

[89] G. E. Bulman, L. W. Cook, and G. E. Stillman, "The effect of electroabsorption on the determination of ionisation coefficients," *Appl. Phys. Lett.*, vol. 39, p. 813, 1984.

[90] G. E. Stillman, L. W. Cook, G. E. Bulman, N. Tabatabaie, R. Chin, and P. D. Dapkus, "Long-wavelength (1.3 to 1.6 μm) detectors for

fiber-optical communications," *IEEE Trans. Electron Devices*, vol. ED-29, p. 1355, 1982.
[91] J. C. Campbell, A. G. Dentai, W. S. Holden, and B. L. Kasper, "High-speed InP/InGaAsP/InGaAs avalanche photodiodes," in *IEDM Tech. Dig.*, p. 464, 1983.
[92] H. Ando, Y. Yamasaki, H. Nakagome, N. Susa, and H. Kanbe, "InGaAs/InP separated absorption and multiplication regions avalanche photodiode using liquid- and vapor-phase epitaxies," *IEEE J. Quantum Electron.*, vol. QE-17, pp. 250–254, 1981.
[93] H. Ando, N. Susa, and H. Kanbe, "Low-temperature Zn- and Cd-diffusion profiles in InP and formation of guard-ring in InP avalanche photodiodes," *IEEE Trans. Electron Devices*, vol. ED-29, pp. 1408–1413, 1982.
[94] M. Ikeda, K. Wakita, S. Hata, S. Kondo, and H. Kanbe, "Planar InP/InGaAs APD with a guard ring formed by Cd diffusion through SiO_2," *Electron. Lett.*, vol. 19, pp. 61–62, 1983.
[95] T. Shirai, F. Osaka, S. Yamasaki, K. Nakajima, and T. Kaneda, "1.3 μm InP/InGaAsP planar avalanche photodiodes," *Electron. Lett.*, vol. 17, pp. 826–827, 1981.
[96] T. Shirai, T. Mikawa, and T. Kaneda, "InGaAs avalanche photodiodes for 1 μm wavelength region," *Electron. Lett.*, vol. 19, pp. 534–535, 1983.
[97] T. Shirai, S. Yamasaki, H. Kawata, K. Nakajima, and T. Kaneda, "A planar InP/InGaAsP heterostructure avalanche photodiode," *IEEE Trans. Electron Devices*, vol. ED-29, pp. 1404–1407, 1982.
[98] K. Yasuda, Y. Kishi, T. Shirai, T. Mikawa, S. Yamasaki, and T. Kaneda, "InP/InGaAs buried structure avalanche photodiode," *Electron. Lett.*, vol. 20, pp. 158–159, 1984.
[99] K. Yasuda, T. Mikawa, T. Kishi, and T. Kaneda, "Multiplication-dependent frequency response of InP/InGaAs avalanche photodiode," *Electron. Lett.*, vol. 20, pp. 373–374, 1984.
[100] H. Melchior and W. T. Lynch, "Signal and noise response of high speed germanium avalanche photodiodes," *IEEE Trans. Electron Devices*, vol. ED-13, p. 829, 1966.
[101] H. Ando, H. Kanbe, T. Kimura, T. Yamaoki and T. Kaneda, "Characteristics of germanium avalanche photodiodes in the wavelength region of 1.0–1.6 μm," *IEEE J. Quantum Electron.*, vol. QE-14, p. 804, 1978.
[102] T. Mikawa, S. Kagawa, T. Kaneda, T. Sakurai, H. Ando and O. Mikami, "A low-noise n^+np germanium avalanche photodiode," *IEEE J. Quantum Electron.*, vol. QE-17, pp. 210–216, 1981.
[103] H. Kanbe, G. Grosskopf, O. Mikami, and S. Machida, "Dark current noise characteristics and temperature dependence in germanium avalanche photodiodes," *IEEE J. Quantum Electron.*, vol. QE-17, p. 1534, 1981.
[104] O. Mikami, H. Ando, H. Kanbe, T. Mikawa, T. Kaneda, and Y. Toyama, "Improved germanium avalanche photodiodes," *IEEE J. Quantum Electron.*, vol. QE-16, pp. 1002–1007, 1980.
[105] H. D. Law, L. R. Tomasetta, K. Nakano, and J. S. Harris, "1.0 μm to 1.4 μm high speed avalanche photodiodes," *Appl. Phys. Lett.*, vol. 33, pp. 416–417, 1978.
[106] F. Capasso, M. B. Panish, S. Sumski, and P. W. Foy, "Very high quantum efficiency GaSb mesa photodetectors between 1.3 and 1.6 μm," *Appl. Phys. Lett.*, vol. 36, pp. 165–166, 1980.
[107] F. Capasso, M. B. Panish, and S. Sumski, "The liquid phase epitaxial growth of low net donor concentration (5×10^{14}–5×10^{15}/cm^3) GaSb for detector applications in the 1.3–1.6 μm region," *IEEE J. Quantum Electron.*, vol. QE-17, p. 273, 1981.
[108] H. D. Law, R. Chin, K. Nakano, and R. A. Milano, "The GaAlAsSb quaternary and GaAlSb ternary alloys and their application to infrared detectors," *IEEE J. Quantum Electron.*, vol. QE-17, p. 275, 1981.
[109] N. Tabatabaie, G. E. Stillman, R. Chin, and P. D. Dapkus, "Tunneling in the reverse dark current of GaAlAsSb avalanche photodiodes," *Appl. Phys. Lett.*, vol. 40, pp. 415–416, 1982.
[110] O. Hildebrand, W. Keubart, W. K. Benz, and M. H. Pilkuhn, "$Ga_{1-x}Al_xSb$ avalanche photodiodes: Resonant impact ionization with very high ratio of ionization coefficients," *IEEE J. Quantum Electron.*, vol. QE-17, p. 284, 1981.
[111] C. Verie, F. Raymond, J. Besson, and T. Nguyen-Duy, "Bandgap spin-orbit splitting resonance effects in $Hg_{1-x}Cd_xTe$ alloys," *J. Crystal Growth*, vol. 59, p. 342, 1982.
[112] J. P. Gordon, R. E. Nahory, M. A. Pollack, and J. M. Warlock, "Low-noise multistage avalanche photodetectors," *Electron. Lett.*, vol. 15, pp. 518–519, 1979.
[113] R. Chin, N. Holonyak, Jr., G. E. Stillman, J. Y. Tang, and K. Hess, "Impact ionisation in multilayer heterojunction structures," *Electron. Lett.*, vol. 16, pp. 467–469, 1980.
[114] F. Capasso, W. T. Tsang, A. L. Hutchinson, and G. F. Williams, "The superlattice photodetector: A new avalanche photodiode with a large ionisation rate ratio," in *IEDM Tech. Dig.*, pp. 184–187, 1981.
[115] ——, "Enhancement of electron impact ionisation in a superlattice: A new avalanche photodiode with a large ionisation rate ratio," *Appl. Phys. Lett.*, vol. 40, pp. 38–40, 1982.
[116] H. Shichijo, R. M. Kolbas, N. Holonyak, Jr., R. D. Dupuis, P. D. Dapkus, "Carrier collection in a semiconductor quantum well," *Solid State Commun.*, vol. 27, pp. 1029–1032, 1978.
[117] H. Kroemer, "Theory of a wide-gap emitter for transistors," *Proc. IRE*, vol. 45 pp. 1535–1537, Nov. 1957.
[118] P. D. Wright, R. J. Nelson, and T. Cella, "High-gain InGaAsP-InP heterojunction phototransistors," *Appl. Phys. Lett.*, vol. 37, no. 2, pp. 192–194, July 15, 1980.
[119] J. C. Campbell, A. G. Dentai, C. A. Burrus, and J. F. Ferguson, "InP/InGaAS heterojunction phototransistors," *IEEE J. Quantum Electron.*, vol. QE-17, no. 2, pp. 264–269, Feb. 1981.
[120] D. Fritzsche, E. Kuphal, and R. Aulbach, "Fast response InP/InGaAsP heterojunction phototransistors," *Electron. Lett.*, vol. 17, no. 5, pp. 178–180, Mar. 5, 1981.
[121] J. C. Campbell, C. A. Burrus, A. G. Dentai, and K. Ogawa, "Small-area high-speed InP/InGaAs phototransistor," *Appl. Phys. Lett.*, vol. 39, no. 10, pp. 820–821, Nov. 15, 1981.
[122] R. A. Milano, P. D. Dapkus, and G. E. Stillman, "An analysis of the performance of heterojunction phototransistors for fiber optic communications," *IEEE Trans. Electron Devices*, vol. ED-29, no. 2, pp. 266–274, Feb. 1982.
[123] K. Tabatabaie-Alavi, C. G. Fonstad, Jr., "Performance comparison of heterojunction phototransistors, p-i-n FET's, and APD-FET's for optical fiber communication systems," *IEEE J. Quantum Electron.*, vol. QE-17, no. 12, pp. 2259–2261, Dec. 1981.
[124] J. C. Campbell and A. G. Dentai, "InP/InGaAs heterojunction phototransistor with integrated light emitting diode," *Appl. Phys. Lett.*, vol. 41, no. 2, pp. 192–193, July 15, 1983.
[125] M. C. Brain and D. R. Smith, "Phototransistors in digital optical communication systems," *IEEE J. Quantum Electron.*, vol. QE-19, no. 6, pp. 1139–1148, June, 1983.
[126] F. H. de la Moneda, E. R. Chenette, and A. van der Ziel, "Noise in phototransistors," *IEEE Trans. Electron Devices*, vol. ED-18, no. 6, pp. 340–346, June 1971.
[127] J. C. Campbell and K. Ogawa, "Heterojunction phototransistors for long-wavelength optical receivers," *J. Appl. Phys.*, vol. 53, no. 2, pp. 1203–1208, Feb. 1982.
[128] C. Y. Chen, Y. M. Pang, K. Alavi, A. Y. Cho, and P. A. Garbinski, "Interdigitated $Al_{0.48}In_{0.52}As/Ga_{0.47}In_{0.53}As$ photoconductive detectors," *Appl. Phys. Lett.*, vol. 44, pp. 99–101, 1984.
[129] C. Y. Chen, B. L. Kasper, and H. M. Cox, "High-sensitivity $Ga_{0.47}In_{0.53}As$ photoconductive detectors prepared by vapor phase epitaxy," *Appl. Phys. Lett.*, vol. 44, pp. 1142–1144, 1984.
[130] C. Y. Chen, B. L. Kasper, H. M. Cox, and J. K. Plourde, "2-Gb/s sensitivity of a $Ga_{0.47}In_{0.53}As$ photoconductive detector/GaAs field effect transistor hybrid photoreceiver," *Appl. Phys. Lett.*, vol. 46, pp. 379–381, 1985.
[131] S. R. Forrest, "The sensitivity of photoconductor receivers for long-wavelength optical communications," *J. Lightwave Technol.*, vol. LT-3, pp. 347–360, 1985.
[132] T. G. Hodgkinson, R. Wyatt, and D. W. Smith, "Experimental assessment of a 140 Mbit/s coherent optical receiver at 1.52 μm," *Electron. Lett.*, vol. 18, no. 12, pp. 523–525, June 10, 1982.
[133] T. G. Hodgkinson, D. W. Smith, and R. Wyatt, "1.5 μm optical heterodyne system operating over 30 km of monomode fiber," *Electron. Lett.*, vol. 18, no. 21, pp. 929–930, Oct. 14, 1982.
[134] T. G. Hodgkinson, R. Wyatt, D. J. Malyon, B. K. Nayar, R. A. Harmon, and D. W. Smith, "Experimental 1.5 μm coherent optical fibre transmission systems," *Ann. Telecomm.*, vol. 38, no. 1–2, pp. 73–81, Jan.–Feb. 1983.
[135] D. J. Malyon, T. G. Hodgkinson, D. W. Smith, R. C. Booth, and B. E. Daymond-John, "PSK homodyne receiver sensitivity measurements at 1.5 μm," *Electron. Lett.*, vol. 19, no. 4, pp. 144–146, Feb. 17, 1983.
[136] R. Wyatt, T. G. Hodgkinson, and D. W. Smith, "1.52 μm PSK heterodyne experiment featuring an external cavity diode laser local oscillator," *Electron. Lett.*, vol. 19, no. 14, pp. 550–552, July 7, 1983.
[137] R. Wyatt, D. W. Smith, T. G. Hodgkinson, R. A. Harmon, and W.

J. Devlin, "140 Mbit/s optical FSK fibre heterodyne experiment at 1.54 μm," *Electron. Lett.*, vol. 20, no. 22, pp. 912–913, Oct. 25, 1984.

[138] S. D. Walker and L. C. Blank, "Long-wavelength transimpedance optical receiver performance enhancement using cooled germanium avalanche photodiodes," *Electron. Lett.*, vol. 20, pp. 16–18, 1984; also "GeAPD/GaAsFET/OP-AMP transimpedance optical receiver design having minimum noise and intersymbol interference characteristic," *Electron. Lett.*, vol. 20, pp. 808–809, 1984.

[139] T. Mikawa, T. Kaneda, H. Nishimoto, M. Motegi, and H. Okushima, "Small-active-area germanium avalanche photodiode for single-mode fibre at 1.3 μm wavelength," *Electron. Lett.*, vol. 19, pp. 452–453, 1983.

[140] L. Bickers, L. C. Blank, and S. D. Walker, "Long-span optical transmission experiment over 222.8 km of commercial monomode fibre at 140 Mbit/s and 1.525 μm," *Electron. Lett.*, vol. 21, no. 7, pp. 267–268, Mar. 28, 1985.

[141] S. R. Forrest, G. F. Williams, O. K. Kim, and R. G. Smith, "Excess noise and receiver sensitivity measurements of $In_{0.53}Ga_{0.47}As/InP$ avalanche photodiodes," *Electron. Lett.*, vol. 17, pp. 917–919, 1981.

[142] Y. Matsushima, N. Seki, S. Akiba, Y. Noda, and Y. Kushivo, "Receiver sensitivity of InGaAs/InP heterostructure avalanche photodiode with InGaAsP buffer layers at 1.5–1.6 μm region," *Electron. Lett.*, vol. 19, pp. 845–846, 1983.

[143] Y. Sugimoto, T. Torikai, K. Makita, H. Ishikara, K. Minemura, and K. Taguchi, "High speed planar-structure InP/InGaAsP/InGaAs avalanche photodiode grown by VPE," *Electron. Lett.*, vol. 20, pp. 653–654, 1984.

[144] B. L. Kasper, R. A. Linke, K. L. Walker, L. G. Cohen, T. L. Koch, T. J. Bridges, E. G. Burkhart, R. A. Logan, R. W. Dawson, and J. C. Campbell, "A 130 km transmission experiment at 2 Gb/s using silica-core fiber and a vapor phase transported DFB laser," presented at the 10th Europ. Conf. Comm., Stuttgart, Germany, Sept. 3–6 1984, post-deadline paper.

[145] A. P. McDonna, D. J. McCartney, and D. B. Mortimore, "1.3 μm bi-directional optical transmission over 31 km of installed single-mode fibre using optical couplers," *Electron. Lett.*, vol. 20, pp. 722–723, 1984.

[146] M. C. Brain, D. R. Smith, P. P. Smyth, and B. R. White, "PIN-FET hybrid optical receivers for Gbit/s transmission systems at 1.1–1.6 μm," presented at the 9th IEEE Int. Semicond. Laser Conf., Rio de Janeiro, Brazil, Aug. 7–10, 1984.

[147] M. Niwa, Y. Abe, Y. Tashiro, K. Minemura, and H. Iwasaki, "Receiver sensitivity temperature dependence of small detectable area Hi-Lo germanium avalanche photodiode," *Electron. Lett.*, vol. 20, pp. 993–994, 1984.

[148] *IEEE Spectrum*, Jan. 1985.

*

Mike Brain was born in England in 1949. He received the B.A. and M.A. degrees in natural sciences from Fitzwilliam College, Cambridge University, Cambridge, England, in 1971 and 1976, respectively.

Since 1971 he has been employed by British Telecom Research Laboratories (formerly the British Post Office Research Department) to study various aspects of optical fiber communications, including propagation in optical fibers, the characterization of photodetectors, and the design and development of optical receivers. He now leads a group concerned with coherent optical systems and measurements.

Mr. Brain is a Member of the Institution of Electrical Engineers.

*

Tien-Pei Lee, photograph and biography not available at time of publication.

Planar Embedded InP/GaInAs p-i-n Photodiode for Very High-Speed Operation

S. MIURA, H. KUWATSUKA, T. MIKAWA, AND O. WADA, MEMBER, IEEE

***Abstract*—Planar embedded InP/GaInAs p-i-n photodiodes have been fabricated by using preferential ion-beam etching for planarizing and embedding the p-i-n photodiode structure in a semi-insulating InP substrate. The stray capacitances caused by a bonding pad and an interconnection have been markedly reduced, which resulted in extremely low capacitance of less than 0.08 pF for a diameter of 20 μm of photosensitive area. It has been demonstrated by an optical heterodyne technique that the photodiode exhibits a maximum cutoff frequency of 14 GHz. This result was analyzed taking the depletion layer thickness into account and has been found to be dominated by the carrier transit time. The demonstrated low capacitance and high-speed response result indicates the suitability of the p-i-n photodiodes not only for a discrete p-i-n photodiode but also for optoelectronic integration.**

I. Introduction

Marvelous advances have been made in this decade in fiber-optic communication systems in both the 0.8–0.9 μm and 1.3–1.6 μm wavelength regions, and these developments have led to practical communication systems operating at a few hundred megabits per second. The development of advanced optical communication systems that operate at several gigabits per second has recently been accelerated in the long-wavelength region, because of the low attenuation as well as the lowest dispersion of optical fiber in this wavelength region. As for photodetectors, avalanche photodiodes (APD's) have been widely used for high-bit-rate systems primarily due to the high signal-to-noise ratio of the APD's. As far as the noise performance is concerned, APD's are considered to be the most attractive photodetectors. However, the response speed of APD's is limited by the avalanche buildup time in the bit rate range above several gigabits per second [1]. On the other hand, the speed of InP/GaInAs p-i-n photodiodes is free from the avalanche buildup time and is limited only by carrier transit time and/or the *RC* time constant. Therefore, considering the application to multigigabit systems, p-i-n photodiodes would be one of the most promising and applicable photodetectors.

In a lightwave receiver, InP/GaInAs p-i-n photodiodes are combined with a front-end such as field-effect transistor (p-i-n/FET) [2]–[5]. The layer thickness of the p-i-n photodiode must therefore be optimized either for discrete p-i-n photodiode or for p-i-n/FET. In a p-i-n/FET configuration, the response speed is most likely governed by the *RC* time constant rather than the transit time because a large load resistance is usually used in order to reduce the thermal noise. Therefore, in this application, the n^- layers must be thicker compared with those for discrete applications, to reduce *C* and thus the *RC* time constant. Moreover, the noise of the p-i-n/FET is strongly dependent on the figure of merit (C^2/g_m, with g_m the transconductance of the front-end FET). Towards the realization of a high-speed and low-noise p-i-n/FET, the reduction of the p-i-n photodiode capacitance must be the most important target.

In constructing InP/GaInAs p-i-n photodiodes, several structures have been developed so far. One is a planar structure using multilayers grown on an n^+-InP substrate [6]. In this structure, a photosensitive area is formed by selective Zn or Cd diffusion or Be ion-implantation by using a dielectric mask such as SiO_2 or Si_3N_4 with local openings. The p^+-region serves as the photosensitive area and also the ohmic contact and bonding pad. A planar structure can provide high reliability because the surface is passivated during fabrication, but large stray capacitances will result from bonding pads. Another structure is a back-illuminated mesa structure with which can reduce parasitic capacitances [7]–[9]. However, these mesa structures would have some disadvantages in diode reliability and in compatibility with electronic elements for optoelectronic integration. So as to accomplish a low capacitance, simultaneously avoiding the reliability problem and realizing applicability for optoelectronic integration, we believe that a planar embedded structure is the best choice. This is because the parasitic capacitances caused by the bonding pads and the interconnections are able to be substantially minimized in this planar embedded structure. However, there have been no reports on photodiodes which realize low capacitance and high-speed response by making use of the full benefit of an embedded structure [10], [11].

The purpose of this paper is to describe the structure, fabrication process, and performance of a planar embedded InP/GaInAs p-i-n photodiode. The present p-i-n photodiode exhibits an extremely low capacitance of as low as 0.08 pF and a very high-speed response of up to 14 GHz. The response characteristics are confirmed by a calculation.

Manuscript received February 9, 1987; revised May 12, 1987.
The authors are with Fujitsu Laboratories, Ltd., 10-1, Morinosato-Wakamiya, Atsugi 243-01, Japan.
IEEE Log Number 8716075.

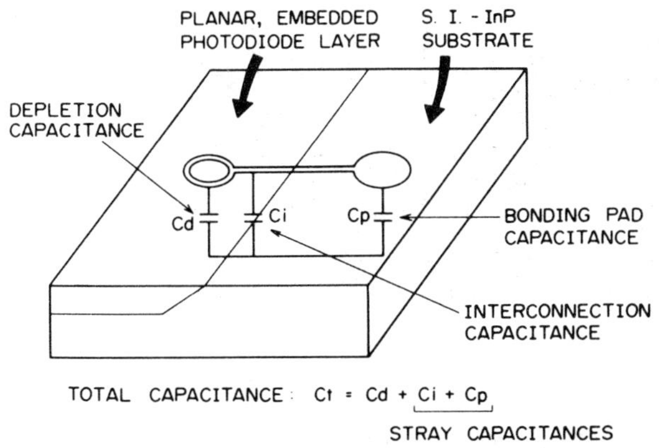

Fig. 1. Simplified schematic illustration of planar embedded p-i-n photodiode. Stray capacitances caused by bonding pad and interconnection are expected to be minimized in this structure.

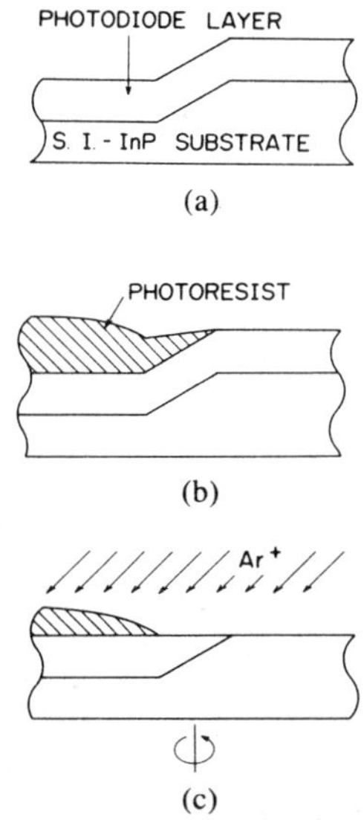

Fig. 2. Key procedure of planarization technique. The well is formed on the substrate by a graded-step process. (a) After epitaxial growth, (b) a double-layer photoresist mask is formed. (c) Ion-beam etching is again performed resulting in a planar embedded structure.

II. Fabrication

In this section, we will first describe the idea of lowering the capacitance in a planar embedded structure. A simplified structure of a planar embedded p-i-n photodiode is schematically illustrated in Fig. 1. In this structure, the photodiode layers are embedded in a well formed on a semi-insulating (SI) InP substrate. The junction area which includes the photosensitive area and the contact area is formed selectively in the photodiode layer. The depletion capacitance of the p-n junction is indicated as C_d in the figure. A bonding pad is formed on the SiN on SI-InP substrate and it adds a bonding pad capacitance C_p. The bonding pad capacitance becomes almost negligible in this structure by the use of a semi-insulating substrate. The most influential source of stray capacitance is considered to be that of the interconnection capacitance C_i between the junction and the bonding pad. Since in the planar embedded structure the substrate surface is planar, a fine geometry interconnection can be easily formed. Therefore, the stray capacitance given by the sum of C_i and C_p is expected to be substantially minimized in this planar embedded structure.

With these considerations in mind, we have developed a technique to fabricate a planar embedded InP/GaInAs p-i-n photodiode [12]. The key procedure of the planarization technique is shown in Fig. 2. First, a stripe geometry well was formed in the $<0\bar{1}\bar{1}>$ direction on a SI-InP substrate by using the graded step process, which utilizes preferential ion-beam etching with a photoresist mask as has been reported previously [13], [14]. After the formation of the well, a short chemical etching was performed to remove damage from the ion-beam etched surface. Then the photodiode layer was grown on the stepped substrate surface by chloride vapor-phase epitaxy (VPE) as shown in Fig. 2(a). After the growth, double photoresist layers (AZ-4620, Shipley) were delineated so as to fill up the well as shown in Fig. 2(b). The first photoresist layer was patterned to coincide with the upper edge of the sidewall. Then, it was baked at 200°C to obtain a smoothly graded pattern on the step region. In the mask formation process, the photoresist thickness was precisely controlled because the surface profile of the photoresist at the step region is transferred to the substrate by ion-beam etching. The second photoresist layer was patterned to protect the bottom of the well. Then the ion-beam etching was performed again to transfer the photoresist pattern on the step region and also to etch away the unnecessary photodiode layers on the upper plane of the substrate surface [15], [16]. In the ion-beam etching procedure, the ion energy and the ion current density were chosen to be 500 eV and 0.41 A/cm^2, respectively. These conditions yield smooth and defect-free surfaces. Moreover, the substrate surface was sufficiently inclined against the ion-beam to reduce defects such as bumps; the ion-beam incident angle was chosen to be 75° [14]. By using this planarization technique, a planar surface was reproducibly obtained.

III. Structure

A schematic cross section of the planar embedded InP/GaInAs p-i-n photodiode is shown in Fig. 3. The photodiode layers were embedded in a well in a SI-InP substrate. Two kinds of photodiode layers were grown in order to optimize the photodiode structure both for low capacitance (diode I) and for high-speed (diode II). A thick n^- layer structure was used for the low-capacitance p-i-n photodiode, while a thin n^- layer structure was used for high-speed applications. The wafers consisted of the following layers; an n^--InP window layer (1.5-μm 1×10^{16} cm^{-3}), an n^--$Ga_{0.47}In_{0.53}As$ light absorption layer (1.8 or 0.5-μm 1.5×10^{15} cm^{-3}), an n^--InP (6 or 3-μm 1×10^{15} cm^{-3}), and an n^+-InP contact layer (2-μm 1×10^{17} cm^{-3}). The photosensitive area was formed by selective Zn diffusion in a closed silica tube at the temperature of 500°C to reach the hetero-interface. The ohmic contacts were formed by using Au/Zn/Au for the p-contact and Au/AuGe for the n-contact, respectively. Anti-

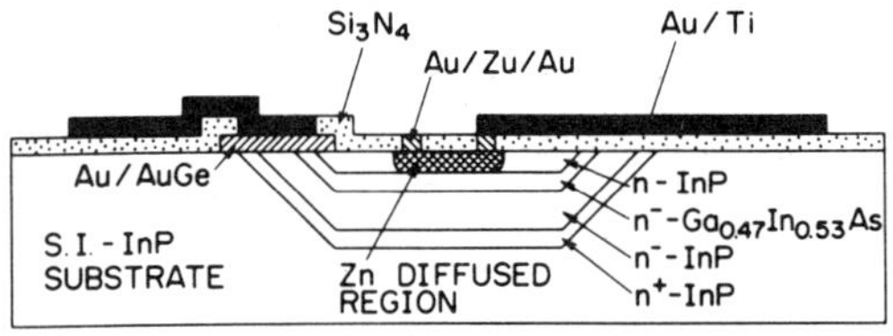

Fig. 3. Schematic cross section of planar embedded InP/GaInAs p-i-n photodiode. Photodiode layer is embedded in a well formed on SI-InP substrate.

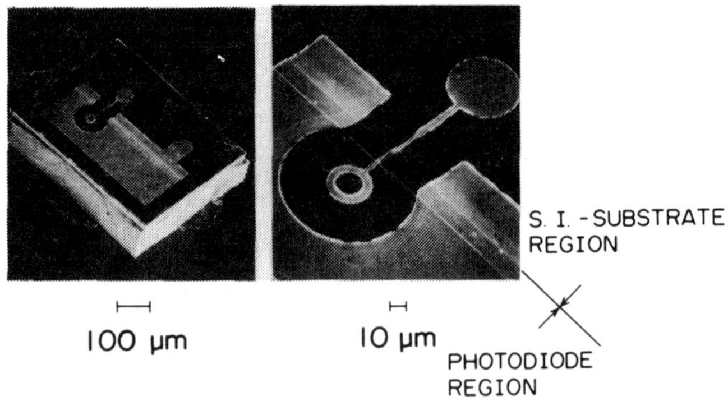

Fig. 4. Scanning electron photomicrographs of planar embedded InP/GaInAs p-i-n photodiode. Photodiode layer is embedded with good planarity, which facilitated the fine line interconnection as narrow as 2.5 μm.

reflection and passivation coatings were performed using plasma-deposited Si_3N_4 films. The interconnection and the bonding pads were formed by patterning a Au/Ti film. The n-contact was formed on the surface where the n^+-InP layer was exposed, and this resulted in a planar structure over the whole chip.

Scanning electron photomicrographs of the photodiode are shown in Fig. 4. The chip size was 500 × 800 μm. The photodiode layer was successfully embedded in a well formed on the SI-InP substrate in the left half of the chip, while the bonding pads were formed on a Si_3N_4 passivation film deposited on the surface of the SI-InP substrate exposed in the right half of the chip. With the benefit of the planar embedded structure, a fine geometry interconnection as narrow as 2.5 μm was realized, which is expected to reduce the parasitic capacitances, as has been explained before.

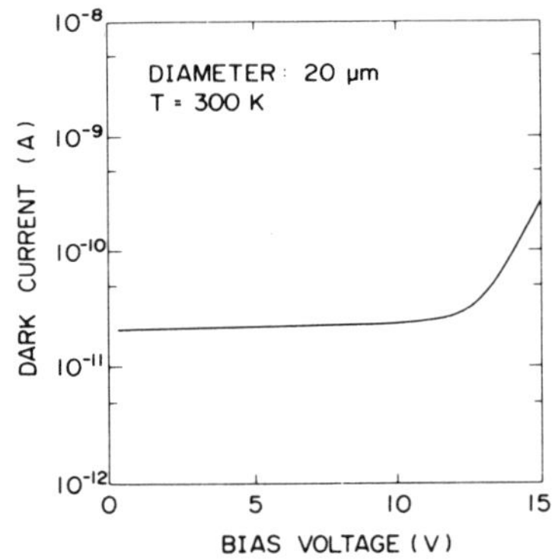

Fig. 5. Current-voltage characteristic of p-i-n photodiode with diameter of photosensitive area of 20 μm. The dark current is as low as 30 pA at 10 V.

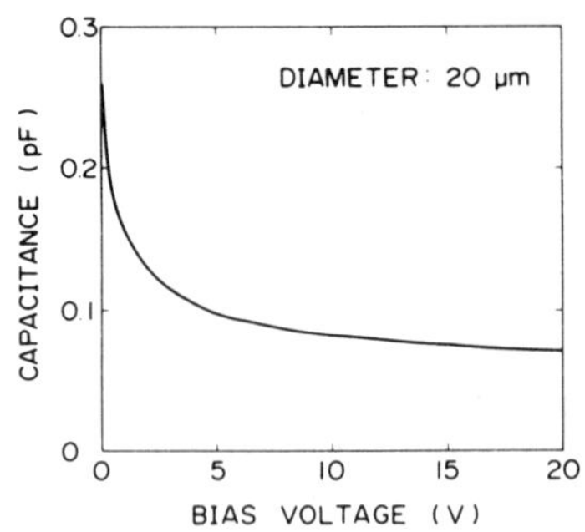

Fig. 6. Capacitance-voltage characteristic of p-i-n photodiode. Extremely low capacitance of 0.08 pF was obtained by marked reduction of stray capacitances.

IV. Device Performance

A. DC Characteristics

We have fabricated photodiodes with different diameters of the photosensitive area. The current-voltage characteristics of the p-i-n photodiode with a 20-μm-diameter photosensitive area and with thicker n^- layers (diode I) is shown in Fig. 5. The leakage current was as low as 30 pA at the bias voltage of 10 V. For bias voltages lower than 10 V, the variation of the dark current with bias voltage is weak and the leakage is thought to be generated by surface generation-recombination current. A sudden increase of the dark current around 13 V is attributed to tunneling current in the $Ga_{0.47}In_{0.53}As$ layer, especially at the periphery of the junction. This was ascertained by the dependence of the dark current on the junction diameter; the dark current above 13 V was proportional to the junction diameter. Therefore, a strong field is considered to be generated at the periphery. The quantum efficiencies of AR coated photodiodes with thicker and thinner n^- layers were 85 and 40 percent, respectively, at 1.3-μm wavelength. The capacitance-voltage characteristic of the p-i-n photodiode with a diameter of the photosensitive area of 20 μm and with thicker n^- layers (diode I) is shown in Fig. 6. The capacitance of these p-i-n photodiodes was as low as 0.07 pF at 20 V. The p-i-n photodiode with thinner n^- layers (diode II) showed a capacitance of 0.08 pF at 20 V. The dependence of the capacitance on the junction diameter is shown in Fig. 7 (V = 10 V). The capacitances of diode I and diode II were almost the same at 10 V because of the undepleted n^- layers. The ideal line shows the calculation without any stray capacitances, so that the line is proportional to the square of a diameter. The present p-i-n photodiodes have bonding pads with a diameter of 50 μm. The capacitance of a conventional p-i-n photodiode having a bonding pad with the same diameter on a Si_3N_4 film on a conductive layer was calculated and is shown in the figure for comparison. We note that the experimental data is very close to the ideal line due to the virtual elimination of the bonding pad capacitance as well as the reduction of the interconnection capacitance. The stray capacitances caused by the bonding pad and the interconnection were measured separately and

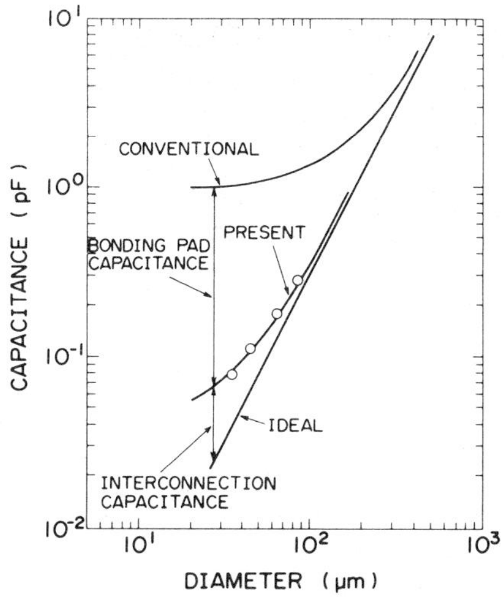

Fig. 7. Dependence of capacitance on diameter. Experimental data is very close to ideal line due to virtual elimination of bonding pad capacitance as well as reduction of interconnection capacitance.

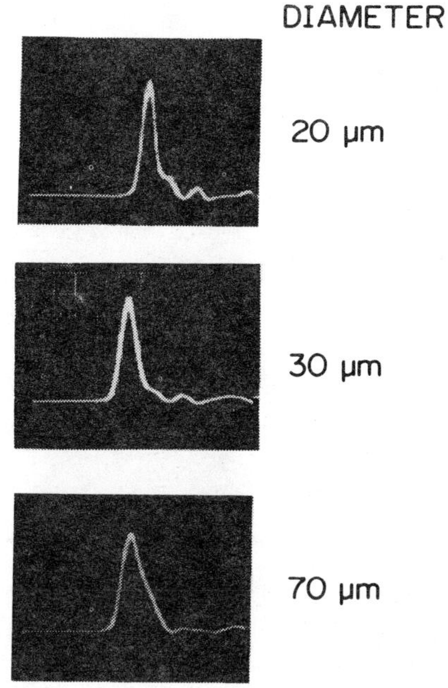

Fig. 8. Pulse response characteristics of p-i-n photodiodes. Values of FWHM are 75, 85, and 120 ps for photodiodes with diameter of photosensitive area of 20, 30, and 70 μm, respectively. Horizontal: 100 ps/div.

were revealed to be a few femtofarads and 0.04 pF, respectively. The advantage of the present structure in reducing the stray capacitance is enhanced for smaller diameters.

B. High-Speed Response Characteristics

For the measurement of the high-speed response characteristics, p-i-n photodiodes having thinner n^- layers were mounted on a stripline. A low-capacitance high-speed InP/GaInAsP DFB laser emitting at 1.3 μm was used as an optical source. The pulse response characteristics of the photodiodes are shown in Fig. 8. The values of full width at half maximum (FWHM) were 75, 85, and 120 ps at the bias voltage of 18 V for the photodiodes with the diameter of the photosensitive area of 20, 30, and 70 μm, respectively. The square of the measured FWHM is simply expressed as the sum of the square of the response time factors including the *RC* and carrier transit time-limited response times, as well as the response times of the laser and the sampling head. The square of the calculated *RC*-limited time and that of the response of the sampling head were subtracted from the square of the measured response time for each photodiode, then the square root of the difference was deduced. The deduced value is a composite of the carrier transit time and the response of the laser so that it must be unique for photodiodes with different diameters. The calculated values were fitted to fulfill this requirement, and the composite response time of 73 ps was obtained. This value is almost the same as the measured FWHM of the 20-μm-diameter photodiode, which means that the carrier transit time or the response time of the laser is dominant in the experiment.

In order to exclude the ambiguity in laser pulse shape determination, we have applied the optical heterodyne technique, in which the lasing wavelength of two DFB lasers emitting around 1.3 μm were precisely controlled by the adjustment of the temperature of the laser heat sinks to a precision of 0.01°C. The light emitted from the two DFB lasers was mixed by a 50/50 optical coupler. The frequency of the beat signal of the two DFB lasers was swept by increasing the dc bias current of one of the lasers. The frequency response of the planar embedded p-i-n photodiode with the diameter of the photosensitive area of 20 μm and with thinner n^- layers (diode II) is shown in Fig. 9. The bias voltage of the photodiode was chosen to be 18 V. The cutoff frequency of the p-i-n photodiode was ascertained to be 14 GHz. This is the best result ever reported, as far as we know, for surface-illuminated InP/GaInAs p-i-n photodiodes.

It is worthwhile analyzing the speed of this p-i-n photodiode in view of its application to the p-i-n/FET configuration. The resistance used for the load is usually large, in order to reduce the thermal noise of the p-i-n/FET photoreceiver. The response speed of the p-i-n/FET is then determined by the *RC* product rather than the transit time of the photo-induced carriers in the p-i-n photodiode. Therefore, considering the application of the p-i-n photodiode for p-i-n/FET photoreceivers, the capacitance of the p-i-n photodiode must be sufficiently reduced to the extent that the response of the photodiode is limited by the carrier transit time. In order to clarify the limiting factor of the response speed, we have calculated the frequency response of the present p-i-n photodiode when terminated by a 50-Ω load, taking both the *RC* time constant and the carrier transit time into consideration [17]. The calculation was consistent with the present experimental result of the −3 dB cutoff frequency of 14 GHz. It was also verified by the calculation that the response speed of the InP/GaInAs p-i-n photodiode is determined by the carrier transit time particularly of electrons, which suggests that the capacitance of the present p-i-n photodiode even with thinner n^- layers (diode II) is

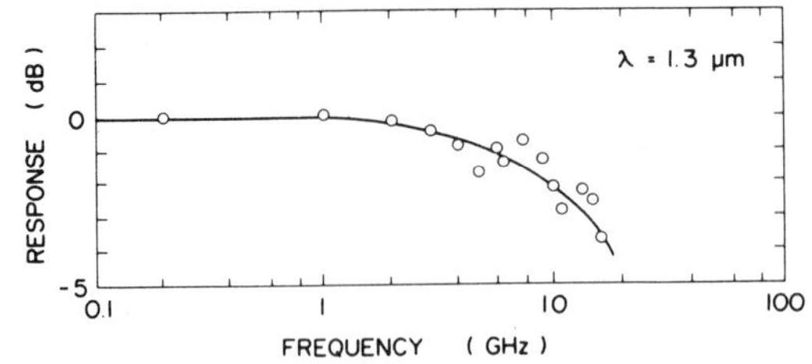

Fig. 9. Frequency response of planar embedded InP/GaInAs p-i-n photodiode. Measurement was performed by using an optical heterodyne technique. A cutoff frequency of 14 GHz was obtained owing to the low capacitance characteristics of the planar embedded InP/GaInAs p-i-n photodiode.

TABLE I
STRUCTURE AND MEASURED CHARACTERISTICS OF TWO TYPES DIODES

Diode	Layer thickness (μm)		Capacitance (pF) (Bias voltage)	Quantum efficiency (%)	Pulse FWHM (ps)	Cut-off frequency (GHz)
	n^--GaInAs	n^--InP				
I	1.8	6	0.07 (20V)	85	-	-
II	0.5	3	0.08 (20V)	40	75	14

sufficiently reduced. The frequency response characteristics of the p-i-n photodiode with thicker n^- layers (diode I) hasn't been measured yet. However, it is expected that the cutoff frequency of these p-i-n photodiodes will be approximately 7 GHz because of the increased transit time.

Table I summarizes the structure and the measured characteristics of two types of diodes. Owing to the low capacitance and high-speed response, the present p-i-n photodiodes should be useful for the p-i-n/FET photoreceiver with high-impedance configuration especially in low-noise, very high-speed applications.

VI. Conclusion

Planar embedded InP/GaInAs p-i-n photodiodes, with two different n^- layer thicknesses for high-speed and low capacitance, were fabricated by a planarization technique which uses a photoresist mask and preferential ion-beam etching. Virtual elimination of bonding pad capacitances and drastic reduction of interconnection capacitances were both realized, which resulted in extremely low capacitances of less than 0.08 pF for both types of photodiodes, with the diameter of the photosensitive area of 20 μm. The p-i-n photodiode with thinner n^- layers was confirmed to exhibit a quantum efficiency of 40 percent and respond up to 14 GHz, and this frequency response was ascertained by a calculation to be limited by the carrier transit time. These results not only indicate the applicability of the planar embedded InP/GaInAs p-i-n photodiodes as high-speed discrete p-i-n photodiodes but also strongly encourage one to use these photodiodes for optoelectronic integration.

Acknowledgment

The authors wish to thank T. Chikama and K. Kamite for their innumerous helpful and useful discussions, and also are deeply indebted to K. Dazai and T. Sakurai for their continuous encouragement and suggestions throughout this work.

References

[1] K. Yasuda, T. Mikawa, Y. Kishi, and T. Kaneda, "Multiplication-dependent frequency responses of InP/InGaAs avalanche photodiode," *Electron. Lett.*, vol. 20, pp. 373–374, Apr. 1984.

[2] R. S. Sussmann, R. M. Ash, A. J. Moseley, and R. C. Goodfellow, "Ultra-low-capacitance flip-chip-bonded GaInAs PIN photodetector for long-wavelength high-data-rate fiber-optic systems," *Electron. Lett.*, vol. 21, pp. 593–595, July 1985.

[3] M. C. Brain, P. P. Smyth, D. R. Smith, B. R. White, and P. J. Chidgey," p-i-nFET hybrid optical receivers for 1.2-Gbit/s transmission systems operating at 1.3 and 1.55 μm wavelength," *Electron. Lett.*, vol. 20, pp. 894–896, Oct. 1984.

[4] D. R. Smith, R. C. Hooper, K. Ahmad, D. Jenkins, A. W. Mabbitt, and R. Nicklin," p-i-n/FET hybrid optical receiver for longer-wavelength optical communication systems," *Electron. Lett.*, vol. 16, pp. 69–71, Jan. 1980.

[5] D. R. Smith, R. C. Hopper, P. P. Smith, and D. Wake, "Experimental comparison of a germanium avalanche photodiode and InGaAs p-i-nFET receiver for longer wavelength optical communication systems," *Electron. Lett.*, vol. 18, pp. 453–454, 1982.

[6] J. Benoit, M. Boulou, and R. Vergnaud," InP based heterojunction photodiodes for high bit rate communication in the 1–1.6 μm wavelength range," in *Tech. Dig. SPIE*, 1985, vol. 587, pp. 138–143.

[7] T. P. Pearsall, R. A. Logan, and C. G. Bethea," GaInAs/InP large bandwidth (>2 GHz) p-i-n detectors," *Electron. Lett.*, vol. 19, pp. 611–612, Aug. 1983.

[8] T. P. Lee, C. A. Burrus, A. G. Dentai, and K. Ogawa, "Small area InGaAs/InP p-i-n photodiodes: fabrication, characteristics, and performance of devices in 274 Mbit/s and 45 Mbits/s lightwave receivers at 1.31-μm wavelength," *Electron. Lett.*, vol. 16, pp. 155–156, Feb. 1980.

[9] T. P. Lee, C. A. Burrus, and A. G. Dentai, "InGaAs/InP p-i-n pho-

todiodes for lightwave communications at the 0.95–1.65 μm wavelength," *IEEE J. Quantum Electron.*, vol. QE-17, pp. 232–238, Feb. 1981.
[10] A. S. H. Liao, T. J. Bridges, E. G. Burkhardt, B. Tell, R. F. Leheny, and E. D. Beebe, "A planar embedded InGaAs photodiode on semi-insulating InP substrate for monolithically integrated PIN-FET receivers, using selective vapor phase epitaxy and ion implantation technique," in *Tech. Dig. IEDM*, 1983, pp. 478–481.
[11] P. J. G. Dawe, D. A. H. Spear, and G. H. B. Thompson, "Embedded GaInAs photodiode and InP-based JFET for optoelectronic integration," in *Tech. Dig. IOOC-ECOC*, 1985, pp. 307–310.
[12] S. Miura, H. Kuwatsuka, T. Mikawa, and O. Wada, "Planar, embedded InP/GaInAs p-i-n photodiode with very high-speed response characteristics," *Appl. Phys. Lett.*, vol. 49, pp. 1522–1524, Dec. 1986.
[13] S. Miura, H. Machida, O. Wada, K. Nakai, and T. Sakurai, "Monolithic integration of a p-i-n photodiode and a field-effect transistor using a new fabrication technique—graded step process," *Appl. Phys. Lett.*, vol. 46, pp. 389-391, Feb. 1985.
[14] O. Wada, S. Miura, H. Machida, K. Nakai, and T. Sakurai, "A new fabrication technique for optoelectronic integrated circuits (OEIC's)—the graded-step process—applied to the fabrication of AlGaAs/GaAs p-i-n/FET and p-i-n/amplifier photoreceivers," *J. Electrochem Soc.*, vol. 132, pp. 1996–2002, Aug. 1985.
[15] S. Miura, O. Wada, M. Makiuchi, and K. Nakai, "Optoelectronic integrated AlGaAs/GaAs p-i-n/field-effect transistor with an embedded, planar p-i-n photodiode," *Appl. Phys. Lett.*, vol. 48, pp. 1461–1463, May 1986.
[16] S. Miura, O. Wada, and K. Nakai, "A novel planarization technique for optoelectronic integrated circuits and its application to a monolithic AlGaAs/GaAs p-i-n FET," *IEEE Trans. Electron Devices*, Feb. 1987.
[17] H. Kuwatsuka, T. Mikawa, S. Miura, and O. Wada, "High-speed response characteristic over 10 GHz of planar embedded InGaAs p-i-n photodiode," unpublished.

*

Shuichi Miura was born in Sakata, Japan, on September 26, 1957. He received the B.S. degree in physics from Tohoku University, Sendai, Japan, in 1981.

In 1981, he joined the Fujitsu Laboratories Ltd., Atsugi, Japan, where he has been engaged in research and development of GaAs-based optoelectronic integrated receivers for applications in the optical fiber communication systems in the 0.8-μm wavelength region.

Mr. Miura is a member of the Japanese Society of Applied Physics.

Haruhiko Kuwatsuka was born in Nobeoka, Japan, on April 16, 1962. He received the B.S. degree in electrical engineering from Tokyo University, Tokyo, Japan, in 1985.

In 1985, he joined the Fujitsu Laboratories, Ltd., Atsugi, Japan, where he has been engaged in research and development of high-speed photodetectors for application in the optical fiber communication systems in the 1-μm wavelength region.

Mr. Kuwatsuka is a member of the Japanese Society of Applied Physics.

*

Takashi Mikawa was born in Tokyo, Japan, on January 10, 1946. He received the B.S. degree in electrical engineering from Keio University, Tokyo, Japan, in 1969.

In 1970 he joined the Fujitsu Laboratories Ltd., Atsugi, Japan. From 1970 to 1974 he was engaged in research on the microwave Gunn device. Since 1975 he has been doing research on avalanche photodiodes (Si, Ge, III-V alloys) for optical communication systems.

Mr. Mikawa is a member of the Institute of Electronic and Communication Engineers of Japan.

*

Osamu Wada (S'70–M'72) was born in Kobe, Japan, on December 7, 1946. He received the B.S. degree in electrical engineering from the Himeji Institute of Technology, Himeji, Japan, in 1969, the M.S. degree in electrical engineering from the University of Kobe, Japan, in 1971, and the Ph.D. degree in electronic and electrical engineering from the University of Sheffield, Sheffield, England, in 1980.

In 1971, he joined the Fujitsu Laboratories, Ltd., Kobe and Kawasaki, where he worked on Gunn-effect logic devices and electron beam lithography. From 1976 to 1978, he was engaged in the research of growth and characterization of InP crystals at the University of Sheffield, on leave from Fujitsu Laboratories, as an Independent Research Worker granted by the U. K. Science Research Council. Since 1978, he has been with Fujitsu Laboratories, where he has been concerned with research on InGaAs/InP light emitting diodes and GaAs-based optoelectronic integrated circuits (OEIC's) incorporating quantum well lasers, metal–semiconductor–metal photodiodes and GaAs field-effect transistors, both for optical communication applications. He is currently Section Manager of a research group for photodetectors for optical communication systems including avalanche photodiodes and OEIC receivers in the Optical Semiconductor Laboratory, Fujitsu Laboratories, Ltd., Atsugi.

Dr. Wada is a member of the Physical Society of Japan and the Japanese Society of Applied Physics.

Multigigabit-per-Second Avalanche Photodiode Lightwave Receivers

BRYON L. KASPER, MEMBER, IEEE, AND JOE C. CAMPBELL, MEMBER, IEEE

(Invited Paper)

Abstract—**High-speed avalanche photodiodes and high-sensitivity receivers are vital components for future multigigabit-per-second lightwave transmission systems. We review theoretical and experimental performance of high-speed III-V avalanche photodiodes, and also that of multigigabit-per-second lightwave receivers using FET and bipolar amplifiers. Particular attention is given to APD gain-bandwidth product, and to its effect on high-speed receiver sensitivity. Comparisons between measured receiver sensitivities and calculated performance are presented for bit rates up to 8 Gbit/s.**

I. Introduction

A NUMBER of recent experiments have demonstrated long-span optical fiber transmission at bit rates of 2, 4, and 8 Gbit/s [1]–[11]. All of these experiments have utilized long-wavelength avalanche photodiodes (APD's) made from either germanium or III-V compounds; the best receiver performance has been obtained by combining these APD's with low-noise GaAs MESFET preamplifiers. It is well known that avalanche photodiodes can provide substantial improvement in receiver sensitivity compared to p-i-n photodiodes; the margin provided by the APD increases with bit rate [12]–[14]. The first generation systems, which operated in the wavelength range of 0.8 to 0.9 μm, utilized high-performance Si APD's [15]. Subsequent development of systems which operate at 1.3 and 1.55 μm has motivated research on APD's for these longer wavelengths. Many of the early long-wavelength lightwave transmission experiments utilized Ge APD's [9]–[11]. However, as a result of their high dark current, unfavorable ratio of ionization coefficients, and relatively low absorption coefficient at 1.55 μm, it is generally accepted that the highest performance can be achieved with APD's fabricated from III-V compounds. The structure that has achieved the best performance to date is the separate absorption and multiplication regions (SAM) APD [16] which was first developed to eliminate the tunneling component of the dark current in $In_{0.53}Ga_{0.47}As$ homojunction APD's [17], [18]. Recently, as lightwave systems have progressed to higher and higher bit rates, the emphasis of the research on these APD's has shifted toward improving the frequency response [19]–[22]. In this paper we will describe the physical effects which determine the frequency response of this type of APD and review the improvements in bandwidth that have been achieved.

We also review noise and sensitivity calculations for FET and bipolar receivers using APD detectors. At very high bit rates, the APD gain-bandwidth product can be a limiting factor; the APD may have to be operated at less than optimum gain in order to have adequate bandwidth for the signal. We present sensitivity calculations for both equalized and nonequalized gain-bandwidth-limited receivers. Finally, we summarize experimental results to date for multigigabit-per-second lightwave receivers, and present comparisons to theory for low-capacitance III-V APD receivers from 420 Mbit/s to 8 Gbit/s. These results include sensitivities of −46.2 dBm (437 photons/bit) at 420 Mbit/s and −42.1 dBm (471 photons/bit) at 1 Gbit/s, which are the best sensitivities to date in terms of photons/bit for long-wavelength direct-detection receivers.

Manuscript received April 2, 1987; revised April 29, 1987.
The authors are with AT&T Bell Laboratories, Holmdel, NJ 07733.
IEEE Log Number 8715759.

II. High-Speed III-V Avalanche Photodiodes

In the initial implementation of the SAM-APD structure the p-n junction and thus the high-field region is located in InP where tunneling is insignificant and absorption occurs in an adjacent layer of $In_{0.53}Ga_{0.47}As$ (referred to henceforth as InGaAs). To improve the frequency response of the SAM-APD a transition region is usually incorporated between the InP multiplication region and the InGaAs absorption region [23]. (The motivation for the transition region will be discussed in Section II-A.) This structure is often referred to as an SAGM-APD [24]. Fig. 1 shows the bandwidth of a back-illuminated InP/InGaAsP/InGaAs mesa-structure SAGM-APD versus the dc avalanche gain M_0. At high gain the frequency response approaches a constant gain-bandwidth limit of 70 GHz. The roll-off at lower gains is due to a combination of three physical effects: the RC time constant, the transit time of carriers through the space-charge region, and charge accumulation at the heterojunction interfaces. An approximate expression for the frequency dependence of the gain which includes these effects is given by [25]

$$\frac{M(\omega)}{M_0} = T(\omega)\left[\left[1 + (\omega/e_h)^2\right] \cdot \left[1 + (\omega RC)^2\right]\left[1 + (\omega\tau_e M_0)^2\right]\right]^{-1/2} \quad (1)$$

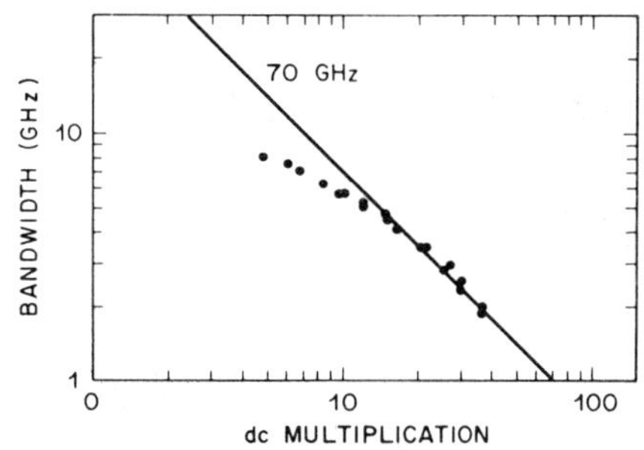

Fig. 1. Bandwidth of a back-illuminated mesa-structure InP/InGaAsP/InGaAs avalanche photodiode versus dc multiplication M_0.

where $T(\omega)$ is the normalized transit-time factor, e_h is the emission rate for holes trapped at the heterojunction interfaces, R is the total resistance (series resistance of the APD plus the load resistance), C is the diode capacitance, and τ_e is the effective transit time through the multiplication region. We now proceed to discuss each of the terms in (1) and their effect on the bandwidth of these APD's.

A. Interface Hole Trapping

The origin of the term containing e_h (second term) is charge accumulation at a heterojunction interface [23]. This effect is illustrated by the valence band diagram in Fig. 2. As the photogenerated holes are swept toward the multiplication region many of them get trapped in the potential well formed by the valence band discontinuity at the heterojunction interface. The number of holes that accumulate in the well and the rate at which they are emitted e_h depend on the height and width of the barrier E_b and W, respectively. Both of these parameters depend, in turn, on the magnitude of the valence band discontinuity E_v and the amount of compositional grading at the heterojunction interface. It has been found that this effect gives rise to a long tail (> 10 ns) in the pulse response of InP/InGaAs SAM-APD's; their bandwidth is typically less than 50 MHz [26], [27].

There is still some uncertainty regarding the nature of the emission process for the trapped holes. Originally, it was assumed that the holes escape by thermionic emission [23]. For this model the emission rate is given by

$$e_h = K \exp(-E_b/kT) \quad (2)$$

where K is a constant over a limited temperature range. A more general model which incorporates tunneling, however, may be more appropriate. Fowler and Nordheim [28] have shown in another context that for sufficiently high barriers or low temperature, tunneling is the dominant emission process. At higher temperatures the strong field emission becomes sensitive to temperature and blends into thermionic emission. Similarly, Padovani and Stratton [29] have demonstrated that transport across Schottky barriers in GaAs can best be described in terms of a thermionic-field emission model which becomes pure thermionic emission or tunneling at high or low temperature, respectively. An extension of this model to interfacing trapping in SAM-APD's is illustrated in Fig. 2. Holes with energy E tunnel through the barrier; higher values of E correspond to a narrower barrier and consequently higher tunneling rates. A second factor which should be considered is that of "hot holes," i.e., the trapped holes may not be in equilibrium with the lattice. This would result in enhanced emission and lower effective barrier heights than would be predicted by conventional thermionic emission. In spite of the remaining questions about the emission process, it is clear that the emission rate is very sensitive to changes in the barrier height and interface width. Fig. 3 shows the time constant ($\tau = 1/e_h$) calculated from (2) for holes trapped at an $InP/In_xGa_{1-x}As_yP_{1-y}$ heterojunction as a function of the valence band discontinuity and different values of the grading parameter L. The total interface width is $2 \times L$. The values of E_b at 50-V bias were obtained from the calculated band structure using the model of Cheung *et al.* [30]. The carrier concentration and thickness of the InP multiplication layer were assumed to be 2×10^{16} cm^{-3} and 1.25 μm, respectively. The endpoint $\Delta E_v = 0.36$ eV corresponds to an $InP/In_{0.53}Ga_{0.47}As$ heterojunction.

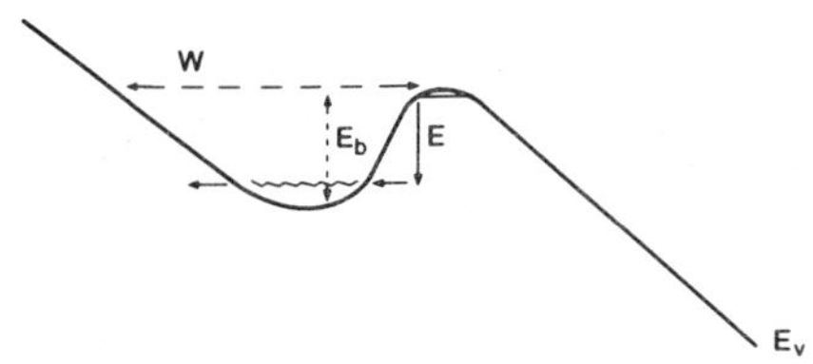

Fig. 2. Valence band at a heterojunction interface.

There have been several solutions proposed to eliminate the problem of charge accumulation at the heterojunction interfaces. Each involves introducing a transition layer between the multiplication and absorption regions. The optimum approach would be to continuously grade the bandgap energy between that of InP and InGaAs. However, it is extremely difficult to grade the crystal composition to provide a uniform change in the bandgap energy and simultaneously maintain the lattice constant. The most sophisticated approach, proposed and demonstrated by Capasso *et al.* [31], utilized a chirped period superlattice to mimic grading of the bandgap. Bandwidths as high as 900 MHz were obtained.

The simplest and most successful solution, to date, has been to do the grading stepwise with a small number of intermediate-bandgap layers [23], [32], the SAGM-APD structure. The "G" denotes the presence of the "grading" layers. The effect of adding these layers is to reduce the valence band discontinuity. Also at each heterojunction interface there is a certain amount of compositional grading. From Fig. 3 it can be seen that both of these effects can have a significant impact on the frequency response. For a grading parameter of 100 Å, which is typical of LPE-grown crystals, a reduction in the valence band discontinuity by less than a factor of 2 can result in a pulse response of less than 200 ps. For more abrupt heterojunctions such as might be obtained by vapor-phase

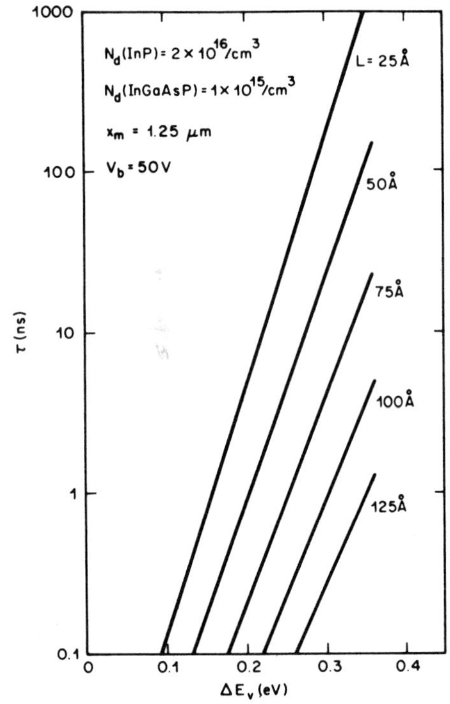

Fig. 3. Calculated time constant versus valence band discontinuity.

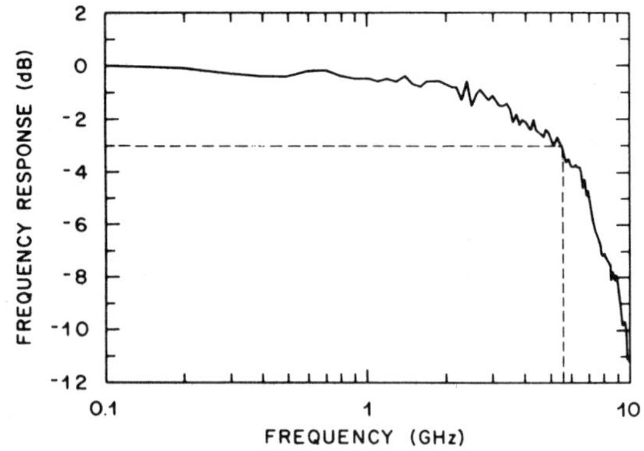

Fig. 4. Frequency response at $M_0 = 4$ of a back-illuminated mesa-structure InP/InGaAsP/InGaAs avalanche photodiode grown by chemical beam epitaxy.

growth techniques such as molecular beam epitaxy (MBE) or metalorganic chemical vapor deposition (MOCVD), even smaller valence band steps are required. Nevertheless, it is clear that a few (two to four) discrete transition layers should be sufficient to achieve bandwidths of several gigahertz.

In the first successful demonstration of an InP/InGaAsP/InGaAs SAGM-APD, Matsushima *et al.* [33], achieved a pulse response <1 ns; the bandwidth was not measured. Most of the early SAGM-APD's utilized a single transition layer [24], [34]–[36]. This greatly reduced but did not eliminate hole trapping. The bandwidths of these APD's is typically 1–2 GHz in the low-gain regime before the onset of the gain-bandwidth limit [25], a significant improvement compared to the InP/InGaAs SAM-structure. Recently, bandwidths greater than 5 GHz have been achieved with LPE- [20] grown wafers and 8 GHz with CBE-grown [37] wafers having two and three intermediate-bandgap transition layers, respectively. The frequency response of a mesa-structure SAGM-APD grown by chemical beam epitaxy (CBE) is shown in Fig. 4. At $M_0 = 9$ the bandwidth is 5.5 GHz.

B. Transit Time of SAGM-APD's

The transit time factor $T(\omega)$ in (1) accounts for the time required for the primary and secondary carriers to drift across the depletion region. The transit time for the SAGM-APD structure will usually be longer than that of a conventional homojunction photodiode because the multiplication and absorption regions are not coincident. This adds length to the drift region. In addition, there will be a transit time associated with the secondary electrons which are injected from the multiplication region into the absorbing layer. In fact, for significant multiplication ($M_0 \geq 2$), these secondary electrons will dominate the transit time.

The transit time factor can be written as a superposition of contributions from the primary electrons and holes and the secondary electrons, $N(\omega)$, $P(\omega)$, and $N_s(\omega)$. Since the absorbing layer is usually much wider than the multiplication region, the drift path of the secondary electrons will be longer than that of the secondary holes. As a result, the contribution of the secondary holes can be ignored and $T(\omega)$ can be written as

$$T(\omega) = \left|P(\omega) + N(\omega) + N_s(\omega)\right| \tag{3}$$

where

$$\begin{aligned} P(\omega) = \frac{N_{ph}}{\sqrt{2\pi}} \Bigg[& \frac{[1 - \exp(-\alpha W_a)]}{jw} + \exp(-jwx_t/v_p) \\ & \times [1 - \exp(-j\omega W_a/v_p - \alpha W_a)] \\ & \cdot [1/(jw + \alpha' v_p) - 1/jw] \Bigg] \end{aligned} \tag{4}$$

$$\begin{aligned} N(\omega) = \frac{N_{ph}}{\sqrt{2\pi}} \Bigg[& \frac{[1 - \exp(-j\omega W_a/v_n)]}{j\omega} - \exp(-\alpha W_a) \\ & \times \frac{[1 - \exp(-j\omega W_a/v_n - \alpha W_a)]}{(j\omega - \alpha v_n)} \Bigg] \end{aligned} \tag{5}$$

and

$$\begin{aligned} N_s(\omega) = \frac{N_{ph}}{\sqrt{2\pi}} (M_0 - 1) \Big[& \exp(-j\omega x_t/v_p) \\ & \cdot [1 - \exp(-j\omega x_a/v_n)] \\ & \times [1 - \exp(-j\omega W_a/v_p)] \\ & \cdot [1/i\omega - 1/(\alpha v_p + j\omega)] \Big]. \end{aligned} \tag{6}$$

In these equations N_{ph} is the number of photogenerated electron-hole pairs, ω is the angular frequency, $W_a = x_a - x_t$ is the width of the absorption region, x_m is the width of the multiplication region, α is the absorption coefficient at the incident wavelength, and v_n and v_p are the saturated electron and hole drift velocities, respectively. for the derivation of $N_s(\omega)$ we have assumed a thin multiplication region with "infinite" gain-bandwidth product at $x = 0$ where the electric field is the highest.

Fig. 5 shows the normalized frequency dependence of $T(\omega)$ for absorbing layer thickness of 2 and 3 μm at $M_0 = 1$ and 10. As might be expected, for $M_0 = 1$ (no avalanche multiplication) the response is very close to that of a p-i-n photodiode with comparable depletion width [38]. At higher gains the transit time of the secondary electrons significantly decreases the bandwidth. The total depletion width of an SAGM-APD is typically about 3 μm of which 2 μm is the absorbing layer. From Fig. 5 we can see that the transit time will usually not be a limiting factor at frequencies less than 10 GHz. Just as for p-i-n photodiodes higher bandwidths can be achieved by decreasing the width of the absorbing layer but at the cost of lower quantum efficiency.

C. RC Time Constant

The *RC* time constant is the origin of the third term in (1). The capacitance of a packaged SAGM-APD is typically in the range of 0.1–0.2 pF. The total resistance is the sum of the 50-Ω load resistance and the series resistance of the APD itself. The latter is usually between 10 and 50 Ω. The minimum and maximum bandwidths associated with these parameters are 8 and 26 GHz.

D. Gain-Bandwidth Product of SAGM-APD's

The fourth term in (1) is due to the avalanche buildup time. Owing to its regenerative nature, the higher the gain the longer it takes the avalanche process to build up. This gives rise to a constant gain-bandwidth product which is often the most restrictive term in (1). One of the factors that determines the avalanche buildup time and hence the gain-bandwidth product is the ratio of the electron (α) and hole (β) ionization coefficients. Emmons [39] has shown that for a given avalanche gain M_0, the more α and β differ the higher the bandwidth that can be attained. This condition also leads to improved receiver sensitivity, c.f., Fig. 7. Higher speed and lower noise have been the driving forces behind the design of super-lattice APD structures with artificially high (or low) β/α ratios, e.g., the "staircase APD" [40]. For the SAGM-APD structure the values of α and β are determined by the bulk crystal properties of the multiplication region. There are, however, two structural modifications that can be incorporated to provide some improvement in the gain-bandwidth product: 1) In InP the convergence of α and β at high electric fields degrades the gain-bandwidth product [41], [42]. Capasso *et al.* [43] have shown that the peak electric field can be reduced by using a low-high doping profile in the multiplication region. 2) Since the avalanche process is regenerative, the effects of carrier feedback can be reduced by decreasing the width of the multiplication region. In the SAGM-APD structure this must be accompanied by an increase in the carrier concentration in order to insure that the performance will not be degraded by carrier diffusion into the drift region or excess dark current. By reducing the multiplication layer thickness from approximately 1.5 μm to less than 0.5 μm, Campbell *et al.* [37] have increased the gain-bandwidth product of back-illuminated mesa-structure InP/InGaAsP/InGaAs SAGM-APD's from 18 to 70 GHz. Recently, Sugimoto *et al.* [21] have demonstrated a planar-structure SAGM-APD with a thin multiplication region; they report a gain-bandwidth product of 50 GHz.

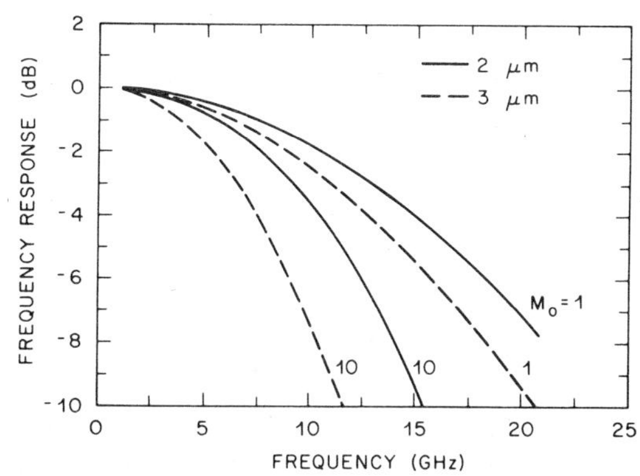

Fig. 5. Normalized transit-time frequency response of an SAGM-APD for absorbing layer thickness of 2 and 3 μm at $M_0 = 1$ and 10.

This section has described the factors which limit the bandwidth and gain-bandwidth product of SAGM-APD's. In addition, we have shown that by optimizing the device parameters to reduce the effects of the transit time, the *RC* time constant, hole trapping, and the avalanche buildup time, these APD's can achieve the high bandwidths required for multigigabit lightwave receivers.

III. Receiver Sensitivity Calculations

The low noise and high gain of silicon avalanche photodiodes allowed high receiver sensitivity to be achieved in first-generation lightwave systems with relative ease. The higher noise and lower optimum gain of long-wavelength APD's has made careful preamplifier design more critical if the best possible sensitivity is desired. This is particularly true at gigabit-per-second rates, as preamplifier noise increases rapidly with bit rate. The following sections discuss the noise characteristics of FET and bipolar receiver front ends, and the sensitivity that can be achieved with APD detectors.

A. Noise Calculations for FET Front Ends

At present, the lowest noise receivers at gigabit-per-second rates are made using GaAs MESFET preamplifiers. The principal noise sources are Johnson noise associated with the FET channel conductance, Johnson noise from the load or feedback resistor, shot noise arising from gate leakage current, and FET $1/f$ noise. The variance of the equivalent input noise current $\langle i^2 \rangle_a$ is given by [13], [14], [44], [45]

$$\langle i^2 \rangle_a = \frac{4kT}{R_L} I_2 B + 2q I_{gate} I_2 B + \frac{4kT\Gamma}{g_m} (2\pi C_T)^2 f_c I_f B^2 + \frac{4kT\Gamma}{g_m} (2\pi C_T)^2 I_3 B^3 \tag{7}$$

where B = bit rate, R_L = load or feedback resistor, I_{gate} = FET gate leakage current, g_m = FET transconductance, C_T = total input capacitance, f_c = FET $1/f$-noise corner frequency, and Γ = FET channel noise factor.

The values of Personick integrals I_2, I_3, and I_f depend on input and output pulse shapes. Expressions for I_2 and I_3 are given in [12] and [13], and the expression for I_f is

$$I_f = \int_0^\infty |H'_T(y)|^2 y \, dy \tag{8}$$

where $H'_T(y)$ is a normalized receiver transfer function and y is normalized frequency. Common pulse shape assumptions are rectangular return-to-zero (RZ) or non-return-to-zero (NRZ) input pulses, and output pulses with a full raised-cosine spectrum which results in zero intersymbol interference. For NRZ input pulses, the corresponding Personick integral values are $I_2 = 0.562$, $I_3 = 0.0868$, and $I_f = 0.184$. For RZ input pulses with a 50-percent duty cycle, the values are $I_2 = 0.403$, $I_3 = 0.0361$, and $I_f = 0.0984$.

FET channel noise is characterized by the numerical factor Γ which accounts for channel Johnson noise and gate-induced noise, plus their correlation [46]–[49]. The value of Γ for GaAs MESFET's is generally taken as 1.1 [13], [50], [51], although it may be as high as 1.75 [52]. The total input capacitance C_T is given by

$$C_T = C_d + C_{gs} + C_{gd} + C_s \tag{9}$$

where C_d is the detector capacitance, C_{gs} is the FET gate–source capacitance, C_{gd} is the FET gate–drain capacitance, and C_s is the stray capacitance (including load resistor capacitance).

The terms in (7) account for the four principal noise sources in the FET preamplifier. The first term arises from thermal noise of the load or feedback resistor R_L. This noise can be minimized by making R_L as large as possible. However, there is generally a tradeoff of sensitivity and dynamic range in the choice of R_L, since the maximum signal level which can be handled by the receiver without saturation decreases as R_L becomes larger. Receivers are often designed with lower-than-optimum values of R_L from a noise perspective in order to provide wider dynamic range.

The second term in (7) arises from FET gate leakage current. This noise source can be insignificant for gigabit-per-second receivers if a FET with low gate leakage current (<100 nA) is used. However, some microwave FET's, particularly HEMT's, have gate leakage currents as high as a few microamps which can be significant noise sources, even at high bit rates.

The third term arises from FET $1/f$ noise. This noise source is a major problem for GaAs amplifiers at low frequencies. It is not well controlled for many GaAs MESFET's or HEMT's, which are optimized mainly for operation at microwave frequencies. Fortunately, the wide bandwidth required for gigabit-per-second receivers makes $1/f$ noise less important than for low-bit-rate applications.

The fourth noise source is Johnson noise associated with the FET channel. This noise is inherent in FET operation, and should ideally be the dominant receiver noise mechanism. The channel noise is minimized by making g_m/C_T^2 as large as possible, which implies an optimization of the FET gate width such that [13]

$$C_{gs} + C_{gd} = C_d + C_s. \tag{10}$$

A figure of merit for optimized FET receivers can be derived in terms of f_T, the FET short-circuit common-source current-gain-bandwidth product, given by

$$f_T = \frac{g_m}{2\pi(C_{gs} + C_{gd})}. \tag{11}$$

For an ideal FET preamplifier in which only channel noise is important, the optimized input noise current is found from (7), (10), and (11) to be

$$\langle i^2 \rangle_{opt} \approx 32\pi kT \frac{\Gamma(C_d + C_s)}{f_T} I_3 B^3. \tag{12}$$

The optimized FET preamplifier figure of merit is then given by

$$\text{FET figure of merit} = \frac{f_T}{\Gamma(C_d + C_s)}. \tag{13}$$

Channel noise can be minimized by using a FET with the highest possible f_T and lowest Γ, by minimizing detector and stray capacitances, and by optimizing gate width to make the FET input capacitance equal to the detector-plus-stray capacitance.

Equation (7) is somewhat idealized because it does not consider noise sources following the input FET [45]. This assumption is more easily justified at low bit rates than at high bit rates, because higher values of drain load resistor can be used at low bit rates leading to higher first-stage gain, thereby lessening the importance of noise from subsequent stages. For high bit rates, the drain load resistor must be smaller to reduce the associated RC time constant, hence the noise from this resistor and subsequent amplifier stages becomes more significant. Noise from a drain resistor R_D can be included in (7) by using an equivalent value of Γ given by

$$\Gamma_{eq} = \Gamma + \frac{1}{g_m R_D}. \tag{14}$$

Since Γ_{eq} depends on g_m, the optimum gate width condition specified in (10) is no longer valid. By setting the derivative of total noise with respect to gate width to be equal to zero, a new optimum value of $C_{gs} + C_{gd}$ is found to be

$$C_{gs} + C_{gd} = (C_d + C_s) \cdot \left[\frac{1}{2} + \left(\frac{1}{4} + \frac{1}{\pi f_T \Gamma R_D (C_d + C_s)}\right)^{1/2}\right]. \qquad (15)$$

As can be seen by comparing this result to (10), the optimum sensitivity is now obtained for a wider gate, giving an FET input capacitance that is larger than the detector-plus-stray capacitance. A wider gate increases the first-stage gain, thereby raising the signal level above noise from R_D. This calculation can easily be extended to include additional subsequent noise sources if desired.

B. Noise Calculations for Bipolar Front Ends

Bipolar junction transistors can also be used in high-bit-rate optical receiver front ends. The noise of current silicon bipolar transistors is somewhat higher than that of GaAs MESFET's, although future III-V bipolar transistors may have performance comparable or superior to that of present FET's.

The principal noise sources in a bipolar front end are shot noise from the base and collector bias currents, thermal noise from the base-spreading resistance $r_{bb'}$, and thermal noise from the load or feedback resistor. The variance of the equivalent input noise current is given by [13], [44]

$$\langle i^2 \rangle_a = \frac{4kT}{R_L} I_2 B + 2qI_b I_2 B + \frac{2qI_c}{g_m^2}(2\pi C_T)^2 I_3 B^3 + 4kTr_{bb'}\left[2\pi(C_d + C_s)\right]^2 I_3 B^3 \qquad (16)$$

where I_b = base bias current, I_c = collector bias current, g_m = transconductance, and $r_{bb'}$ = base-spreading resistance. The transconductance is given by

$$g_m = I_c / V_T$$

where $V_T = kT/q$, and the total input capacitance C_T is given by

$$C_T = C_d + C_s + C_{b'e} + C_{b'c} \qquad (17)$$

where $C_{b'e}$ and $C_{b'c}$ are the small-signal hybrid-π-model capacitances of the transistor.

The emitter capacitance $C_{b'e}$ is the sum of a space-charge capacitance C_{je} and a diffusion capacitance which is proportional to the collector current, such that

$$C_{b'e} = C_{je} + \frac{I_c}{2\pi V_T f_T} \qquad (18)$$

where f_T is the short-circuit common-emitter current-gain-bandwidth product at high I_c, where the diffusion capacitance dominates. Muoi [44] has calculated the optimum bias current when this variation of emitter capacitance is taken into account. The optimum collector bias current can be written as

$$I_c = 2\pi C_0 f_T V_T \psi(B) \qquad (19)$$

where C_0 = total input capacitance for zero bias current $= C_d + C_s + C_{b'c} + C_{je}$. The dependence of the optimum I_c on bit rate is contained in the term $\psi(B)$ which is given by

$$\psi(B) = \left[1 + \frac{I_2 f_T^2}{\beta I_3 B^2}\right]^{-1/2} \qquad (20)$$

where β = current gain = I_c/I_b. At low bit rates, the factor $\psi(B)$ can be approximated by

$$\psi(B) \approx \frac{B}{f_T}\left[\frac{\beta I_3}{I_2}\right]^{1/2}. \qquad (21)$$

The optimum bias current at low bit rates is therefore proportional to B, and agrees with the result of an analysis by Smith and Personick [13] which assumed that C_T is constant. This assumption is valid for low bit rates, as using (17), (18), and (19), C_T is given by

$$C_T = C_0[1 + \psi(B)] \qquad (22)$$

and for low bit rates $\psi(B) \ll 1$.

At very high bit rates, the value of $\psi(B)$ approaches 1, hence the optimum collector current reaches a constant value and no longer depends on B. At this point, the total capacitance reaches a maximum value of $2C_0$. The crossover point between these high- and low-bit-rate regimes occurs when $B = f_T[I_2/\beta I_3]^{1/2}$, which for current microwave bipolar transistors with $f_T \approx 10$ GHz and $\beta \approx 100$ occurs at $B \approx 3$ Gbit/s.

Combining (16) and (19), the input noise-current variance for an optimized bipolar preamplifier is found to be

$$\langle i^2 \rangle_a = \frac{4kT}{R_L} I_2 B + \frac{4\pi kTC_0 f_T}{\beta}\psi(B) I_2 B + \frac{4\pi kTC_0}{f_T}\frac{\left[1 + \psi(B)\right]^2}{\psi(B)} I_3 B^3 + 4kTr_{bb'}\left[2\pi(C_d + C_s)\right]^2 I_3 B^3. \qquad (23)$$

At low bit rates, the approximation of (21) can be used to obtain

$$\langle i^2 \rangle_a = \frac{4kT}{R_L} I_2 B + 8\pi kTC_0\left[\frac{I_2 I_3}{\beta}\right]^{1/2} B^2 + 4kTr_{bb'}\left[2\pi(C_d + C_s)\right]^2 I_3 B^3 \qquad (24)$$

which agrees with the minimum bipolar noise result of Smith and Personick [13]. At very high bit rates where $\psi(B) \approx 1$, the second term of (16) which accounts for base-current shot noise can be neglected and the expression simplifies to

$$\langle i^2 \rangle_a = \frac{4kT}{R_L} I_2 B + \frac{16\pi k T C_0}{f_T} I_3 B^3 + 4kTr_{bb'}\left[2\pi(C_d + C_s)\right]^2 I_3 B^3. \quad (25)$$

This expression differs from the result of Muoi [44] by a factor of 2 in the second term. The difference arises from an approximation made in Muoi's derivation which leads to underestimation of noise at very high bit rates; (25) above is a more exact result.

The noise in bipolar preamplifiers, as for FET preamplifiers, is minimized by choosing a device with the highest possible f_T, by minimizing detector and stray capacitance, and by making the load or feedback resistor as large as possible. Additional considerations for bipolar amplifiers are the minimization of parasitic contributions to base resistance, collector–base capacitance, and emitter–base capacitance.

C. Noise Comparison of GaAs MESFET and Si Bipolar Transistors

A comparison of equivalent input-noise-current variance for state-of-the-art GaAs MESFET and Si bipolar transistors is shown in Fig. 6. Parameters for the MESFET are $f_T = 20$ GHz, $\Gamma = 1.1$, $f_c = 50$ MHz, and $I_{\text{gate}} = 100$ nA. Parameters for the bipolar transistor are $\beta = 100$, $f_T = 10$ GHz, $r_{bb'} = 20\ \Omega$, $C_{b'c} = 0.2$ pF, and $C_{je} = 0.8$ pF. NRZ input-pulse format is assumed, and both amplifiers are optimized for a detector-plus-stray capacitance of 0.5 pF. The load resistor R_L is assumed to be large enough that its noise can be neglected, and noise from sources following the input transistor is also neglected.

The slope of the bipolar noise curve changes from 6 dB/octave (i.e., $\sim B^2$) at 100 Mbit/s to 9 dB/octave (i.e., $\sim B^3$) at 20 Gbit/s as a result of the saturation of the optimum collector current caused by the increase in emitter diffusion capacitance with I_c. The slope of the FET curve is a constant 9 dB/octave, except at the low-frequency end where $1/f$ noise and gate leakage current cause it to bend upwards.

With parameter values as assumed, the electrical signal-to-noise ratio for the MESFET is larger than that for the bipolar transistor by from 10 dB at 100 Mbit/s to 5 dB at 20 Gbit/s. This translates into a 2.5–5 dB optical sensitivity advantage for the MESFET if one employs a p-i-n photodetector. With an APD, the difference in sensitivity is less, with the exact difference depending on the excess noise factor of the APD, as can be calculated as described in Section III-E.

D. p-i-n Sensitivity Calculations

For a p-i-n photodetector, the only additional source of noise to be considered is detector dark current. The noise from the dark current I_d can be added to the amplifier noise to obtain a total circuit noise given by

$$\langle i^2 \rangle_c = \langle i^2 \rangle_a + 2qI_d I_2 B. \quad (26)$$

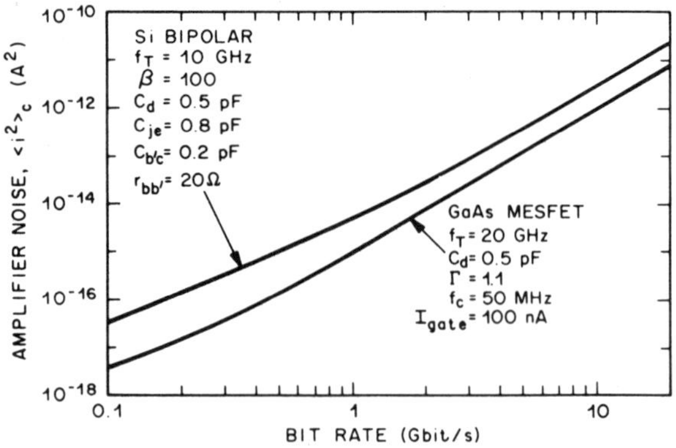

Fig. 6. Calculated equivalent input-noise-current variance $\langle i^2 \rangle_c$ versus bit rate for optimized silicon bipolar preamplifier and optimized GaAs MESFET preamplifier.

The average detected optical power $\eta \bar{P}$ needed for a given error rate is then obtained from [13]

$$\eta \bar{P} = \left(\frac{h\nu}{q}\right) Q \langle i^2 \rangle_c^{1/2} \quad (27)$$

where η = detector quantum efficiency, h = Planck's constant, ν = optical frequency, and Q = signal-to-noise ratio ($Q = 6$ for BER $= 1 \times 10^{-9}$).

E. APD Sensitivity Calculations

If the multiplied dark current I_{dm} in an APD is sufficiently low that its noise can be neglected, then the sensitivity of an APD receiver for an avalanche gain M is given by [13]

$$\eta \bar{P} = \left(\frac{h\nu}{q}\right) Q \left[\frac{\langle i^2 \rangle_c^{1/2}}{M} + qBI_1 QF(M)\right]. \quad (28)$$

Personick integral I_1 is a normalized noise-bandwidth integral that has a value of 0.548 for NRZ input pulses and full raised-cosine-spectrum output pulses, and a value of 0.500 for 50-percent duty-cycle RZ input pulses. $F(M)$ is McIntyre's excess noise factor given by [53]

$$F(M) = kM + (1 - k)(2 - 1/M) \quad (29)$$

where k is the electron and hole ionization coefficient ratio. The optimum gain M_{opt} for the case of $I_{dm} = 0$ is given by

$$M_{\text{opt}} = \frac{1}{k^{1/2}} \left[\frac{\langle i^2 \rangle_c^{1/2}}{qBI_1 Q} + k - 1\right]^{1/2}. \quad (30)$$

If I_{dm} cannot be neglected, then the receiver sensitivity as a function of M is found from [44], [45]

$$\eta \bar{P} = \left(\frac{h\nu}{q}\right) Q \left\{ QqBI_1 F(M) + \left[\frac{\langle i^2 \rangle_c}{M^2} + 2qI_{dm} F(M) BI_2\right]^{1/2} \right\}. \quad (31)$$

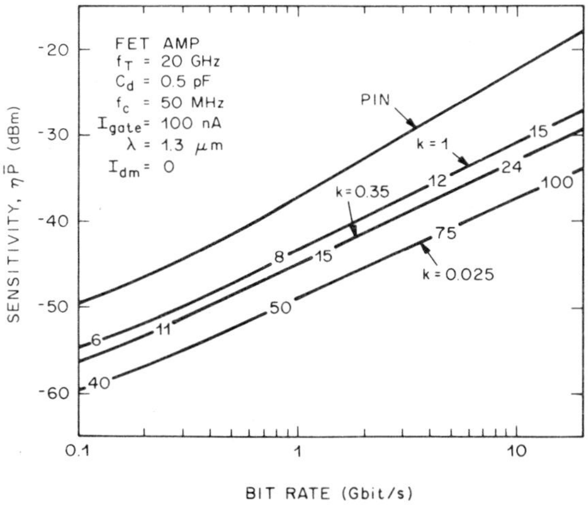

Fig. 7. Calculated receiver sensitivities versus bit rate for an optimized GaAs MESFET preamplifier and a p-i-n detector or an APD with $k = 1$, 0.35, or 0.025. Optimum APD gains are indicated at points along the curves.

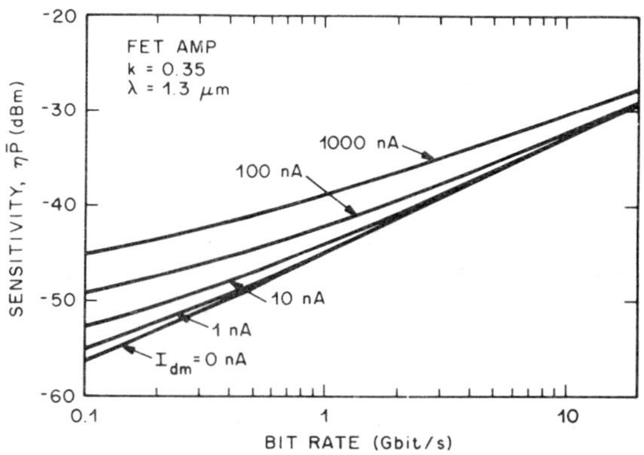

Fig. 8. Calculated receiver sensitivities versus bit rate for an optimized GaAs MESFET preamplifier and an APD with $k = 0.35$ and primary-multiplied-dark-current values of $I_{dm} = 0$, 1, 10, 100, and 1000 nA.

The optimum gain which minimizes (31) is most easily found either numerically or graphically.

An example of theoretical APD or p-i-n sensitivities for an optimized FET receiver at bit rates from 100 Mbit/s to 20 Gbit/s is shown in Fig. 7. The sensitivities are for a bit error rate of 10^{-9}, and NRZ input-pulse format is assumed. The calculations are done for a wavelength of 1.3 μm using (7), (27), and (28). The APD gain is optimized as in (30), and the FET gate width is optimized as in (10). The load resistor is assumed to be very large so its noise can be neglected. The detector capacitance is 0.5 pF, and stray capacitance is assumed to be zero. FET parameters are $f_T = 20$ GHz, $\Gamma = 1.1$, $I_{\text{gate}} = 100$ nA, and $f_c = 50$ MHz. Curves are shown for ionization rate ratios of $k = 1$ corresponding to a Ge APD, $k = 0.35$ corresponding to an InGaAs SAM or SAGM APD, and 0.025 corresponding to possible future long-wavelength APD's with silicon-like excess-noise factors. Optimum avalanche gains are indicated at points along the curves.

From Fig. 7, InGaAs APD's are seen to have a 7–11 dB sensitivity advantage over p-i-n's and a 1.5–2 dB advantage over Ge APD's. The advantage over Ge would be larger at 1.55-μm wavelength where the speed of Ge APD's is reduced as a result of diffusion tails in the pulse response [9]. A silicon-like APD would give a 3–5 dB sensitivity improvement over an InGaAs APD.

The above calculations assume zero dark current and an infinite gain-bandwidth product. The effect of limited gain-bandwidth is dealt with in the next section. To illustrate the effect of multiplied dark current, Fig. 8 shows sensitivities for an InGaAs APD with primary multiplied dark current ranging from 1 to 1000 nA. The same FET receiver as in Fig. 7 is assumed. The sensitivity penalty is 1 dB for a 1 nA dark current at 100 Mbit/s, whereas at 20 Gbit/s a dark current of 100 nA produces a penalty of only about 0.5 dB.

IV. Effect of APD Gain-Bandwidth Limitation on Receiver Sensitivity

At very high bit rates, the limited gain-bandwidth product of an avalanche photodiode can prevent operation at the optimum gain in (30), causing the receiver sensitivity to be reduced [54]. The receiver could be operated either with no equalization for the limited APD bandwidth, in which case the gain must be reduced sufficiently that no intersymbol interference is generated, or with equalization of the APD response, in which case intersymbol interference is removed by the equalization and the APD gain can be higher, but a penalty is incurred due to increased high-frequency receiver noise. The analyses below assume that the gain-bandwidth product is the only bandwidth-limiting factor, and that other potentially limiting effects such as interface hole trapping, transit time, and *RC* time constant can be ignored.

A. No Equalization of APD Frequency Response

The first approach, in which no equalization is used, has been described by Forrest [54]. To find the receiver sensitivity when an APD has a limited gain-bandwidth product denoted by f_{GB}, one must first find the usable avalanche gain which is given by the *smaller* of either the optimum gain from (30) or by the maximum gain for no intersymbol interference given by

$$M_{\max} = f_{GB}/B. \tag{32}$$

It is assumed here that the APD gain is high enough that the gain-bandwidth-limited regime applies (i.e., $M > 1/k$) [39], and that a 3-dB-bandwidth equal to the bit rate is sufficient to produce negligible intersymbol interference. Using the smaller of the two gains found above, one can substitute into (28) to calculate sensitivity. An example calculation is included in Section IV-B.

Before discussing equalization of gain-bandwidth-limited APD's, we would like to mention that we could not reproduce Forrest's result that APD's with an ionization rate ratio of $k = 0.5$ are slower in optimized receivers than APD's with $k = 1$. A discussion of this point is contained in Appendix A.

B. Equalization of APD Frequency Response

The effects of equalization of detector bandwidth limitations on the sensitivity of APD receivers has been discussed by Ando and Kanbe [55]. Our method is similar to theirs, although we consider only avalanche build-up time (gain-bandwidth product) and do not include transit time limitations, which are included in their results.

To calculate the effect of equalization on the sensitivity of a gain-bandwidth-limited APD receiver, it can be assumed that the APD impulse response has a simple exponential shape. Its complex frequency response is then given by

$$M(\omega) = \frac{M_0}{1 + j\omega M_0 \tau_e}. \tag{33}$$

The magnitude of this complex response gives the gain-bandwidth term of (1). Such an exponential impulse response can be equalized by a simple RC circuit.

In calculating the effect of this equalization on sensitivity, it is first necessary to know its effect on APD noise. It is well known that APD noise and detected signals are filtered identically by the limited APD bandwidth. In fact, measurements of APD noise spectra have been used to derive both frequency response and temporal response [56]–[58]. Hence, if the APD frequency response is equalized to be flat, then equalized APD noise can be assumed to have a uniform spectrum and can be modeled as white Gaussian noise. The signal-to-noise ratio before or after equalization is identical to that for an APD with no bandwidth limitation, as assumed in sensitivity calculations such as (28). Another assumption which simplifies the analysis is that all of the APD noise, including excess noise, occurs at the time photons are detected, hence after equalization this noise is present for the duration of the received pulse and ceases after the received light pulse is gone. This assumption is not strictly true, as due to the slow buildup and decay time of the avalanche process and its regenerative nature for $k > 0$, excess noise continues to be generated by daughter electron-hole pairs after the light pulse ceases. Such temporal evolution of APD noise has not been analyzed, and will not be considered here.

With the above assumptions, equalization corrects for the limited detector bandwidth and does not affect APD signal-to-noise ratio or change the APD-related expressions for noise and sensitivity. It is only necessary to consider the effect of equalization on receiver noise, which will have its spectral content enhanced at frequencies above the corner frequency of the equalizer.

An RC equalizer that corrects for the APD roll-off has a frequency response $H(\omega)$ which is the inverse of the frequency-dependent portion of the APD response, hence

$$H(\omega) = 1 + j\frac{\omega M_0}{2\pi f_{GB}} \tag{34}$$

where f_{GB} is related to the effective avalanche-region transit time by

$$f_{GB} = 1/2\pi\tau_e. \tag{35}$$

In calculating the effect of this equalizer on receiver noise, we can include it in an analysis similar to that presented by Smith and Personick [13], where the receiver transfer function $Z_T(\omega)$ includes the additional term $H(\omega)$ above. It is easily shown that this analysis leads to modified Personick integrals I_2' and I_3' of the form

$$\begin{aligned} I_2' &= \int_0^\infty \left|H_T'(y)\right|^2 \left[1 + \frac{M_0^2 B^2}{f_{GB}^2} y^2\right] dy \\ &= I_2 + \frac{M_0^2 B^2}{f_{GB}^2} I_3 \end{aligned} \tag{36}$$

$$\begin{aligned} I_3' &= \int_0^\infty \left|H_T'(y)\right|^2 \left[1 + \frac{M_0^2 B^2}{f_{GB}^2} y^2\right] y^2\, dy \\ &= I_3 + \frac{M_0^2 B^2}{f_{GB}^2} I_4 \end{aligned} \tag{37}$$

where

$$I_4 = \int_0^\infty \left|H_T'(y)\right|^2 y^4\, dy. \tag{38}$$

The value of I_4 for a full-raised-cosine-spectrum output pulse and a rectangular NRZ input pulse is 0.0304, and for a 50-percent-duty-cycle RZ input pulse is 0.00826.

To calculate the sensitivity of an equalized APD receiver, the modified Personick integrals above can be substituted for I_2 and I_3 in either (7) for an FET amplifier or (23) for a bipolar amplifier, and the input-noise-current variance of the amplifier can be found. This amplifier noise is now a function of M_0 because the corner frequency of the equalizer must be adjusted to match the 3-dB bandwidth of the APD, which depends in turn on M_0. The amplifier noise for a given value of M_0 can be substituted into (31), and the sensitivity can be found. As in the cases of nonzero dark current or nonzero extinction ratio [12], [13], the optimum gain for highest sensitivity is most easily found either graphically or numerically.

The shunt and series receiver noise terms of Ando and Kanbe [55] do not show the above dependence of the receiver noise on avalanche gain, although we do agree with their overall conclusion that high-impedance receivers are less affected by limited APD gain-bandwidth product than low-impedance receivers.

As an example of the effect of limited APD gain-bandwidth product, Fig. 9 shows calculated receiver sensitivities for an APD with $k = 0.35$ and f_{GB} values of 20, 50, 100 GHz, and infinity. The calculations assume zero dark current and an FET preamplifier with parameters identical to those used for Fig. 7. Equalization of the APD frequency response is assumed, except for $f_{GB} = 20$ GHz where an additional dashed curve is included to indicate

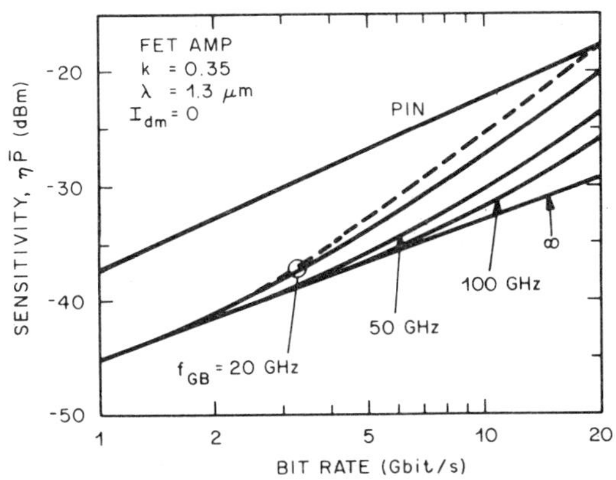

Fig. 9. Receiver sensitivity versus bit rate for an optimized GaAs MESFET preamplifier and a p-i-n detector or APD's with $k = 0.35$ and gain-bandwidth products of 20, 50, and 100 GHz, and infinity. Solid curves assume equalization of the APD frequency response, whereas dashed curve for $f_{GB} = 20$ GHz is for no equalization.

sensitivity when no equalization is employed and M_0 is limited to the value given by (32).

We note from Fig. 9 that for $f_{GB} = 20$ GHz, the sensitivity begins to be reduced above 2 Gbit/s, whereas for $f_{GB} = 50$ GHz a reduction is not present until 4 Gbit/s, and for $f_{GB} = 100$ GHz, reduced sensitivity does not occur until above 6 Gbit/s. The onset frequency of gain-bandwidth limitation does not scale linearly with f_{GB} because the APD optimum gain increases with bit rate in approximate proportion to $B^{0.25}$, as a result of increasing receiver noise. Also, from the $f_{GB} = 20$ GHz curves it can be seen that equalization of the APD response can improve the sensitivity by a few decibels when operating in the gain-bandwidth-limited regime.

The curves in Fig. 9 are for a very good optimized FET preamplifier. If other, less sensitive preamplifiers were employed, the penalties from gain-bandwidth would be larger because more receiver noise necessitates higher optimum gain, which would be precluded by limited APD bandwidth.

V. Experimental Results

A summary of experimental receiver sensitivities as reported in the literature for bit rates up to 8 Gbit/s is contained in Table I. These experimental receivers have used either Ge or III-V APD's; one p-i-n result at 1.2 Gbit/s is also shown. The preamplifiers are mainly GaAs MESFET's, however four bipolar amplifiers and one HEMT amplifier are also included.

The best experimental sensitivities in this table have been obtained with InGaAs APD's and high-impedance GaAs MESFET preamplifiers. These best results are lower than the theoretical sensitivities for an APD with $k = 0.35$ in Fig. 7 by 3 dB at 1 Gbit/s and 8 dB at 8 Gbit/s. The larger difference at the higher bit rate reflects both limited APD gain-bandwidth product and, perhaps more significantly, the greater difficulty in achieving near-ideal operation of the many electronic components comprising an optical regenerator at very high bit rates. Electronic components for bit rates near 1 Gbit/s, including amplifiers, decision circuits, multiplexers and demultiplexers, are now quite mature and have excellent performance, whereas at 4 and 8 Gbit/s all components are still experimental and will require considerable development to attain the performance presently available at lower speeds.

Experimental and theoretical bit-error-rate curves for entries 1, 5, and 15 of Table I are shown in Fig. 10 for bit rates of 420 Mbits, and 1, 2, 4, and 8 Gbit/s. The theoretical BER curves, assuming no equalization of the APD response, were derived using (7) and (31), with the APD gain limited to the maximum value specified by (32) to account for gain-bandwidth restrictions. The experimental 8-Gbit/s receiver used an SAGM-APD with a gain-bandwidth product of 60 GHz [20] and a high-impedance GaAs FET preamplifier, and is described in detail in [59]. Other parameters used in the theoretical calculation are $C_T = 0.71$ pF, $g_m = 40$ mS, $\Gamma = 1.1$, $R_L = 1.3$ kΩ, $R_D = 68$ Ω, $I_{\text{gate}} = 0$, $f_c = 0$, $k = 0.35$, $I_{dm} = 52$ nA, and $\eta = 62$ percent. The experimental optimum gain of 7.4 agrees with the theoretical maximum gain of 7.5 for no equalization as predicted by (32). The sensitivity difference of 3.7 dB between theory and experiment is a result of the transmitter extinction ratio of ~ 10 : 1 which contributes a penalty of about 1.4 dB, and of nonidealities in the 8-Gbit/s electronics leading to intersymbol interference and imperfect receiver equalization. The APD in this receiver had uniform gain and low series resistance of about 30 Ω.

The 2 and 4 Gbit/s results were obtained using an SAGM-APD with a gain-bandwidth product of 18 GHz, and a high-impedance GaAs FET preamplifier with a C_T of 0.63 pF [61]. Other parameters used for the theoretical calculations are $g_m = 40$ mS, $\Gamma = 1.1$, $R_L = 22$ kΩ, $R_D = 100$ Ω, $I_{\text{gate}} = 70$ nA, $f_c = 0$, $k = 0.35$, $I_{dm} = 4$ nA, and $\eta = 71$ percent. The experimental optimum gain of 8.5 at 2 Gbit/s is very close to the theoretical maximum gain of 9 found from (32). At 4 Gbit/s, however, the experimental optimum gain of 7.8 is higher than the theoretical $M_{\max}$ of 4.5, suggesting that some equalization of the APD response occurs due to receiver bandwidth not being tightly filtered. Estimates of penalties contributing to the differences of about 2.5–3.5 dB between experiment and theory at 2 and 4 Gbit/s are 0.7 dB from the transmitter extinction ratio of 20 : 1, and 0.4 dB due to a high APD series resistance of 820 Ω as a result of low substrate doping level. The remaining differences are attributed to imperfect filtering and to nonuniformity of the APD gain profile which leads to additional excess noise [67].

The 1-Gbit/s and 420-Mbit/s results in Fig. 10 are previously unpublished measurements obtained with the same receiver as [61], but with a different APD. This APD was another 18-GHz gain-bandwidth device, but had a very low dark current of $I_{dm} = 2$ nA and a very good antireflection coating giving a unity-gain quantum efficiency of 95 percent. Other theory parameters are the same as for 2

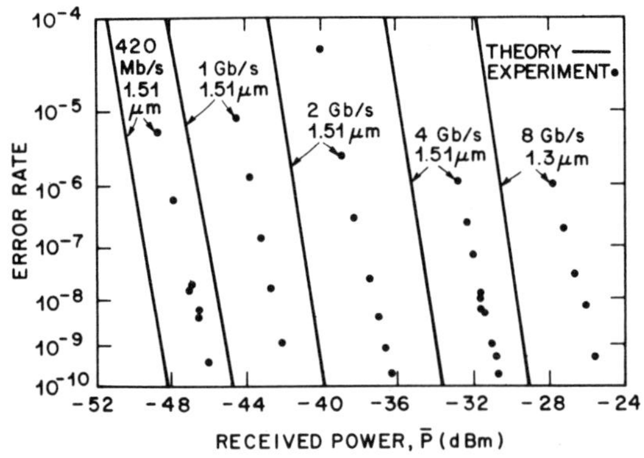

Fig. 10. Theoretical and experimental error rate versus received power for receivers using SAGM APD's and high-impedance GaAs MESFET preamplifiers.

TABLE I
STATE-OF-THE-ART RECEIVER SENSITIVITIES

No.	B (Mb/s)	λ (μm)	$\bar{P}$ [1] (dBm)	Photons[1] /Bit	Detector[2]	Amplifier[3]	Pulse Format	Ref.
1	8000	1.3	-25.8	2150	IGA APD	GA FET-HZ	NRZ	[59]
2	8000	1.3	-15.5	23040	IGA APD	GA FET-LZ	NRZ	[2]
3	4800	1.3	-23	6828	Ge APD	HEMT-TZ	NRZ	[60]
4	4000	1.54	-32.4	1114	IGA APD	GA FET-HZ	RZ	[3]
	4000	1.3	-31.5	1157	"	"	"	
5	4000	1.51	-31.2	1440	IGA APD	GA FET-HZ	NRZ	[61]
	2000	1.51	-36.6	847	"	"	"	
6	4000	1.3	-29.1	2011	IGA APD	GA FET-TZ	RZ	[6]
7	4000	1.3	(-20.5)	(14570)	Ge APD	GA FET-TZ	RZ	[62]
8	2240	1.3	-27	5825	Ge APD	Si Bipolar-TZ	RZ	[11]
9	2000	1.54	-37.4	705	IGA APD	GA FET-HZ	RZ	[7]
	1200	1.54	-40.0	646	"	"	"	
	565	1.54	-42.9	703	"	GA FET-TZ	"	
10	2000	1.3	-33.4	1494	Ge APD	Si Bipolar-LZ	RZ	[9]
	"	1.55	-32.0	2460	"	"	"	
	1600	1.3	-34.5	1450	"	"	"	
	"	1.55	-33.0	2442	"	"	"	
	1200	1.3	-35.9	1401	"	"	"	
	"	1.55	-34.4	2359	"	"	"	
	800	1.3	-37.9	1326	"	"	"	
	"	1.55	-36.2	2338	"	"	"	
	400	1.3	-40.2	1561	"	"	"	
	"	1.55	-39.0	2454	"	"	"	
11	1800	1.3	(-33.0)	(1821)	Ge APD	Si Bipolar-TZ	RZ	[63]
12	1600	1.55	-31.6	3371	IGA APD	GA FET-TZ	NRZ	[64]
13	1600	1.3	(-30.5)	(3642)	Ge APD	Si Bipolar-TZ	RZ	[65]
14	1200	1.53	-36.5	1436	IGA PIN	GA FET-HZ	NRZ	[66]
15	1000	1.51	-42.1	471	IGA APD	GA FET-HZ	NRZ	†
	420	1.51	-46.2	437	"	"	"	
16	1000	1.55	-38.0	1235	IGA APD	GA FET-HZ	NRZ	[24]
	"	1.3	-37.5	1162	"	"	"	
	420	1.55	-43.0	930	"	"	"	
	"	1.3	-41.5	1102	"	"	"	

1. $\bar{P}$ for 10^{-9} bit-error rate. Parentheses indicate $\bar{P}$ for 10^{-11} bit-error rate.
2. IGA - InGaAs.
3. GA - GaAs. HZ - High-impedance. TZ - Transimpedance. LZ - Low-impedance.
† This work.

and 4 Gbit/s, except that $C_T = 0.69$ pF because of higher APD capacitance. Calculated penalties include 0.7 dB from the transmitter extinction ratio of 20:1, and 0.4 dB from the APD series resistance of 740 Ω. Remaining differences between theory and experiment are due to imperfect filtering and APD nonuniformity. The experimental optimum gains of 9.0 and 8.0 in this case are lower than the theoretical values of 12.1 and 13.1 at 420 Mbit/s and 1 Gbit/s, respectively, because of nonuniformity in the gain profile of the particular APD employed.

The above receivers use high values of load resistor in order to maximize receiver sensitivity, and as such are not optimized for wide dynamic range. The front end for 420 Mbit/s to 4 Gbit/s was partially a transimpedance design, as feedback was used around the first FET to reduce the amount of equalization required. We have not measured the dynamic range of this receiver, but have conducted dynamic range measurements on the 8 Gbit/s receiver, which is a true high-impedance design with a load resistor of 1.3 kΩ. As our amplifiers all have fixed gain, automatic gain control was simulated by the insertion of attenuators between amplifiers as the optical signal level was increased. Using this means of gain control, we were able to accomodate signal levels over a range from −13.8 dBm to −25.8 dBm while maintaining an error rate less than 10^{-9}, for a dynamic range of 12 dB through amplifier gain control. An additional 8 dB of dynamic range was gained through reduction of APD gain. The overall dynamic range of the 8-Gbit/s receiver is therefore 20 dB.

The measured sensitivities of −42.1 dBm (471 photons/bit) at 1 Gbit/s and −46.2 dBm (437 photons/bit) at 420 Mbit/s are the highest noncoherent sensitivities reported to date at these bit rates, and are believed to be the best results in terms of photons per bit for any long-wavelength direct-detection receiver. As a comparison, the best results for short-wavelength silicon APD receivers are $\bar{P}$ = 110 photon/bit at 325 Mbit/s using NRZ data format [68], and $\eta\bar{P}$ = 175 photon/bit at 250 Mbit/s using Manchester coding [69]. The InGaAs SAGM APD results above are therefore within 4–6 dB of the best Si APD results.

VI. Conclusions

Very-high-speed avalanche photodiodes and high-sensitivity lightwave receivers have been realized in the laboratory, and are expected to have considerable impact on future gigabit-per-second lightwave systems. We have described the physical mechanisms that determine the bandwidth of high-speed III-V SAGM avalanche photodiodes and have reviewed recent progress towards increasing APD speed. We have also outlined factors determining the noise of FET and bipolar amplifiers, and the sensitivity of APD receivers using these devices. The effect of APD gain-bandwidth product on multigigabit-per-second receiver sensitivity has been discussed. Experimental receiver performance has also been reviewed, and comparisons to theory have been presented.

Appendix A

Forrest [54] has done a calculation of the 3-dB bit rate $B_{1/2}$ of a nonequalized APD receiver versus the ionization rate ratio k in which he claims that $k = 0.5$ gives a lower value of $B_{1/2}$ than $k = 1$. Using the same equations and parameter values, we have not been able to reproduce this result.

$B_{1/2}$ is defined as the bit rate B at which the APD 3-dB bandwidth is equal to $B/2$. The expression for $B_{1/2}$ is

$$B_{1/2} = \frac{\eta\bar{P}_{\text{p-i-n}}}{2Q^2(hc/\lambda)I_1}\frac{1}{(1-k)} \cdot \left\{1 - \left[1 - (1-k)k\left(\frac{Q^2(hc/\lambda)}{\pi\tau_e(\eta\bar{P}_{\text{p-i-n}})}\right)^2\right]^{1/2}\right\} \quad \text{(A1)}$$

where $\eta\bar{P}_{\text{p-i-n}}$ is the receiver sensitivity that would be obtained with a p-i-n detector rather than an APD, and other parameters are as defined previously. The effective transit time τ_e as a function of k has been expressed by Emmons [39] as

$$\tau_e = Nk\tau_0 \quad \text{(A2)}$$

where τ_0 is the mean transit time of carriers across the gain region and N is a number which varies between 1/3 and 2, depending on k.

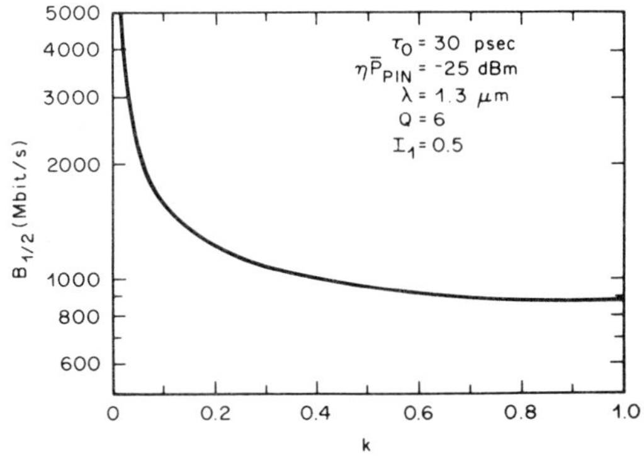

Fig. 11. Bit rate $B_{1/2}$ at which APD bandwidth is equal to $B/2$ versus ionization rate ratio k for a p-i-n receiver sensitivity of $\eta\bar{P}_{\text{p-i-n}} = -25$ dBm, τ_0 = 30 ps, and zero APD dark current.

We have calculated $B_{1/2}$ versus k using the above equations for parameter values identical to those used by Forrest: τ_0 = 30 ps, $\eta\bar{P}_{\text{p-i-n}} = -25$ dBm, $Q = 6$, $\lambda = 1.3$ μm, and $I_1 = 0.5$. The only parameter not specified by Forrest is N as a function of k, which we have derived from [39, fig. 4] to be N = 0.33, 0.45, 0.63, 0.79, 0.93, 1.14, and 1.23 for k = 1, 0.5, 0.2, 0.1, 0.05, 0.02, and 0.01, respectively. Using these parameter values, we obtained a plot of $B_{1/2}$ versus k as shown in Fig. 11. The curve is different than Forrest's, which had a minimum at $k = 0.4$. Our curve has a minimum $B_{1/2}$ at $k = 1$ and a somewhat higher value at $k = 0.5$, showing that the slowest response speed in optimized APD receivers occurs for an ionization-rate ratio of $k = 1$ and not for $k = 0.5$. This result seems more agreeable intuitively, and in any case there can be no doubt that the higher gain-bandwidth product and lower noise of an APD with $k = 0.5$ will always provide better sensitivity at high bit rates than would an APD with $k = 1$.

References

[1] A. H. Gnauck, S. K. Korotky, B. L. Kasper, J. C. Campbell, J. R. Talman, J. J. Veselka, and A. R. McCormick, "Information-bandwidth-limited transmission at 8 Gbit/s over 68.3 km of single-mode optical fiber," in *Tech. Dig. OFC'86* (Atlanta, GA), 1986 (postdeadline).

[2] A. H. Gnauck, J. E. Bowers, and J. C. Campbell, "8-Gbit/s transmission over 30 km of optical fiber," *Electron Lett.*, vol. 22, pp. 600–602, 1986.

[3] S. Fujita, I. Takano, N. Henmi, S. Shikada, I. Mito, K. Taguchi, and K. Minemura, "4-Gbit/s long span transmission experiments employing high speed DFB-LD's and InGaAs-APD's," in *Proc. 12th Eur. Conf. Optical Commun.* (Barcelona, Spain), 1986.

[4] S. K. Korotky, G. Eisenstein, A. H. Gnauck, B. L. Kasper, J. J. Veselka, R. C. Alferness, L. L. Buhl, C. A. Burrus, T. C. D. Huo, L. W. Stulz, K. Ciemiecki Nelson, L. G. Cohen, R. W. Dawson, and J. C. Campbell, "4-Gbit/s transmission experiment over 117 km of optical fiber using a Ti-$LiNbO_3$ external modulator," *J. Lightwave Technol.*, vol. LT-3, pp. 1027–1031, 1985.

[5] A. H. Gnauck, B. L. Kasper, R. A. Linke, R. W. Dawson, T. L. Koch, T. J. Bridges, E. G. Burkhardt, R. T. Yen, D. P. Wilt, J. C. Campbell, K. Ciemiecki Nelson, and L. G. Cohen, "4-Gbit/s transmission experiment over 103 km of optical fiber using a novel electronic multiplexer/demultiplexer," *J. Lightwave Technol.*, vol. LT-3, pp. 1032–1035, 1985.

[6] I. Takano, I. Mito, K. Taguchi, T. Watanabe, and B. Hirosaki, "A 4-Gbit/s optical regenerative repeater for its fiber transmission experiments," in *Proc. 6th Int. Conf. Integrated Opt. and Opt. Fiber Commun.* (Reno, NV), 1987 (postdeadline).

[7] M. Shikada, S. Fujita, I. Takano, N. Henmi, I. Mito, K. Taguchi, and K. Minemura, "1.5-μm high-bit-rate long-span transmission experiments employing a high-power DFB-DC-PBH laser diode," in *Proc. 5th Int. Conf. Integrated Opt. and Opt. Fiber Commun.* (Venice, Italy), 1985.

[8] B. L. Kasper, R. A. Linke, K. L. Walker, L. G. Cohen, T. L. Koch, T. J. Bridges, E. G. Burkhardt, R. A. Logan, R. W. Dawson, and J. C. Campbell, "A 130-km transmission experiment at 2 Gbit/s using silica-core fiber and a vapor phase transported DFB laser," in *Proc. 10th Eur. Conf. Opt. Commun.* (Stuttgart, W. Germany), 1984 (postdeadline).

[9] J. Yamada, A. Kawana, T. Miya, J. Nagai, and T. Kimura, "Gigabit/s optical receiver sensitivity and zero-dispersion single-mode fiber transmission at 1.55 μm," *IEEE J. Quantum Electron.*, vol. QE-18, pp. 1537–1546, 1982.

[10] J. I. Yamada, S. Machida, and T. Kimura, "2-Gbit/s optical transmission experiments at 1.3 μm with 44-km single-mode fiber," *Electron. Lett.*, vol. 17, pp. 479–480, 1981.

[11] W. Albrecht, G. Elze, B. Enning, G. Walf, and G. Wenke, "Experiences with an optical long-haul 2.24-Gbit/s transmission system at a wavelength of 1.3 μm," *Electron Lett.*, vol. 18, pp. 746–748, 1982.

[12] S. D. Personick, "Receiver design for digital fiber-optic communication systems, Part I and II," *Bell Syst. Tech. J.*, vol. 52, pp. 843–886, 1973.

[13] R. G. Smith and S. D. Personick, "Receiver design for optical fiber communication systems," in *Semiconductor Devices for Optical Communication.* New York: Springer-Verlag, 1980, ch. 4.

[14] S. R. Forrest, "Sensitivity of avalanche photodetector receivers for high-bit-rate long-wavelength optical communication systems," in *Semiconductors and Semimetals*, vol. 22: *Lightwave Communications Technology.* Orlando, FL: Academic, 1985, ch. 4.

[15] H. Melchior, A. R. Hartman, D. P. Schinke, and T. E. Seidel, "Planar epitaxial silicon avalanche photodiode," *Bell Syst. Tech. J.*, vol. 57, pp. 1791–1807, 1978.

[16] K. Nishida, K. Taguchi, and Y. Matsumoto, "InGaAsP heterojunction avalanche photodiodes with high avalanche gain," *Appl. Phys. Lett.*, vol. 35, pp. 251–253, 1979.

[17] S. R. Forrest, M. DiDomenico, Jr., R. G. Smith, and H. J. Stocker, "Evidence of tunneling in reverse-bias III-V photodetector diodes," *Appl. Phys. Lett.*, vol. 36, pp. 580–582, 1980.

[18] H. Ando, H. Kanbe, M. Ito, and T. Kaneda, "Tunneling current in InGaAsP and optimum design for InGaAs/InP avalanche photodiodes," *Japan J. Appl. Phys.*, vol. 19, pp. 1277–1280, 1980.

[19] W. S. Holden, J. C. Campbell, J. F. Ferguson, A. G. Dentai, and Y. K. Jhee, "Improved frequency response of InP/InGaAsP/InGaAs avalanche photodiodes with separate absorption, grading, and multiplication regions," *Electron. Lett.*, vol. 21, pp. 886–887, 1985.

[20] J. C. Campbell, W. S. Holden, J. F. Ferguson, A. G. Dentai, and Y. K. Jhee, "A high-speed InP/InGaAsP/InGaAs avalanche photodiode exhibiting a gain-bandwidth product of 60 GHz," in *Proc. 5th Int. Conf. Integrated Opt. and Opt. Fiber Commun.* (Venice, Italy), 1985 (postdeadline).

[21] Y. Sugimoto, T. Torikai, K. Makita, H. Ishihara, and K. Taguchi, "Gain-bandwidth improvement in planar InP/InGaAsP/InGaAs avalanche photodiodes," in *Proc. 5th Int. Conf. Integrated Opt. and Opt. Fiber Commun.* (Venice, Italy), 1985.

[22] M. Kobayashi, H. Machida, T. Shirai, Y. Kishi, N. Takagi, and T. Kaneda, "Optimized GaInAs avalanche photodiode with low noise and large gain-bandwidth product," in *Proc. 6th Int. Conf. Integrated Opt. and Opt. Fiber Commun.* (Reno, NV), 1987.

[23] S. R. Forrest, O. K. Kim, and R. G. Smith, "Optical response time of $In_{0.53}Ga_{0.47}As$ avalanche photodiodes," *Appl. Phys. Lett.*, vol. 41, pp. 95–98, 1982.

[24] J. C. Campbell, A. G. Dentai, W. S. Holden, and B. L. Kasper, "High-performance avalanche photodiode with separate absorption, grading, and multiplication regions," *Electron. Lett.*, vol. 18, pp. 818–820, 1983.

[25] J. C. Campbell, W. S. Holden, G. J. Qua, and A. G. Dentai, "Frequency response of InP/InGaAsP/InGaAs avalanche photodiodes with separate absorption, grading, and multiplication regions," *IEEE J. Quantum Electron.*, vol. QE-21, pp. 1743–1746, 1985.

[26] S. R. Forrest, G. F. Williams, O. K. Kim, and R. G. Smith, "Excess-noise and receiver sensitivity measurement of $In_{0.53}Ga_{0.47}As$/InP avalanche photodiodes," *Electron. Lett.*, vol. 17, pp. 917–919, 1981.

[27] H. Ando, Y. Yamauchi, and N. Susa, "Reach-through type planar InGaAs/InP avalanche photodiode fabricated by continuous vapor phase epitaxy," *IEEE J. Quantum Electron.*, vol. QE-20, pp. 256–264, 1984.

[28] R. H. Fowler and L. Nordheim, "Electron emission in intense electric fields," *Proc. Roy. Soc. (London)*, vol. 119, pp. 173–181, 1928.

[29] F. A. Padovani and R. Stratton, "Field and thermionic-field emission in Schottky barriers," *Solid-State Electron.*, vol. 9, pp. 695–707, 1966.

[30] D. T. Cheung, S. Y. Chiang, and G. L. Pearson, "A simplified model for graded-gap heterostructures," *Solid-State Electron.*, vol. 18, pp. 263–266, 1975.

[31] F. Capasso, H. M. Cox, A. L. Hutchinson, N. A. Olsson, and S. G. Hummel, "Psuedo-quaternary GaInAsP semiconductors: A new $Ga_{0.47}In_{0.53}As$/InP graded gap superlattice and its applications to avalanche photodiodes," *Appl. Phys. Lett.*, vol. 45, pp. 1193–1195, 1984.

[32] Y. Matsushima, K. Sakai, and Y. Noda, "New type InGaAs/InP heterostructure avalanche photodiode with buffer layer," *IEEE Electron. Dev. Lett.*, vol. EDL-2, pp. 179–181, 1981.

[33] Y. Matsushima, A. Akiba, K. Sakai, K. Kushiro, Y. Noda, and K. Utaka, "High-speed response InGaAs/InP heterostructure avalanche photodiode with InGaAsP buffer layers," *Electron. Lett.*, vol. 18, pp. 945–946, 1982.

[34] Y. Sugimoto, T. Torikai, K. Makita, H. Ishihara, K. Minemura, and K. Taguchi, "High-speed planar-structure InP/InGaAsP/InGaAs avalanche photodiode grown by VPE," *Electron. Lett.*, vol. 20, pp. 653–654, 1984.

[35] K. Yasuda, T. Mikawa, Y. Kishi, and T. Kaneda, "Multiplication-dependent frequency responses of InP/InGaAs avalanche photodiode," *Electron. Lett.*, vol. 20, pp. 373–374, 1984.

[36] R. Trommer and L. Hoffman, "Time-response of InGaAs/InP avalanche photodiodes," in *Proc. 11th Int. Symp. GaAs and Related Compounds* (Biaritz, France), 1984.

[37] J. C. Campbell, W. T. Tsang, G. J. Qua, and B. C. Johnson, "High-speed InP/InGaAsP/InGaAs avalanche photodiodes grown by chemical beam epitaxy," *IEEE J. Quantum Electron.*, to be published.

[38] J. E. Bowers, C. A. Burrus, and R. J. McCoy, "InGaAs photodetectors with modulation response to millimeter wavelengths," *Electron. Lett.*, vol. 21, pp. 812–814, 1985.

[39] E. B. Emmons, "Avalanche-photodiode frequency response," *J. Appl. Phys.*, vol. 38, pp. 3705–3714, 1967.

[40] G. F. Williams, F. Capasso, and W. T. Tsang, "The graded bandgap multilayer avalanche photodiode: A new low-noise detector," *IEEE Electron. Dev. Lett.*, vol. EDL-3, pp. 71–73, 1982.

[41] G. E. Stillman, N. Tabatabaie, V. M. Robbins, and J. A. Aguilar, "III-V compound semiconductor optical detectors," in *Proc. Int. Symp. GaAs and Related Compounds* (Biaritz, France), 1984.

[42] I. Umeba, A. N. M. M. Choudhury, and P. N. Robson, "Ionization coefficients measured in abrupt InP junctions," *Appl. Phys. Lett.*, vol. 36, pp. 302–303, 1980.

[43] F. Capasso, A. Y. Cho, and P. W. Foy, "Low-dark-current low-voltage 1.3–1.6 μm avalanche photodiode with high–low electric field profile and separate absorption and multiplication regions by molecular beam epitaxy," *Electron. Lett.*, vol. 20, pp. 635–637, 1984.

[44] T. V. Muoi, "Receiver design for high-speed optical-fiber systems," *J. Lightwave Technol.*, vol. LT-2, pp. 243–267, 1984.

[45] M. Brain and T. P. Lee, "Optical receivers for lightwave communications systems," *J. Lightwave Technol.*, vol. LT-3, pp. 1281–1300, 1985.

[46] A. van der Ziel, "Thermal noise in field-effect transistors," *Proc. IRE*, vol. 50, pp. 1808–1812, 1962.

[47] A. van der Ziel, "Gate noise in field effect transistors at moderately high frequencies," *Proc. IEEE*, vol. 57, pp. 461–467, 1963.

[48] F. N. H. Robinson, *Noise and Fluctuations in Electronic Devices and Circuits.* Oxford, England: Clarendon, 1974.

[49] M. J. Buckingham, *Noise in Electronic Devices and Systems.* Chichester, England: Ellis Horwood, 1983.

[50] W. Baechtold, "Noise behavior of GaAs field-effect transistors with short gate lengths," *IEEE Trans. Electron Devices*, vol. ED-19, pp. 674–680, 1972.

[51] J. E. Goell, "Input amplifiers for optical PCM repeaters," *Bell Syst. Tech. J.*, vol. 53, pp. 1771–1793, 1974.

[52] K. Ogawa, "Noise caused by GaAs MESFET's in optical receivers," *Bell Syst. Tech. J.*, vol. 60, pp. 923–928, 1981.

[53] R. J. McIntyre, "Multiplication noise in uniform avalanche diodes," *IEEE Trans. Electron Devices*, vol. ED-13, pp. 164–168, 1966.

[54] S. R. Forrest, "Gain-bandwidth-limited response in long-wavelength avalanche photodiodes," *J. Lightwave Technol.*, vol. LT-2, pp. 34–39, 1984.

[55] H. Ando and H. Kanbe, "Effect of avalanche build-up time on avalanche photodiode sensitivity," *IEEE J. Quantum Electron.*, vol. QE-21, pp. 251–255, 1985.

[56] T. Kaneda, H. Takanashi, H. Matsumoto, and T. Yamaoka, "Avalanche buildup time of silicon reach-through photodiodes," *J. Appl. Phys.*, vol. 47, pp. 4960–4963, 1976.

[57] T. Andersson, A. R. Johnston, and H. Eklund, "Temporal and frequency response of avalanche photodiodes from noise measurements," *Appl. Opt.*, vol. 19, pp. 3496–3499, 1980.

[58] Y. K. Jhee, J. C. Campbell, J. F. Ferguson, A. G. Dentai, and W. S. Holden, "Avalanche buildup time of InP/InGaAsP/InGaAs avalanche photodiodes with separate absorption, grading, and multiplication regions," *IEEE J. Quantum Electron.*, vol. QE-22, pp. 753–755, 1986.

[59] B. L. Kasper, J. C. Campbell, J. R. Talman, A. H. Gnauck, J. E. Bowers, and W. S. Holden, "An APD/FET optical receiver operating at 8 Gbit/s," *J. Lightwave Technol.*, vol. LT-5, pp. 344–347, 1987.

[60] L. C. Blank, J. M. Boggis, R. A. Garnham, and S. D. Walker, "High-electron-mobility transistor lightwave receiver for broadband optical transmission system application," in *Proc. 6th Int. Conf. Integrated Opt. and Opt. Fiber Commun.* (Reno, NV), 1987.

[61] B. L. Kasper, J. C. Campbell, A. H. Gnauck, A. G. Dentai, and J. R. Talman, "SAGM avalanche photodiode receiver for 2 and 4 Gbit/s," *Electron. Lett.*, vol. 21, pp. 982–984, 1985.

[62] I. Takano, T. Iwakami, I. Mito, and Y. Tashiro, "4-Gbit/s modulation experiments for long-wavelength region optical fiber transmission," in *Tech. Dig. OFC'85* (San Diego, CA), 1985.

[63] K. Iwashita, N. Ohta, Y. Akazawa, N. Ishihara, and S. Konaka, "Monolithic integrated amplifiers for a gigabit optical repeater," *Electron. Lett.*, vol. 20, pp. 470–471, 1984.

[64] S. Lumish and F. T. Stone, "Cabled-fiber 1.6-Gbit/s 1550-nm transmission experiment," *Electron. Lett.*, vol. 22, pp. 1247–1248, 1986.

[65] M. Ohara, Y. Akazawa, N. Ishihara, and S. Konaka, "Bipolar monolithic amplifiers for a gigabit optical repeater," *IEEE J. Solid-State Circuits*, vol. SC-19, pp. 491–497, 1984.

[66] M. C. Brain, P. P. Smyth, D. R. Smith, B. R. White, and P. J. Chidgey, "PINFET hybrid optical receivers for 1.2-Gbit/s transmission systems at 1.3- and 1.55-μm wavelength," *Electron. Lett.*, vol. 20, pp. 894–896, 1984.

[67] Y. K. Jhee, J. C. Campbell, W. S. Holden, A. G. Dentai, and J. K. Plourde, "The effect of nonuniform gain on the multiplication noise of InP/InGaAsP/InGaAs avalanche photodiodes," *IEEE J. Quantum Electron.*, vol. QE-21, pp. 1858–1861, 1985.

[68] T. V. Muoi, "Extremely sensitive direct detection receiver for laser communication," in *Tech. Dig. CLEO'87* (Baltimore, MD), 1987.

[69] T. V. Muoi, "Receiver design for digital fiber optic transmission systems using Manchester (biphase) coding," *IEEE Trans. Commun.*, vol. COM-31, pp. 608–619, 1983.

*

Bryon L. Kasper (S'71–M'73) received the B.Sc. and Ph.D. degrees in electrical engineering from the University of Alberta, Edmonton, Canada, in 1973 and 1981, respectively.

From 1973 to 1975 he worked for Alberta Government Telephones in mobile radio and microwave radio systems. In 1981, he joined the Lightwave Systems Research Department of AT&T Bell Laboratories, Crawford Hill Lab, Holmdel, NJ, and is currently engaged in research on high-speed optical receivers and transmission systems. He has designed high sensitivity avalanche photodiode receivers and has collaborated in long-span transmission experiments at bit rates from 400 Mbit/s to 8 Gbit/s, and has also designed receivers and collaborated in coherent transmission experiments to 2 Gbit/s.

*

Joe C. Campbell (S'73–M'74) was born in Gorman, TX, on January 11, 1947. He received the B.S. degree in physics from the University of Texas at Austin in 1969, the M.S. and Ph.D. degrees in physics from the University of Illinois at Urbana-Champaign in 1971 and 1973, respectively.

From 1974 to 1976 he was employed by Texas Instruments where he worked on integrated optics in GaAs/AlGaAs. Since 1976 he has been at the Crawford Hill Laboratory of Bell Laboratories in Holmdel, NJ, where he has worked on electrooptic devices for fiber-optics applications.

Dr. Campbell is a member of Sigma Xi, the American Physical Society, and the Optical Society of America.

Chapter 5
Multiplexing Components and Techniques

Multiplexing increases the usable information capacity of a fiber through the simultaneous transmission of more than one channel over that fiber. A variety of methods can be used: incoherent or coherent carrier-frequency (or *wavelength division*) multiplexing, subcarrier-frequency multiplexing, and time-division multiplexing. Multiplexing is finding increased application in fiber optics systems, but no single technique is recognized as clearly superior at this stage. The papers in this chapter were selected to give an indication of the various multiplexing possibilities: **"Review and Status of Wavelength-Division Multiplexing Technology and Its Application,"** by **H. Ishio, I. Minowa, and K. Nosu; "60-Channel FM Video Subcarrier Multiplexed Optical Communication System,"** by **R. Olshansky and V.A. Lanzisera; "All Single-Mode Optical Fibre Wavelength-Tunable WDM with Very Narrow Pass/Stopband Separations,"** by **S.P. Shipley and A. El Fatatry; "Optical Time-Division Multiplexed Transmission System Experiment at 8 Gbit/s,"** by **G. Eisenstein, R.S. Tucker, S.K. Korotky, G. Raybon, J.J. Veselka, L.L. Buhl, A.H. Gnauck, B.L. Kasper, and R.C. Alferness; "5 GHz-Spaced, Eight-Channel Guided-Wave Tunable Multi/Demultiplexer for Optical FDM Transmission Systems," by H. Toba, K. Oda, N. Takato, and K. Nosu; "5 GHz-Spaced, Eight-Channel Optical FDM Transmission Experiment using Guided-Wave Tunable Demultiplexer," by H. Toba, K. Oda, K. Nosu, N. Takato, and H. Miyazawa;** and **"Densely Spaced WDM Coherent Optical Star Network,"** by **B. Glance, K. Pollock, C.A. Burrus, B.L. Kasper, G. Eisenstein, and L. W. Stulz.**

Review and Status of Wavelength-Division-Multiplexing Technology and Its Application

(Invited Overview)

HIDEKI ISHIO, JUNICHIRO MINOWA, AND KIYOSHI NOSU, MEMBER, IEEE

***Abstract*—Wavelength-division-multiplexing (WDM) technology is now recognized as one of the key technologies in optical communications systems. This is because it has great potential to enhance system design and flexibility.**

This paper reviews state-of-the-art optical multi/demultiplexers (MUX/DEMUX) and WDM system design. Various system applications are also summarized.

I. INTRODUCTION

WAVELENGTH-DIVISION-MULTIPLEXING (WDM) technology, by which multiple optical channels can be simultaneously transmitted at different wavelengths through a single optical fiber, is a useful means of making full use of the low-loss characteristics of optical fibers over a wide-wavelength region.

This concept first appeared in 1970 according to published reports [1]. However, it was the middle of 1977 before fundamental research on WDM technology actually started. This research focused on practical applications to communications systems [2]-[4]. Since then, research has intensified along with dramatic progress in optical fibers, light sources, and detectors. Particularly, optical multi/demultiplexers, which are key devices in the WDM transmission systems, are currently among the most popular research and development items. Many types of multi/demultiplexers have been proposed and fabricated to data.

Now, although WDM technology is not fully matured at present, it has been gradually introduced into practical systems in some countries. It is also expected that this technology will play a major role in future communications. This is primarily due to the excellent applicability of WDM technology, which is based on utilization of the wide range of the low-loss spectrum region in optical fibers.

This paper reviews WDM technology and its application, with emphasis on practical use. Section II presents an outline of WDM systems, and the state-of-the-art of related devices such as light sources and detectors. Section III presents optical multiplexing devices. Section IV gives an outline of WDM system design methods. Section V describes some applications of WDM technology.

Wavelength-division-multiplexing technology is currently one of the most popular research and development items and many review papers have already been published [5], [6]. Therefore, in this paper overlapping descriptions of well-known data and technology are minimized.

II. OUTLINE OF WDM SYSTEMS

There are two basic configurations for WDM transmission systems, i.e., one-way and two-way transmission, as shown in Fig. 1. Either requires a single optical fiber, several light sources and photodetectors, and optical multi/demultiplexing devices. The one-way transmission system requires a multiplexer (MUX) and a demultiplexer (DEMUX), while two-way transmission system requires a multi/demultiplexer (MUX/DEMUX) on each end.

The advantages of WDM systems are: transmission capacity increase per fiber, system cost reduction, simultaneous transmission of different modulation-scheme signals, and service channel expandability after fiber installation. These are the reason WDM technology is expected to be widely applied to systems in various fields of communications.

In WDM system design, performance of optical multi/demultiplexers (MUX, DEMUX, and MUX/DEMUX) should be the primarily consideration, together with fibers, light sources and photodetectors. Since optical multi/demultiplexers are examined in the following sections, the latter part of this section will focus on the status of other optical devices and components. The performance of optical devices and components, some of which are shown in Fig. 2, depend on wavelength. The items necessary for WDM system design are the following.

A. Optical Fibers

Loss characteristics of silica fiber now approximately meet the theoretical values in the 0.8- to 1.7-μm regions, except in the 0.95- and 1.38-μm regions, where OH absorption occurs. Therefore, silica fiber can be applied to WDM transmission at almost all wavelengths from 0.8 to 1.7 μm.

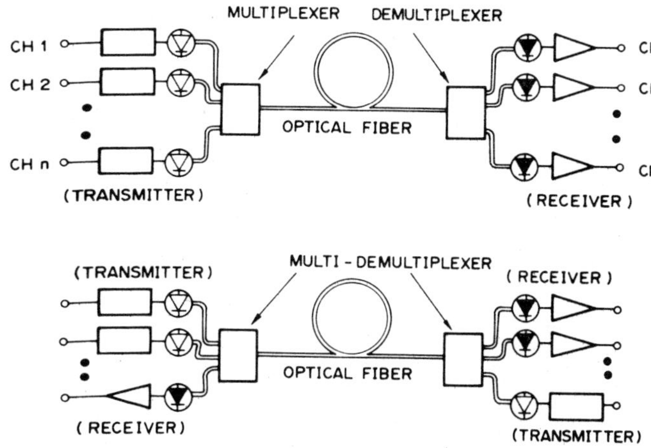

Fig. 1. Fundamental configuration for WDM transmission.

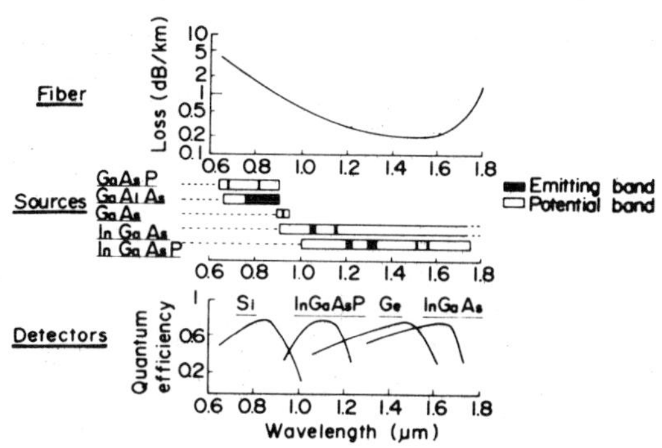

Fig. 2. Low-loss wavelength region of fiber, emission wavelengths for semiconductor light service, and wavelength response of semiconductor photodetectors.

B. Light Sources

GaAlAs LD's and LED's are now fully developed and have a considerably long lifetime of around 10^5–10^6 h in the 0.8-μm region. InGaAsP LD's are also gradually being employed in the long-wavelength regions: the 1.2-, 1.3-, and 1.5-μm regions.

When these devices are used in WDM systems, wavelength deviation is one of the most important design factors. Therefore, as an example, let us consider oscillating wavelength deviation for an LD. Wavelength deviations result from various factors including deviation from target wavelength during production (wavelength reproducibility), temperature fluctuation, feedback effect due to reflected light.

In addition, the effective spectrum width of an LD should be considered. Wavelength deviation due to these factors and the effective spectrum width are evaluated in Table I [7], [8]. Thus broadly speaking, an LD has an oscillating wavelength deviation range of $\lambda_0 \pm 125$ Å (λ_0: nominal wavelength) in the 0.8-μm region and an oscillating wavelength deviation range of $\lambda_0 \pm 250$ Å in the 1.3-μm region. Here, it is assumed that the yield of laser diodes is maintained at a reasonable value and no measures are taken to stabilize temperature fluctuation. That is, if wavelenght deviation due to production and temperature fluctuation were reduced by strictly specifying the oscillating wavelength of laser diodes (thus decreasing the yield) and using Peltier effect devices, then, the bandwidth requirement for multiplexing devices could be relaxed. However, that

TABLE I
FACTORS AND VALUES OF WAVELENGTH VARIATION IN LD'S

Factor	GaAlAs-LD	InGaAsP-LD
Wavelength reproducibility	-5 ∼ +5 (nm)	-10 ∼ +10 (nm)
Temperature fluctuation (30 ±20°C)	-5 ∼ +5 (nm) (2.5 Å/°C)	-10 ∼ +10 (nm) (5 Å/°C)
Feedback reflected light	-1 ∼ +1 (nm)	-3 ∼ +3 (nm)
Effective spectral width	-1.5 ∼ +1.5 (nm)	-2 ∼ +2 (nm)
Total wavelength variation	-12.5 ∼ +12.5 (nm)	-25 ∼ +25 (nm)

would still leave many problems such as high LD cost, high power consumption and maintenance of Peltier devices.

Thus an LD oscillating wavelength deviation range of about $\lambda_0 \pm 125$ Å and $\lambda_0 \pm 250$ Å is desired in the 0.8- and 1.3-μm regions, respectively.

C. Photodetectors

In the short-wavelength region, Si-p-i-n's and Si-APD's have been developed, as well as Ge-p-i-n's and Ge-APD's in the long-wavelength region. In addition, InGaAs photodetectors are being developed in the long-wavelength region.

III. OPTICAL MULTI/DEMULTIPLEXING DEVICES

Optical multi/demultiplexing devices are primarily classified into two types, wavelength selective and wavelength nonselective, as shown in Fig. 3. The wavelength selective type is further classified into two types, passive and active. The passive type is further classified into four types, angularly dispersive, dielectric thin-film filter (DTF), hybrid, and planar waveguide. The active type includes multiwavelength light sources (LD, LED) and multiwavelength photodiodes. The wavelength selective type is primarily important because the WDM system involves principles based on the combination and selection of different wavelength light. In a few cases, however, the wavelength nonselective type is applicable for MUX-only use because MUX does not always demand a wavelength selective element.

In the following subsections, these devices are discussed with respect to their features, configurations, and performance characteristics. Here, subsections *A*, *B*, *C*, *D*, and *E* correspond to the symbols *A*, *B*, *C*, *D*, and *E* written on the shoulders of the rectangles in Fig. 3.

A. Angularly Dispersive Devices

Angularly dispersive devices have the advantage of compactness for application to a large number of channels. This is because they can multiplex or demultiplex many channels using only one angularly dispersive element, such as a prism or grating. However, it is not easy to obtain the required passband width. The reason for this is that, in order to obtain the required passband width, the output fiber core must be much larger than the input fiber core. For MUX, since the output fiber is the transmission fiber, the input fiber core must have a diameter of less than 50 μm, which is the standard fiber core diameter. This thin core significantly degrades coupling efficiency between the light source and fiber. Such degradation is more severe in MUX for a single-mode WDM system. Therefore, a prism or grating may be applicable to DEMUX, but is very difficult to apply to MUX or MUX/DEMUX. Angularly

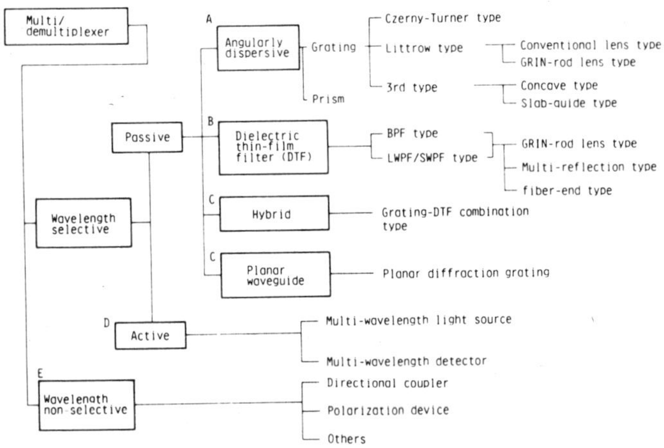

Fig. 3. Classification of multi/demultiplexers.

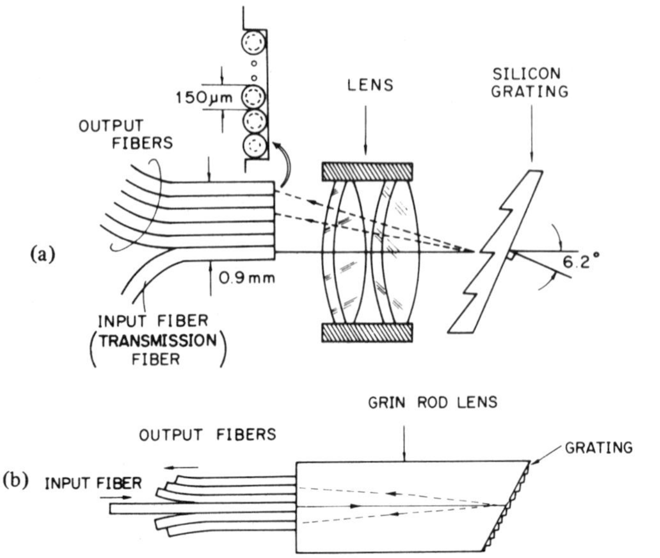

Fig. 4. Typical structure for grating-Littrow type DEMUX. (a) Conventional lens type. (b) GRIN-rod lens type.

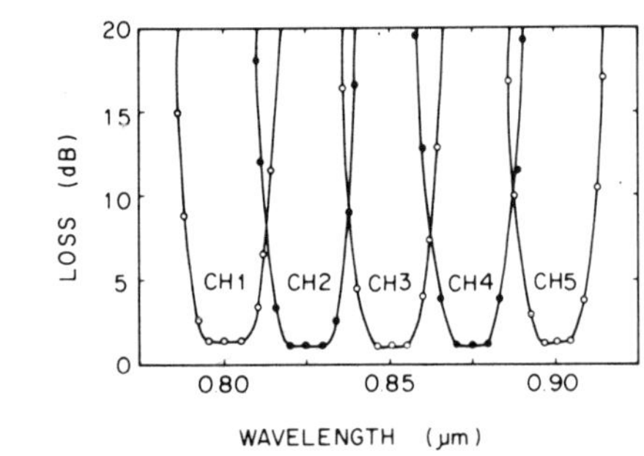

Fig. 5. Demultiplexing characteristic of the Littrow-type DEMUX shown in Fig. 4(a).

dispersive devices also have a disadvantage in that they cannot achieve low insertion loss in both the 0.8- and 1.3-μm regions simultaneously, due to the prism or grating characteristics.

In the initial stage, prisms were employed as an element in the development of angularly dispersive devices [9]-[11], but a practical device was not realized due to the material constraints of prisms, i.e., low angular dispersion. At present, the diffraction grating is primarily being used as the angularly dispersive element.

For such a grating, a blazed grating that can efficiently diffract light into a specific diffraction order is widely used. There are two types of blazed gratings. One is fabricated by conventional mechanical ruling and the other by anisotropic etching of single-crystal silicon [13] (silicon grating). Since the silicon grating provides design freedom (i.e., blaze wavelength and grating constant), high efficiency, and a stable surface [13], it is superior to the conventional mechanically ruled grating for practical use.

With respect to structure, there are three types: the Czerny-Turner type, which has different lenses for input and output, the Littrow type, which has one common lens, and a third type which has no lens. The Littrow and third types have been more extensively studied than the Czerny-Turner type due to their compact structures.

1) Littrow Type: Examples of the Littrow-type DEMUX [13], [17] are shown in Fig. 4(a) (conventional lens type) and (b) (GRIN-rod lens type). The GRIN-rod lens type is superior to the conventional lens type in compactness and ease of alignment. Examples of the conventional lens type and GRIN-rod lens type are listed in [12]-[15] and [16]-[19], respectively.

Now, let us examine the operating principle and performance for the Littrow-type DEMUX, referring to the structure shown in Fig. 4(a) as an example. An input fiber and multiple output fibers are arranged on the focal plane of the lens. Wavelength multiplexed light from the input fiber is collimated by the lens and reaches the diffraction grating. The light is angularly dispersed, according to different wavelengths, and reflected at the same time. Then, the different wavelengths pass through the lens and are focused to their corresponding output fibers. Each wavelength is fed to one individual detector through its output fiber. Performance for this DEMUX is shown in Fig. 5. Insertion lossses for five channels are 1.4, 1.1, 1.1, 1.1, and 1.2 dB. Crosstalk levels are less than −30 dB. Interchannel wavelength spacing is equal to the 25-nm designed value. The 0.1-dB bandwidths are all 11 nm. Such wide passband width can be obtained due to the difference between the core diameter Δd of the input fiber (core diameter: 60 μm) and the output fibers (core diameters: 130 μm). Bandwidth B_W can be estimated from

$$B_W = \frac{\Delta d \cdot \Delta\lambda}{\Delta Z} \tag{3.1}$$

where $\Delta\lambda/\Delta Z$ is linear dispersion in the focal plane, which is 25 nm/150 μm in this device. Since Δd is 70 μm, B_W is 11.7 nm from (3.1). This value is nearly equal to the experimental value of 11 nm. This wide bandwidth is necessary in a DEMUX to prevent insertion loss being influenced by wavelength variation in the LD's.

The performance of Littrow-type DEMUX's listed in the references includes:

number of channels	3 ~ 20;
insertion loss	1 ~ 4 dB;
wavelength spacing	20 ~ 40 nm; and
crosstalk attenuation	20 ~ 30 dB.

2) Third Type: As previously stated, third-type devices have no lens. They are further classified into two types, one, using only a concave grating [20] (concave grating type, the other) using a combination of slab waveguide and concave grating [21]-[23] or a combination of slab waveguide and plane chirp-grating [24] (slab-waveguide type). Slab-waveguide types are constructed with different materials between the slab waveguide and grating. These waveguides do not function as single-mode waveguides.

A typical structure is shown in Fig. 6 [23]. The operating principle of the device is similar to that for the Littrow-type DEMUX, except that the concave grating functions as both the lens and the plane grating in the case of the Littrow-type DEMUX. The slab waveguide confines the multiplexed light from the input fiber to a two-dimensional guiding layer. Light from the guiding layer hits the concave grating, is dispersed into individual beams, and is reflected back to the guiding layer. These light beams are then focused toward the output fibers.

Performance of this device is close to that of the Littrow-type device, but it does not have adequate crosstalk attenuation or passband loss. This type is very attractive since it is compact and has no lens. It is expected that slab-waveguide loss, aberration, and efficiency of the concave grating will be improved.

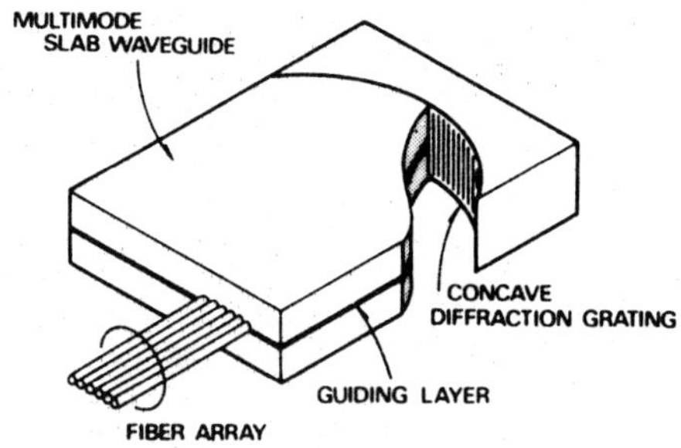

Fig. 6. DEMUX structure using a concave grating and slab waveguide (Slab-guide type).

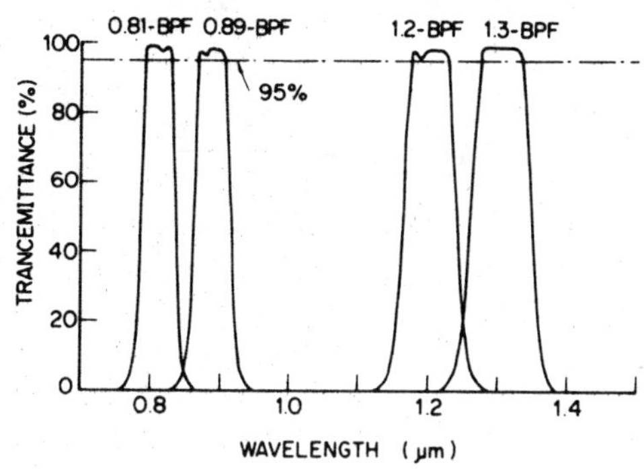

Fig. 7. Spectral responses of bandpass-type DTF's developed by ECL.

B. Dielectric Thin-Film Filter Type

A multiplexing device that makes use of the wavelength-selective characteristic of a dielectric thin-film filter (DTF) has been realized. This device contains high and low refractive-index dielectric films in alternating layers. Each layer has an optical thickness of approximately one quarter or half wavelength. In this type of device, only the DTF element has a wavelength-selective function. Therefore, it can be designed almost independent of the fiber parameters used. This is quite different from angularly dispersive devices. In other words, by applying a DTF designed and manufactured according to the system requirements (light source wavelength allocation, passband width, crosstalk attenuation in adjacent channels, etc.), the desired MUX and DEMUX can easily be realized. A MUX/DEMUX for two-way transmission, which demands high near-end crosstalk attenuation, can also be fabricated using a DTF cascade-connection technique. Here, the DTF characteristics prove to be very important.

On the other hand, such devices require DTF's with center wavelengths that differ from one another according to the number of optical channels multiplexed. In addition, total insertion loss tends to increase with the number of optical channels.

1) DTF Element: As just described, the characteristics of DTF type devices depend on DTF performance to a great extent. Therefore, the design and performance of the DTF are very important. DTF's are classified into two types, the band pass-filter (BPF) type and the edge-filter type. The edge filter type includes the long-wavelength pass-filter (LWPF) type and the short-wavelength pass-filter (SWPF) type. The BPF type is more advantageous than the edge-filter type because it has rejection bands on both sides of the passband. In the initial stage, these DTF's consisted of alternate layers of chiolite (low refractive-index material: $n = 1.35$) and ZnS (high refractive index: $n = 2.3$). Such filters, however, were unsuitable for practical use due to the moisture instability of chiolite and ZnS. Recently, DTF's have been realized by employing an electron beam that evaporates high-stability materials, i.e., SiO_2 (low refractive-index material: $n = 1.46$) and TiO_2 (high refractive-index material: $n = 2.3$) [25], [26].

Here, let us consider DTF design. The parameters of primary importance are passband loss, passband width, and rejection band attenuation. In general, passband loss must be low (typically, less than 0.3 dB, including ripples), passband width must be wide (typically, wider than oscillating wavelength deviation in the light source), and rejection band attenuation must be large (typically, more than 20 ~ 30 dB). Thus simulations should be performed to ensure that the DTF characteristics satisfy the above requirements, since DTF characteristics are largely dependent on total layer number and layer structure. The matrix method [27] is applicable for such simulation. One particularly significant point concerning DTF fabrication is how to precisely control each film thickness during vacuum evaporation. In the case of a BPF, less than 1-percent deviation from the theoretical value is demanded [26]. For example, BPF's of 0.81, 0.89, 1.2, and 1.3 μm have been developed by ECL, NTT, as shown in Fig. 7. These are 23-layer 3-cavity BPF's composed of SiO_2 and TiO_2 films. Their characteristics are:

passband loss	less than 0.2 dB;
passband width	400 Å (0.8-μm region) and 600 Å (1.3-μm region); and
rejection band attenuation	25 ~ 30 dB.

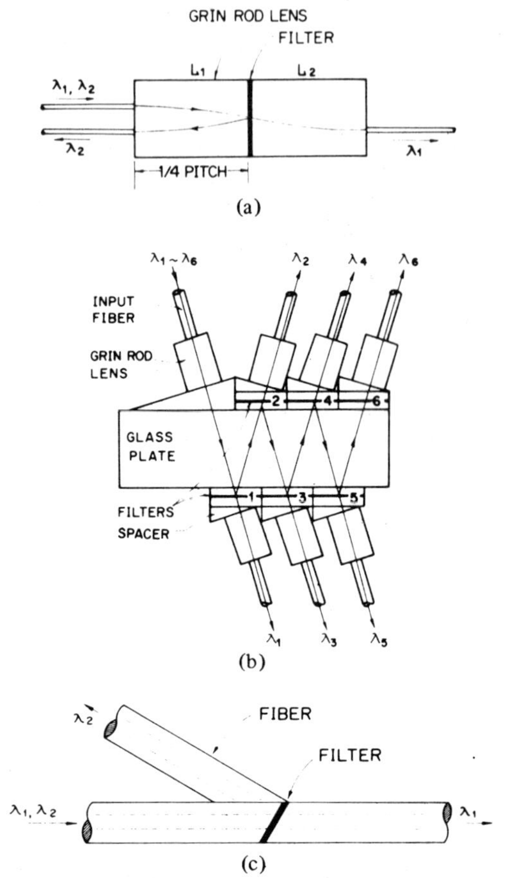

Fig. 8. Typical structure of DTF-type DEMUX. (a) GRIN-rod-lens type. (b) Multi-reflection type. (c) Fiber-end type.

2) DTF Types: Typical examples of the DTF device types are shown in Fig. 8. They are the Fig. 8(a) GRIN-rod lens type [28]-[31], (b) multireflection type [32]-[34], and (c) fiber-end type [35], [36]. While these examples are shown as DEMUX configurations, they can also be modified for MUX or MUX/DEMUX configurations by reversing the light input at the ports.

Let us consider the operation of this type of device by referring to the multireflection type shown in Fig. 8(b). The filters are all of the BPF type. Their passband wavelengths are all different. For example, the center wavelength of filter 1 is λ_1. That is, a light beam of wavelength λ_1 passes through filter 1; however, wavelength λ_2-λ_6 are sufficiently different from λ_1 and are reflected by filter 1. Similarly, the center wavelengths of filters 2-6 are λ_2-λ_6, respectively, and each filter reflects wavelengths sufficiently different from its particular center wavelength. When wavelengths in the range λ_1-λ_6 are applied to the GRIN-rod lens from the input fiber; they are collimated by the lens to form parallel beams and enter filter 1. Due to the characteristics of filter 1, wavelength λ_1 is transmitted through the filter, while the others are reflected and enter filter 2 through the common glass plate. Wavelength λ_1, which is transmitted by filter 1, passes through a spacer, is focused by the GRIN-rod lens and enters the output fiber. Among the light beams reflected by filter 1, only those of wavelength λ_2 are transmitted by filter 2; wavelengths λ_3-λ_6 are reflected and enter filter 3 through the common glass plate. In the same way, λ_3, λ_4, λ_5, and λ_6 are separated in sequence through filters 3, 4, 5, and 6, respectively. Although six filters are illustrated in the figure, more light waves can be demultiplexed by arranging additional filters, each with a different center wavelength.

Now, let us compare the features of the three types. The GRIN-rod-lens type has the advantage of compactness, but it is not easy to fabricate MUXs or DEMUX with more than three channels, because this would require GRIN-rod lenses of more than three elements with off-axis cascade connection between elements, and such off-axis cascade connection has very critical alignment. The multireflection type has a more complicated structure and has an even more difficult alignment. However, it is applicable to MUX or DEMUX with more than six channels. The fiber-end type has the simplest structure, but it is difficult to fabricate for MUX or DEMUX with more than three channels. This is because it tends to increase insertion loss based on alignment, since it has no collimating lens.

DTF-type devices are now available with the following performance characteristics:

number of channels	$2 \sim 6$ channels;
insertion loss	$1 \sim 5$ dB;
crosstalk attenuation	$20 \sim 70$ dB; and
channel spacing	$30 \sim 100$ nm (or very wide spacing, for example, 0.8 and 1.3 μm).

A practical, two-channel (1.2- and 1.3-μm) two-way MUX/DEMUX using SiO_2/TiO_2 BPF's of the GRIN-rod-lens type has been fabricated with the configuration shown in Fig. 9 [31]. In this device, cascade connection of BPF's is employed to obtain high crosstalk attenuation. Its spectral response is shown in Fig. 10. Here, insertion loss is 1.5 dB or less and crosstalk attenuation is 58 dB or more in the optical level. This device has proven to be very stable (insertion-loss variation of less than 0.3 dB) during stability tests, i.e., vibration, temperature-cycle, shock, and damp-heat tests.

C. Hybrid- and Planar-Waveguide-Type Devices

1) Hybrid Type: Hybrid-type devices [37], [38] are constructed by incorporating a grating and DTF, as shown in Fig. 11. This device makes use of the special features of both DTF's and gratings, i.e., efficient multiplexing by DTF's and demultiplexing of a large number of channels by gratings. Thus this device is expected to be useful for two-way large-scale WDM transmission in the future.

2) Planar-Waveguide Type: These devices are constructed by incorporating a planar waveguide and planar diffraction grating formed on the same substrate surface. They are applicable only to single-mode WDM system because they have a planar waveguide for single-mode operation. This is different from the slab-waveguide type previously described. The planar-waveguide types include the chirped-grating types [39], the contradirectional-Bragg-coupler type [40], the Bragg-grating type [41], [42], and the selective cascaded-mode-converter type [43]. All are far inferior to such bulk WDM devices as the bulk-grating or DTF types mentioned previously, with respect to insertion loss, passband width, and crosstalk attenu-

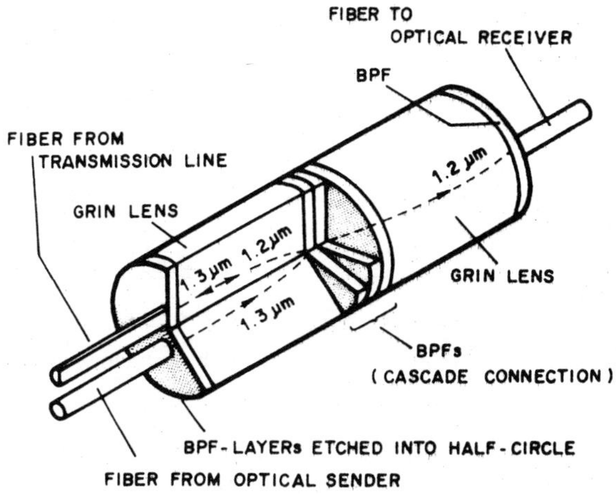

Fig. 9. Two-channel DTF-type MUX/DEMUX structure.

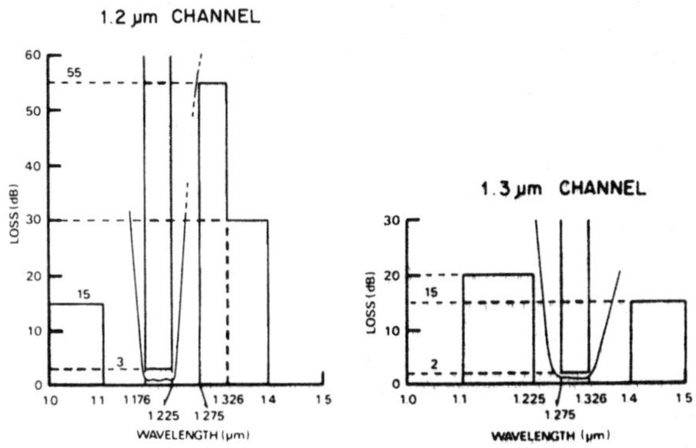

Fig. 10. MUX/DEMUX characteristics for the DTF type shown in Fig. 9.

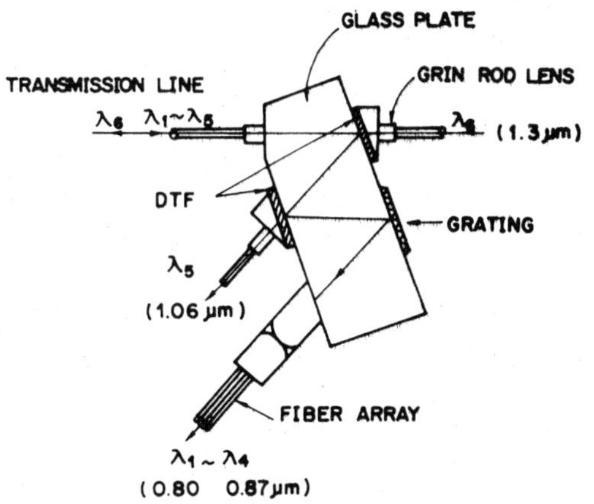

Fig. 11. Typical hybrid-type MUX/DEMUX structure.

ation. However, these configurations have the potential advantages of planar technology, such as reproducibility, batch fabrication, incorporation of complex optical circuits into photomasks, and ease of alignment with connecting fibers. Therefore, they are expected to become more advanced, although they are, at present, still in the fundamental stage of development.

D. Active WDM Devices

Multiwavelength light sources and multiwavelength photodetectors as active WDM devices have been studied. A multiwavelength light source performs both as a light source and a conventional passive multiplexer. A multiwavelength photodetector can also perform as both a detector and a passive demultiplexer. If these active devices function satisfactorily, WDM system configuration can be dramatically simplified.

To date, LD and LED type multiwavelength light sources for 2-4-channel multiplexing have been studied [44]-[47]. Additionally, various types of multiwavelength photodetectors for 2- or 3-channel demultiplexing have also been studied [48], [49].

These active WDM devices, however, present some problems. For multiwavelength light sources these problems include the following.

1) Total output power is restricted. Therefore, individual-channel power is much less than that for a conventional single-wavelength light source.
2) Coupling efficiency to the fiber is much lower than that for a conventional single-wavelength light source.
3) Stability and lifetime have not been clearly determined.

Problems with respect to multiwavelength photodetectors include the following.

1) dull wavelength selectivity; and
2) unsatisfactory degree of crosstalk suppression.

Since great emphasis has been put on these active WDM devices, it is important to solve these problems. At present, however, the level of development of passive WDM devices is higher than that of active WDM devices.

E. Wavelength Nonselective Devices

In WDM systems, although DEMUX principally requires wavelength selective elements (e.g., grating of DTF), MUX doesn't always require such elements. However, there is a problem with wavelength nonselective MUX for practical use, that is, wavelength nonselective MUX produces greater insertion loss, except for certain special types (see, for example, the polarization type). Also, DEMUX in a WDM system having wavelength nonselective MUX must have much greater crosstalk attenuation than that in a system using wavelength selective MUX.

Under these restrictions, directional and butt-joint couplers, in principle, can be used in MUX. In addition, the polarization [50] and double GRIN-rod-lens type devices [51], [52] should be applicable in nonselective MUX. The polarization-type device is based on enabling ordinary and extraordinary rays to coincide in a birefringent crystal (i.e., calcite). Here, the ordinary and extraordinary rays are of different wavelengths. They are combined into one ray at the output end of the crystal. The double GRIN-rod lens device is based on beam-diameter transformation, and consists of a set of small diameter first-stage quarter-pitch GRIN-rod lenses followed by a large-diameter, second-stage quarter-pitch GRIN-rod lens.

Such wavelength nonselective devices may be used only in special WDM systems.

TABLE II
WDM SYSTEM DESIGN PARAMETERS

Device/Item	Parameter
Light Source	◦ Emission Spectrum ◦ Wavelength dependence on temperature
Photo detector	◦ Sensitivity dependence on wavelength
Multi/demulti-plexer	◦ Loss ◦ Crosstalk ◦ Modal Noise ◦ Signal distortion by optical filters ◦ Reflection from a connector to LD
Fiber	◦ Loss and bandwidth dependence on wavelength ◦ Raman scattering
Signal	◦ Tolerable interchannel crosstalk ◦ minimum received level

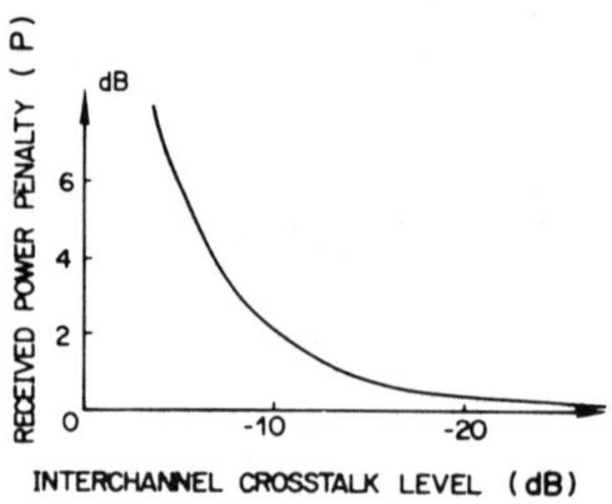

Fig. 12. Received-power penalty due to interchannel crosstalk.

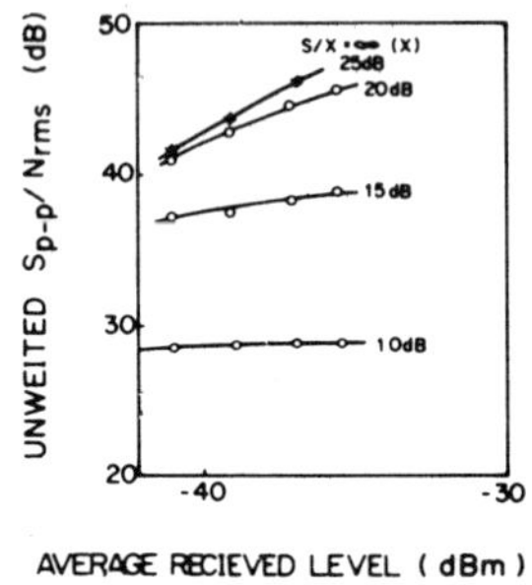

Fig. 13. SNR degradation due to interchannel crosstalk.

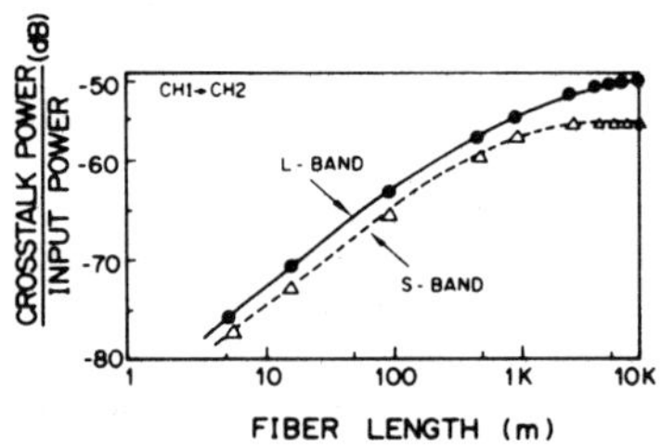

Fig. 14. Interchannel crosstalk caused by Raman scattering.

IV. WDM SYSTEM DESIGN

The previous section examined optical multi/demultiplexers (MUX, DEMUX, and MUX/DEMUX) for WDM transmission. This section discusses WDM system design and related problems.

Section IV-A reviews the items taken into account in designing a system; Section IV-B rather precisely describes multi/demultiplexing system design; and Section IV-C provides some examples of WDM transmission system designs.

A. WDM System Design Parameters

In comparison with conventional fiber-optic transmission systems, a WDM transmission system needs additional optical multi/demultiplexers. When multi/demultiplexer insertion loss is considered, the loss budget design for a WDM transmission system is basically almost the same as that for conventional fiber optic transmission systems. However, other factors must be taken into account [53]-[57].

The major parameters for WDM transmission system design are shown in Table II; these parameters are closely related to each other because a number of optical signals having different wavelengths share a single fiber whose loss and bandwidth depend on wavelength. Among these parameters, crosstalk is the most important when considering system design.

1) Interchannel Crosstalk: Since, multiple optical signals traveling in the same direction, or in opposite directions, share a common single fiber, optical interchannel crosstalk is a significant problem in WDM systems. In optical multi/demultiplexers, such crosstalk is related to light-source spectrum width, photodetector sensitivity (which is dependent on wavelength) and so on. The tolerable crosstalk levels are determined by the nature of the transmission signal.

Error-rate degradation in an 8-Mb/s digital transmission system [58] is shown in Fig. 12. When the optial crosstalk level is less than minus 15-20 dB, error-rate degradation is not significant. SNR degradation in a 4-MHz analog baseband transmission is shown in Fig. 13. For optical crosstalk levels of less than -30 dB, SNR degradation is also not significant [59]. Multi/demultiplexer crosstalk problems will be discussed in connection with multi/demultiplexer design in Section IV-B. Here, we concentrate on crosstalk caused by the transmission fiber.

In multimode fiber systems, as well as in single-mode fiber systems, optical interchannel crosstalk due to Raman scattering along a transmission fiber can affect WDM transmission performance [60], [61] (see Fig. 13). Crosstalk dependence on fiber length in a two-channel multimode fiber system in which 0.813- (CH.1) and 0.885-μm (CH.2) LD's are utilized is shown in Fig. 14.

One inherent characteristic of Raman scattering is that crosstalk occurs only between a shorter wavelength channel and a longer wavelength one (CH.1 → CH.2). This Raman crosstalk level increases as channel wavelength separation decrease.

A similar phenomena occurs in single-mode fibers, where stimulated Raman scattering is significant. However, from the practical point of view, Raman scattering is not significant for multimode fiber digital transmission systems or for single-mode fiber digital transmission systems. However, analog or analog-digital hybrid transmission system designs may require attention to this problem.

2) Modal Noise: In some multimode fiber transmission systems, such as in an analog transmission system using laser diodes or in a very high speed digital transmission system, an optical multi/demultiplexer causes significant SNR, differential gain (DG), and differential phase (DP) degradation [62]. This is primarily due to fiber propagation mode filtering in

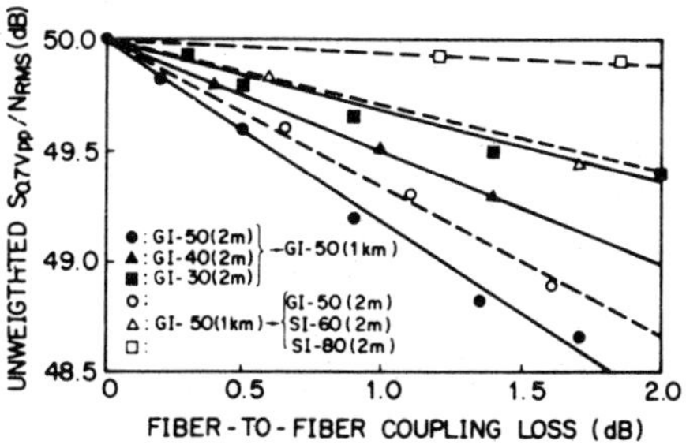

Fig. 15. SNR degradation versus fiber-to-fiber coupling loss.

TABLE III
TOTAL SYSTEM POWER PENALTY

CORE DIAMETER		50 μm	40 μm	30 μm
LD TO FIBER COUPLING LOSS **		0.0	0.5	1.2
CONNECTOR LOSS**		0.0	0.0	0.3
FIBER-TO-FIBER COUPLING LOSS	5 μm*	0.5	0.15	0.03
	10 μm*	1.4	0.9	0.4
AVERAGE RECEIVED POWER PENALTY	5 μm*	0.45	0.1	0.0
	10 μm*	1.2	0.45	0.2
TOTAL POWER PENALTY	5 μm*	0.95	0.75	1.53
	10 μm*	2.6	1.85	2.1

* : AXIAL DISPLACEMENT. (dB)
** : NORMALIZED BY 50 μm VALUE

fiber-to-fiber coupling through the lenses in the optical multi/demultiplexers [63], [64].

SNR degradation as a function of fiber-to-fiber coupling loss is shown in Fig. 15. In general, when an input fiber diameter is smaller than the output core diameter, the modal noise can easily be reduced. Since the photodetector sensing area diameter is more than 100 μm, it is easy for DEMUX output ports to use large-core-diameter fibers. However, for MUX input ports, core-diameter reduction decreases the LD-to-fiber coupling. The comparison between power penalty due to modal noise and LD-to-fiber coupling loss is shown in Table III. Here, the modal-noise power penalty is defined as the power required to maintain 50-dB unweighted SNR. From Table III, for a 50-μm core-diameter graded-index fiber transmission line, the optimum core diameter for a MUX input port is 40 μm.

B. Multi/Demultiplexer Design

This section examines multi/demultiplexing loss, interchannel crosstalk and channel spacing for given wavelength-selective element performance and light-source characteristics.

The following terminology for multi/demultiplexer design will be used hereafter:

1) L_0 sum of multiplexing loss and demultiplexing loss (dB);
2) L_0' demultiplexing loss (dB);
3) S/X optical crosstalk from adjacent optical channels (dB); and
4) L_e effective optical multi/demultiplexing loss for digital transmission systems (dB).

L_e is obtained as

$$L_e = L_0 + \Delta P, \qquad \Delta P = \Delta\,\mathrm{SNR} \cdot (2 + x)/(2 + 2x)$$
$$\Delta\,\mathrm{SNR} = -20 \log (1 - 2 \times 10^{-(S/X)/10}) \quad \text{(dB)} \tag{4.1}$$

where x is the excess noise factor for an avalanche photodiode. Hereafter, $x = 0.4$ is used. Δ SNR is the signal-to-noise ratio deterioration caused by crosstalk and ΔP is the loss penalty due to crosstalk [65].

In order to simplify the argument, the reflection and aberration losses in optical systems are neglected in the following discussion. It is also assumed that the LED and LD emitting spectrums are expressed by [53], [58]

$$\text{LED: } \Gamma_{\mathrm{LED}}(\lambda) = \frac{1}{1 + \left| \dfrac{2(\lambda - \lambda_0)}{\Delta\lambda_0} \right|^3}$$

and

$$\text{LD: } \Gamma_{\mathrm{LD}}(\lambda) = \delta(\lambda - \lambda_0) \tag{4.2}$$

where λ_0 and $\Delta\lambda_0$ are the center wavelength and the emitting spectrum halfwidth, respectively.

When fiber baseband width is sufficient for signal bandwidth, the digital system repeater span l (km) is deteremined by

$$\left.\begin{aligned} \alpha l &= P_o - P_r - L_e \\ L_e &= L_0 + \Delta P \end{aligned}\right\} \tag{4.3}$$

where P_o is an optical power coupled to a fiber (dBm), P_r is the minimum received power (dBm), and α is fiber loss (dB/km). Therefore, multi/demultiplexers should be designed so that the effective multi/demultiplexer loss L_e is kept to a minimum.

In analog transmission system design, the multi/demultiplexer configuration should be chosen so that L_o is the smallest among multi/demultiplexer configurations in which the crosstalk level is less than the desired level, for example, -30 dB for a 4-MHz baseband analog system.

Practical WDM systems take various configurations, such as one-way or two-way systems, analog-digital hybrids, and so on. Within the limits of this paper, it is difficult to cover all possible systems. Therefore, we will focus on one-way digital WDM systems and one-way analog systems. However, the basic design principle described here—namely the balance between crosstalk and multi/demultiplexing loss—can be applied to other system configurations.

1) DTF Multiplexers: Fig. 16 shows the structure for a multireflection type MUX that uses DTF's [65]. It includes optical bandpass filters with different reflectance/transmittance wavelength dependencies. A light signal having a specific wavelength passes through its channel filter and is reflected to the different channel filters. The channel 3 wave ($\lambda = \lambda_3$), for example, passes through filter 3 and is reflected to filters 4 and 5; it then couples to the output fiber. In the same way, channels 1, 2, 4, and 5 waves couple to the output fiber by a zigzag path. This MUX also functions as a DEMUX when it is used inversely.

Wide LED emission spectrums or LD oscillation wavelength deviations due to temperature changes create interchannel

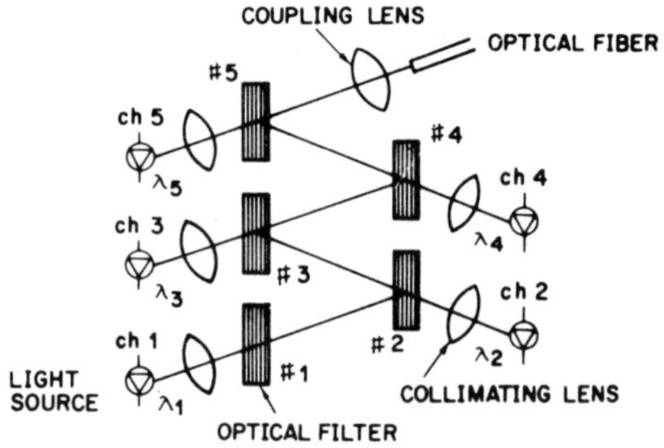

Fig. 16. Schematic structure of multireflection-type multi/demultiplexer.

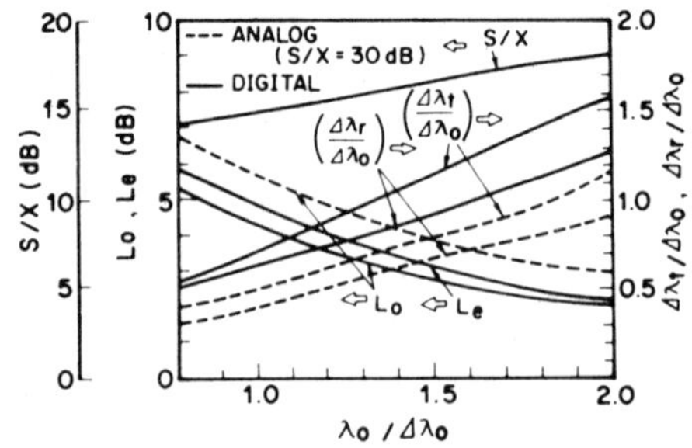

Fig. 17. Relation between channel spacing (λ_s) optimum widths and crosstalk for LED-WDM transmission systems.

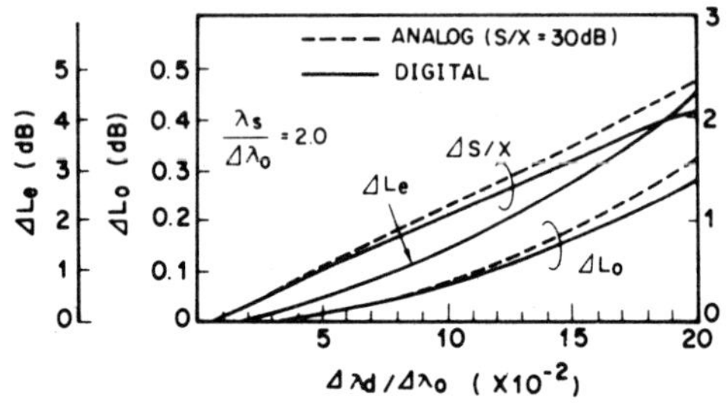

Fig. 18. Relation between wavelength deviation, degraded values of crosstalk and multi/demultiplexing loss.

crosstalk. When the bandwidth of the bandpass filter becomes narrow, crosstalk decreases but MUX loss increases.

a) LED systems: When LED's are utilized in a WDM transmission system using DTF-type MUX, filtering loss L_0 and crosstalk S/X are given as

$$\left.\begin{aligned} L_0 &= 10 \log [S]/[O] \\ S/X &= 10 \log [S]/[O] \end{aligned}\right\} \tag{4.4}$$

$$[O] = \int_0^{\infty} \Gamma_{\text{LED}}(\lambda)\, d\lambda \tag{4.5}$$

$$\begin{aligned} [S] = \int_0^{\infty} & \Gamma_{\text{LED}}(\lambda)\, T_t(\lambda)\, [1 - T_t(\lambda + \lambda_s)] \\ & \cdot [1 - T_t(\lambda + 2\lambda_s)]\; [1 - T_r(\lambda + 2\lambda_s)] \\ & \cdot [1 - T_r(\lambda + \lambda_s)]\; T_r(\lambda)\, d\lambda \end{aligned} \tag{4.6}$$

$$\begin{aligned} [X] = \int_0^{\infty} & \Gamma_{\text{LED}}(\lambda - \lambda_s)\, T_t(\lambda - \lambda_s)\, [1 - T_t(\lambda)] \\ & \cdot [1 - T_r(\lambda + \lambda_s)]\; (1 - T_t(\lambda + 2\lambda_s)] \\ & \cdot [1 - T_r(\lambda + \lambda_s)]\; T_r(\lambda)\, d\lambda \\ + \int_0^{\infty} & \Gamma_{\text{LED}}(\lambda + \lambda_s)\, T_t(\lambda + \lambda_s) \\ & \cdot [1 - T_t(\lambda + 2\lambda_s)]\; [1 - T_r(\lambda + 2\lambda_s)] \\ & \cdot (1 - T_r(\lambda + \lambda_s)\; T_r(\lambda)\, d\lambda \end{aligned} \tag{4.7}$$

$$T_t(\lambda) = \exp\, [-2 \ln 2 |(\lambda - \lambda_0)/\Delta\lambda_t|^2] \tag{4.8}$$

$$T_r(\lambda) = \exp\, [-2 \ln 2 |(\lambda - \lambda_0)/\Delta\lambda_r|^2] \tag{4.9}$$

where λ_s, $\Delta\lambda_t$, and $\Delta\lambda_r$ are channel-spacing filter bandwidths of the multiplexing device and of the demultiplexing device, respectively. $T_t(\lambda)$ and $T_r(\lambda)$ express the transmittance of a filter in the multiplexing device and in the demultiplexing device, respectively.

In designing analog transmission systems, $\Delta\lambda_t$ and $\Delta\lambda_r$ should be chosen to minimize filtering loss among pairs of $\Delta\lambda_r$ and $\Delta\lambda_t$ to make the crosstalk less than a given level, for example, S/X = 30 dB. For digital transmission systems, $\Delta\lambda_t$ and $\Delta\lambda_r$ should be chosen so that effective loss L_e is minimized.

Optimum filter bandwidth $\Delta\lambda_t$ and $\Delta\lambda_r$; and L_0, S/X and L_e, for analog and digital transmission systems, both using LED's, respectively, are shown in Fig. 17.

LED center-emission wavelength shift due to temperature change results in increased loss and crosstalk degradation. Loss increases ΔL_0 and ΔL_e, and crosstalk degradation $\Delta(S/X)$, all as a function of wavelength deviation, are shown in Fig. 18.

b) LD system: When using LD's, it is necessary to consider wavelength deviation due to temperature change and/or wavelength setting error. Let us assume that LD wavelength shifts $\Delta\lambda_d$ from the given wavelength λ_0 while, at a longer wavelength, adjacent channel LD oscillates at $\lambda = \lambda_0 + \lambda_s - \Delta\lambda_d$ and, in another adjacent channel, LD wavelength is $\lambda = \lambda_0 - \lambda_s + \Delta\lambda_d$. This assumption produces the worst case for the given wavelength deviation $\Delta\lambda_d$. In this case, L_0 and S/X are obtained by replacing $\Gamma_{\text{LED}}(\lambda)$ and $\Gamma_{\text{LED}}(\lambda \pm \lambda_s)$ in (4.4) ~ (4.6) by $\delta(\lambda + \Delta\lambda_d)$ and $\delta(\lambda \pm (\lambda_s - \Delta\lambda_d))$, respectively. Fig. 19 shows optimum filter bandwidths $\Delta\lambda_t$ and $\Delta\lambda_r$ and L_0, S/X, and L_e for analog and digital transmission systems.

1) Angular Dispersion Demultiplexers: Fig. 20 shows the configuration of a DEMUX using an angular dispersive element such as a grating or a prism. The light signals from the input fiber are collimated. After striking the dispersive element, their propagation angles correspond to their wavelengths. They are then focused on the receiver plane by a Fourier transform lens having focal length f_2. Collimated beam diameter D is

$$D = 2f_1 \frac{(\text{NA})}{n_M} \tag{4.10}$$

where (NA) is the numerical aperture of the input fiber, n_M is the reflective index of the medium between the dispersive

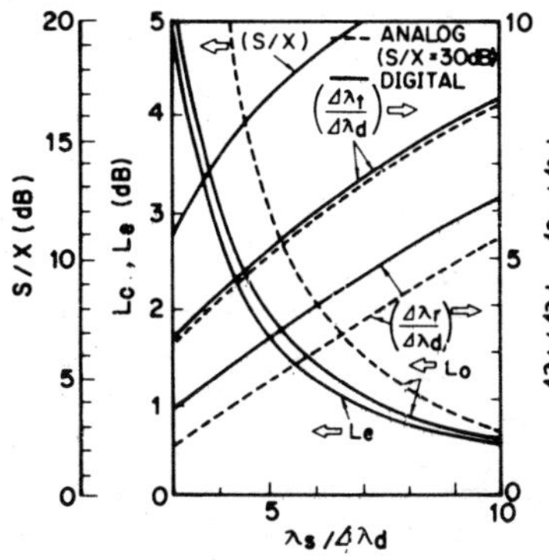

Fig. 19. Relation between channel spacing optimum fiber width, crosstalk and multi/demultiplexing loss for LD-WDM transmission systems.

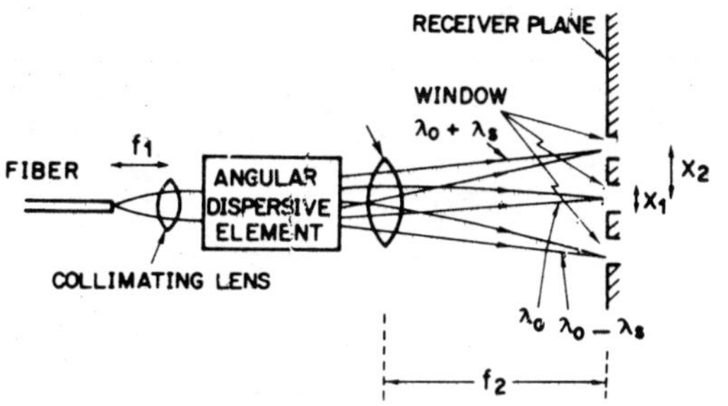

Fig. 20. Schematic structure of an angular-dispersive optical demultiplexing circuit.

element and the lens, and f_1 is the focal length of the collimating lens. The angular dispersion for a Littrow-mounted grating is given by

$$\frac{d\theta}{d\lambda} = 2\frac{\tan\theta}{\lambda} \cong 2\frac{\tan\theta_B}{\lambda} \tag{4.11}$$

where θ_B is the blaze angle. Prism angular dispersion for minimum deviation angle is given by

$$\frac{d\theta}{d\lambda} = \frac{B}{n_M D}\frac{dn}{d\lambda} \tag{4.12}$$

where n, B, and D are the refractive index, length of the base of the prism, and incident-beam diameter, respectively [53], [54].

A grating- or prism-type MUX needs sophisticated optical systems for low insertion loss [66]. Therefore, this discussion will focus on DEMUX's.

From (4.10), positional dispersion for a Littrow-type grating DEMUX is given by

$$\Delta x = \int_{\lambda_0}^{\lambda_0+\delta\lambda} f_2\frac{\tan\theta}{\lambda}\,d\lambda \cong 2f_2\tan\theta_B\int_{\lambda_0}^{\lambda_0+\delta\lambda}\frac{1}{\lambda}\,d\lambda$$

$$= 2f_2\tan\theta_B\ln\left(1+\frac{\delta\lambda}{\lambda_s}\right). \tag{4.13}$$

From (4.10) and (4.12) positional dispersion for a prism-type DEMUX is given by

$$\Delta x = \frac{Bf_2}{2f_1(\mathrm{NA})}\int_{\lambda_0}^{\lambda_0+\delta\lambda}\frac{dn}{d\lambda}\,d\lambda \tag{4.14}$$

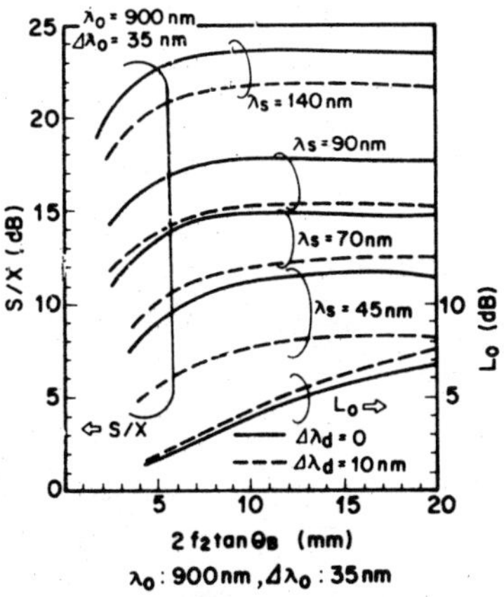

Fig. 21. Relation between the structural parameter for a grating type demultiplexing circuit, demultiplexing loss and crosstalk, as a function of channel spacing.

In order to simplify discussion of demultiplexing loss and crosstalk, it is assumed that a light signal having a fixed wavelength is incident to the center of its channel window—a photodiode or a fiber—in the receiving plane, and any light signals incident to the channel window are completely converted to an electrical signal. Thus light signals incident to an adjacent channel window create crosstalk. It is also assumed that, as indicated in Fig. 20, the window dimension is x_1 and the interval between windows is x_2.

a) LED systems: When LED's are used as light sources, the wide emission spectrum creates crosstalk. The demultiplexing loss and crosstalk for a grating type DEMUX are obtained as

$$L_0' = -10\log\frac{[S]}{[O]} \tag{4.15}$$

$$S/X = 20\log\frac{[S]}{[X]} \tag{4.16}$$

$$[O] = \int_0^{\infty}\Gamma_{\mathrm{LED}}(\lambda)\,d\lambda \tag{4.17}$$

$$[S] = \int_{\lambda_0-\delta\lambda}^{\lambda_0+\delta\lambda}\Gamma_{\mathrm{LED}}(\lambda)\,d\lambda \tag{4.18}$$

$$[X] = \int_{\lambda_0-\delta\lambda}^{\lambda_0+\delta\lambda}\Gamma_{\mathrm{LED}}(\lambda-\lambda_s)+\Gamma_{\mathrm{LED}}(\lambda+\lambda_s)\,d\lambda \tag{4.19}$$

$$\delta\lambda = \lambda_0\exp\left[\frac{X_1(\mathrm{NA})}{2Dn_M\tan\theta_B}\right] - \lambda_0. \tag{4.20}$$

Loss and crosstalk as a function of grating parameter $2f_2 \cdot \tan\theta_B$ are shown in Fig. 21. The broken lines denote values for wavelength deviation of 10 nm. As a LED emitting spectrum spreads widely, channel spacing should be larger, about 70–100 nm, or so. A prism-type DEMUX presumably requires similar channel spacing.

b) LD system: When using LD's, an angular dispersive DEMUX improves performance. It becomes demultiplexing-loss free and crosstalk free if proper channel spacing λ_s and

TABLE IV
WAVELENGTH ALLOCATION

TRANSMISSION SCHEME	LIGHT SOURCE	NUMBER OF CHANNELS	WAVELENGTH SPACING (nm)	WAVELENGTH BAND (μm)	MULTIPLEXER	DEMULTIPLEXER	Lo (dB)	S/X (dB)
DIGITAL	LED $\Delta\lambda_e = 30$nm	3	60	0.90 ~ 0.78	DTF TYPE $\Delta\lambda_t = 47$ nm	DTF TYPE $\Delta\lambda_r = 38$ nm	2.0	18.2
		5	30	0.90 ~ 0.75	DTF TYPE $\Delta\lambda_t = 16$ nm	DTF TYPE $\Delta\lambda_r = 15$nm	5.1	14.2
	LD $\Delta\lambda d = \pm 2.5$ nm	10	10	0.89 ~ 0.80	DTF TYPE $\Delta\lambda_t = 10$ nm	DTF TYPE $\Delta\lambda_r = 5$nm	<2.7	<14.6
					DTF TYPE $\Delta\lambda_t = 7$nm	ANGULAR DISPERSIVE TYPE	<1.5	$\simeq \infty$
ANALOG	LED $\Delta\lambda_0 = 100$nm	3	200	1.5 ~ 1.1	DTF TYPE $\Delta\lambda_t = 114$ nm	DTF TYPE $\Delta\lambda_r = 91$nm	2.8	30
	LED $\Delta\lambda_0 = 30$nm	3	60	0.90 ~ 0.78	DTF TYPE $\Delta\lambda_t = 34$ nm	DTF TYPE $\Delta\lambda_r = 27$ nm	2.8	30

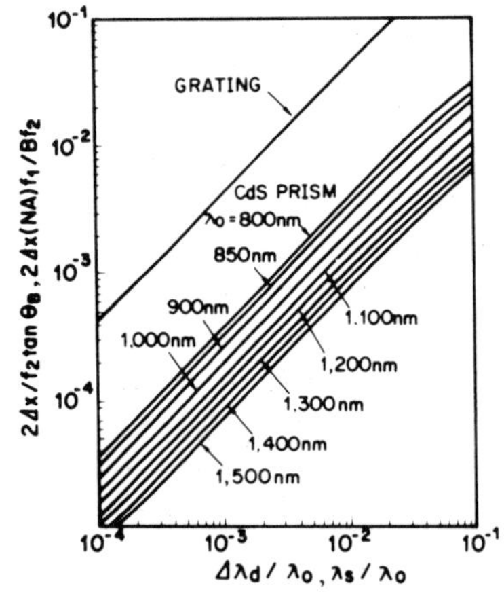

Fig. 22. Relation between wavelength deviation in an LD and a grating or CdS prism demultiplexing circuit.

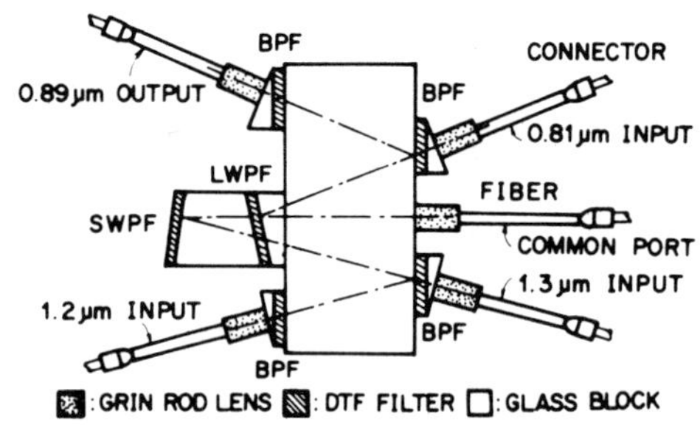

Fig. 23. Structure of MUX-DEMUX for a business use subscriber system.

TABLE V
NTT FIELD-TRIAL SUBSCRIBER-LOOP SYSTEM

SYSTEM	ADVANCED BUSINESS SYSTEM	
SIGNAL	DIGITAL (6.3Mb/s)	CTV WITH voice
CABLE LENGTH	5 km	
FIBER	GI . 4dB/km AT 0.85 μm	
SIGNAL FORMAT	DIGITAL IN BASEBAND	ANALOG IN BASEBAND
NUMBE OF WDM	4 1.2μm (down) 0.81μm (up)	1.3μm (down) 0.89μm (up)
FILTER	DIELECTRIC MULTILAYER THIN FILM FILTER	

sufficient window dimensions x_1 are chosen to receive a light signal deviating within a certain wavelength range.

From (4.13) and (4.14), we can obtain the device parameters. Replacing $\delta\lambda$ by the tolerable wavelength deviation $\Delta\lambda_d$, $2 \times \Delta X$ gives the window interval x_2. Fig. 22 shows device parameters as a function of channel spacing and wavelength deviation. Taking $\Delta\lambda_d = 2.5$ nm, $\lambda_s = 10$ nm, and $\lambda_0 = 850$ nm, for example, we obtain $x_1/f_2 \tan\theta_B = 1.2 \times 10^{-2}$ and $x_2/f_2 \tan\theta_B = 2.3 \times 10^{-2}$ for a grating-type DEMUX, and $f_1(\mathrm{NA})\,x_1/f_2B = 9.0 \times 10^{-4}$ and $f_1(\mathrm{NA})\,x_2/f_2B = 1.8 \times 10^{-3}$ for a prism-type DEMUX. When using a grating having $\lambda_B =$ 800 nm and $\Lambda = 1/600$ nm, an input fiber having (NA) = 0.17, and a lens having $f_1(=f_2) = 30$ mm, we obtain $x_1 = 90$ μm, $x_2 = 170$ μm, and $D = 10.2$ mm for n_M 1.0. When a prism having $B \simeq 15$ mm and lenses having $f_1 = f_2 = 30$ mm are used, we obtain $x_1 = 80$ μm, $x_2 = 154$ μm, and $D = 10.2$ mm for $n_M = 1.0$.

C. WDM System Design Examples [67]-[69]

Wavelength allocation design examples obtained from the above design maps (Figs. 17-22) are shown in Table IV. In this table the influence of LD spontaneous emission and multi/demultiplexer losses caused by optical element imperfections such as lens aberration, etc., are not considered.

The previous discussion dealt with simple models. More complicated systems, including two-way WDM systems, require more sophisticated considerations, such as fiber loss and bandwidth dependence on wavelengths, minimum received levels of different modulation scheme signals, and so on. A design precedure for such systems may be difficult to show in simple form. Rather heuristic design methods are used for such systems, although the basic loss and crosstalk design procedure is valid. Fig. 23 shows a MUX/DEMUX structure for the business use fiber-optic subscriber transmission system whose major system parameters are shown in Table V. To suppress modal noise, the multiplexing input port uses a 40-μm ϕ core fiber, while the demultiplexing output port uses an 80-μm ϕ core fiber. Oblique polished fiber optic connectors are utilized to obtain low reflection return losses. The MUX/DEMUX characteristics, including design objectives, are shown in Fig. 24 and Table VI.

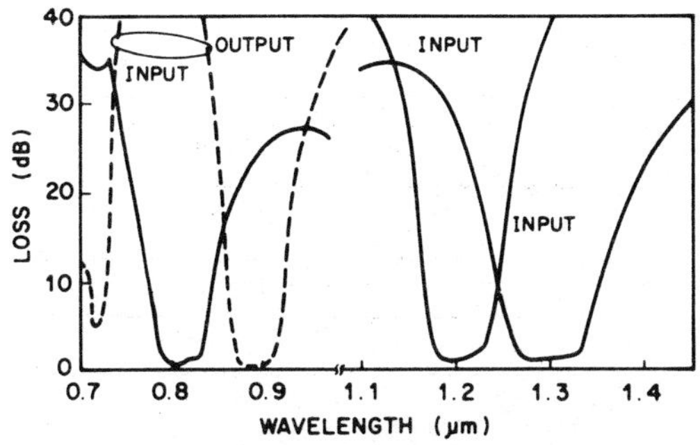

Fig. 24. Measured branching characteristics of the MUX/DEMUX shown in Fig. 23.

TABLE VI
MUX/DEMUX CHARACTERISTICS

OUTPUT / INPUT	0.81 µm	0.89 µm	1.2 µm	1.3 µm
0.81 µm	2.0	>83	73.8	>83
	< 3.4	> 76.6	>50.0	>50.0
0.89 µm	68.1	1.9	>79	>79
	>55.8	< 3.9	>75.0	>75.0
1.2 µm	24.4	43.9	2.4	>80
	>11.9	>36.9	< 5.1	> 40.6
1.3 µm	23.2	48.4	30.5	3.0
	>11.8	> 36.8	> 12.1	< 6.7
RETURN LOSS	27	29	23	25

Upper : measured . below : specification (dB)

V. System Applications

WDM transmission has many attractive advantages, e.g., greater transmission capacity, two-way transmission, simultaneous transmission of various types of signals such as digital and analog, and easy system expansion. That is to say, WDM technology has two aspects. First, it is a technology that makes it feasible to transmit as much information as possible through a single optical fiber. Second, it is a technology that enhances flexibility in system design [9].

For example, in long-haul and/or large-capacity transmission systems, fiber cable cost is one of the dominant factors in system cost. Therefore, system cost can be significantly reduced by applying WDM technology. This is the first aspect of WDM technology and it can be said that it more or less applies to almost all trunk transmission systems, from short-haul to long-haul.

On the other hand, the second aspect seems to be more attractive for subscriber systems and on-premises applications, that is, it is foreseen that in the near future a wide variety of information services must be available at any place and at any time due to diversification of social activities. These information services will be provided by broad-band fibered subscriber networks in conjunction with large-capacity optical trunk transmission systems. WDM technology could be effectively applied in these areas because it makes it possible to transmit various types of signals, such as video and data or analog and digital, through a single optical fiber and to expand systems without changing underground facilities.

In the following subsections, various system applications are described with emphasis on practical applications, including field trials.

TABLE VII
MAIN PARAMETERS FOR A SMALL-CAPACITY WDM TRANSMISSION SYSTEM

System	Type A	Type B
Bit rate /wavelength	6.312 Mb/s	
Line code	CMI	
Repeater spacing	10 kilometers	17 kilometers
Optical fiber	graded-index multimode fiber	
Number of wavlengths	2	
Nominal wavelengths	0.81 µm, 0.89 µm	1.2 µm, 1.3 µm
Light source	GaAlAs-LD	InGaAsP-LD
Optical detector	Si-APD	Ge-APD
Optical multiplexer	dielectric thin-film filter type	

A. *Trunk Transmission System Applications*

In the early stage of WDM technology research, application to trunk transmission systems was the most attractive research item because this technology was expected to greatly improve system economy. Since then, numerous experiments, such as laboratory tests or field trials, have been carried out on trunk transmission system applications for both short- and long-haul systems [70]-[73].

Today, such applications seem to be shifting from the field-experiment stage to the commercial-use stage. In the following, two typical WDM trunk transmission systems that are now being used in existing communications networks are briefly described.

In Japan, 32- and 100-Mb/s digital transmission systems using 1.3-µm laser diodes have been widely employed in short-haul inter-city trunk lines [74]. Furthermore, large-capacity 400-Mb/s digital transmission systems using single-mode fibers have been installed in long-haul intercity trunk lines [75].

On the other hand, in order to introduce optical fibers into intracity trunks, it is necessary to design low bit-rate transmission systems that are more economical than conventionally paired cable systems. A 6.3-Mb/s digital transmission system has been developed using WDM technology, taking into account the above requirements. Fundamental system parameters are shown in Table VII [8].

Two optical channels with wavelengths of 1.2 and 1.3 µm are simultaneously transmitted, either one-way or two-way. DTF's are used in the multiplexing devices, as shown in Fig. 9.

Another example is the FT3C System in the United States [30]. In this type of system, 3-channel one-way transmission is planned using three laser diodes (0.825, 0.875, and 1.3 µm). DTF's will also be used for multiplexing devices in this system.

B. *Subscriber System Applications*

It is foreseen that all information services, including broadcasting and advanced telecommunications services, will be provided by a single telecommunication network in the future. In this network, information will be transmitted to all subscribers by broad-band subscriber networks in conjunction with large-capacity trunk-transmission systems.

Optical-fiber transmission is expected to be one of the key

TABLE VIII
FIELD TRIALS OF FIBER-OPTIC SUBSCRIBER SYSTEMS USING WDM TECHNOLOGY

	Canada	France (SAFO)	FRG (BIGFON)	Japan
Place Number of subscribers	Elie, Manitoba 150 - 200	-------- --------	Berlin,Darmstadt - 100	Yokosuka - 2
Test year	1981.09 -	--------	1980 - 1983	1980 -1981
Network configuration	star	star	star	star
Information provided	telephone data (56kbps) FM radio TV	telephone data TV	telephone data FM stereo TV	telephone data TV hi-definition TV
Modulation scheme	analog FDM	digital	digital	analog FDM
Optical fibers	GI(50/125)	--------	GI	GI(50/125)
Number of wavelength multiplexed	2 (two-way)	4 to 6 (two-way)	2 to 3 (two-way)	2, 4 (two-way)
Wavelength used	840nm, 930nm (down), (up)	3 ch in 0.8μm 1 ch in 1.3μm	850nm, 1,300nm (up) (down)	800nm,830nm,860nm 890nm,1150nm,1300nm
Light sources	LED, LD	--------	LD	LED, LD
Multiplexing devices	optical coupler	grating type	DTF type	DTF type grating type hybrid type

technologies leading to realization of broad-band subscriber networks. This is because of the optical fiber's capability to transmit a large volume of information, as well as its physical superiority, e.g., small size, light weight, and immunity to electromagnetic interference.

However, the most significant problems in applying optical fiber to subscriber applications are in the area of cost effectiveness and in technical difficulties with bilateral transmission of a wide variety of types of information, i.e., narrow-band and wide-band signals, as well as analog and digital signals.

The use of wavelength multiplexing technology in conjunction with other multiplexing methods, such as frequency division multiplexing (FDM) and time division multiplexing (TDM), is expected to solve these problems.

With these backgrounds in mind, experiments on fiber-optic subscriber systems using WDM technology have been vigorously carried out all over the world during the last few years. Relatively large-scale experiments, including those still in the planning stages, as summarized in Table VIII.

The Bigfon system has been tested in several cities in the Federal Republic of Germany. This system employs 2-channel two-way WDM transmission using two different laser diodes (0.85 and 1.3 μm) that transmit FM stereo, video, and telephone in digital formats [76], [77].

Elie experiments in Canada have been going on in rural field environments since the end of 1981. Two wavelength (0.84 and 0.93 μm) are used for two-way transmission of FM stereo, video, telephone, and data signals [78].

The French SAFO system, now in the planning stage, is attempting to use grating devices for 4–6-channel WDM transmission. However, details of this work have not been provided [79].

The Netherlands TTOSS system is in the laboratory test stage and uses 3-channel WDM transmission [80].

The Electrical Communication Laboratories (ECL's), which are the research and development arm of the Nippon Telegraph and Telephone Public Corporation (NTT), have carried out their first field trials on fiber-optic subscriber systems [81]. These experiments were carried out from April 1980 to May 1981, with the objectives of confirming the technical feasibility of the systems and to find any unexpected problems under actual operating conditions.

In these experiments, four types of subscriber systems were tested and the multiplexing devices shown in Figs. 6, 8(b), and 11 were developed.

Based upon these experiments, a larger scale field trial of a 'home-use' subscriber system and a "business-use" subscriber system has been going on in the Mitake area near Tokyo since September 1982. Main parameters of the systems are listed in Table IX.

The "home-use" subscriber system has a star network in which one optical fiber provides a single subscriber with two-way transmission. Various types of services, such as broadcast TV, center-to-end video, data, and telephones are multiplexed at the telephone office and provided to each subscriber. Each subscriber also has a wavelength-multiplexing device and can generate upstream channels.

On the other hand, in the "business-use" system, high-speed data and video signals are transmitted bilaterally to provide data and TV-conference services between business offices.

In both systems, DTF-type multiplexing devices are employed for practical and economic reasons [26], [62].

As these systems are now under field trial, detailed system performance and device characteristics will be presented at a later date. However, it should be noted that in designing, full consideration should be given to analog video degradation resulting from internal reflection and modal noise in optical multi/demultiplexers [62].

TABLE IX
SYSTEM PARAMETERS FOR TWO TYPES OF SUBSCRIBER SYSTEMS IN MITAKA FIELD TRIALS

	Business Use System		Home Use System			
			Type A		Type B	
Service provided	high-speed data & TV-conference service		CATV & center-to-end video services			
Signals transmitted	digital data (two-way)	color TV (two-way)	color TV (7 channels) (one-way)	color TV control sig. (two-way)	color TV (2 channels) (one-way)	control sig. (one-way)
Transmission distance	5 kilometers		2 kilometers		5 kilometers	
Optical fiber	graded-index multimode fiber (50 μm/125 μm)					
Signal format	CMI	baseband analog	VHF-FDM	baseband analog	baseband analog	CMI
Number of wave-lengths	4		3		3	
Light sources	InGaAsP-LD GaAlAs-LED	InGaAsP-LD GaAlAs-LD	GaAlAs-LD	InGaAsP-LD GaAlAs-LD	GaAlAs-LD InGaAsP-LD	GaAlAs-LED
Nominal wavelength	1.2μm (down) 0.81μm (up)	1.3μm (down) 0.89μm (up)	0.81μm(down)	1.3μm (down) 0.89μm (up)	0.89μm(down) 1.3μm (down)	0.81μm (up)
Optical detectors	Si-APD & Ge-APD		Si-APD	Ge-APD Si-APD	Si-APD Ge-APD	Si-APD
Optical multiplexer	dielectric thin-film filter type					

C. *Other Applications*

In addition to trunk-transmission system and subscriber system applications, an optical fiber has been applied to such areas as short-link video transmission, local area network (LNA) for on-premises use, and so on [82].

IV. CONCLUSION

State-of-the-art WDM technology and its application to optical communications systems has been reviewed. The impact of this new technology on communications systems is that it will enhance system and design flexibility, as well as system cost-effectiveness. Thus this multiplexing technology is expected to play a major role in advancing communications systems, in conjunction with such conventional multiplexing technologies as FDM and TDM. Particularly, the application to multiservice broad-band subscriber networks is most promising.

Although steady progress has been made in optical multiplexing devices, light sources, and detectors, this is still the first stage of development with respect to the technology itself, as well as to system applications. Therefore, more efforts should be directed toward system applications as well as to research on refining WDM technology. In particular, keen attention should be paid to cost reduction of WDM devices to increase the application areas.

ACKNOWLEDGMENT

The authors would like to express their gratitude to Dr. E. Iwahashi, Dr. S. Shimada, and Dr. M. Koyama of Yokosuka Electrical Communication Laboratory (ECL) for their encouragement. Thanks are also due to the members of the Optical Transmission Section at the Yokosuka ECL for their helpful discussions.

REFERENCES

[1] O. E. Delange, "Wideband optical communication systems, Part II–Frequency division multiplexing," *Proc. IEEE*, vol. 58, p. 1683, Oct. 1970.

[2] W. Tomlinson, "Wavelength multiplexing in multimode optical fibers," *Appl. Opt.*, vol. 18, no. 8, pp. 2180–2194, Aug. 1977.

[3] H. Ishio and T. Miki, "A preliminary experiment on wavelength division multiplexing transmission using LED," in *Proc. IOOC'77*, vol. C7-3 (Tokyo, Japan), July 1977, also S. Sugimoto *et al.*, "High speed digital signal transmission experiments by optical wavelength division multiplexing," in *Proc. IOOC'77*, vol. C7-4 (Tokyo, Japan), July 1977.

[4] K. Nosu and H. Ishio, "A design of optical multi/demultiplexers for optical wavelength-division multiplexing transmission," *Trans. IECE*, vol. 62-B, pp. 1030–1036, Nov. 1979.

[5] A. Nicia, "Wavelength multiplexing and demultiplexing systems for single mode and multimode fibers," presented at 7th Europ. Conf. Opt. Commun., vol. 8.1 (Copenhagen, Denmark), Sept. 1981.

[6] G. Winzer, "Wavelength division multiplexing-status and trends," presented at 8th Europ. Conf. Opt. Commun., Commun. AIX-1 (Cannes, France), Sept. 1982.

[7] T. Kitami, Y. Okano, and T. Kanada, "6.3 Mb/s optical transmission systems using wavelength-division-multiplexing technology," presented at Tech. Group, IECE Japan), Sept. 1982, paper CS82-60, (in Japanese).

[8] T. Kanada, Y. Okano, K. Aoyama, and T. Kitami, "Design and performance of WDM transmission systems at 6.3 MB/s," *IEEE Trans. Commun.*, vol. COM-31, pp. 1095–1102, Sept. 1983.

[9] T. Miki and H. Ishio, "Viabilities of wavelength-division-multiplexing transmission system over an optical fiber cable," *IEEE Commun.*, vol. COM-26, pp. 1082–1087, July 1978.

[10] S. Hata, "An optical channel filter for fiber transmission systems," *Opt. Quantum Electron.*, vol. 9, pp. 265–267, 1977.

[11] S. Sugimoto *et al.*, "High-speed digital-signal transmission experiments by optical wavelength-division multiplexing," *Electron. Lett.*, vol. 13, pp. 680–681, 1977.

[12] K. Aoyama and J. Minowa, "Low-loss optical demultiplexer for WDM systems in the 0.8-μm wavelength region," *Appl. Opt.*, vol. 18, no. 16, pp. 2834–2836, Aug. 1979.

[13] Y. Fujii, K. Aoyama, and J. Minowa, "Optical demultiplexer using a silicon echelette grating," *IEEE J. Quantum Electron.*, vol. QE-16, pp. 165–169, Feb. 1980.

[14] R. Watanabe, K. Nosu, and Y. Fujii, "Optical grating multiplexer in the 1.1–1.5-μm wavelength region," *Electron. Lett.*, vol. 16, no. 3, pp. 108–109, Jan. 1980.

[15] N. Nicia, "Wavelength multiplexing and demultiplexing systems for single mode and multimode fibers," presented at 7th Europ. Conf. Opt. Commun., vol. 8.1 (Copenhagen, Denmark), Sept. 1981.

[16] W. Tomlinson and G. Aumiller, "Optical multiplexer for multimode fiber transmission systems," *Appl. Phys. Lett.*, vol. 31, no. 3, pp. 169–171, Aug. 1977.

[17] K. Kobayashi and M. Seki, "Microoptic grating multiplexers and optical isolators for fiber-optic communications," *IEEE J. Quantum Electron.*, vol. QE-16, pp. 11–12, Jan. 1980.

[18] M. Seki *et al.*, "20-channel micro-optic grating demultiplexer for 1.1 ~ 1.6 μm band using a small focusing parameter graded-index rod lens," *Electron. Lett.*, vol. 18, no. 6, pp. 257–258, Mar. 1982.

[19] B. Metcalf and J. Providakes, "High-capacity wavelength demultiplexer with a large-diameter GRIN rod lens," *Appl. Opt.*, vol. 21, no. 5, pp. 794–796, Mar. 1982.

[20] R. Watanabe and K. Nosu, "Optical demultiplexer using concave grating in 0.7–0.9 μm wavelength region," *Electron. Lett.*, vol. 16, no. 3, pp. 106–108, Jan. 1980.

[21] ——, "Slab waveguide demultiplexer for multi-mode optical transmission in the 1.0–1.4 μm wavelength region," *Appl. Opt.*, vol. 19, no. 21, pp. 3588–3590, Nov. 1980.

[22] H. Yen *et al.*, "Planar Rowland Spectrometer for fiber-optic wavelength demultiplexer," *Opt. Lett.*, vol. 6, no. 12, pp. 639–641, Dec. 1981.

[23] Y. Fujii and J. Minowa, "Optical demultiplexer using a silicon concave diffraction grating," *Appl. Opt.*, vol. 22, no. 7, pp. 974–978, Apr. 1983.

[24] T. Suhara, J. Viljanen, and M. Leppihalme, "Integrated-optic waveguide multi- and demultiplexers using a chirped grating and an ion-exchanged waveguide," *Appl. Opt.*, vol. 21, no. 12, pp. 2195–2198, June 1982.

[25] H. Mahlein, "A high performance edge filter for wavelength-division multi-demultiplexer units," *Opt. Laser Technology*, pp. 13–20, Feb. 1981.

[26] J. Minowa and Y. Fujii, "Dielectric multilayer thin film filters for WDM transmission systems," *J. Lightwave Technol.*, vol. LT-1, pp. 116–121, Mar. 1983. ——, "High performance bandpass filter for WDM transmission," *Appl. Opt.*, vol. 23, no. 2, p. 193, Jan. 15, 1984.

[27] A. Thelen, "Design of multilayer interference filters," in *Physics of Thin Films*, vol. 5, G. Hass and R. E. Thun, Eds. New York: Academic, 1969, pp. 47–86.

[28] S. Sugimoto *et al.*, "Wavelength division two-way fiber-optic transmission experiments using micro-optic duplexer," *Electron. Lett.*, vol. 14, no. 1, pp. 15–17, Jan. 1978.

[29] T. Tanaka, S. Kishi, and T. Tsutsumi, "Fiber-optic multifunction device using a single GRIN-rod lens for WDM transmission systems," *Appl. Opt.*, vol. 21, no. 19, pp. 3423–3429, Oct. 1982.

[30] C. Schwizer, "A three channel wavelength multiplexing device," presented at 8th Europ. Conf. Opt. Commun., vol. C-6 (Cannes, France), Sept. 1982.

[31] Y. Fujii, J. Minowa, and T. Kanada, "Practical two-wavelength multi-demultiplexer: Design and performance," *Appl. Opt.*, vol. 22, no. 19, pp. 3090–3097, Oct. 1, 1983.

[32] K. Nosu, H. Ishio, and K. Hashimoto, "Multireflection optical multi/demultiplexer using interference filters," *Electron. Lett.*, vol. 15, pp. 414–415, July 1979.

[33] R. Watanabe, K. Sano, and J. Minowa, "Design and performance of multi/demultiplexer for subscriber loop system," presented at IOOC '83, 30Cl-2 (Tokyo, Japan), June 1983.

[34] S. Ishikawa, K. Takahashi, and K. Doi, "Multi-reflection wavelength-division multiplexer/demultiplexers with the wavelength tunability," presented at 6th Europ. Conf. Opt. Commun., University of York, United Kingdom, Sept. 1980, pp. 298–301.

[35] Miyauchi *et al.*, "Compact wavelength multiplexer using optical fiber pieces," *Opt. Lett.*, vol. 5, no. 7, pp. 321–322, July 1980.

[36] G. Winzer, H. Mahlein, and A. Reichlet, "Single-mode and multimode all-viber directional couplers for WDM," *Appl. Opt.*, vol. 20, no. 18, pp. 3128–3135, Sept. 1981.

[37] R. Watanabe, Y. Fujii, K. Nosu, and J. Minowa, "Optical multi/demultiplexers for single-mode fiber transmission," *IEEE J. Quantum Electron.*, vol. QE-17, pp. 974–981, June 1981.

[38] A. Koonen, "A compact wavelength demultiplexer using both interference filters and a diffraction grating," presented at 7th Europ. Conf. Opt. Commun., vol. 8.5 (Copenhagen, Denmark), Sept. 1981.

[39] A. Livanos, K. Katzir, A. Yariv, and C. Hong, "Chirped-grating demultiplexers in dielectric waveguides," *Appl. Phys. Lett.*, vol. 30, no. 10, pp. 519–521, May 1977.

[40] P. Yhe and F. Taylor, "Contradirectional frequency-selective couplers for guided-wave optics," *Appl. Opt.*, vol. 19, no. 16, pp. 2848–2855, Aug. 1980.

[41] T. Handa, T. Suhara, H. Nishihara, and J. Koyama, "Microgratings for high-efficiency guided-beam deflection fabricated by electron-beam direct-writing techniques," *Appl. Opt.*, vol. 19, no. 16, pp. 2842–2847, Aug. 1980.

[42] T. Suhara, Y. Handa, H. Nishihara, and J. Koyama, "Monolithic integration microgratings and photodiodes for wavelength demultiplexer," *Appl. Phys. Lett.*, vol. 40, no. 2, pp. 120–122, Jan. 1982.

[43] R. Alferness and L. Buhl, "Electro-optic waveguide TE ⟷ TM mode converter with low drive voltage," *Opt. Lett.*, vol. 5, no. 11, pp. 473–475, Nov. 1980.

[44] K. Aiki, M. Nakamura, and J. Umeda, "A frequency-multiplexing light sources with monolithically integrated distributed-feedback diode lasers," *IEEE J. Quantum Electron.*, vol. QE-13, pp. 220–223, Apr. 1977.

[45] W. Tsang, "CW multiwavelength transverse-junction-stripe lasers grown by molecular beam epitaxy operating predominantly in single-longitudinal modes," *Appl. Phys. Lett.*, vol. 36, no. 6, pp. 411–443, Mar. 1980.

[46] S. Sakaki, T. Aoki, and M. Umeno, "InGaAsP/InP dual wavelength lasers," *Electron. Lett.*, vol. 18, no. 1, pp. 17–18, Jan. 1982.

[47] N. Bouadma, J. Bouley, and J. Riou, "Dual-wavelength (GaAl)As laser," *Electron. Lett.*, vol. 18, no. 20, pp. 871–873, Sept. 1982.

[48] J. Cambell *et al.*, "Improved two-wavelength demultiplexing InGaAsP photodetector," *IEEE J. Quantum Electron.*, vol. QE-16, pp. 601–602, 1980.

[49] T. Lee, "LEDs and photodetectors for wavelength-division-multiplexed light-wave systems," *Opt. Laser Technol.*, pp. 15–20, Feb. 1982.

[50] R. Watanabe and K. Nosu, "New 2-channel multiplexer using polarization," *Electron. Lett.*, vol. 17, no. 3, pp. 134–135, Feb. 1981.

[51] S. Sugimoto *et al.*, "High-speed digital-signal transmission experiments by optical wavelength-division multiplexing," *Electron. Lett.*, vol. 13, no. 12, pp. 680–681, Oct. 1977.

[52] B. Metcalf and L. Jou, "Dual GRIN lens wavelength multiplexer," *Appl. Opt.*, vol. 22, no. 3, pp. 455–459, Feb. 1983.

[53] K. Nosu and H. Ishio, "A design of multiplexers for optical wavelength-division-multiplexing transmission via a single fiber," in *Proc. 1979 Int. Symp. Circuits Syst.*, vol. R20.3, July 1978.

[54] W. J. Tomlinson, "Wavelength multiplexing in multimode fibers," *Appl. Opt.*, vol. 16, no. 8, pp. 2186–2194, Aug. 1977.

[55] H. Ishio, "WDM technology," in *Proc. 5th Europ. Conf. Opt. Commun.*, vol. 11.1 (Amsterdam, The Netherlands), Sept. 1979.

[56] A. M. Hill, "Practical constraints on wavelength multiplexing of multimode, digital systems in 850 nm region," in *Proc. 5th Europ. Conf. Opt. Commun.*, vol. 21.6 (Amsterdam, The Netherlands), Sept. 1979.

[57] N. Nakamura *et al.*, "Nonlinear distortion in WDM analog transmission systems," IECE Japan, Paper of Tech. Group Commun., CS80-4, Apr. 1980 (in Japanese).

[58] H. Ishio and K. Nosu, "On channel spacing and interchannel crosstalk in WDM transmission," in *Proc. Nat. Conv. IECE Japan*, no. 2002, Mar. 1978 (in Japanese).

[59] K. Nosu *et al.*, "Optical access loop network design," IECE Japan, Paper Tech. Group Commun., CS80-4, Apr. 1980 (in Japanese).

[60] K. Aoyama, E. Yanai, and T. Kanada, "Crosstalk due to spontaneous Raman scattering in WDM transmission systems," in *Proc. IOOC-83*, vol. 27C2-3, June 1983.

[61] A. Tomita, "Crosstalk caused by stimulated Raman scattering in single mode wavelength division multiplexed systems," *Opt. Lett.*, vol. 8, no. 7, pp. 412–414, July 1983.

[62] R. Watanabe, A. Sano, and J. Minowa, "Design and performance of multi/demultiplexer for subscriber loop system," in *Proc. IOOC'83*, vol. 30C1-2, June 1983.

[63] R. E. Epworth, "The phenomenon of modal noise in analog and

digital optical fiber system," *4th Europ. Conf. Opt. Commun.* (Genoa, Italy), Sept. 1978.

[64] K. Peterman and G. Arnold, "Noise and distortion characteristics of semiconductor lasers in optical fiber communication systems," *IEEE J. Quantum Electron.*, vol. QE-18, pp. 543–555, Apr. 1982.

[65] K. Nakagawa, H. Ishio, and T. Miki, "On loss margin for a fiber optic repeater," in *Proc. 1978 Nat. Conv. IECE Japan*, no. 2010, Mar. 1978 (in Japanese).

[66] K. Nosu, H. Ishio, and K. Hashimoto, "Multireflection optical multi/demultiplexer using interference filters," *Electron. Lett.*, vol. 15, no. 4, pp. 414, July 1979.

[67] E. Iwahashi, "Optical fiber systems for public telecommunication," in *Proc. IOOC'83*, vol. 28C1-1 (Tokyo, Japan), June 1983.

[68] Y. Fujii, J. Minowa, and T. Kanada, "Practical two-wavelength multi-multiplexer: Design and performance," *Appl. Opt.*, vol. 22, no. 19, pp. 3090–3097, Oct. 1, 1983.

[69] T. Kanada, Y. Okano, K. Aoyama, and T. Kitami, "Design and performance of WDM transmission systems at 6.3 Mbit/s," *IEEE Trans. Commun.*, vol. COM-31, pp. 1095–1102, Sept. 1983.

[70] I. Kobayashi *et al.*, "A 4 channel wavelength-division multiplexing transmission system," in *Proc. FOC'80* (San Francisco), Sept. 1980, pp. 27–30.

[71] K. Nakagawa, T. Ito, and K. Aoyama, "400 Mb/s two-way 2 wavelengths multiplexing transmission experiment," in *Nat. Conv. Rec. IECE Japan*, vol. S4-6, Mar. 1980 (in Japanese).

[72] K. Minemura *et al.*, "Optical multiplexing transmission in 1 μm wavelength region using single mode fibers," in *Nat. Conv. Rec. IECE Japan*, vol. S4-7, Mar. 1980.

[73] P. Kaiser *et al.*, "Two-channel single-mode WDM in the 1.3 micron wavelength region," in *Proc. IOOC'83*, vol. 27CL-4 (Tokyo, Japan), June 1983.

[74] H. Ishio, "Japanese field trials and applications in telephony," *IEEE Trans. Commun.*, vol. SAC-1, pp. 404–412, Apr. 1983.

[75] E. Iwahashi, "Trends in long-wavelength single-mode transmission systems and demonstrations in Japan," *IEEE J. Quantum Electron.*, vol. QE-17, pp. 890–896, June 1981.

[76] D. Kneisel, "Goals and strategies for the introduction of broadband optical communication systems into the Deutsche Bundespost telecommunication networks," in *ICC'83 Conf. Rec.*, vol. 4D.1 (Philadelphia, PA), June 1982.

[77] U. Horn and F. Sporlender, "Service-integrated optical subscriber loop systems research and industrial prototype projects guided by the Deutsche Bundespost," in *ISSLS 82 Proc.* (Toronto, Canada), Sept. 1982, pp. 214–219.

[78] G. A. Tough, K. B. Harries, K. Y. Chang, C. I. Nisbet, and J. F. Chalmers, "An integrated fiber optics distribution field trial in Elie, Manitoba," in *ICC'82 Conf. Rec.*, vol. 4D.4 (Philadelphia, PA), June 1982.

[79] J. Popovics, "Optical fiber multiservice subscriber connection system: SAFO," *IEEE Trans. Commun.*, vol. COM-30, pp. 2215–2220, Sept. 1982.

[80] E. Roza *et al.*, "TV systems with remote selection for subscriber glass fiber networks," in *Proc. 8th ECOC'82*, vol. AXV-4 (Cannes, France), Sept. 1982.

[81] K. Asatani, R. Watanabe, K. Nosu, and T. Matsumoto, "A field trial of fiber optic subscriber loop systems utilizing wavelength-division multiplexers," *IEEE Trans. Commun.*, vol. COM-30, pp. 2172–2184, Sept. 1982.

[82] M. Saruwatari, K. Nosu, T. Miki, and K. Asatani, "Optically accessing loop (OPAL) network using wavelength-division-multiplexing technology," presented at the *Int. Conf. Commun.* (Seattle, WA), June 1980, Paper 28.6.

Hideki Ishio was born in Osaka, Japan, in 1943. He received the B.S. degree in 1966 and the Ph.D. degree in 1980, both from Kyoto University, Kyoto, Japan. Since 1966, he has been with the Electrical Communication Laboratories, Nippon Telegraph and Telephone Public Corporation.

He was involved in the research and development on high-speed modulation and demodulation technologies for guided-millimeter-wave transmission systems. From 1975, he has been engaged in research and development on optical-fiber transmission systems. He is currently head of Electronic Components Section, Electronic Equipment Development Division, Musashino ECL, Musashino, Japan.

Dr. Ishio is a member of the Institute of Electronics and Communication Engineers of Japan (IECEJ). He received the Yonezawa Memorial Scholarship Award and the Paper Awards for his papers from the IECEJ.

*

Junichiro Minowa was born in Nagano, Japan, on March 3, 1943. He received the B.S. degree from Yamanashi University in 1966, and the M.S. and Ph.D. degrees in electronics engineering from Tohoku University, Sendai, Japan, in 1968 and 1972, respectively.

In 1972 he joined the Yokosuka Electrical Communication Laboratory, Nippon Telegraph and Telephone Public Corporation, Kanagawa, Japan. He has been engaged in research and development of surface-acoustic-wave devices. Currently, he is working on optical passive devices such as multiplexers, thin-film filters, and switches.

Dr. Minowa is a member of the Institute of Electronics and Communication Engineers of Japan and the Acoustic Society of Japan. He received the Young Engineers Award from the IECEJ.

*

Kiyoshi Nosu (S'73–M'77) was born in Gifu, Japan, on June 19, 1949. He received B.S. degree in electric engineering from Keio University, Tokyo, Japan, in 1972, and M.S. and Ph.D. degrees in electronics engineering from Tokyo University, Tokyo, Japan, in 1974 and 1977, respectively.

He joined the Yokosuka Electrical Communication Laboratory, Nippon Telegraph and Telephone Public Corporation, Yokosuka, Kanagawa, Japan, in 1977, and has been engaged in research on fiber-optic communication systems, in particular, wavelength-division-multiplexing transmission systems and fiber-optic local area networks. In 1982 and 1983, he stayed at Naval Research Laboratory, Washington, DC, as a Visiting Scientist.

Dr. Kiyoshi Nosu is a member of the Institute of Electronics and Communication Engineers of Japan, the Japanese Society of Applied Physics, and Japan Operations Research Society. He received the Young Engineers Award from the IECEJ.

60-CHANNEL FM VIDEO SUBCARRIER MULTIPLEXED OPTICAL COMMUNICATION SYSTEM

Indexing terms: Optical communications, Optical fibres, Video

A subcarrier multiplexed (SCM) optical communication system is described which transmits 60 frequency-modulated (FM) video channels over 18 km of single-mode fibre with a 56 dB weighted signal/noise ratio.

The distribution of voice, data and video services to the subscriber is one of the next major steps in the evolution of optical fibre communication systems. Although completely switched, time division multiplexed broadband subscriber networks have been proposed, at present the high cost of the digital electronics and the large data rates required for each video channel make such systems difficult to justify economically. In this letter it is proposed that subcarrier multiplexing[1-3] of video signals at microwave carrier frequencies provides an attractive, alternative method for distribution of video signals. In particular, the use of frequency-modulated (FM) microwave subcarriers[4,5] in the 2–8 GHz band is very attractive because the required electronics have been previously developed for microwave systems, and much of it is relatively inexpensive.

SCM systems have other advantages. With the recent progress in high-speed lasers and photodiodes, the total available bandwidth can exceed 10 GHz. SCM systems have sufficient bandwidth to provide a multiplicity of services, including analogue and digital video, high-definition video and digital audio. The great capacity of SCM systems and their flexibility in the allocation of bandwidth can provide many new opportunities for optical fibre distribution networks.

The possibility of using FM microwave carriers for subscriber loop transmission has been motivated by several recently published studies. In one,[4] five FM video channels were successfully transmitted using the 150–300 MHz band. In another,[5] a C-band satellite signal (4·9–5·2 GHz) carrying ten 30 MHz FM video channels was transmitted over 35 km of single-mode fibre. In addition, several SCM digital experiments have been reported. Three 44 Mbit/s signals were transmitted[1] using FSK-modulated carriers in the 3·5–4·0 GHz band, and a 2 Gbit/s signal was transmitted[2] using PSK-modulated carriers in the 3–16 GHz band.

In this letter we report the transmission of 60 FM video channels over 18 km of single-mode fibre using the 2·7–5·2 GHz band. A block diagram of a 60-channel subcarrier multiplexed (SCM) system is shown in Fig. 1. Sixty FM microwave carriers are electrically combined and used to intensity-modulate a 1·3 μm GaInAsP high-speed, vapour-phase-regrown, buried heterostructure (VPR-BH) laser.[6] At 5 mW bias the laser bandwidth is 11 GHz and the relative intensity noise (RIN) is less than −135 dB/Hz.

The 60-channel video source is obtained by taking the signal from a C-band satellite (3·7–4·2 GHz) carrying 11 channels and adding a locally generated signal, for a total of 12 channels. As shown in Fig. 1, the 12-channel microwave signal is then up- and downconverted by two balanced mixers with

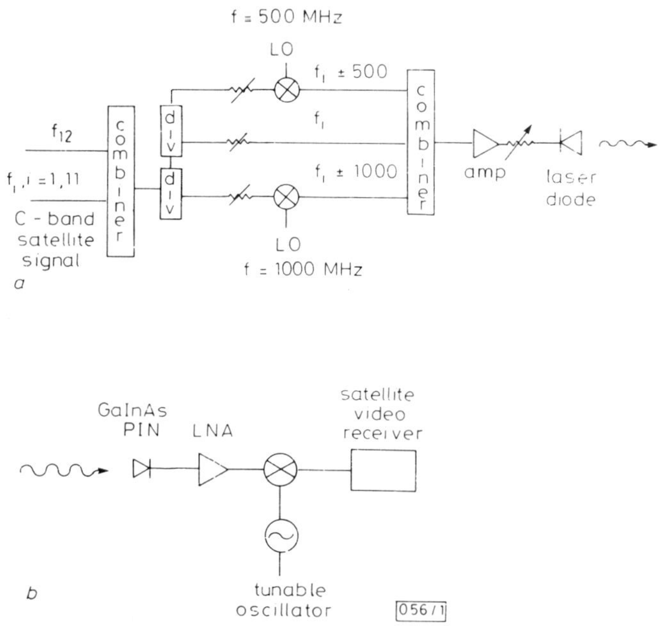

Fig. 1 *System diagram for 60-channel subcarrier multiplexed optical communication system*

a 60-channel video source
b 60-channel SCM receiver

500 MHz and 1000 MHz local oscillators. The output of the mixers and the original 12-channel signal are then combined to provide a 60-channel source occupying the 2·7–5·2 GHz band.

The locally generated signal is obtained by directly modulating a voltage-controlled oscillator with a pre-emphasised baseband video signal. The presence of the locally generated signal permits the direct measurement of the carrier/noise ratio (CNR) for the unmodulated carrier and of the weighted signal/noise ratio (SNR) using locally generated test signals.

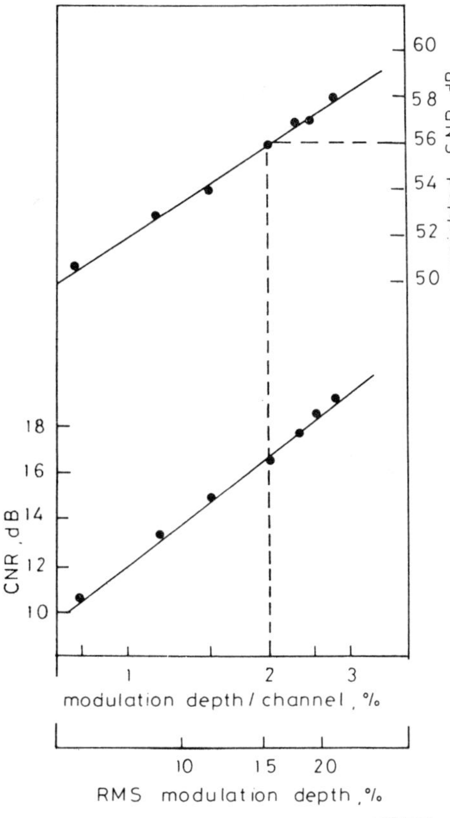

Fig. 2 *Measured CNR and weighted SNR against modulation depth for 60-channel FM/SCM transmission over 18 km of single-mode fibre*

$I_{pd} = 30\,\mu A$

The optical signal is transmitted over 18 km of single-mode fibre, including four GTE elastomeric splices, and having a total optical loss of 8 dB. The fibre-coupled power is 700 μW and the system operates with 4 dB of margin. The signal is detected with a high-speed GaInAs PIN diode with a bandwidth of 20 GHz[7] and a responsivity of 0·7 A/W. An optical attenuator is used to set the direct photocurrent at 30 μA, a value appropriate for a subscriber distribution link having a total optical loss budget of 8 dB and 4 dB margin.

The signal from the photodiode is amplified with a 2–8 GHz low-noise amplifier with a noise figure of 3 dB. The signal is downconverted to a 950 MHz IF using a mixer and a tunable oscillator. The FM signal is demodulated using a low-cost, 12-channel satellite video receiver with a 30 MHz video bandwidth.

The CNR for a 30 MHz noise bandwidth is measured for the unmodulated carrier using a spectrum analyser, and the result is plotted in Fig. 2 as a function of modulation depth. The weighted SNR is determined using a Tektronix video signal generator and NTSC Vectorscope. The weighted SNR against modulation depth is also shown in Fig. 2.

For the 60-channel signal transmitted over 18 km of single-mode fibre, a CNR of 16·5 dB produces a 56 dB weighted SNR, in agreement with theory.[5] The optical modulation depth is determined by using a 10 dB tap to measure the RF signal output of the low-noise amplifier, and then correcting for the measured net gain of the cable, amplifier and tap. The 56 dB weighted SNR is achieved with a modulation depth m of 2·0%/channel. The RMS modulation depth for all N channels is $m\sqrt{N}$, and is measured to be 15·8%.

The system noise is dominated by the thermal noise of the amplifier. There is no evidence for noise contributions from third-order intermodulation products, and only a small contribution (less than 1 dB) from laser intensity noise (RIN). The CNR is thus approximately given as

$$CNR = \frac{(mI_{dc})^2 R}{2\,NF\,kTB} \tag{1}$$

where I_{dc} is the direct photocurrent (30 μA), R is the 50 Ω load, NF is the amplifier noise figure (3 dB) and the receiver bandwidth $B = 30$ MHz. Eqn. 1 predicts $m = 2{\cdot}2\%$ for a 16·5 dB CNR, in very good agreement with the measured value.

The amplifier's thermal noise floor ($NF \times kT$) is at −171 dBm/Hz. The noise power due to the laser RIN is $(I_{dc}^2 R) \times RIN$. For the operating conditions described, the RIN is thus less than −179 dBm/Hz and adds at most 0·8 dB to the system noise. The shot noise at −183 dBm/Hz does not contribute significantly to the total noise.

At the centre of the transmission band there are approximately $N^2/2$ third-order intermodulation products (IMPs).[8] The ratio of the optical power in the IMP to that in the carrier can be written as

$$P_{IM3}/P_c = m^2 IM_3 \tag{2}$$

From either theory or measurements[9] one can set the bound $IM_3 \leq 0{\cdot}1$ for dominant IMPs of the form $f_i = f_j \pm f_k \pm f_l$.[8] Each IMP introduces a noise power N_{IM3}, given as

$$N_{IM3} = (m^2 IM_3)^2 (mI_{dc})^2 R/2 \tag{3}$$

Assuming the $N^2/2$ IMPs add incoherently and are spread evenly over the 30 MHz transmission band, the average contribution of IMP noise is −210 dBm/Hz. Indeed, no evidence for third-order intermodulation products is observed as the modulation depth is varied from 0·09 to 2·2%/channel. No second-order IMPs are present because the transmission band is less than one octave.

Finally, it should be noted that a total optical modulation depth might be defined as mN, where N is the number of channels. For the 60-channel system at the operating point which gives a 56 dB SNR, the product $mN = 1{\cdot}2$. The signals from the N FM channels, however, add incoherently, and thus the RMS modulation depth is $m\sqrt{N}$ or only 15·8%. No degradation in measured SNR has been observed until the value of mN exceeds about 1·4. Thus it appears possible to design SCM systems with a modulation depth/channel exceeding $1/N$.

Summary: Sixty FM video channels have been transmitted over 18 km of single-mode fibre with an optical system margin of 4 dB. A weighted SNR of 56 dB is achieved for a 16·5 dB CNR at a modulation depth of 2·0%/channel, and an RMS modulation depth of 15·8% for all 60 channels. The 60 channels occupy the band from 2·7 to 5·2 GHz. Such analogue systems appear to be very promising for the distribution of video signals in the subscriber loop. The great flexibility of SCM systems in allocating bandwidth makes them extremely attractive for providing voice, data, video, high-definition video and other services to the subscriber.

Acknowledgments: We thank Dr. R. B. Lauer for his strong support, and the many colleagues of GTE Laboratories whose efforts contributed to this work in a multitude of ways.

R. OLSHANSKY
V. A. LANZISERA

GTE Laboratories
40 Sylvan Road
Waltham, MA 02254, USA

15th September 1987

References

1 DARCIE, T. E., DIXON, M. E., KASPER, B. L., and BURRUS, C. A.: 'Lightwave system using microwave subcarrier multiplexing', *Electron. Lett.*, 1986, **22**, pp. 774–775

2 BOWERS, J. E.: 'Optical transmission using PSK-modulated subcarriers at frequencies to 16 GHz', *ibid.*, 1986, **22**, pp. 1119–1121

3 DARCIE, T. E.: 'Subcarrier multiplexing for multiple-access lightwave networks', *J. Lightwave Technol.*, 1987, **LT-5**, pp. 1103–1110

4 TJHUNG, T. T., SELVAN, B., HALDAR, M. K., and MENDIS, F. V. C.: 'Single-mode 1·3 μm optical fibre link for analogue multichannel television', *Electron. Lett.*, 1986, **22**, pp. 507–509

5 WAY, W. I., KRAIN, M., and WOLFF, R. S.: 'A 35 km satellite earth station 1·3 mm fiber optic microwave multicarrier transmission system'. Optical fiber communication conference (OFC/IOOC '87), Reno, NV, 19–22 Jan. 1987, Postdeadline paper PD-18

6 OLSHANSKY, R., HILL, P., LANZISERA, V., and POWAZINIK, W.: 'Frequency response of 1·3 μm InGaAsP high speed lasers', *IEEE J. Quantum Electron.*, 1987, **QE-23**, pp. 1410–1418

7 SCHLAFER, J., SU, C. B., POWAZINIK, W., and LAUER, R. B.: '20 GHz bandwidth InGaAs photodetector for long-wavelength microwave optical links', *Electron. Lett.*, 1985, **21**, pp. 469–471

8 DALY, J. C.: 'Fiber optic intermodulation distortion', *IEEE Trans.*, 1982, **COM-30**, pp. 1954–1958

9 DARCIE, T. E., TUCKER, R. S., and SULLIVAN, G. J.: 'Intermodulation and harmonic distortion in InGaAsP lasers', *Electron. Lett.*, 1985, **21**, pp. 665–666

ALL SINGLE-MODE OPTICAL FIBRE WAVELENGTH-TUNABLE WDM WITH VERY NARROW PASS/STOPBAND SEPARATIONS

Indexing terms: Optical fibres, Optical multiplexers, Optical couplers

A single-mode all-fibre wavelength-division multiplexer (periodic wavelength filter) exhibiting very narrow separations (~1 nm) between adjacent pass- and stopbands has been fabricated. The wavelengths at which this device has stop- and passbands may be varied by means of an external electrical control while maintaining a nearly constant band separation. Stopband extinctions in excess of 20 dB are reported for band separations of ~30 nm.

Introduction: A number of authors have reported the use of strongly overcoupled fused biconic couplers as WDM/filter devices.[1,2] The suitability of such devices for this purpose arises from the periodic wavelength dependence of the coupling strength and the ease with which this period may be selected by adjusting the coupling length. The coupling coefficient of such fused biconic devices is dependent on the refractive index of the material surrounding the coupling region.[3] This dependence is usually minimised by controlling the cross-sectional shape of the fused region during the fabrication process. Such devices may then be used as conventional WDMs.

The device reported in this letter, however, makes use of this dependence to achieve wavelength tunability. This is done by encapsulating the coupling region in a material with a temperature-dependent refractive index.[4]

Such devices may be used in multichannel WDM or coherent optical systems, whereby the tuning facility will enable matching of the operational wavelength bands with the transmission frequencies.[5]

Alternatively, the wavelength tunability of narrow pass/stopband filters provides a useful function for the characterisation of the mode structure of laser sources.[4]

Device fabrication: The fused biconic coupler is fabricated by tapering and fusing two single-mode fibres placed side-by-side in an oxybutane flame.[2] The narrowband couplers reported in this letter were fabricated using this process whereby the fibres were tapered and stretched, ensuring the occurrence of many power transfers and hence a strong wavelength selectivity. The length of the resultant coupling region was 23 mm, giving a total coupler length (bare of primary coating) of approximately 50 mm.

The WDM coupler, expoxy-bonded on to a silicon substrate for mechanical support, is encapsulated in a silicone compound to protect it against vibration and shock as well as to provide a surrounding with a temperature-dependent refractive index.

The coupler on the substrate is placed on to a thermoelectric Peltier cooler and the complete assembly is housed in a small metal package (75 mm long). A thermistor probe is potted in the silica compound in close proximity to the coupling region. Fig. 1 illustrates a schematic diagram of the complete device.

The WDM coupler is in good thermal contact with one side of the thermoelectric cooler, with the other side secured firmly to the metal package to ensure good thermal contact.

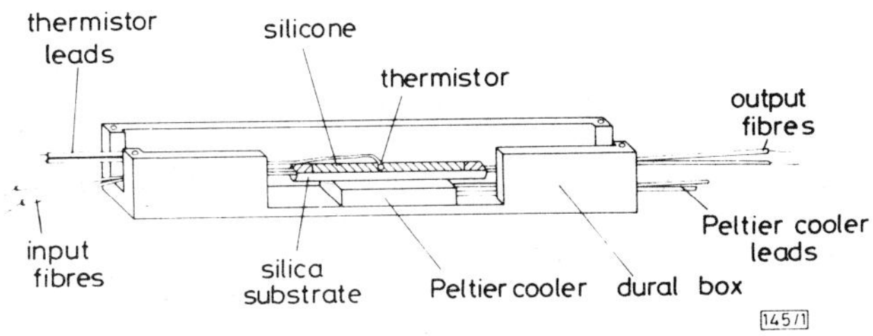

Fig. 1

Device performance: The coupler was measured to have a wavelength periodicity of approximately 60 nm (pass/stopband separation of 30 nm). Fig. 2 illustrates the shape of the measured spectral response for this coupler.

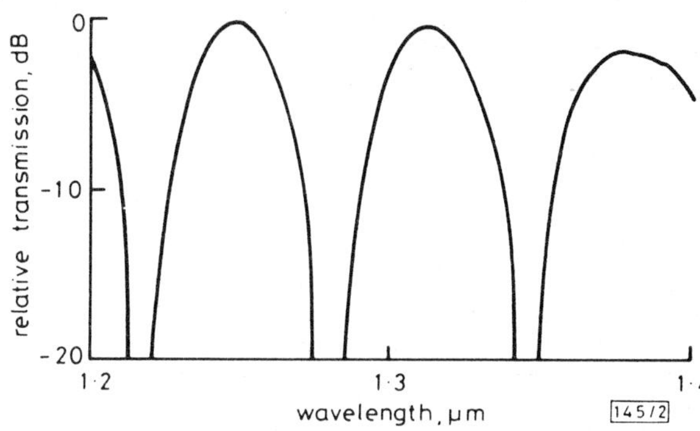

Fig. 2

The temperature of the silicone compound housing the coupler can be controlled by varying the current through the thermoelectric cooler. The temperature is monitored and may be stabilised using the thermistor probe and Peltier cooler contained in the package in a feedback-controlled system.

In this manner, it was possible to maintain the device at a constant temperature within the range 10–35°C. The spectral response of the coupler from one of the output ports was measured at various temperatures within this range, from which the variation of the position of the stopband centre nearest 1·3 μm and the positions of the two adjacent passband centres were determined. The variations in position of these pass- and stopbands are illustrated in Fig. 3 as a function of temperature. The results show that the stopband centre wavelength varies from 1·27 to 1·345 μm within the temperature range 11–35°C. This shift of 75 nm is equivalent to just over one period of the wavelength response. Fig. 3 also shows that, while the shorter-wavelength passband remains a constant 30 nm from the stopband, the adjacent passband at longer wavelength, however, varies in separation from the stopband from 25 to 36 nm.

The device was a 10 kΩ (at 25°C) thermistor probe and a 9A Peltier cooler yielding a maximum measured power consumption of ~2·5 W when operating at room temperature.

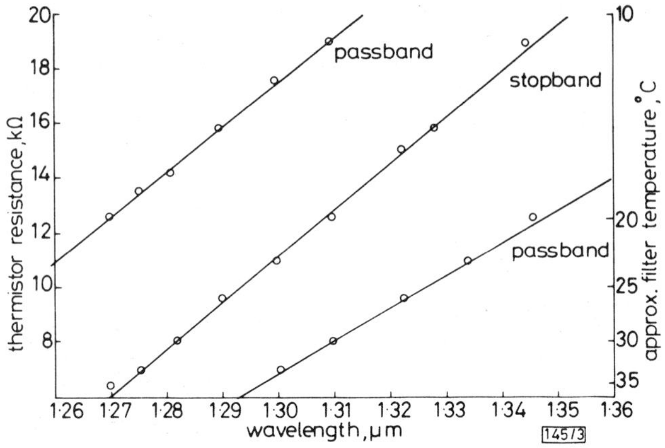

Fig. 3

Conclusions: An all-fibre WDM with a narrow tunable pass/stop wavelength band was reported. The device demonstrates a 30 nm pass/stopband separation which may be tuned over the wavelength range 1270–1345 nm. This device may be used in WDM optical systems for optimally matching the source and channel wavelengths.

S. P. SHIPLEY
A. EL FATATRY

GEC Research Limited
Hirst Research Centre
East Lane
Wembley, Middx. HA9 7PP, United Kingdom

26th March 1987

References

1 YATAKI, M. S., PAYNE, D. N., and VARNHAM, M. P.: 'All-fibre wavelength filters using concatenated fused-taper couplers', *Electron Lett.*, 1985, **21**, pp. 248–249

2 GEORGIOU, G., and BOUCOUVALAS, A. C.: 'High-isolation single-mode wavelength-division multiplexer/demultiplexer', *ibid.*, 1986, **22**, pp. 62–63

3 PAYNE, F. P., FINEGAN, T., YATAKI, M. S., MEARS, R. J., and HUSSEY, C. D.: 'Dependence of fused taper couplers on external refractive index', *ibid.*, 1986, **22**, pp. 1207–1209

4 BAKER, N.: 'A novel laser spectral noise measurement technique'. ECOC '86, 1986, p. 109

5 FUSSGAENGER, K., KOISTER, W., SALLER, H. D., and VOLLIMEER, T.: '4λ × 560 Mbit/s WDM system using 3 wavelength selective fused single mode fibre couplers as multiplexer'. ECOC '86, 1986, p. 447

OPTICAL TIME-DIVISION MULTIPLEXED TRANSMISSION SYSTEM EXPERIMENT AT 8 Gbit/s

Indexing terms: Optical communications, Optical transmission

We describe the design and performance of an experimental two-channel optical time-division multiplexed fibre transmission system operating at a wavelength of 1·3 μm and a bit rate of 8 Gbit/s. Transmission over 8 km of single-mode fibre with low crosstalk and transmission error rates as low as 10^{-10} is demonstrated.

High-bit-rate optical transmission systems are usually driven electronically by digital bit streams that are generated by electronic time-division multiplexing. Optical transmission based on electronic multiplexing has been reported at data rates as high as 8 Gbit/s,[1] but at these bit rates the electronic circuits are difficult to implement. Optical time-division multiplexing (OTDM)[2-5] extends time multiplexing into the optical regime, where the availability of short optical pulses and fast electro-optical switches enables very high bit rate operation while employing moderate bandwidth electronics.[2,4] We report here on recent advances in the design and characterisation of an experimental multigigabit per second optical time-division multiplexed and demultiplexed fibre transmission system. In particular, we describe a new system experiment at a bit rate of 8 Gbit/s which is twice the bit rate of our previous experiments.[4] Details of the system design as well as measurements of crosstalk and other characteristics which affect system performance are discussed.

The 8 Gbit/s system is illustrated in Fig. 1. It is similar in structure to the 4 Gbit/s system described in Reference 4, but most components operate at higher bit rates. The system uses two 1·3 μm optical transmitters[4] which consist of mode-locked semiconductor lasers for the generation of short (15 ps) optical pulses and Ti:$LiNbO_3$ waveguide external modulators for

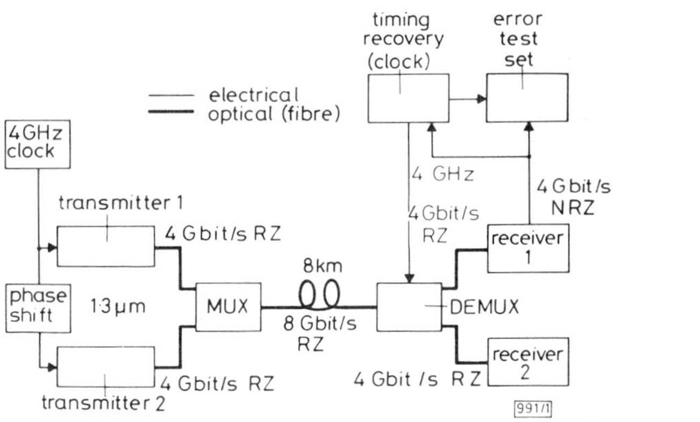

Fig. 1 *Schematic diagram of 8 Gbit/s optical time-division multiplexed system*

data encoding. Both transmitters are driven by a common 4 GHz clock, and produce return-to-zero (RZ) data streams at 4 Gbit/s. The 4 Gbit/s bit streams are multiplexed in a 3 dB fibre directional coupler, and the resulting multiplexed 8 Gbit/s RZ signal is launched into an 8 km length of fibre. After transmission, the 8 Gbit/s signal is demultiplexed to its two 4 Gbit/s baseband components using a Ti:$LiNbO_3$ directional coupler switch (DEMUX).[3,4] The demultiplexed 4 Gbit/s signal is detected using an avalanche photodiode and GaAs FET amplifier.[6] The bandwidth of this receiver (2·6 GHz) has been minimised for high-sensitivity operation, and the associated integrating characteristic converts the baseband signal to the desired non-return-to-zero (NRZ) format. A 4 GHz phase-locked loop at the output of receiver 1 provides timing signals both for demultiplexing and error detection. Note that timing information is recovered from a demultiplexed (4 Gbit/s) data stream rather than the 8 Gbit/s data. This is a unique requirement of optical time-division multiplexed systems because there is no 8 Gbit/s electrical signal available.

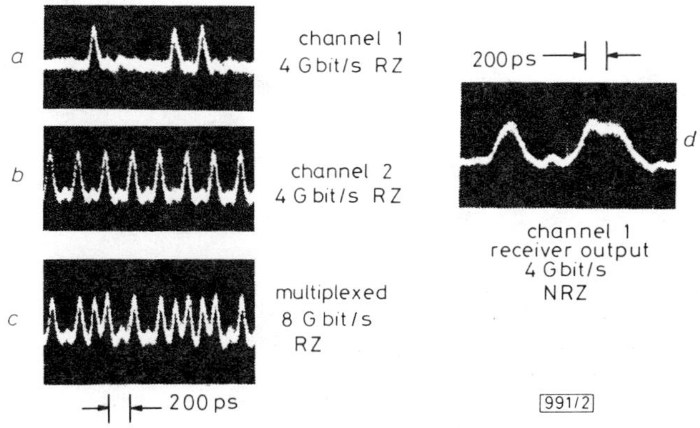

Fig. 2 *Sample bit patterns*
a Transmitter 1
b Transmitter 2
c Multiplexed 8 Gbit/s stream
d Receiver 1

Operation of the two-channel 8 Gbit/s system is illustrated in Fig. 2 with examples of measured bit patterns. Fig. 2*a* is the RZ bit pattern '0100110' at the output of transmitter 1, while Fig. 2*b* is the RZ bit pattern '11111111' at the output of transmitter 2. Fig. 2*c* is the corresponding multiplexed 8 Gbit/s RZ signal at the input to the transmission fibre and Fig. 2*d* is the demultiplexed 4 Gbit/s NRZ '0100110' output from receiver 1.

Low-error-rate operation of the system depends on low system crosstalk. Crosstalk, in the form of intersymbol interference, can arise from poor on/off ratio of the mode-locked pulses, improper switch or data timing, pulse overlap due to fibre dispersion, and from incomplete switch extinction. The effects of crosstalk are illustrated in Fig. 3. Fig. 3*a* shows the eye pattern of an 8 Gbit/s RZ multiplexed signal, made up from a pseudorandom bit sequence on channel 1, and a '101010 ...' bit pattern on channel 2. When converted to NRZ by the receiver, this latter bit pattern is close to a 2 GHz sinusoid. Crosstalk from channel 2 to channel 1 therefore

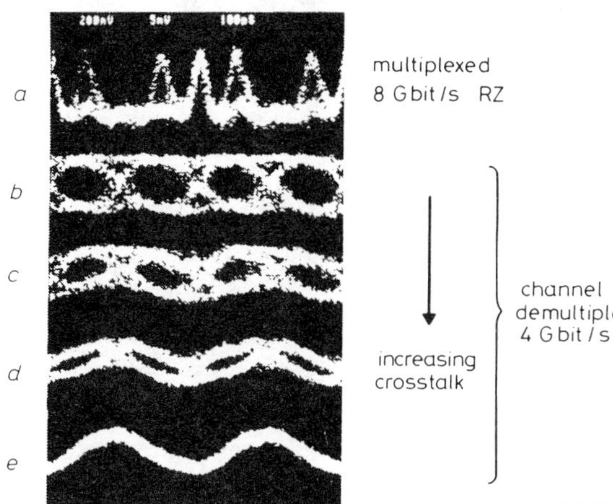

Fig. 3 *Effect of intentionally induced crosstalk*

results in a regular 2 GHz skewing of the 4 Gbit/s NRZ eye pattern at the output of receiver 1. This skewing is illustrated in Figs. 3*b–e* for different levels of intentionally induced crosstalk. For the transmission experiments, the operating conditions of the DEMUX switch were adjusted for the lowest crosstalk (Fig. 3*b*). An upper limit for the system crosstalk was determined using a microwave spectrum analyser to compare the 2 GHz components at the outputs of channel 1 and channel 2. The crosstalk was less than −23 dB (electrical), i.e. −11·5 dB (optical).

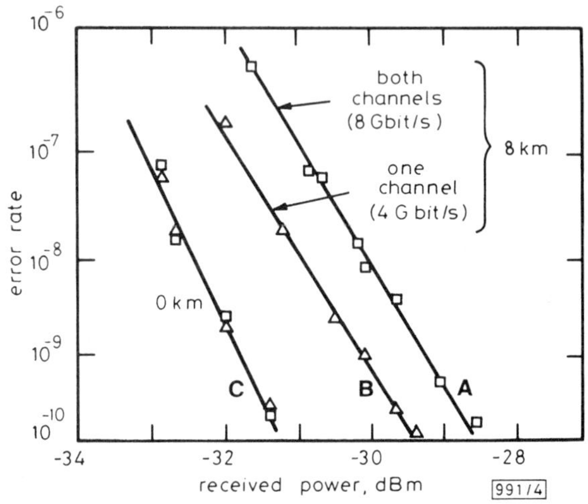

Fig. 4 *Error rate against received power for one 4 Gbit/s channel*

Bit error rate (BER) measurements were carried out using a pseudorandom bit sequence in channel 1, and with the bit patterns '111111 ...' and '101010 ...' in channel 2. The bit error rate in the pseudorandom 4 Gbit/s channel is plotted in Fig. 4 as a function of received power level at 4 Gbit/s. Three curves are shown in Fig. 4: two with the 8 km length of fibre included, and one with the fibre removed. The base line, curve C, characterises the system with no transmission fibre. The symbols represent the BER for the 8 Gbit/s system with both channels on (squares) and with channel 2 turned off (triangles). The two sets of points overlap, independent of the bit pattern in channel 2, indicating negligible system penalty due to crosstalk in the absence of the transmission fibre. The receiver senitivity with no fibre was −31·8 dB. The system performance with the 8 km fibre included is illustrated by curves A and B. The addition of the fibre resulted in a penalty of 2·5 dB with both channels on. This penalty decreased by 0·9 dB with channel 2 turned off. These penalties are caused by pulse spreading and shifting due to fibre dispersion, combined with a small difference in wavelength and the relatively wide spectral width of our mode-locked lasers. Reduced penalties are expected with improvements in the spectral characteristics of the mode-locked lasers.

G. EISENSTEIN
R. S. TUCKER
S. K. KOROTKY
G. RAYBON
J. J. VESELKA
L. L. BUHL
A. H. GNAUCK
B. L. KASPER
R. C. ALFERNESS

AT&T Bell Laboratories
Crawford Hill Laboratory
Holmdel, NJ 07733, USA

1st September 1987

References

1 GNAUCK, A. H., KOROTKY, S. K., KASPER, B. L., CAMPBELL, J. C., TALMAN, J. R., VESELKA, J. J., and MCCORMICK, A. R.: 'Information bandwidth limited transmission at 8 Gbit/s over 68·3 km of single mode optical fiber'. Digest of conf. on optical fiber comm., Atlanta, 1986, Paper PDP9

2 KINSEL, T. S.: 'Wide-band optical communication systems: Part I—time division multiplexing', *Proc. IEEE*, 1970, **58,** pp. 1666–1683

3 KOROTKY, S. K., EISENSTEIN, G., ALFERNESS, R. C., VESELKA, J. J., BUHL, L. L., HARVEY, G. T., and READ, P. H.: 'Fully connectorized high-speed Ti : $LiNbO_3$ switch/modulator for time-division multiplexing and data encoding', *J. Lightwave Technol.*, 1985, **LT-3,** pp. 1–6

4 TUCKER, R. S., EISENSTEIN, G., KOROTKY, S. K., KOREN, U., RAYBON, G., VESELKA, J. J., BUHL, L. L., KASPER, B. L., and ALFERNESS, R. C.: 'Optical time-division multiplexing and demultiplexing in a multigigabit/second fibre transmission system', *Electron. Lett.*, 1987, **23,** pp. 208–209

5 PRUCNAL, P. R., BLUMENTHAL, D. J., and SANTORO, M. A.: '12·5 Gbit/s fibre-optic network using all optical processing', *ibid.*, 1987, **23,** pp. 629–630

6 KASPER, B. L., CAMPBELL, J. C., GNAUCK, A. H., DENTAI, A. G., and TALMAN, J. R.: 'SAGM avalanche photodiode optical receiver for 2 Gbit/s and 4 Gbit/s', *ibid.*, 1985, **21,** pp. 982–984

5 GHz-SPACED, EIGHT-CHANNEL, GUIDED-WAVE TUNABLE MULTI/DEMULTIPLEXER FOR OPTICAL FDM TRANSMISSION SYSTEMS

Indexing terms: Optical communications, Optical multi/demultiplexers, Optical transmission

A 5 GHz-spaced, eight-channel, guided-wave tunable multi/demultiplexer for optical frequency-division-multiplexing (FDM) transmission systems is demonstrated. It is constructed with high-silica channel waveguides and thin-film heaters loaded on the waveguides for frequency tuning.

Introduction: Optical periodic filter-type multi/demultiplexers constructed with antisymmetric Mach–Zender interferometers are useful for optical FDM transmission systems.[1,2] Guided-wave periodic filters have various advantages such as small size, stability, accurate frequency spacing and suitability for multiple integration on one chip.[3] This letter describes a tunable multi/demultiplexer consisting of guided-wave periodic filters.

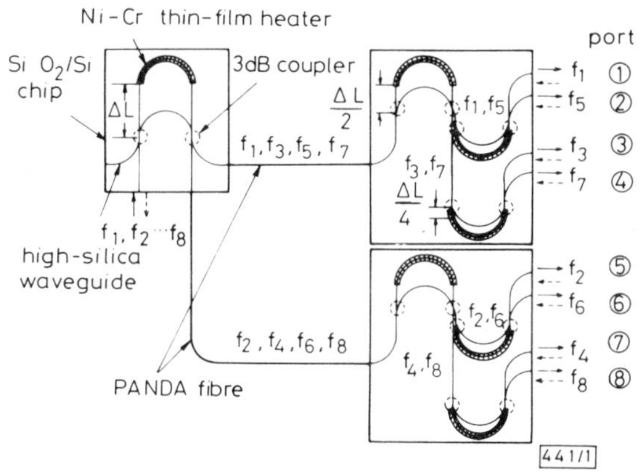

Fig. 1 *Configuration of eight-channel, tunable optical multi/demultiplexer*

Tunable multi/demultiplexer configuration: The configuration of the eight-channel tunable multi/demultiplexer is shown in Fig. 1. It consists of seven cascaded periodic filters. Construction of the periodic filter is the same as that of the Mach–Zender interferometer. It consists of two directional couplers with a 50% divided ratio and two waveguide arms whose lengths are different. The output power transmittance of the periodic filter shows a sinusoidal variation with frequency.[1] The frequency spacing Δf of a periodic filter corresponds to the frequency spacing between the maximum and minimum transmittance. Δf is determined by $c/(2\ \Delta L)$, where c is the light velocity in vacuum and $2\ \Delta L$ is the difference in length between the two waveguide arms of the Mach–Zender interferometer; n is the effective refractive index of the waveguide. For $n = 1{\cdot}47$, multi/demultiplexing of two input lightwaves spaced at 5 GHz requires $2\ \Delta L = 2{\cdot}04$ cm. A frequency spacing of 5 GHz corresponds to a wavelength spacing of 0·04 nm in the 1·5 μm wavelength region. An eight-channel demultiplexer has been formed with cascading periodic filters whose arm length differences are $2\ \Delta L$, ΔL and $\Delta L/2$, which correspond to frequency spacings of Δf, $2\ \Delta f$ and $4\ \Delta f$. The centre frequency of the multi/demultiplexer is shifted by changing the phase delay of one of the arms. The phase delay change is caused by changing the effective optical path length of the arm. As a result, frequency tuning is performed by the phase shifter, such as a thin-film heater loaded on the waveguide.[4]

The waveguide multi/demultiplexer was fabricated on a silicon substrate by flame hydrolysis deposition of high-silica glass and reactive ion etching.[3] As shown in Fig. 1, four-channel demultiplexers with a frequency spacing of 10 GHz are integrated on one chip. They are connected to a 5 GHz-spaced, two-channel periodic filter chip through PANDA fibres. The polarisation of the input lightwave should be either TE or TM mode because of the birefringence of the waveguide.[5] The refractive index difference between the core and cladding was 0·8%, and the core cross-section was 7·5 × 7·5 μm. The radius of the bending was 5 mm and no bending loss appeared. Ni–Cr thin-film heaters with a width of 20 μm and a thickness of 1000 Å were loaded on one of the arms of each filter for tuning.

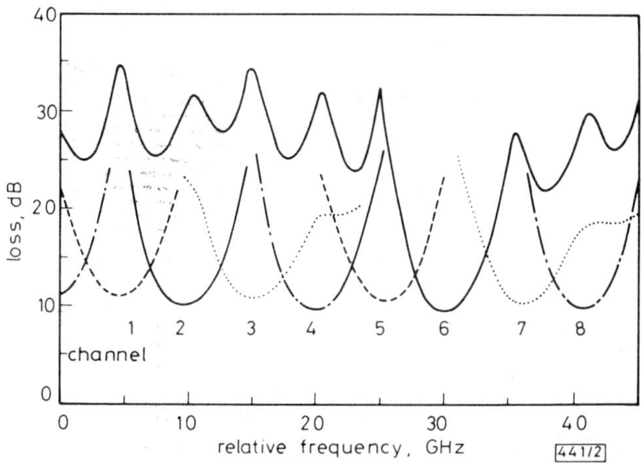

Fig. 2 *Demultiplexing characteristics of 5 GHz-spaced, eight-channel demultiplexer*

Experimental results: Demultiplexing characteristics of the eight-channel demultiplexer are shown in Fig. 2. A linearly polarised lightwave from a 1·53 μm DFB-LD was the light source, and it was coupled to the demultiplexer in TM mode through a PANDA fibre. The output power of the demultiplexer is again coupled to a PANDA fibre. The temperature of the demultiplexer was stabilised to within 0·02 deg. The excess losses from the input fibre to the output fibre were 9·8 ~ 11·2 dB (average of 10·4 dB). The crosstalk between channels was −8·3 ~ −25·8 dB (average of −16·9 dB). The lowest crosstalk of −17·8 ~ −25·2 dB was measured for channel 6. Degradation of the crosstalk is caused by slight deviations in the equal power-splitting ratio for the periodic filters.[6]

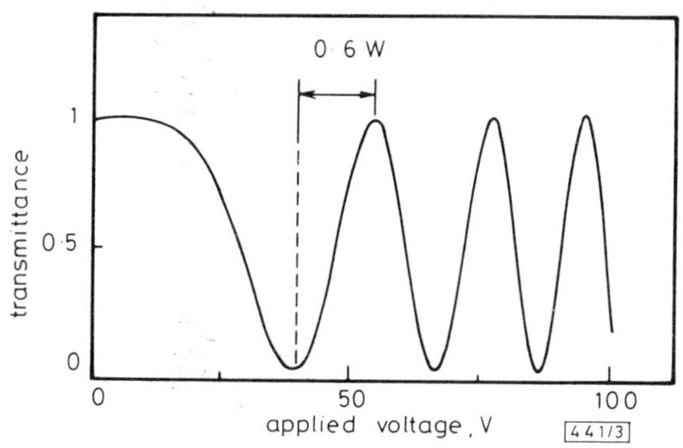

Fig. 3 *Transmittance of periodic filter with Δf of 20 GHz against voltage applied to Ni–Cr thin-film heater*

Resistance of heater $R = 2{\cdot}5\,k\Omega$

Tuning characteristics of the 20 GHz-spacing periodic filter against heater voltage are shown in Fig. 3. The heater resistance was 2·5 kΩ. The power required to change the transmittance from maximum to minimum (or vice versa) was 0·6 W at every point. When the heater changes the transmittance from minimum to maximum, the demultiplexer works as a frequency selection switch (FS-SW) which selects a single arbitrary frequency from densely packed multiplexed frequencies.[2] The pulse response characteristic of the FS-SW is shown in Fig. 4. The transit time for switching was 0·8 ms.

Conclusion: A 5 GHz-spaced, eight-channel, guided-wave tunable multi/demultiplexer was investigated. The average loss was 10·4 dB and the average crosstalk was −16·9 dB. Frequency tuning was performed by means of Ni–Cr thin-film heaters loaded on the waveguides. When the demultiplexer was used as an FS-SW, the applied power to the heater for switching was 0·6 W and the switching time was 0·8 ms. The

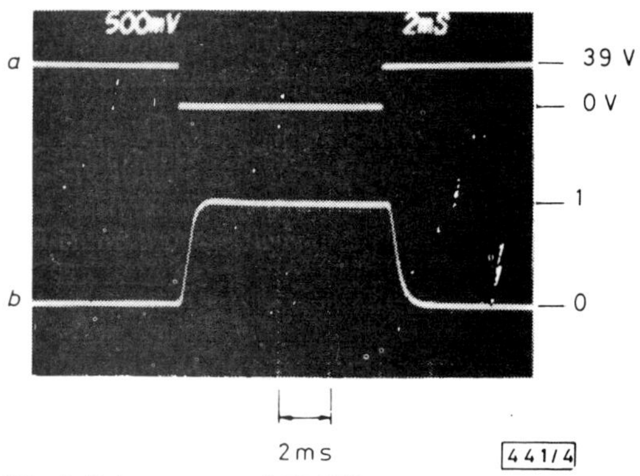

Fig. 4 *Pulse response of FS-SW output*

a Drive voltage *b* Optical output

FS-SW will be a key device in the receivers of optical FDM distribution systems using the conventional intensity modulation/direct-detection scheme.

Acknowledgment: The authors wish to thank T. Miyashita and M. Kawachi for their encouragement, and K. Inoue for his participation in useful discussions.

H. TOBA
K. ODA
N. TAKATO
K. NOSU

NTT Electrical Communications Laboratories
Take 1-2356, Yokosuka-shi, Kanagawa-ken 238-03, Japan

26th May 1987

References

1 INOUE, K., TOBA, H., and NOSU, K.: 'Tunable optical multi/demultiplexer for optical FDM transmission system', *Electron. Lett.*, 1985, **21**, pp. 387–389

2 TOBA, H., INOUE, K., and NOSU, K.: 'A conceptual design on optical frequency-division-multiplexing distribution systems with optical tunable filters', *IEEE J. Sel. Areas Commun.*, 1986, **SAC-4,** pp. 1458–1467

3 TAKATO, N., INOUE, K., JINGUJI, K., YASU, M., TOBA, H., and KAWACHI, M.: 'Guided-wave multi/demultiplexer for optical FDM transmission'. Proc. 12th ECOC '86, 1986, pp. 443–446

4 HARUNA, M., and KOYAMA, J.: 'Thermo-optic waveguide interferometric modulator/switch in glass', *IEE Proc. H, Microwaves, Opt. & Antennas*, 1984, **131,** pp. 322–324

5 KAWACHI, M., TAKATO, N., JINGUJI, K., and YASU, M.: 'Birefringence control in high-silica single-mode channel waveguides on silicon'. Proc. OFC/IOOC '87, 1987, p. 125

6 INOUE, K., TAKATO, N., TOBA, H., and KAWACHI, M.: 'A 4 channel optical waveguide multi/demultiplexer for 5 GHz spaced optical FDM transmission', *J. Lightwave Technol.*, to be published

5 GHz-SPACED, EIGHT-CHANNEL OPTICAL FDM TRANSMISSION EXPERIMENT USING GUIDED-WAVE TUNABLE DEMULTIPLEXER

Indexing terms: Optical communications, Multiplexing, Optical multi/demultiplexers

A 5 GHz-spaced, eight-channel optical FDM transmission experiment using a guided-wave tunable demultiplexer is presented. 400 Mbit/s intensity-modulated optical carriers are multiplexed and transmitted through 13 km single-mode fibre. One of the carriers is selected by the tunable demultiplexer and directly detected.

Introduction: Optical FDM systems permit the transmission of a large number of carriers. Heterodyne detection is often used at receivers to select one of the multiplexed carriers.[1,2] When an optical demultiplexer is used for carrier selection at the receiver, a conventional IM-DD scheme can be constructed.[3] The authors have previously presented an eight-channel guided-wave demultiplexer with frequency spacing of 5 GHz.[4] In this demultiplexer frequencies are tuned by changing the waveguide refractive index. In this way, it works as a frequency-selection switch (FS-SW), which is utilised in FDM information distribution systems.

This letter reports a 5 GHz-spaced, eight-channel FDM transmission experiment using the tunable guided-wave demultiplexer.

Experimental system configuration: A schematic diagram of the optical FDM transmission experiment is shown in Fig. 1. Eight DFB LDs with wavelengths of around 1·53 μm were used as optical sources. Their temperature was stabilised within 0·01°C and frequencies were maintained at 5 GHz intervals. Each output is launched to a PANDA fibre through

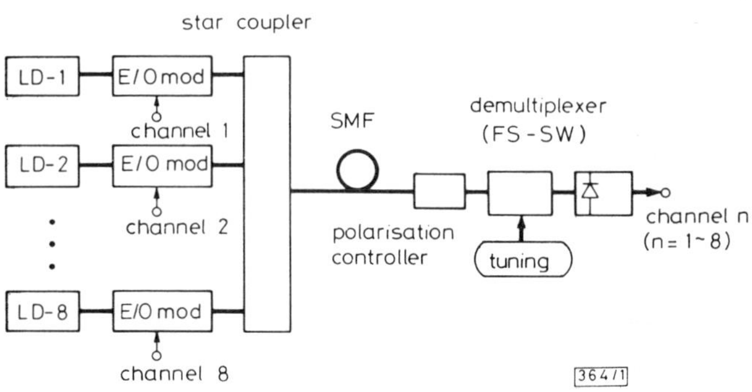

Fig. 1 *Schematic diagram of optical FDM distribution transmission experiment*

the isolator with approximately 50 dB isolation. Each fibre output is coupled to an $LiNbO_3$ intensity modulator and modulated by a 400 Mbit/s NRZ signal. Then lightwaves are multiplexed by a star coupler and coupled to a transmission fibre. The polarisation states of all output signals are aligned by rotating connectors at the input of the star coupler. A 13 km single-mode fibre is used as the transmission fibre. At the fibre output, the polarisation state is regulated by a polarisation controller, and the lightwave is coupled to the demultiplexer. The polarisation controller is used to convert the polarisation state of the lightwave to the TM mode of the demultiplexer because of stress-induced birefringence in the waveguides.[5]

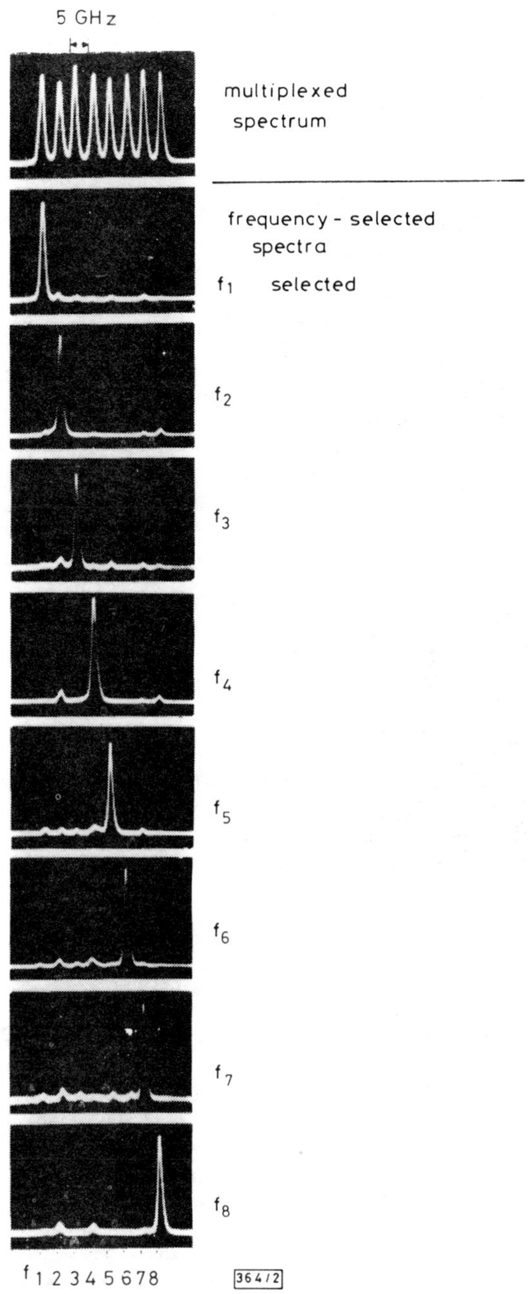

Fig. 2 *Frequency selection characteristics measured by Fabry–Perot interferometer*

The demultiplexer consists of three-step cascaded periodic filters in an asymmetrical Mach–Zehnder interferometer configuration.[6] It was fabricated on a silicon substrate by flame hydrolysis deposition of high-silica glass and reactive ion etching.[7] In this experiment, the demultiplexer works as an FS-SW, utilising its frequency tuning characteristics. Thus, only one output port was used. Excess loss of the FS-SW was 10·3 dB and crosstalk between channels was −17 to −25 dB.[4] Frequency tuning was performed by changing the voltage applied to an Ni–Cr thin-film heater loaded on the waveguides which make up the periodic filters, and thereby effecting a phase shift. The output signal of the FS-SW was directly detected by a GaInAs APD followed by an FET preamplifier.

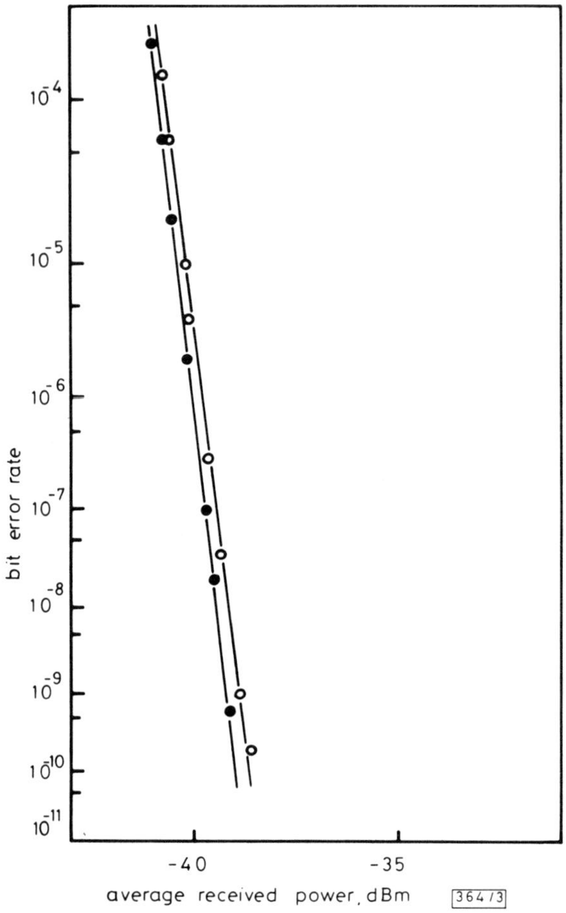

Fig. 3 *Bit error rate characteristics*

○ with interference ● without interference

Experimental results: A multiplexed spectrum and frequency-selected spectra were measured using the Fabry–Perot interferometer and are shown in Fig. 2. The top trace shows the multiplexed spectrum. The lower traces show the output as the FS-SW was successively tuned from f_1 to f_8. This corresponds to a 35 GHz tuning range. Bit error rate characteristics for the 400 Mbit/s NRZ pseudorandom signal after 13 km transmission, for f_6, are shown in Fig. 3. Open circles show the error rate characteristics when all frequencies were transmitted. Closed circles show those when only f_6 was transmitted. A bit error rate of 10^{-9} was obtained at a received power level of $-39{\cdot}2$ dBm when only f_6 was transmitted. It was degraded by 0·3 dB when all frequencies were transmitted while the

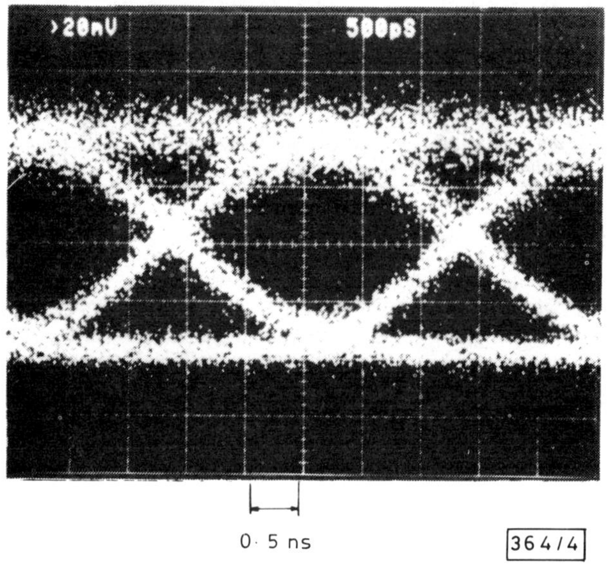

Fig. 4 *Received eye pattern after 13 km transmission with interference power*

total crosstalk level was 12 dB below the signal level. The received eye pattern after 13 km transmission with interference power is shown in Fig. 4.

Conclusions: An eight-channel optical FDM transmission experiment with channel spacing of 5 GHz was carried out using a tunable, periodic filter type, guided-wave demultiplexer. The demultiplexer's tunable range over 35 GHz was confirmed. Receiver sensitivity degradation of 0·3 dB appeared for 400 Mbit/s NRZ IM signal transmission through a 13 km single-mode fibre because of the crosstalk between channels.

Acknowledgment: We thank Dr. G. Motosugi for providing DFB LDs, and Dr. M. Kawachi for his encouragement.

H. TOBA
K. ODA
K. NOSU

NTT Laboratories
1-2356 Take, Yokosuka-shi
Kanagawa-ken 238-03, Japan

N. TAKATO

NTT Laboratories
Tokai, Ibaraki-ken 319-11, Japan

H. MIYAZAWA

NTT Laboratories
3-9-11, Midori-cho, Musashino-shi
Tokyo 180, Japan

5th November 1987

References

1 BACHUS, E. J., BRAUN, R. P., CASPER, C., GROSSMANN, E., FOISEL, H., HEIMES, K., LAMPING, H., STREBEL, B., and WESTPHAL, F. J.: 'Ten-channel coherent optical fibre transmission', *Electron. Lett.*, 1986, **22,** pp. 1002–1003

2 GLANCE, B., POLLOCK, K., BURRUS, C. A., KASPER, B. L., EISENSTEIN, G., and STULZ, L. W.: 'Densely-spaced WDM coherent star network', *ibid.*, 1987, **23,** pp. 875–876

3 TOBA, H., INOUE, K., and NOSU, K.: 'A conceptional design on frequency-division-multiplexing distribution systems with optical tunable filters', *IEEE J. Sel. Areas Commun.*, 1986, **SAC-4,** pp. 1458–1467

4 TOBA, H., ODA, K., and NOSU, K.: '5 GHz-spaced, eight-channel, guided-wave tunable multi/demultiplexer for optical FDM transmission systems', *Electron. Lett.*, 1987, **23,** pp. 788–789

5 KAWACHI, M., TAKATO, N., JINGUJI, N., and YASU, M.: 'Birefringence control in high-silica single-mode channel waveguides on silicon'. Proc. OFC/IOOC '87, 1987, p. 125

6 INOUE, K., TOBA, H., and NOSU, K.: 'Tunable optical multi/demultiplexer for optical FDM transmission systems', *Electron. Lett.*, 1985, **21,** pp. 387–389

7 TAKATO, N., INOUE, K., JINGUJI, K., YASU, M., TOBA, H., and KAWACHI, M.: 'Guided-wave multi/demultiplexer for optical FDM transmission'. Proc. 12th ECOC '86, 1986, pp. 443–446

DENSELY SPACED WDM COHERENT OPTICAL STAR NETWORK

Indexing terms: Optics, Star couplers

An optical WDM star network consisting of three lasers transmitting at about 234 000 GHz, spaced by 300 MHz, has been used to demonstrate dense packing of WDM signals. The three optical signals, FSK-modulated at 45 Mbit/s, are multiplexed by a 4 × 4 star coupler and demultiplexed by a balanced heterodyned receiver. Receiver sensitivity is −61 dBm for a BER of 10^{-9}, or 113 photons/bit, which is 4·5 dB from the shot noise limit and represents the best sensitivity yet reported for FSK modulation. The results indicate that 100 000 users in a 10 km radius could be interconnected with such a system.

A coherent detection, wavelength-division-multiplexed, optically star-coupled system offers several advantages in interconnecting large numbers of users:[1–3] (i) WDM allows use of the wide available optical bandwidth;[4] (ii) coherent detection provides both high receiver sensitivity and high frequency selectivity; and (iii) the star coupler distributes power efficiently.[5] We demonstrate experimentally that an optical network with large throughput (number of users × bit rate) can result from combining these advantages. The demonstration system consists of three optical channels spaced by the minimum frequency difference necessary to prevent cochannel interference in the IF domain. Our results indicate that 100 000 users within a radius of 10 km can be interconnected with such a system, where injection-laser sources are FSK-modulated at 45 Mbit/s, the channels are spaced by 300 MHz and a received optical power of −61 dBm (113 photons/bit) is sufficient for a BER of 10^{-9}. This represents a receiver sensitivity 4·5 dB above the shot noise limit, the best sensitivity yet reported for FSK modulation.

The experimental optical sources are fast frequency-tunable external-cavity lasers.[6] They provide a narrow-line single-frequency output at 1·28 μm that can be modulated up to 100 MHz. Three such lasers, FSK-modulated at 45 Mbit/s (index about one) with a random NRZ bit stream, were used to generate three optical signals spaced by 300 MHz. The three signals were multiplexed by a 4 × 4 star coupler. Each of the four output fibres transmitted the three WDM signals. The WDM signals from one of these fibres were combined (3 dB coupler) with the LO signal from a conventional external-cavity laser[7] (Fig. 1), and the polarisation of each of the three WDM signals was adjusted to match that of the LO. The combined signals from the 3 dB coupler outputs were then fed to a balanced receiver[8] that heterodyned the received signals to an IF frequency of 225 MHz. This IF signal was amplified, filtered (60 MHz BW), demodulated by a frequency discriminator and subjected to a lowpass filter (3 dB cutoff at 35 MHz). Channel selection was accomplished by tuning the LO frequency to place the channel within the IF filter bandwidth. Finally, the resulting IF frequency was maintained by an AFC circuit on the LO.

In this arrangement, the desired optical channel is heterodyned by the LO signal frequency-tuned at the edge of an adjacent optical channel. Thus the channel selected by the IF filter is the second lowest IF channel, an arrangement that minimises the frequency spacing required between the optical channels and separates the demodulated baseband signal from its IF signal. For 45 Mbit/s modulation the minimum optical frequency spacing without cochannel interference in the IF domain is about 260 MHz. The channel spacing in the experiments is 300 MHz, so that the spacing between the interleaved

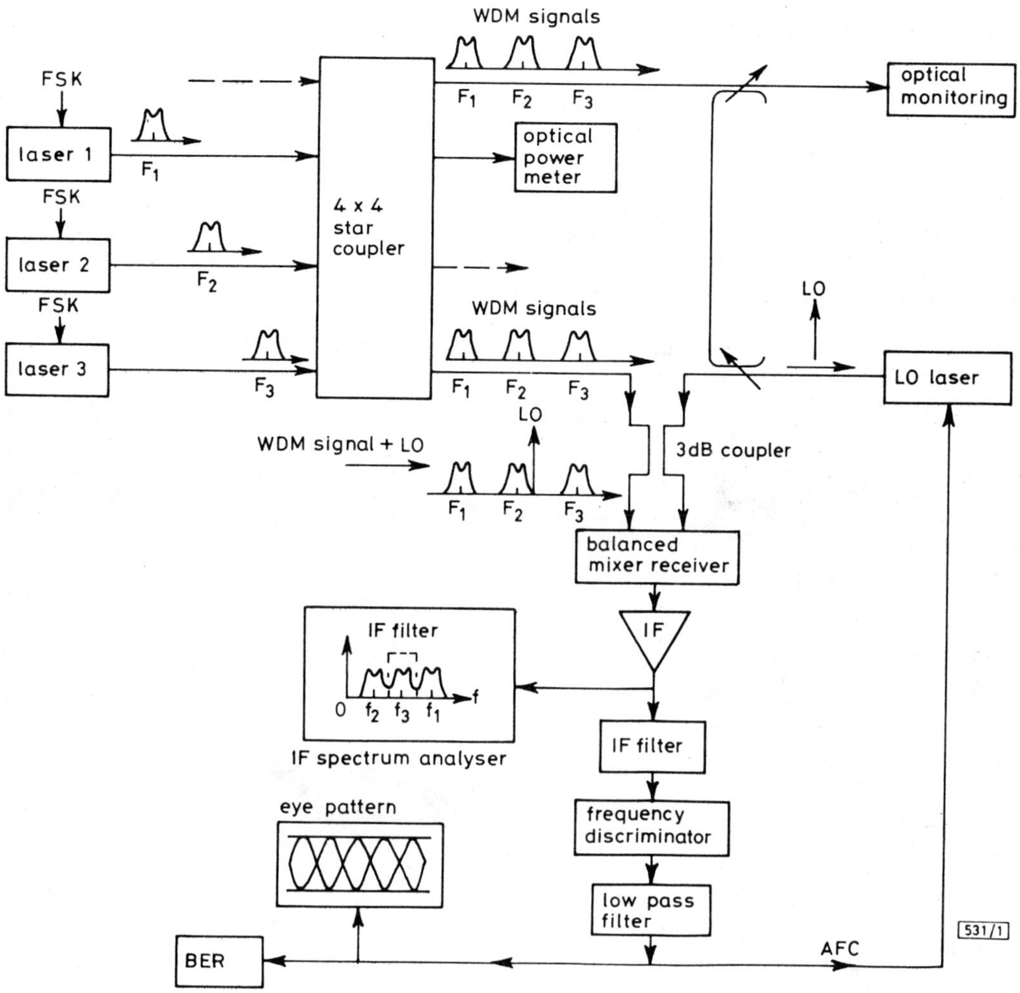

Fig. 1 *Schematic diagram of WDM network used for making the measurements*

IF channels is 150 MHz. The results are illustrated in Fig. 2, which shows the IF power spectra of the three channels, heterodyned from an optical frequency of about 234 000 GHz and selected successively by the IF filter centred at 225 MHz. The first three cases show the selection, respectively, of channel 1, channel 2 and channel 3, where channel 3 is interleaved between two adjacent channels spaced by 150 MHz, as would occur in a system having a large number of channels. The fourth case shows the selection of channel 3 for an LO frequency yielding an IF channel distribution reversed relative to that given by the selection of channel 1. Shown beside each IF channel spectrum distribution is the corresponding optical signal power spectrum and the frequency position of the LO signal yielding the resulting IF spectra.

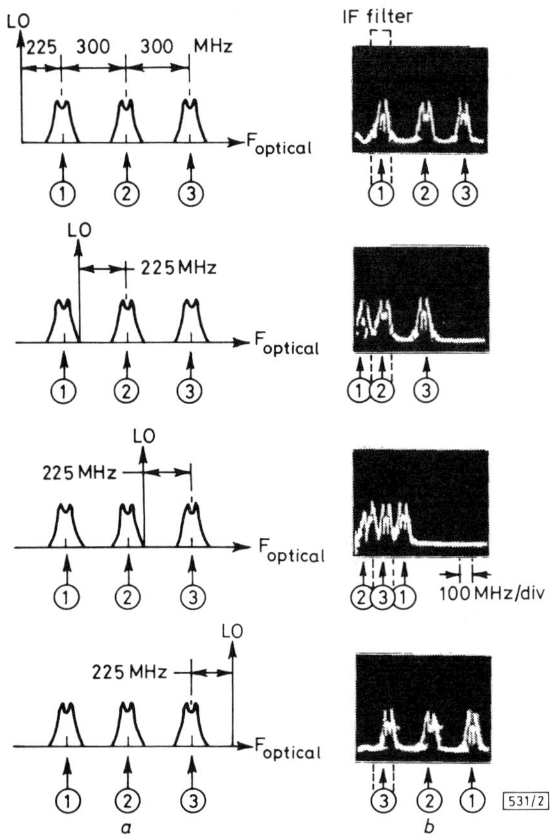

Fig. 2 *Power spectra of the three IF channels corresponding to the LO frequency tuning illustrated with the measured IF spectra*

a Optical domain; 234 000 GHz
b IF domain; IF = 225 MHz, 45 Mbit/s

For one channel, the minimum received optical power needed to obtain a BER of 10^{-8} is $-61{\cdot}6$ dBm (Fig. 3). Extrapolating this result to a BER of 10^{-9} (actual measurement of which requires an excessive counting time of 45 Mbit/s), yields a received optical power of -61 dBm, corresponding to the optical power at the output of the 4×4 star coupler.

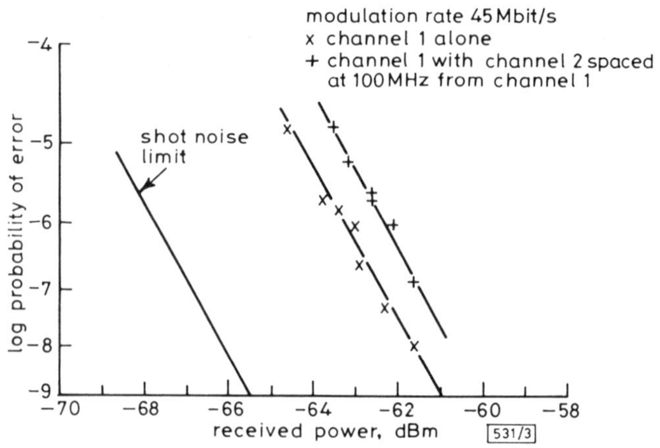

Fig. 3 *BER against received optical power for one channel alone, and for one channel in presence of an interfering channel spaced by 100 MHz in IF domain*

However, it does not include the additional excess loss due to the 3 dB coupler and the fibre connectors (totalling about 0·2 dB). Thus, in terms of the average number of photons/bit, the receiver sensitivity is 113 photons/bit, about 4·5 dB above the shot noise limit of an ideal FSK optical receiver. The power penalty from cochannel interference is 1 dB when the selected channel is within 100 MHz of an adjacent channel,

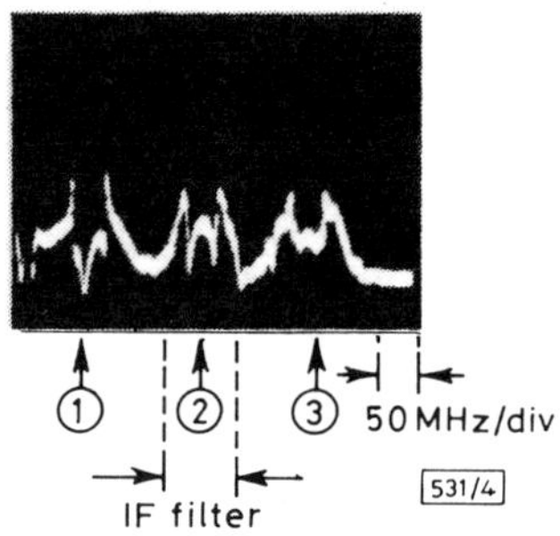

Fig. 4 *IF power spectrum of three interleaved channels, each received with $-61{\cdot}6$ dBm, modulated at 45 Mbit/s*

IF = 225 MHz, optical power/channel = $-61{\cdot}6$ dBm, 45 Mbit/s

has equal power, and the modulation rate is 45 Mbit/s. This result is illustrated in Fig. 3. No cochannel degradation is observed when the selected channel is interleaved between two adjacent channels spaced by 150 MHz. Fig. 4 shows the measured IF spectrum for this case, with the three channels at $-61{\cdot}6$ dBm and modulated at 45 Mbit/s. From the above results the throughput of a system operating at 1·55 μm wavelength can be estimated. We assume that a user transmits 0 dBm of optical power into a fibre connected to a star coupler interconnecting 2^{17} subscribers, and that the excess loss of the signal propagating through the coupler is proportional to the 17 stages of 3 dB couplers constituting the star (commercial versions of which have about 0·1 dB excess loss each). To the resulting 2 dB excess loss of the coupler, we add 4 dB of fibre loss for a system with a 10 km radius, to obtain a power level of −57·2 dBm optical power received by each subscriber. This value is 3·8 dB larger than necessary to obtain a BER of 10^{-9} at 45 Mbit/s. Thus 100 000 subscribers can be interconnected within a radius of 10 km by the WDM star network.

B. GLANCE
K. POLLOCK
C. A. BURRUS
B. L. KASPER
G. EISENSTEIN
L. W. STULZ

AT&T Bell Laboratories
Crawford Hill Laboratory
Holmdel, NJ 07733, USA

11th June 1987

References

1 BACHUS, E. J., BRAUN, R. P., CASPAR, C., GROSSMANN, E., FOISED, H., HEIMES, K., LAMPING, H., STREBEL, B., and WESTPHAL, F. J.: 'Ten-channel coherent optical fibre transmission', *Electron. Lett.*, 1986, **22,** pp. 1002–1003
2 CHRAPLYVY, A. R., and TKACH, R. W.: 'Narrowband tunable optical filter for channel selection in densely packed WDM systems', *ibid.*, 1986, **22,** pp. 1084–1085
3 LINKE, R. A.: 'Beyond gigabit-per-second transmission rates'. OFC sixth international conference, Reno, 19–22 Jan. 1987, Tech. digest, p. 184
4 HENRY, P. S., LINKE, R. A., and GNAUCK, A. A.: 'Introduction to lightwave systems', *in* MILLER, S. E., and KAMINOW, I. P. (Eds): 'Optical fiber telecommunications II' (Academic Press, NY, 1987), Chap. 21
5 SALEH, A. A. M., and KOGELNIK, H.: 'Reflective, single-mode, fiber optic, passive star couplers'. TM 11344-861031-14
6 GLANCE, B., BURRUS, C. A., and STULZ, L. W.: 'Fast frequency-tunable external-cavity laser', *Electron. Lett.*, 1987, **23,** pp. 98–99
7 WYATT, R., and DELVIN W. J. JR.: '10 kHz linewidth 1·5 μm InGaAsP external cavity laser with 55 mm tuning range', *ibid.*, 1983, **19,** pp. 110–112
8 KASPER, B. L., BURRUS, C. A., TALMAN, J. R., and HALL, K. L.: 'Balanced dual-detector receiver for optical heterodyne communication at Gbit/s rates', *ibid.*, 1986, **22,** pp. 413–415

Chapter 6
Integrated Optics and Optoelectronics

Integrated optics and integrated optoelectronics are technologies which have received considerable attention in research laboratories, but have not yet had much of an effect on fiber optics systems. However, these technologies have considerable potential for future application (for example, for cost reduction in high–data–rate systems). Integrated optics is concerned with the development of optical waveguide components on planar substrates for performing functions such as directional coupling, modulation, and switching. The most commonly used substrate material is lithium niobate ($LiNbO_3$). The current technology is reviewed by **E. Voges and A. Neyer** in **"Integrated-Optic Devices on $LiNbO_3$ for Optical Communication"**, and application in a communication experiment is described by **S.K. Korotky, A.H. Gnauck, B.L. Kasper, J.C. Campbell, J.J. Veselka, J.R. Talman, and A.R. McCormick** in **"8-Gbit/s Transmission Experiment over 68 km of Optical Fiber using a Ti:$LiNbo_3$ External Modulator"**. Integrated optoelectronics is concerned with combining optical components (such as lasers and photodetectors) with electronic elements (such as amplifiers and multiplexers) on a single semiconductor substrate. Gallium arsenide (GaAs) and indium phosphide (InP) are commonly used substrate materials. The current developments in this field are described in the next two papers: **"Research Aiming for Future Optoelectronic Integration: The Optoelectronics Joint Research Laboratory"** by **I. Hayashi,** and **"InGaAsP/InP Long Wavelength Optoelectronic Integrated Circuits (OEICs) for High-Speed Optical Fiber Communication Systems,"** by **A. Suzuki, K. Kasahara, and M. Shikada.**

Integrated-Optic Devices on $LiNbO_3$ for Optical Communication

EDGAR VOGES AND ANDREAS NEYER

(Invited Paper)

***Abstract*—Integrated-optic components on $LiNbO_3$ for fiber-optical communication systems are reviewed with respect to technology, fiber coupling, and device configurations. Typical devices for communication applications, such as modulators/switches, multiplexers, switch arrays, and heterodyne receiver chips rely on electrooptically controlled waveguide structures. Their operation mechanisms and performance characteristics are described to highlight the present status of development and of system applications.**

I. Introduction

THE EVOLUTION of fiber-optical communications with its emphasis on high data rate single-mode fiber networks in the 1.3–1.6 μm wavelength region, and on sophisticated optical systems has stimulated the application of guided wave devices. Integrated optics which basically employ single-mode guided wave techniques are expected to contribute to rugged, cost-effective, and batch-fabricated components such as modulators/switches, multiplexers, tunable filters, frequency shifters, polarization controllers, signal routing, and timing devices.

Such functions can be realized by electrooptic waveguide components. At present the performance of $LiNbO_3$ devices is superior to comparable semiconductor electrooptic waveguide components. For about one decade, the research efforts in electrooptic $LiNbO_3$ waveguide components have resulted in a high status of development with regard to technology, device performance, and fiber coupling. This may be characterized by intensity modulators with a beyond-10-GHz modulation bandwidth and about 100-mW drive power [1]–[4], high-speed low-insertion-loss PSK-modulators for coherent systems [5], 8 × 8 nonblocking switch arrays [6], multichannel wave length multiplexers [7], [8], and 0.5-dB insertion loss of fiber-coupled 1.3-μm $LiNbO_3$ channel waveguides [9].

The reproducible fabrication of stable waveguide devices on $LiNbO_3$ by standardized technologies, and an efficient waveguide to fiber coupling are essential for system applications. The common indiffusion of titanium stripes for channel waveguides on $LiNbO_3$ ($Ti:LiNbO_3$) is first discussed with respect to technological tolerances and reproducible fabrication conditions. The attachment of fibers to $LiNbO_3$ devices requires special means if polarization preserving fibers are demanded. Generally, the electrodes of electrooptic devices must be decoupled from the evanescent field of waveguide modes. The electrode fabrication crucially influences the device stability. We discuss several electrode designs (underlying dielectric buffer layers, transparent indium-tin-oxide (ITO) electrodes, and buried $Ti:LiNbO_3$ waveguides by an additional MgO indiffusion).

Electrooptic guided wave devices such as intensity modulators, PSK-modulators, and multiplexers are utilized for terminal functions, whereas switched branching structures and switch arrays are examples of network functions. Polarization control or polarization independent properties are required for network components and receiver terminal components if standard communication fibers are utilized. Application areas and state-of-the-art performance characteristics of electrooptic $LiNbO_3$ devices are illustrated by several selected examples.

II. Device Fabrication and Fiber Coupling

Electrooptic waveguide devices on $LiNbO_3$ are fabricated by what are in principle simple technologies. Commonly, thin titanium films ($\tau \approx$ 50–150 nm) are evaporated or sputtered and photolithographically delineated on the $LiNbO_3$ wafer. Thermal indiffusion at temperatures $T \sim$ 980–1050°C leads to a desired channel waveguide configuration [10]–[12]. Thereafter, metal electrodes are deposited on top or beside the waveguides to generate the modulating electric fields for electrooptic functions such as phase shifting, switching, and polarization conversion.

For a reproducible fabrication of uniform and high performance devices, several problems have to be overcome, and a number of critical tolerances have to be met.

The indiffusion of titanium stripes is often accompanied by a parasitic film waveguide at the surface of the crystal due to an increase of the extraordinary index n_e. This is attributed to the outdiffusion of O_2/Li_2O [13], [14], and various means for a suppression of this parasitic film waveguide have been proposed [14]–[18]. The addition of water vapor [17], [18] to the gas flow is now a common technique in many laboratories. The immediate surrounding of the $LiNbO_3$ sample is of critical importance, also [14]. We achieved reproducible results by placing the $LiNbO_3$ substrate in a closed platinum box (Fig. 1) [19]. The platinum box provides a chemically clean surrounding for the crystal, and yields an excellent uniformity of the indiffusion process due to a high temperature homo-

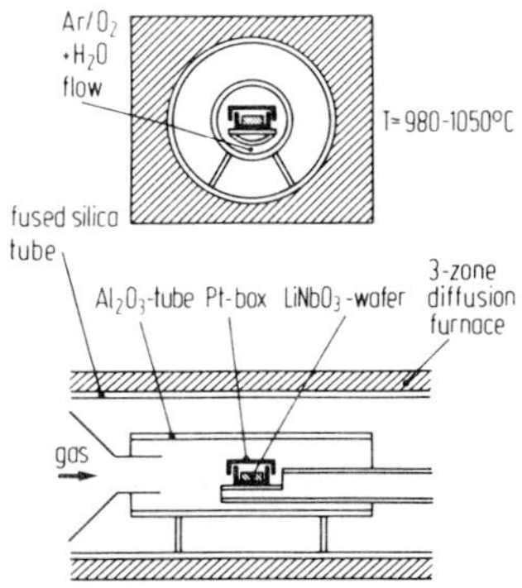

Fig. 1. Schematic arrangement of titanium diffusion with the $LiNbO_3$ wafer surrounded by a platinum box.

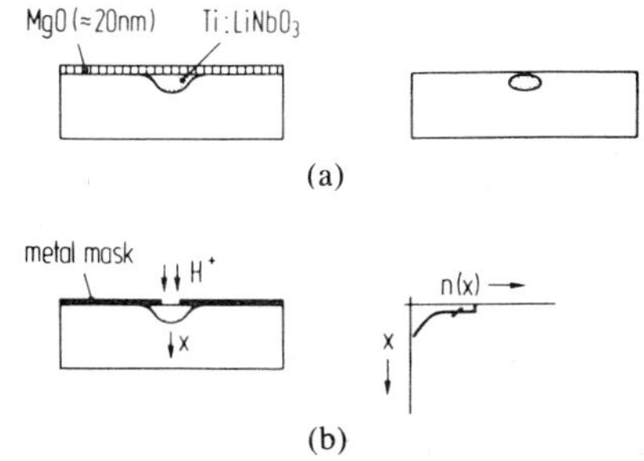

Fig. 2. Waveguide tuning. (a) Indiffusion of MgO ($\tau \approx$ 10–30 nm, 900°C, 2-h diffusion time, $\Delta n < 0$) leads to buried channel guides with reduced losses due to surface scattering and to an almost circular near-field pattern (b) Proton exchange strongly increases n_e near the surface and changes, for example, the dispersion, the birefringence, and the near field pattern of the channel guide. ($\Delta n_e/n_e \approx +6$ percent, $\Delta n_o \approx -2$ percent.)

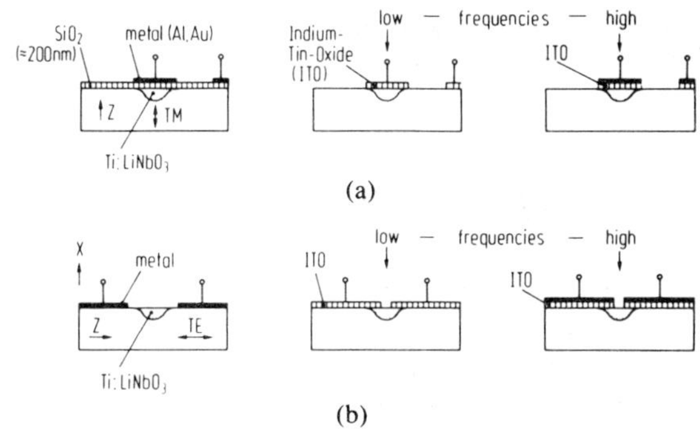

Fig. 3. Electrode configurations and electrode materials. (a) Electrooptic modulation on Z-cut $LiNbO_3$ with SiO_2 buffer layers or transparent ITO electrodes. (b) Electrooptic modulation on X-cut $LiNbO_3$ with metal electrodes beside the channel waveguide or with ITO-electrodes.

geneity. (No outdiffused waveguide is observed at 0.63-μm wavelength even without water vapor.)

For uniform results, tight photomask (<0.25 μm) and alignment (<0.5 μm) tolerances are required. The evaporation vacuum should be better than 10^{-6} torr for a reproducible titanium mass density, and the diffusion temperature must be controlled within better than ± 0.5°C. The control of the thickness and uniformity of the titanium film imposes a severe problem, and tolerances must be below 1 percent [20]–[22]. Fluctuations of the titanium stripe dimensions and variations of the Li/Nb stoichiometry influence for example the coupling length of directional couplers, too [20], [21].

The photorefractive effect [23], [24] severely limits the optical power at shorter wavelength (e.g., $\lambda \approx 0.8$ μm) to several microwatts, only. At $\lambda \gtrsim 1.3$ μm, however, no significant degradation of electrooptic devices is predicted for power levels up to about 5 mW [25]. Therefore, $Ti:LiNbO_3$ devices can well handle typical transmitter powers in the wavelength region of 1.3–1.6 μm.

Titanium-indiffused ($Ti:LiNbO_3$) channel waveguides can be tuned by an additional indiffusion of MgO [9], [26] or by an additional proton exchange [27], [28] (see Fig. 2). The MgO indiffusion buries the $Ti:LiNbO_3$ channel waveguide below the surface and shapes its index profile. This reduces the waveguide attenuation due to surface scattering, enhances the coupling efficiency to fibers because of an improved mode matching [9], and decouples metal electrodes from the evanescent fields of the waveguide modes. A proton exchange strongly increases the extraordinary index n_e ($\Delta n_e/n_e \approx 6$ percent) and reduces the ordinary index n_0 by $\Delta n_0/n_0 \approx -2$ percent. Therefore, the combination of a titanium indiffusion and a subsequent low-temperature proton exchange in so-called TIPE-waveguides [29] can be utilized to tune the birefringence or the waveguide dispersion [30], [31] of $LiNbO_3$ channel waveguides. Furthermore, efficient polarizers on Z-cut $LiNbO_3$ can be fabricated by this technique [32].

The electrode configurations and the electrode materials strongly affect the performance and stability (drift) of electrooptic $Ti:LiNbO_3$ devices. Electrode designs for devices on Z-cut and X-cut $LiNbO_3$ are depicted in Fig. 3. (Devices on Y-cut $LiNbO_3$ are not included since they are inferior to devices on X-cut material because of their low coupling efficiency to fibers due to a strongly elliptical near field pattern, and because of their stronger tendency to exhibit outdiffused waveguides. Y-cut $LiNbO_3$ is useful in case of SAW acoustooptic interactions [33].)

Devices on Z-cut $LiNbO_3$ (see Fig. 3(a)) show excellent mode matching to connecting fibers, and allow a simple push–pull operation of interferometer modulators and of directional coupler switches [34]–[35]. TM-polarization and the vertical electrical field yield the strongest electrooptic effect. The evanescent field of the waveguide mode must be decoupled from the overlying metal electrodes by an intermediate low-index buffer layer to avoid light absorption by the metal. Usually, SiO_2 buffer layers are sputter-desposited, or CVD-fabricated. The properties of the buffer layer and the substrate/film interface critically influence the short- and long-term stability of electrooptic devices [36]. A reproducible fabrication of drift-free devices still imposes difficulties. Transparent indium-tin-oxide (ITO) electrodes [37]–[39] do not require buffer layers. They improve the device stability and lower the modulation voltage.

The near field distribution of $Ti:LiNbO_3$ channel waveguides on X-cut material can also be sufficiently matched to the mode field of connecting fibers. Again, the modulating electric field and the polarization of the optical wave (TE) should both be parallel to the Z-axis to use the largest electrooptic coefficient r_{33}. In this case, the electrodes are positioned along the waveguide. If the electrode gap is sufficiently large, no buffer layer is necessary. Otherwise, ITO electrodes can be used. In both cases, the fabrication is simplified and drift effects are reduced.

The choice of the crystal orientation, *Z*-cut or *X*-cut, depends on the particular device. For example, directional coupler switches are advantageous on *Z*-cut material, whereas phase modulators or *X*-switches [40] are better fabricated on *X*-cut material. The electrical-optical overlap factor, which determines the required modulation voltage, must also be evaluated carefully [41], [42].

Recently, electrooptic components on *X*- or *Y*-cut material with wave propagation along the *Z*-axis have been of interest [43]–[46]. They require higher voltages due to the relatively small electrooptic coefficients r_{12}, r_{22}. They allow, however, the construction of optically broad-band polarization transformers and polarization-independent couplers.

Communication applications of integrated-optic devices necessitate practical techniques for a permanent attachment of single-mode fibers. Low insertion losses of fiber-coupled devices are of fundamental importance. The following mechanisms contribute to the insertion loss of fiber-coupled $LiNbO_3$ devices:

- waveguide attenuation which is now below 0.2 dB/cm at 1.3–1.6 μm wavelengths,
- device excess losses due to bends, *Y*-branches, or electrodes amounting to typically 0.1–0.5 dB per device,
- Fresnel reflection losses of about 0.15 dB/coupling without antireflection coating. Return loss of 35 dB is achieved for antireflection coated devices with a single Y_2O_3 layer [47],
- mode mismatch.

Near 100-percent coupling efficiencies demand a close matching of the channel waveguide mode to the fiber mode and submicrometer alignment tolerances. Fig. 4(a) shows the array coupling of single-mode fibers. Standard communication fibers are aligned in closely spaced, anisotropically etched silicon V-grooves [48]–[51]. Fiber arrays with up to 18 fibers have been coupled to $LiNbO_3$ switch arrays with an average excess loss of about 0.3 dB/interface [52]. Flattened cladding polarization maintaining or polarizing fibers [53], [54] automatically orient the polarization axis in rectangular grooves (Fig. 4(b)). An average excess insertion loss of <0.3 dB has been achieved in this case [55]. Ti : $LiNbO_3$ channel waveguides on *Z*- and *X*-cut material have a slightly elliptical near field pattern with a depth-to-width-ratio of roughly 1 : 1.1 to 1 : 1.5. This mode distribution can be well matched to the elliptical mode distribution of polarization-maintaining fibers.

Low-loss array coupling requires highly concentric fibers. Otherwise the fibers must be attached one by one with UV-curing epoxy (see Fig. 4(c)). Insertion losses as low as 0.5 dB have been obtained for fiber-coupled channel waveguides of 2-cm length [9]. Fig. 5 demonstrates that insertion losses below 1.5 dB can be achieved over a broad wavelength range for Ti : $LiNbO_3$ channel waveguides coupled to polarization maintaining fibers [56].

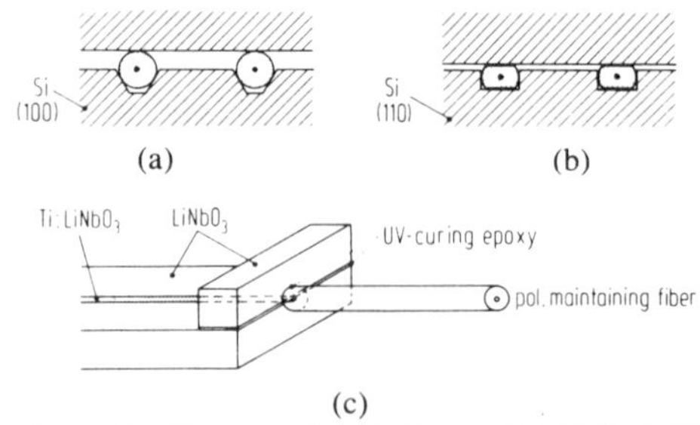

Fig. 4. Techniques for fiber coupling to integrated-optic waveguides. (a) Array coupling of standard communication fibers. (b) Array coupling of flattened cladding polarization maintaining fibers. (c) Bonding of single fibers with UV-curing epoxy.

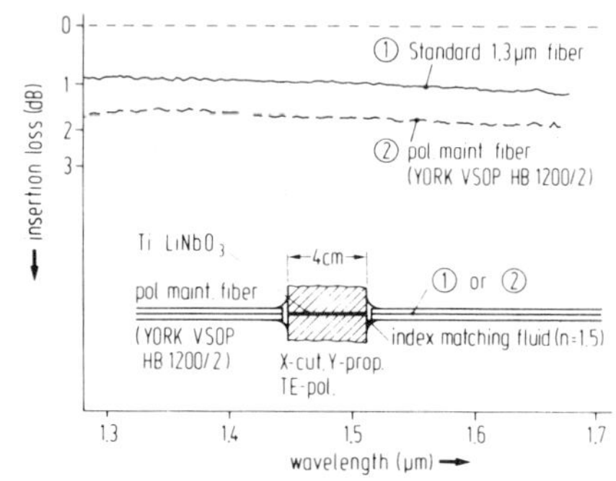

Fig. 5. Insertion loss of a fiber-coupled Ti : $LINbO_3$ channel guide (7-μm width of the Ti-stripe, τ = 100 nm as a function of wavelength.

III. Ti : $LiNbO_3$ Devices For Communication Systems

Terminal and network functions of electrooptic $LiNbO_3$ waveguide devices are demonstrated for an assessment of their present abilities to provide high-speed operation, low insertion loss, and high-performance characteristics of complex electrooptic waveguide configurations.

A. High-Speed Modulators

External $LiNbO_3$ intensity modulators for laser transmitters of direct detection fiber-optic systems offer the advantages of: large signal gigahertz-modulation-bandwidth with low modulation voltage and high on–off ratio, elimination of dynamic laser chirping, and phase modulation in coherent systems. In general, devices for transmitter terminal functions require—above all—low insertion loss and a high modulation bandwidth to compete with electronic and direct laser modulation techniques.

Phase-shift-keying (PSK) yields the highest signal-to-noise ratio in coherent systems. Here, external $LiNbO_3$ phase modulators are particularly attractive since they provide a clean phase modulation. Efficient and low-loss $LiNbO_3$ PSK-modulators are presently used in coherent system experiments [5]. Fig. 6 demonstrates the electrode configuration of a fiber-pigtailed phase modulator for 1.5-μm wavelength [5]. All electrooptic modulators with a

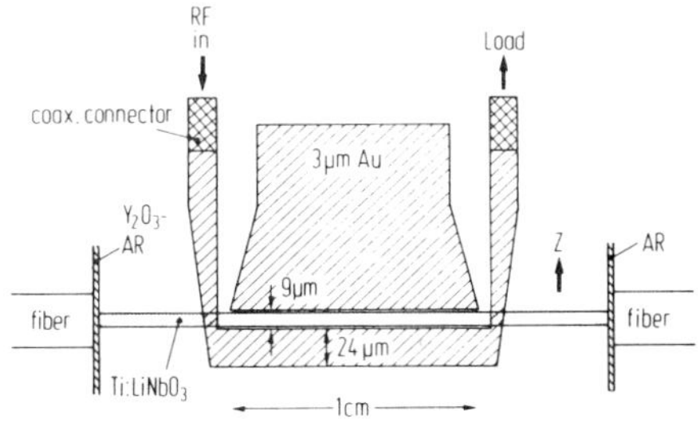

Fig. 6. Electrode configuration of a traveling-wave phase modulator on X-cut $LiNbO_3$ for 1.5-μm wavelength coherent systems [5] (1.8-dB insertion loss of fiber-pigtailed device, U_π = 8-V half-wave voltage, ≈ 5-GHz modulation bandwidth).

modulation bandwidth in excess of 1.5–2 GHz and low voltages require traveling-wave electrodes. Coplanar transmission line electrodes of 2–3 μm thickness show the best RF-performance. The coplanar lines are tapered for a connection to 50-Ω microstrip or coaxial lines. Their wave impedance is commonly chosen lower than 50 Ω to allow for broader electrode stripes with less attenuation. Alternatively, capacitively loaded 50-Ω coplanar lines are used to improve, for example, the electrical-optical overlap [58].

High-speed intensity modulators employing Mach–Zehnder interferometers [1]–[4], directional couplers (DC's) (see, for example, [59]–[63]), X-switches [64], and cutoff modulation [65] have been demonstrated on $LiNbO_3$. Integrated Mach–Zehnder interferometers yield the largest modulation bandwidth. Fig. 7(a) and (b) shows the electrode designs of interferometer modulators on Z-cut and X-cut $LiNbO_3$ for $\lambda \approx 0.8$-μm wavelength [2], [3]. One recognizes that push–pull operation is achieved with a simpler electrode configuration on Z-cut material. DC-modulators/switches have the advantage of being a four-port devices. The extra port can be used for monitoring the laser power. High-speed low-insertion-loss switches for λ = 1.3 μm and λ = 1.56 μm have been reported [61]–[63] and have been used in system experiments [63]. Fig. 8 shows a schematic view of a packaged and fiber-coupled DC-switch on Z-cut $LiNbO_3$ [63]. At λ = 1.56-μm operating wavelength, about 4-GHz RF bandwidth, 8.1-V switching voltage, 25-dB extinction ratio, and only 2.9-dB insertion loss of the connectorized device have been achieved. Low-DC-switches with a uniform electrode demand extremely tight fabrication tolerances. DC-switches with a low crosstalk are achieved by applying the $\Delta\beta$-reversal principle [66], where electrode segments are driven by voltages of opposite polarity. The concepts of $\Delta\beta$-reversal and of travelling-wave electrodes are combined by connecting the electrode segments to a coplanar transmission line as shown in Fig. 9 [61], [67].

Electrooptic modulators operate most effectively for fixed linear polarizations (TE or TM, depending on the crystal orientation). Network and receiver terminal components demand a polarization-insensitive behavior or additional polarization controllers when connected to standard communication fibers. The polarization dependence of modulators arises primarily from the different elec-

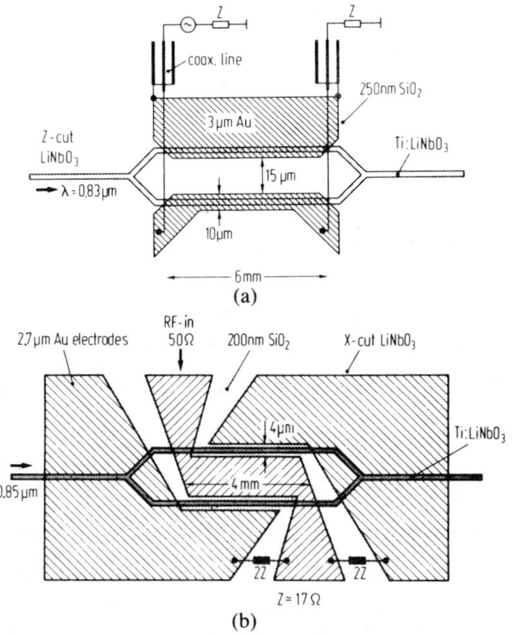

Fig. 7. (a) Schematical arrangement of a Mach–Zehnder interferometric modulator with Z = 50 Ω asymmetric coplanar line traveling-wave electrodes on Z-cut $LiNbO_3$ (± 3.5-V switching voltage, 17-GHz 3-dB modulation bandwidth, ≈ 7-dB insertion loss, λ = 0.83 μm) [2]. (b) Traveling-wave Mach–Zehnder modulator on X-cut $LiNb0_3$ with coplanar electrodes (± 2.25-V switching voltage, 16-GHz 3-dB modulation bandwidth, λ = 0.85 μm) [3].

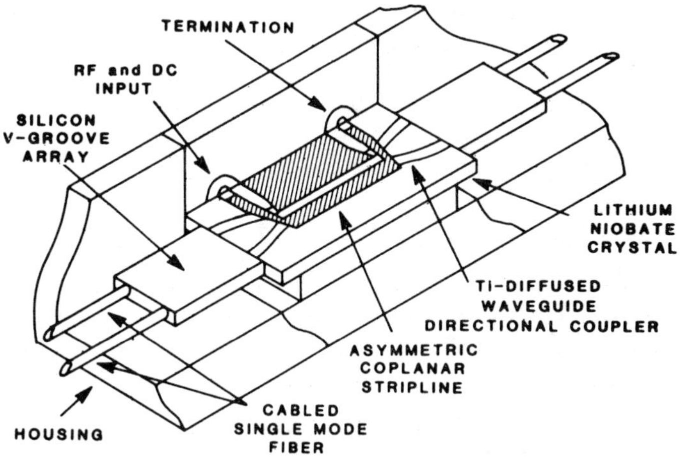

Fig. 8. Schematic arrangement of fiber-coupled packaged directional coupler switch for 1.56-μm wavelength (8.1-V switching voltage, > 4-Gbit/s switching speed, 2.9-dB insertion loss) [63].

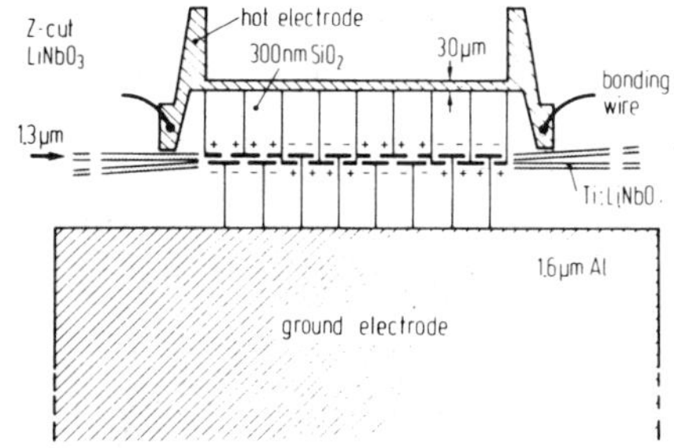

Fig. 9. Schematic drawing of a $\Delta\beta$-reversal traveling-wave directional coupler switch for 1.3-μm wavelength (± 8-V switching voltage, > 6-Gbit/s switching speed, 2-dB insertion loss) [61].

trooptic coefficients for TE- and TM-modes. Polarization insensitive Mach–Zehnder modulators on *X*-, *Y*-, and *Z*-cut $LiNbO_3$ are designed in a straightforward way by utilizing two electrode systems which almost independently influence TE- and TM-polarizations [68] (Fig. 10(a)). For a particular set of applied voltages the off state is achieved. The on state is not fully achievable and causes the device to have additional loss. Since the lower electrooptic coefficient $r_{13}(r_{13} \simeq r_{33}/3)$ is used in one electrode system, long electrodes are necessary for low-voltage operation.

The principle of weighted coupling is applied for polarization insensitive directional coupler switches [69], [70] (Fig. 10(b)). Switches for 1.3-μm wavelengths have been developed by careful design and well-controlled technology [71]. They are applied, for example, in polarization insensitive 1×16 switched branching devices [52].

Generally, electrooptic $LiNbO_3$ devices exhibit better performance characteristics for linear polarization. For applications one should take into consideration the better performance of modulators/switches for linear polarization against the enlarged device complexity when adding polarization controllers (see below).

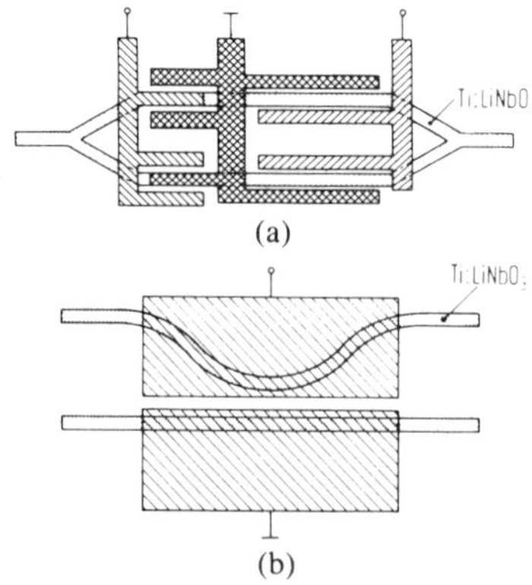

Fig. 10. (a) Electrode configuration for polarization independent Mach–Zehnder interferometric modulators. (b) Polarization insensitive directional coupler switch with weighted coupling.

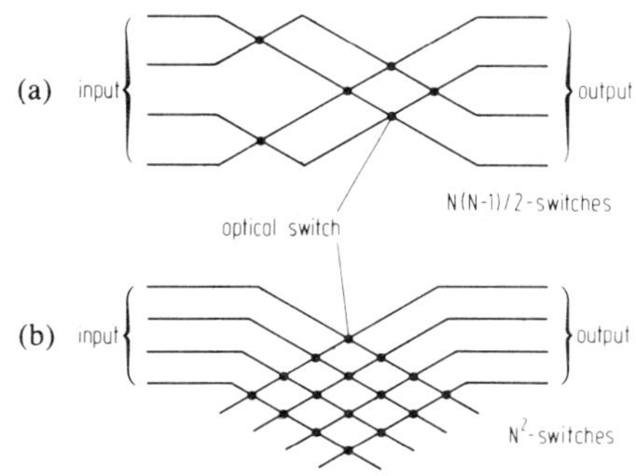

Fig. 11. Examples of architectures of nonblocking $N \times N$ switch arrays for $N = 4$. (a) Statically nonblocking rearrangeable matrix. (b) Strictly nonblocking busbar structure.

B. Switch Arrays

Future wide-band communication systems may require optical switching networks, especially for routing high-speed digital signals and for frequency-multiplexed channels in coherent communication systems. Direct switching of the optical signals—without E/O conversion—by using optical waveguide switches may be preferable because of the high switching speed, lack of limitation in signal bit-rate, bidirectional switching, and conservation of optical wavelength. Optical switching networks have been proposed and demonstrated for switched branching applications [52], time division switching [72], and $N \times N$ space-division switching. As an example of the integration capability of $LiNbO_3$ devices, we discuss $N \times N$ switch arrays.

Among others [73], there are mainly two concepts of $N \times N$ matrix architectures: the blocking or statically nonblocking matrix of the rearrangeable type [74]–[76] and the nonblocking full crosspoint array, the so-called busbar structure [6], [77] (Fig. 11). The rearrangeable matrix has the advantage that the number of switch points $N(N_2 - 1)/2$ is considerably reduced in comparison to the busbar structure with N^2 switches, where N is the number of input and output ports. However, since several switches are involved in establishing an interconnection, some information is lost during the time required for the rearrangement of the signal path, in the case of changing an interconnection. In the busbar structure, the activation of only one switch is sufficient to build up an interconnection from any of the inputs to any of the outputs. Therefore, this structure is most suitable for the application in high data rate transmission systems.

All of the recently reported switch arrays have been realized by using directional couplers [6], [55] or *X*-switches [77]. As an alternative to the well-known DC, the *X*-switch has been developed especially as a simple switch of compact size for the integration in large switching networks. Its operation principle is based on the electrooptically controlled two-mode interference in the intersection region [78]–[80]. Typical dimensions and experimental results of both switches implemented in switch arrays are given in Fig. 12. Although the voltage-length product of *X*-switches on *X*-cut $LiNbO_3$ is lower than that of DC-switches on *Z*-cut $LiNbO_3$ [40], the absolute switching voltage of *X*-switches is relatively high due to the extremely short length of the *X*-switch of only 700 μm. An advantage of DC-switches is the lower sensitivity to wavelength changes, whereas the *X*-switch has a simpler electrode configuration which allows a less sophisticated electrode design and electronic control of switch arrays. Recently, excellent performance characteristics of realized switch arrays have been reported: one 8×8 matrix of the busbar type with 6.8-dB maximum insertion loss and −30.50dB average single-switch crosstalk [6], several 4×4 switch arrays of the busbar structure [55], [77] and of the rearrangeable type [76] with single-switch crosstalks lower than −35 dB [55], and total insertion losses in the order of 5 dB [55], [76].

Fig. 13 shows the schematic configuration of the realized 4×4 matrix using *X*-switches [77] and a table of experimentally obtained single switch crosstalk values at $\lambda = 1.3$ μm.

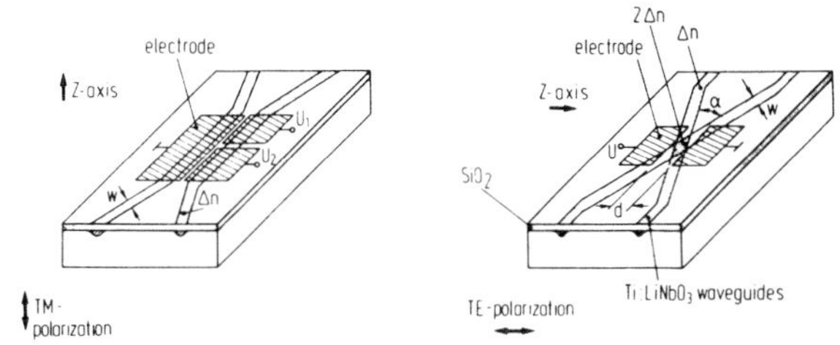

(a) (b)

Fig. 12. Configuration and experimental results of: (a) directional coupler switch [6] (Interaction length: 2 mm, electrode length: 3 mm; Switching voltages: barstate: $U_1 = U_2 = +18.4$ V, cross state: $U_1 = +26.4$ V, $U_2 = -26.4$ V, crosstalk: < -35 dB, excess loss: 0.1 dB), and (b) *X*-switch [77] (Intersection length: 0.7 mm, electrode length: 1 mm; Switching voltages: barstate: +25 V, cross state: −35 V, crosstalk: <35 dB, excess loss: 0.5 dB.

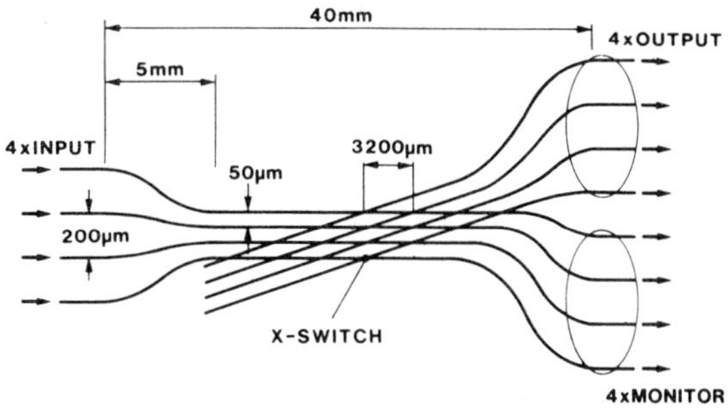

(a)

	1	2	3	4
1	-27,4	-35	-32,9	-31,5
2	-35	-32,9	-32,3	-29,2
3	-28,4	-35	-31,6	-27,1
4	-29,2	-24,3	-35	-35

(b)

Fig. 13. (a) Schematic configuration of 4 × 4 switch array using *X*-switches. (b) Table of single switch crosstalk values (in decibels).

The maximum size of an optical switching matrix is limited by the tolerable insertion loss and the total crosstalk. In a $N \times N$ matrix of the busbar structure, optical crosstalk is added at each switch element passed by the optical signal. Therefore, the output signal includes, in the worst case, a total crosstalk $C_{tot} \simeq 10 \log (N - 1) + C_0$, where C_0 is the single-switch crosstalk. In Fig. 14 C_{tot} is plotted as a function of C_0, with N as parameter. For example, if a total crosstalk of −20 dB could be tolerated, a 16 × 16 matrix would require a single-switch crosstalk greater than −32 dB, which is in the range of the above-mentioned experimental results. Furthermore, estimating the physical length of a matrix from the single-switch dimensions, $N = 16$ could be the maximum size to be realized on commercially available 75-mm-long substrates. An enhancement of the overall crosstalk of a matrix can be achieved in a double-switch arrangement [81].

Since all reported switch arrays are designed for a single polarization, the suppression of the unwanted cross polarization is of great importance for the total crosstalk of a switch matrix. Therefore, the coupling to the switch should be accomplished by polarization-maintaining fibers, which exhibit a cross polarization in the order of −20 to −30 dB. If lower crosstalk levels are required, additional polarizers have to be implemented, for example, by using high efficient resonant absorbing polarizers on the same $LiNbO_3$-substrate [82]. Integrated-optics-polarizers are especially useful for switch arrays built up with *X*-switches, since these switches utilize generally TE-modes. For the TM-modes in directional couplers, other means for the suppression of the unwanted TE-modes have to be investigated, e.g., taking advantage of the polarizing properties of proton-exchanged waveguides [83], [84]. The polarization problem in switch arrays may be overcome by using polarization-independent switches or polarization controllers at the input ports of the single-polarization switches. Optically narrow-band polarization transformers could be integrated together with switch arrays fabricated on *X*- or *Y*-cut substrates [85], [86], whereas broad-band polarization controllers require optical propagation in *Z*-direction [43], [44], [46] which impedes the utilization of the largest electrooptic coefficient r_{33} for switching. Therefore, chip-to-chip coupling of single-polarization switch arrays on *X*- or *Z*-cut $LiNbO_3$ and of *Z*-propagating polarization controllers could be an attractive combination of both device characteristics.

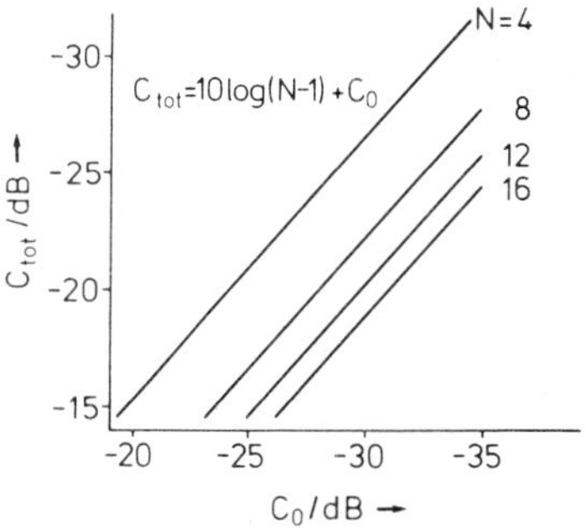

Fig. 14. Total crosstalk C_{tot} of $N \times N$ switch arrays as a function of single-switch crosstalk C_0.

C. *Wavelength Multiplexers*

Wavelength-division-multiplexing (WDM) increases the transmission capacity of fiber links and adds to network flexibility [87], [88]. Effective multichannel WDM devices have already been demonstrated for single-mode fiber systems [89], [90]. They employ microoptic or fiberoptic components, gratings, or interference filters as wavelength-selective elements. Single-mode WDM devices are particularly sensitive to alignment tolerances and fabrication costs [87]. Integrated-optic $LiNbO_3$ WDM-devices are attractive because of their channel waveguide configuration, ruggedness, and electrooptic tunability. Several wavelength-dependent operational schemes for guided-wave devices have been conceived and tested [34], [91].

Here, we consider two-mode interference for low-loss tunable WDM [91]. For a simple dual-channel multiplexer, a two-moded waveguide is coupled to single-mode input and output waveguides by *Y*-junctions of low ($\approx 1°$) branching angle (Fig. 15). An input waveguide mode is adiabatically coupled to the fundamental and first-order lateral modes of the two-moded waveguide. They propagate with phase constants $\beta_{00}(\lambda)$ and $\beta_{01}(\lambda)$. The light powers in the output waveguides depend on the phase difference between both modes at the end of the two-moded waveguides. This phase difference varies almost linearly with wavelength, leading to $\sin^2\{\pi(\lambda - \lambda_0)/\Delta\lambda\}$ transmission characteristics with a channel separation $\Delta\lambda$. The wavelength-selective branching is visualized by a beam propagation simulation [92] (Fig. 16). The Ti : $LiNbO_3$ waveguide structure is modelled by an equivalent effective index distribution [92]. The two-moded waveguide is obtained by an increased Ti-thickness $\tau_1 + \tau_2$. Fig. 16 demonstrates the interference of the modes within the two-moded section, and the adiabatic coupling and mode excitation within the *Y*-branches. Fig. 17(a) shows the configuration of a tunable MUX/DEMUX. The measured transmission characteristic is shown in Fig. 17(b). A white source/monochromator unit and a liquid nitrogen cooled Ge-detector are used in a computer-controlled measuring setup [93].

The fiber-coupled device has an insertion loss below 3 dB and about 30-nm channel separation. (Channel separations down to 10 nm are achieved.) The center channel wavelength can be tuned electrooptically by 2.5 nm/V in both directions.

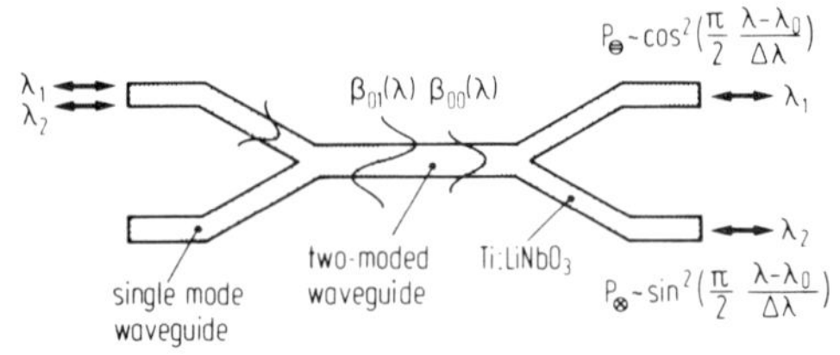

Fig. 15. Dual-channel MUX/DEMUX operation by two-mode-interference.

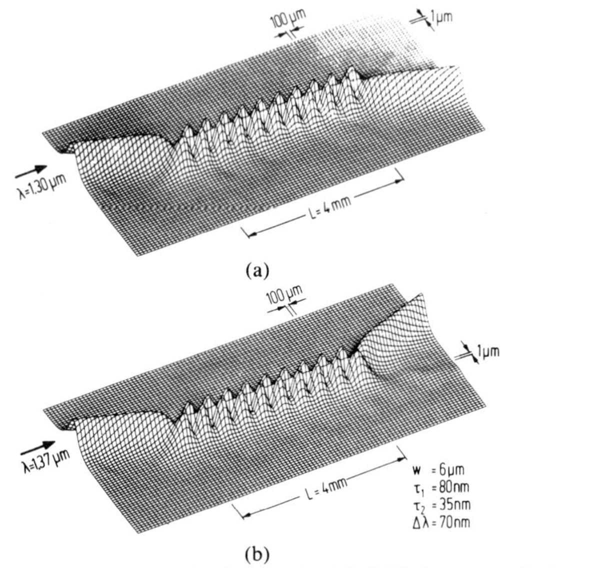

Fig. 16. Beam propagation simulation of a Ti : $LiNbO_3$ two-mode-interference MUX/DEMUX for the wavelengths (a) $\lambda = 1.30$ μm and (b) 1.37 μm.

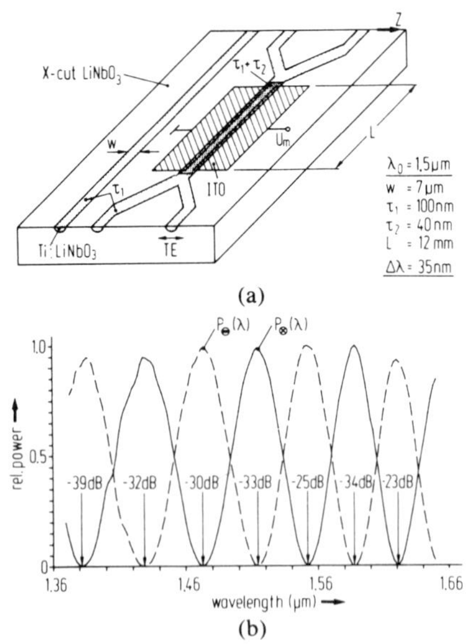

Fig. 17. (a) Configuration of a tunable Ti : $LiNbO_3$ 1.5-μm MUX/DEMUX. (b) Measured transmission characteristic.

D. Devices for Coherent Systems

Fiber-optic transmission using heterodyne or homodyne coherent detection offers an up to about 20-dB gain in receiver sensitivity when compared to direct detection. Moreover, the transmission capacity can be increased significantly by a closely spaced frequency multiplexing. $LiNbO_3$ devices can provide efficient transmitter and receiver terminal components with several functions integrated on a common $LiNbO_3$ substrate [94], [95]. We already addressed coherent systems when discussing Ti : $LiNbO_3$ phase modulators.

A minimum Ti : $LiNbO_3$ integrated coherent receiver chip contains a polarization controller and a power combiner. A phase shifter must be added for homodyne detection. Fig. 18 shows a recent example [46]. The device is built on *X*-cut *Z*-propagating $LiNbO_3$. The phase shifters and TE–TM converters required for polarization control with unlimited range are distributed in the local oscillator (LO) and the signal arms. The components are optically broad band due to the use of *Z*-propagation. The directional coupler is tuned electrooptically and it is polarization independent. The electrooptic coefficients which influence TE- and TM-polarizations are indicated in Fig. 18.

The further integration of a frequency translator would provide a precise frequency control and a potentially compact frequency multiplexer/demultiplexer. The corresponding device should preferably provide a single-line output spectrum and gigahertz frequency shifts. A variety of methods can be used for frequency shifting (see, for example, [33], [96]–[100]). As shown above, a multigigahertz modulation bandwidth and low drive power are most easily achieved with Mach–Zehnder modulators.

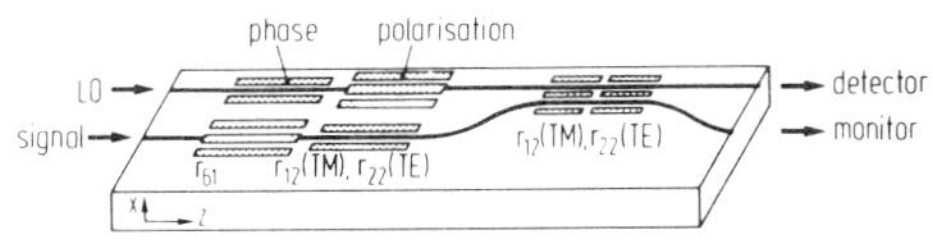

Fig. 18. Schematic arrangement of a broadband coherent receiver chip on X-cut Z-propagating $LiNbO_3$. The monitor output provides control signals for the polarization and coupling adjustment [46].

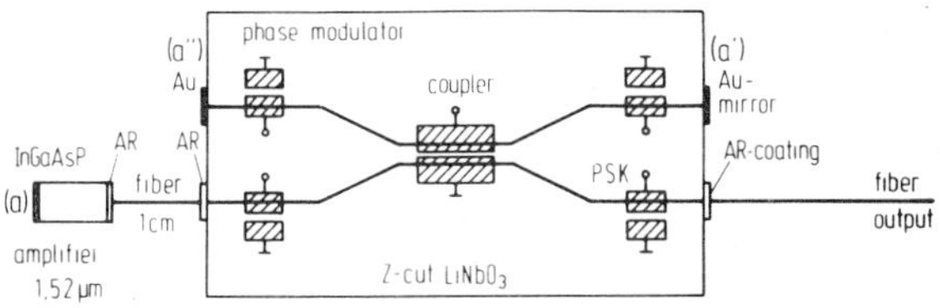

Fig. 19. Electrooptically tuned PSK-modulated 1.52-μm transmitter [101].

They allow low-distortion frequency multiplexing for a proper choice of the modulation bandwidth [100]. No convincing solution, however, has been developed for true fequency translators with gigahertz frequency range.

The integration capabilities and the performance characteristics of $Ti:LiNbO_3$ devices for transmitters are demonstrated by Fig. 19. The hybrid integrated-optic laser consists of an InGaAsP amplifier coupled to a $Ti:LiNbO_3$ waveguide circuit. The main laser resonator indicated by the mirror (a), (a') is parallel coupled to a second waveguide resonator with mirrors (a'), (a''). The coupling factor and the resonator tuning are accomplished electrooptically. In addition, the output can be PSK-modulated. A linewidth of less than 100 GHz, 0.25-mW fiber-coupled output power, and 1-GHz phase modulation have been achieved [101].

IV. Conclusion

The progress in $LiNbO_3$ waveguide technology and the development of high-performance electrooptic devices with respect to insertion loss, bandwidth, and drive power, allows the system applications of specialized discrete devices, such as modulators and switches. Full advantage of $LiNbO_3$ integrated optics can be taken by integrating a number of components on a single substrate. The integration of up to 64 devices in switch arrays has been demonstrated. $Ti:LiNbO_3$ devices performing complex signal processing operations have been used in communication system experiments.

References

[1] T. Sueta and M. Izutsu, "High speed guided-wave optical modulators," *J. Opt. Commun.*, vol. 3, pp. 52–58, 1982.

[2] C. M. Gee, G. D. Thurmond, and H. W. Yen, "17-GHz bandwith electrooptic modulator," *Appl. Phys. Lett.*, vol. 43, pp. 998–1000, 1983.

[3] R. A. Becker, "Traveling-wave electrooptic modulator with maximum bandwith-length product," *Appl. Phys. Lett.*, vol. 45, pp. 1168–1170, 1984.

[4] H. Haga, M. Izutsu, and T. Sueta, "$LiNb0_3$ traveling-wave light modulator switch with an etched groove," *IEEE J. Quantum Electron.*, vol. QE-22, pp. 902–906, 1986.

[5] R. C. Alferness, L. L. Buhl, M. D. Divino, S. K. Korotky, and L. W. Stulz, "Low-loss broad band $Ti:LiNbO_3$ waveguide phase modulators for coherent systems," *Electron. Lett.*, vol. 22, pp. 309–310, 1986.

[6] P. Granestrand, B. Stoltz, L. Thylen, K. Bergvall, W. Döldissen, H. Heidrich, and D. Hoffmann, "Strictly nonblocking 8 × 8 integrated optical switch matrix," *Electron. Lett.*, vol. 22, pp. 816–817, 1986.

[7] R. C. Alferness and R. V. Schmidt, "Turnable optical waveguide directional coupler filter," *Appl. Phys. Lett.*, vol. 33, pp. 161–163, 1978.

[8] A. Neyer, "Integrated-optical multichannel wavelength multiplexer for monomode systems," *Electron. Lett.*, vol. 20, pp. 744–746, 1984.

[9] K. Komatsu, M. Kondo, and Y. Ohta, "Titanium/magnesium double diffusion method for efficient fiber-$LiNbO_3$ waveguide coupling," presented at Topical Meet. Integrated and Guided Wave Optics, Atlanta, GA, 1986, pap. PDP 2-1.

[10] R. V. Schmidt and I. P. Kaminow, "Metal diffused optical waveguides in $LiNbO_3$," *Appl. Phys. Lett.*, vol. 15, pp. 458–460, 1974.

[11] M. Minakata, S. Saito, M. Shibata, and S. Miyazawa, "Precise determination of refractive-index changes in Ti-diffused $LiNbO_3$ waveguides," *J. Appl. Phys.*, vol. 49, pp. 4677–4682, 1978.

[12] G. J. Griffiths and R. J. Esdaile, "Analysis of titanium diffused planar optical waveguides in lithium niobate." *IEEE J. Quantum Electron.*, vol. QE-20, pp. 149–159, 1984.

[13] L. P. Kaminow and J. R. Carruthers, "Optical waveguiding layers in $LiNbO_3$ and $LiTaO_3$," *Appl. Phys. Lett.*, vol. 22, pp. 326–328, 1973.

[14] R. J. Esdaile, "Closed-tube control of out-diffusion during fabrication of optical waveguides in $LiNbO_3$," *Appl. Phys. Lett.*, vol. 33, pp. 733–734, 1978.

[15] B. Chen and A. C. Pastor, "Elimination of Li_2O out-diffusion waveguide in $LiNbO_3$," *Appl. Phys. Lett.*, vol. 30, pp. 570–571, 1977.

[16] T. R. Ranganath and S. Wang, "Suppression of Li_2O out-diffusion from Ti-diffused $LiNbO_3$ optical waveguides," *Appl. Phys. Lett.*, vol. 30, pp. 376–379, 1977.

[17] J. L. Jackel, V. Ramaswamy, and S. P. Lyman, "Elimination of out-diffused surface guiding in titanium-diffused $LiNb0_3$," *Appl. Phys. Lett.*, vol. 38, pp. 509–511, 1981.

[18] J. L. Jackel, "Suppression of outdiffusion in titanium diffused $LiNbO_3$: a review," *J. Opt. Commun.* vol. 3, pp. 82–85, 1982.

[19] A. Neyer, to be published.

[20] L. McCaughan and K. D. Choguette, "Origin and solution of finite extinction ratios in Ti :$LiNbO_3$ directional coupler switches," presented at Topical Meet. Integrated and Guided Wave Optics, Atlanta, GA, 1986, pap. THBB6.

[21] R. A. Becker, "Impact of material parameters on $LiNbO_3$ waveguide devices," presented at Topical Meet. Integrated and Guided Wave Optics, Atlanta, GA, 1986, pap. FFA1.

[22] W. Döldissen, H. Heidrich, and D. Hoffmann, "Microfabrication of planar waveguide devices on $LiNbO_3$," *Proc. SPIE*, vol. 651, pp. 15–19, 1986.

[23] A. Askin, G. D. Boyd, J. M. Dziedzic, R. G. Smith, A. A. Ballman, J. J. Levinstein, and K. Nassau, "Optically induced refractive index inhomogeneities in $LiNbO_3$ and $LiTaO_3$," *Appl. Phys. Lett.*, vol. 9, pp. 72–74, 1966.

[24] A. M. Glass, "The photorefractive effect," *Opt. Eng.*, vol. 17, pp. 470–479, 1978.

[25] G. T. Harvey, G. Astfalk, A. Y. Feldblum, and B. Kassahun, "The photorefractive effect in titanium indiffused lithium niobate optical directional couplers at 1.3 μm," *IEEE J. Quantum Electron.*, vol. QE-22, pp. 939–946, 1986.

[26] J. Noda, M. Fukuma, and S. Saito, "Effect of Mg diffusion on Ti-diffused $LiNbO_3$ waveguides," *J. Appl. Phys.*, vol. 49, pp. 3150–3154, 1978.

[27] J. L. Jackel, C. E. Rice, and J. J. Veselka, "Proton exchange for high-index waveguides in $LiNbO_3$," *Appl. Phys. Lett.*, vol. 47, pp. 607–608, 1982.

[28] M. Goodwin and C. Stewart, "Proton-exchanged optical waveguides in Y-cut lithium niobate," *Electron. Lett.*, vol. 19, pp. 223–225, 1983.

[29] M. De Micheli, J. Botineau, P. Sibillot, D. B. Ostrowsky, and M. Papuchon, "Fabrication and characterization of titanium indiffused proton exchanged (TIPE) waveguides in lithium niobate," *Opt. Commun.*, vol. 42, pp. 101–103, 1982.

[30] V. Hinkov and E. Ise, "Control of birefringence in $Ti:LiNbO_3$ optical waveguides by proton exchange of lithium ions," *J. Lightwave Tech.*, vol. LT-4, pp. 444–448, 1986.

[31] A. C. G. Nutt, E. Sudo, and K. Nishizawa, "Control of Ti: diffused $LiNbO_3$ waveguide profile and propagation characteristics by the proton exchange of lithium ions," presented at Conf. Optical Fiber Communication/International Conf. on Integrated Optics and Optical Fiber Commun., Reno, NV, 1987, pap. TuH1.

[32] J. J. Veselka and G. A. Bogart, "Low-loss TM-pass polarizer fabricated by proton exchange for *Z*-cut Ti : $LiNbO_3$ waveguides," *Electron. Lett.*, vol. 23, pp. 37–38, 1987.

[33] C. S. Tsai, "Guided-wave acoustooptic Bragg modulator for wide-band integrated optic communications and signal processing," *IEEE Trans. Circuits Syst.*, vol. CAS-26, pp. 1072–1098, 1979.

[34] R. C. Alferness, "Guided-wave devices for optical communication," *IEEE J. Quantum Electron.*, vol. QE-17, pp. 946–959, 1981.

[35] R. C. Alferness, "Waveguide electrooptic modulators," *IEEE J. Trans. Microwave Theory Tech.*, vol. MTT-30, pp. 1121–1137, 1982.

[36] S. Yamada and M. Minakata, "DC drift phenomena in $LiNbO_3$ optical waveguides devices," *Japan. J. Phys.*, vol. 20, pp. 733–737, 1981.

[37] C. M. Gee, G. D. Thurmond, H. Blauvelt, and H. W. Yen, "Minimizing dc drift in $LiNbO_3$ waveguide devices," *Appl. Phys. Lett.*, vol. 47, pp. 211–213, 1985.

[38] K. Sreenivas, T. Sudersena Rao, A. Mansingh, and S. Chandra, "Preparation and characterization of RF sputtered indium tin oxide films," *J. Appl. Phys.*, vol. 57, pp. 384–392, 1985.

[39] H. Heidrich, C. H. von Helmolt, O. Hoffmann, H.-J. Hensel, and A. Kleinwaechter, *Electron. Lett.*, to be published, 1987.

[40] J. Ctyroky, "Voltage-length product of *X* and *Z*-cut Ti : $LiNbO_3$ directional coupler and BOA switches : A comparison,"*J. Opt. Commun.*, vol. 7, pp. 139–143, 1986.

[41] D. Marcuse, "Optimal electrode design for integrated optics modulators," *IEEE J. Quantum Electron.*, vol. QE-18, pp. 393–398, 1982.

[42] F. Auracher, "Design trade-offs for high-speed directional coupler modulators with $\Delta\beta$-reversal in $LiNbO_3$," *J. Opt. Commun.*, vol. 5, pp. 2–6, 1984.

[43] C. Mariller and M. Papuchon, "Simple and wide optical bandwidth TE/TM mode converter using Z-propagating $LiNbO_3$ waveguides," in *Proc. 3rd Eur. Conf. Integrated Optics*, (Berlin, W. Germany), 1985, pp. 174–176.

[44] S. Thaniyavarn, "Wavelength-independent, optical-damage-immune $LiNb0_3$ TE-TM mode convertor," *Opti. Lett.*, vol. 11, pp. 39–41, 1986.

[45] A. J. Cook and M. J. Wale, "Integrated optical modulator with high extinction ratio employing resonant absorption," in *Tech. Dig. 12th Eur. Conf. Optical Commun.* (Barcelona, Spain), 1986, vol. III, pp. 45–47.

[46] H. Heidrich, D. Hoffmann, P. Nolting, H. Ahlers, and A. Kleinwaechter, "Polarization transformer with unlimited range on Ti : $LiNbO_3$," in *Tech. Dig. 12th Eur. Conf. Optical Commun.* (Barcelona, Spain), 1986, vol. I, pp. 411–414.

[47] G. Eisenstein, S. K. Korotky, L. W. Stulz, J. J. Veselka, R. M. Jopson, and K. L. Hall, "Antireflection coatings on lithium niobate waveguide devices using electron beam evaporated yttrium oxide," *Electron. Lett.*, vol. 21, pp. 363–364, 1985.

[48] K. E. Petersen, "Silicon as a mechanical material," *Proc. IEEE*, vol. 70, pp. 420–454, 1982.

[49] O. G. Ramer, C. C. Nelson, and C. M. Mohr, "Experimental integrated optic circuit losses and fiber pigtailing of chips," *IEEE J. Quantum Electron.*, vol. QE-17, pp. 970–974, 1981.

[50] E. J. Murphy and T. C. Rice, "Low loss coupling of multiple fiber arrays to single-mode waveguides," *J. Lightwave Technol.*, vol. LT-1, pp. 479–481, 1983.

[51] E. J. Murphy, T. C. Rice, L. McCaughan, G. T. Harvey, and P. H. Read, "Permanent attachment of single-mode fiber arrays to waveguides," *J. Lightwave Technol.*, vol. LT-3, pp. 795–798, 1985.

[52] J. E. Watson, M. A. Milbrodt, and T. C. Rice, "A polarization-independent 1 × 16 guide-wave optical switch integrated on lithium niobate," *J. Lightwave Technol.*, vol. LT-4, pp. 1717–1721, 1986.

[53] R. H. Stolen, W. Pleitzel, and J. Simpson, "High-birefringence optical fibers by preform deformation," *J. Lightwave Technol.*, vol. LT-2, pp. 639–641, 1984.

[54] K. Okamoto, T. Hosaka, and J. Noda, "High-birefringence, polarizing fiber with flat cladding," *J. Lightwave Technol.*, vol. LT-3, pp. 758–767, 1985.

[55] G. A. Bogert, E. J. Murphy, and R. T. Ku, "Low crosstalk 4 × 4 Ti : $LiNb0_3$ optical switch with permanently attached polarization matintaining fiber array," *J. Lightwave Technol.*, vol. LT-4, pp. 1542–1545, 1986.

[56] F. Rottmann and E. Voges, to be published.

[57] R. A. Linke, B. K. Kasper, N. A. Olsson, and R. C. Alferness, "Coherent lightwaves transmission over 150-km fiber length at 400 Mbit/s and 1 Gbit/s data rates using phase modulation," *Electron. Lett.*, vol. 22, pp. 30–31, 1986.

[58] F. Auracher and D. Imhof, "Design tradeoffs for high-speed guided-wave phase modulators in $LiNbO_3$," *Siemens Forsch.- u. Entwickl. Ber.*, vol. 15, pp. 19–22, 1986.

[59] K. Kubota, J. Noda, and O. Mikami, "Traveling-wave optical modulator using directional coupler Ti : $LiNbO_3$ waveguide," *IEEE J. Quantum Electron.*, vol. QE-16, pp. 754–760, 1981.

[60] R. C. Alferness, S. K. Korotky, L. L. Buhl, and M. D. Divino, "High-speed, low loss, low drive power traveling-wave optical switch/modulator for $\lambda = 1.32\ \mu m$," *Electron. Lett.*, vol. 20, pp. 354–355, 1984.

[61] F. Auracher, D. Schicketanz, and K.-H. Zeitler, "High-speed $\Delta\beta$-reversal directional coupler modulator with low insertion loss for 1.3 μm in $LiNb0_3$," *J. Opt. Commun.*, vol. 5, pp. 7–9, 1984.

[62] S. K. Korotky, G. Eisenstein, R. C. Alferness, J. J. Veselka, L. L. Buhl, G. T. Harvey, and H. Read, "Fully connectorized high-speed Ti : $LiNb0_3$ switch/modulator for time-division multiplexing and data encoding," *J. Lightwave Technol.*, vol. LT-3, pp. 1–6, 1985.

[63] S. K. Korotky, G. Eisenstein, A. H. Gnauck, B. L. Kasper, J. J. Veselka, R. C. Alferness, L. L. Buhl, C. A. Burreis, T. C. D. Huo, L. W. Stulz, K. C. Nelson, L. G. Cohen, R. W. Dawson, and J. C. Campbell, "4-Gibt/s transmission experiment over 117 km of optical fiber using a Ti : $LiNb0_3$ external modulator," *J. Lightwave Technol.*, vol. LT-3, pp. 1027–1030, 1985.

[64] C. S. Tsai, C. C. Lee, and P. Le, "A 8.5-GHz bandwidth single-mode crossed channel TIR modulator and switch in $LiNbO_3$," presented at Top. Meet. Integrated and Guided-Wave Optics, Kissimee, FL, 1984, pap. PD5.

[65] A. Neyer and W. Sohler, "High-speed cutoff modulator using a Ti-diffused $LiNbO_3$ channel waveguide," *Appl. Phys. Lett.*, vol. 35, pp. 256–258, 1979.

[66] H. Kogelnik and R. V. Schmidt, "Electrooptically switched couplers with alternating $\Delta\beta$," *IEEE J. Quantum Electron.*, vol. QE-12, pp. 396–401, 1976.

[67] L. Thylen, A. Djupsjöbacka, B. Lagerström, and P. Svensson, "High speed $LiNb0_3$ integrated optics modulators and switches," *SPIE Int. Opt. Circuit Eng.*, vol. 517, pp. 249–257, 1984.

[68] W. K. Burns, T. G. Giallorenzi, R. P. Moeller, and E. J. West, "Interferometric waveguide modulator with polarization independent operation," *Appl. Phys. Lett.*, vol. 33, pp. 944–947, 1978.

[69] R. C. Alferness, "Optical directional couplers with weighted coupling," *Appl. Phys. Lett.*, vol. 35, pp. 160–262, 1979.

[70] O. G. Ramer, C. Mohr, and J. Pikulski, "Polarization-independent optical switch with multiple sections of $\Delta\beta$ reversal and Gaussian taper functions," *IEEE J. Quantum Electron.*, vol. QE-18, pp. 1760–1767, 1982.

[71] L. McCaughan, "Low-loss polarization-independent electrooptical switsches at $\lambda = 1.3\ \mu m$," *J. Lightwave Technol.*, vol. LT-2, pp. 51–55, 1984.

[72] H. Haga, M. Izutsu, and T. Sueta, "An integrated 1 × 4 high-speed optical switch and its applications to a time demultiplexer," *J. Lightwave Technol.*, vol. LT-3, pp. 116–120, 1985.

[73] R. A. Spanke, "Architectures for large nonblocking optical space switches," *IEEE J. Quantum Electron.*, vol. QE-22, pp. 964–967, 1986.

[74] H. F. Taylor, "Optical-waveguide connection networks," *Electron. Lett.*, vol. 10, pp. 41–43, 1974.

[75] R. V. Schmidt and L. L. Buhl, "Experimental 4 × 4 optical switching network," *Electron. Lett.*, vol. 12, pp. 575–577, 1976.

[76] D. Hoffmann, H. Heidrich, W. Döldissen, H. Ahlers, A. Kleinwaechter, and R. Schmidt, "Low-loss rearrangeably nonblocking 4 × 4 switch matrix module," in *Tech. Dig. 12th Eur. Conf. Optical Commun.* (Barcelona, Spain), 1986, vol. I, pp. 167–170.

[77] A. Neyer, W. Mevenkamp, and B. Kretschmann, "Nonblocking 4 × 4 switch array with sixteen *X*-switches in Ti : $LiNb0_3$, in *Tech. Dig. Top. Meet. Integrated and Guided-Wave Optics* (Atlanta, GA), 1986, pap. WAA2.

[78] A. Neyer, "Electrooptic *X*-switch using single-mode Ti:$LiNbO_3$ channel waveguides," *Electron. Lett.*, vol. 19, pp. 553–554, 1983.

[79] A. Neyer, W. Mevenkamp, and B. Kretschmann, "Optimization of *X*-switches for integrated optical switching networks," in *Tech. Dig. 5th Int. Conf. Integrated Optics and Optical Fibre Commun./11th*

Europ. Conf. Optical Commun. (Venezia, Italy), 1985, vol. I, pp. 369–372.
[80] A. Neyer, W. Mevenkamp, L. Thylen, and B. Lagerström, "A beam propagation method analysis of active and passive waveguide crossing," *J. Lightwave Technol.*, vol. LT-3, pp. 635–642, 1985.
[81] M. Kondo, N. Takado, K. Komatsu, and Y. Ohta, "32 switch-elements integrated low-crosstalk $LiNb0_3$ 4 × 4 optical matrix switch," in *Tech. Dig. 5th Int. Conf. Integrated Optics and Optical Fibre Commun./11th Europ. Conf. Optical Commun.* (Venezia, Italy), 1985, vol. I, pp. 361–364.
[82] J. Ctyroky and H.-J. Henning, "Thin-film polarizer for Ti:$LiNbO_3$ waveguides at $\lambda = 1.3\ \mu m$," *Electron. Lett.*, 1986, vol. 22, pp. 756–757.
[83] D. F. Clark, A. C. G. Nutt, K. K. Wong, P. J. R. Laybourn, and R. M. De La Rue, "Characterization of proton-exchange slab optical waveguides in Z-cut $LiNbO_3$," *J. Appl. Phys.*, vol. 54, pp. 6218–6220, 1983.
[84] J. J. Veselka and G. A. Bogert, "Low-loss proton exchange channel waveguides and TM-pass polarizers in Z-cut $LiNbO_3$," presented at Optical Fiber Commun. Conf. (Reno, NV), 1987, pap. TuH3.
[85] R. C. Alferness, "Electrooptic guided-wave device for general polarization transformations," *IEEE J. Quantum Electon.*, vol. QE-17, pp. 2225–2227, 1981.
[86] R. C. Alferness and L. L. Buhl, "New low-loss waveguide electrooptic polarization controller for $\lambda = 1.32\ \mu m$" in *Proc. 4th Int. Conf. Integrated Opt. Optical Fiber Commun.* (Tokyo, Japan), 1983, pp. 38–39.
[87] G. Winzer, "Wavelength multiplexing components—A review of single-mode devices and their applications," *J. Lightwave Technol.*, 1984, vol. LT-2, pp. 369–378.
[88] H. Ishio, J. Minowa, and K. Nosu, "Review and status of wavelength -division- multiplexing technology and its application," *J. Lightwave Technol.*, vol. LT-2, pp. 448–463, 1984.
[89] N. A. Olsson, J. Hegarty, R. A. Logan, L. F. Johnson, K. L. Walker, and L. G. Cohen, "68.3-km transmission with 1.37-Tbit/s capacity using wavelength division multiplexing on ten single-frequency lasers at $1.5\mu m$," *Electron. Lett.*, vol. 21, pp. 105–106, 1985.
[90] J. Lipson, W. J. Minford, E. J. Murphy, T. C. Rice, R. A. Linke, and G. T. Harvey, "A six-channel wavelength multiplexer and demultiplexer for single mode systems," *J. Lightwave Technol.*, vol. LT-3, pp. 1159–1162, 1985.
[91] A. Neyer, "Integrated-optical multichannel wavelength multiplexer for monomode systems," *Electron. Lett.*, vol. 20, pp. 744–746, 1984.
[92] W. Mevenkamp and E. Voges, "Modelling and beam propagation analysis of integrated electro-optic devices," *Arch. Elek. Übertragung*, vol. 40, pp. 289–296, 1986.
[93] F. Rottmann and E. Voges, to be published.
[94] W. A. Stallard, A. R. Beaumont, and R. C. Booth, "Integrated optic devices for coherent transmission," *J. Lightwave Technol.*, vol. LT-4, pp. 852–857, 1986.
[95] W. A. Stallard, T. G. Hodgkinson, K. R. Preston, and R. C. Booth, "A novel $LiNbO_3$-integrated optic component for coherent optical heterodyne detection," *Electron. Lett.*, vol. 21, pp. 1077–1079, 1985.
[96] B. Culshaw and M. G. F. Wilson, "Integrated optic frequency shifter modulator," *Electron. Lett.*, vol. 17, pp. 135–136, 1981.
[97] M. Izutsu, S. Shikama, and T. Sueta, "Integrated optical SSB modulator/frequency shifter," *IEEE J. Quantum Electron.*, vol. QE-17, pp. 965–969, 1981.
[98] F. Heisman and R. Ulrich, "Integrated-optical frequency translator with stripe waveguide," *Appl. Phys. Lett.*, vol. 45, pp. 490–492, 1984.
[99] W. A. Stallard, B. E. Daymond-John, and R. C. Booth, "$LiNbO_3$ optical frequency translator for coherent optical-fiber systems," in *Proc. 3rd Eur. Conf. Integrated Optics* (Berlin, W. Germany), 1985, pp. 164–168.
[100] G. R. Hill, R. A. Lobbett, and W. A. Stallard, "Frequency multiplexing in an optical system using a Mach–Zehnder waveguide frequency translator," *Electron. Lett.*, vol. 21, pp. 1214–1215, 1985.
[101] S. K. Korotky, G. Eisenstein, F. Heismann, J. J. Veselka, and E. A. J. Marcatili, "Narrow linewidth, electrooptically tuned parallel coupled waveguide laser," presented at Top. Meet. Integrated and Guided-Wave Optics, Atlanta, GA, 1986, pap. PDP 10-1.

*

Edgar Voges was born in Braunschweig, Germany, on August 26, 1941. He received the Diplom-Physiker degree in physics and the Doktor-Ingenieur degree from the Technische Universität Braunschweig, Braunschweig, W. Germany, in 1967 and 1970, respectively.

From 1967 to 1974 he was an Assistant at the Institut für Hochfrequenztechnik of the Universität Braunschweig, where he was engaged in investigations on microwave semiconductor devices and acoustic surface waves. From 1974 to 1977 he was a Professor with the Department of Electrical Engineering of the Universität Dortmund. From 1977 to 1982 he was a Professor of Communication Engineering at the FernUniversität, Hagen, W. Germany. Since 1982 he has again been with the Department of Electrical Engineering of the Universität Dortmund, Chair of High-Frequency Techniques. His current interests are in optical communications, integrated optics, and high-freqency integrated circuits.

*

Andreas Neyer was born in Ibbenbüren, W. Germany, on April 14, 1950. He received the Dipl.-Phys. degree in physics from the Universitaet Dortmund, Dortmund, W. Germany, in 1978 and the Dr.-Ing. degree in electrical engineering from the FernUniversitaet, Hagen, W. Germany, in 1982.

From 1978 to 1982 he was an Assistant at the Lehrstuhl fuer Nachrichtentechnik of the Fern-Universitaet, where he worked on electrooptic modulators and bistable devices. Since 1982 he has been an Assistant at the Lehrstuhl fuer Hochfrequenztechnik of the University Dortmund, where he has been engaged in the development of the $LiNbO_3$-technology for integrated optics devices such as electrooptic switches, switch arrays, and wavelength multiplexers.

8-Gbit/s Transmission Experiment over 68 km of Optical Fiber using a Ti:$LiNbO_3$ External Modulator

S. K. KOROTKY, A. H. GNAUCK, B. L. KASPER, J. C. CAMPBELL, J. J. VESELKA, J. R. TALMAN, AND A. R. MCCORMICK

Abstract—**We describe the performance of an experimental 1.5-μm lightwave transmission system operating at 8 Gbit/s over 68.3 km of single-mode fiber. The dispersion penalty is limited to 1 dB through the use of external modulation and is attributable to the intrinsic information bandwidth.**

Introduction

Although the fundamental and practical limits of characteristics such as laser power, transmitter/receiver speed, transmitter spectral purity, receiver sensitivity, fiber dispersion, and fiber loss can be probed at the device level, the efficiency with which a given component trades off one characteristic against another is best evaluated in the system environment. One figure of merit that measures the degree of optimization is the bit rate times distance product. Another is the efficiency of the use of the total fiber bandwidth. At present, the spectral width of directly modulated high-speed injection lasers is a limiting factor for both of these measures [1]. Here we describe an experiment [2] that makes use of an electronic multiplexer and a Ti:$LiNbO_3$ external modulator to encode data in optical form at 8 Gbit/s. Using the external modulator, the spectral width of the transmitted signal is essentially that minimum value determined by the information bandwidth. Thus, the effect of chromatic dispersion is minimized and the use of the fiber capacity is maximized. In addition, we report on a sensitive 8-Gbit/s receiver and an 8-Gbit/s timing recovery circuit. With these devices we have transmitted 8 Gbit/s at 1.53-μm wavelength over 68.3 km of single-mode fiber with zero dispersion at 1.3 μm. The corresponding bit rate (B) times distance (L) product, $BL \sim 0.55$ Tbit · km/s, represents a new record for a single-wavelength fiber span.

Experiment

The arrangement of the transmission system experiment is similar to that used in a previous experiment at 4 Gbit/s [3] and is shown in Fig. 1. Bit-error-rate (BER) measurements at 8 Gbit/s are made possible by a recently demonstrated electronic multiplexer/demultiplexer [1]. Four 2-Gbit/s pseudorandom digital data streams of length $2^{15} - 1$, which are obtained from a commercially available 2-Gbit/s BER test set, are combined to form the 8-Gbit/s non-return-to-zero (NRZ) signal using circuits built from discrete GaAs FET's. The electrical multiplexing from 2 to 8 Gbit/s is accomplished in two binary stages. In each stage the input NRZ data is first converted to a RZ (return-to-zero) format by sampling each bit for one-half the bit period. This is done using a grounded-source dual-gate FET circuit which exhibits a −3-dB bandwidth of 13 GHz. The NRZ data is applied to one gate and a sinusoidal clock signal with a frequency corresponding to the input data rate (making one full cycle each bit period) is applied to the other gate. Only when both inputs are at the on level does the transistor turn on, thereby making the inverted RZ data available at the drain. The RZ data stream is split by a microwave resistive power divider into two identical streams and one of these is delayed in a coaxial cable for an integral number of half bit periods with respect to the other. The two streams are then interleaved in a circuit comprised of two single-gate FET's with coupled drains (−3-dB bandwidth of 8 GHz). Each RZ data stream controls one transistor; therefore, NRZ data at twice the original bit rate is available at the transistors' drains.

Our transmitter is a hybrid combination of an InGaAsP/InP injection laser and a Ti:$LiNbO_3$ directional coupler modulator. The laser is of the channel-substrate, buried heterostructure design [4] and was cleaved to form two coupled resonators. The drive to the laser is *not* modulated. In this manner, the laser provides a high power ($\sim$10 dBm into air per facet), single frequency, and relatively narrow linewidth emission at 1.53-μm wavelength. Light from the laser is coupled into a microlensed fiber that is in turn connected to one port of the directional coupler modulator. The crossover waveguide of the Ti:$LiNbO_3$ coupler serves as the output and is connected to the fiber span. Data is encoded by switching the Ti:$LiNbO_3$ modulator between the port attached to the fiber span and the second output port using the amplified signal from the 8-Gbit/s multiplexer. The design of the Ti:$LiNbO_3$ waveguide switch is summarized elsewhere [3]. The relevant characteristics of the present modulator are an active length of 1 cm, a switching voltage of 10.6 V, dc extinction ratio of >28 dB, a total optical insertion loss of 4 dB, an optical reflectivity of ~ -26 dB, and a small-signal 3-dB electrical bandwidth of $\sim$4–5 GHz. The overall performance with respect to these characteristics

Manuscript received February 17, 1987; revised May 6, 1987.
The authors are with AT&T Bell Laboratories, Holmdel, NJ 07733.
IEEE Log Number 8715758.

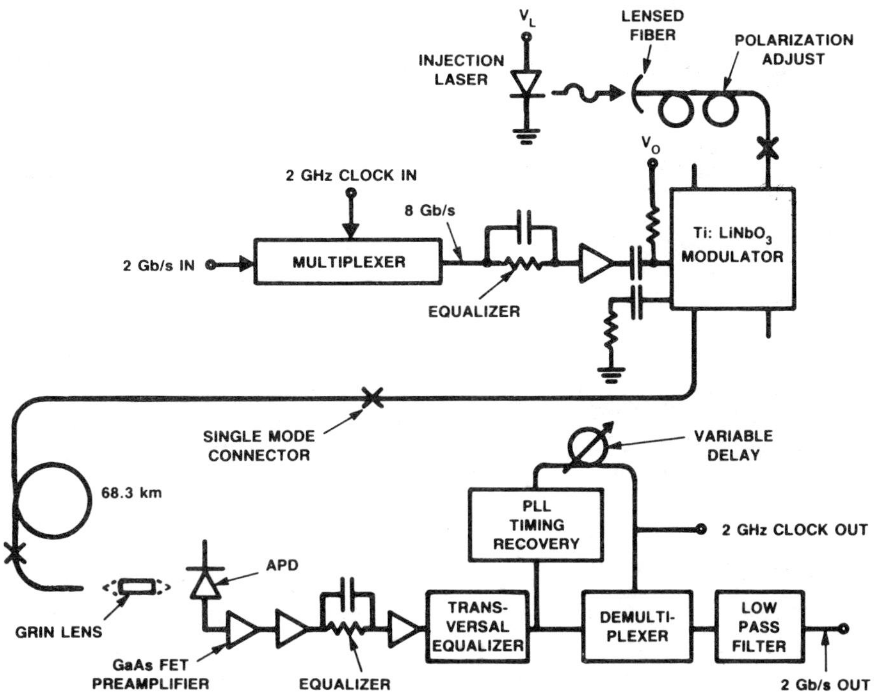

Fig. 1. Schematic diagram of the transmission experiment.

was somewhat better for our earlier device [3]; however, for the present device special attention was placed on attaining a smooth frequency response to ~ 10 GHz. In Fig. 2(a) we show the eye pattern at the output of the modulator, as observed with a fast InGaAs p-i-n detector.

The maximum power that could be coupled from the laser into the microlensed fiber was +6.5 dBm. Because the antireflection coating was not optimal for this particular device, it was necessary to decouple from this maximum value by 4 dB to reduce the level of the feedback to the laser below −32 dB. (We note that AR coatings providing reflectivities below −35 dB have been demonstrated [3], [5]). Because the electrical drive amplifier was unable to provide the necessary full modulation voltage swing, the modulator was not adjusted to the center of its switching curve. Rather, the bias was adjusted to obtain good extinction (> 10 dB) and therefore nearly optimal receiver sensitivity. The incomplete modulator swing resulted in a 2-dB loss penalty. As a consequence the average modulated power launched into the transmission fiber was ~ −6.5 dBm. In principle, this power can be increased by 7 dB through tighter control of the modulator fabrication to ensure optimal AR-coatings, drive voltage, and insertion loss [3]. The length of the single-mode, almost-pure-silica core [6] fiber span is 68.3 km with a total loss of 14.8 dB and dispersion of 19 ps/km/nm at 1.53-μm wavelength.

The 8-Gbit/s receiver used a high-speed InGaAs SAGM (separate absorption, grading, and multiplication) avalanche photodiode [7] combined with a high-impedance GaAs MESFET amplifier. The APD had a gain-bandwidth product of 60 GHz and a maximum bandwidth at low gain of 6 GHz. It was fabricated as a mesa device with a diameter of ~50 μm, and exhibited a quantum efficiency of 62 percent without antireflection coating. The total input capacitance of the front end was 0.71 pF, of which 0.23 pF was contributed by the APD, 0.42 pF came from the 0.3 μm gate-length MESFET, and 0.06 pF came from the 1.3-kΩ load resistor. The preamplifier consisted of a two-stage direct-coupled common-source amplifier with the first stage providing a voltage gain of about 2, and the second stage having unity gain but very good impedance matching to 50 Ω.

The preamplifier was followed by a linear channel consisting of a number of commercial wide-band amplifiers with frequency response from 2 kHz to 7 GHz. The linear channel included an *RC* equalizer to correct for integration of the signal which occurs in the high-impedance front end (the 3-dB bandwidth of which was 170 MHz before equalization). After equalization, the front-end bandwidth is extended to >6 GHz. The linear channel also included a 3-tap transversal equalizer built from commercial microwave components, including phase shifters, wide-band amplifiers, attenuators, and a splitter, inverter, and combiner. The purpose of this adjustable equalizer was to improve on the frequency response and bandwidth obtained when a number of the wide-band linear-channel amplifiers were cascaded together. With the help of this equalizer, an overall receiver bandwidth of 6.9 GHz was obtained. The sensitivity for 10^{-9} BER at 1.3-μm wavelength as measured with a very fast directly-modulated laser source was −25.8 dBm for the best of the demultiplexed 2-Gbit/s output channels and −23.4 dBm for the worst 2-Gbit/s channel. Additional details and measurements of this receiver's performance have been previously published [8].

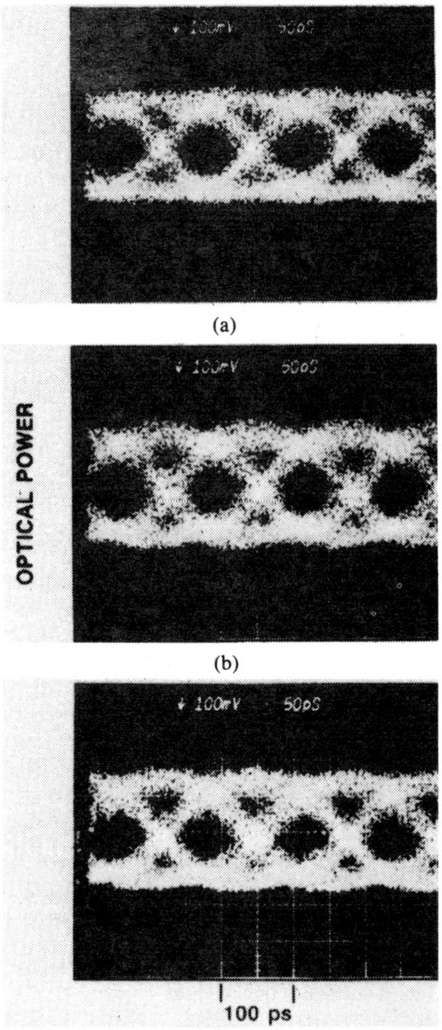

(a)

(b)

(c)

Fig. 2. 8-Gbit/s eye patterns: (a) transmitted signal from the Ti:$LiNbO_3$ modulator, (b) received baseline signal through short length of fiber, (c) received signal through 68.3 km of fiber.

Clock recovery was achieved with an 8-GHz phase-locked loop built from commercial microwave components. The phase-locked loop consists of a transition marker generation circuit (to produce marker pulses synchronized with bit transitions in the receiver data), a phase comparator, a dc amplifier, and an 8-GHz voltage-controlled oscillator (VCO). Marker pulses were derived from NRZ data at the receiver linear-channel output by splitting the signal, delaying one branch by one-half bit, and applying both signals to the inputs of a balanced microwave mixer. The mixer output provided transition marker pulses with a strong spectral component at the bit rate. This signal and the 8-GHz VCO output are applied to another balanced mixer acting as a phase comparator, providing a dc output signal proportional to the phase difference between the VCO and the marker pulses. This low frequency signal was amplified and filtered by a dc-coupled amplifier with a bandwidth of ~200 kHz, then applied to the frequency control input of the VCO. The VCO is a 2-GHz crystal-controlled oscillator with a voltage-controlled tuning range of 400 kHz. Its output is doubled twice in order to obtain the 8 GHz sinewave which goes to phase comparator. The fundamental 2-GHz VCO output is used as the clock to derive sampling pulses needed for the demultiplexer, described below, that extracts 2-Gbit/s data from the 8-Gbit/s receiver output. Measured performance characteristics of the 8-GHz phase-locked loop include a pull-in range of ±75 kHz and a hold-in range of ±200 kHz.

Any one of the four 2-Gbit/s streams can be separately recovered from the received 8-Gbit/s NRZ stream using a sampling circuit. This circuit consists of a dual-gate FET identical to the one used in the mulitplexer and a commercially available 2-GHz comb generator (step recovery diode) capable of producing 50 ps (full width at half maximum) pulses. The 8-Gbit/s data is applied to one gate of the FET and the sampling pulses are applied to the other gate. Coarse phase synchronization is accomplished with an adjustable coaxial delay-line phase shifter and active synchronization is maintained by the timing recovery circuit described above. Every fourth bit is sampled by the comb generator to recover one of the 2-Gbit/s streams. The sampled output is low-pass filtered with a 1300-MHz filter, amplified, and then checked for errors in the commercial 2-Gbit/s BER test set.

Results/Discussion

Received eye patterns for transmission through a baseline fiber length of ~3 m and the 68.3-km length are shown in Fig. 2(b) and (c), respectively. The bit-error rate was measured as a function of the received optical power for each of the 2-Gbit/s demultiplexed data streams. The transversal equalizer was adjusted for minimum error rate. In Fig. 3 we compare the BER versus received power for the 3-m baseline and the 68.3-km length for the best 2-Gbit/s streams. The baseline sensitivity for 10^{-9} BER was −22.7 dBm and for 68.3 km the sensitivity was −21.7 dBm, giving a dispersion penalty of 1.0 dB. The lowest error rate measured through 68.3 km was 3×10^{-10} at a received power of −21.3 dBm. Best sensitivity was obtained with an avalanche gain of $\langle M \rangle = 7$. There is no bending of the BER curve at low error rates, hence no evidence of an error rate floor. The remaining three 2 Gbit/s data streams all gave BER measurements that were within 0.5 dB of the best results shown in Fig. 3.

The baseline receiver sensitivity obtained in this experiment is slightly lower than that previously obtained with a fast laser [8], with the difference being the result of bandwidth limitations in the transmitter. The high-power amplifier used to drive the external modulator had a frequency response that was down by 2.5 dB at 5 GHz. An *RC* equalizer was used prior to this amplifier in an attempt to improve its eye pattern; however, the equalizer's effectiveness was limited by saturation of the amplifier output. Under optimal conditions the amplifier provided a peak-to-peak swing of ~8 V, although single bits in the pattern

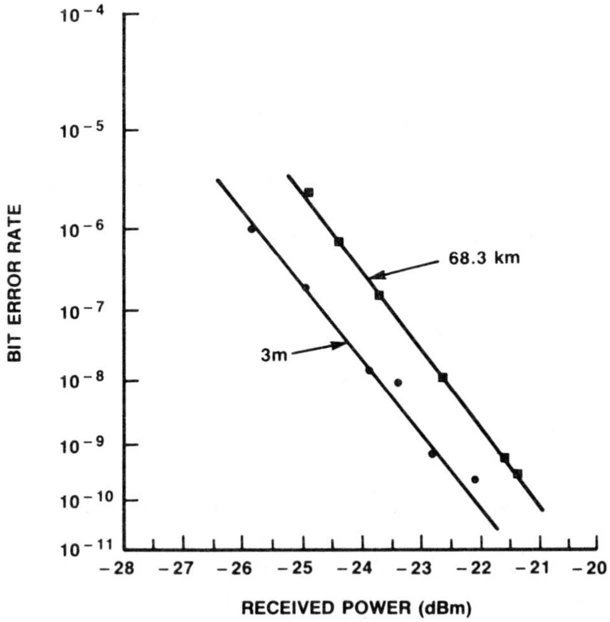

Fig. 3. Bit-error-rate curves.

saturated at 6 V_{p-p}. This driver voltage limitation in combination with the modulator's own rolloff caused a reduced bandwidth of the transmitted signal, leading to a receiver sensitivity reduction of about 1.2 dB for the worst 2-Gbit/s channel. Other factors affecting the receiver sensitivity include the limited gain-bandwidth product of the APD which reduces the avalanche gain that can be employed, and intersymbol interference from nonideal linear channel amplifiers which could not be perfectly equalized.

As in our previous external modulation experiment at 4 Gbit/s, the present dispersion penalty is intrinsic to the information bandwidth and fiber dispersion at 1.53 μm. Indeed, measurement of the time-averaged transmitted optical spectrum using a Fabry–Perot interferometer revealed that essentially all of the optical power is contained within a spectral range of ~ 16 GHz. The observed dispersion penalty of 1 dB and the B^2L product of 4400 $(\text{Gbit/s})^2 \cdot$ km attained are in good agreement with the estimated dispersion limit [9]. We note that systems may be operated beyond the 1-dB dispersion limit as the dispersion penalty increases quasilinearly with distance for penalties up to ~ 2 dB. Thus longer transmission distances at 8 Gbit/s are possible if the loss budget is increased by improved modulator performance as discussed earlier or through the use of higher power lasers.

SUMMARY

In conclusion, we have demonstrated 8-Gbit/s digital transmission over 68.3 km of single-mode optical fiber with the use of an external modulator. This is the longest transmission distance reported at this rate and is the largest bit rate times distance product for a single-wavelength fiber span. The dispersion penalty of 1 dB and B^2L product of 4400 $(\text{Gbit/s})^2 \cdot$ km are in good agreement with the estimated theoretical dispersion limit.

ACKNOWLEDGMENT

The authors thank T. C. D. Huo and C. A. Burrus for the C^3 laser, G. Eisenstein and L. W. Stulz for the AR coatings, K. Ciemiecki Nelson, L. G. Cohen, and V. Mazurczyk for the fiber, K. L. Hall for the optical reflectivity measurements, A. R. Chraplyvy and R. W. Tkach for the Fabry–Perot measurements, and R. A. Linke and R. C. Alferness for stimulating discussions.

REFERENCES

[1] A. H. Gnauck, J. E. Bowers, and J. C. Campbell, "8-Gbit/s transmission over 30 km of optical fiber," *Electron. Lett.*, vol. 22, pp. 600–602, 1986.

[2] A. H. Gnauck, S. K. Korotky, B. L. Kasper, J. C. Campbell, J. R. Talman, J. J. Veselka, and A. R. McCormick, "Information-bandwidth-limited transmission at 8 Gbit/s over 68.3 km of single-mode optical fiber," in *Tech. Dig. Conf. Optical Fiber Commun.* (Atlanta, GA), 1986, post-deadline pap. PDP9.

[3] S. K. Korotky, G. Eisenstein, A. H. Gnauck, B. L. Kasper, J. J. Veselka, R. C. Alferness, L. L. Buhl, C. A. Burrus, T. C. D. Huo, L. W. Stulz, K. Ciemiecki Nelson, L. G. Cohen, R. W. Dawson, and J. C. Campbell, "4-Gbit/s transmission experiment over 117 km of optical fiber using a Ti : $LiNbO_3$ external modulator," *J. Lightwave Technol.*, vol. LT-3, pp. 1027–1031, 1985.

[4] D. P. Wilt, R. F. Karlicek, K. E. Strege, W. C. Dautremont-Smith, N. K. Dutta, E. J. Flynn, W. D. Johnston, Jr., and R. J. Nelson, "Channeled-substrate buried heterostructure InGaAsP/InP lasers with vapor phase epitaxial base structure and liquid phase epitaxial regrowth," *J. Appl. Phys.*, vol. 56, pp. 710–712, 1984.

[5] G. Eisenstein, S. K. Korotky, L. W. Stulz, J. J. Veselka, R. M. Jopson, and K. L. Hall, "Anti-reflection coatings on lithium niobate waveguide devices using electron beam evaporated yttrium oxide," *Electron. Lett.*, vol. 21, pp. 363–364, 1985.

[6] K. Ciemiecki Nelson, D. L. Brownlow, L. G. Cohen, F. V. DiMarcello, R. G. Huff, J. T. Krause, W. A. Reed, D. S. Shenk, E. A. Sigety, J. R. Simpson, and K. L. Walker, "The fabrication and performance of long lengths of silica core fiber," *J. Lightwave Technol.*, vol. LT-3, pp. 935–941, 1985.

[7] W. S. Holden, J. C. Campbell, J. F. Ferguson, A. G. Dentai, and Y. K. Jhee, "Improved frequency response of InP/InGaAsP/InGaAs avalanche photodiodes with separate absorption, grading, and multiplication regions," *Electron. Lett.*, vol. 21, pp. 886–887, 1985.

[8] B. L. Kasper, J. C. Campbell, J. R. Talman, A. H. Gnauck, J. E. Bowers, and W. S. Holden, "An APD/FET optical receiver operating at 8 Gbit/s," *J. Lightwave Technol.*, vol. LT-5, pp. 344–347, 1987.

[9] P. S. Henry, "Lightwave primer," *IEEE J. Quantum Electron.*, vol. QE-21, pp. 1862–1879, 1985.

*

Steven K. Korotky was born in Paterson, NJ, on February 13, 1953. He received the B.A. degree in physics from Rutgers University, New Brunswick, NJ, in 1975, and the M.S. and Ph.D. degrees in physics from Yale University, New Haven, CT, in 1976 and 1980. His thesis work was on nuclear molecular states formed in heavy-ion collisions.

In 1980 he joined the Photonics Circuits Research Department of AT&T Bell Laboratories, Holmdel, NJ. His research there has been focused on integrated-optic devices and experimental lightwave systems.

Dr. Korotky is a member of the American Physical Society and the Institute of Electrical and Electronic Engineers.

Alan H. Gnauck was born in Teaneck, NJ, on August 30, 1952. He received the B.A. degree in physics from Rutgers University, Newark, NJ, in 1975 and the M.S. degree in electrical engineering from Rutgers University, Piscataway, NJ, in 1986.

From 1976 to 1982 he was employed by Exxon Research and Engineering Company, where his work was in the area of laser chemistry. In 1982 he joined AT&T Bell Laboratories, Holmdel, NJ, where he is currently working on high-speed optical communications systems.

*

Bryon L. Kasper (S'71–M'73) received the B.Sc. and Ph.D. degrees in electrical engineering from the University of Alberta, Edmonton, Canada, in 1973 and 1981, respectively.

From 1973 to 1975 he worked for Alberta Government Telephones in mobile radio and microwave radio systems. In 1981, he joined the Lightwave Systems Research Department of AT&T Bell Laboratories, Crawford Hill Lab., Holmdel, NJ, and is currently engaged in research on high-speed optical receivers and transmission systems. He has designed high sensitivity avalanche photodiode receivers and has collaborated in long-span transmission experiments at bit rates from 400 Mbit/s to 8 Gbit/s, and has also designed receivers and collaborated in coherent transmission experiments to 2 Gbit/s.

*

Joe C. Campbell (S'73–M'74) was born in Gorman, TX, on January 11, 1947. He received the B.S. degree in physics from the University of Texas at Austin in 1969, and the M.S. and Ph.D. degrees in physics from the University of Illinois at Urbana-Champaign in 1971 and 1973, respectively.

From 1974 to 1976 he was employed by Texas Instruments where he worked on integrated optics in GaAs/AlGaAs. Since 1976 he has been at the Crawford Hill Laboratory of Bell Laboratories in Holmdel, NJ, where he has worked on electrooptic devices for fiber optics applications.

Dr. Campbell is a member of Sigma Xi, the American Physical Society, and the Optical Society.

John J. Veselka received the B.A. and M.A. degrees in physics from the State University of New York, Binghamton, in 1974 and 1983, respectively.

He joined AT&T Bell Laboratories in 1981 where he has been engaged in research on optical guided-wave devices.

*

J. R. Talman was born in Pasadena, CA, on February 19, 1961. He received the A.B. degree in physics from Cornell University, Ithaca, NY, in 1984 and is currently studying electronics engineering at Rutgers University, New Brunswick, NJ.

He joined AT&T Bell Labs. in 1984 where he has engaged in research into high-speed low-noise fiber-optic receivers.

*

Alfred McCormick was born in Washington, NJ, on May 21, 1940. He received the B.S. degree in electrical engineering from Monmouth College in 1983.

He joined Bell Labs in 1972, as a Technical Associate, where he has been working in lightwave systems research. He is presently interested in high sensitivity receivers.

Research aiming for future optoelectronic integration

The Optoelectronics Joint Research Laboratory

Izuo Hayashi

Indexing terms: *Optical fibres, Optoelectronics*

Abstract: Progress in semiconductor optoelectronic devices in the past 20 years has been significant. With the progress of optical fibres, a new era of communication has been established by overcoming major difficulties in device technologies. However, if one considers the present status of optoelectronic technologies, one finds that a large potential advantage of semiconductor technologies, that of 'integration', has been left unrevealed. The monolithic integration of optical and electrical devices, called optoelectronic integrated circuit (OEIC), will result in a big improvement in optoelectronic technology. The Optoelectronics Joint Research Laboratory (OJL) has been established to develop new generic technologies for optoelectronic integration. Significant progress in the material process technologies has been obtained in bulk GaAs crystal growth, a new maskless dry process using focused ion beam doping, MBE and dry etching for GaAs. The future evolution of optoelectronic technology and OEIC markets is also discussed.

1 Introduction

The progress of optoelectronic semiconductor devices in the past twenty years has been significant. There were no such devices for practical use before that time and it was even difficult to foresee what kind of market would develop for these devices. The developments that actually occurred during that time, both in device technologies and in market opportunities, have been overwhelming. An entirely new field of communication has been established using semiconductor optoelectronic devices together with the ultra low loss glass fibres. The possibilities of loss reduction was predicted in 1966 [1], and this new technology has made significant progress in the past 20 years.

The realisation of a semiconductor laser in 1962 was the start of semiconductor light emitters [2]. In 1970 there was a major breakthrough in semiconductor laser technology, the double heterostructure (DH) [3]. Room temperature operation of semiconductor lasers using the DH structure [3] coincided with the realisation of the low loss glass fibre [4]; these developments showed the feasibility of optical communication. Efforts aimed at the development of practical devices for optical communication started. There were many difficult problems to overcome to reach the goal, in particular, degradation and mode control. After several years of struggle, AlGaAs DH lasers with a lifetime and mode control sufficient for communication were obtained [3]. As the minimum loss wavelength of quartz glass fibres had been shifted into a longer wavelength, lasers with another III-V material, InGaAsP, were developed [5]. Since then both of these lasers, as well as high brightness LEDs made of these materials, have been installed in optical fibre communication systems all over the world.

In another field of optical device application, semiconductor optoelectronic devices, visible light emitting diodes (LEDs) and photodetectors, began to be developed in the late 1960s and established a large market for display lamps and many other applications [6]; this has grown to almost a billion dollar market per year. Compactness, ruggedness and low prices, due to mass production of semiconductor devices, are fully utilised. The structures and characteristics of these devices are much simpler than those used for communication, and will not be considered in this article.

However, the background in optoelectronic technology around the end of the 1970s and new possibilities of optoelectronic integration will be discussed. The reasons for choosing OEIC technology as the target of the Optoelectronics Joint Research Laboratory (OJL) will be explained. It will be emphasised that new technological breakthroughs will be needed for the OEIC, which would lead to a new era in optoelectronic technology. Research subjects of OJL aiming at the OEIC and their achievements will be presented. Finally, the future prospect of optoelectronic technology and expectations for the OEIC market will be discussed.

2 Background for the Optoelectronics Joint Research Laboratory (OJL): optoelectronic technology in the late 1970s and new possibilities for optoelectronic integration.

As described in the introduction, in the 1970s the progress of optical devices had been significant. Although more improvements were needed for these devices, no fundamental problems were apparent in the optical fibre communication system applications, the major market for these devices.

When Sakurai, Tanaka and others started to discuss the OJL project [7], serious questions were asked concerning future developments of optoelectronic devices. Optoelectronic devices for communications, semiconductor lasers, high radiance LEDs or detectors, were high performance devices produced in small quantities. High performance, such as fast response, single-mode operation or reliability, was the major concern of the system, and one specific aspect of these semiconductor devices, high performance, including compactness, efficiency, and reliability, had been emphasised. Whereas little attention had been paid to other potential aspects, represented symbolically by silicon LSIs, monolithic integration and mass production.

Silicon technology provided an important model for compound semiconductor development: the great impact of Silicon devices arose from the high degree of integration possible and the consequent reduction of cost to unanticipated levels. This was based on the significant progress in fabrication technology. The older technologies, such as those used for 'alloy' transistors, had been replaced by new technologies, such as 'planar' design, and 'dry' processing.

One could have asked what the advantage would be of optical devices integration, as these devices were mainly used individually, and would there be any market for these devices?

I believed, as well as some others, that, even for such

optical devices, 'integration' was clearly the direction of evolution and a large market would be developed [8–10].

2.1 Optoelectronic integrated circuit, OEIC

Let me first discuss the optoelectronic integrated circuit (OEIC) (Fig. 1).

Optical devices such as lasers or photodetectors are essentially devices which convert electrical signals to optical or vice versa, utilising the basic characteristics of semiconductors. Therefore optical devices are always connected with electronic devices, as we see, for example, in optical transmitters or repeaters. Therefore, if one considers integration of optical devices, that integration will include electrical devices as well. Integrated transmitters or repeaters are good examples of OEICs.

High performance and low cost will be the general benefits of optoelectronic integration.

Higher speed can be expected with an OEIC than with a hybrid circuit with discrete devices, in which optical devices are wire bonded with their electrical counterpart, because monolithic integration will minimise parasitic impedances between optical and electrical devices. Faster transmitters or receivers can be obtained reproducibly with integration rather than with the hybrid approach. Such improvements will be vital for gigabit optical communication in the future.

In short-distance optical-fibre networks with many interconnections, OEIC will be useful for a higher system performance as well as for lowering the system cost. This was, in fact, the first argument for choosing OEIC technology as the target of the OJL. Many OEICs located at the intersection points of the network will serve as distributed exchange nodes of the system.

The unique aspect of the OEIC is the fact that it combines two different kinds of device, one suitable for signal processing and the other for signal transmission. Optical signal transmission is superior to electrical transmission over both long and short distances. For short-distance optical signal transmission in electrical systems, a concept expressed as 'optical interconnection', which is like excess signal delay in conventional electrical interconnections, can be avoided, as a result of the charging time of stray capacitance. This will become the limiting factor for high-speed high-density electrical systems and appears between consoles or LSIs. Increased charging time delay due to increased 'fan out' in electrical systems will also be avoided with optical interconnections.

Free-space light propagation can also be used in optical interconnections, which eliminates the complex network of wiring usually found in electrical systems [10]. In addition to high speed, the optical interconnection will provide electrical voltage isolation (DC isolation) between electrical circuits, and immunity for electrical disturbances. This DC isolation will be quite useful in many electrical systems, because circuits of different DC levels can therefore be safely connected, which results in additional freedom in the system design.

Thus, the use of OEICs instead of conventional electrical ICs provides all the benefits of optical interconnection to the system. Systems with many interconnections, such as those of computer systems, will benefit by the use of OEICs.

However, because of many difficulties in fabrication, we are not able to enjoy all the benefits of OEIC described above. Compound semiconductor technologies, particularly those for optical devices which were available at the end of the 1970s, were far from those required to make OEICs.

For example, 'cleaving' had been the standard method for fabricating cavity mirror facets of semiconductor lasers, and this could not be applied to integrated lasers. 'Etching' had to be used instead. Laser stripe fabrication was another problem requiring new technology for integration. Most of the stripe processes for high-performance lasers had been complex, often needing two epitaxy steps plus wet chemical etching. These were generally delicate procedures and were very difficult to control. Fabrication yield was low, even for discrete devices, and much lower than the value needed for integrated devices.

Molecular beam epitaxy (MBE) had been developed, and it had achieved pretty good control for the thin layers needed for optical devices [11]. However there was no appropriate technique available for controlling the dimensions of optical devices parallel to the epitaxial layer plane. Hence we needed to make a stripe structure with simple but precise submicron definition. Some kind of 'planar' technology was needed for a new generation lasers suitable for integration.

Elsewhere the status of compound semiconductor electronic devices, another important part of the OEIC, had taken a significant step forward. For GaAs field effect transistors (FETs) development had already started in fabrication technologies for integration. The use of ion implantation, instead of epitaxy, to fabricate active layers was initiated by a group at Rockwell in the 1970s and was becoming a standard technology for GaAs FETs [12]. The

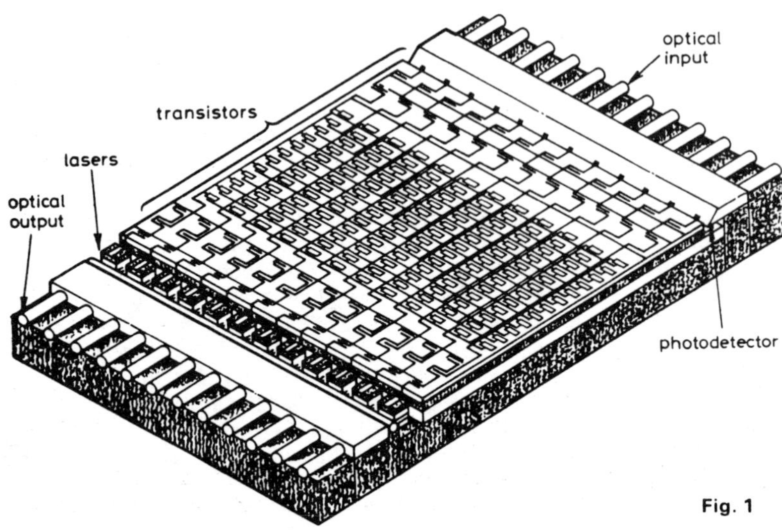

Fig. 1 *Schematic diagram of an optoelectronic integrated circuit*

self-aligned gate technique, which was used in conjunction with ion implantation [13], was becoming recognised as a technology well suited for integration. Using these technologies, developments aimed at large scale GaAs FET integration had just started.

Before these new 'planar' technologies were introduced, GaAs FETs had used old methods which were only suitable for discrete devices. This situation was quite similar to that of optical devices. A very serious obstacle still remained, however, which prevented GaAs FET integration (GaAs IC): the very poor quality of GaAs substrates.

To obtain OEICs, several breakthroughs in process technologies were therefore required, particularly those for optical devices, as described above. New innovative processes were required to solve these problems.

3 Optoelectronics Joint Research Laboratory (OJL)

To fabricate OEICs, high-grade materials and high-process technologies would be needed.

The Optoelectronics Joint Research Laboratory was established in 1981, as part of the optical measurement and control systems†. A total of about 50 researchers have been temporarily transferred from nine major electronics companies and the Japanese Electrotechnical Laboratory*.

Several key OEIC processes were chosen as targets for the OJL, because they required large investments of apparatus and man power, making it difficult for individual companies to pursue them. These targets were as follows [7,8]:

(*a*) High quality bulk GaAs crystal growth for OEICs and characterisation of grown crystals.

(*b*) A focused ion beam (FIB) doping system and a new maskless process system for OEICs

(*c*) A dry etching system

(*d*) Epitaxial growth with MBE and OMVPE

(*e*) Surface and Interface characterization and their application.

GaAs and AlGaAs were chosen as materials for all the processes mentioned above because GaAs based OEICs would have many applications for short-distance communications and for the optical interconnections in electronic systems. Details of each research target will be described later.

3.1 High quality bulk GaAs crystal growth

The quality of LEC GaAs crystals produced when OJL started was so poor that it was difficult to satisfy either optical or electrical device requirements for OEICs. Substrate crystals must be semi-insulating and defect free. Semi-insulating LEC GaAs crystals available at the time had a high density of dislocations (10^4–$10^5/cm^2$) and were homogeneous. In order not to degrade light emitting devices, dislocation-free crystal was mandatory. In addition, the crystal inhomogeneity resulted in large fluctuations in GaAs FET characteristics, so that large scale integration was impossible. It was believed necessary to control the stoichiometry of this binary crystal to minimise native defects. However, it was not entirely clear whether any technique would be available to solve these problems. Shortly after the laboratory was open in January 1982, a newly developed automatic diameter control system [14] was placed into operation. Control within 1% was obtained for 2 inch diameter GaAs crystal. Most of the prior to LEC growth of GaAs had been carried out manually, and the result was poor.

Next, a new purification method was tried and proved to be effective, including a new *in situ* monitoring [14].

However, the real challenge was suppression of dislocations which were introduced primarily by thermal stress in the furnace.

After two years of struggle, nearly dislocation-free undoped GaAs crystals were obtained in 1985, using a specially designed apparatus [15]. These crystals were grown in an arsenic atmosphere, so that an extremely low temperature gradient could be used during growth. Arsenic evaporation from the GaAs ingot grown was suppressed by the As atmosphere. Otherwise the evaporation would have eroded the crystal severely. The arsenic pressure was maintained by a container placed inside the high pressure chamber. A new X-ray viewing apparatus had to be developed to monitor the growth in this double container [15] (Fig. 2). Average dislocation densities of hundreds per cm^2 undoped LEC GaAs crystals about two inches in diameter

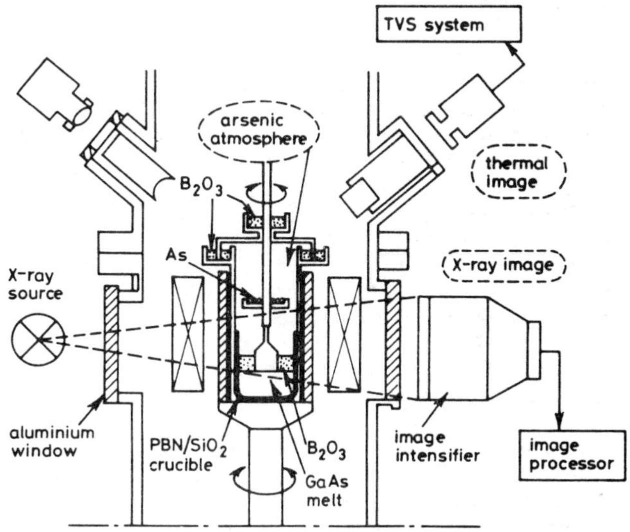

Fig. 2 *Schematic diagram of As atmosphere LEC growth system for undoped dislocation-free GaAs crystals [16]*

were obtained for the first time. Although dislocation-free crystals had been obtained with high In doping [16], undoped crystals would be far more advantageous, leading to higher uniformity and yield.

Another large obstacle for controlling GaAs crystals was the nonstoichiometry problem. A small difference between Ga and As composition, of the order of one part in ten thousand or less, created excessive native defects, such as host atom vacancies or interstitials. Principal among these was the arsenic antisite (As_{Ga}), where an arsenic atom misplaced into a gallium site [17], created the midgap level, usually referred to as the EL2, which was responsible for the semi-insulating behaviour. The As_{Ga} was therefore a useful defect, but it was necessary to control its generation during the crystal growth [18]. The As_{Ga} density depended sensitively on crystal non-stoichiometry.

Adjustment of the melt composition was the first step towards the stoichiometry control. The melt composition

† OJRL is a Japanese national project and is sponsored by the Agency of Industrial Science and Technology, an agency of the Ministry of International Trade and Industry

* The director of the OJL is Takashi Iizuka and the nine companies are Fujitsu, Furukawa, Hitachi, Matsushita, Mitsubishi, NEC, Oki, Sumitomo and Toshiba. The OJL will operate from 1981 to 1986 and the total budget will be approximately forty million US dollars, including the companies' expenditure, which is about 30% of the total budget

was changed by arsenic evaporation during the growth, making it quite difficult to maintain a constant melt composition throughout the crystal growth.

A new technique, 'As injection', was developed and found to be an excellent technique to control stoichiometry [18]. By injecting As before and during growth, the melt composition was stabilised quite effectively throughout the growth [18]. For the first time crystals with uniform characteristics in the plane of the wafer, as well as in the growth direction, have been achieved consistently [18].

A magnetic field was applied to LEC growth and proved to be effective for suppressing striations [19]. It was also found that the midgap level EL2 could be controlled better by the magnetic field [19].

3.2 Bulk GaAs characterisation

The characteristics of these LEC grown GaAs wafers have been evaluated by a variety of electrical and optical techniques, including TEM, X-rays, SIMS, DLTS, photoluminescence, infrared absorption, EPR etc. However, evaluation was difficult because these techniques were not established for crystal defects, particularly for point defects. Large dislocations were relatively easy to observe, but atomic defects, such as vacancies, interstitials or antisites, which often controlled the optical or electrical characteristics, were very difficult to measure. These atomic defects existed in small quantities (parts per 10^6 or less), yet they played a crucial role for device characteristics. The information obtained was indirect and insufficient to identify the atomic origin and to understand the behaviours of these defects. The arsenic, antisite As_{Ga}, had received great attention, yet it had not clearly identified.

The amount of nonstoichiometry, which should be very important for point-defect generation, was almost impossible to measure. Recently a few approaches have been made for obtaining information correlated with the amount of nonstoichiometry [20–22].

At OJL, measurements have been carried out to obtain correlation between the growth conditions and crystal characteristics for establishing optimum growth conditions for OEIC applications.

Efforts have been concentrated to contribute better control of crystal growth and to achieve a consistent picture of the semi-insulating GaAs characteristics [23–26].

The tests of these GaAs crystals as substrates for device fabrication have been performed in collaboration with member company laboratories in the research association. Crystals were evaluated by the characteristics of a group of MESFET devices fabricated on them. Stoichiometry controlled crystals gave much smaller FET threshold voltage distributions than those made on conventional LEC crystals [18]. An LEC GaAs semi-insulating crystal was tested for the first time as a substrate for semiconductor lasers. The result was satisfactory [27]. These results indicate that LEC crystals of satisfactory quality can be made for electrical devices as well as for optical devices, which in turn indicates feasibility for OEICs.

3.3 Focused ion beam (FIB) and the maskless process system

The FIB doping and the maskless process activities were exploratory research aiming at a totally new process technology, which would open up possibilities for new device structures needed for OEIC construction.

It had been known that ordinary (non-clean) interfaces or bare surfaces of GaAs contain many defects, which trap carriers causing electrical or optical deficiencies [28]. One of my important findings in double heterostructure (DH) research was the clean defect-free interface between GaAs and AlGaAs, which could be obtained by successive growth using liquid phase epitaxy (LPE) [3]. This clean interface was the basis of a successful DH-laser structure.

A clean as-grown surface of GaAs is defectless. However, as soon as it is exposed to air, an environment which contains oxygen or H_2O, defects are created instantly, which trap carriers [28]. For this reason, an interface made by any epitaxial or deposition process on an exposed GaAs (AlGaAs) surface always contains defects. One important example is the process of photolithographic masking, which always creates defects on the GaAs crystal surfaces. The FIB ion doping, on the other hand, does not need any masking to define doping patterns for device structures. If the FIB doping apparatus is connected to an MBE machine through an ultra-high vacuum duct, the combination of doping and epitaxy can be carried out without degrading the surface.

The FIB apparatus at OJL utilised a liquid-metal ion source with an extremely small diameter (hundreds of Å), so that an accelerated and focused ion beam of less than 0.1 μm in diameter could be obtained [29]. The acceleration voltage was 100 to 40 kV. An ion source metal containing both *n*-type (Si) and *p*-type (Be) elements, (Au-Si-Be) alloy was developed [29]. By switching a mass separator bias, Si or Be ions with the desired energy could be extracted.

The ion source has been working stably for several hundred hours, thus establishing the usefulness of this apparatus. A computer system to control this FIB apparatus was developed, and a variety of operating modes, such as scanning, positioning and doping, can be carried out with appropriate software [30]. Focused ion doping characteristics were tested and found to be as good as conventional nonfocused ion implantation, both in electrical and optical characteristics, even though the ion density of the FIB (~ 1 A/cm^2) was several orders of magnitude higher than is conventional [29]. Special attention was paid to the optical characteristics of doped layers, which would be important for OEIC fabrication.

Testing has been carried out on the maskless system (Fig. 3). No interface traps, either electrical or optical, were produced with this *in situ* process of multiple epitaxy and doping [30]. No trace of oxygen or carbon, which are commonly found on air-exposed and ion-implanted surfaces, was detected at the interface. A demonstration to fabricate a stripe laser was successful. The first AlGaAs layer in the DH structure was semi-insulating, and an FIB doping stripe provided a current path into the active layer [30]. It was shown that the optical quality of the GaAs/AlGaAs interface, which was exposed in a vacuum of 5×10^{-10} Torr for three hours, was equally good as a similar interface made in the MBE without growth interruption [30]. Positioning of FIB dopings in different epilayers was feasible, using an etched marking on the substrate and detecting secondary electrons, within a few tenth of microns [30]. A new 'self aligning' process can be carried out with this technique, which will be quite versatile for device fabrication.

A new expansion of this maskless system, by adding an etching and an annealing chamber, is under way. The system capability will be increased greatly.

Structures suitable for OEIC integration, i.e. a variety of 'planar structures' can be made with this technique. A greater ability to reproduce structures with submicron dimension will bring optical device technology from a handcraft to engineering.

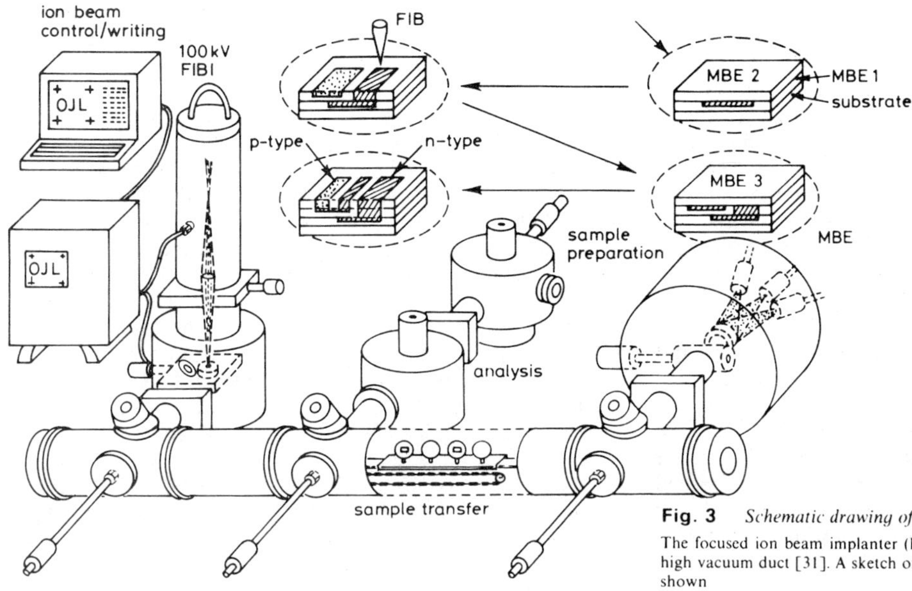

Fig. 3 *Schematic drawing of the maskless process system*

The focused ion beam implanter (FIBI) is connected to an MBE through an ultra high vacuum duct [31]. A sketch of the pattern doped multilayer growth progress is shown

under proper conditions, one can achieve etching with the desired shape and with the minimum amount of damage in GaAs. Such control of the etching is essential for GaAs processing.

A demonstration of vertical mirror etching was performed with a GaAs DH laser structure [34]. Laser characteristics, threshold and efficiency were identical with those made by the cleaving technique [34] (Fig. 4B). Another handcraft was eliminated by the new engineering.

3.4 Reactive ion beam etching (RIBE)

An etching process suitable for GaAs devices has been developed using Cl_2 gas RIBE. Study showed that the conventional RIE apparatus with parallel plates, developed mainly for Si, produced damage on etched GaAs surfaces. Therefore a new experimental etching apparatus with microwave discharge under a magnetic field was constructed. Ions as well as radicals were extracted from the discharge chamber on to a GaAs wafer placed in a high vacuum chamber [32].

It was found that chlorine radicals enhanced the etching rate with a given amount of ions, and reduced etching damage at the same time [31]. The beam intensity ratio between radicals and ions could be adjusted by the discharge condition. Over ten Ga and As atoms were removed per one incoming ion, which was one to two orders of magnitude larger than that which can be obtained by argon ions. A 'vertical' etching with submicron definition was obtained with a shadowing mask, as shown in Fig. 4A [32]. It was also shown that the chlorine radicals alone could etch GaAs or AlGaAs, if the substrate temperature was raised. In this case, the etching was purely chemical and no damage was produced [33]. It is then clear that with a combination of radicals and ions

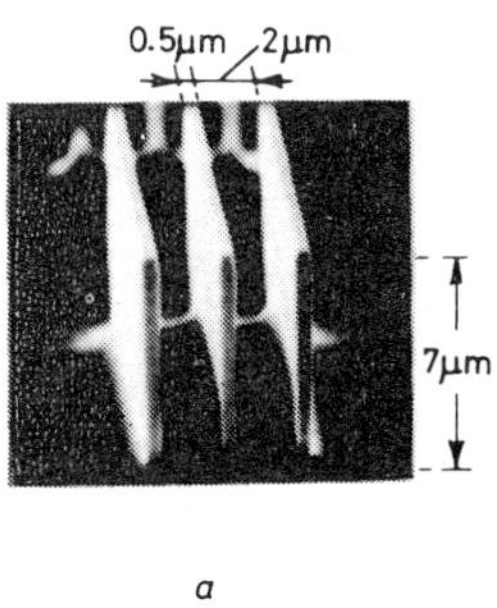

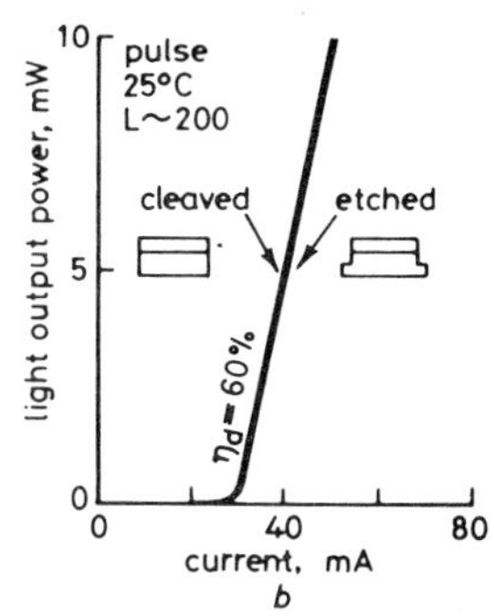

Fig. 4A *SEM photograph of vertically etched GaAs structures*

Etching was carried out with a hard baked mask at CL_2 gas pressure of 8×10^{-4} Torr, ion extraction voltage was 500 V etching rate 0.2 μm/min

Fig. 4B *Comparison of etched and cleaved laser characteristics made from the same wafer* [35]

3.5 New surface process and evaluation techniques

A group has been working to develop new surface interface evaluation techniques and surface related processes.

A special XPS apparatus equipped with a small MBE chamber was built, with which a fresh (100) surface could be provided *in situ*. Reactions of metals deposited on the (100) clean GaAs surfaces was studied [35]. A new investigation on LaB_6 for feasible Schottky electrodes on GaAs is underway, and gives promising characteristics [36] for self-aligning MESFET gate electrodes.

A Rutherford backscattering apparatus (RBS) equipped with an MBE and an X-ray analysing system (PIXE) has been developed. Investigation of the lattice location of impurities in GaAs crystals [37] and studies on interfaces between GaAs and deposited metals and insulators were performed [35].

To develop a new buried heterostructure (BH) process, disordering [38] of GaAs/AlGaAs superlattices had been investigated. The BH stripe structure lasers have been known to provide an ideal lateral confinement, but their fabrication procedure is complicated and delicate. It is a typical handcraft process which is still used extensively. If the 'disordering', smoothing process of the GaAs/AlGaAs superlattice structure [38] can be applied effectively to bury regions outside a stripe, a BH stripe structure will be obtained by a well-controlled procedure in one step. Diffusion of some impurities can cause the disordering; thus a BH stripe can be made by selective impurity diffusion. A laser with a disordered BH stripe (DBH) by Zn diffusion, worked successfully [39]. Another disordering experiment using Si implantation, is under investigation [40]. Combining with the FIB doping technique, it is also hoped to achieve a disordering process with fine definition, such as the DFB laser structure. These are examples of 'planar' structure laser process needed for OEICs. Better ability to reproduce, higher yields and higher reliabilities will be obtained by such planar processes.

3.6 *Exploring new epitaxial processes for OEICs*

Molecular beam epitaxy (MBE) was known to provide excellent epilayer thickness control, which was difficult with liquid phase epitaxy (LPE), a common technology for optical devices. However, detailed study on the MBE in OJL revealed that the optical quality and luminescent efficiency of AlGaAs layers grown by MBE was generally inferior to that of LPE AlGaAs [41]. Studies aiming for high luminescent AlGaAs are underway using MBE and OMVPE techniques.

Low pressure OMVPE growth, 10^{-2}–10^{2} Torr, has been studied extensively at OJL. Layers of GaAs and AlGaAs with high luminescent efficiency have been obtained. It was also found that in the low pressure MOCVD, a highly selective planar epitaxy of GaAs and AlGaAs could be obtained, which was difficult at higher pressures [42]. An embedded local epitaxy was obtained by this technique, which will be very useful for optical device epitaxy on an OEIC substrate without producing a step height (Fig. 5) [42]. Studies on gas reaction in this

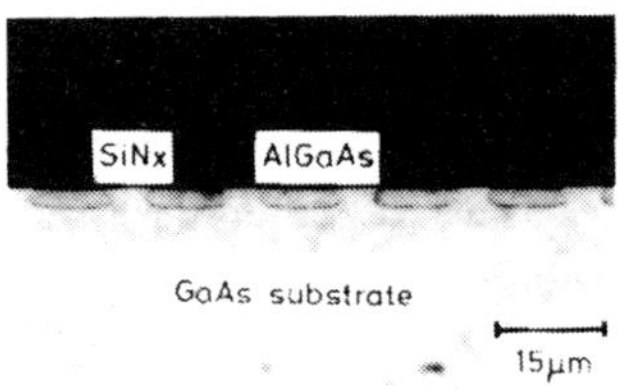

Fig. 5 *$Al_{0.35}Ga_{0.65}As$ selective epitaxial layers grown into etched grooves*

Areas outside the groove were covered by SiN_x. TMA, TMG and AsH_3 were used at 10 Torr [43]

system are under way, in the hope of clarifying the kinetics in the OMVPE process.

Research on OMVPE in a high vacuum has started, which hopes to combine the benefits of OMVPE and MBE. Ultraviolet laser stimulation is combined, in order to obtain high quality epitaxies at lower temperatures with spacial selectivity. These characteristics will be quite effective for the process interference problem between electrical and optical devices in OEIC processes.

4 Summary and future development of OEIC technology

GaAs technology, and that of some other III–V materials, has entered a stage of revolution. Until the late 1970s the III–V devices, optical as well as electrical, had been utilised extensively in a variety of areas, such as optical communications, microwaves, displays and optical instrumentation. Their characteristics were unique, and obtainable only with III–Vs. These compounds had been used mainly for discrete devices, and the device structures had been rather complex, except for those for simple display use. The devices had been fabricated by sophisticated 'handcraft-like' techniques, made by skilled workers. As a result, the devices were difficult to make and therefore expensive. However, some of the electronic devices, GaAs MESFETs, have recently taken the first step towards monolithic integration. New 'planar' technologies, such as ion implantation [12] or self-aligning gates [13] have been developed.

In 1981 the Optoelectronics Joint Research Laboratory (OJL) started research aimed at the technological breakthroughs needed for optoelectronic integrated circuits (OEICs). Several generic technologies, which had been missing for OEIC fabrication, were selected as targets. Four years later, significant progress has been made. Several breakthroughs have been achieved in GaAs substrate growth [14, 15, 18–19], focused ion maskless implantation [29, 30], dry etching [31–34] and other process technologies [35–37, 39–42].

The quality of LEC GaAs material has been greatly improved, owing to efforts at OJL, as well as at many other laboratories [43]. Control of defects, dislocation, impurities and deep levels has progressed significantly. The progress has eliminated an important bottleneck in GaAs IC technology. Because of the improved substrate quality, large scale integration of GaAs LSI is becoming feasible. However, there are still many other technical problems left unsolved, such as surface related treatments.

It is expected that GaAs ICs will be used for practical applications in the near future, starting from small-scale microwave circuits. For large-scale high-speed digital integrated circuits, competition with Si LSIs will be a big challenge. Cost reduction will be the major system demand, which will require a total upgrading of the fabrication technology.

For optical devices, a new application of semiconductor lasers, that of optical memory, has recently emerged [44]. Although this is another application involving discrete devices, the emphasis has shifted from quality to quantity, from performance to cost. More than several million devices per year are now being produced, which is orders of magnitude larger than the number for communication [44]. Lower price will induce more applications. However, for further development, new high-yield low-price technologies, which are essentially the same as OEIC technologies, will be required.

Efforts on small-scale OEICs have already started in many laboratories. Various combinations, such as a laser and driving transistors [45], a photodetector and amplifiers [46], are reported. Some use etched mirrors to form the laser cavity. However, these early attempts of integration primarily use conventional technologies, which are still difficult to apply for large scale integration. The experience gained by this early work on integration made it clear that an altogether new technology was needed. For example, a step height made by epitaxy for optical devices should be eliminated. For OEICs, problems are expected due to the coexistence of optical and electrical devices' 'process interference' problems. New generation process technologies, such as those developed at OJL, will be useful for further development of this field.

It was demonstrated at OJL that high quality laser facets can be fabricated by dry etching [31–34]. This is a significant breakthrough in semiconductor laser technology, and it makes possible OEIC integration, as well as mass production for low price discrete devices.

The superlattice disordering is another technique which will be useful for integrated laser structures. A high quality BH stripe laser structure can be made by this 'planar' technique [39], and the use of ion implantation is particularly attractive [40]. The combination of disordering and dry etching should make it possible to fabricate extremely low threshold lasers, which will be compatible with FETs on the same OEIC.

For optical detectors, planar structures using ion implantation have been used successfully [47]. Problems for detectors are similar to those for electrical devices.

The focused ion beam doping combined with epitaxy and etching in ultra-high vacuum systems will be extremely useful for further advancement of OEIC technology [29, 30]. New optical and electrical device structures will be made possible by these techniques. However, for further advancement of OEIC technology, selective etching, epitaxy and deposition without conventional masking and other techniques, such as high quality 'passivation', will be necessary.

Combining the technologies described above, a new generation of OEIC devices could be produced in the near future, which correspond to silicon 'planar' devices. Devices should be made with submicron precision, with techniques which are easy to reproduce, and with high quality materials, to guarantee high yields.

These new 'planar' technologies for OEICs will also have a large impact on discrete devices. Drastic improvement in productivity and price reduction will expand their market greatly.

4.1 OEIC market

The first application of OEIC will be in high-speed optical communication, where improvements in system characteristics will be the major concern. As the requirements are for small quantities and little integration, 'artistic' and high-priced fabrication techniques will be allowed. InP-based small-scale OEIC will be used. For larger scale integration, InP-based optical devices monolithically integrated with GaAs-based transistors will be the solution‡. On the other hand, for applications for short distance fibre networks, higher integration and lower price will be most important, where GaAs OEIC with extended planar technologies will be needed. A larger market is expected for GaAs-based OEICs, which include the application of optical interconnection of electrical systems to be described shortly.

There is another benefit of OEIC technology, which can be introduced in the relatively short term. The monolithic integration of optical devices will provide a high degree of geometrical reproducibility between them. For instance, the relative position of a laser and a monitoring detector can be reproduced with micrometer precision. A phased laser diode array can be made only by a monolithic integration, as has already been demonstrated [48]. It will minimise the troublesome assembling process of an optical system, such as that in an optical memory equipment. There will be many such applications which require relatively simple techniques.

It is likely that the first use of OEICs for electrical systems will be for ultra-high-speed digital systems, where GaAs-based FETs will play the major role, and optical interconnection will assist them to speed up the system performance, and thus guarantee their superiority over silicon systems.

Accompanying such progress in fabrication technology and cost reduction, more and more optical interconnects will be introducced in electrical systems between consoles, between LSIs, or even between some individual devices inside LSIs, in the future. Electrical isolation, immunity from electrical disturbances, or a smaller volume of interconnection, will become the major reasons for the introduction, rather than the speed requirement.

Eventually, silicon-based conventional LSIs will be converted into OEICs for these purposes. Recent developments of high quality GaAs epitaxy on Si have made this feasible [49]. For Si-based OEICs, only small area islands of GaAs will be needed for light emitters.

It will take time before the technology of III–Vs becomes compatible with silicon technology. Even though III–Vs are very difficult materials to handle, it should be possible to achieve a high level of integration, once the technologies are established.

For optical interconnection between LSIs, new optical waveguiding systems will be needed. Some sort of 'optical printed board' replacing metallic wire printed boards will be used. 'Optical wiring' will be needed to interconnect devices on a chip. Another scheme involves the free-space interconnection [10]. The increased flexibility introduced by such interconnections will be useful for future computers.

It will be extremely important to have the freedom to choose either light or electricity in a given system, whichever is most suitable. Such choices will become possible with the development of OEIC technology. Eventually, systems which can take best advantage of both photons and electrons will find the most extensive use. I believe this approach should give birth to true 'optical' computers, and OEIC technology will lead the way.

‡ High quality InP epitaxy on GaAs should be possible using a similar technique with GaAs on Si [48].

5 Acknowledgment

The author gratefully acknowledges the assistance of J.L. Merz, T. Iizuka and M. Mihara for critically reading the manuscript and valuable discussions, and H. Hashimoto for providing valuable suggestions related to the GaAs FET IC technology development.

The author wishes to thank members of the OJL for providing material included in this paper.

Work at OJL described in this paper is supported by the Agency of Industrial Science and Technology, Ministry of International Trade and Industry, Japan.

6 References

1 KAO, K.C., and HOCKHAM, G.A.: 'Dielectric-fibre surface waveguides for optical frequencies', *Proc. IEE* 1966, **113**, (7), pp. 1151–1158

2 HALL, R.N.: 'Injection lasers', *IEEE Trans.*, 1976, **ED–23**, pp. 700–704

3 HAYASHI, I.: 'Heterostructure lasers', *ibid.*, 1984, **ED–31** (11), pp. 1630–1642

4 KAPRON, F.P., KECK, D.B., and MAURER, R.D.: 'Radiation losses in glass optical waveguides', *Appl. Phys. Lett.*, 1970, **17**, pp. 423–425

5 HSIEH, J.J., ROSSI, J.A., and DONNELLY, J.P.: 'Room temperature cw operation of (Ga, In) (As, P)/InP double-heterostructure diode lasers emitting at 1.1 μm', *ibid.*, 1976, **28**, pp. 709–711

6 BERGH, A.A., and DEAN, P.J.: 'Light emitting diodes' (Clarendon Press, 1976)

7 SAKURAI, K. 'The optoelectronic industry in Japan: development and prospects', *Optics News*, 1985, pp. 6–14

8 HAYASHI, I.: 'Development of optoelectronics IC's in Japan — Activities of the Optoelectronics Joint Research Laboratory', *Jpn. Semiconductor Technology News*, 1984, **3**, (1), pp. 32–38

9 YARIV, A.: 'The beginning of integrated optoelectronics circuits', *IEEE Trans.*, 1984, **ED–31**, (11), pp. 1656–1661

10 GOODMAN, J.W., LEONBERGER, F.I., KUNG, S.Y., and ATHALE, R.A.: 'Optical interconnection for VLSI systems', *Proc. IEEE*, 1984, **72**, (7), pp. 850–866

11 BARNHAM, R.D., and SCIFRES, D.R.: 'Integrated optical devices fabricated by MBE', Molecular Beam Epitaxy, Progress Crystal Growth Charact., Vol. 2, pp. 95–113, Pergamon Press Ltd., 1979

12 EDEN, R.C., WELCH, B.M., and ZUCCA, R.: 'Planar GaAs IC technology: Applications for digital LSI', *IEEE J.*, 1978 **SC–13**, (4), pp. 419–426

13 YOKOYAMA, N., OHNISHI, T., ODANI, K., ONODERA, H., and ABE, M. 'TiW silicide gate self-alignment technology for ultra-high-speed GaAs MESFET LSI/VLSIs', *IEEE Trans.*, 1982, **ED–29**, (10), pp. 1541–1547

14 FUKUDA, T.: 'Growth of undoped semi-insulating GaAs single crystal', *Jpn. J. Appl. Phys.*, 1983, **22**, Suppl. 22–1, pp. 413–418

15 OZAWA, S., MIYAIRI, H., NAKAJIMA, M., and FUKUDA, T.: 'Growth and characterization of undoped dislocation free GaAs by arsenic ambient controlled LEC with the aid of X-ray imaging system', Proc. Gallium Arsenide and Related Compounds Conf., 1985, Karuizawa, Japan

16 Jacob, G., DUSEAUX, M., FARGES, J.P., VAN DEN BOOM, M.M.B., and ROKSNOER, P.J.: 'Dislocation-free GaAs and InP crystals by isoelectronic doping', *J. Cryst. Growth*, 1983, **61**, pp. 417–427

17 MAKRAM-EBEID, S., LANGLADE, P., and MARTIN, G.M.: 'Nature of EL2: the main native midgap electron trap in VPE and bulk GaAs', Semi-insulating III–V Materials, Kah-nee-ta, pp. 184–203 (Shiva Publishing Ltd., 1984)

18 INADA, T., SATO, T., ISHIDA, K., and FUKUDA, T.: 'Homogeneous semi-insulating GaAs crystal growth by new LEC technique with As injection into melt', *J. Electron. Mater.* (to be published)

19 TERASHIMA, K., and FUKUDA, T.: 'Growth and characterization of undoped GaAs single crystals by a new magnetic field applied LEC technique', Proc. of the Symposium on III–V Optoelectronics epitaxy and device related process', *Electrochem. Soc. Proc.*, 1983, **83–13**, pp. 79–82

20 FUJIMOTO, I.: 'X-ray diffraction analysis of stoichiometry of GaAs crystal', Extended Abstracts of the 16th (1984 International) Conference on Solid State Devics and Materials, Kobe, 1984, pp. 165–168

21 TERASHIMA, K., WASHIZUKA, S., NISHIO, J., OKADO, A., YASUAMI, S. and WATANABE, M.: 'Non-stoichiometry effects on crystal properties of undoped LEC GaAs', 12th International Symposium on GaAs and related Compounds, September 1985, Karuizawa, Japan, pp. 37–42

22 TAKANO, Y., ISHIBA, T., MATSUNAGA, N., and HASHIMOTO, N.: 'Measurements of compositional change in semi-insulating GaAs single crystals by precise lattice parameter measurements', *Jpn. J. Appl. Phys.*, 1985, **24**, (4), pp. L239–L240

23 YAHATA, A., SATO, T., KIKUTA, T., and ISHIDA, K.: 'Detailed analysis of DLTS signal of EL2 in LEC n-GaAs crystals', Mat. Res. Soc. Symp. Proc., **46**, pp. 179–184

24 ISHIDA, K., YAHATA, A., and KIKUTA, T.: 'Distinction between midgap levels in LEC n-GaAs determined by DLTS and optical absorption at 1.1 um', *Jpn. J. Appl. Phys.*, 1985, **24**, (4), pp. (4), pp. L250–L252

25 TSUKADA, N., KIKUTA, T., and ISHIDA, K.: 'Photo-electron paramagnetic resonance study of As_{Ga} antisite defect in as-grown GaAs crystals of different stoichimetry', *Jpn. J. Appl. Phys.*, 1985, **(9)**, pp. L689–L692

26 TANIO, H., and TAJI, A.M.: 'Radiative recombination process of deep levels in GaAs manifested by photoluminescence under high pressure', 12th International Symposium on GaAs and related compounds, September 1985, Karuizawa, Japan, *Inst. Phys. Conf. Ser.*, 1986, **79**

27 OKADA, H., KATSUMATA, T., OBOKATA, T., and FUKUDA, T.: 'LEC growth of undoped SI GaAs single crystal for optoelectronic application', *J. Crystal Growth* (to be published)

28 LINDAU, I., and SPICER, W.E.: 'The use of synchrotron radiation to study oxygen and metal overlayers on GaAs (110) surfaces', in BRUNDLE, C.R., and BAKER, A.D. (Eds.): 'Electron spectroscopy, theory, techniques and application, Vol. 4' (Academic Press, 1981) pp. 197–276

29 HASHIMOTO, H., and MIYAUCHI, E.: 'Focused ion implantation for optoelectronic integrated circuits', Extended Abstracts of the 16th Conference on Solid State Devices and Materials, Kobe, Japan, 1984, pp. 121–124

30 MIYAUCH, E., and HASHIMOTO, H., 'Application of FIB technology to maskless ion implantation in an MBE-grown GaAs or AlGaAs epitaxial layer for three-dimensional pattern doping crystal growth', 32nd National Symposium American Vacuum Society, November 1985, *Vac. Science & Technol.*, 1986 (to be published)

31 SUGATA, S., and ASAKAWA, K.: 'Characterization of reactive ion beam etching-induced damage on GaAs using schottky electrode technique', Proceedings of Symposium on Dry Process, Tokyo, Japan, October 1985, pp. 144–148

32 ASAKAWA, K., and SUGATA, S.: 'GaAs and AlGaAs anisotropic fine pattern etching using a new reactive ion beam etching system', *J. Vac. Sci. & Technol.*, 1985, **B3**, pp. 402–405

33 SUGATA, S., and ASAKAWA, K.: 'GaAs and AlGaAs crystallographic etching with low pressure chlorine radical in ultra-high-vacuum RIBE system', 7th International Symposium on Plasma Chemistry, Eindhoven, Netherlands, 1985, Symposium proceedings, pp. 977–982

34 YUASA, T., ASAKAWA, K., MANNOH, M., SHINOZAKI, K., and ISHII, M. 'Reactive-ion-beam-etched cavity GaAs-AlGaAs multi-quantum well lasers', 5th International Conference on Integrated Optics and Optical Fibre Communication (IOOC–ECOC), Venice, Italy, October 1985, Technical Digest, pp. 505–508. (Identical characteristics for etched and cleaved mirror lasers was presented aurally at the conference.)

35 KOBAYASHI, K.L.I., WATANABE, N., NARUSAWA, T., and NAKASHIMA, H.: 'Formation and thermal stability of Au-GaAs (001) interface studied by angle resolved XPS and Rutherford back-scattering', Extended Abstracts of the 15th Conference on Solid-State Device and Materials, Tokyo, Japan, 1983, pp. 165–168

36 YOKOTSUKA, T., UCHIDA, Y., KOBAYASHI, K.L.I., NARUSAWA, T., and NAKASHIMA, H.: 'New self-aligned gate material; LaB_6 for GaAs IC', Extended abstracts of the 17th Conference on Solid-state Devices and Materials, Tokyo, Japan, 1985, pp. 437–440

37 NAKASHIMA, H., NARUSAWA, T., SEMURA, S., and KOBAYASHI, K.L.I.: 'Lattice location of Zn in Zn-diffused GaAlAs using particle-induced X-ray emission', Proceedings of the 13th International Conference on Defects in Semiconductors, August 1984, Coronado, California, USA, pp. 463–469

38 LAIDIG, W.D., HOLONYAK, N., CAMRAS, M.D., HESS, K., COLEMAN, J.J., DAPKUS, P.D., BARDEEN, J.: 'Disorder of an AlAs-GaAs superlattice by impurity diffusion', *Appl. Phys. Lett.*, 1981, **38**, (10), pp. 776–778

39 NAKASHIMA, H., SEMURA, S., OHTA, T., UCHIDA, Y., SAITO, H., FUKUZAWA, T., KURODA, T., and KOBAYASHI, K.L.I.: 'Device characteristics of GaAlAs buried-multiquantum-well lasers fabricated by Zn-diffusion-induced disordering', *IEEE J. Quantum Electron.*, 1985, **QE–21**, (6), pp. 629–633

40 FUKUNAGA, T., ISHIDA, K., KURODA, T., MATSUI, K., NARUSAWA, T., MORITA, T., MIYAUCHI, E., HASHIMOTO, H., and NAKASHIMA, H.: 'Enhanced disordering of AlGaAs-GaAs multiquantum well by focused Si ion beam implantation', 12th International Symposium on GaAs and Related Compounds, Karuizawa, Japan, September 1985, pp. 439–444

41 NOMURA, Y., MANNOH, M., MIHARA, M., NARITSUKA, S., YAMANAKA, K., YUASA, T., and ISHII, M.: 'Effect of group V/III flux ratio on lightly Si-doped $Al_xGa_{1-x}As$ grown by molecular beam eptaxy', *J. Electrochem. Soc.*, 1984, **131**, pp. 2630–2633

42 KAMON, K., TAKAGISHI, S., and MORI, H.: 'Selective embedded growth of $Al_xGa_{1-x}As$ by low-pressure organometallic vapor phase epitaxy', *Jpn. J. Appl. Phys., Part 2 (Letters)*, 1986, **25**, pp. L10–L12

43 KIRKPATRICK, C.G., CHEN, R.T., HOLMES, D.E., ASBECK, P.M., ELLIOTT, K.R., FAIRMAN, R.D., and OLIVER, J.R.: 'LEC GaAs for integrated circuit applications', *Semiconductors and Semimetals*, 1984, **20**, pp. 159–231

44 'Compact Disc market', Nikkei Electronics, 1985, October 21, p. 328 (in Japanese).

45 For Example, WADA, O., SANADA, T., YAMAKOSHI, S., HAMAGUCHI, H., FUJII, T., HORIMATSU, T., and SAKURAI, T.: 'Monolithic integration of a quantum-well laser and a driver circuit on a GaAs substrate', International Electron Device Meeting, December 1984, pp. 520–523

46 For Example MAKIUCHI, M., WADA, O., MIURA, S., HAMAGUCHI, H., MACHIDA, H., NAKAI, K., HORIMATSU, H., and SAKURAI, T.: 'Fast-responce PIN/preamplifier photoreceiver monolithically integrated on a GaAs substrate', International Electron Device Meeting, December 1984, pp. 862–863

47 SHIRAI, T., YAMAZAKI, S., KAWATA, H., NAKAJIMA, K., and KANEDA, T.: 'A Planar InP/InGaAsP heterostructure avalanche photodiode', *IEEE Trans.*, 1982, **ED–29**, (9), pp. 1404–1407

48 SCIFRES, D., BURNHAM, R.D., STREIFER, W., and BERNSTEIN, M.: 'Lateral beam collimation of a phased array semiconductor laser', *Appl. Phys. Lett.*, 1982, **41**, (7) pp. 614–616

49 NISHI, S., INOMATA, H., AKIYAMA, M., and KAMINISHI, K.: 'Growth of single domain GaAs on 2-inch Si (100) substrate by molecular beam epitaxy', *Jpn. J. Appl. Phys. Part 2*, 1985, **24**, (6), pp. L391–L393

InGaAsP/InP Long Wavelength Optoelectronic Integrated Circuits (OEIC's) for High-Speed Optical Fiber Communication Systems

AKIRA SUZUKI, KENICHI KASAHARA, AND MINORU SHIKADA

(*Invited Paper*)

***Abstract*—This paper describes long wavelength InP-based OEIC's, in which optical components and electrical components are monolithically integrated on a single chip. Recent progress in the InP-based OEIC technologies is reviewed. Results obtained from transmission experiments using long wavelength OEIC's are described and applications for optical systems are discussed.**

I. Introduction

IN RECENT YEARS, marked progress has been made in optical fiber communication systems. The optical fiber communication system applications scope has expanded from trunk line systems to local area network, subscriber, and intraoffice systems. Optoelectronic devices with higher performance and increased functions are required for those optical fiber communication systems. The optoelectronic integrated circuit (OEIC) is one of the most important key elements. The OEIC was first proposed by Yariv in 1973 [1]. It has been actively studied by many research groups [2]–[23]. The monolithic integration of optical components and electrical components is a natural trend of the semiconductor technology. The potential advantages inherent in OEIC's are drawn from Si or GaAs electronic integrated circuits. The advantages of monolithic integration are recognized as an improvement in compactness and reliability, a reduction in cost, an improvement in performance, and an increase in functionality over a similar hybrid circuit.

Such potential advantages have led OEIC research in two major directions. One research direction is to improve optoelectronic device performance for use in high-speed optical fiber communication systems by monolithic integration of electronic circuits. Wire bonding among the discrete components, involving bonding wires and bonding pads, leads to an increase in parasitic reactance. The parasitic reactance considerably deteriorates the speed and the sensitivity performances for optoelectronic devices in the region above 1 Gbit/s. Assembling discrete components also leads to performance variation as well as to higher construction cost. The main research target for this direction is the one-chip optical transmitter/receiver OEIC, easy to handle and with high-speed and high sensitivity performance. The other research direction is to improve electronic circuit performance by introducing optical signal interconnection. Optical signal interconnection has such advantages as freedom from stray capacitance problems in electrical interconnection wires, electrical isolation, and small size. Because of these advantages, optical signal interconnections will be extremely useful for signal transmission between pieces of apparatus, LSI mounting boards, LSI chips, and even within a VLSI chip.

The feasibility of fabricating OEIC's has been demonstrated by the following device combinations: laser diode (LD)/FET or HBT, light emitting diode (LED)/FET, photodiode (PD)/FET, and so on. Fig. 1 shows dates for experimental work on OEIC's reported so far. The total number of integrated optical and electronic components are plotted as function of a year. In the earlier stages, most OEIC's consisted of only one optical component and one electronic component. In the past several years, however, the number of integrated components has been rapidly increased. Four channel receiver OEIC's were reported with fairly good results regarding crosstalk [14].

A major obstacle to the successful integration of diverse components, such as optical components and electronic components, is the coordination of material parameters, fabrication process and device structure. Technological problems are as follows:

a) Interconnection between optical component and electronic component. This problem arises from the fact that electronic components are essentially surface devices, while optical components are bulk devices. For example, optical components, such as LD's and PD's, require more than 5-μm thickness and electronic components, such as FET's, need less than 0.2-μm thickness. Interconnection yield deteriorates due to the geometrical step formed betweeen those components, through the imperfect lithography near the step edges.
b) Fine lithography applicable over several micrometer high steps. High resolution photolithography

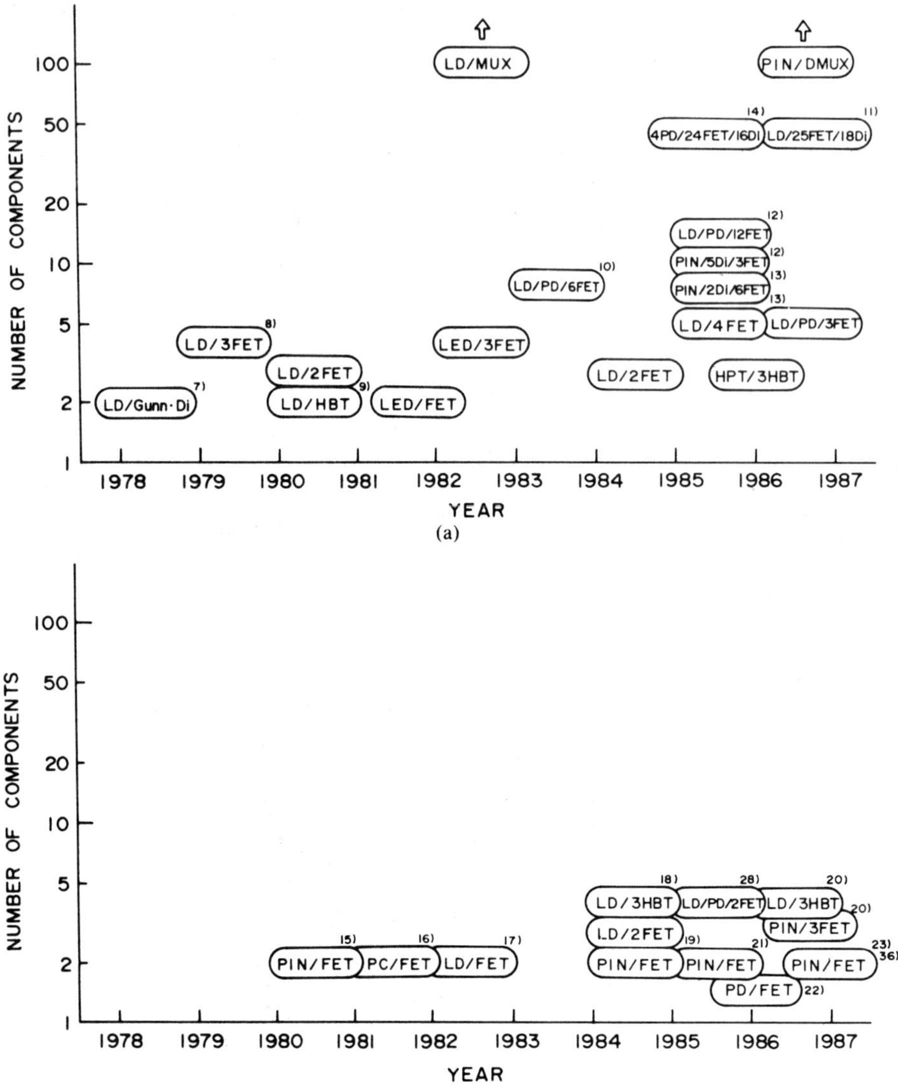

Fig. 1. Experimental work on OEIC's, (a) short wavelength OEIC's, (b) long wavelength OEIC's.

must be obtained over the processed wafer, where several micrometers surface steps are common, for high performance FET's. The steps often make fine lithography difficult due to the nonuniformity of the spincoated photoresist thickness and the exposure pattern being out of focus.

c) Independent optimization of optical components and electrical components. The generally contradictory structure parameters, such as composition, thickness, and doping concentration, for optical components and electrical components, requires that the component regions be segregrated onto different areas of the chip and that they be optimized independently.

d) Suitable design for components layout. For example, the substrate cleavage required for obtaining the Fabry–Perot LD cavity restricts the electrical circuit layout. The alternative noncleaved laser cavity is desired from the circuit layout freedom viewpoint.

Overcoming these problems will allow OEIC's to compete with and successfully replace hybrid optoelectronic components.

The work on OEIC's reported so far is divided into two categories; GaAs-based OEIC's for short wavelength optical systems and InP-based OEIC's for long wavelength optical systems. Short wavelength GaAs-based OEIC's are useful for short-haul (< 1 km) systems and are mainly aimed at the application to optical signal interconnections. Long wavelength InP-based OEIC's are most important for high-speed optical fiber communication systems.

InP-based OEIC's with complex integration, comparable to GaAs-based OEIC's, have not yet been demonstrated due to the relative immaturity of InP-based technology as compared with Ga-As-based technology. InP-based OEIC's seem to be still in a preliminary stage in spite of their importance for optical fiber communication use. Basic combinations, however, have already been investigated, as shown in Fig. 1 [15]–[23].

This paper is focused on InP-based OEIC's for high-speed optical fiber communication systems. First, the basic components available for InP-based OEIC's will be discussed in Section II. Next, the InP-based transmitter/receiver OEIC fabricated by the authors will be described in Section III. Section IV will discuss optical fiber transmission systems using long wavelength InP-based transmitter/receiver OEIC's. Finally, conclusions will be presented in Section V.

II. Components for Long Wavelength OEIC's

Recent progress in optical components and electronic components for the InP-based OEIC's is reviewed.

A. Laser Diodes

Laser diodes have to satisfy several requirements for them to be successfully integrated with electronic components. First of all, the power consumption has to be reduced. The LD has a p-side-up mounting configuration to allow interconnecting with the electronic circuit within a chip. The p-side-up configuration requires low threshold current and high efficiency. From this point of view, buried-heterostructure (BH) lasers are commonly used for InP-based transmitter OEIC's. The InGaAsP active region is fully embedded in an InP cladding region. Suitable design and fabrication techniques for the active region and the injection current confinement are most important. Typical characteristics, 10–20 mA threshold current and 50–60 percent external quantum efficiency, were achieved in 1.3-μm wavelength DC-PBH LD's [24]. Quantum well (QW) structures have also been investigated to reduce the threshold current density. The effective reduction in threshold current, however, has not yet been obvious in the InP-based material system. Regarding this point, InP-based materials are different from GaAs-based materials and more elaborate investigations concerning both crystal growth and materials are still needed.

The second important requirement concerns the LD modulation capability. The high-speed modulation characteristics of LD's have been a subject of recent active research. Fast frequency response can be obtained by driving the LD at high current level to increase the photon density, by shortening the LD cavity length to decrease the photon lifetime or by using materials having a larger differential gain coefficient, such as quantum well structures or highly doped InGaAsP materials [25], [26]. Also, LD parasitic elements have to be taken into account. Principal parasitic elements are shunt capacitance and series resistance inherent in the LD structure. These elements shunt the injected high frequency current and severely limit the LD operating speed. Efforts toward reducing parasitic capacitance by optimizing the LD structure have successfully led to a modulation bandwidth exceeding 10 GHz. In these LD's, the shunt capacitance is reduced to 1–2 pF [26], [27].

The last requirement for integrated LD's concerns noncleaved cavity facets. Since conventional wafer cleavage restricts the chip size according to the cavity length and limits the integrated electronic circuit scale, mirror cleaving is not compatible with OEIC technology. In addition, noncleaved facets are needed for monitoring PD integration. Distributed feedback (DFB) structures and etched mirrors are the most important technological bases, both from a practical standpoint and in the realization of full OEIC potential. DFB structures don't employ cleaved mirrors for optical feedback. Moreover, they make it possible to integrate the LD with the monitoring PD with high optical coupling efficiency. These features are very attractive for a transmitter OEIC including a monitoring PD. An integrated DFB-DC-PBH LD with two FET's and a monitoring PD has been reported by the authors [28]. Reactive ion etched mirrors are also important. The angled RIE process and the Br_2-RIE process have been reported [29], [30]. The etched mirror LD's, however, were found to exhibit an increase in threshold current of 20–30 percent or a decrease in external quantum efficiency of about a third. Further investigations on the RIE process for InP-based materials is necessary to achieve high performance devices.

B. Photodiodes

Photodiodes suitable for InP-based OEIC's are p-i-n-PD's, because of their simple structures, stability, and low operating voltage. The most important requirement for the high-speed and high sensitivity receiver OEIC is a reduction in the total input capacitance. Planarization approaches, such as a selective VPE approach [31] and a graded-step approach [32], not only minimize parasitic capacitance but also greatly simplify the interconnection between the PD and the electronic circuit. Recently, a planar embedded InGaAs p-i-n-PD, fabricated by using the graded-step process, has been reported by Miura *et al.* Fig. 2 shows a cross section of the planar p-i-n-PD. A capacitance of 0.08 pF and high-speed operation exceeding 10 GHz have been successfully demonstrated.

InGaAs p-i-n-PD's, grown on GaAs substrates by strained heteroepitaxy, are worthy of notice. This technology makes it possible to combine InP-based optical components with the maturing technology for GaAs integrated circuits. An InGaAs/InP p-i-n-PD, grown on a GaAs substrate, has been fabricated by Dentai *et al.* [33]. InGaAs/InP epitaxial layers were directly grown by MOVPE. The p-i-n-PD's reported exhibit relatively low dark current and high quantum efficiency, in spite of the highly strained material system.

C. Electronic Components

Since the small Schottky barrier height in the InP-based materials prevents the formation of metal-semiconductor FET's (MESFET's), alternatively, metal-insulator-semiconductor FET's (MISFET's) and junction FET's (JFET's) are used for the long wavelength OEIC's. Recently, InP-based JFET's have been shown to be suitable for OEIC's and the self-aligned gate formation process has been demonstrated by Cheng *et al.* [34], Wake *et al.* [35], and the authors [36]. Cheng *et al.* developed a high-

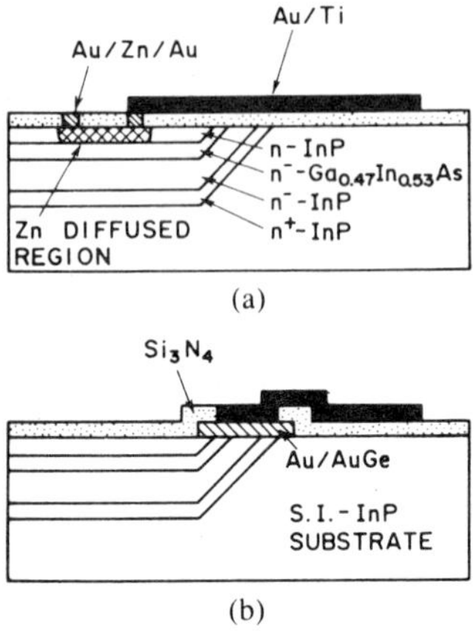

Fig. 2. Planar p-i-n-PD, (a) p-side contact area (b) n-side contact area.

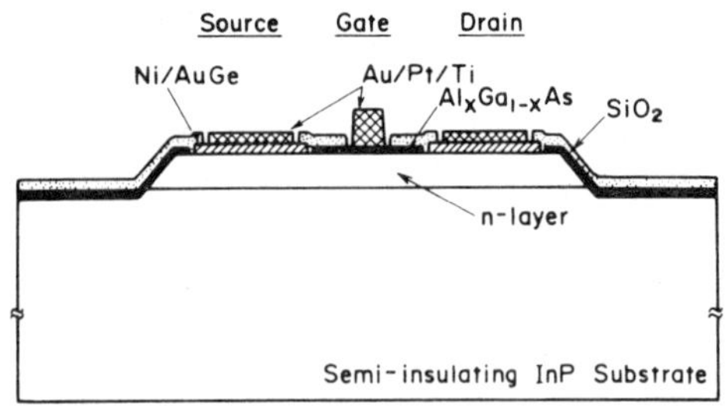

Fig. 3. AlGaAs hetero-MIS gate InP FET.

speed device ($f_T \sim 20$ GHz) and built a monolithic integrated direct-coupled JFET amplifier.

JFET's, however, have disadvantages, such as small tolerances in the device structure and the fabrication process. MISFET's have a simple gate structure and can eliminate these disadvantages. More recently, AlGaAs hetero-MIS gate InP FET's, shown in Fig. 3, have been successfully fabricated. Stable dc characteristics and good microwave performance have been demonstrated by Itoh *et al.* [37]. A strained-epitaxial AlGaAs gate insulator was grown by MBE. The AlGaAs hetero-MIS gate InP FET's exhibit no drift in drain current. Such drift is a serious problem to be solved for conventional InP MISFET's. The AlGaAs hetero-MIS gate InP FET is an attractive candidate for InP-based electronic components.

Heterojunction bipolar transistors (HBT's) have a multilayered structure, similar to those for LD's. From the viewpoint of structure coordination between optical components and electrical components, monolithic integration of HBT's with a BH-LD has been investigated by Shibata *et al.* [18] and the authors [38].

III. Long Wavelength OEIC's

This section describes the long wavelength InP-based transmitter/receiver OEIC's fabricated by the authors.

A. Transmitter OEIC's

The transmitter OEIC consists of an LD and a driving circuit. Fig. 4 shows the structure of the LD-HBT transmitter OEIC fabricated by the authors [12], [30]. The LD has an improved DC-PBH structure, whose buried layers are thinner than in a conventional DC-PBH structure. It is mesa-striped 40-μm-wide and mirror facets are made by the conventional cleavage method. The LD driving circuit consists of a CML and a buffer amplifier. HBT's were used from the viewpoint of their high current drivability. The HBT has a double heterostructure, which has an n-InGaAsP spacing layer between base layer and collector layer to improve the base transport efficiency. The p^+-InGaAsP base layer is 0.2 μm thick and Cd-doped to 5×10^{18} cm^{-3}. The n-InP emitter and collector layers are Sn-doped to 5×10^{17} cm^{-3} and 2×10^{17} cm^{-3}, respectively.

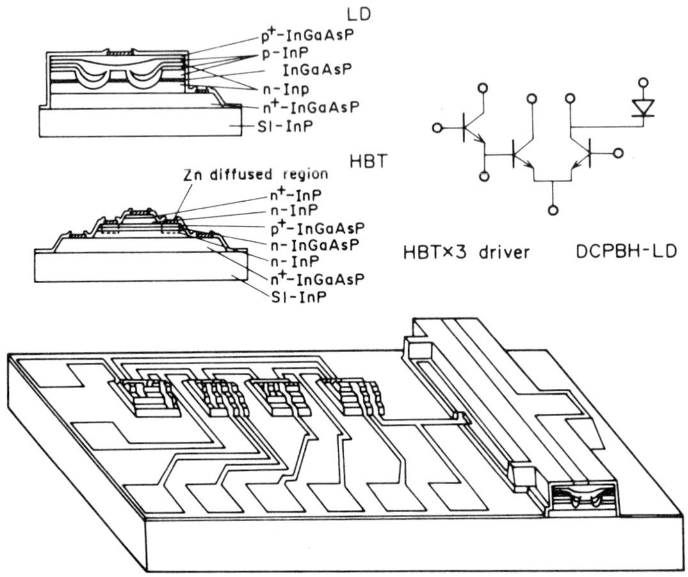

Fig. 4. Long wavelength LD-HBT's transmitter OEIC.

These concentrations are optimized for high-frequency operation, considering the junction capacitance, Kirk effect [39] and the Cd-dopant auto-diffusion during the fabrication process. N-contact layers for LD and HBT's are highly Te-doped to 1×10^{19} cm^{-3} to reduce bulk and contact resistance. Base electrodes are located on both sides of the striped emitter to reduce base resistance. Emitter and collector sizes are 11×34 μm and 34×30 μm, respectively.

Fabrication includes three key processes. First is selective LPE. This process enables optimizing the LD structure and the HBT structure independently. Second is preferential mesa etching. For example, using SiO_2 or InGaAsP as a mesa-etching mask, side-etching hardly occurs in InP preferential etching. Making use of these controlled side-etching characteristics, a fine mesa structure can be easily fabricated. Third is inerconnection using selective Au plating [38]. This process enables reproducible interconnection on mesa-shaped and step-formed LD and HBT's. It has 100-percent yield within 7-μm step height.

The wafers, grown by three-step LPE, were preferentially mesa-etched to isolate LD and HBT's from each other and to reveal electrode contact layers. Zn diffusion in the external base was carried out and a SiO_2 film was deposited by CVD. Electrodes were patterned by a conventional liftoff technique. Bonding pads and interconnections were patterned by a selective Au plating technique. In the last stage, the wafers were polished to about 100-μm thickness.

A 7.5-GHz cutoff frequency f_T, 200 current gain h_{FE}, and more than 200-mA maximum collector current were obtained for the HBT. These characteristics indicate that this HBT has a high enough current drivability to operate light sources at high-speed and high output operation. A 23-mA threshold current, 38-percent external quantum efficiency, and more than 20-mW (dc at room temperature) maximum output power were obtained for the LD. High-speed operation to 2 Gbit/s and 20 $mW_{p\text{-}p}$ high output power operation were successfully demonstrated for the fabricated transmitter OEIC. Input voltage conversion to light output power was achieved with a ratio as large as 10 mW/facet/V.

The wafer cleavage restricts the chip size and limits the integration scale. Using noncleaved DFB LD's increases the circuit layout freedom. They are attractive for transmitter OEIC's as described in Section II. The first transmitter OEIC, consisting of a DFB LD, two MISFET's and a monitoring PD, has been fabricated [20]. MISFET's were used from a viewpoint of their processing simplicity. The device structure is shown in Fig. 5. The fabrication procedures are as follows. First, the n^+-InGaAsP layer and the n-InP layer were grown on the step-formed substrate by LPE. Next, a first-order corrugation was formed by holographic photolithography and the DFB-DC-PBH LD structure was fabricated. The LD structure with a 20-μm mesa width was defined by selectively removing the layers until the first epitaxial n^+-InGaAsP layer was exposed. MISFET's with self-aligned gates [40] were fabricated. The gate insulator was a 0.1-μm-thick CVD-SiO_2 film deposited at 350°C. Gate length was 1 μm. Gate widths were 150 μm for the first stage and 300 μm for the second stage. After forming the electrode, the LD mesa was separated into two sections by chemical etching. One of the two sections was used as a monitoring PD.

Transconductance g_m and cutoff frequency f_T values were typically 65 mS/mm and 7 GHz for the MISFET, respectively. The CW threshold current was 53 mA and the external quantum efficiency was 22 percent for the DFB-DC-PBH LD. The 3-dB roll-off frequency for the LD was 2.8 GHz. Operation at 2 Gbit/s was successfully demonstrated for the transmitter OEIC using the DFB-DC-PBH LD. Input gate voltage conversion ratio to light output power was about 1 mW/facet/V. The isolation resistance between the DFB-DC-PBH LD and the monitoring PD was over 100 kΩ. A responsivity of 0.6 mA/mW was achieved for the integrated monitoring PD.

B. Receiver OEIC's

The receiver OEIC consists of a p-i-n-PD and a preamplifier circuit. Fig. 6 shows the p-i-n-JFET receiver OEIC structure [20]. JFET's were used from the viewpoint of their low noise performance. A p-i-n-PD, an amplifier, and two stage buffer amplifiers were formed in a common n-InGaAsP layer on a semi-insulating InP sub-

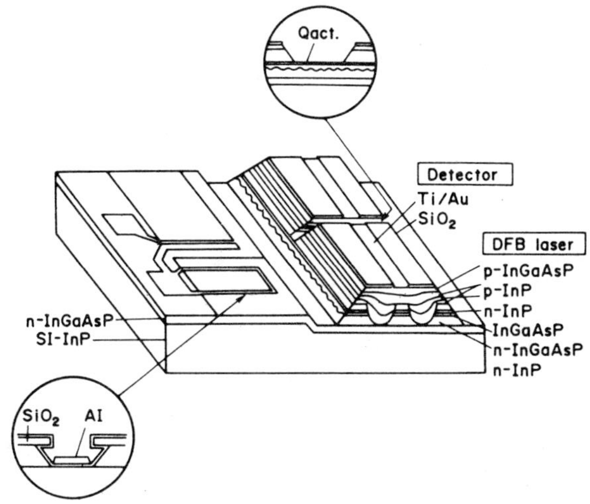

Fig. 5. Long wavelength DFB LD-PD-FET's transmitter OEIC.

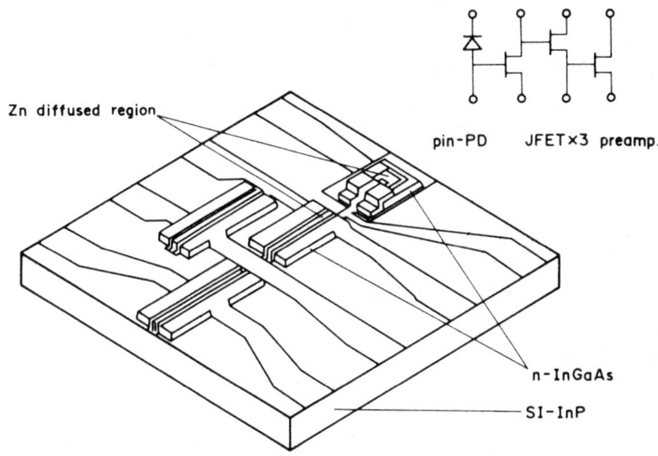

Fig. 6. Long wavelength p-i-n-JFET's receiver OEIC.

strate. The PD size was 40 × 40 μm. The JFET gate length was about 2 μm and gate widths were 150 and 200 μm for the first stage and the buffer stages, respectively. A cutoff frequency f_T, of about 3 GHz, and a transconductance g_m, of about 65 mS/mm, were obtained. PD quantum efficiency and total input capacitance were about 30 percent and about 2.5 pF, respectively. Receiver sensitivities of −25.5 dBm for 565 Mbit/s and −14.2 dBm for 1.2 Gbit/s were obtained for a 10^{-9} bit-error rate.

More recently, the p-i-n-JFET receiver performance has been improved [36]. A new self-aligned JFET process, consisting of T-shaped gate structure formation and self-aligned drain and source electrode metallization, was developed. Fig. 7 shows a cross-sectional view of an improved p-i-n-FET receiver OEIC. JFET and p-i-n-PD layers were successively grown on a step-formed semi-insulating InP substrate by LPE. Owing to the Cr/p-InGaAsP/n-InP structure, p-InGaAsP gate layer side-etching could be precisely controlled under the Cr gate metal and a T-shaped gate was reproducibly fabricated. After T-shaped gate formation, drain and source electrodes, which were cut by the T-shaped structure overhang, were formed. This new self-aligned process enables making a less than 1-μm gate length and a less than 0.5-

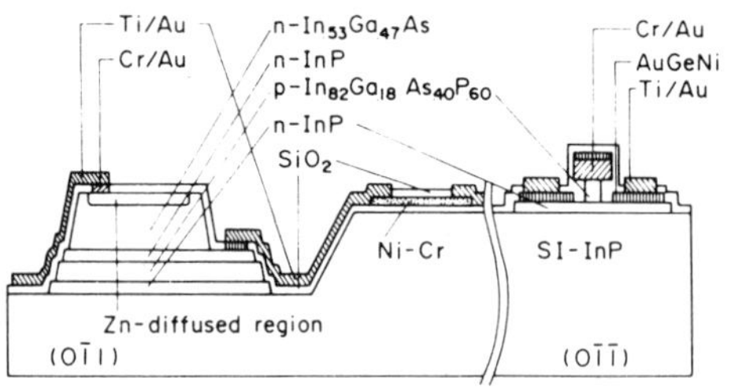

Fig. 7. Improved long wavelength p-i-n-FET receiver OEIC.

μm gate–drain source spacing. A 5-kΩ PD load resistance was also integrated monolithically. The fabricated JFET exhibited 13-GHz cutoff frequency f_T, and 110 mS/mm transconductance g_m. PD quantum efficiency and total input capacitance were 43 percent and 0.8 pF, respectively. High receiver sensitivity, −22.8 dBm for 1.2 Gbit/s, was achieved. The sensitivity values for the improved receiver OEIC's are still 10 dB lower than those for the well designed hybrid p-i-n-FET receiver. Total input capacitance reduction, FET transconductance increase, and PD quantum efficiency improvement are required to improve the receiver sensitivity.

IV. Long Wavelength OEIC's Applications

Long wavelength OEIC's will be applied to every field in optical fiber systems, such as trunk line systems, subscriber systems, local area network systems, and so on. The first half of this section describes transmission experimental results obtained using long wavelength OEIC's. The experiments were performed by the authors to demonstrate the feasibility of applying InP-based OEIC's to high-speed optical fiber communication systems. The LD-HBT transmitter OEIC and the p-i-n-JFET receiver OEIC described in the previous section were used in the experiments.

The experimental setup is shown in Fig. 8. A 1.2-Gbit/s or 565-Mbit/s non-return-to-zero (NRZ) signal from a pseudorandom pulse pattern generator was directly supplied to the LD-HBT transmitter OEIC. The transmitter OEIC output signal waveforms are shown in Fig. 9. A spherical-ended gradient index rod lens (coupling loss: 3.5–4 dB) was used for coupling between the LD-HBT OEIC and the single-mode optical fiber. The optical fiber was a conventional 1.3-μm zero dispersion one, with 0.46-dB/km transmission loss at the 1.3-μm wavelength. The transmitted signal was demodulated by the p-i-n-JFET receiver OEIC and amplifiers with an equalizing circuit. The demodulated signal bit error rate was measured by an error detector. The signal waveforms are shown in Fig. 10.

The LD-HBT transmitter OEIC consists of three HBT's and a 1.28-μm DC-PBH LD, whose structure is shown in Fig. 4. The transmitter OEIC operated up to 2 Gbit/s and the input voltage conversion ratio to light output was 10 mW/facet/V.

The receiver circuits developed were high impedance receiver circuits with the p-i-n-JFET OEIC and equalizing circuits. The authors have developed two receiver OEIC categories. One consists of a PD and three JFET's, monolithically integrated on the InP substrate, as shown in Fig. 6 (PD-JFET 1). A 1-kΩ PD load resistance was hybridly integrated with the PD-JFET's chip. The other receiver OEIC consists of a PD, a JFET, and a 5-kΩ PD load resistance, monolithically integrated on a substrate

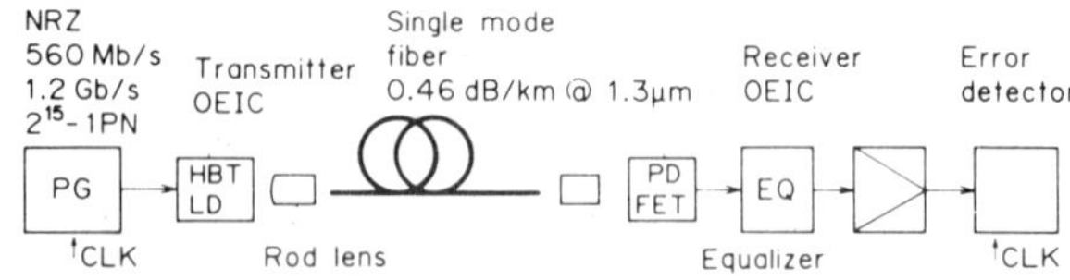

Fig. 8. Transmission experimental setup using long wavelength OEIC's.

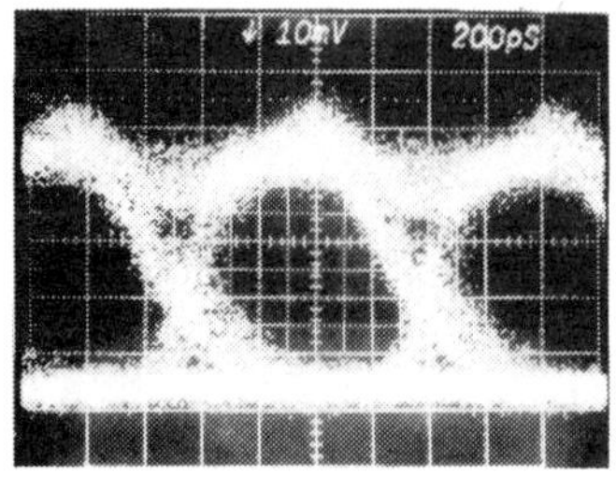

(a)

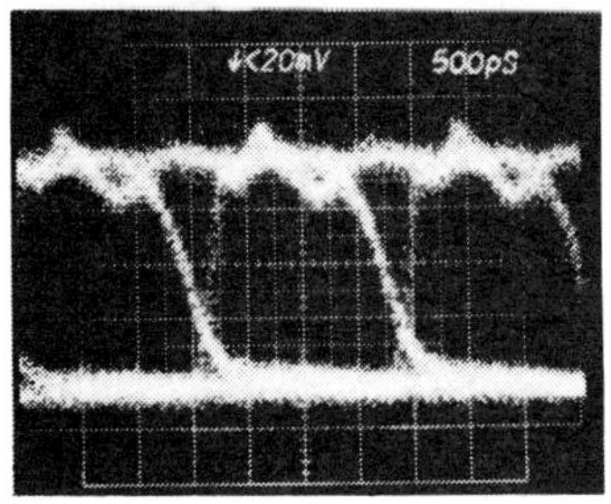

(b)

Fig. 9. Transmission OEIC output signal waveforms, (a) 1.2 Gbit/s, (b) 565 Mbit/s.

(a)

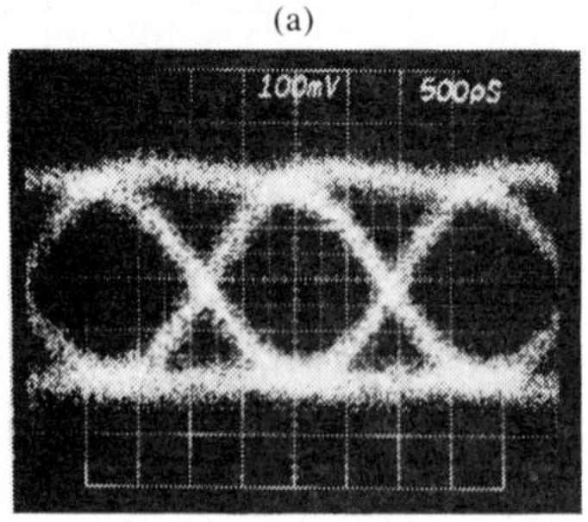

(b)

Fig. 10. Demodulated signal waveforms, (a) 1.2-Gbit/s 11.8-km transmission, (b) 565-Mbit/s 22-km transmission.

as shown in Fig. 7 (PD-JFET 2). Receiver sensitivities, measured at 1.2-Gbit/s and 565-Mbit/s for PD-JFET 1, were −14.2 dBm and −21.5 dBm (BER = 10^{-9}), respectively. The receiver sensitivity at 1.2-Gbit/s was improved to −22.8 dBm for PD-JFET 2.

Transmission experiments were carried out using the PD-JFET 1 receiver OEIC. The bit-error-rate characteristics, obtained as a result of the optical fiber transmission experiment, are shown in Fig. 11. Experimental results are summarized in Table I. For the 1.2-Gbit/s transmission, 13.7-dB allowable transmission line loss was obtained. A transmission experiment over an 11.8-km span length has been carried out, achieving a 7.7-dB level margin. About 1-dB power penalty was observed after transmission, due to chromatic dispersion from the LD wavelength (λ = 1.28 μm) difference from the optical fiber zero-dispersion wavelength. A 2.2-dB power penalty was also observed in the 1.2-Gbit/s · 22-km transmission experiment, which gave no level margin. A 565-Mbit/s transmission experiment was also carried out over a 22-km span with 9.9-dB level margin. Power penalty was not observed at this bit rate. The feasibility of long wavelength InP-based OEIC application to high-speed optical fiber communication systems was successfully demonstrated.

In the latter half of this section, long wavelength OEIC applications for various optical systems are discussed. Various application fields are shown in Fig. 12.

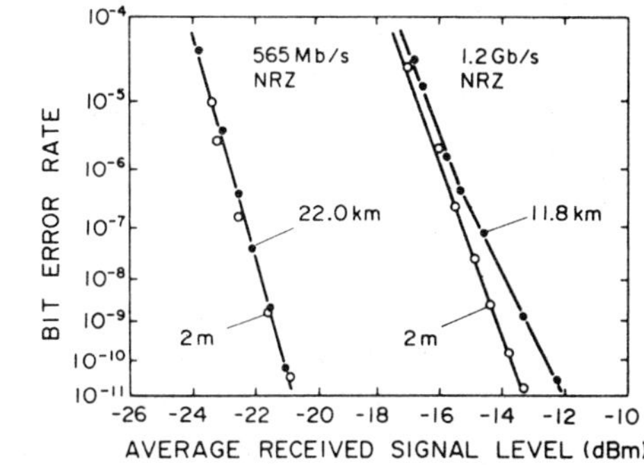

Fig. 11. Bit-error-rate characteristics obtained as result of optical fiber transmission experiment.

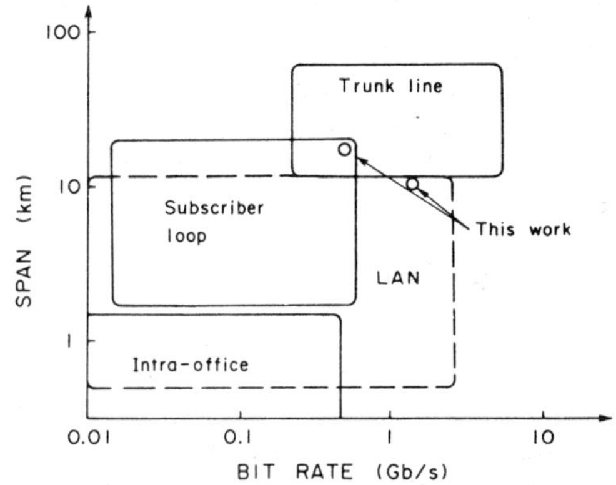

Fig. 12. Long wavelength OEIC applications.

TABLE I
EXPERIMENTAL RESULTS

BIT RATE	565Mb/s NRZ	1.2Gb/s NRZ
LD-HBTs OUTPUT POWER*	+3.0dBm	+3.0dBm
FIBER INPUT LEVEL*	-1 dBm	-0.5 dBm
PIN-JFETs SENSITIVITY**	-21.5dBm	-14.2dBm
ALLOWABLE TRANSMISSION LINE LOSS	20.5dB	13.7 dB
TRANSMISSION LINE LENGTH	22.0 km	11.8 km
TRANSMISSION LINE LOSS	10.6 dB	5 dB
POWER PENALTY BY DISPERSION	0 dB	1 dB
LEVEL MARGIN	9.9 dB	7.7 dB

* AVERAGE POWER LEVEL
** B.E.R. = 10^{-9}

A. General Applications

OEIC compactness and capability of high-speed operation make them suitable for use in trunk lines, subscriber loops, high-speed local/wide area networks, intra-office links, and so on.

1. Trunk Line Systems: Long wavelength gigabit-per-second range OEIC's are promising for trunk line systems. Package stray capacitances and lead wire inductances degrade gigabit-per-second pulse waveforms in commercially available optical and electrical devices. OEIC modules, with low stray and parasitic reactance, will fit into this high speed application field. Distributed feedback laser diodes have to be integrated on these OEIC's and receiver sensitivity has to be improved for longer span transmission systems.

2. Subscriber Loops: The maximum signal bit rate reported for subscriber loops is around 600 Mbit/s. OEIC high-speed modulation/demodulation capabilities are very important. Compact and low cost long wavelength OEIC modules, together with GaAs or Si SSI/SI's in them, will be suitable for this application. Bidirectional transmission modules are also important for subscriber loops.

3. Local Area Networks: Long wavelength OEIC applications for LAN's are important. The transmission bit rate for LAN systems will increase to the gigabit-per-second range in the near future. A 1.2-Gbit/s experimental LAN system has already been reported. Compact, low cost and high-speed OEIC modules will be required for the system.

4. Intraoffice Links: One of the most fascinating OEIC applications for intraoffice links will be multichannel transmission using a single OEIC chip. Especially for computer links, in which coaxial cables are used for interconnection between CPU, memory, and remote terminals, application of multichannel OEIC's will enable space reduction, high-speed signal transfer and will solve various problems concerning electrical crosstalk and grounding. Ultra-high-speed multichannel optical fiber links will facilitate the development of supercomputers. Thermal isolation, electrical crosstalk, and coupling between an optical fiber array and an OEIC will be key issues in these multichannel OEIC's.

B. New and/or Future Applications

The OEIC's multifunction and multichannel capabilities are very important when applying these devices for signal processing as well as for signal transmission. Time domain and wavelength domain switching and multiplexing functions are suitable for optical exchange in optical fiber transmission networks. One of the authors reported an optical time division switching system which utilized 1.3-μm wavelength bistable laser diodes as an optical flip-flop (optical memory) [41]. Multichannel bistable LD OEIC's will enable the time division switching system to become more compact and reliable. An optical wavelength division switching system also needs multichannel LD array OEIC's operable at different wavelengths.

It has also been reported that OEIC's will be used, over the long term, for optical interconnection applications between LSI/VLSI chips and their mounting boards. In high-speed computer systems, it has already been seen that the interconnection delay limits the operation speed. OEIC's can furnish wireless optical interconnections, which are free from stray capacitance, grounding problems and transmission line problems. The increasing freedom in optical interconnection of OEIC's will be extremely useful in future optical systems.

V. Conclusion

The first attempt to achieve long wavelength InP-based OEIC's has been successful. The promising features are expected to be useful in introducing various system applications and in furnishing a bright future for optical systems. Although their fundamentals and attractive feasibility have been demonstrated, there are many problems still to be solved in achieving their expected full potentials, as described above. Among them is development of the integrable InP-based electronic technology. When these problems are overcome, monolithic integration of high performance long wavelength optoelectronic devices and electronic circuits will be accelerated. Future optical systems may be replaced by a small InP-based OEIC chip. A large amount of research effort will be required to realize high performance InP-based OEIC's.

Acknowledgment

The authors are grateful to T. Uchida, M. Sakaguchi, K. Kobayashi, and K. Minemura for their encouragement.

References

[1] A. Yariv, "The beginning of integrated optoelectronic circuit," *IEEE Trans. Electron. Devices*, vol. ED-31, pp. 1656, 1984.

[2] N. Bar-Chaim, S. Margalit, A. Yariv, and I. Ury, "GaAs integrated optoelectronics," *IEEE Trans. Electron Devices*, vol. Ed-29, pp. 1372–1381, 1982.

[3] I. Hayashi, "OEIC: Its concepts and prospects," in *Tech. Dig. 4th IOOC* (Tokyo, Japan), 1983, pp. 170–173.

[4] U. Koren, S. Margalit, T. R. Chen, K. L. Yu, A. Yariv, N. Bar-Chaim, K. Y. Lan, and I. Ury, "Recent developments in monolithic integration of InGaAsP/InP optoelectronic devices," *IEEE J. Quantum Electron*, vol. QE-10, pp. 1653–1662, 1982.

[5] S. R. Forrest, "Monolithic optoelectronic integration: A new component technology for lightwave communication," *J. Lightwave Technol.*, vol. LT-3, pp. 1248–1263, 1985.

[6] O. Wada, T. Sakurai, and T. Nakagami, "Recent progress in optoelectronic integrated circuits (OEIC's)," *IEEE J. Quantum Electron.*, vol. QE-22, pp. 805–821, 1986.

[7] C. P. Lee, S. Margalit, I. Ury, and A. Yariv, "Integration of an injection laser with a Gunn oscillator on a semi-insulating GaAs substrate," *Appl. Phys. Lett.*, vol. 32, pp. 806–807, 1978.

[8] M. Yust, N. Bar-Chaim, S. H. Izadpanah, S. Margalit, I. Ury, D. Wilt, and A. Yariv, "A monolithically integrated optical repeater," *Appl. Phys. Lett.*, vol. 35, pp. 795–797, 1979.

[9] J. Katz, N. Bar-Chaim, P. C. Chen, S. Margalit, I. Ury, D. Wilt, M. Yust, and A. Yariv, "A monolithic integration of GaAs/GaAlAs bipolar transistor and heterostructure laser," *Appl. Phys. Lett.*, vol. 37, pp. 211–213, 1980.

[10] H. Matsueda, S. Sasaki, and M. Nakamura, "GaAs optoelectronic integrated circuits," *J. Lightwave Technol.*, vol. LT-1, pp. 261–269, 1983.

[11] K. Hamada, N. Yoshikawa, H. Shimizu, T. Otsuki, A. Shimano, K. Itoh, G. Kano, and I. Teramoto, "Monolithic integration of high-power buried stripe (GaAl)As laser with high frequency modulator MESFET circuit," in *Proc. 18th Conf. Solid-State Devices and Materials* (Tokyo, Japan), 1986, pp. 181–184.

[12] H. Matsueda, M. Hirao, T. P. Tanaka, H. Kodera, and M. Nakamura, "Integration of optical devices with electronic circuits for high speed communications," in *Proc. Int. Symp. GaAs and Related Compunds* (Karuizawa, Japan), 1985, pp. 655–660.

[13] T. Horimatsu, T. Iwama, Y. Oikawa, T. Touge, M. Makiuchi, O. Wada, and T. Nakagami, "Compact transmitter and receiver modules with optoelectronic-integrated circuits for optical LAN's," *J. Lightwave Technol.*, vol. LT-4, pp. 680–688, 1986.

[14] O. Wada, H. Hamaguchi, M. Makiuchi, T. Kumai, M. Ito, K. Nakai, T. Horimatsu, and T. Sakurai, "Monolithic four-channel photodiode/amplifier receiver array integrated on a GaAs substrate," *J. Lightwave Technol.*, vol. LT-4, pp. 1694–1703, 1986.

[15] R. F. Leheny, R. E. Nahory, M. A. Pollack, A. A. Ballman, E. D. Beebe, T. C. Dewinter, and R. J. Martin, "Integrated InGaAs p-i-n FET photoreceiver," *Electron. Lett.*, vol. 16, pp. 353–355, 1980.

[16] J. Barnard, H. Ohno, E. C. Wood, and L. F. Eastman, "Integrated heterostructure GaInAs photoreceiver with automatic gain control," *IEEE Trans. Electron. Devices Lett.*, vol. EDL-2, pp. 7–9, 1981.

[17] U. Koren, K. L. Yu, T. R. Chen, N. Bar-Chaim, S. Margalit, and A. Yariv, "Monolithic integration of very low threshold GaInAsP laser and MISFET on semi-insulating InP," *Appl. Phys. Lett.*, vol. 40, pp. 643–645, 1982.

[18] J. Shibata, I. Nakao, Y. Sasai, S. Kimura, N. Hase, and H. Serizawa, "Monolithic integration of an InGaAs/InP laser diode with heterojunction bipolar transistors," *Appl. Phys. Lett.*, vol. 45, pp. 191–193, 1984.

[19] K. Ohnaka, K. Inoue, T. Uno, K. Hasegawa, N. Hase, and H. Serizawa, "A planar InGaAs p-i-n/JFET fiber-optic detector," *IEEE J. Quantum Electron.*, vol. QE-21, pp. 1236–1240, 1985.

[20] K. Kasahara, A. Suzuki, S. Fujita, Y. Inomoto, T. Terakado, and M. Shikada, "InGaAsP/InP long wavelength transmitter and receiver OEIC's for high speed optical transmission systems," in *12th Eur. Conf. Optical Fiber Commun. Tech. Dig.* (Barcelona, Spain), 1986, pp. 119–122.

[21] K. Ohtsuka, H. Sugimoto, Y. Abe, T. Matsui, and H. Ogata, "Monolithic integration of InGaAs/In pin PD with MISFET on stepless substrate," *Electron. Lett.*, vol. 22, pp. 652–653, 1986.

[22] S. Hata, H. Yasaka, K. Nakashima, H. Asahi, and S. Uehara, "Monolithic integration of a Schottky photodiode and MESFET using an InAlAs/InGaAs heterostructure," in *12th Eur. Conf. Optical Fiber Commun. Tech. Dig.* (Barcelona, Spain), 1986, pp. 123–126.

[23] T. Mikawa, S. Miura, H. Kuwatsuka, T. Fujii, and O. Wada, "A planar monolithic p-i-n/FET fabricated by using an embedded structure InP/GaInAs p-i-n photodiode and an AlInAs/GaInAs field-effect transistor," in *Optical Fiber Commun. Conf. Tech. Dig.* (Reno, NV), 1987, pap. WG2.

[24] I. Mito, M. Kitamura, Ke. Kobayashi, S. Murata, M. Seki, Y. Odagiri, H. Nishimoto, M. Yamaguchi, and Ko. Kobayashi, "InGaAsP double-channel-planar-buried-heterostructure laser diodes (DC-PBH LD) with effective current confinement," *J. Lightwave Technol.*, vol. LT-1, pp. 195–202, 1983.

[25] K. Y. Lau and A. Yariv, "Ultra-high speed semiconductor lasers," *IEEE J. Quantum Electron.*, vol. QE-21, pp. 121–138, 1985.

[26] C. B. Su, V. Lanzisera, R. Olshansky, J. Schlafer, W. Powazinik,

E. Meland, and R. B. Lauer, "15-GHz direct modulation bandwidth of vapor phase regrown 1.3-μm InGaAsP buried heterostructure laser under CW operation at room temperature," *Electron. Lett.*, vol. 2, pp. 577, 1985.

[27] J. E. Bowers, B. R. Hemenway, A. H. Gnauck, T. J. Bridges, E. G. Burkhardt, D. P. Wilt, and S. Maynard, "High frequency constricted mesa laser," *Appl. Phys. Lett.*, vol. 47, pp. 78, 1985.

[28] K. Kasahara, T. Terakado, A. Suzuki, and S. Murata, "Monolithically integrated high-speed light source using 1.3-μm wavelength DFB-DC-PBH laser," *J. Lightwave Technol.*, vol. LT-4, pp. 908–912, 1986.

[29] T. Terashige, K. Ohnaka, J. Shibata, and T. Kajiwara, "An etched facet InGaAsP/InP buried-heterostructure laser fabricated by a Br_2-reactive-ion-etching," in *First Optoelectronics Conf. Post-Deadline Pap. Tech. Dig.* (Tokyo, Japan), 1986, pap. B11-1.

[30] H. Saito, Y. Noguchi, and H. Nagai, "Low-threshold InGaAsP/InP 1.3-μm doubly buried-heterostructure lasers with a reactive-ion-etched facet," *Electron. Lett.*, vol. 22, pp. 36–38, 1986.

[31] A. S. H. Liao, T. J. Bridges, E. G. Burkhardt, B. Tell, R. F. Leheny, and E. D. Beebe, "A planar embedded InGaAs photodiode on semi-insulating InP substrate for monolithically integrated p-i-n-FET receivers, using selective vapor phase epitaxy and ion implantation technique," in *IEDM83 Tech. Dig.*, 1983, pp. 478.

[32] S. Miura, H. Kuwatsuka, T. Mikawa, and O. Wada, "Planar, embedded InP/GaInAs p-i-n photodiode with very high-speed response characteristics," *Appl. Phys. Lett.*, vol. 49, pp. 1522–1524, 1986.

[33] A. G. Dentai, J. C. Campbell, C. H. Joyner, and G. J. Qua, "InGaAs p-i-n photodiodes grown on GaAs substrates by metal organic vapour phase epitaxy," *Electron. Lett.*, vol. 23, pp. 38–39, 1987.

[34] J. Cheng, G. Guth, M. Washington, S. R. Forrest, and R. Wunder, "Monolithically integrated InGaAs/InP direct-coupled junction field-effect transistor amplifier," *IEEE Electron Device Lett.*, vol. EDL-7, pp. 225–228, 1986.

[35] D. Wake, A. W. Livingstone, D. A. Andrews, and G. J. Davies, "A self-aligned InGaAs junciton field-effect transistor grown by molecular beam epitaxy," *IEEE Electron Device Lett.*, vol. EDL-5, pp. 285–287, 1984.

[36] T. Terakado, S. Fujita and A. Suzuki, "High sensitivity long wavelength p-i-n-FET OEIC receiver with self-aligned junction FET," in *CLEO'87 Tech. Dig.*

[37] T. Itoh, K. Kasahara, T. Ozawa, and K. Ohata, "Microwave $Al_xGa_{1-x}As$ hetero-MIS gate depletion-mode InP field effect transistors," in *IEDM'86 Tech. Dig.*

[38] Y. Inomoto, T. Terakado, and A. Suzuki, "High performance InGaAsP/InP field LD-HBT's light source OEIC," in *First Optoelectronics Conf. Tech. Dig.* (Tokyo, Japan), 1986, pap. A6-4.

[39] C. T. Kirk, "A theory of transistor cutoff frequency (f_T) falloff at high current densities," *IRE Trans. Electron Devices*, vol. ED-9, pp. 164–174, 1962.

[40] T. Itoh and K. Ohata, "X-band self-aligned gate enhancement-node InP MISFET'S," *IEEE Trans. Electron Devices*, vol. ED-30, pp. 811–815, 1983.

[41] S. Suzuki, T. Terakado, K. Komatsu, K. Nagashima, A. Suzuki, and M. Kondo, "An experiment on high-speed optical time-division switching," *J. Lightwave Technol.*, vol. LT-4, pp. 894–899, 1986.

Akira Suzuki was born in Nagoya, Japan, on March 21, 1953. He received the B.E., M.E., and Ph.D. degrees in electronics engineering from the University of Tokyo, Tokyo, Japan, in 1975, 1977, and 1980, respectively.

In 1980, he joined the Opto-Electronics Research Laboratories, NEC Corporation, Kawasaki, Japan, where he has been engaged in research on semiconductor lasers, light emitting diodes, and integrated optics.

Dr. Suzuki is a member of the Japan Society of Applied Physics and the Institute of Electronics and Communication Engineers of Japan.

*

Kenichi Kasahara was born in Nagano, Japan, on February 3, 1953. He received the B.E., M.E., and Ph.D. degrees from the University of Tokyo, Japan, in 1976, 1978, and 1981, respectively.

In 1981, he joined the NEC Corporation, where he has been engaged in research on integrated optics.

Dr. Kasahara is a member of the Institute of Electronics and Communication Engineers of Japan and the Japan Society of Applied Physics.

*

Minoru Shikada was born in Gunma Prefecture, Japan, on March 7, 1951. He received the B.E. degree in electronics from Tohoku University, Sendai, Japan, in 1973.

In 1973, he joined Central Research Laboratories, NEC Corporation, where he worked on microoptic devices for optical fiber communications. Since 1981 he has been engaged in research and development on optical fiber transmission systems in OptoElectronics Research Laboratories, NEC Corporation, Kawasaki, Japan.

Mr. Shikada is a member of the Institute of Electronics, Information and Communication Engineers of Japan.

Chapter 7
Analog Signal Transmission

Most of the development effort in fiber optics has focused on digital transmission, but progress in high–speed lasers and photodetectors has made analog applications increasingly attractive. We can now transmit signals with bandwidths in excess of 20 GHz over long distances through fibers. This opens the way for use of fibers as an alternative to metal–walled waveguides or coaxial cable for signal transmission over a wide range of microwave frequencies. The papers in this chapter describe experiments directed toward such applications: **"1.3-μm 35-km Fibre-Optic Microwave Multicarrier Transmission System for Satellite Earth Stations,"** by **W.I. Way, M. Krain, and R.S. Wolff; "Long Distance Fiber-Optic Transmission of C-Band Microwave Signals to and from a Satellite Antenna,"** by **J.E. Bowers, A.C. Chipaloski, S. Boodaghians, and J.W. Carlin;** and **"11-GHz Fiber Optic Links for Airborne Satellite Communications Applications,"** by **C.M. Gee, G.D. Thurmond, I.L. Newberg, H.W. Yen, D. Lafaw, and B. Spink.**

1·3 μm 35 km FIBRE-OPTIC MICROWAVE MULTICARRIER TRANSMISSION SYSTEM FOR SATELLITE EARTH STATIONS

Indexing terms: Optical fibres, Optical transmission, Microwave circuits and systems, Satellite links

A field experiment that transmits the full bandwidth of C-band satellite downlink signals over 35 km of 1·3 μm single-mode fibre was demonstrated. Analytical results of the fibre-optic link match well with the measured data.

The high capacity and low attenuation of optical fibre make it ideal for extending the reach of existing satellite-based networks, serving as a link between earth station and terrestrial networks. Alternatives include coaxial cable, which has limited range, or terrestrial microwaves, which are vulnerable to interference and may be difficult to use in cities that are already crisscrossed with many microwave routes. In this letter we report a field experiment of a C-band satellite earth station fibre-optic multichannel transmission system which has been running since early November 1986. The system transports a multicarrier ensemble of video, data and voice signals in a full 500 MHz bandwidth (3·7–4·2 GHz) from a satellite-receiving earth station to a remote indoor receiver. A preliminary report of our experimental results has been given earlier.[1]

The complete system configuration is shown in Fig. 1. The downlink satellite signal from an earth station antenna is amplified and used to directly modulate a high-speed multilongitudinal-mode (MLM) InGaAsP laser diode (LD) which has a 6 GHz bandwidth at a bias level of $2{\cdot}33 I_{th}$ (threshold current $I_{th} = 15$ mA at room temperature). The centre wavelength and the spectral width of the LD are 1298·5 nm and 2 nm, respectively. The modulated optical signal then passes through a 500 ft (152·4 m) steel-centred fibre cable which was installed between the earth station and our Navesink indoor laboratory. In the laboratory, 34·7 km more fibre on spools was added. The optical signal was detected by an InGaAs PIN diode which has a measured bandwidth of 6 GHz and negligible dark current. The satellite microwave signals from the PIN diode were then amplified by a C-band low-noise amplifier (LNA). No signal processing at the earth station other than amplification is needed.

Considering major noise terms in the system, we can write the inverse of carrier/noise ratio (CNR^{-1}) per unit bandwidth as

$$(CNR)^{-1} = \frac{2(F-1)kT}{m^2 I_D^2 R_D} + \frac{2RIN_T}{m^2} + \frac{2P_{IM}}{R_D G(I_D m)^2} \qquad (1)$$

where F is the noise figure of the LNA (connected after the PIN diode), k is Boltzmann's constant, R_D and G are the input resistance and power gain of the LNA, respectively, T is the absolute temperature in degrees Kelvin, m is the optical modulation index (OMI) per channel, I_D is the detected DC photocurrent, RIN_T is the relative intensity noise (including reflection-induced noise) of the laser diode[2,3] and P_{IM} is the intermodulation noise per hertz caused by system nonlinearity.[4,5]

The measured light output power and RIN_T values for different bias levels are shown in Table 1. The measured OMI values and linearity as functions of input C-band signal power, for two different bias levels, are shown in Fig. 2. System linearity is characterised in terms of the power ratio of the intermodulation product and the fundamental carrier (IM/C). Cases of minimum and high intermodulation products (IM/C $\simeq -20$ dB) are shown in Figs. 3*a* and *b*, respectively. From Fig. 2, higher values of OMI can be obtained with minimum nonlinear distortion for a lower bias such as 35 mA, but the RIN_T is also high. For a higher bias such as 60 mA, the RIN_T is lower, but the obtainable OMI with minimum nonlinear distortion is lower. Test results are given in this letter for a bias level of 35 mA and input microwave

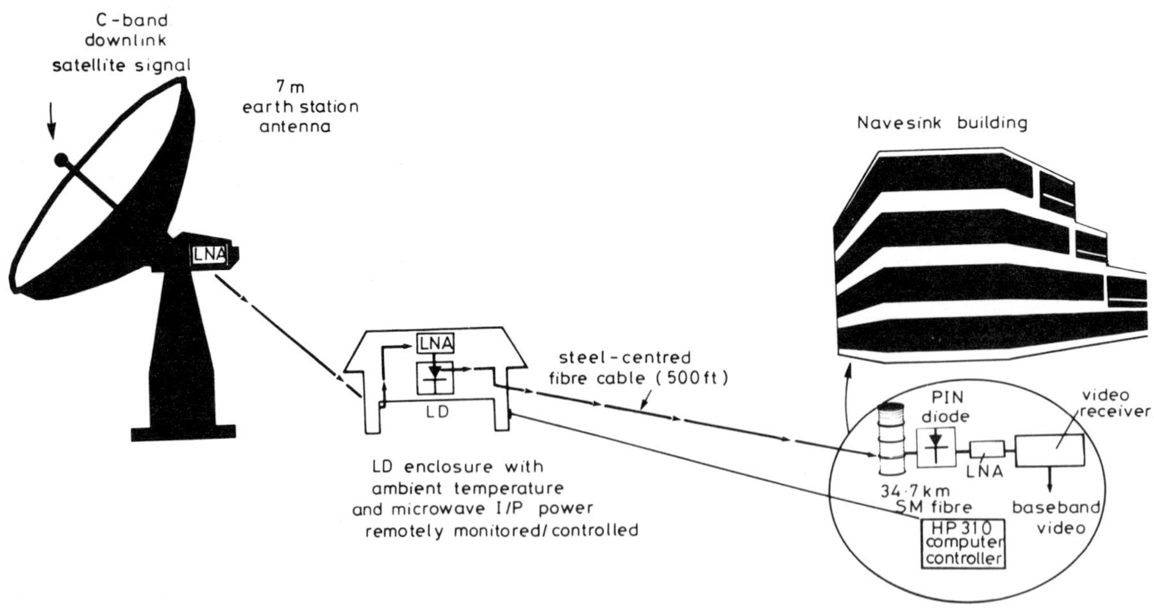

Fig. 1 *Experimental set-up*

power of 4·5 dBm. Under this operating condition, $RIN_T = -129$ dB/Hz, OMI per channel = 0·027, and P_{IM} is minimum (so that the third term in eqn. 1 can be neglected).

Table 1

Bias	RIN_T	Light output from pigtail
mA	dB/Hz	mW
35	−129	1·2
40	−131	1·5
45	−132	1·8
50	−134	2·1
55	−135	2·4
60	−135	2·6
65	−136	2·9

$I_{th} = 15$ mA

Assessment of the fibre link performance was based on measurement of video signal quality, as satellite transmission is predominantly used for video services. For FM-TV, the baseband signal/noise ratio (SNR) is related to C-band CNR by[6]

$$SNR = CNR + 39{\cdot}5 \qquad (2)$$

for a 525-line NTSC system and a 36 MHz satellite transponder channel. The standard EIA minimum requirement on SNR for a satellite system is 56 dB.[7] Therefore from eqn. 2, the CNR in eqn. 1 must not be less than 16·5 dB. The measured C-band CNR (per 36 MHz bandwidth) after transmission over 0 and 35 km of fibre are shown in curves *c* and *e* in Fig. 4, respectively. Both curves are closely matched to the calculated curve *d* in Fig. 4. Curve *d* was obtained by using the measured RIN_T and OMI values and eqn. 1. The measured SNRs of an FM-TV signal on Telstar 301 at 3840 MHz for 0 and 35 km transmission distances are shown in curves *a* and *b*, respectively. When the CNR is 16·5 dB, the measured SNRs are close to the predicted value of 56 dB. The measured video differential gain (DG) and differential phase (DP) after 35 km transmission as functions of received optical power are shown in curves *f* and *g*, respectively. From the measured SNR, DG and DP after 35 km of transmission ($I_D \simeq 17\,\mu$A), it is assured that the received signals meet the EIA end-to-end standard for NTSC video. Note that the total light power loss of the 35 km fibre link is around 16 dB. The system performance for this link length is power-limited and not dispersion-limited. Limitation of the system performance by dispersion was observed in terms of a severely degraded video quality when only a 2·6 km dispersion-shifted fibre on spool was used (nominal dispersion 21 ± 1 ps/nm km). The quality of the downlink satellite signal which was characterised right after the antenna LNA can be represented by the following parameters: C-band CNR = 21 ± 1 dB, baseband SNR = 61 ± 2 dB, DG = (6 ± 1·5)% and DP = (3 ± 1)°.

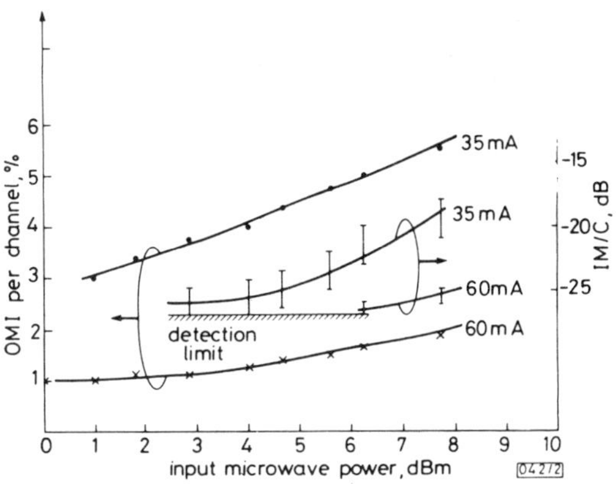

Fig. 2 *OMI per channel and IM/C against input microwave power*

The minimum value of CNR obtained for an FM video signal channel, 16·5 dB, is also satisfactory for several other satellite downlink applications that transport digital QPSK signals. Some of the downlink satellite digital data on BER performance via a fibre link are given in Reference 8.

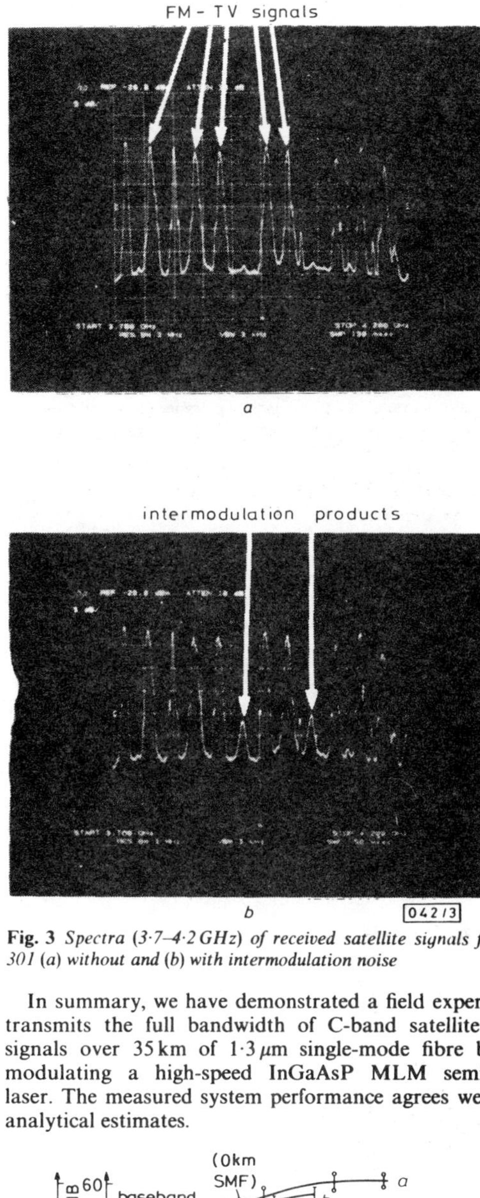

Fig. 3 *Spectra (3·7–4·2 GHz) of received satellite signals from Telstar 301 (a) without and (b) with intermodulation noise*

In summary, we have demonstrated a field experiment that transmits the full bandwidth of C-band satellite downlink signals over 35 km of 1·3 μm single-mode fibre by directly modulating a high-speed InGaAsP MLM semiconductor laser. The measured system performance agrees well with the analytical estimates.

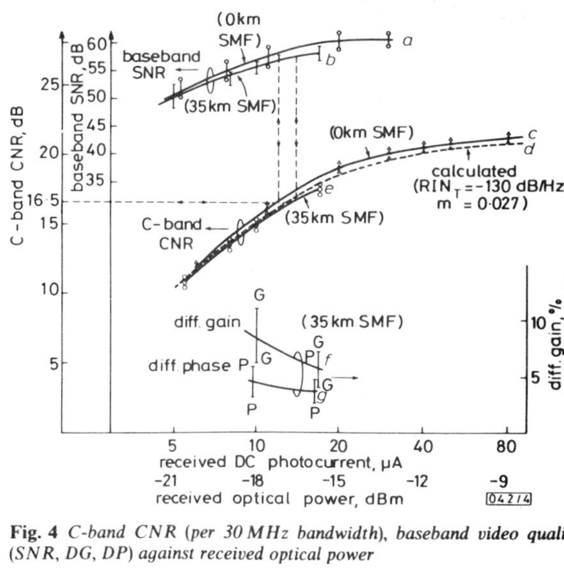

Fig. 4 *C-band CNR (per 30 MHz bandwidth), baseband video quality (SNR, DG, DP) against received optical power*

CNR and SNR degradations in (*b*) and (*e*), as compared to (*a*) and (*c*), respectively, are due to the additional optical reflections caused by eight more pairs of biconic connectors, and fibre bandwidth limitation

The authors thank R. Spicer, L. Curtis, F. E. Placek and M. B. Romeiser for installing biconic connectors and assembling termination boxes. They also wish to thank D. C. Cox for encouragement during the course of this project, which started in late 1984.

W. I. WAY
M. KRAIN
R. S. WOLFF

Bell Communications Research
331 Newman Springs Road
Red Bank, NJ 07701, USA

5th March 1987

References

1 WAY, W. I., KRAIN, M., and WOLFF, R. S.: 'A 35 km satellite earth station 1·3 μm fiber optic microwave multicarrier transmission system'. OFC '87, Reno, NA, 1987, Postdeadline paper PDP18

2 WAY, W. I., and ELREFAIE, A. F.: 'Fiber optic transmission of microwave 16-ary quadrature-amplitude-modulated signal at 1·3 μm wavelength region'. Tech. digest, ICC '87, Seattle, WA, 1987

3 WAY, W. I., and GARDINA, M. F.: 'Fiber optic transmission of microwave 8 phase-PSK signal'. Tech. digest of SPIE '87 southeast conf. on optics/electro-optics, Orlando, FL, 1987

4 WAY, W. I.: 'Large signal nonlinear distortion prediction for a single-mode laser diode under microwave intensity modulation', *J. Lightwave Technol.*, March 1987, **LT-5**

5 WAY, W. I., and AFRASHTEH, A.: 'Linearity characterization of connectorized laser diodes under microwave intensity modulation by AM/AM and AM/PM measurements'. Tech. digest IEEE international microwave symposium, Baltimore, MD, 1986, pp. 659–662

6 SIOCOS, C. A.: 'Satellite FM television channel bandwidth', *Space Commun. & Broadcast.*, 1984, **2**, pp. 363–369

7 'Electrical performance standards for television relay'. Electronics Industries Association (EIA) report RS-250B, 1976, p. 30

8 BOWERS, J. E., CHIPALOSKI, A. C., BOODAGHIANS, S., and CARLIN, J. W.: 'Direct fibre-optic transmission of entire microwave satellite antenna signals', *Electron. Lett.*, 1987, **23**, pp. 185–187

Long Distance Fiber-Optic Transmission of C-Band Microwave Signals to and from a Satellite Antenna

JOHN E. BOWERS, SENIOR MEMBER, IEEE, A. C. CHIPALOSKI, S. BOODAGHIANS, AND J. W. CARLIN

Abstract—**We demonstrate here the use of high-speed semiconductor lasers and detectors with low loss optical fiber for the transmission of the 500-MHz C-band microwave signal spectrum to and from a satellite antenna. The optical system has low attenuation and large bandwidth, so the microwave signals can be transmitted directly at the microwave frequency (4 or 6 GHz) over 20 km of fiber without regeneration or qualization. The optical system introduces ≈ 1 dB or less of degradation for both low spectral density QPSK signals and high spectral density FM video signals present in typical satellite transmission systems. For the video signals, the signal to noise after AM conversion was reduced from 56 to 55 dB. The microwave drive level to the laser must be significantly larger (typically > −10 dBm) than the intensity noise of the laser, yet small enough (typically < 10 dBm) to reduce the intermodulation distortion signals to an acceptable level. The laser bias current must be several times threshold so that the resonance frequency is beyond the frequency band of interest (4–6 GHz in this case). A flat modulation response is then obtained, and the intensity noise and intermodulation levels are low.**

I. Introduction

SATELLITE communication systems provide a cost effective means for obtaining wide-band video and data communication lines. Coaxial cable or waveguide is usually used to interconnect a C-band (6/4 GHz) or Ku-band (14/11 GHz) antenna to terminal equipment located in a nearby equipment shelter or building. Losses in the cable or waveguide limit the interconnection distance to 100 m or less if the transmission frequencies are above 4 GHz. If the signal spectrum is frequency translated to around 1 GHz, interconnect lengths of several hundred meters are feasible. If only a small portion of the 500-MHz signal spectrum is used at the site, the narrow-band signal can be transmitted at 70 MHz over lengths of a few kilometers. Optical fiber has two or more orders of magnitude less loss than microwave cable, and the possible transmission distances extend to 50 km or more, provided efficient, broad-band lasers and detectors are used. Other advantages of optical fiber over coaxial cable and waveguide include greater flexibility, smaller size and weight, and lower cost. The fiber-optic cable cost advantage is offset by the cost of lasers, detectors, and fiber connectors. However, these optical component costs are dropping rapidly. Fiber-optic transmission also offers immunity to electrical interference from power lines, lightning strokes, or electromagnetic pulses from upper atmosphere nuclear bursts.

We describe here the use of recently demonstrated high speed lasers [1] and detectors [2], [3] to transmit the entire satellite signal directly as the microwave transmission frequency over as much as 50 km of fiber. This advance opens up many new applications. For instance, finding a site for C-band antennas may be difficult because of interference from terrestrial microwave links at the same transmission frequencies. With the advance described here, the transmission signal can now be multiplexed and frequency translated at a convenient location (local office or customer premises) and then transmitted over a fiber-optic link to a remote antenna site, which can be selected on the basis of interference, availability, cost, or other criterion. Fiber-optic transmission should also be of interest for Ku-band systems, which are currently undergoing significant commercial growth. Satellite transponders at these frequencies provide more power and the antennas are much smaller and cheaper, allowing the location of antennas on customer premises. High-speed optical devices and fiber-optic links can be used to transmit the signal directly at Ku-band frequencies or at a lower down converted L-band frequency. The optical link would allow for distribution of the full 500-MHz satellite signal spectrum throughout a building without repeated amplification. Other applications of optical links include antenna diversity to reduce the possibility of rain outage. Presently, hard rainfall can reduce the signal-to-noise level to unacceptable levels. Using the optical techniques described above, two antennas can be separated by 40 km, and the probability of both sites suffering simultaneous outages due to rain induced fades is greatly reduced. The optical system also has applications in long baseline (50-km) radio astronomy arrays or radars.

In the following sections, we describe the experimental arrangement for characterizing the transmission systems and describe the device design tradeoffs. We then describe measurements of the optical system performance as a function of laser bias, detector bias, microwave drive level, reflection level, and fiber length. We examine the transmission characteristics of both high spectral density FM video signals and low spectral density QPSK signals and quantify the degradation that the optical system imposes on the earth station up- and down-links [4]. Finally, we examine the usefulness of this system at L, Ku, and K bands.

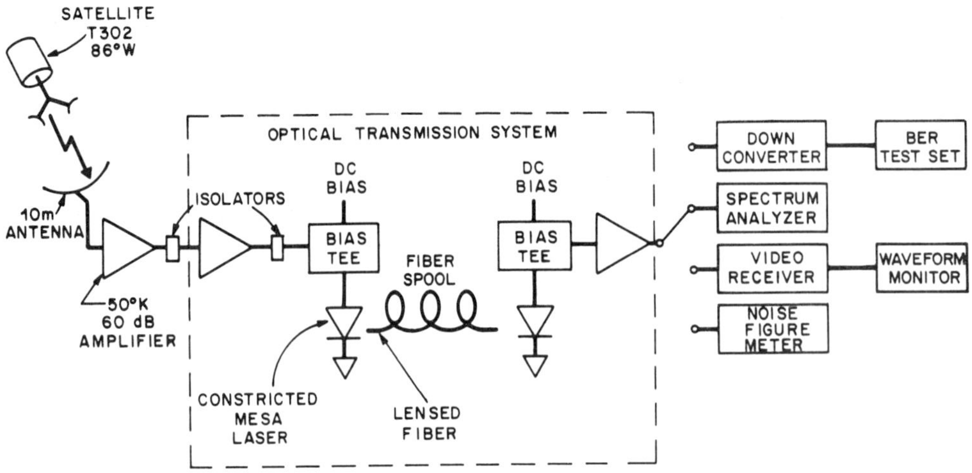

Fig. 1. Schematic diagram of a downlink fiber-optic system and diagnostic equipment.

II. Experimental Arrangement

A schematic drawing of the downlink experiment is shown in Fig. 1. A 10-m antenna was used to receive the 4-GHz signal from the AT&T Telstar T-302 satellite. A typical received spectrum is shown in Fig. 2. The signal is composed of 12 channels in each polarity, each 40 MHz wide. Video, digital, and telemetry signals are marked on the figure. Existing nonencoded video signals were used to measure the video performance using a commercial video waveform analyzer. The digital transmission quality was measured in terms of bit error rate (BER) performance of the link versus the signal bit energy to noise density ratio (E_B/N_0). The baseband digital data drives a QPSK satellite modem at 1.544 Msymbol/s. The output is up converted to 6.385 GHz and transmitted to the satellite. The entire 500-MHz-wide 4-GHz-downlink signal spectrum was transmitted over the optical link.

The laser was a multilongitudinal mode 1.3-μm constricted mesa laser [1] with a 3-dB bandwidth of 14.6 GHz at a bias of 100 mA. The external differential quantum efficiency of the laser was 0.2 mW/mA at threshold and decreased to 0.15 mW/mA at an operating current of 100 mA and an output power per facet of 12 mW. The peak laser power in the experiments below was at most 14 mW, and was usually below 10 mW. The frequency response of the laser is shown in Fig. 3(a) for several bias levels. The response is typically flat up to a resonance frequency, commonly called the relaxation-oscillation frequency. The dependence of the relaxation-oscillation frequency on the bias current is shown in Fig. 3(b). The laser should be biased so that the relaxation-oscillation frequency is beyond the desired transmission band. The amplitude and phase transfer characteristics of the laser are then flat, and the laser intensity noise and intermodulation distortion levels, which are a maximum at the resonance frequency, are at a level well below the desired signal level.

Light from the laser was coupled with 50-percent efficiency into single-mode optical fiber using a lensed tip on the fiber. We also replaced the lensed fiber with two mi-

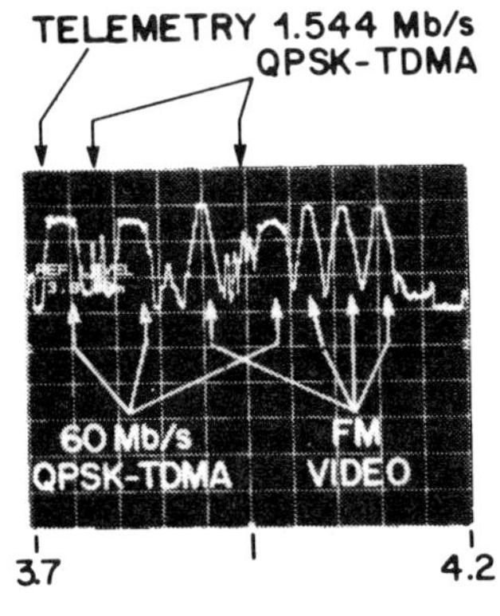

Fig. 2. Typical spectrum of the received satellite signal showing the frequencies of FM video, QPSK digital, and telemetry signals.

croscope objectives and an optical isolator [5] between the objectives to reduce optical reflections back to the laser and thus reduce intensity noise [6]. The system performance was better without the isolator and objectives since they introduced 9-dB (electrical power[1]) additional loss, which was greater than any increase in intensity noise due to optical reflections. The effect of optical reflections back to the laser is greatly reduced by including a polarization controller [7] on the fiber close to the lens. The controller can be used to adjust the dominant polarization of the reflected light to be orthogonal to the laser output and thus minimize the effect of reflections.

The optical fiber had a loss of 0.45 dB (optical)/km at 1.3 μm. The zero dispersion wavelength (~ 1.30 μm) was close to the center wavelength of the lasers used, but no

[1] Note that the optical power from the laser is proportional to the electrical drive *current*. The detected *photocurrent* is proportional to optical power. Consequently, the entire system is linear, but the value in decibels calculated from electrical power (10 log (ratio of electrical power) = 20 log (ratio of electrical current)) is twice the value in decibels calculated from optical power (10 log (ratio of optical power) = 10 log (ratio of electrical current)). Unless otherwise indicated, all decibels used here are calculated from electrical power.

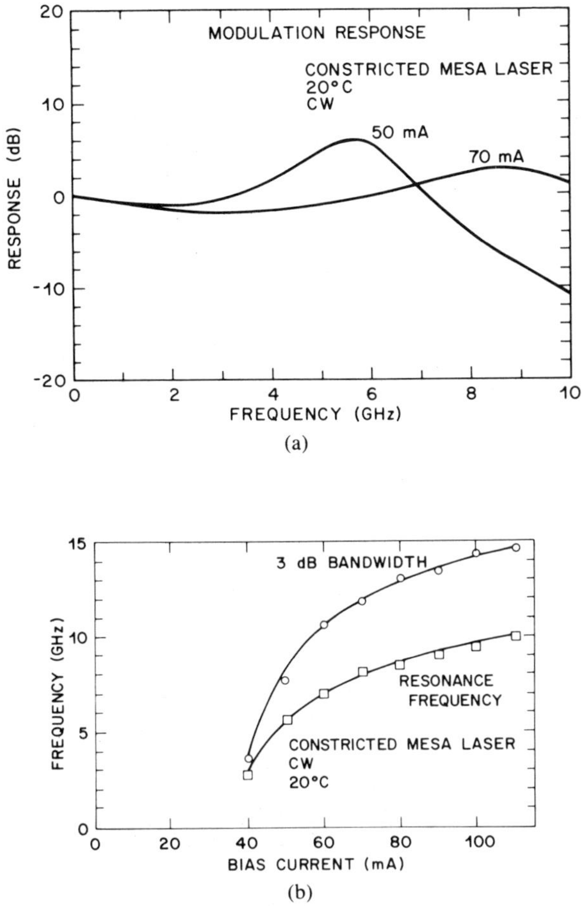

Fig. 3. (a) Frequency response of the constricted mesa laser used in these measurements. (b) Dependence of resonance frequency and 3-dB bandwidth on the dc bias current.

attempt was made to tune the wavelength of the lasers to the zero of dispersion. The lasers typically had 3 to 4 longitudinal modes, spaced 9-Å apart.

Two photodetectors were used. We initially used a p-i-n InGaAs back-illuminated detector [3] with a 3-dB bandwidth of 22 GHz, a quantum efficiency of 50 percent, and a leakage current of 10 nA at the bias voltage of 4 V. This detector had a flat response across the 500-MHz band. We also used an avalanche photodetector (APD) [2], [8] with a gain bandwidth product of 45 GHz and a maximum 3-dB bandwidth (at low gain) of 7 GHz. The total microwave to optical to microwave conversion loss of the fiber-optic system for a short length of fiber (60 m) was 45 dB for an uncoated constricted-mesa laser, a lensed fiber and a p-i-n detector. Using an APD decreased the loss to as low as 25 dB for 45-V bias. Using high reflectivity coatings on one facet of the laser and low reflecting coatings on the other facet reduces the loss by >6 dB. Using other, higher efficiency lasers with facet coatings, we have achieved conversion losses as low as 4 dB for 3-dB bandwidths of 1 GHz (limited by the gain bandwidth product of the APD). Adding an optical isolator adds typically 2 dB (1 dB optical) of loss, and the use of microscope objectives rather than a lensed fiber increased the loss by an additional 7 dB.

III. Fiber-Optic System Characterization

Two important limitations in the use of fiber-optic systems to transmit microwave signals are 1) the laser intensity noise, which typically is 20 dB greater than thermal noise, and 2) the intermodulation products generated by nonlinearities in the optical system, particularly in the laser. The first factor is minimized by driving the laser with as large a microwave power as possible; the second factor is minimized by decreasing the microwave drive level. The important parameter is how much does the insertion of the fiber-optic link decrease the system signal to noise level? We shall see below that this parameter is very similar to the degradation measured with digital BER measurements or with video signal to noise analyzers. However, this measurement is essentially a real-time measurement, and allows easy optimization of system parameters such as laser and detector bias, connectors, and polarization controller position. BER measurements on the other hand, take much longer to make, and are best used to confirm a system meets specifications rather than optimizing performance.

The noise measurement system is shown in Fig. 4(a). By including attenuators and amplifiers before and after the optical transmission system, the total loss of the "link" is zero. The microwave drive level at the laser is adjusted to a level (−3 dBm) which gives acceptable intermodulation distortion products −35 to −40 dBc. We use a commercial noise figure meter to measure the degradation in the signal caused by insertion of the 20-km link.

The noise from a noise diode (Fig. 4(a)) is amplified by 50 dB with a 3.7–4.2 GHz amplifier whose noise figure (1.7 dB) determines the minimum achievable noise figure (any or all of the subsequent amplifiers, attenuators, isolators or optical components can be removed, and the minimum noise figure was 1.7 dB). The broad-band amplified noise is transmitted over 20 km of fiber, amplified, and then downconverted to pass within the dc to 1-GHz passband of the noise figure meter. Fig. 4(b) shows the resultant gain and noise figure as a function of APD bias voltage. The microwave gain at 4 GHz monotonically increased with bias. The noise figure decreases with APD voltage as the detected signal becomes larger and larger than the thermal noise of the preamp following the detector. However, after 40-V detector bias, the noise figure increases with increasing voltage as the excess avalanche noise increases faster than the signal and dominates over the thermal noise of the preamp. Lower noise figures could be obtained with a high input impedance integrating receiver rather than using an APD driving a 50-Ω amplifier. Typically, about 20-dB (optical) increase in sensitivity is obtained in this way, which translates into a 40-km additional transmission distance (ignoring fiber dispersion) for the same noise figure. If a transmission distance of 1 km or less of fiber is used with the present receiver, the minimum noise figure drops from 2.7 dB (Fig. 4(b)) to the 1.7-dB limit of the amplifier following the noise diode.

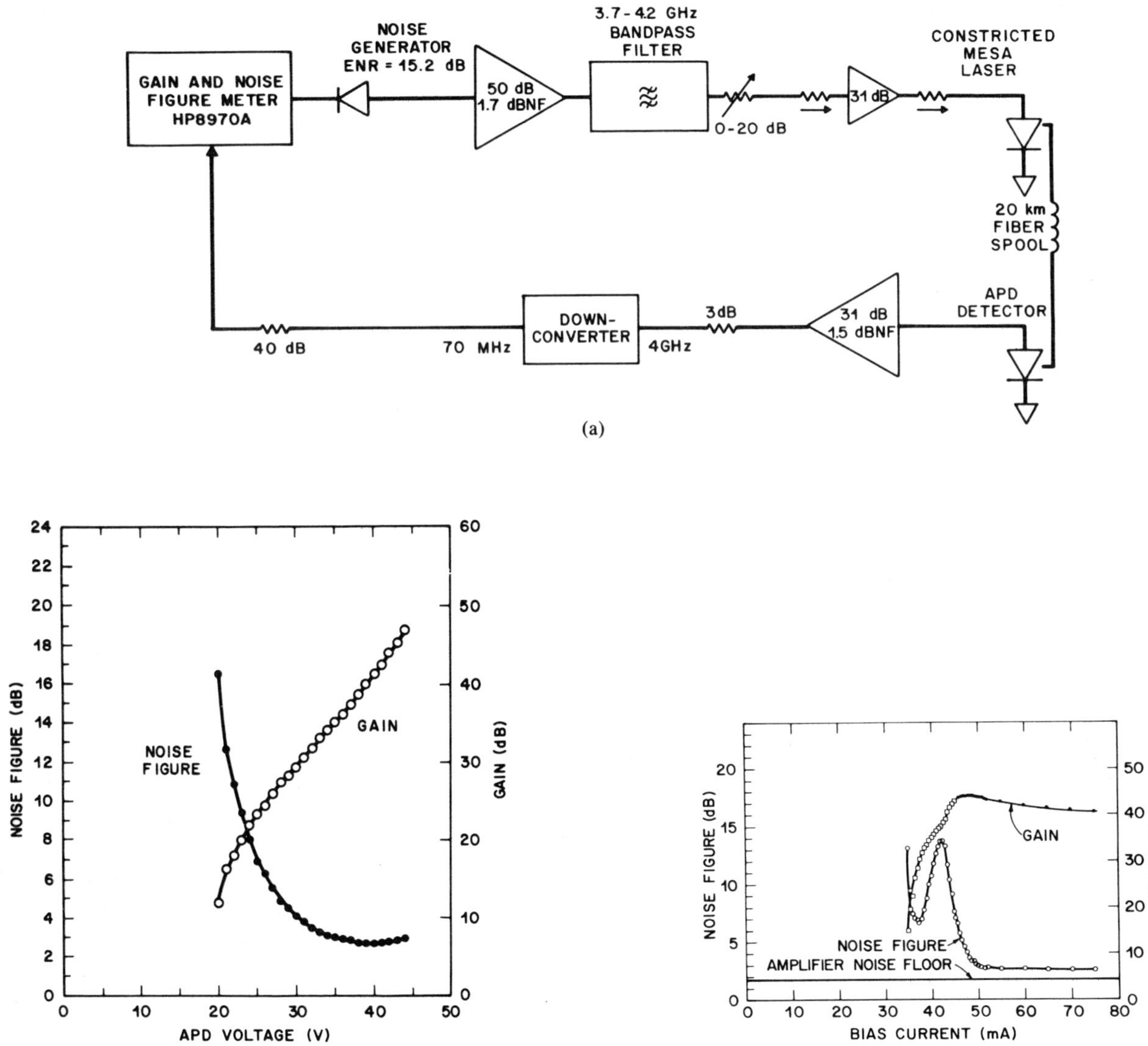

Fig. 4. (a) Schematic diagram of noise figure measurement. (b) Dependence of noise figure and gain on APD bias voltage for 20 km of fiber, a 70-mA bias on the laser, and −3-dBm microwave power at the laser. (c) Dependence of noise figure and gain on laser bias for 20 km of fiber, an APD voltage of 39 V, and −3 dBm microwave power at the laser.

The dependence of noise figure and gain on the laser bias current is shown in Fig. 4(c). Near threshold (35 mA) the gain is low because the laser response severely rolls off above 1 GHz (Fig. 3(a)). The gain increases as the current is increased because the resonance frequency is increasing and the resonance frequency eventually surpasses the measurement band. The gain gradually rolls off (4 dB) as the bias current is further increased from 50–80 mA because the power-current dependence of the laser is not completely linear, and slightly curves at high bias currents. The peak in the noise figure at 42-mA bias occurs when the peak in the intensity noise spectrum is centered on the measurement band; at higher bias, the position of the peak in the noise spectrum continues increasing up to about 10 GHz. The height of the noise intensity peak continually decreases with increasing bias [6], [9]. Over a broad current range, 50–80 mA, the noise figure is low and the gain is relatively constant. Most of the subsequent experiments were done at 70-mA laser bias and 39-V detector bias, where the noise figure is minimized.

Another operational parameter to consider is the microwave drive level. For 70-mA bias current, the laser is 100-percent intensity modulated for a drive level of 15 dBm. In general, the use of a large modulation signal puts the signal level far above the laser intensity noise. However, large drive levels produce a large amount of intermodulation distortion [10], [11].

The intermodulation distortion level was measured using the combined output of two microwave oscillators spaced 200 kHz apart, and varying the power of both of the oscillators. The minimum intermodulation level that could be measured using an APD detector was − 55 dBc, limited by the intensity noise on the laser. The minimum level that could be measured using a p-i-n detector was −43 dBc, limited by the thermal noise of the preamp following the detector. We compare the IMD levels for a p-

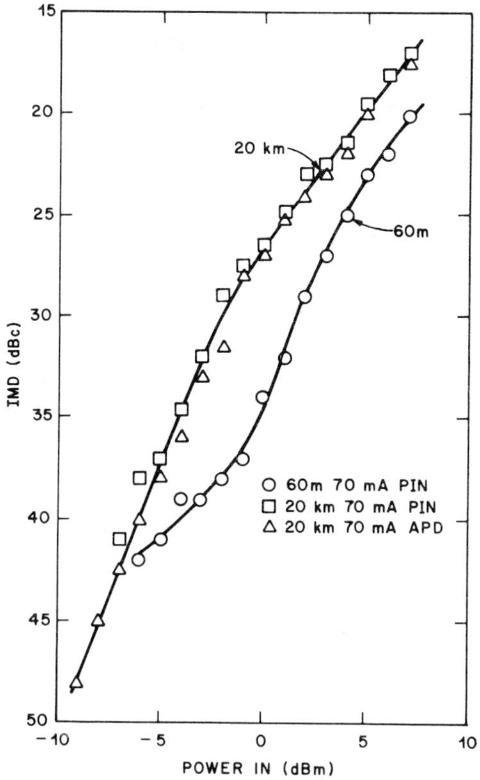

Fig. 5. Dependence of IMD on microwave drive power for different lengths of fiber and APD or p-i-n detectors.

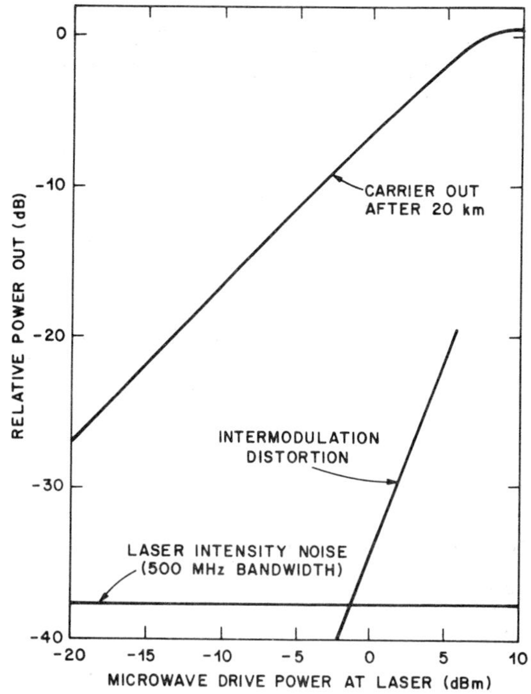

Fig. 6. Power levels of optical system output, IMD, and laser intensity noise.

i-n and an APD detector in Fig. 5 for the same conditions, (70-mA laser bias and 20 km of fiber). No difference was seen, indicating the intermodulation distortion level of the detector is less than that caused by the laser and fiber. The laser IMD level (−42 dBc for −5-dBm drive) is similar to that seen on other lasers [10]–[12]. The power (P) dependence of the IMD varies between $P^{2.5}$ and $P^{1.5}$. We also compare the IMD levels with and without 20 km of fiber. The fiber adds approximately 4 dB of IMD. This is not due to optical reflections from connectors or Rayleigh scattering coming back to the laser. We determined that reflections were not the cause by inserting an optical isolator between the laser and fiber. The IMD level was the same with or without the isolator. Rather, the IMD increase with fiber appears to be due to FM to AM conversion in the fiber since AM modulation of the laser also causes modulation of its optical frequency. This frequency modulation is converted to AM by the fiber dispersion [13], [14].

The optical system linearity, IMD level, and laser intensity noise level are shown in Fig. 6. Because the IMD level increases faster than the linear output level, there is an optimum drive level (0 to −10 dBm) that gives the largest dynamic range. For a drive power of −5 dBm, the IMD level is 37 dB below the carrier. This level is acceptable and is 11 dB above the IMD level of the ultralinear amplifiers used in this system.

We now describe several experiments utilizing actual satellite-transmitted signals to determine in practice how much degradation the optical system introduces. The first experiment examines the optical system effect on large bandwidth FM video signals after transmission over the satellite (i.e., downlink at 4 GHz). These signals occupy the entire 36-MHz bandwidth of one transponder (see Fig. 2). The second experiment examines the optical system effect on the low level QPSK digital signals after transmission over the satellite. The downlink received power spectral density is approximately 18 dB lower than the saturated video spectral density (Fig. 2) and the digital signals are particularly susceptible to IMD (2A-B) interference products of the saturated FM video signals which fall into the 500-MHz received signal spectrum. The final set of experiments examines the problems when the fiber-optic system is inserted on the uplink (i.e., before transmission at 6 GHz up to the satellite).

The video noise level was measured with a waveform monitor. The dependence of signal to noise ratio (SNR) on laser bias (Fig. 7(a)) is similar to the dependence of noise figure on bias current (Fig. 4(b)). Commercial network quality video signals have a weighted SNR of 57 dB. The signal received by the Holmdel earth station had a SNR of 56 dB. If 40 dB of attenuation (equivalent to an "ideal" but lossy, fiber system) and amplifiers with 40 dB of gain were added, then the SNR decreased to 55 dB. When the attenuators were replaced by a constricted-mesa laser, 20 km of fiber and an APD, the SNR was the same, 55 dB, indicating no additional degradation other than that required to make up for the fiber loss and conversion loss.

The dependence of SNR on transmission distance is shown in Fig. 7(b) for p-i-n and APD detectors. For 50-km transmission with a p-i-n detector, the fiber loss, 45 dB (electrical), was so large that the thermal noise of the preamp following the detector dominated the noise level.

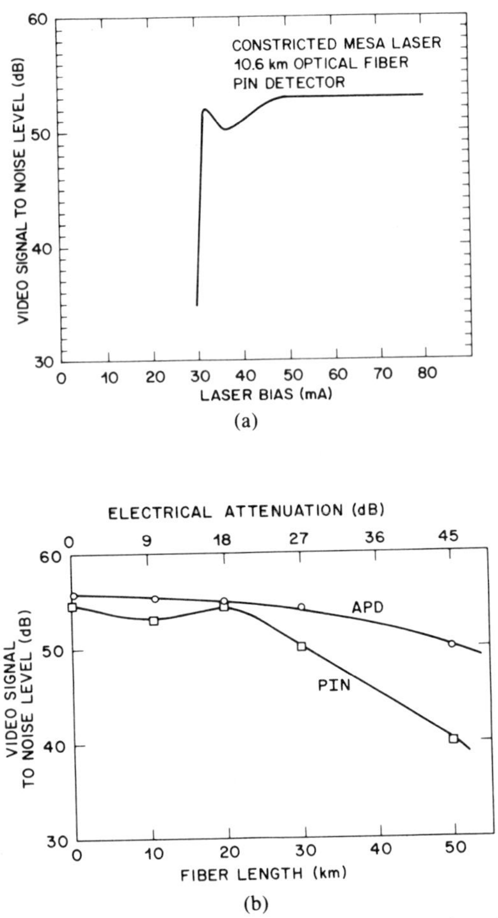

Fig. 7. Signal-to-noise level of FM video signal as a function of (a) laser bias and (b) different fiber lengths for p-i-n and APD photodetectors.

Since the APD has 20 dB more gain than the p-i-n, the SNR was greatly improved by using an APD for the longest distances. As discussed earlier, even longer distances could have been traversed with less noise if integrating high impedance receivers [8] had been used.

The effect on the low density QPSK signal of adding the fiber system before or after satellite transmission was measured in terms of additional bit error (BER) degradation of the satellite link (Fig. 8). Unlike most fiber-optical system characterizations where the BER is measured as a function of received signal level, satellite transmission is usually characterized by measuring the BER as a function of bit energy to noise density ratio (E_B/N_0) by adding noise to the received signal. Note that the normal operating E_B/N_0, or the available over the satellite E_B/N_0, is measured without the fiber system in the link and when the noise attenuator in Fig. 8 is 60 dB or higher. E_B/N_0 is determined by the RF noise density level due to the earth station antenna and low noise amplifier and the signal strength which depends on the satellite traveling wave tube (TWT) amplifier operating point. A realistic operating point for a multicarrier 12-1.544 Msymbols/s QPSK frequency division multiple access (FDMA) service is that each carrier is approximately 15 to 20 dB below the satellite TWT amplifier saturation level. This figure can vary depending on the modulation bandwidth and the number of carriers per satellite transponder. The number of carriers and the power of each carrier is determined in such a way to minimize the total degradation due to the thermal noise and IMD levels in the multicarrier QPSK-FDMA satellite transmission system. The system BER versus E_B/N_0 curve is obtained by measuring the BER count for a number of E_B/N_0 values. To obtain several E_B/N_0 values, the noise attenuator shown in Fig. 8 is varied to obtain a filtered noise power output corresponding to the desired N_0 level. BER degradation can occur in two ways, as a reduction in E_B/N_0 or as a change in the dependence of BER on E_B/N_0. Therefore, by placing the fiber system in the down path or up path of the satellite link and measuring the new available E_B/N_0, we can assess the E_B/N_0 reduction due to the fiber system. To evaluate the change in BER due to the addition of the fiber system to the satellite link, we can plot and compare the obtained BER versus E_B/N_0 curve to a baseline curve obtained in the absence of the fiber system.

Typical results are shown in Fig. 9. For white noise, the theoretical probability of bit errors has a complementary error function dependence on E_B/N_0 [15]. Also shown in Fig. 9 is the performance of the modems back to back (i.e., no satellite transmission), and the performance for satellite transmission without subsequent optical transmission. The available E_B/N_0 in the latter case was 14.6 dB when the uplink power was adjusted to be 22 dB below the satellite TWT saturation level. All of the following experiments used this transmission power. When a short (60-m) optical link was added, the available E_B/N_0 decreased typically 0.5 dB, but the shape of the BER curve did not change. When 20 km of fiber was added, the available E_B/N_0 decreased by an additional 0.7 dB (13.4 dB). Note that 20 km of fiber adds 18-dB additional loss, but the available E_B/N_0 only slightly changes since the noise is limited by the laser intensity noise and IMD levels, and these levels decrease just as rapidly as the signal with additional fiber. With 20 km of fiber, as with 60 m of fiber, no change in the BER curve was seen (Fig. 9).

The results in Fig. 9 were obtained with a drive level of -3 dBm. The dependence of available E_B/N_0 at the QPSK modem demodulator input is plotted in Fig. 10 as a function of the microwave drive level. As expected from the IMD measurements presented earlier, there is a broad range (-10 to 0 dBm) over which the system can be operated without significant degradation.

All of the above experiments involved inserting the fiber-optic system after reception of the satellite signal. These experiments were at the downlink frequency band (3.7–4.2 GHz). We now discuss inserting the fiber system on the uplink at 5.9–6.4 GHz before transmission to the satellite. Whereas for downlink transmission the available E_B/N_0 was small both without (14.6 dB) and with (13.4 dB) the fiber link, now on the uplink, the input E_B/N_0 is large (>30 dB) both before and after the fiber-

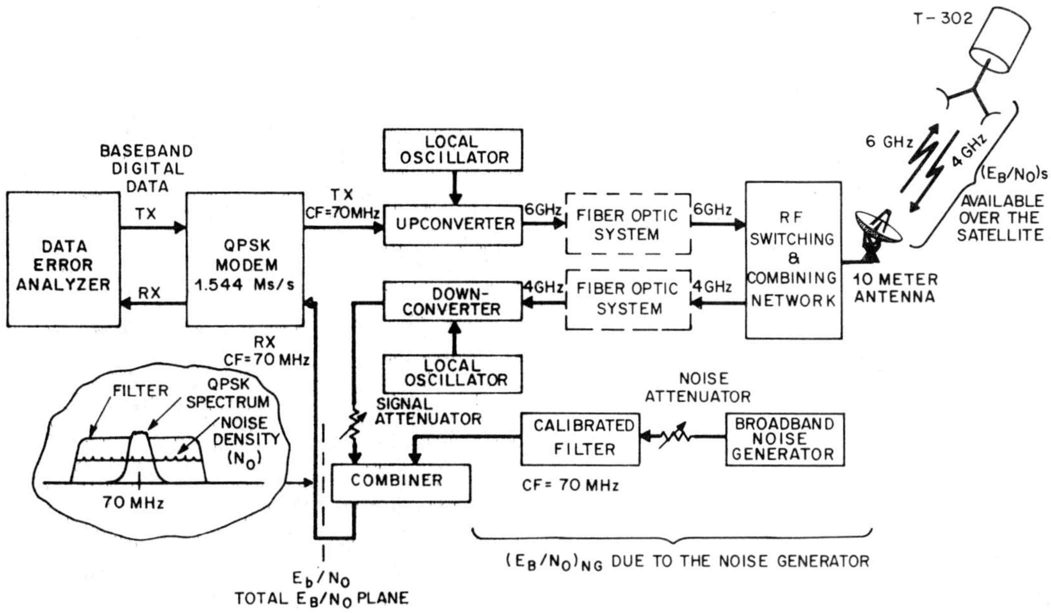

Fig. 8. Bit error rate measurement schematic.

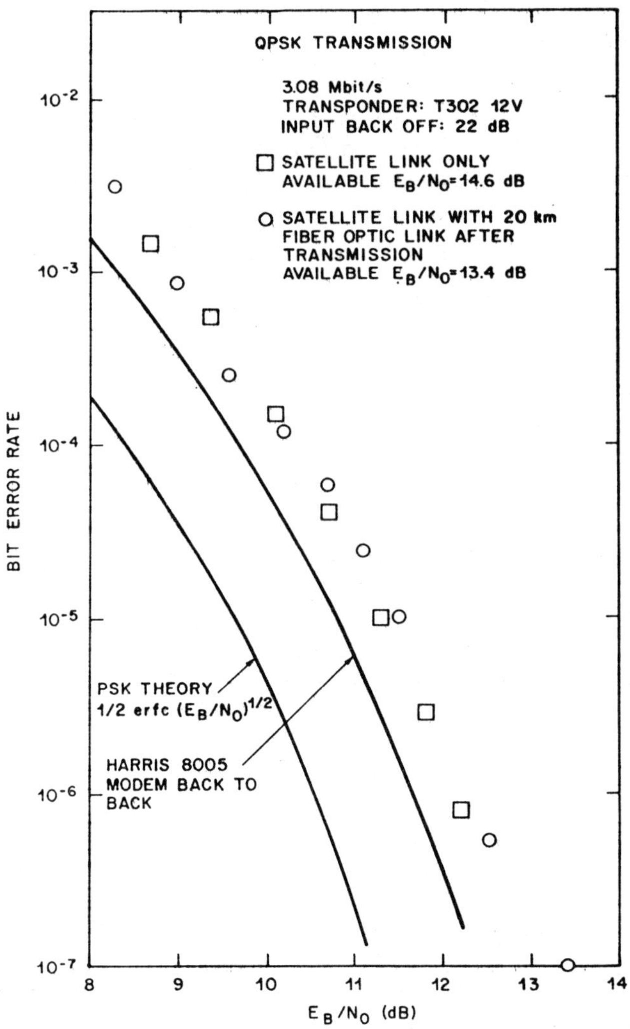

Fig. 9. Dependence of bit error rate on signal-to-noise level for 3.08-Mbit/s QPSK signals on the 12-V channel with the uplink backed off 22 dB from saturation. Curves are shown for the modem only, satellite transmission, and satellite transmission followed by a 20-km fiber link containing the entire 500-MHz signal.

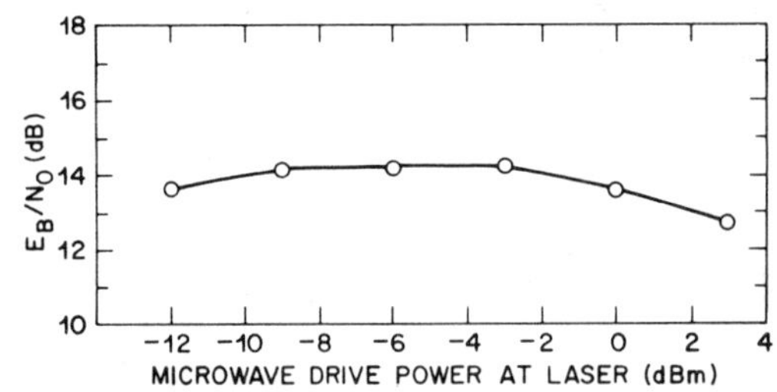

Fig. 10. Dependence of final signal-to-noise level (E_B/N_0) on the total microwave drive power at the laser input. The entire 500-MHz signal from the satellite was transmitted over a 20-km fiber-optic link, and the error rate of one 3.08 Mbit/s QPSK signal was measured. (Fiber-optic system at downlink 4 GHz, 60 m of fiber).

optic link. After transmission over the satellite, the E_B/N_0 was again reduced (due to the tremendous attenuation over $2 \times 35\ 000$ km of free space) and was 14.6 dB whether or not the 20-km fiber link was installed in the uplink. The BER curves for this case are shown in Fig. 11 and are essentially the same for 70 m of fiber, 20 km of fiber, or no optical system at all. Note that the total microwave drive level to the laser was −3 dBm, as on the downlink, but because there is only one signal (the 3.088-Mbit/s QPSK signal) on the uplink, its relative level is higher, and IMD from other frequencies is not present. Note also that the earth station uplink power amplifier follows the fiber system and is at the antenna uplink input port.

Many scenarios can be envisioned for using optical fiber systems on an uplink. The two extremes for the case measured here, an FDMA system, would be 1) ten separate 3.088-Mbit/s uplinks or 2) one fiber link carring all ten 3.088-Mbit/s signals. In the first case, the satellite sees the broad-band noise and intermodulation deterioration from ten fiber systems, not just one system. While we did not do such an experiment, we can estimate its performance. The uplink signal to noise level (primarily due to the laser intensity noise) would decrease from

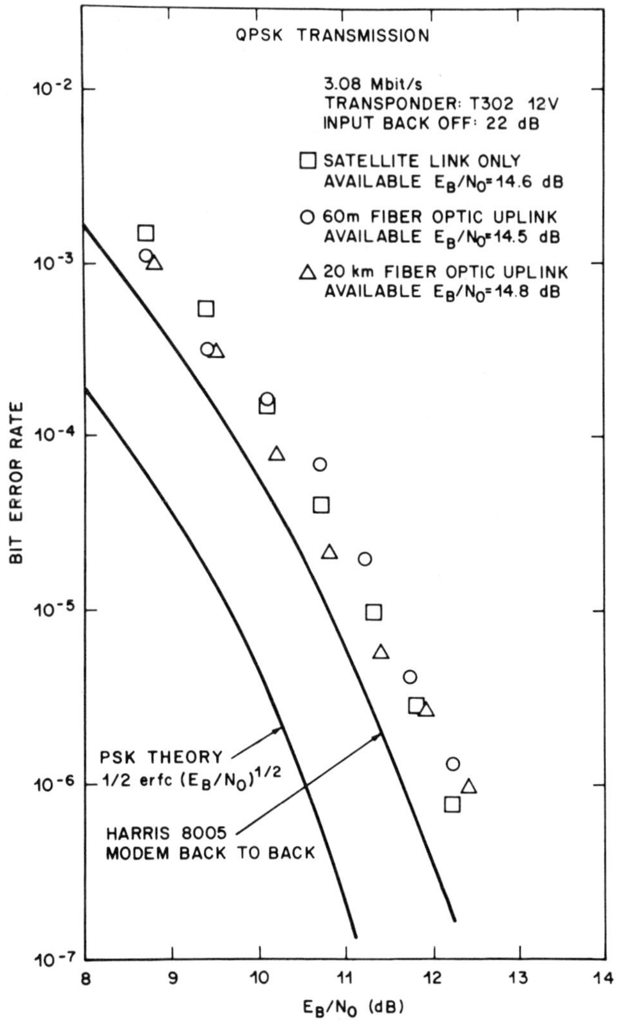

Fig. 11. Dependence of bit error rate on signal to noise level for satellite transmission, and for 60-m and 20-km optical system before transmission over the satellite.

greater than 30-dB to greater than 20-dB. The intermodulation levels would not significantly degrade from the levels measured here because the different frequency signals are combined after going through the laser system. Hence, we expect to have significantly more dynamic range than required for 10^{-8} BER (15 dB). For the other extreme of 10 carrier transmission over one laser and fiber system, the signal-to-noise level would be less (a reduction from 30 to 20 dB), because there is less power per carrier (-3 dBm for all ten carriers instead of just one carrier). The intermodulation levels for the 10 carrier transmission over a single fiber would also be higher than the measured two-tone value of -32 dBc for -3-dBm drive. This increase is on the order of 6–8 dB in a system for which the IM levels are primarily third order [16], [17]. Thus the expected IM levels for 10 carrier transmission are expected to be at the -25-dBc level for -3-dBm drive. In summary, for both of the limiting scenarios, the uplink carrier to total noise level will be greater than 20 dB, which is sufficient for such uplink applications.

IV. Conclusions

Satellite signals composed of typically 3–4 video signals, 1–2 sets of 60-Mbit/s QPSK signals, and several sets of 3.088-Mbit/s QPSK signals can be transmitted over distances of up to 20 km with 1 dB or less of degradation in the signal-to-noise ratio for both the video and digital signals. For most earth station applications where the fiber length is 100 m or less, no degradation in the earth station gain over the system noise temperature (G/T) was observed. While we did not examine the transmission of the 60-Mbit/s QPSK signals in this paper, we have previously [9] examined the transmission of a much higher bit rate signal (2 Gbit/s) at carrier frequencies from 3 to 16 GHz and found that 10^{-9} BER could be obtained for transmission over 34 km of fiber. Hence, we believe the 60-Mbit/s signals will not be degraded any worse than the digital or video signals examined here. Transmission over 50 km of fiber is possible with a 6-dB reduction in video signal to noise level.

The laser intensity noise and intermodulation distortion determine lower and upper limits, respectively, on the possible microwave drive powers, with an acceptable variation in total drive power of about 10 dB. While it would be preferable to have all of the frequency multiplexed signals at the same drive level because of IMD limits, the 20-dB variation between the FM video and QPSK signal levels encountered in typical satellite transmission is acceptable.

References

[1] J. E. Bowers, B. R. Hemenway, A. H. Gnauck, and D. P. Wilt, "High-speed InGaAsP constricted mesa lasers," *J. Quantum Electron.*, vol. QE-22, pp. 833–844, 1986.

[2] J. C. Campbell, W. S. Holden, J. F. Ferguson, A. G. Dentai, and Y. K. Jhee, "A high-speed InP/InGaAsP/InGaAs avalanche photodiode exhibiting a gain bandwidth product of 60 GHz," in *Tech. Dig. IOOC-ECOC'85* (Venice, Italy), pp. 66–68, Oct. 1985.

[3] J. E. Bowers, C. A. Burrus, and R. J. McCoy, "InGaAs p-i-n photodetectors with modulation response to millimeter wavelengths," *Electron. Lett.*, vol. 21, pp. 812–814, 1985.

[4] J. E. Bowers, A. C. Chipaloski, S. Boodaghians, and J. W. Carlin, "Direct fiber-optic transmission of entire microwave satellite antenna signals," *Electron. Lett.*, vol. 23, 1987.

[5] R. M. Jopson, G. Eisenstein, H. E. Earl, and K. L. Hall, "Bulk optical isolator tunable from 1.2 to 2 μm," *Electron. Lett.*, vol. 21, p. 783, 1985.

[6] Y. Yamamoto, "AM and FM quantum noise in semiconductor lasers—Part I: Theoretical analysis," *IEEE J. Quantum Electron.*, vol. QE-19, no. 34, 1983.

[7] B. G. Koehler and J. E. Bowers, "In-line single-mode fiber polarization controllers at 1.55, 1.30, and 0.63 μm," *Appl. Opt.*, vol. 25, pp. 349–353, 1985.

[8] B. L. Kasper, J. C. Campbell, J. R. Talman, A. H. Gnauck, J. E. Bowers, and W. S. Holden, "An APD/FET optical receiver operating at 8 Gbits," to be published.

[9] J. E. Bowers, "Optical transmission using PSK modulated subcarriers at frequencies to 16 GHz," *Electron. Lett.*, vol. 22, p. 1119, 1986.

[10] K. Y. Lau and A. Yariv, "Intermodulation distortion in directly modulated semiconductor lasers," *Appl. Phys. Lett.*, vol. 45, pp. 1034–1037, 1985.

[11] T. E. Darcie, R. S. Tucker, and G. J. Sullivan, ''Intermodulation and harmonic distortion in InGaAsP lasers,'' *Electron. Lett.*, vol. 21, pp. 665–666, 1985.
[12] W. I. Way and A. Afrashteh, ''Linearity characterization of connectorized laser diodes under microwave intensity modulation by AM/AM and AM/PM measurements,'' in *1986 IEEE MTT-S Int. Microwave Symp. Tech. Dig.*, pp. 659–662.
[13] P. Brosson, C. Ribeiro, and J. Ripper, ''Frequency demodulation of a semiconductor laser by FM-AM conversion in an optical fiber,'' *IEEE J. Quantum Electron.*, vol. QE-17, p. 689, 1981.
[14] A. R. Craplyvy, R. W. Tkach, L. L. Buhl, and R. C. Alferness, ''Phase modulation to amplitude modulation conversion of CW laser light in optical fibers,'' *Electron. Lett.*, vol. 22, pp. 409–411, 1986.
[15] S. Haykin, *Communication Systems.* New York: Wiley, 1983, p. 559.
[16] E. D. Sunde, *Communication Systems Engineering Theory.* New York: Wiley, 1969, p. 351.
[17] J. Boyhar, private communication.

*

John E. Bowers (SM'86) received the M.S. and Ph.D. degrees in applied physics from Stanford University. His dissertation involved the design, fabrication, and demonstration of SAW signal processing devices.

He was a Research Associate at Stanford University conducting research on single-mode fiber-optic devices, sensors, and signal processors. He was employed by Honeywell for one year in 1979, where he developed an LPE system for the large area multilayer growth of HgCdTe. Since 1982, he has worked at Bell Laboratories, Holmdel, NJ, investigating quaternary lasers and photodetectors. He is primarily concerned with high frequency devices, and their physics application to high-speed digital and analog communications systems.

Dr. Bowers is a member of Phi Beta Kappa, Tau Beta Pi, and the American Physical Society. He is a recipient of Sigma Xi's Thomas F. Andrew's prize for undergraduate research.

*

A. C. Chipaloski, photograph and biography not available at time of publication.

*

S. Boodaghians, photograph and biography not available at time of publication.

*

J. W Carlin, photograph and biography not available at time of publication.

11-GHz FIBER OPTIC LINKS FOR AIRBORNE SATELLITE COMMUNICATIONS APPLICATIONS

C. M. Gee, G. D. Thurmond, I. L. Newberg, and H. W. Yen
Hughes Research Laboratories
Malibu, California 90265

D. Lafaw and B. Spink
Rome Air Development Center
Griffiss AFB, New York

KEY TERMS

Wideband fiber optic systems, microwave and optical waveguide applications, high-speed modulation techniques, X-band fiber optics links, integrated optics

ABSTRACT

We report the first implementation of fiber optics at 11 GHz in an experimental satellite communication EHF/SHF research terminal for airborne applications. Signal-to-noise and bit-error rate were measured to characterize the system after a 100-m-long fiber optic link replaced a 4-ft-long coaxial cable to enable X-band signal remoting.

Future satellite communication systems will be operating with transmit and receive frequencies above 20 GHz (K-band). In addition, signals in the X-band frequency range (10 GHz) are required at remotely located antennas for signal modulation and demodulation. Currently, the antenna, synthesizers, mixers, and other terminal equipment are located close together to keep coaxial cable runs short. Longer coaxial cable lengths significantly increase insertion loss and incur interference signals that can degrade signal-to-noise (S/N). Consequently, supporting hardware placement is inflexible in generally cramped airframes.

High-frequency analog fiber optic communication is developing at a rapid rate and can be important for future avionic systems. Fiber optics offer the aircraft and satellite communication system designer flexibility and advantages over coaxial cable and waveguide. A fiber optic link allows the X-band signal to be transmitted directly between the antenna and the signal generation and processing equipment with virtually no signal degradation. Thus, the antenna can then be remotely positioned relative to the terminal and processing equipment. In addition, the fiber optic cable is light, compact, EMI resistant, and flexible.

Figure 1 shows the functional block diagram of the experimental satellite communication (SATCOM) EHF/SHF research terminal (SERT) at the Air Force Rome Air Development Center (RADC). The SATCOM system provides communication in which a microwave frequency hopped carrier that is digitally coded is sent between a 44-GHz airborne uplink terminal and a 22-GHz satellite downlink terminal. In the research terminal, both the 22-GHz downlink and 44-GHz uplink are colocated for test purposes. The information is digitally coded in the modem/processor which provides eight tone modulation and demodulation for a spread spectrum frequency shift keyed (FSK) process. Programmable frequency synthesizers and other peripheral equipment provide the FSK X-band signals for both the downlink local oscillator mixing tc 1.53 GHz and for the transmitter oscillator signal that is upconverted to 44 GHz. The digital information is demodulated by further mixing to 70 MHz.

Using a fiber optic link, the processing equipment can be remotely located from the antenna, the upconverter mixer, and the downconverter mixer. In this program, we replaced the coaxial cable connecting the frequency synthesizer to the subharmonic downconverter mixer in the downlink chain with a 100-m-long fiber optic link. The fiber optic link supplies the 11-GHz local oscillator (LO) input to the mixer. Another fiber optic link at 1.53 GHz replaced the coaxial cable connecting the subharmonic mixer with the second IF downconverter mixer.

The fiber optic link, illustrated in Figure 2, consists of a directly modulated semiconducter laser, 100 m of optical fiber, a high-speed semiconductor photodetector, and low-noise rf amplifiers. In this technique, an rf signal amplitude modulates (AM) the laser, producing a coherent AM optical signal which is coupled into single-mode fiber. The optical output is detected with a photodiode and demodulated to produce a coherent replica of the input rf signal.

The fiber optic links are required to be transparent to the signals passing through them without introducing degradation. Since the fiber optic links contain active optoelectronic components, they could add noise to the SATCOM system relative to a short passive coaxial cable. If the fiber optic link noise is considerably less than the SATCOM system noise, then the effects are negligible. Of course, fiber optics provide remoting capability not available with the present SATCOM research terminal system. Nevertheless, designing a low-noise fiber optic link is most important. The performance of fiber optics for this application is primarily determined by the laser and detector frequency response and by laser noise.

Present lasers and detectors operate well near 1.53 GHz, but few lasers are commercially available near 11 GHz. Be-

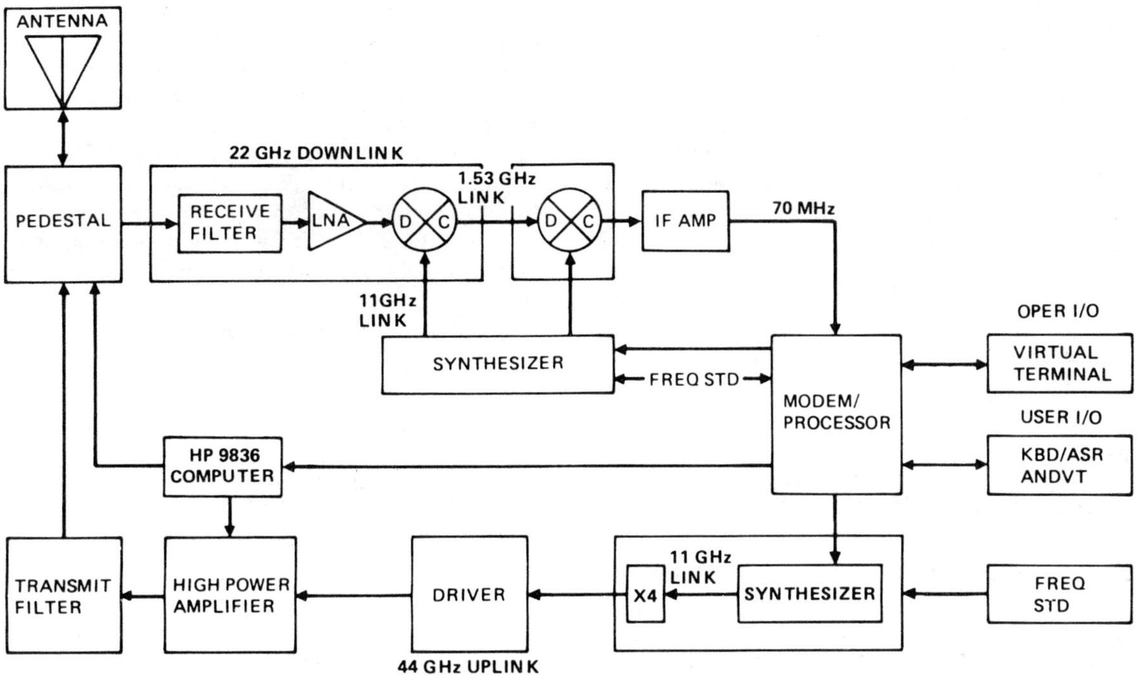

Figure 1 Functional block diagram of RADC SATCOM EHF/SHF research terminal (SERT) showing 11- and 1.53-GHz fiber optic links for remoting the antenna from the processing equipment

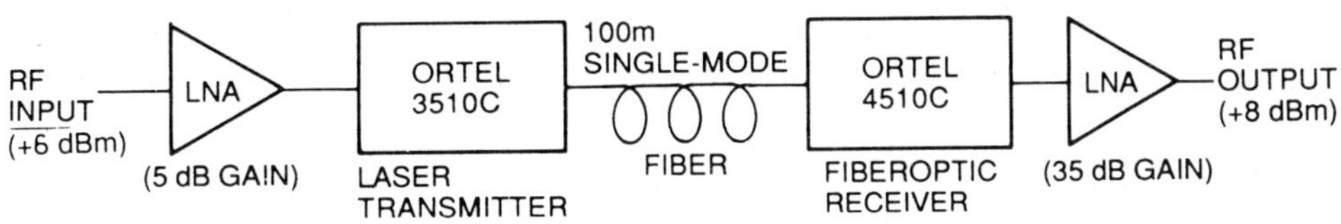

Figure 2 Fiber optic link, consisting of a directly modulated semiconductor laser, 100-m-long optical fiber, and fiberoptic receiver, operated at 11 GHz

sides frequency response, laser noise must be lower than the phase noise of the LO synthesizer. We have extensively characterized several commercial lasers and detectors. The best performance at 11 GHz for this application are the Ortel Corporation[1] model 3510C fiber optic transmitter and the model 4501C fiber optic receiver. Table 1 summarizes their performance at 11 GHz and compares them to the LO synthesizer output. The fiber optic link must be lower in noise than the synthesizer in order not to degrade the SATCOM system. Also, although the SATCOM synthesizers have very low AM noise far from the carrier, the fiber optic link contains very broadband laser AM at a level of −127 dBc/Hz. Specifically, laser noise AM translated to the 1.53-GHz IF can couple into the mixer output and affect the system noise figure. Consequently, the limiting amplifier, bandpass filter, and the balanced dual diode design of the subharmonic mixer are essential in preventing broadband fiber optic link AM noise at a level of −121 dBm/Hz (including amplification) from passing through the 1.53-GHz output of the mixer.

The primary criteria for evaluating the performance of the SATCOM system is bit-error-rate (BER) measurements. The 100-m (≃ 330-ft) fiber optic links were compared to a 4-ft-long Gore-Tex type 3 microwave coaxial cable. Figure 3 displays the BER of the cable and the 11-GHz fiber optic link. It appears that the fiber optic link did not significantly degrade the SATCOM system. Overall S/N, stability, repeatability, and statistical measurement error of the experimental SATCOM system also contribute to any apparent discrepancy between the BER performance of the two links. The SATCOM system performance can also be evaluated by measuring the S/N of the system at the 70-MHz IF. S/N is within 1 dB of each other for the coaxial cable and the fiber optic links.

Because lasers and detectors operating at a few GHz are readily available, the requirements of the 1.53-GHz link between the subharmonic mixer and IF downconverter can be easily satisfied with a fiber optic link. The fiber optic link has a frequency flatness of ±0.05 dB/MHz with a variation of 0.1 dB across the 10-MHz passband at 1.53 GHz, well within the requirements of the SERT system. In addition, as evident from a comparison of the S/N of the SERT 70-MHz signal for the 4-ft coaxial cable with the 100-m-long fiber optic link at 1.53 GHz (Figure 4), the fiber optic link did not degrade the system performance.

In conclusion, we have demonstrated for the first time that insertion of 100-m-long fiber optic links in a satellite communication terminal for airborne applications did not signifi-

TABLE 1 Performance of 100-m-long Fiber Optic Link at 11 GHz[a]

SPECIFICATION	SYNTHESIZER	FIBER OPTIC LINK
SERT FREQUENCY	10.865–11.375 GHz	10.865–11.375 GHz
FREQUENCY FLATNESS		±0.4 dB
RF POWER OUTPUT	6 dBm	8 dBm
VSWR	< 1.5:1	< 1.5:1
INTERMODULATION DISTORTION		−52 dBc AT 10 dBm INPUT
SPURIOUS RESPONSE	< −72 dBc	< −72 dBc
PHASE NOISE AT 10 MHz	−122 dBc/Hz	−130 dBc/Hz
AM NOISE AT 100 kHz		−127 dBc/Hz
LENGTH		100 m

[a] Ortel model 3510C laser transmitter and model 4510C fiber optic receiver.

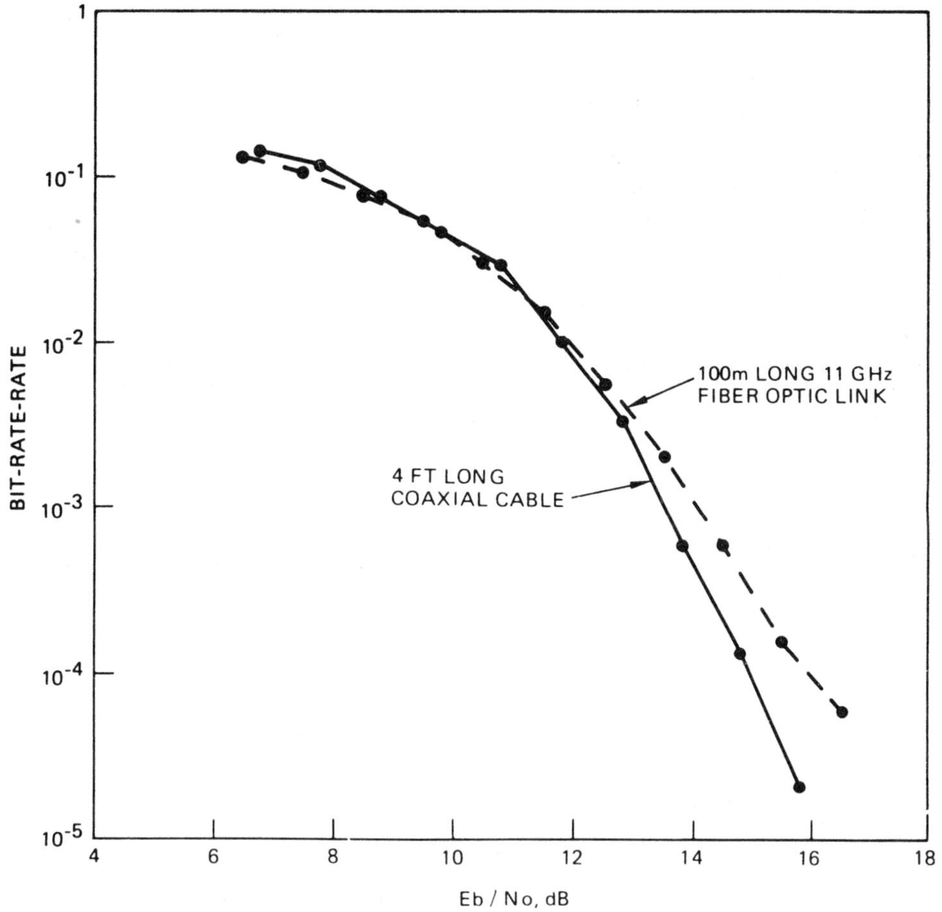

Figure 3 Bit-error-rate as a function of energy per bit for the SERT system with (1) a 4-ft-long coaxial cable and (2) a 100-m-long fiber optic link between the 11 GHz synthesizer output and the LO to the subharmonic mixer

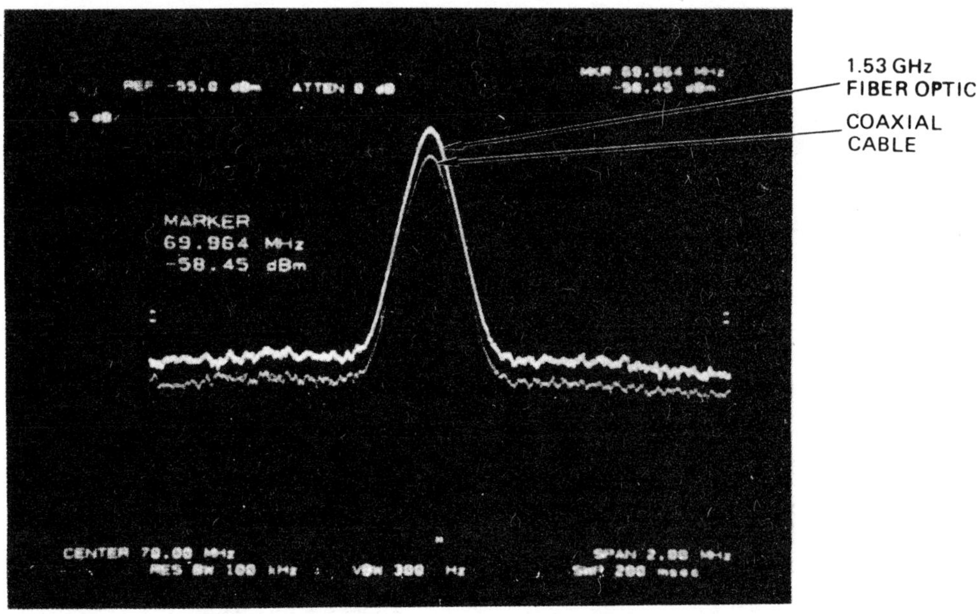

Figure 4 SATCOM CW signal-to-noise with (1) a 4-ft coaxial cable and (2) a 100-m-long fiber optic link between the 1.53-GHz subharmonic mixer and the IF downconverter mixer.

cantly degrade system performance. The links provide lightweight long distance remoting that is not practical at high frequencies with coaxial cable or waveguides. The BER results demonstrate that fiber optic links can meet the low-noise, high-frequency, and flat frequency response required, and that they could enhance the flexibility and performance of SATCOM terminals for MILSTAR applications.

The authors would like to acknowledge J. Dodge and P. Heim of Mitre Corporation for providing technical support and useful discussions for the system measurements and to Ortel Corporation for supplying the prototype fiber optic link. This work was supported by the Air Force Rome Air Development Center contract no. F30602-86-C-0156 under the direction of Brian Hendrickson.

REFERENCE

1. Ortel Corporation, 2015 West Chestnut, Alhambra, California 91803.

Chapter 8
Long-Distance Digital Transmission Systems

Throughout the two decades during which fiber optical communication has been seriously pursued, the desire to provide high-data-rate links between two terminal locations at a low cost has been the main impetus behind the technological progress. When the cost of optical transmission of information over long distances became favorable in comparison with coaxial cable and microwave links, the rapid expansion in fiber optics began. The cost-effectiveness of fiber optics became firmly established in the early 1980s, and since then technological progress has continued in both commercial systems and research laboratories. Commercial long-haul systems are now operating above 1 Gbit/s, and data-rate-length products in excess of 500 Gbit-km/s have been achieved in the laboratory.

The first paper of this chapter, **"Design Considerations for Long-Haul Lightwave Systems"** by **I. Jacobs,** discusses the trade-offs to be considered in planning a fiber communication system. A review of wideband system results is given by **S.D. Walker and L.C. Blank** in **"Progress in Long-Span, High-Capacity Optical Transmission System Research."** Two high-data-rate commercial systems are described in **"1.13-Gbit/Lightwave Transmission System,"** by **K.Y. Maxham, J.M. Dugan, M.A. McDonald, and C.R. Hogge**, and by **V. Nagarajan and G. Nordqvist** in **"Field Performance of 565 Mbit/s Optical Fibre Line System."** The transatlantic undersea cable system is detailed by **R.L. Smith and R. Whittington** in **"TAT-8: An Overview."** Recent laboratory results are presented in the next three papers: **"313 km Transmission Experiment at 1 Gbit/s using Optical Amplifiers and a Low-Chirp Laser,"** by **M.G. Oberg, N.A. Olsson, L.A. Koszi, and G.J. Przybylek; "5 Gbit/s Transmission System Experiment over 111 km of Optical Fibre,"** by **R. Heidemann, U. Scholz, and B. Wedding**; and **"Transmission at 1 Gbit/s over 7.4 km of Single-Mode Fibre Using an Edge-Emitting LED and PINFET Receiver,"** by **M. Stern, D.J. Millicker, A.E. Elrefaie, and K. Runge.** Finally, an important system–related noise effect is described in **"Effects of Dynamic Spectral Behavior and Mode-Partitioning of 1550 nm Distributed Feedback Lasers on Gbit/s Transmission Systems,"** by **S. Sasaki, M.M. Choy, and N.K. Cheung.**

Design Considerations for Long-Haul Lightwave Systems

IRA JACOBS, FELLOW, IEEE

Abstract—**The following design considerations for long-haul lightwave systems are reviewed: outage and error rate requirements, terminal architecture, regenerator span design and performance specification, capacity upgrade alternatives, and broad-gauge economic factors.**

I. Introduction

LIGHTWAVE transmission systems were initially applied [1]–[4] largely for metropolitan trunking where there was already extensive application of digital transmission on wire-pair carrier systems. In the United States, early metropolitan lightwave systems [5] transmitted at 45 Mbits/s with a regenerator spacing of 7 km. Such systems afforded conduit space and repeater savings relative to wire-pair T1 systems transmitting at 1.5 Mbits/s with a regenerator spacing of 2 km.

It was difficult initially to economically prove-in lightwave systems for long haul intercity transmission where existing analog microwave radio and coaxial cable installations could be upgraded in capacity. However, the advent and wide deployment of toll digital switching afforded considerable terminal cost savings when interfacing with digital rather than analog transmission facilities. Lightwave systems became the economic choice on high growth routes, particularly when the distance between terminal cities was relatively short. A prime example was the AT&T Northeast Corridor System between New York and Washington, DC, construction approval for which was sought in January 1980 [6], and which began service in January 1983.

On the Northeast corridor, information is transmitted at 90 Mbits/s at a wavelength of 825 nm on multimode fiber [7].[1] Regenerators are spaced 7 km apart, utilizing existing repeater stations from an old coaxial cable system. The rapidly advancing nature of lightwave technology was recognized by the FCC at the time of the Northeast Corridor application approval.

"Continuing improvements in fiber technology, like most other new technologies, should produce innovations and increased capacity in fiber systems and reduce future per circuit fiber system costs relative to existing technology. We believe, however, that the experience to be gained from this project outweighs whatever theoretical benefits might arise from delay in its implementation. This experience is necessary to foster the technological developments that will lead to 'learning curve' decreases in cost. Without experience in the early stages of development, we may never see some of the advances that fiber technology promises to bring to telecommunication users. Thus, we find that Applicants should be authorized to begin implementation of this new technology in the long haul network at this time" [8].

Rapid technology improvements have been and continue to be made, and long-haul systems presently being installed in the United States all utilize single mode fiber, and typically have transmission rates in the range of 400–600 Mbits/s with regenerator spacings of about 40 km. Field experiments [9] have been performed at 1.7 Gbits/s and laboratory experiments [10] have been performed at 8 Gbits/s.

The applications arena is also dynamic, with a large number of long-haul carriers planning lightwave networks in the United States. Similarly, a large number of manufacturers (North American, Japanese, and European) are competing for these applications. Design considerations for long-haul lightwave transmission systems are, of necessity, influenced by this highly competitive environment.

This article includes requirements, architecture, and performance considerations for terrestrial long-haul lightwave systems. Emphasis is given to the U.S. network but some similarities and differences for European and Japanese applications are noted. Technology capability is the subject of other articles in this issue. However, the dependence on technology of system architecture and performance is considered briefly here. Also, since design choices are strongly influenced by economics, some very broad-gauge cost considerations are included. The intent is to be illustrative rather than definitive.

II. System Performance Requirements

There are two types of performance requirements: system requirements (e.g., cross-connect specifications, outage, and error performance) and component subsystem requirements (e.g., the performance of a regenerator). The

Manuscript received May 1, 1986.
The author is with AT&T Bell Laboratories, Holmdel, NJ 07733.
IEEE Log Number 8610779.

[1]The system is being upgraded to three times its original capacity, but somewhat differently than originally planned. A second channel, transmitting 180 Mbits/s at 1310 nm. is being added (wavelength-division multiplexed with the original channel). Regenerators for the 180 Mbit/s channel are twice as far apart as on the original channel.

former (typically contained in standards documents [11]) are intended to assure proper interworking with other facilities and appropriate performance for subscriber services (e.g., data services have more demanding error performance requirements than voice services). The latter (subsystem requirements) are imposed by the designer (and, not infrequently, by the purchaser) to assure that the end-to-end requirements are met.

This section considers only two of the many performance criteria for long-haul lightwave systems—viz., outage and error performance.[2]

A. Outage

An outage is said to occur "when the bit error rate (BER) in each second is worse than 10^{-3} for a period of ten consecutive seconds" [12]. In the U.S., short haul systems (up to 400 km in length) are required to have an outage (measured at the 1.544 Mbit/s DSX-1) of no more than 0.02 percent. 75 percent of the outage is allocated to the DSX-3 to DSX-3 transmission link and is prorated by length; viz., 0.0000375 percent/km.

For long-haul systems, the same fractional outage, 0.02 percent, is used for a 6400 km system. This full outage is allocated to the DSX-3 to DSX-3 transmission link, and is prorated by length giving 0.000003 percent/km, a more stringent requirement than for short-haul systems.

More stringent requirements imposed on long-haul than on short-haul facilities are likely a carry-over from the time when long-haul transmission facilities were of higher quality than short-haul facilities (e.g., coaxial cable compared to paired cable). Such distinctions may not be meaningful for lightwave facilities.

High-capacity lightwave systems invariably employ automatic protection switching to meet outage requirements. This is the case even if the outage requirements could be theoretically met without automatic protection, since the consequences of an outage at high capacity are so severe, and the cost of protection is relatively low.

If there are N working lines and M protection lines in a protection group, then the fractional outage per working line is given by

$$\text{Outage} = \frac{(N + M)!\, P^{M+1}}{N!\,(M + 1)!} \tag{1}$$

where P is the outage in the absence of protection:

$$P = \frac{\text{Mean time to repair (MTTR)}}{\text{Mean time before failure (MTBF)}}. \tag{2}$$

Typically, $M = 1$, in which case (1) and (2) reduce to

$$\text{Outage} = \left[\frac{N + 1}{2}\right]\left[\frac{\text{MTTR}}{\text{MTBF}}\right]^2. \tag{3}$$

[2]Other performance criteria include protection switching performance (switching thresholds and switching times), transient response (recovery times), and jitter generation and accommodation. These criteria are associated with a level of design detail outside the scope of this paper.

Equation (3) provides the requirements on the reliability (MTBF) of the equipment in a protection span to achieve a given outage for a given protection ratio (N) and a given MTTR (typically assumed to be 4 h). In Fig. 1, the MTBF to achieve an outage of 10^{-5} is shown. This outage corresponds to the requirements for a 320 km protection span in a long haul system. Assuming 40 km spacing between regenerators, there are 16 regenerators in the two-way span. If half of the outage is allocated to the regenerators, then for a protection ratio of 7, each regenerator should have an MTBF of at least 80 000 h (no more than 12 500 FITS). Although this is adequate to meet the outage requirement, higher reliability is desired to minimize repair activity.

B. Error Performance

There are three common measures of error performance:

i) long term BER;
ii) percent error-free seconds (EFS);
iii) severe errored seconds (SES).

In the U.S., both long-haul (6400 km) and short-haul (400 km) BER requirements are typically $< 10^{-8}$. These are allocated by mileage; thus, a 40 km short-haul system would have a BER requirement $< 10^{-9}$. Assuming randomly distributed errors, it makes no difference whether the BER requirement is applied to a tributary input rate or to the line rate.

In the case of EFS, the requirement is, of necessity, dependent on the bit rate at which it is measured. For example, a 1 Gbit/s system operating at a 10^{-8} BER would have essentially zero EFS measured at the line rate, since each second would contain, on average, ten errors. Requirements on EFS typically apply to end-to-end digital services, measured at 1.5 Mbits/s or below. Establishment of similar requirements at 45 Mbits/s is currently under study. An errored second at 45 Mbits/s (DS3) can affect from 1 to 28 DS1's. If only one DS1 is affected (isolated random errors), then the DS3 EFS requirement may be relaxed by a factor of 28 relative to the DS1 requirement.[3] On the other hand, if all 28 DS1's are affected (burst error), then the DS3 EFS requirement is the same as the DS1 requirement.

A severe errored second is generally defined as a second in which there are more than 100 bit errors. A less precise, but somewhat more meaningful, definition is that an SES contains a sufficiently large burst of errors such that *all* tributaries are affected and are considered severely errored. Thus, by definition severe errored seconds are independent of whether they are measured at the line or tributary rates.

SES requirements are typically in the range of, on the average, no more than 4–8 SES per day, where the lower figure corresponds to short-haul (400 km) systems, and the higher to long-haul systems (6400 km).

[3]I.e., the percent errored seconds could be a factor of 28 higher.

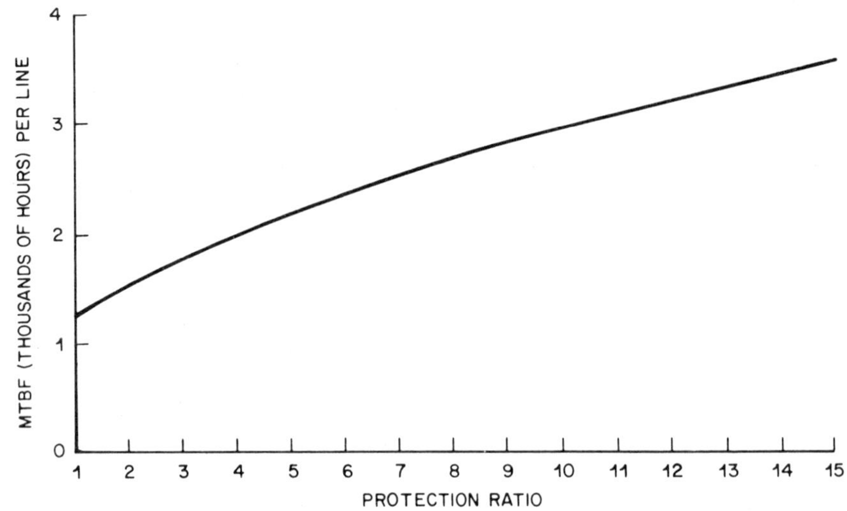

Fig. 1. Mean time before failure (MTBF) of a two-way line between switch points to achieve 10^{-5} outage probability as a function of the protection ratio. (Assumes two-way protection switching and 4 h MTTR.)

How do the above requirements impact the design of a long-haul lightwave system? Clearly, individual repeater sections have to be designed for BER considerably below 10^{-8} (typically $< 10^{-11}$). Also, the design should assure that events causing severe burst errors (e.g., external interference or internal intermittents such as amplifier "dropouts") occur less than of the order of one per day. Lightwave systems should operate essentially error free.

III. System Architecture (Terminal Design)

A simplified functional block diagram of a lightwave system terminal is shown in Fig. 2. Each element of this block diagram (with the exception of the regenerator which is discussed separately in Section IV) is discussed below.

A. *Cross-Connect*

Long-haul lightwave systems interface at a standard cross-connect (DSX). In the U.S. this has been almost universally at the DSX-3 (44.736 Mbit/s third level of the North American digital heirarchy). In Europe the interface is at the CEPT-4 (139.264 Mbit/s) level,[4] and in Japan at the 397.20 Mbit/s fifth level of the Japanese hierarchy. (See Table I for comparison of the North American, European, and Japanese hierarchies.)

It is interesting that the North American DSX-4 has not been used for lightwave systems (except for an early system in Canada). For both economic and technology reasons, the DSX-4 cross-connect has largely been abandoned, and it it not clear whether a cross-connect beyond the DSX-3 will be required in the future.

B. *Low-Speed Interface*

The principal functions of the low-speed interface are to provide the coding and decoding (codec) for the cross-connect interface,[5] synchronization and desynchronization (syndes) for the subsequent multiplexing and demultiplexing, and the addition (and removal) of bits for performance monitoring and order wire and telemetry channels. The advantage of adding these bits at the low-speed interface rather than at the line rate[6] is that it reduces the amount of high-speed signal processing required and facilitates upgrade to higher capacity. The

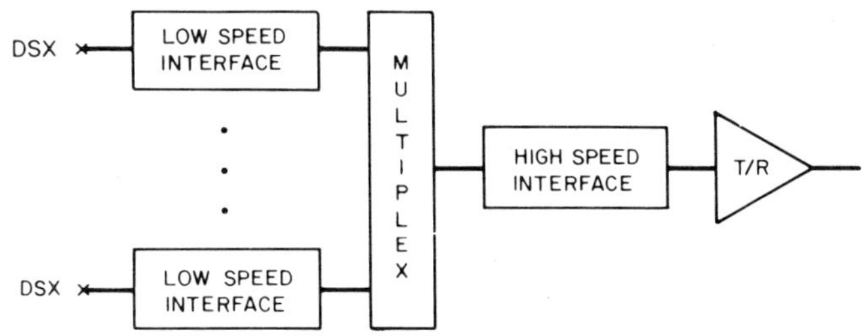

Fig. 2. Lightwave system transmission terminal architecture.

TABLE I
DIGITAL HIERARCHIES

		North American	European (CEPT)	Japanese
Level 1	Bit rate (Mbits/s)	1.544	2.048	1.544
	Voice circuits	24	30	24
Level 2	Bit rate (Mbits/s)	6.312	8.448	6.312
	Voice circuits	96	120	96
Level 3	Bit rate (Mbits/s)	44.736	34.368	32.064
	Voice circuits	672	480	480
Level 4	Bit rate (Mbits/s)	274.176	139.264	97.728
	Voice circuits	4032	1920	1440
Level 5	Bit rate (Mbits/s)	Not defined	565.148	397.20
	Voice circuits		7680	5760

[4]Some European systems have been applied in the U.S., utilizing a separate multiplexer (factor of three) between the U.S. DSX-3 and the CEPT-4 levels.

[5]The DSX-3 utilizes a B3ZS code.

[6]In some systems the bits are added at a rate intermediate between the interface and line rates.

disadvantage is that it is somewhat more wasteful of bandwidth, and it requires demultiplexing of the line signal to access these bits.

C. Multiplexing

If the signal processing is all performed at the low-speed input level, then the multiplexing and demultiplexing can be simply bit interleaving and separation. Proper identification of the tributaries may be achieved by complementing one of the tributaries.

D. High-Speed Interface

The function of the high-speed interface is to introduce appropriate statistics into the line signal. This function may, in principle, be implemented by appropriate parallel processing of the input tributaries, prior to multiplexing, rather than serial processing after multiplexing. However, this may necessitate aligning the frames in each of the tributaries which makes the dropping and inserting of tributaries more difficult.

Most systems in the United States use unrestricted binary transmission, with scrambling to guard against low-frequency patterns occurring with probability greater than predicted by random statistics. This format has been used by all of AT&T's high-speed (45 Mbit/s and above) lightwave systems, as well as a previous high-speed coaxial cable system [13].

In Europe, balanced block codes are generally used [14]. This allows a higher low-frequency cutoff in the regenerator circuits, and facilitates timing recovery and performance monitoring. However, the use of block codes with binary transmission results in a higher line rate. For example, the 5B6B code (translating a 5 bit binary input sequence to a 6 bit binary line signal) requires a 678 Mbit/s line rate for a 565 Mbit/s line capacity. This puts an added burden on high-speed regenerator circuitry.

Ternary transmission, commonly used on metallic systems, has not generally been used on lightwave systems owing to the higher required signal-to-noise ratio,[7] and the greater sensitivity to eye degradations. However, one such European system has been announced [15], utilizing a 4B3T code. This allows achieving a 565 Mbit/s capacity using the same 423 Mbaud transmission employed in coaxial systems. However, it does reduce eye margins in half (see the following section).

In Japan, long haul systems use a 10B1C code in which a complementary bit (to the previous bit) is added after every 10 information bits [16]. This assures no more than 11 bits in a row without transitions.

There does not appear to be convincing arguments in favor of any particular coding arrangement.

IV. Regenerator Performance

All of the key technologies for a lightwave system are contained within the regenerator: single mode connectors and fiber pig-tails, the optical receiver (photodetector and preamplifier), wide-band AGC amplifier, timing recovery circuit, decision circuit, and the optical transmitter (laser source, drive and control circuitry). All of these technologies impact on regenerator design and capability (bit rate and spacing). The emphasis here will not be on the technologies, but rather on the methodology for designing regenerator spans and for specifying performance.

[7] On metallic systems, reduced bandwidth offers the advantage of lower cable loss. This does not occur in lightwave systems.

TABLE II
Regenerator Spacing Calculation; Representative Example for FT-Series G

$P_T = -3$ dBm
$P_R = -33$ dBm (PIN receiver, BER $= 10^{-9}$)
$L_c + L_x = 8$ dB (combined in a statistical analysis)
$L_1 = 22$ dB
$S = 44$ km
$M = 23$ (2 km between splices)
$\mu_f = 0.4$ dB/km $\quad \sigma_f = 0.1$ dB/km
$\mu_s = 0.1$ dB $\quad \sigma_s = 0.1$ dB

A. Regenerator Spacing

The loss budget for a regenerator span is calculated by taking the difference between the transmitter power P_T and receiver sensitivity P_R, subtracting for connector losses (L_C), and allowing for various degradations and margin (L_X).

$$L_1 = P_T - P_R - L_C - L_X. \qquad (4)$$

The loss of the connectors within the regenerator are usually included in the definitions of P_T and P_R; L_C is intended to accommodate additional connectors (and jumpers) within terminal or repeater sites. For example, an optical patch panel may be desired to permit flexible interconnection and rearrangement, and also serves as an appropriate point for the installation of wavelength division multiplexing (WDM) filters.

This maximum loss allowance L_1 is then allocated to the installed fiber cable loss. Since there are statistical variations in cabled fiber loss and in splice loss, characterization of these variations allows a statistical treatment that is less pessimistic than a worst case analysis. The 2σ outside plant loss is given by [17].

$$L_2 = \mu_f S + \mu_s M = 2\left[\frac{\sigma_f^2 S^2}{M-1} + \sigma_s^2 M\right]^{1/2} \qquad (5)$$

where μ_f and μ_s are mean fiber and splice losses (including any temperature effects), σ_f and σ_s are the corresponding standard deviations, S is the regenerator spacing, and M is the number of splices including one at each end of the span. Thus, $M - 1$ is the number of statistically independent fiber sections.

The outside plant loss L_2 must be less than or equal to the loss allowance L_1. A representative calculation for the AT&T FT-Series G system, with a loss allowance of 22 dB leading to a repeater spacing of 44 km, is provided in Table II.

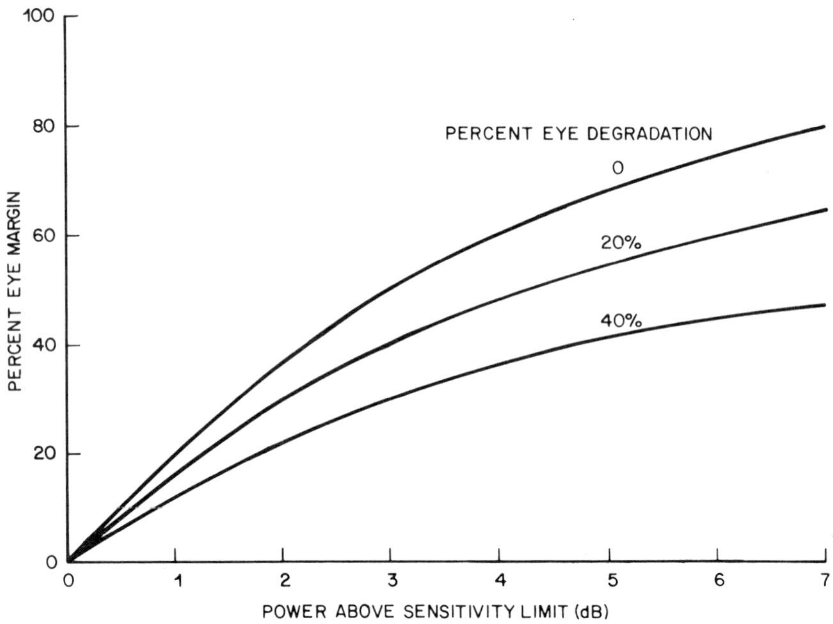

Fig. 3. Percent eye margin resulting from increases in power above the sensitivity limit, with initial eye degradation as a parameter. (The sensitivity limit is also dependent on the initial degradation.)

B. Degradation Allowances and Requirements

There are several phenomena that must be considered in the degradation allowance L_x. Although it is customary to express the allowance as a power penalty L_x, in general the effects are more properly described as a reduction in "eye opening" resulting from phenomena such as intersymbol interference, signal dependent noise, decision threshold offsets, timing offsets, and degradations in laser on–off ratios.

Consider a binary system in which the ideal eye height is unity, in which there is an initial eye degradation D_0, and in which we wish to allow for a subsequent degradation D. The additional power margin required to accommodate an eye margin D is given by

$$L_x = 10 \log_{10} \left[\frac{1 - D_0}{1 - D - D_0} \right]. \qquad (6)$$

The relation between eye margin and power margin is shown in Fig. 3. The power margin is dependent on the initial eye degradation. Thus, if it is desired to have a 50 percent eye margin for subsequent degradations, the power margin needs to be 3 dB if there are no initial degradations, but needs to be 5 dB if there are initial degradations totaling 25 percent. Eye margin, rather than power margin, is a more appropriate measure of regenerator performance. The degradation allowance, L_x in (4), should thus be based on an analysis of eye degradations, and on a *measured curve* of power margin. Regenerator dynamic range should be specified as the range of input power over which the eye margin exceeds a preassigned threshold.

One source of eye degradation that impacts on subsystem specifications is chromatic dispersion, which results in both intersymbol interference and in mode partition noise [18]. To limit the equivalent eye degradation from dispersion to no more than 10 percent requires that the dispersion be less than 0.1 of a pulse period [19]. For systems operating at the wavelength of minimum chromatic dispersion (λ_0), this sets requirements on both the laser spectral width, and on the offset of the laser center wavelength from λ_0. For operation far from λ_0 (e.g., at 1.55 μm on non-dispersion-shifted fiber) there are much more stringent requirements on laser spectral width (see Table III).

TABLE III
SPECTRAL REQUIREMENTS RESULTING FROM CHROMATIC DISPERSION AND MODE PARTITION NOISE FOR SINGLE MODE FIBER (MINIMUM DISPERSION AT 1.3 μm) AND 40 km REPEATER SPACING

$\Delta\lambda$ is the rms spectral width of the laser, and $\Delta\lambda_0$ is the difference between the laser center wavelength and the minimum dispersion wavelength of the fiber. The mode partition noise k-parameter is assumed to be 0.5 [18], [19].

Bit rate	λ = 1.3 μm	λ = 1.55 μm
417 Mbits/s	$(\Delta\lambda)(\Delta\lambda_0)$ = 40 nm^2	$\Delta\lambda$ = 0.2 nm
1668 Mbits/s	$(\Delta\lambda)(\Delta\lambda_0)$ = 10 nm^2	$\Delta\lambda$ = 0.05 nm

V. APPLICATION CONSIDERATIONS

A. Relative Costs

There are three principal materials cost components of a lightwave system—the fiber cable, the terminals, and the regenerators. A highly simplified example, depicted in Fig. 4, illustrates the relative costs of these components as a function of system capacity and length. For the purpose of this example, we have assumed the cost of two fibers to be \$30 000 per repeater section (e.g., 40 km repeater spacing, and cabled fiber price of about \$0.40 per meter), terminal cost (exclusive of regenerators) to be

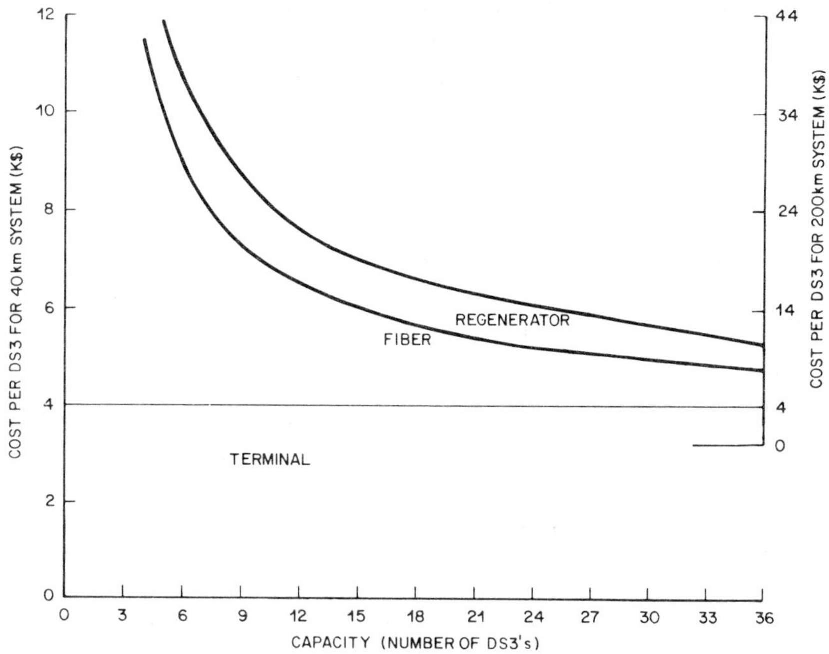

Fig. 4. Illustrative material costs for high-speed lightwave systems. The scale on the left corresponds to an unrepeatered 40 km system. The scale on the right corresponds to a 200 km span between terminals (four intermediate repeaters) in a long-haul system. Cost assumptions are given in the text.

\$2000 per DS3 independent of the capacity of the system, and regenerator cost to be \$2000 $\sqrt{n}$ where n is the capacity of the system in DS3's.[8] For simplicity, we have assumed repeater spacing to be independent of n. This requires either increased transmitter power or improved receiver sensitivity (smaller number of photons per bit) as the bit rate increases. This is one of the reasons for the assumed $\sqrt{n}$ dependence of regenerator cost.

It follows from Fig. 4 that for capacities below 9 DS3, fiber rapidly becomes the dominant system cost element. At capacities in the gigabit range, fiber and regenerator costs are comparable. Also, as capacities increase, terminal costs become a larger fraction of system costs. It should be noted, however, that much of the terminal costs may be deferred until the circuits are implemented, whereas all of the fiber costs and a larger fraction of the regenerator costs are paid up front.

B. Upgrade

There are two principal techniques for extending the capacity of a lightwave system upon exhaust—either increasing the capacity by higher speed serial transmission or by wavelength division multiplexing (WDM).

The economics favor higher speed serial transmission, since the cost of a single high-speed regenerator is generally less than that of multiple lower speed regenerators and WDM filters. There is a point, however, at which speed limitations on both regenerator and terminal circuitry are likely to make WDM a more viable option for upgrading capacity than continuing to increase the serial transmission speed.

C. Trends

The two principal trends in lightwave systems, higher capacity transmission and longer repeater spacing, have particularly benefited long-haul transmission systems. It should be noted, however, that the route capacities of installed systems have not changed nearly as much. For example, 144 fibers with 45 Mbit/s transmission have the same capacity as 16 fibers with 405 Mbit/s transmission. However, the economics has improved considerably—resulting from a nine-fold reduction in the quantity of fiber and about a 50-fold reduction in the number of regenerators from that of early short wavelength systems. This, in turn, has greatly increased the number of routes in which lightwave systems have economically proved-in.

Further increases in capacity and distance are dependent upon both technology and application considerations. Lightwave systems may be upgraded in capacity at fixed repeater spacing—the power budget needs to accommodate the additional noise bandwidth of the higher speed transmission or the attenuation of the WDM filters. For example, in AT&T's FT-Series G system, in converting from 417 Mbits/s to 1.7 Gbits/s, the additional noise bandwidth is accommodated by the use of an APD receiver. Additional expansions in the future may be accommodated by increased transmitter power and/or further improvements in receiver sensitivity.

Further advances in repeater spacing are also anticipated. This may be more significant for new routes than

[8]Note that all costs have been intentionally chosen to one significant figure.

for existing routes where repeater spacings are quantized by existing terminal and repeater sites. Of itself, 1.5 μm technology does not allow a doubling of repeater spacing, but may well do so combined with other advances.

Technology advances (resulting in lower costs) have been a spur to increased application, which in turn are a spur both to further technology advances and to increased usage. The existence of long-haul nationwide fiber facilities, interconnected to both metropolitan and local fiber networks, provides not only the capability of meeting today's communication needs, but offers the opportunity of shaping tomorrow's needs.

References

[1] I. Jacobs, "Fiberguide transmission system capability and needs," presented at IEEE Int. Conf. Commun., June 1976.

[2] J. S. Cook and O. I. Szentisi, "North American field trials and early applications in telephony," *IEEE J. Select. Areas Commun.*, vol. SAC-1, pp. 393-397, Apr. 1983.

[3] A. Moncalvo and F. Tosco, "European field trials and early applications in telephony," *IEEE J. Select. Areas Commun.*, vol. SAC-1, pp. 398-403, Apr. 1983.

[4] H. Ishio, "Japanese field trials and applications in telephony," *IEEE J. Select. Areas Commun.*, vol. SAC-1, pp. 404-412, Apr. 1983.

[5] I. Jacobs and J. R. Stauffer, "FT3-A metropolitan trunk lightwave system," *Proc. IEEE*, vol. 68, pp. 1286-1290, Oct. 1980.

[6] Filing by AT&T before Fed. Commun. Commission, File W-P-C-3071, Jan. 23, 1980.

[7] J. R. Stauffer, "FT3C—A lightwave system for metropolitan and intercity applications," *IEEE J. Select. Areas Commun.*, vol. SAC-1, pp. 413-19, Apr. 1983.

[8] Fed. Commun. Commission, File W-P-C-3071, Memo. Opinion, Order, and Authorization, adopted Nov. 25, 1980, released Jan. 27, 1981.

[9] D. A. Fishman *et al.*, "1.7 Gb/s lightwave transmission field experiment," in *Proc. OFC'86*, Feb. 1986, paper PDP-11.

[10] A. H. Gnauck *et al.*, "Information-bandwidth-limited transmission at 8 Gb/s over 68.3 km of single-mode optical fiber," in *Proc. OFC'86*, Feb. 1986, paper PDP-9.

[11] "Single-mode interoffice digital fiber optic systems," Bell Commun. Res., Inc., Tech. Advisory TA-TSY-000038, Issue 2, July 1985.

[12] "Error performance of an international digital connection forming part of an integrated services digital network," CCITT Rec. G. 821.

[13] P. E. Rubin, "The T4 digital transmission system—Overview," presented at IEEE Int. Conf. Commun., June 1975.

[14] G. G. Windus, A. Jessop, and A. D. Hemsworth, "Considerations in the design of a 565 Mbit/s optical fibre systems," in *Proc. 3rd Int. Conf. Telecommun. Transmiss.*, IEE, London, England, Mar. 1985, pp. 146-150.

[15] J. Drupsteen, "A high-capacity 565 Mb/s optical system bridges long distances," *Philips Telecommun. Rev.*, vol. 43, pp. 31-42, Apr. 1985.

[16] E. Iwahashi and H. Fukutomi, "F-400M system overview," *Rev. Elec. Commun. Lab.*, vol. 31, pp. 237-243, May 1983.

[17] D. J. Meskell, Jr., "Fiber optic system engineering," *Proc. Nat. Commun. Forum*, vol. 38, pp. 142-148, 1984.

[18] K. Ogawa, "Considerations for single-mode fiber systems," *Bell Syst. Tech. J.*, vol. 61, pp. 1919-1931, Oct. 1982.

[19] S. Lumish, private communication.

Ira Jacobs (M'57-SM'62-F'81) received the B.S. degree from the City College of New York in 1950, and the M.S. and Ph.D. degree in physics from Purdue University, West Lafayette, IN, in 1952 and 1955, respectively.

He is currently Director of the Transmission Technology Laboratory at AT&T Bell Laboratories, Holmdel, NJ, having responsibility for digital transmission and signal processing technology. Until recently he had responsibility for design and development of digital multiplexers and fiber optic transmission systems. He joined Bell Laboratories in 1955. From 1955 until 1967, he worked in electromagnetic and communication theory and had responsibility for a number of military satellite and space communications projects. Since 1967, he has been involved in transmission systems engineering and development and has directed much of AT&T's development of fiber optic transmission systems.

In 1981, Dr. Jacobs was elected a Fellow of the IEEE for contributions and leadership in the development and application of fiber optic systems.

PROGRESS IN LONG-SPAN, HIGH-CAPACITY OPTICAL TRANSMISSION SYSTEM RESEARCH

S.D.Walker and L.C.Blank

British Telecom Research Laboratories
Martlesham Heath
Ipswich
Suffolk
U.K.

Abstract

Direct detection single mode optical fibre systems have evolved rapidly during recent years such that multigabit transmission has now been demonstrated over 100 km spans. At lower line rates, unrepeatered optical path lengths in excess of 250 km have been reported. This progress is largely due to technical advances such as narrow spectral line lasers, low-loss step-index and dispersion-shifted fibre together with high-sensitivity, broad-band receivers. The high speed system developments have placed increased emphasis on the associated signal processing functions such as timing extraction and regeneration, this latter function being provided with gallium arsenide logic devices.

This paper will review some aspects of the high-speed and long-span transmission experiments carried out at BTRL. These include an ultra-long-span, 34 Mbit/s system and a 2.4 Gbit/s fully-regenerated high-speed optical transmission test over an installed 32 km fibre.

1. Introduction

A number of research laboratories are currently investigating the potential of long-span, high-speed optical transmission systems. Progress has been rapid, such that within the U.K., a 565 Mbit/s trunk optical transmission network is being installed [1,2]. Although the cost and complexity of the electronic circuitry employed in the system terminals is significant, the overall costs are dominated by the optical fibre and electro-optic components. Further enhancements in bit-rate may therefore be introduced with reduced costs per bit. This economic feature may pave the way for a wide variety of new broadband services.

In the submarine systems context, the availability of very-low-loss monomode fibre has allowed a series of long-span unrepeatered system experiments to be carried out, a span in excess of 250 km being achieved at 34 Mbit/s [3]. A number of prospective submarine routes have lesser spans and thus offer clear potential for unrepeatered operation. Step-index monomode fibre with 0.20 dB/km attenuation is now commercially available, 250 km spans having spliced losses in the range 50-60 dB [3,4].

High laser launch power and strongly monochromatic emission is required for long-span, high bit-rate systems in the 1.5 um wavelength region. The development of single-line lasers, such as distributed feedback (DFB) devices [5], has central importance in this area.

In the following sections, the basic requirements for high-performance optical system design are described with reference to the transmitter, optical channel and receiver. Subsequently, the trend towards higher performance is illustrated by reference to optical transmission system experiments ranging from 34 Mbit/s to 2.4 Gbit/s.

2. Optical Components.

The basic requirements for long-span, high-speed optical transmission systems can, in broad terms, be summarized as: high transmitter launch power, low fibre loss and high receiver sensitivity. Operation at 1.55 um wavelength requires additional attention to the spectral width of the transmitter and consequent dispersive characteristics of the fibre. Subsequent regeneration and associated timing extraction functions must be carried out with negligible penalty.

2.1. Transmitter

The following properties determine the suitability of lasers for high-performance optical transmission systems.

* high average launch power into single mode fibre
* 1.5 um wavelength operation
* narrow modulated spectral width
* large external differential quantum efficiency
* operational features such as small size and long life.

Suitable semiconductor injection lasers are the conventional multi-longitudinal-mode Fabry-Perot devices and the more recent single spectral line lasers such as DFB devices. At the present time, the DFB laser is preferred amongst other single line lasers since it requires only a single injection current (unlike cleaved-coupled-cavity types [6]).

Table 1 gives some measured parameters for a Fabry-Perot and DFB device.

Table 1

Laser Characteristics

Laser Type	Wavelength (micron)	Modulated Spectral width (nm)	Efficiency (mA/mW)	Threshold (mA)
Double Channel Planar Buried Heterostructure (DC-PBH)	1.55	3.5 (rms)	0.125	40
Distributed feedback (DFB)	1.52	0.2 (FWHM)	0.225	35

Figures 1 and 2 show the typical modulated spectral characteristics of DFB and Fabry-Perot lasers respectively. The DFB laser has considerably reduced spectral width. However, the modulated spectral width is still sufficient to introduce dispersion penalties at high data rates [7]. In consequence, dynamic linewidth studies have assumed great importance as "Chirping" effects can set an upper limit to achievable system spans [8].

Laser drive may be provided by Silicon bipolar transistors or, at gigabit speeds, GaAs FET's. Modulation currents approaching 120 mA peak to peak are required for the highest laser output powers. With laboratory lens couplers (based on confocal objective lenses [9]), launch powers in excess of 0 dBm can be achieved.

2.2 Low-loss Monomode Fibre

Unrepeatered transmission over spans in excess of 200 km requires high-grade monomode fibre with closely specified parameters since excessive tolerances can lead to large design margins and limited achievable span. Attenuation at the 1.55 um operating wavelength is the most important parameter, whilst further system design constraints arise from fibre dispersion and microbending sensitivity. All these factors can limit the line rate, restrict the choice of laser and prescribe fibre and cabling designs.

At the present time, three fibre types have potential applications in high performance optical transmission systems. Of these, standard step-index monomode fibre [10] is widely used for 1.3 um operation and has its lowest attenuation at approximately 1.55 um. Typical parameters are :- zero dispersion wavelength 1.3 um, dispersion 18 ps/(nm.km) at 1.55 um, cutoff wavelength 1.2 um. This short cutoff wavelength can lead to some microbending sensitivity in the 1.55 um wavelength region. Considerable improvements may be made with a shift to longer cutoff wavelengths. An increase to 1.4 um has given improved microbending performance[11].

A third fibre category is the dispersion-shifted type [12]. This design has a more complex refractive index profile which increases the waveguide dispersion such that the overall fibre zero dispersion wavelength is shifted to 1.55 um. At the present time, the fibre attenuation is marginally higher than that of step-index type. Additionally, reduced microbending sensitivity has been reported [13].

Standard step-index monomode fibre (manufactured by Optical Fibres Ltd., U.K.) and dispersion-shifted fibre (from Corning Glassworks, USA) were available for laboratory experiments. Typical characteristics are:-

Table II

Step-index Monomode Fibre

Attenuation	:	0.20 dB/km	@ 1.55 um
	:	0.21 dB/km	@ 1.52 um
Dispersion	:	18 ps/(nm.km)	@ 1.55 um
Concentricity	:	0.5 um	

Dispersion-shifted Monomode Fibre

Attenuation	:	0.215 dB/km	@ 1.55 um
	:	0.23 dB/km	@ 1.52 um
Dispersion	:	dispersion-zero	@ 1.55 um
Dispersion Slope	:	Max. 0.075ps/(nm^2.km)	

2.3 Receiver Design

The design and performance of receivers for use in optical transmission systems has been discussed extensively by many authors [e.g., 14,15]. In the present context, the discussion will centre on receivers developed at BTRL for use in long-span, high-capacity optical transmission system experiments.

A good optical receiver performance can only be obtained if careful attention is paid to the electrical noise characteristics of the post-detection circuitry. In principle, the lowest noise is obtained with the high-impedance GaAs FET input stage assuming the total input capacitance is minimised. However, accurate compensation for the low-pass filtering action at the gate node must be provided. An alternative configuration is the broad-band transimpedance arrangement. Although this design is typically more noisy than its high impedance counterpart, it offers advantages such as large dynamic range and greatly simplified equalization.

Recently, a number of laboratories have reported good results with both Germanium [16] and III-V [17] avalanche photodiodes (APD's). It has been found that very-high-performance optical receivers can be realised with packaged components and conventional printed circuit board construction techniques. In our work, the achievement of good sensitivity whilst maintaining ease of construction has depended on the commercial availability of small-area Germanium p^+n avalanche photodiodes. These devices have been included in a range of receiver designs ranging from 34 Mbit/s to 2.4 Gbit/s.

Practical Receivers . The optimum input device choice depends on a number of factors. At bit rates of 34 Mbit/s and above either a Silicon bipolar or GaAs FET based preamplifier can give good performance. Figure 3 shows the relative noise performance of these devices as a function of total input capacitance at 1 Gbit/s and 2 Gbit/s.

At bit rates upto 140 Mbit/s, an op-amp based transimpedance APD receiver design has given good performance [18]. The schematic circuit is shown in figure 4 . Standard printed circuit board constructional techniques were used together with commercially-available 30 um diameter p^+n Ge APD's. A variety of low-noise GaAs FET's together with a wide-band linear phase op-amp gave highly reproducible receiver characteristics. At bit rates of 34 Mbit/s and 140 Mbit/s, sensitivities of -55.8 dBm and -49.3 dBm respectively were achieved. In both cases, a fast 1.52 um laser was used as the optical source.

Future developments are likely to include the use of III-V APD's together with more advanced preamplifier designs. In this respect, the availability of photodetectors and GaAs FET's in chip form should allow significant improvements in receiver performance.

2.4 Regenerators

In order to restore the magnitude, shape and timing of the data pulses after degradation in the optical transmission channel, a regenerator is required. The regeneration process is controlled by a clock signal derived from the incoming data stream. Clock information is carried within the non-return-to-zero (NRZ) data stream, which can be extracted by passing the incoming signal through a non-linear element. This process introduces a spectral line at the clock frequency which can be filtered out and amplified to retime the decision circuit. Retiming signal phase adjustment is provided by inserting the correct delay in the clock or data path.

A number of component options exist to implement the regenerator function. Commercial Silicon bipolar emitter-coupled logic (ECL) covers the range 34-750 Mbit/s, with custom ECL usable upto 1 Gbit/s [19]. For higher bit rates a GaAs Logic family is required [20].

Filtering elements can either be used directly or as part of an injection-locked oscillator/phase-locked-loop (PLL) arrangement. Crystal filters are suitable for bit rates upto approximately 140 Mbit/s with surface acoustic wave or lumped component devices delivering improved performance at higher bit rates.

3. System Experiments

The previous section outlined the basic features of recent long-span and high bit rate laboratory developments at BTRL. This section will describe a range of experimental optical transmission systems at bit rates from 34 Mbit/s to 2.4 Gbit/s with spans upto 251.6 km. Both monomode step-index and dispersion-shifted fibre types feature in the results reported.

3.1 34 Mbit/s, 251.6 km Laboratory Transmission System [3]

A 1.525 um BTRL DFB laser was used as the optical source. As described in section 2.1, the dynamic linewidth of 0.2 nm introduced negligible dispersion penalty into a transmission system with 8.5 Gbit.km span-bit rate product (a limit is expected at approximately 100 Gbit.km). A mean launch power of +4 dBm was achieved with a laser mean output power of +8 dBm. This output level was obtained with a high power GaAs FET drive circuit which provided 150 mA peak-to-peak modulation current from threshold. A single stage Peltier cooler maintained the laser temperature at 17 ^{o}C.

The 251.6 km step-index monomode fibre span contained 43 low-loss fusion splices with a total span loss of 56.6 dB at the operating wavelength of 1.525 um. A receiver sensitivity of -55.3 dBm was obtained at 34 Mbit/s and 1.525 um with a design similar to that shown in figure 4. No discernable dispersion penalty was observed as was expected.

3.2 Fully-regenerated 2 Gbit/s optical transmission system [21]

A schematic diagram of the experimental system is shown in figure 5. The transmitter used was a stud-mounted, ridge-waveguide, DFB laser emitting at 1.535 um. Light was coupled into the fibre by means of an objective lens arrangement with 40 % coupling efficiency, allowing a mean launch power of -1.0 dBm with the laser biased at threshold and a 2 Gbit/s , 60 mA peak-to-peak modulation current provided by a GaAs MESFET pulse amplifier. Particular care was taken to minimise reflection noise effects in the DFB/launch optics interface. The fibre span consisted of 90.5 km of low-loss, dispersion-shifted fibre on 14 reels joined by fusion splices with an average splice loss of 0.1 dB. Total path attenuation at 1.535 um was 22.0 dB. Negative dispersion of approximately 1ps/(nm.km) was introduced by operation at 1.535 um wavelength.

At the receive end of the transmission path, a monomode fibre tailed 30 um diameter, p^+n Ge APD was butt-coupled to the dispersion-shifted system fibre. The receiver circuitry consisted of a surface-mounted Si bipolar transimpedance arrangement constructed from microwave transistors and monolithic integrated circuits. After amplification and equalisation, the detected signal was resistively split and passed directly to a GaAs 'D' flip-flop and timing extraction circuitry. As shown in figure 5, the recovered clock signal was passed to the flip-flop and data test receiver clock inputs. Both regeneration and retiming functions were therefore performed in a single stage. Differential clock port drive was necessary in order to achieve 2 Gbit/s operation with the commercially available flip-flop used. Additionally, DC bias adjustement at the 'D' input was employed to perform decision threshold optimisation as required with avalanche photodetectors. The output of the flip-flop was connected directly to the data test set for error-rate measurements.

D-type flip-flop characterisation experiments with 2 Gbit/s, NRZ, p.r.b. data indicated a decision threshold ambiguity level of 13 mV and a clock phase margin of 149°. Further tests at 1.6 Gbit/s produced figures of 2.7 mV and 198° respectively. A negligible system penalty is therefore expected if signal levels appropriate to emitter-coupled logic and a low jitter clock are available. Stable regenerator performance was obtained with on-board regulators providing power supply, bias and output pulldown voltages. Figure 6 shows the received "eye" diagram whilst figure 7 shows the overall electrical characteristics of the 2 Gbit/s optical transmission channel. Table III below summarises the power budget obtained :-

Table III

Fully-regenerated 2 Gbit/s, 90 km optical transmission system

Bit-rate	2.0 Gbit/s
Launch Power	-1.0 dBm
Fibre Span	90.5 km
Total loss	22.0 dB
Receiver Butt Coupling loss	0.4 dB
Receiver Sensitivity	-24.8 dBm
System Margin over 10^{-9} BER	1.4 dB

3.3 Bit-interleaved 2.4 Gbit/s, 71 km optical transmission system [22]

A block diagram of the experimental 2.4 Gbit/s system is shown in figure 8. The 2.4 Gbit/s, NRZ, p.r.b.s was assembled by a straightforward bit-interleaving process featuring a standard data-generator, RZ converter module and a hybrid combiner. This signal was fed to a 3.5 GHz bandwidth, single-stage GaAs FET amplifier with 100 mA peak-to-peak output capability into a 50 ohm load.

A stud-mounted, impedance matched 1.52 um ridge-waveguide DFB laser was driven from threshold with 30 mA amplitude current pulses. A mean launch power of -7.8 dBm at 2.4 Gbit/s into triangular core dispersion-shifted fibre was obtained with a launch arrangement similar to the 2 Gbit/s experiment. Launch power was limited by reflection noise considerations. At 1.52 um, the total fibre loss over 71 km was 17.5 dB (0.245 dB/km) with an estimated dispersion of -2.2 ps/(nm.km) at 1.52 um.

The 2.4 Gbit/s receiver was fabricated using surface mounting techniques similar to the 2 Gbit/s design. A sensitivity of -27.0 dBm was obtained. After amplification and equalisation, the detected signal was split and passed to a GaAs 'D' type flip-flop as in the 2 Gbit/s case and narrow-band PLL timing extraction circuitry. As shown in figure 8, the extracted 1.2 GHz clock signal was connected to the differential clock-input of the flip-flop and the data test-set such that BER measurements were performed on the disinterleaved data. Retiming, regeneration and disinterleaving functions were therefore performed in a single stage. Clock phase adjustment allowed the BER of each disinterleaved data stream to be assessed. Results were within a factor of 2 of each other. Figure 9 shows the received pulses after 71 km whilst table IV below summarises the power budget obtained:-

Table IV

2.4 Gbit/s, 71 km transmission experiment

Bit Rate	2.4 Gbit/s {2*1.2 Gbit/s}
Launch Power	-7.8 dBm
Fibre Span	71.2 km
Total Loss	17.5 dB
Receiver Butt Coupling Loss	0.4 dB
Receiver Sensitivity	-27.0 dBm
Margin Over 10^{-9} BER	1.3 dB

3.4 2 Gbit/s and 2.4 Gbit/s optical transmission field trial over a 32 km installed route [23]

Recently, the laboratory techniques described in the preceding paragraphs were given a field test over a 32 km installed monomode fibre path between the Midland cities of Birmingham and Tamworth in the UK. This span represents the maximum repeater spacing in the UK network to date. At an operating wavelength of 1.52 um, the overall path loss was 10 dB, including a number of connections and a fibre-to-fibre coupling at the receive end. The dispersion of the 125 um monomode fibre was approximately 15 ps/(nm.km). Satisfactory transmission channel performance was obtained at both 2 Gbit/s and 2.4 Gbit/s. Error ratio maesurements at 2 Gbit/s showed values approaching 10^{-14}, exact assessments being limited by the available measurement time. Table V summarises the results obtained :-

Table V

Power Budget of 2.0 Gbit/s and 2.4 Gbit/s field trial experiment

Bit rate (Gbit/s)	2.0	2.4
Launch power (dBm)	-4.5	-4.5
Path loss (dB)	10.0	10.0
Receiver sensitivity (dBm for 10^{-9} BER	-30.3	-26.5
System margin (dB)	15.8	12.0

4. Review and Conclusions

This paper has reviewed selected direct detection optical transmission experiments which have been conducted at BTRL in recent years. Steady progress has been made from early laboratory trials at 34 Mbit/s to gigabit field trials over installed fibre systems. Further developments are likely to include increasing use of monolithic Silicon and GaAs components and the transfer of today's laboratory technology into tomorrow's installed optical transmission systems.

5. Acknowledgements

The authors would like to thank all their colleagues at BTRL who contributed to this work. The dispersion-shifted fibre was obtained from Corning Glassworks Inc through Optical Fibres Ltd (UK). Acknowledgement is also made to the Director of Technology Applications at British Telecom for permission to publish this paper.

6. References

1. Larner, D.S., Design Features of a 565 Mbit/s Optical Transmission System , Proc. I.E.E. Third Int. Conf. on Telecommunication Transmission, pp. 151-154, 18-21 March 1985.
2. Brooks, R.M., et al., British Telecom 565 Mbit/s Lightline System , Br. Telecom Technol J. Vol. 3, no. 2, April 1985.
3. Blank, L.C., et al., Long Span Optical Transmission Experiments at 34 Mbit/s and 140 Mbit/s . Jnl. Lightwave Tech., Vol. LT-3, no. 5, Oct. 1985, pp. 1017-1025.
4. Marzurcyk, V.J., 202 km Transmission Spans at 1550 nm with multilongitudinal mode lasers, OFC '84 Tech. Dig., 1984, paper WJ8.
5. Westbrook, L.D., et al., Continuous Wave Operation of 1.5 um Distributed-feedback ridge-waveguide lasers , Elect. lett., Vol. 20, no. 6, March 1984, pp.225-226.
6. Kaminow, I.P., et al, High Speed 1.55 um Single-Longitudinal-mode Ridge Waveguide C^3 Laser ,ibid, Vol. 19, no. 19, Sept., 1984, pp. 784-785.
7. Kasper, B.L., et al, A 130 km Transmission Experiment at 2 Gbit/s Using Silica Core Fiber and a Vapour Phase Transported DFB Laser , 10 th ECOC, Stuttgart, Sept. 3 rd-6th 1984, post-deadline paper.
8. Frisch, D.A. et al, Effect of Laser Chirp on Optical Systems-Initial Tests using a 1480 nm DFB Laser , Elect. Lett., Vol. 20, no. 20, July 1984, pp.631-633.
9. Saruwatari, M. et al, Efficient Laser-Diode/Single-Mode Fibre Coupling using Two Confocal Lenses , ibid, Vol. 16, no. 25, Dec. 1980, pp. 955-956.
10. Ainslie. B.J., et al, The Design and Fabrication of Monomode Optical Fibre , I.E.E.E., Jnl. Quant. Elect., Vol. QE-18, no. 4, April 1982, pp. 514-523.
11. Kitayama, K., et al, Ultralow-loss 220 km-long Single-Mode-Fibre Cable Link in 1.5 um Region , Elect. Lett., Vol. 20, no. 19, Sept. 1984, pp. 774-775.
12. Ainslie, B.J., et al, Monomode Fibre with Ultralow Loss and Minimum Dispersion at 1.55 um , ibid, Vol.18, no. 19, Sept. 1982, pp. 842-844.
13. Bhagavatula, V.A., et al, Bend-Optimised Dispersion-Shifted Segmented Core Designs for Specialised 1550 nm Operation , OFC '85 Tech. Digest, pp. 94-96.
14. Smith, D.R., et al, A Simplified Approach to Digital Optical Receiver Design , Opt. Quant., Elect., Vol. 10, no. 3, May 1978, pp. 211-221.
15. Muoi, T.V., Receiver Design For High Speed Optical-Fiber Systems , Jnl. Lightwave Tech., Vol. LT-2, no. 3, June 1984, pp.243-267.
16. Kagawa, S., et al, Germanium Avalanche Photodiodes in the 1.3 um Region , Fujitsu Sci., Tech. Jnl., Vol. 18, no. 3, Sept. 1982, pp. 397-418.
17. Shirai, T., et al, A Planar InP/InGaAsP Heterostructure Avalanche Photodiode , I.E.E.E. Trans. Elect. Dev., Vol. ED-29, no. 9, Sept. 1982, pp. 1404-1407.
18. Walker, S.D. et al, Ge APD/GaAs FET/OP-Amp Transimpedance Optical Receiver Design Having Minimum Noise and Intersymbol Interference Characteristics , Elect. Lett., Vol. 20, no. 20, Sept. 1984, pp. 808-809.
19. Walker, S.D., et al, 1 Gbit/s Optical Transmission Experiment over 83 km using a 1.478 um DFB Laser , 10 th ECOC Stuttgart, Sept. 3rd-6th 1984, paper 11B3.
20. Rocchi M., et al, GaAs Digital Dynamic IC's for Applications up to 10 GHz , Jnl. Solid State Circuits, Vol. SC-18, no. 3, June 1983, pp. 369-376.
21. Bickers, L. et al. Fully-Regenerated 2 Gbit/s, 90 km Optical Transmission System using a Commercial GaAs Master-slave D-type Flip-Flop , 11th ECOC, Venice, Sept., 1985
22. Walker, S.D. et al, Fully-Regenerated 2.4 Gbit/s Lightwave System Experiments Featuring Dispersion-Shifted Fibre and Commercially-available Signal Processing Components , OFC '86, Tech. Digest, Atlanta, Feb. 1986.
23. Blank, L.C. et al, 2 Gbit/s and 2.4 Gbit/s Optical Transmission Field Trial over a 32 km Installed Route , Elect. Lett., Vol. 22, no. 1, Jan. 1986, pp.33-35.

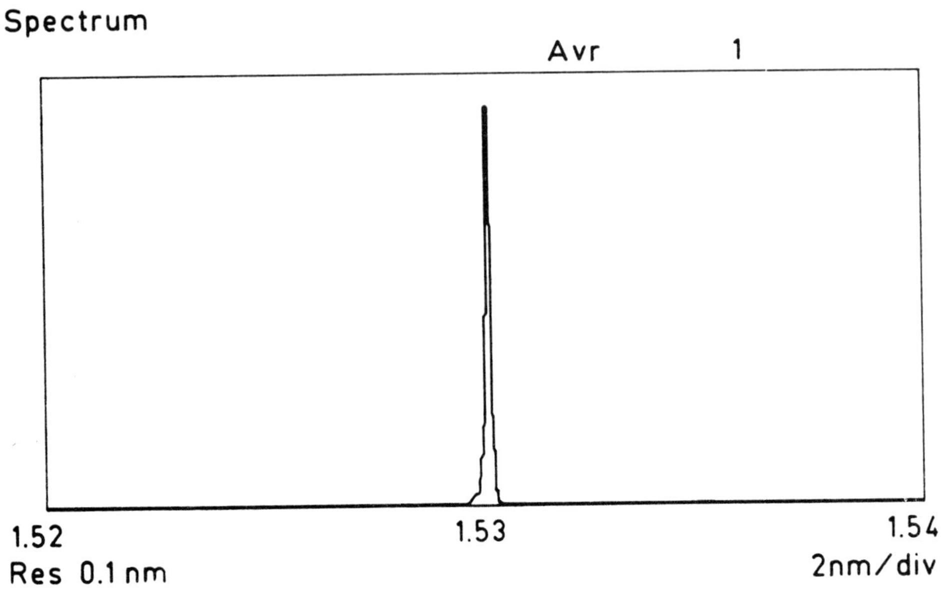

FIG. 1 DFB LASER MODULATED SPECTRAL CHARACTERISTICS

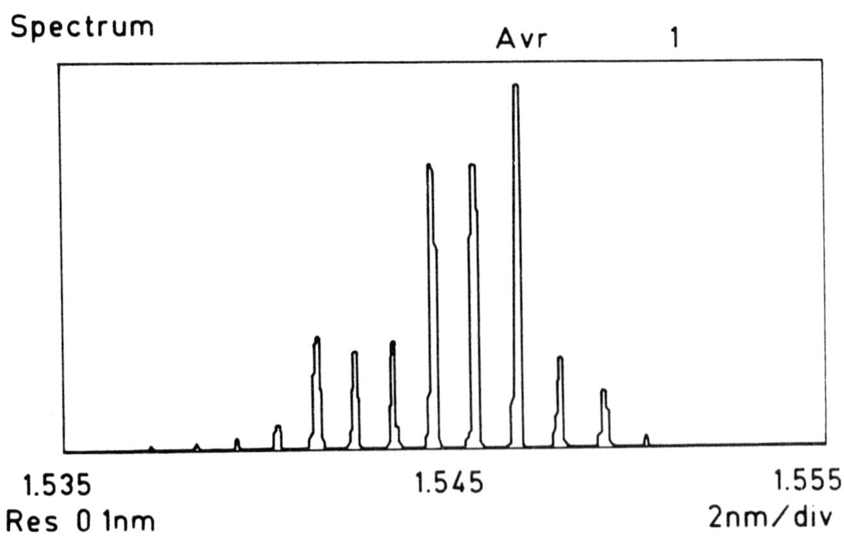

FIG. 2 FABRY-PEROT LASER MODULATED SPECTRAL CHARATERISTICS

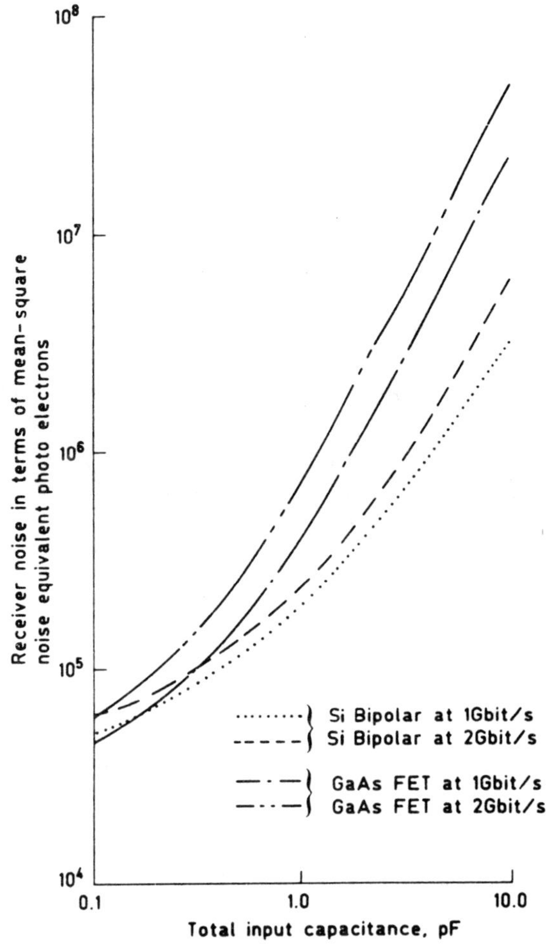

FIG. 3 Si BIPOLAR AND GaAs FET FRONT-END NOISE PERFORMANCE AS A FUNCTION OF TOTAL INPUT CAPACITANCE AT 1Gbit/s AND 2Gbit/s

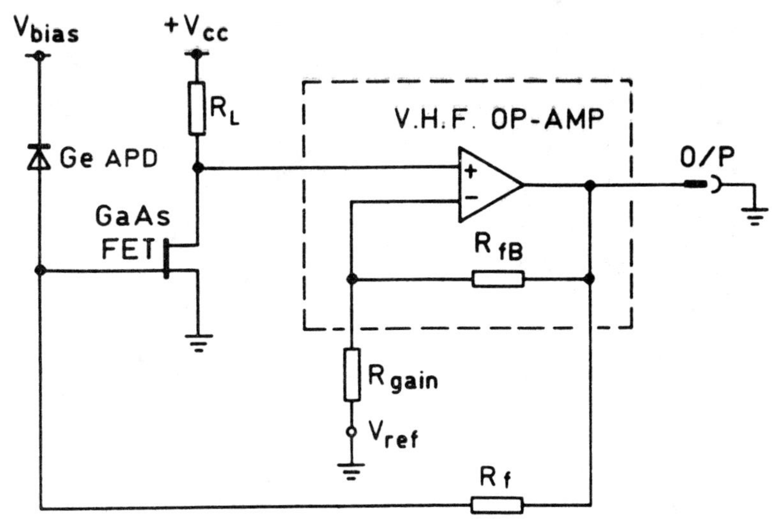

FIG. 4 APD/GaAs – FET/Op – AMP TRANSIMPEDANCE CIRCUIT

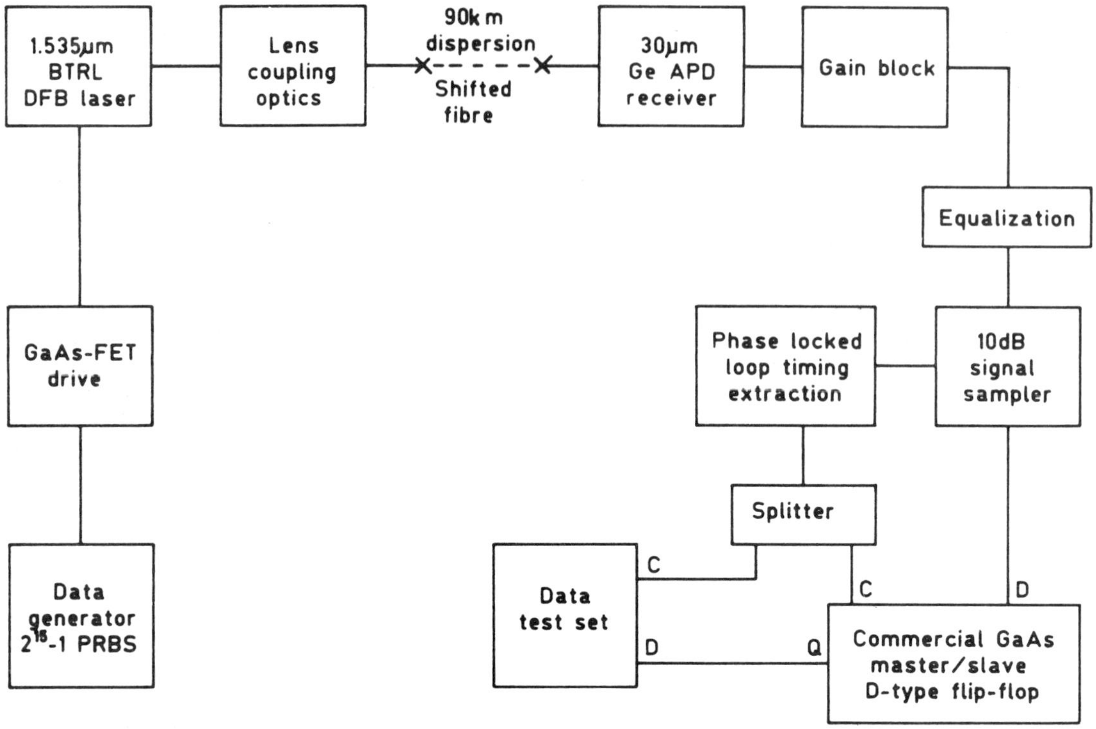

FIG. 5 SCHEMATIC DIAGRAM OF A 2Gbit/s, 90km FULLY-REGENERATED OPTICAL TRANSMISSION SYSTEM

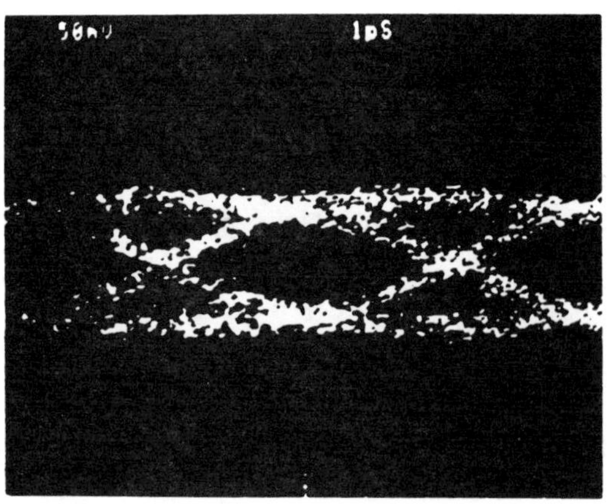

FIG. 6 RECEIVED "EYE" DIAGRAM IN A REFLECTION-NOISE-FREE 2Gbit/s OPTICAL SYSTEM

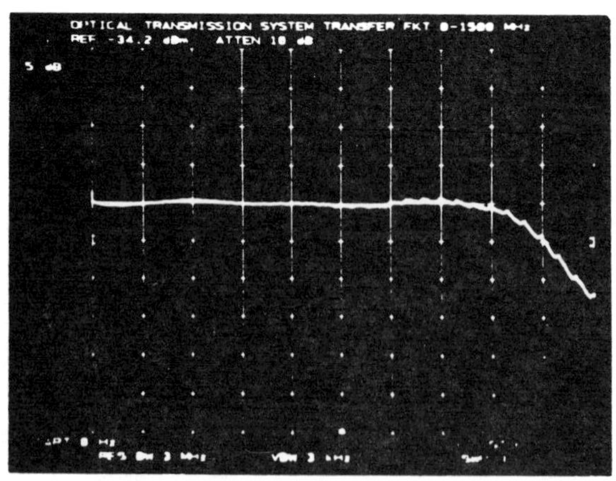

FIG.7 CORRESPONDING SYSTEM SWEPT-FREQUENCY RESPONSE

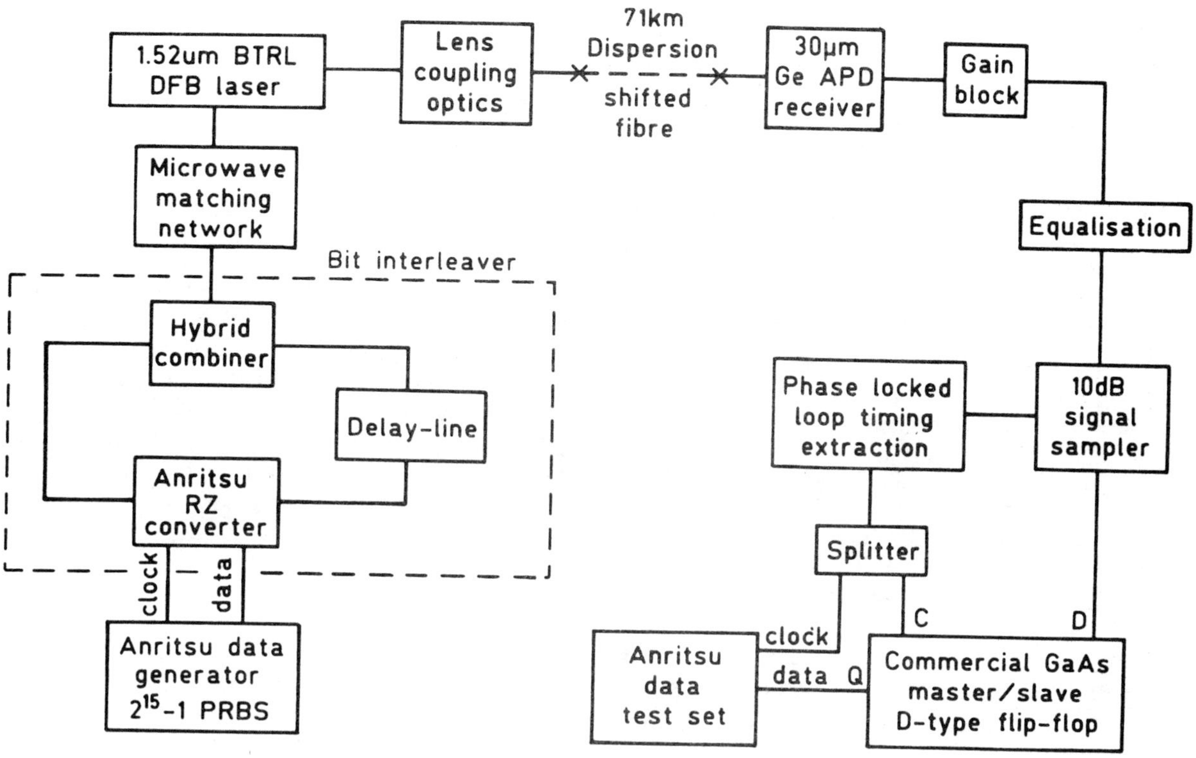

FIG. 8 SCHEMATIC DIAGRAM OF A 2.4 Gbit/s, 71 km BIT-INTERLEAVED OPTICAL TRANSMISSION SYSTEM

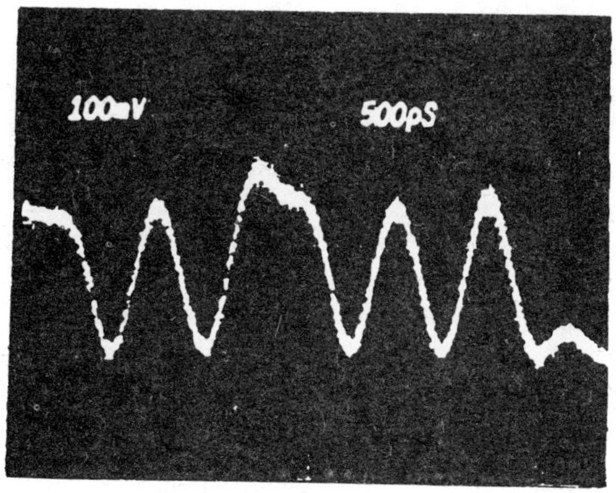

FIG. 9 RECEIVED PULSES AFTER 71km

1.13-Gbit/Lightwave Transmission System

KENNETH Y. MAXHAM, J. MICHAEL DUGAN, MARK A. MCDONALD, MEMBER, IEEE, AND CHARLES R. HOGGE, JR., SENIOR MEMBER, IEEE

Abstract—**A new high capacity lightwave transmission system has been developed using GaAs semicustom logic arrays and a DFB single-mode laser, and is presently in production. The architecture of this product is designed for in-service upgrade of a 565-Mbit/s product. This paper reviews the technical characteristics and design considerations of the Rockwell LTS-21130 lightwave transmission system.**

INTRODUCTION

THE LTS-21130 is a digital transmission system with a line rate of 1.13 Gbit/s, and carries 24 North American standard DS3 signals per optical channel. Significant features in this system are: the use of GaAs standard cell semicustom IC's in the high speed multiplexer and demultiplexer to achieve 1.13-Gbit/s performance in a production system; the use of a 1310-nm wavelength DFB single-mode laser diode to overcome dispersion limitation in long transmission spans; an efficient architecture that enables the upgrade of the LTS-1565 (565-Mbit/s transmission rate) to be performed without interrupting service. With shipments starting in April 1987, the Rockwell LTS-21130 is the first production lightwave transmission system operating in excess of 1 Gbit/s to be introduced in North America. A summary of key features is given in Table I.

Typical receiver sensitivity at a bit error rate of 1×10^{-9} is -36 dBm, and transmitter output is not less than -4 dBm. With margin allowances, the cable loss budget is 27.5 dB, with no additional allowances needed for dispersion. An important feature is that the guaranteed cable loss budget is the same as for the 565-Mbit/s product, enabling upgrade of existing installed 565-Mbit/s systems, without requiring additional margin allowance in the installation of the 565-bit/systems.

During the design planning phase, there was concern that dielectric losses in conventional epoxy-glass circuit board material would be too high, and a more exotic material such as a teflon-based substrate would have to be used for the 1-GHz signal handling circuits. However, laboratory evaluation indicated that the losses were manageable with a) careful layout design, b) use of coax for high frequency interconnections, and c) GaAs devices for much of the high speed signal processing. The VCO in the clock recovery circuit was the only circuit where a teflon substrate is used.

Manuscript received January 15, 1987; revised April 14, 1987.
The authors are with Rockwell International Lightwave Systems Division, Dallas, TX 75207.
IEEE Log Number 8715755.

SYSTEM OVERVIEW

Fig. 1 shows a block diagram of the transmit terminal shelf. There are 24 DS3 Input conditioner modules, a 12-wide 2 : 1 multiplexer module that also inserts overhead signaling bits, a transmitter module that performs the 12 : 1 final multiplexing and optical signal generation, a DS3 protection switching module, a channel monitor module that monitors status and alarms and communicates with other parts of the system, a DS1 interface module for interface to the overhead channels, and two power supply modules. A second shelf performs the receive terminal functions.

UPGRADE FROM 565-MBIT/S SYSTEM

A key design requirement in the planning of this system was that it must be easy for a customer to upgrade a 565-Mbit/s system to the 1.13-Gbit/s configuration. For this to happen, the planning for the 1.13-Gbit system began with the design of the 565-Mbit system. The system is designed with 1 : N protection switched redundancy. The design allows for upgrade by combining two 565-Mbit channels into one 1.13-Gbit channel. The upgrade proceeds in steps: the first step is to convert the protection channel to 1.13-Gbit/s (a space must be reserved for doubling the capacity of the protection channel); thereafter, traffic is switched to the protection channel while the working channel or pair of channels is being upgraded.

A block diagram of the LTS-1565 terminal shelf is shown in Fig. 2. One shelf of equipment contains one full duplex channel, consisting of 12 DS3 input conditioners, 12 DS3 output conditioners, a transmitter, a receiver, monitoring and protection switching modules, and power supplies. The first 565-Mbit shelf is converted to a transmit shelf by replacing the 12 output conditioners with 12 additional input conditioners, and replacing the 565-Mbit transmit and receive cards with the 1.13-Gbit multiplexer and transmit cards. The second 565 shelf is converted to a receive shelf by replacing the 12 input conditioners with 12 additional output containers, and replacing the 565-Mbit transmit and receive cards with the 1.13 Gbit receive and demultiplexer cards. Finally, the DS3 and associated cables are reconfigured to route the signals to the appropriate inputs and outputs.

DISPERSION CONSIDERATIONS

As part of the upgrade philosophy, a design goal for the LTS-21130 was to provide the same maximum span length

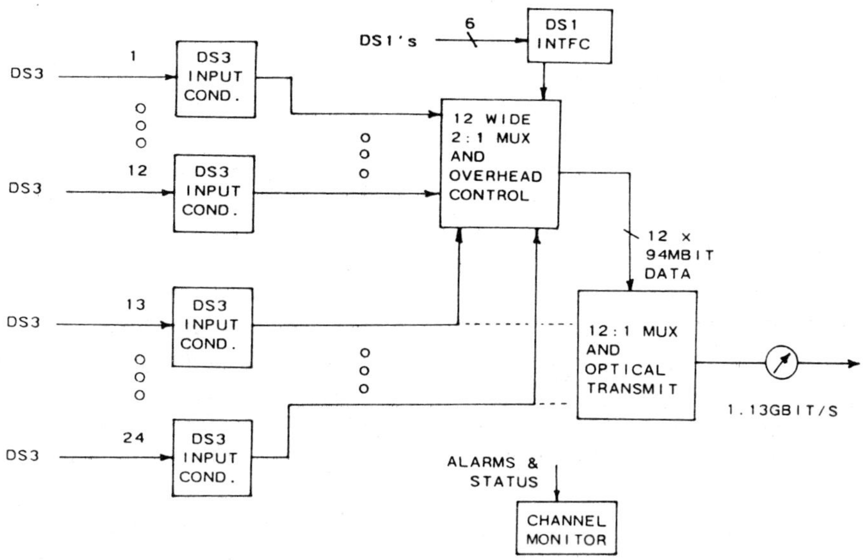

Fig. 1. Block diagram of the 1.13-Gbit/s transmit terminal shelf. The receive configuration is the same block diagram with inverse functions.

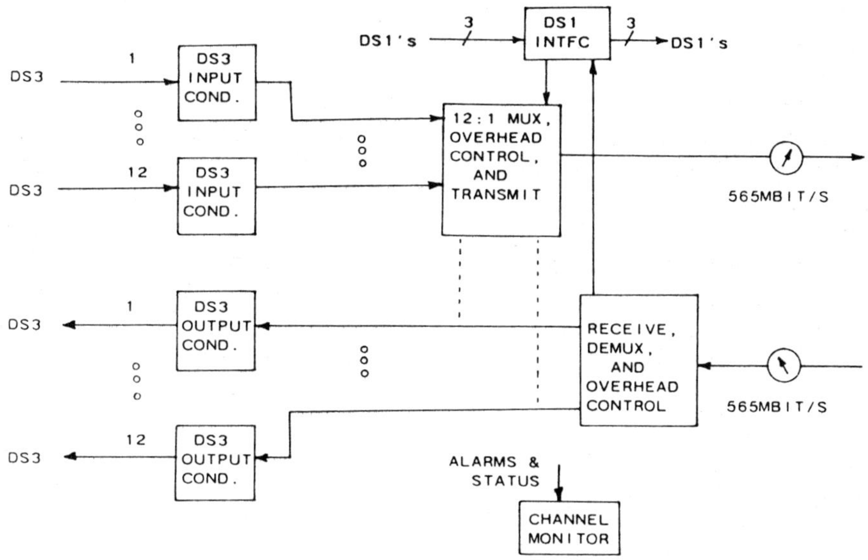

Fig. 2. Block diagram of the 565-Mbit/s duplex terminal shelf.

TABLE I
SUMMARY OF KEY FEATURES

Optical Source	DFB Laser
Center wavelength	1310 nm
Optical line rate	1129.984 Mbit/s
Line code	RZ, scrambled
Fiber span loss, for BER less than 10^{-11}	26.0 dB, guaranteed
Repeater spacing	>40 km
Primary electrical interface	24 DS3 channels
Overhead and supervisory signalling	Time division multiplexed with traffic data stream
Overhead channels	6 DS1's plus 2 Mbit/s supervisory channel

as LTS-1565, i.e., about 40 km. The analysis used to determine maximum span length is shown in Table II.

In the 565-Mbit/s product, the design allowed up to one-fourth of a bit interval of total dispersion in the transmission fiber, which is 440 ps. For this amount of dispersion, our system evaluation indicated there will be less than 1 dB of optical power penalty. Here, the total fiber dispersion can be defined as follows:

$$Dt = Df \times \Delta\lambda$$

where Dt is the total dispersion in picoseconds, Df is the fiber dispersion in picoseconds per nanometer of spectral

TABLE II
MAXIMUM SPAN LENGTH CALCULATION

Receiver sensitivity, typical	-36.0 dBm
Transmitter output, minimum	- 4.0 dBm
Typical system gain	32.0 dB
Connector losses	1.5 dB
Equipment margins	4.5 dB
Available system gain, guaranteed	26.0 dB
System Allowances:	
Optical patch panel @ 1.0 dB / end =	2.0 dB
Unallocated margin	3.0 dB
System gain available for span loss	21.0 dB
Fiber loss	0.4 dB/km
Splice losses:	
installation splices	0.2 dB/ 4 km
repair splice allowance	0.3 dB/ 8 km
Installed fiber span loss allowance =	0.4875 dB/ km
Maximum span length = 21.0 dB / 0.4875 dB/ km =	43.1 km

width, and $\Delta\lambda$ is the full width half maximum spectral width of laser diode as determined by statistical method.

By extrapolation, ignoring increased difficulty in the electronics, the allowable dispersion at 1.13 Gbit/s would be up to 220 ps, reducing maximum span length to half. However, at 1.13 Gbit/s, the typical laser spectral width increases somewhat compared to the width at 565 Mbit/s. Thus the maximum span length is reduced to about 14 km, unless the maximum laser spectral width could be reduced or the laser wavelength could be matched to fiber zero-dispersion wavelength. This latter method was calculated to be feasible but inconvenient and costly to both manufacturer and customer in terms of testing fibers and lasers for wavelength and stocking several different wavelength transmitter modules. The decision was made to use a single-mode DFB laser. With a typical spectral width of much less than 1 nm, the DFB laser theoretically has negligible dispersion for data rates even beyond 2 Gbit/s. A bit error rate curve for the system is shown in Fig. 3. There is no measurable performance difference for back to back through an attenuator only or through a dispersion shifted fiber with 225 ps/nm of dispersion at 1300 nm.

TRANSMITTER MODULE

The transmitter module block diagram is shown in Fig. 4. The DFB laser, laser drive circuit, rear-beam photodetector, thermistor, and thermoelectric heat pump (THP) are hermetically sealed in a hybrid microelectronic package. Optical output is via a lensed single-mode fiber aligned to the laser. The thermistor and THP, together with a temperature feedback control circuit, control the cooling and heating of the laser, regulating it to about 22°C. The circuit design techniques for driving the DFB single-mode laser and regulating its bias current proved to be essentially the same as for the multimode Fabry-Perot type laser used in the 565-Mbit/s product. For the 1.13-Gbit design, the laser bias control circuits were integrated into a linear LSI circuit device, saving space and cost.

The laser is driven with RZ coded data. Although this requires somewhat greater bandwidth in the laser and drive circuits, performance advantages outweighed the disadvantages. RZ coding reduces intersymbol interference at the receiver, reduces pattern dependent jitter in both the transmitter and the receiver, provides a better extinction ratio in the transmitter, and reduces problems related to connector reflections back into the laser. Additionally, since the APD has more noise in the logic 1 state than in the logic 0 state, the shorter duration of the data pulse reduces noise in the receiver [1].

RECEIVER MODULE

The receiver module block diagram is shown in Fig. 5. The Ge avalanche photodiode (APD), a four stage GaAs FET preamplifier, the THP, and the thermistor are housed in a hermetic hybrid microelectronic package. Optical input is via a multimode optical fiber aligned to the APD. The avalanche gain of the APD is approximately 5. The APD is cooled by a thermoelectric heat pump, and the temperature control circuit maintains the APD temperature at about 22°C. Because of this cooling, the APD dark current is kept low, and module ambient temperatures as high as 70°C have negligible effect on receiver sensitivity. Without this cooling, a 4 dB or more sensitivity loss would be incurred at high temperature. The sensitive receiver circuits are housed in a cast aluminum cavity, which provides heat sinking for the THP in the detector hybrid, stage-to-stage isolation for the amplifier chain, and shielding for the low-level portions of the amplifier. The AGC circuits provide about 64 dB of gain control. The output of the receiver subassembly is a 1-V peak-to-peak 50-Ω baseband signal that connects to the demultiplexer subassembly.

CLOCK RECOVERY

The clock recovery circuit uses a phase-locked loop with a voltage controlled oscillator, and employs a patented phase detector circuit, as shown in Fig. 6 [2]. The phase detector includes the same flip-flop that synchronously detects the incoming data. The phase-regulating property of the phase-locked loop is thus used to stabilize the timing of the recovered clock with the recovered data at the data detection flip-flop.

The phase detector portion of the clock recovery circuit is implemented within the GaAs demultiplexer IC. As mentioned in [2], this circuit is capable of operating in either a phase-locked loop with a VCO or to control the phase shifter of a SAW filter clock extractor. This circuit allows a wide variety of choices to be considered in that it can function with either NRZ or RZ waveforms at the input to the decision flip-flop, it requires no special encoding of the data, and it can be used to stabilize several types of clock sources.

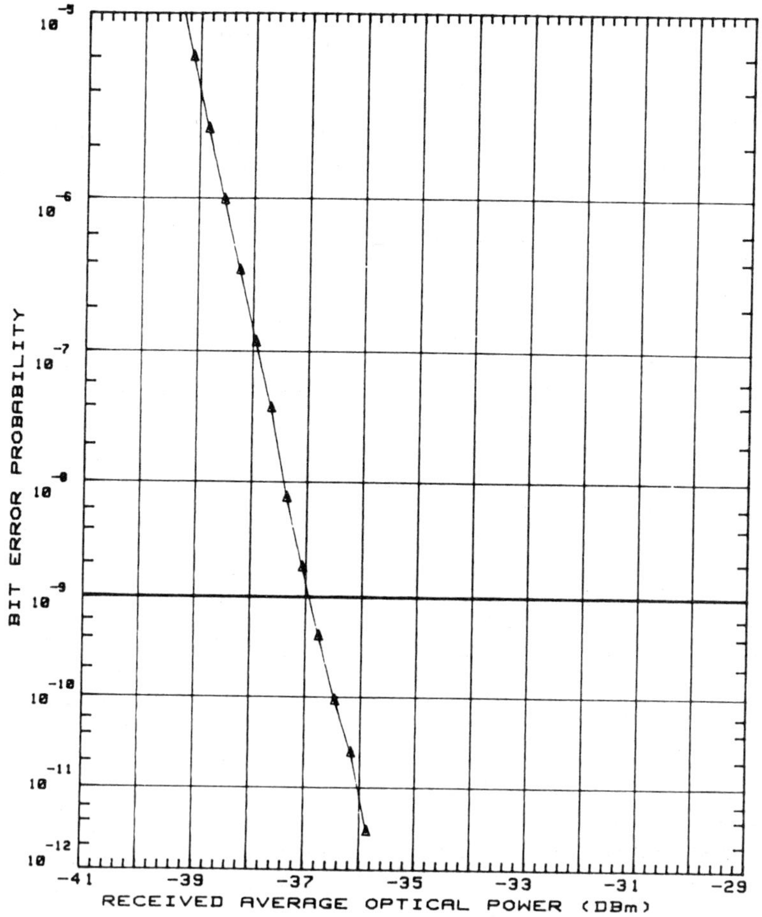

Fig. 3. System bit error rate curve. Performance is the same through attenuator only and through dispersion shifted fiber with 225 ps/nm of dispersion at 1300 nm.

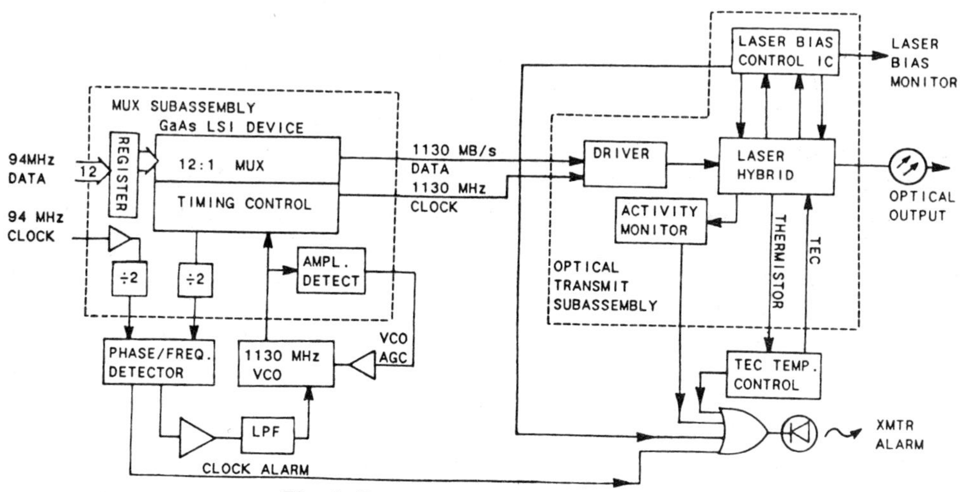

Fig. 4. Transmitter module block diagram.

The SAW filter approach was discarded for several reasons. 1) A SAW filter device is difficult to build at 1.13 GHz. 2) This approach requires waveform manipulation to ensure proper operation. 3) SAW devices have a temperature drift that would have to be compensated for or temperature stabilized. 4) Phase alignment would require circuitry similar to a phase-locked loop. 5) The SAW is a bandpass device with a fixed center frequency and Q (bandwidth), whereas a phase locked loop allows independent control of capture range and loop bandwidth (effective Q).

Other clock recovery options included: a) 141-MHz VCXO followed by a $\times 8$ frequency multiplier or a 282 MHz VCXO followed by a $\times 4$ multiplier; b) 565-MHz

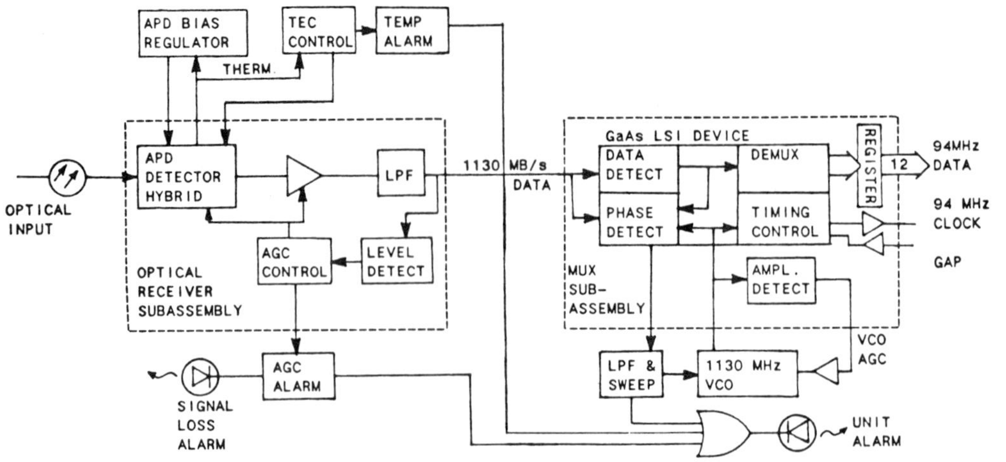

Fig. 5. Receiver module block diagram.

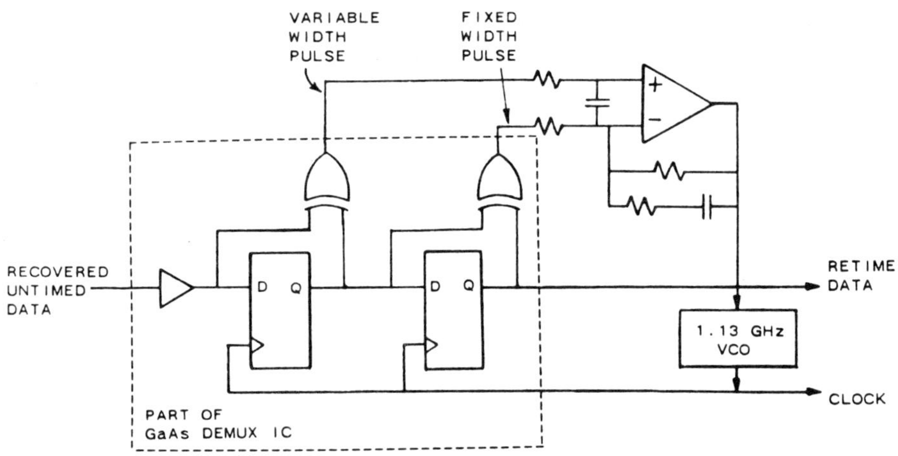

Fig. 6. Phase detector and data retiming circuit.

SAW resonator stabilized VCO followed by a doubler: and c) a fundamental frequency 1.13-GHz VCO using a moderately high Q microstrip tank circuit and a novel (patent pending) circuit to assure rapid acquistion by the phase-locked loop. Option c) was chosen as the most cost effective and most easily manufactured approach. The same VCO is used as the clock source in the transmitter, where its output is frequency divided down to 94 MHz and is phase locked to a 94-MHz crystal reference. The 1.13-GHz VCO is a four stage transistor circuit which is constructed on ceramic loaded teflon based low loss microstrip circuit board material. The 4 transistors are low noise devices with typical gain-bandwidth products of 10 GHz. A plot of the phase noise of the VCO is shown in Fig. 7.

The clock recovery circuit has the properties of a type 2 second order loop. The loop is heavily damped so as to minimize peaking [3]. The characteristics of the clock recovery jitter transfer function are shown in Fig. 8. The graph shows the measured transfer function for a single transmitter-receiver (TR) pair and 10 TR's in cascade. The peaking for the 10 TR case is about 1.0 dB at approximately 25 kHz.

GaAs IC Design Characteristics

The high speed multiplexing and demultiplexing functions of the LTS-21130 are implemented using a pair of GaAs standard cell devices. Features not normally associated with "standard" multiplexers and demultiplexers are included in this pair of parts to ease their integration into the system. The multiplexer includes an edge triggered register to store the low speed data input, and a high speed clock output that is synchronous with the data output. The demultiplexer features differential inputs for both the high speed clock and data and an integral phase detector whose operation is frequency independent. A detailed description of these devices can be found in [3]. A photo of the subassembly containing the mux device is shown in Fig. 9.

The twenty four DS3 input lines are received by the input conditioners. In the mux module, overhead bits are added to each DS3 data stream by bit stuffing up to the rate of 47.083 Mbit/s, and then the data streams are multiplexed two-to-one into twelve rails at 94.166 Mbit/s. (see Fig. 1) The twelve 94.166-Mbit/s rails and a 94.166-MHz reference clock are fed to the multiplexer subassem-

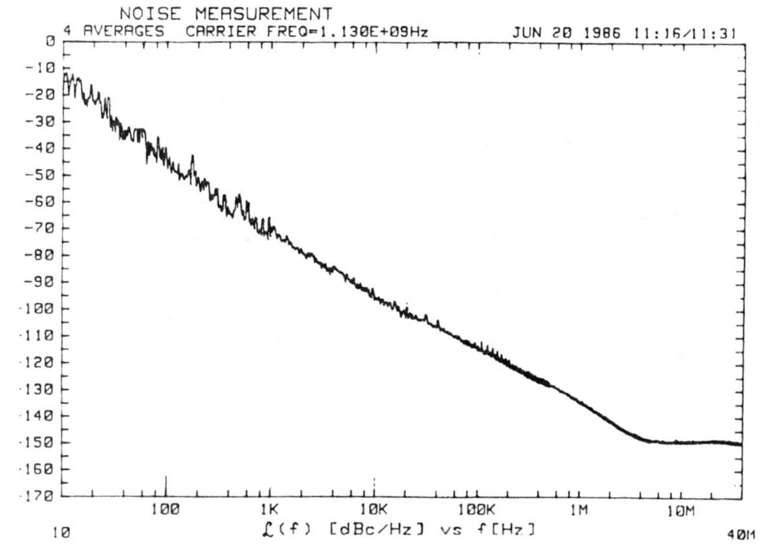

Fig. 7. Graph of VCO phase noise.

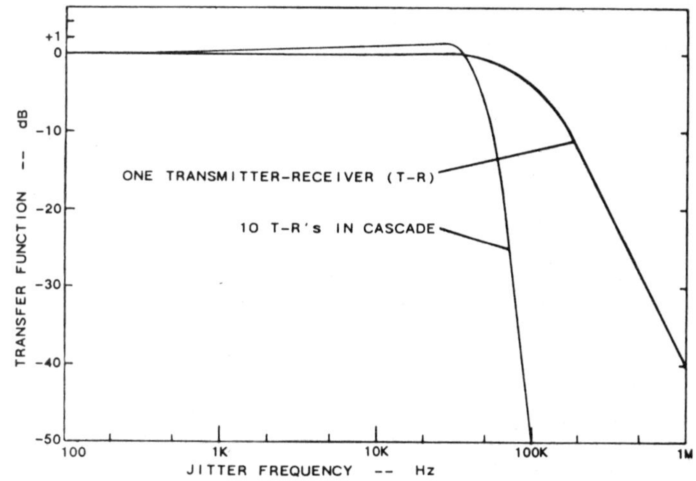

Fig. 8. Graph of regenerator jitter transfer function.

Fig. 9. Photograph of multiplexer subassembly with GaAs semicustom IC.

bly in the optical transmitter module where they are buffered and then input to the GaAs twelve-to-one multiplexer. The multiplexer outputs 1.13-Gbit/s NRZ data and a 1.13-GHz clock. These outputs are AND'ed together to produce an RZ data signal. The RZ data output goes to the optical transmitter subassembly where it is amplified and applied to the laser.

The GaAs demultiplexer receives the recovered baseband data from the optical receiver. Internal to the GaAs demultiplexer is a comparator, a decision flip-flop, a phase detector, and a one-to-twelve demultiplexer. Having both the decision circuit and the phase detector on the same die minimizes timing alignment and clock extraction problems encountered with discrete implementations.

The GaAs demultiplexer outputs twelve tributaries at 94.166 Mbit/s and a 94.166-MHz reference clock to the DEMUX low speed processing module where the twelve tributaries are demultiplexed into the original twenty four data rails.

Using monolithic GaAs circuits for the high speed digital functions has yielded several advantages in manufacturing. Placing almost all of the high speed digital circuitry into one IC (one for the mux and one for the demux) eliminated the need for expensive low loss substrate materials. The GaAs devices are mounted on standard, low cost, glass/epoxy board materials. The high speed interconnects required to go distances greater than two inches are routed on subminiature coaxial cables.

The monolithic GaAs approach has reduced manufacturing and test time. These IC's mean fewer parts to assemble and test. Furthermore, the performance of the GaAs IC's is verified by the vendor using a high temperature functional test. This reduces the amount of the functional testing required after assembly. Since most of the high speed critical circuitry is contained in a single IC there is very little alignment or adjustment required.

One concern with the monolithic GaAs approach is the long term reliability of GaAs IC's. Test results beginning to come in [5] indicate that this concern may not be as big an issue as once thought. A reliability analysis for the LTS-1565 system indicated that going to the GaAs monolithic approach over a discrete ECL implementation caused an increase in the failure rate of 306 FITS (from 660 to 966 FITS) [4] for the mux and demux subassemblies at 40° ambient temperature. However, the impact on the system availability/reliability was negligible.

Final Comments

The 1.13-Gbit/s system described achieves high performance, economical construction, an efficient architecture, and a means to easily upgrade from a 565-Mbit/s system. Upgrade is achieved without penalty to the 565-Mbit/s installation in terms of span length or traffic interruption during upgrade. New technology used in this system includes a DFB laser and the use of semicustom standard cell GaAs devices. Novel circuits in the clock recovery area take advantage of the GaAs performance and minimize timing error in the data detection. Careful system design and PC board layout allow the use of standard glass/epoxy board materials and other mature manufacturing methods. The BER performance has been verified using dispersion-shifted fiber with 225 ps/nm of dispersion. Production shipments began in April 1987.

References

[1] S. D. Personick, "Receiver design for digital fiber-optic communications systems," *Bell Syst. Tech. J.*, vol. 52, pp. 843–874, July–Aug. 1973.

[2] C. Hogge, Jr., "A self correcting clock recovery circuit," *J. Lightwave Technol.*, vol. LT-3, no. 6, pp. 1312–1314, Dec. 1985.

[3] H. Yamamoto and S. Kubo, "Performance of timing recovery circuits with phase-locked loop in a long chain of regenerative repeaters," *Electron. Commun. Japan*, vol. 59-B, no. 3, pp. 87–95, 1976.

[4] M. McDonald and G. McCormack, "A 12:1 multiplexer and demultiplexer chip set for use in a fiber-optic communication system," in *Tech. Dig. 1986 GaAs IC Symp.*, pp. 229–232.

[5] R. Venkataraman and J. Schaefer, "GaAs IC reliability and quality assurance handbook," GigaBit Logic Reliability Rep., Aug. 1, 1986.

Kenneth Y. Maxham received the B.S. degree from Rice University, Houston, TX, in 1966, and the M.S. degree from Southern Methodist University, Dallas, TX, in 1970, both in electrical engineering.

He joined Collins Radio Company, Dallas, TX, in 1966 where he was responsible for developing a variety of analog and digital circuits and systems. From 1970 to 1973 he was an Electronic Project Engineer with Boeing Company, Baggage Handling Division, Dallas, TX. From 1974 to 1975 he was Project Engineer with Sands Measurement Corp., Dallas, responsible for the development of an automatic electronic weighing system for railroad cars. In 1975, he joined Rockwell International, Dallas, where he served as Project Engineer and Systems Engineer on a variety of development programs for their message and circuit switching products. In 1981, he transferred to the Lightwave Systems Division, Rockwell, Dallas, where he has been a Project Engineer involved in the development of several lightwave transmission products.

Mr. Maxham is a Registered Professional Engineer in the state of Texas. He has been granted three patents.

*

J. Michael Dugan received the B.S. degree in electrical engineering from the University of Houston, Houston, TX, in 1975, and the M.S. degree in physics from the University of Texas at Dallas, Richardson, TX, in 1982.

He joined Rockwell International, Dallas, TX, in 1977, where he designed phase-locked circuitry for use in frequency synthesis and clock recovery applications. Since 1979, he has been involved in the design and characterization of lightwave components, and is currently the Project Manager of gigabit transmitters and receivers. He has been granted four patents.

In 1986, Mr. Dugan was a recipient of the Rockwell International Engineer or the Year Award.

*

Mark A. McDonald (S'77–M'80) received the B.S.E.E. degree from Oklahoma State University in 1980, and the Master of Engineering Degree in electrical engineering from Cornell University in 1981.

He spent two years at Bell Laboratories in Holmdel, NJ, working on satellite and fiber-optic transmission systems. He joined the Lightwave Systems Division of Rockwell International in 1983, where he is currently a Project Manager working on the development of multigigabit lightwave transmission products.

Mr. McDonald is a member of Tau Beta Pi, Eta Kappa Nu, and Phi Kappa Phi.

*

Charles R. Hogge, Jr., (S'52–A'53–M-58–SM'86) received the B.S. degree from the Virginia Military Institute in 1952, and took graduate courses at Johns Hopkins University and North Carolina State University.

He began his career with Westinghouse in 1952, working on radar antenna design. He was co-inventor of a distributed feed system for a three-dimensional radar. In 1958 he began work in Microwave Telecommunications and worked on waveguide filters, microwave mixers and up convertors, solid-state RF amplifiers, varactor multipliers, solid-stae sources, and FDMFM systems at RCA, ITT, Sanders Associates, and Raytheon Data Systems. In 1974 he joined Rockwell International in Dallas, TX. He developed a Dual-MSK modulator and demodulator and the demodulator for the 90-Mbit/s 8PSK digital radio product line. In 1981 he transferred to the Lightwave Systems Division where he had project responsibility for developing the LTS-135 receiver, the LTS-3139 transmitter and receiver, and the clock recovery/data detector and other RF circuits for the LTS-1565 transmitter, receiver, and regenerator, and the clock recovery circuitry for the LTS-21130 equipment.

Mr. Hogge has published a paper in the *IEEE Transactions on Communications*, a paper in the JOURNAL OF LIGHTWAVE TECHNOLOGY, and was a co-author of another paper in the JOURNAL OF LIGHTWAVE TECHNOLOGY. He holds seven U.S. patents. In 1987, he was a recipient of the Rockwell International Engineer of the Year Award.

Field Performance of 565 Mbit/s Optical Fibre Line System

Venkataraman Nagarajan and Glenn Nordqvist

Ericsson's optical fibre high-speed system ZAM 565-1 has been installed between St. Paul and Dallas and made operational during the first quarter of 1986. The system has a capacity of 8 064 telephone channels and uses single mode fibre as the transmission medium at a wavelength of 1 300 nm. It has henceforth become one of the very first of this kind to carry normal traffic, providing the long-haul telecommunication services for US SPRINT. Elaborate tests conducted on the fibre and the equipment during the installation and operational phases of the 2 400 km long fibre optic route confirmed the practical viability of this high speed digital optical system.

The authors present the system and cable performance results from the installation.

optical fibres
optical links
installation
telephone lines
electronic equipment testing

The trend towards an all digital/all fibre network is fast emerging all over the world, thanks to the technical/economical advantages of optical fibre transmission systems. Some of the interexchange telecommunication carriers in the United States caught up with this pace at a very early stage, by installing single mode fibre optic cables along the railroads and highways at low costs. In order to exploit the enormous information bandwith potential of the single mode fibres already installed, they required a high capacity system to meet their service demands at a reasonable cost.

In the above context, the economic advantage of installing a high speed 565 Mbit/s optical fibre system compared with a number of low-order 140 Mbit/s systems or a 405 Mbit/s system decided the final selection. Hence, currently in the majority of the US long-haul network applications, 565 Mbit/s systems are widely preferred as the optimum choice for providing high bit-rate service to the customers. Improved reliability and maintenance aspects further justify the use of 565 Mbit/s system. The ever-increasing traffic demands call for consideration of system upgrading possibilities in the near future, using the existing single mode fibre. In this perspective also, the 565 Mbit/s system has emerged out as a suitable building block amenable for such future upgrading plans.

Realizing the above urgent and huge demand in the US market, Ericsson developed this high-technology equipment and delivered to US SPRINT, one of its first customers. The system had been designed with stringent performance requirements and in connection with the first longest installation of

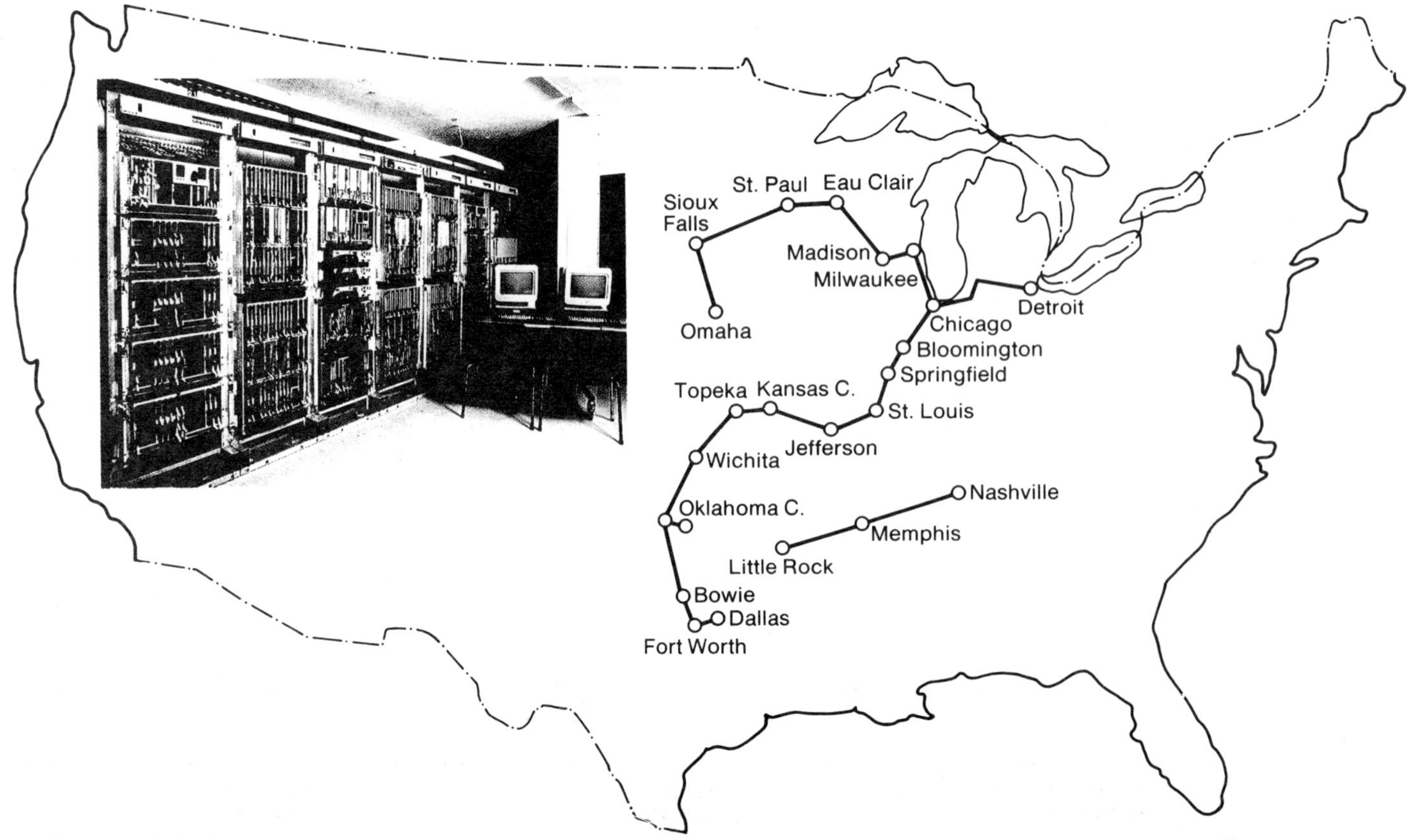

Fig. 1
The 565 Mbit/s optical fibre route between St. Paul and Dallas, USA
The total route length is approximately 2400 km

2400 km system length (St. Paul – Dallas), fig 1, field performance of the system and the cable was verified. Also any possible degradations for long.term operation over long chains of cascaded repeaters were studied and the results were found quite satisfactory.

The most important parameters for an optical fibre line system, which determine the maximum transmission span, are:
- the attenuation and dispersion of the fibre
- the optical output power and spectral width of the transmitter
- the sensitivity of the receiver
- the bit error rate
- the jitter.

System overview

The basic version of the 565 Mbit/s line system[1] consists of terminal equipment and line repeaters, fig. 2. The terminal equipment has a DSX 3 electrical interface (45 Mbit/s) and a 600 Mbaud optical interface. A 600 Mbaud electrical interface is also available. It will be suitable for upgrading to future systems with higher transmission capacity. The muldex version, with an interface for the 140 Mbit/s in the CEPT hierarchy, has been designed with due consideration paid to cable-TV applications. The muldex versions accordingly multiplexes and demultiplexes twelve 45 Mbit/s or four 140 Mbit/s signals.

In addition to the muldex unit the terminal equipment contains a terminal repeater. Its main purpose is to convert the electric 600 Mbaud signal to an optical signal in the send direction and vice versa in the receive direction. The repeater also includes an interface for the connection of a fault location signal and service telephone.

The line repeater contains an optical transmitter and receiver for each direction of transmission, a fault detector unit and an interface for a service telephone. Information regarding faults is transmitted via the fibre carrying the traffic and is processed in the fault location magazine. Fault location can thus be carried out during operation.

Operational reliability and supervision

A system must not only have good performance but also good operational reliability. Components and printed board assembly designs are selected with this in mind. The system reliability can be improved further through the use of a standby line system designated Protection Switch Line. Between one and eighteen systems can be connected via the standby system. It also offers the possibility of setting priority levels for the systems that have access to the standby. The switchover is initiated by an alarm and can be carried out manually or automatically. It can also be controlled from an operator terminal. An alarm is given in the case of a fault in the multiplexed 600 Mbaud stream or the 45 Mbit/s streams. The safety level can be set in accordance with the customer's requirements. Switchover takes place for complex faults of 50 ms. The standby system is designed with reliable components. It can never reduce the operational reliability of the ordinary systems since the latter are restored if a fault should occur in the standby system.

Fig. 2
The equipment in the 565 Mbit/s project
PS1:N Switchover equipment
PROT Standby line

Muldex and terminal repeaters

Line repeaters

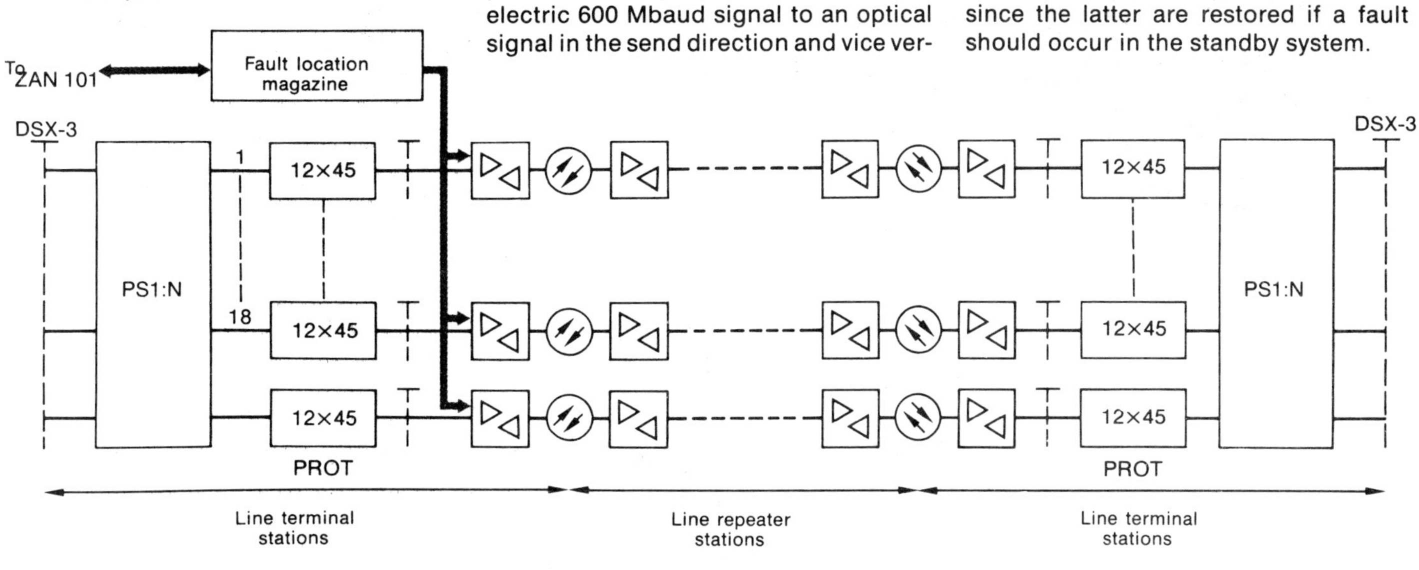

Fig. 3
Bit error rate, BER, as a function of the optical input power with temperature as the parameter

T = −20°C

T = +25°C

T = +70°C

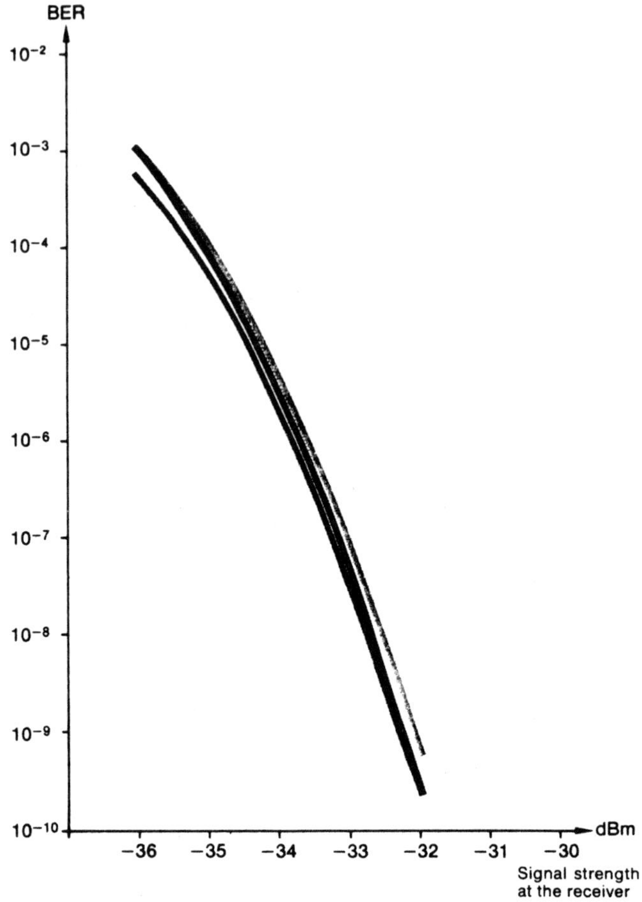

Fig. 4
Bit error rate, BER, as a function of the optical input power with dispersion as the parameter

● Dispersion = 0 ps/nm

● Dispersion = 119 ps/nm

● Dispersion = 157 ps/nm

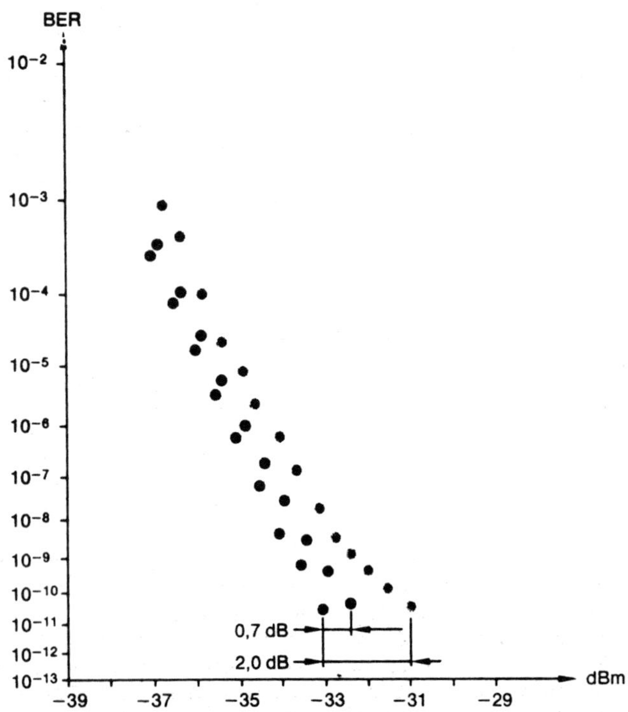

Fault location system ZAN 201 can be used to supervise transmission systems which include ZAM 565-1. Transmission supervision system ZAN 101 offers centralized supervision and alarm assembly.[2, 3]

Maximum permissible cable section attenuation

The system specification states that the attenuation of a cable section may amount to 24 dB without the bit error rate, BER, exceeding 1×10^{-11} The attenuation value includes contributions from the cable and splices. Due consideration has been given to the deterioration caused by the dispersion of a 40 km long line, since cable sections are of that length in most long-distance networks. The margins allowed for ageing, temperature dependence and dispersion are shown in the adjoining figures.

Measurements have shown that the deterioration caused by temperature variations is considerably less than the allowance, see fig. 3.

A fibre causes dispersion of the pulses, the size of which is dependent on the wavelength and spectral content of the transmitted light. The dispersion results in higher BER, fig. 4, and may, if not kept

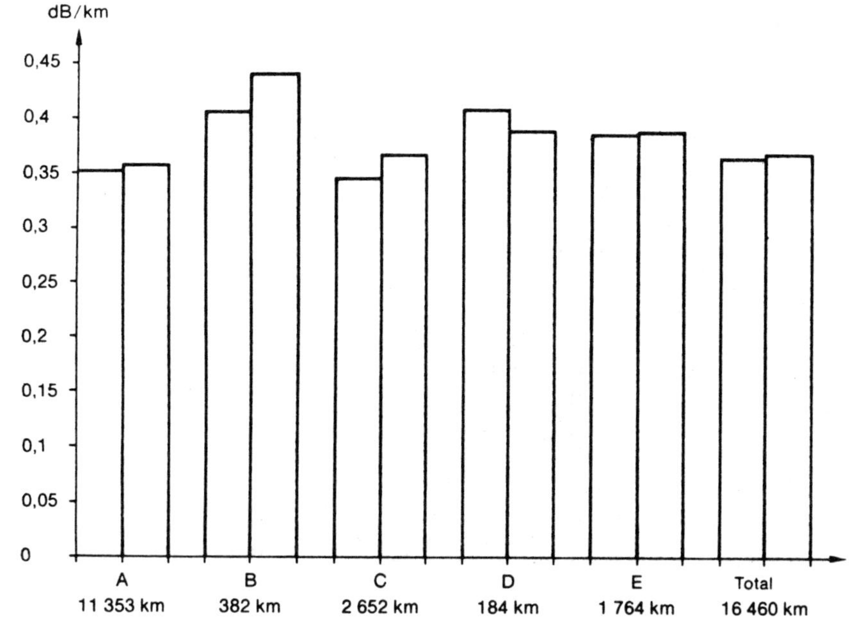

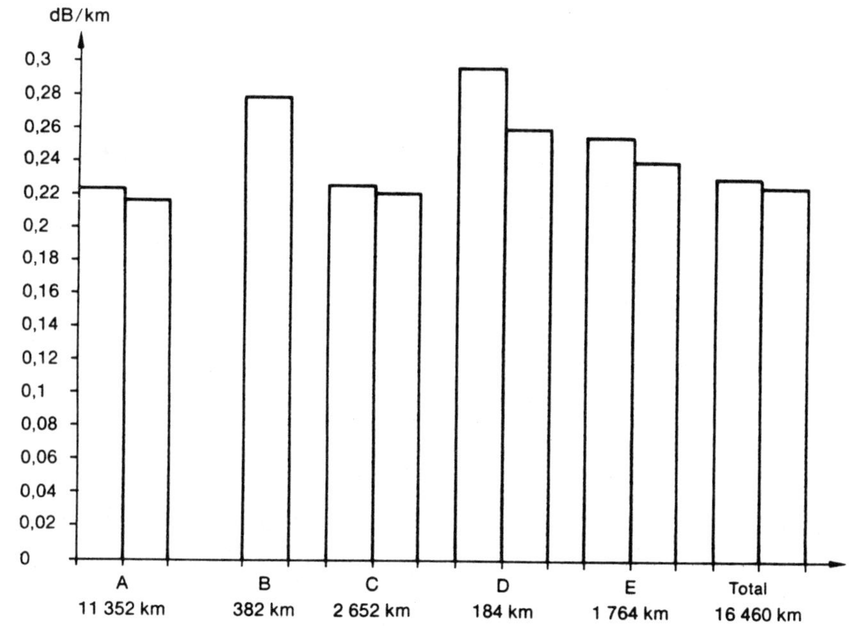

Fig. 5a, top
Fibre attenuation with different cable lengths, measured at 1310 nm

Buried

On a drum

Fig. 5b, bottom
Fibre attenuation with different cable lengths, measured at 1550 nm

under control, lead to the system being unable to operate without errors regardless of the input power. The system has been designed to tolerate a dispersion of 100 ps/nm, at the wavelengths 1290 and 1320 nm. This interval is suitable for present-day single-mode fibres, which have zero dispersion at approximately 1310 nm.

Network structure

The first installation of Ericsson's 565 Mbit/s system in the US SPRINT network consists of systems in normal operation and standby systems (protection switch lines). The network now in operation is intended both as the backbone of a future more extensive long-distance network and as local networks in cities. Some lines are also leased to private networks.

During 1985 US SPRINT laid almost 6400 km (4000 miles) of buried cable and put 5960 km (over 3700 miles) of it into operation. Ultimately the network will comprise 37 000 km (23 000 miles) of buried optical fibre cable.

Measurements

Extensive measurements have been carried out in order to determine the attenuation and dispersion of fibres and the accumulated effect of jitter and bit errors over a number of cascade-coupled amplifiers.

A large amount of data has been obtained through measurements on cable on drums. Measurements were made on 3630 cable drums with an average of 4.22 km of cable each. The fibre cables contain between 10 and 70 fibres. Altogether the drums held over 15 million metres of fibre of different manufactures. Hitherto over 31 000 fibre splices have been made, with a mean attenuation of 0.13 dB. Data for 10 068 cable sections has been recorded. On average the cable sections contain 24 fibres and have a length of 33.8 km.

The optical fibres now being installed are intended for use with both the present and the next generation of transmission systems. At present light having a wavelength of 1310 nm is used, but when suitable lasers and detectors are available the 1 550 nm wavelength will be used. US SPRINT has specified fibre data for both 1310 and 1550 nm. Attenuation measurements have been made at both wavelengths.

Figs. 5a and 5b show the attenuation per km of fibres from different manufacturers. Measurements were made on the cables both on drums and after burial, using an optical time domain reflectometer (OTDR) connected to a Hewlett Packard personal computer. Measurement data from the newly laid cable constitutes an important reference for future maintenance. Fig. 6 shows the attenuation of all fibres in the route between Fort Worth, Texas, and El Reno, Oklahoma, a distance of 394 km. The results show that the route meets the set requirement, 0.5 db/km, satisfactorily.

Fig. 6
Overall attenuation of different fibres on the route between Fort Worth and El Reno, measured at 1310 nm

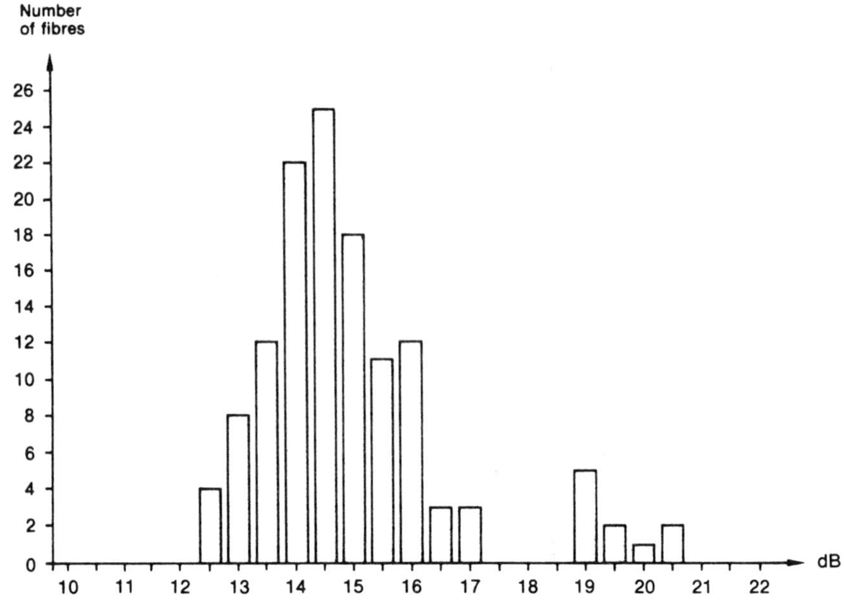

Fig. 7
OTDR curve showing the change in signal strength along a 35 km long route. The measurement was made at 1550 nm

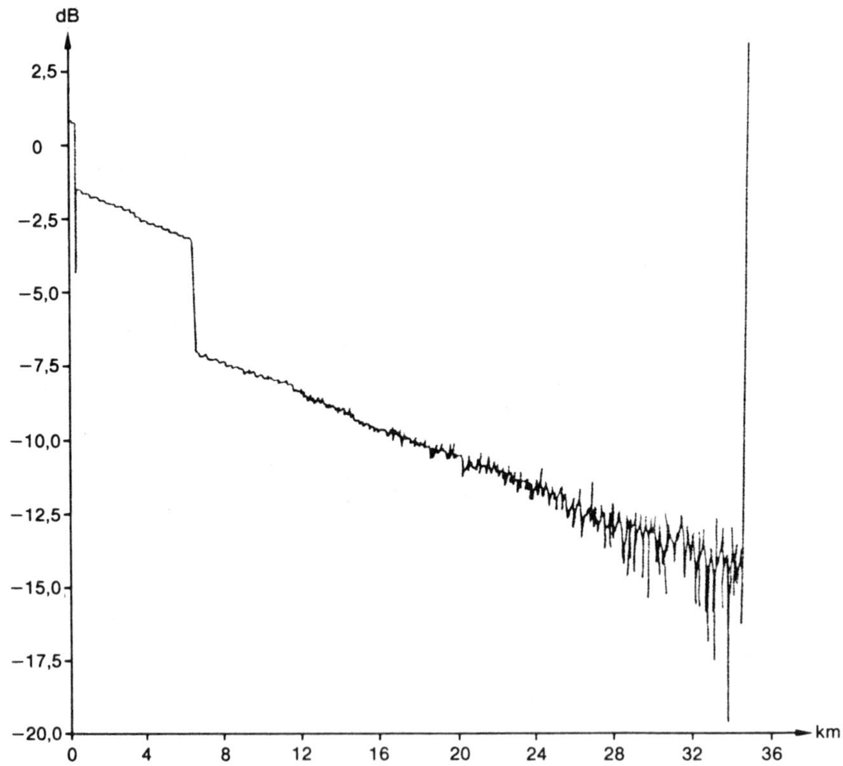

Fig. 8
Dispersion as a function of the light wavelength, measured over a 35 km long fibre. The dispersion is zero at a wavelength of 1310 nm. At 1550 nm the dispersion is 17 ps/nm km

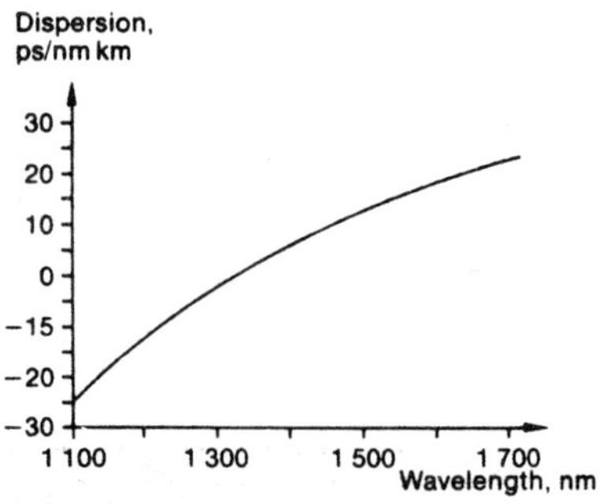

Fig. 7 shows the change in the signal level over a distance of 35 km, measured with OTDR. Abnormal conditions are easily spotted since the measurement was made at 1550 nm, where the fibre attenuation is low. For example, the increase in attenuation at approximately 6 km shows a fault that occurred during the cable laying.

Fig. 8 gives the results of dispersion measurements made at a wavelength of 1550 nm.

From the point of view of maintenance, cables with a loose buffer jacket are preferrable to cables with a tight jacket since they seem to be less prone to microbending problems.

Bit error rate

The bit error rate, BER, of the system is the most important measure of performance. Ageing factors, reduced output power, spectral changes in the laser and reduced receiver sensitivity, may have a negative effect on BER. In one measure-

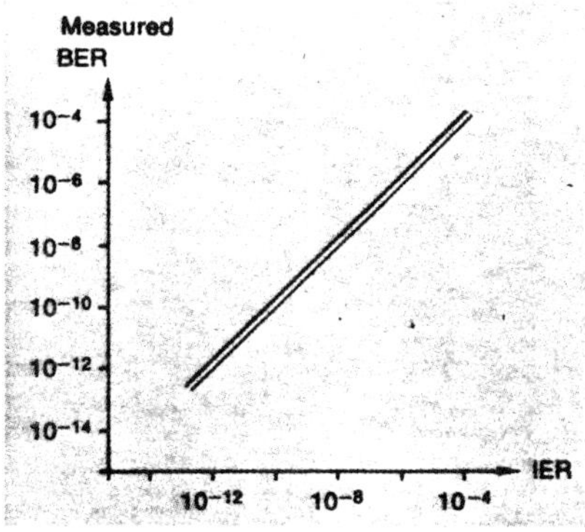

Fig. 9, right
Bit error accumulation after transmission over 25 and 50 cascade-coupled line repeaters respectively

- - - - Data after 25 line repeaters

—— Data after 50 line repeaters

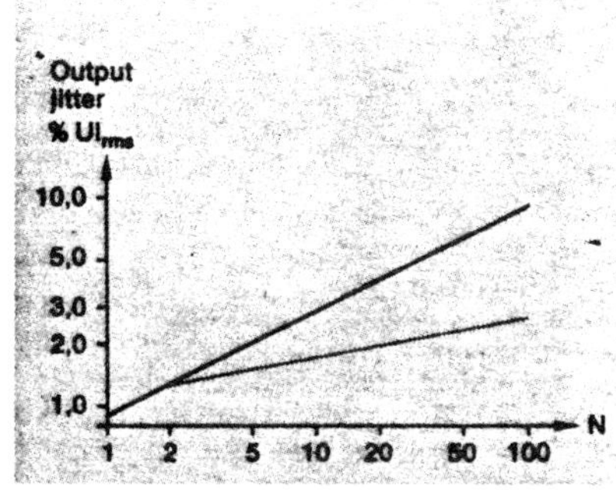

Fig. 10, far right
Jitter accumulation over 66 line repeaters, measured on the 600 Mbaud signal
N = number of line repeaters

—— $N^{0.5}$ addition law

- - - - Measured values

Jitter

In a perfect system all pulse edges arrive exactly at predetermined moments, but in real systems the edges are displaced slightly. This time displacement is called jitter. The jitter is caused partly by systematic and partly by statistically varying factors. A complete description of the jitter therefore requires a distribution diagram. The data provided for the size of the jitter can either be the maximum displacement peak-to-peak or the r.m.s. value of the distribution function. The jitter could be given as a time, but in order to convey a better idea of the effect of the jitter on the system it is given as a fraction of the time it takes to send a bit in the system. This time is designated as unit interval, UI. The UI for the 600 Mbaud signal is 1.67 ns and for the 45 Mbit/s signal 22.35 ns. Jitter in line systems causes bit errors because of faulty sampling and is particularly serious in long, cascade-coupled repeater chains.

ment the system was monitored for 48 hours, during which time no bit errors occurred. This corresponds to a BER of less than 1×10^{-14}. Measurements were also made in order to investigate whether bit errors generate new errors when the signal is sent via a long chain of intermediate repeaters. Errors were injected at the start point at a rate IER (injected error rate). BER was measured when the signal had passed 25 and 50 intermediate repeaters. The results show that there was no error accumulation for the 50 repeaters, fig. 9.

Jitter accumulation and tolerance

The jitter measurements were made on the installed system with 66 cascade-coupled line repeaters. The output jitter was measured on the 600 MHz clock in the terminal. The jitter accumulated in the chain of line repeaters was calculated on the assumption that each repeater made the same contribution to the jitter. The systematic jitter addition was assumed to follow the $N^{0.5}$ law, where N is the number of line repeaters. The measurements at the 600 Mbaud interface indicated that the addition was systematic, but in a more favourable way than the $N^{0.5}$ law, fig. 10. After 66 line repeaters the accumulated line jitter was 3% UI_{rms} (approximately 0.30 UI_{p-p}) at the 600 Mbaud interface and well below 0.1 UI_{p-p} at the 45 Mbit/s interface, fig. 11. The measurements showed that the set requirements were met with a good margin, which was very satisfying as jitter was one of the greatest problems in the development of ZAM 565.

The output signal from the muldex is free from jitter because the bit streams coming in to the muldex unit are stored in a buffer store and then sent out under the control of a free-running high-stability oscillator. Hence jitter is not transmitted from one line section to another.

The requirements regarding the system tolerance to incoming jitter on the 45 Mbit/s streams are given in the DSX 3 specification. Fig. 12 shows the results of the measurements. The system tolerates incoming jitter to an extent well above the requirements.

System recovery

After a signal failure of one minute or more the system with 66 line repeaters is restored within 16 s. This is well below the specified limit of 5 s for 10 line repeaters.

Optical output power and receiver sensitivity

The optical output power of the transmitters and the sensitivity of the receivers have been measured in order to verify important repeater data after manufacture, fig. 13. Typical values are −4.5 dBm for the output power and −33.5 dBm for the sensitivity (BER = 1×10^{-9}). The line repeaters in the system had margins of 10–15 dB to the sensitivity limits, which means that large deterioration can be allowed during their service life.

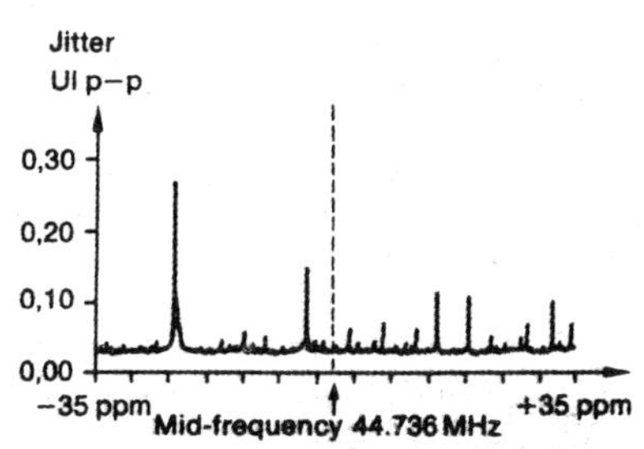

Fig. 11, right
Output jitter on the 45 Mbit/s signal from the muldex after looping of twelve 45 Mbit/s tributaries

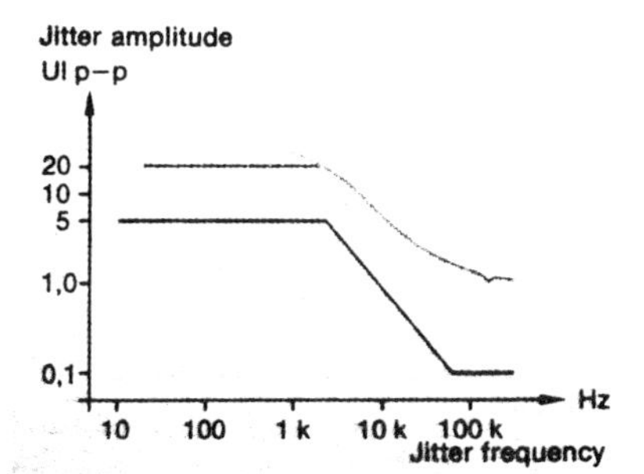

Fig. 12, far right
The ability of the system to tolerate jitter on the 45 Mbit/s input signal to the muldex

—— Requirement

- - - - Measured

—— Limit of the instrument measuring range

Fig. 13a, right
Measurement of the optical output power of approximately 300 laser transmitters.
−5.5 dBm is a guaranteed value

Fig 13b, far right
The sensitivity of approximately 60 working receivers at a BER = 1 x 10⁻⁹ and with fibre lengths varying between 30 and 35 km

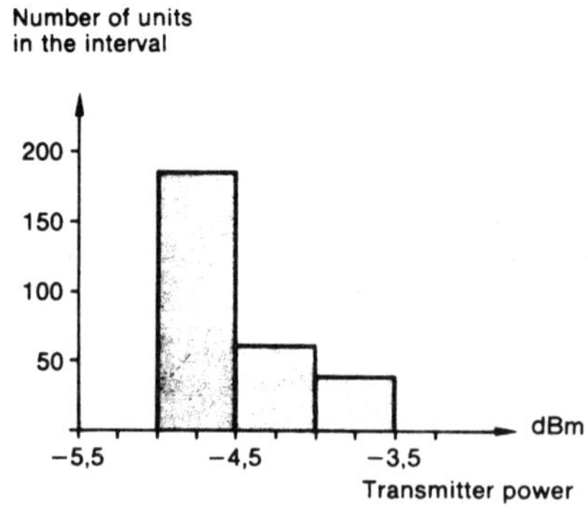

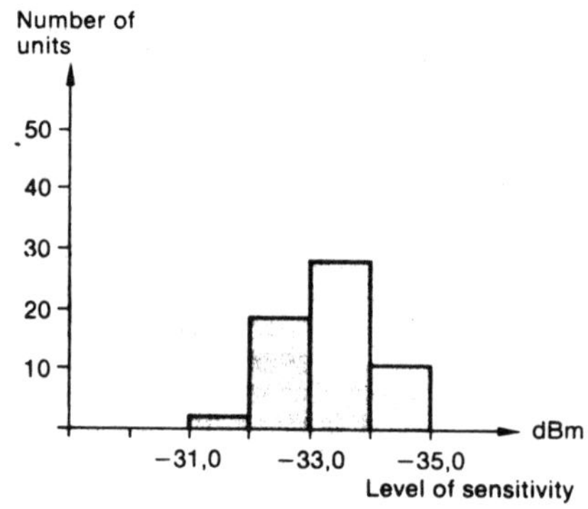

Summary

An optical fibre line system for 565 Mbit/s in operation in a long-haul network has been described. It works with good margins with repeater section lengths of over 40 km. This has been achieved thanks to the optimised design of the terminal and line repeaters. Very little jitter accumulation was obtained even with cascade-coupling of many repeaters, making the system ideally suitable for really long-haul sections between two terminals. The performance spreads of a large number of repeaters lie well within the typical specifications, illustrating further the quality standards of the system.

The results of the cable measurements illustrated in the article were presented by Terry Yake, US SPRINT, in a paper at the Stockholm Maintenance Conference, 1986.

Power budget

Transmitter	
Output power	>−4.5 dBm
Temperature dependence, ageing	1.0 dB
Guaranteed value	−5.5 dBm
Receiver	
Sensitivity at BER = 1 x 10^{-11}	<−31.5 dBm
Margin for dispersion (Dispersion < 100 ps/nm at 1290 and 1320 nm)	1.0 dB
Temperature dependence, ageing	1.0 dB
Guaranteed value	−29.5 dBm

The guaranteed values given above permit a maximum attenuation of 24 dB for a cable section. The system in the present version has been improved through the introduction of low-loss connectors. The section attenuation now amounts to 25 dB.

References

1 Johansen, B. and Stjernlöf, B.: *565 Mbit/s Optical Fibre Line System.* Ericsson Review *62* (1985):3, pp. 130–137.
2 Silvergran, U. and Woldegiorgis, K.: *Fault Location System ZAN 201.* Ericsson Review *61* (1984):4, pp. 162–169.
3 Eneborg, M. and Johansen, B.: *Transmission Maintenance System ZAN 101.* Ericsson Review *61* (1984):1, pp. 18–25.

TAT-8: An Overview

R. L. SMITH, C.ENG., M.I.E.R.E.,† and R. WHITTINGTON*

UDC 621.315.28 : 621.391.63

The eighth transatlantic telephone cable (TAT-8) will be the first system employing optical-fibre digital technology on this prestigious major route. British Telecom International has played a major role in planning the facility, in which it is a major investor. This article describes the planning processes, including the problems faced with the introduction of the new technology, and briefly describes the system and reports on the progress to date.

INTRODUCTION

In order to maintain close contact for planning transatlantic telecommunication facilities, potential European and American telecommunications investors set up a series of consultative meetings together with the USA Federal Communications Commission (FCC). This forum received the information in 1980 that it would be possible to lay an optical transatlantic cable, incorporating multiple landings, for use in 1988. Interested parties then started the detailed planning of such a facility—to become known as *TAT-8*. Twenty-nine different telecommunication entities in North America and in Europe notified their intention to participate in this cable. The American Telephone and Telegraph Company (AT&T) was the major investor with about 35%, followed by British Telecom (BT) with 15·5%; the cable was equally owned by entities on the eastern and western side of the Atlantic.

PRE-CONTRACT PLANNING ARRANGEMENTS

The first tangible step on the road to the implementation of the TAT-8 project was a meeting of potential owners of the system held in Bournemouth, UK, in September 1981. The early start was in recognition of the work to be done and to allow the owners to revert to the procurement of an analogue system in the event of the optical system being shown not to be viable. At this meeting, the owners decided that a number of specific problems relating to an optical system needed to be addressed, and set up specialist groups to study them.

The first problem was to identify the shape of the system to be provided; in other words, to study the traffic flow-predictions and assess which system configuration would most efficiently meet the owners' needs. This was especially necessary for an optical system because a system having multiple landing points by means of submerged branching units was now a possibility. The group set up to do this was called the *configuration working party* (CWP).

The second problem related to the technology and whether in fact a viable system meeting the owners' requirements could be made available in 1988. Associated with this task were studies into capacity, procurement and interconnection of the different digital hierarchies on either side of the Atlantic. A *technical working party* (TWP) was established to consider these problems.

A third group, called the *investment method working party* (IMWP), was set up to assess how the traditional methods of buying circuits in a cable needed to be modified in view of the digital optical-fibre technology proposed for TAT-8, particularly because of the prospect of multiple landings.

The outcome of all these groups resulted in the definition of TAT-8 as it is now being manufactured. The groups themselves worked closely together and made use of sub-groups to study particular aspects of the work. In addition to the other groups, a working party to draft the *construction and maintenance agreement* (C&MA WP) was formed to prepare for the formalisation of the agreement to proceed. Fig. 1 shows the structure of the committees formed to deal with the system planning stage.

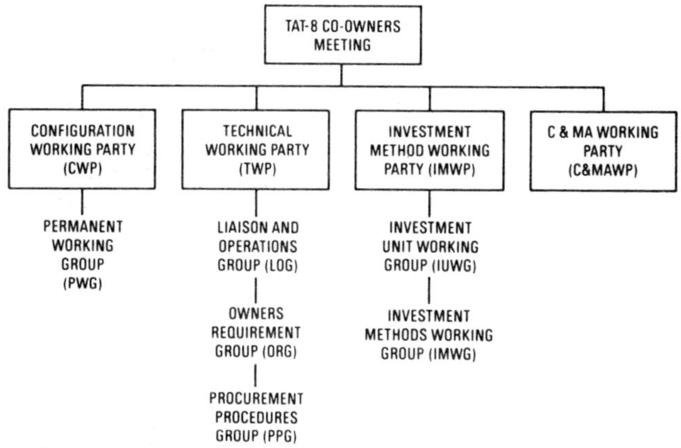

Fig. 1—TAT-8 co-owner arrangements for pre-contract planning

CONFIGURATION

The discussion on configuration, led by the Italian company Italcable, was given a wider scope than usual because of the possibility of using branching units on optical systems. This led to a large number of potential configurations with up to three landing points on either side of the ocean. By December 1982, this had been resolved to five main variants for the purposes of tendering (Fig. 2).

Eventually it was decided to proceed on the basis of configuration A (see Fig. 3) largely in recognition of the economics of adding longer spurs which would not carry high densities of traffic. It was, however, agreed that an Iberian branch could be added, if circumstances justified it later.

INVESTMENT METHOD AND UNIT OF INVESTMENT

The fundamental unit of capacity was identified as the bit per second in each direction. A unit of ownership designated as the minimum practicable unit of ownership and called the *minimum assignable unit of ownership* (MAUO) was agreed to be 73 684·656 bit/s. This comprises 64 kbit/s of usable data and the necessary additional bits to provide for multiplexing etc.; that is, equivalent to a voice channel.

For the simple configuration chosen, the owners agreed to use the overall average unit costing method whereby all intercontinental units would cost the same and the ownership would be equally split between eastward and westward co-

† Satellite and Lines Executive, British Telecom International
* Marine Services, British Telecom International

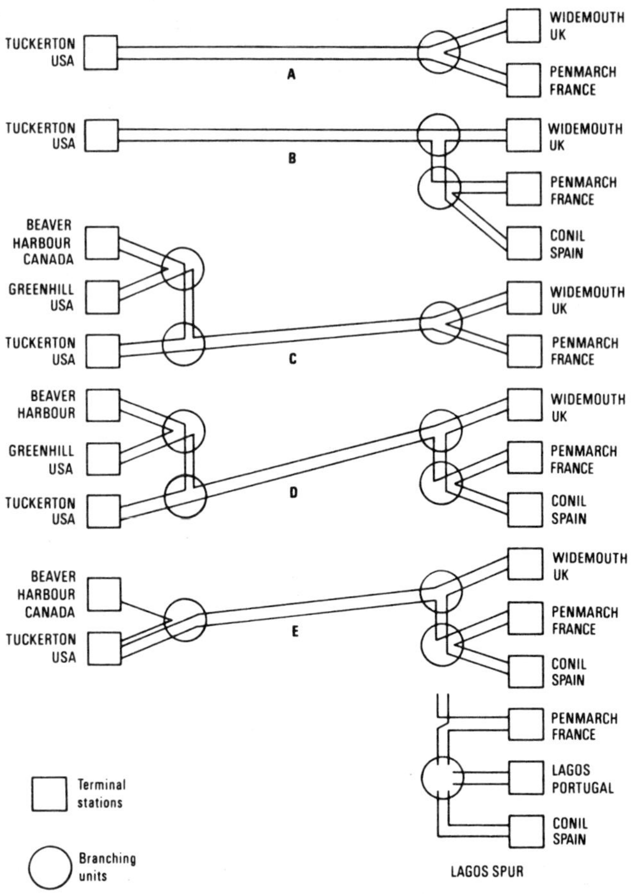

Fig. 2—TAT-8 possible engineering solutions

owners. Several other investment methods were identified, and may well be more appropriate to configurations having multiple landing points, spurs of uneven length and capacity demand. For example, a much longer spur may have to be financed by using an individual branching costing method where each branch is paid for by the carriers using it.

As well as the overall average unit costing method, which is based on the premise of 50/50 cost sharing between eastern and western partners and in which each owner pays the same amount for units of intercontinental traffic, several other methods were identified.

The individual branch costing method is useful for some owners who require longer branches or more lightly-loaded branches. They would finance that branch without cross-subsidisation from owners of a shorter branch.

Also, with the unitary arrangement of an optical-fibre cable system, it is possible to share ownership of pairs of fibre within the cable, each fibre pair taking its share of the common cable and repeater structure. This method was dubbed the *individual fibre pair costing method.*

Methods dealing with the incremental cost of adding fibre pairs to an already agreed configuration were also discussed, but these methods have the disadvantage of not being welcomed by parties owning the basic cable structure because they could be seen as cross-subsidising other users.

TECHNICAL AND PROCUREMENT ARRANGEMENTS

One of the early problems to be considered was the means of interconnecting the different digital hierarchies that had established themselves in the USA and Europe. The European system is based on the 2·048 Mbit/s 30-channel primary unit using A-law encoding, whereas the USA system is based on 1·544 Mbit/s 24-channel primary unit using μ-law encoding. After studies of the frame structure, signalling etc., a special hierarchy was agreed for use on TAT-8. It comprises intermediate levels at 2–45–140 Mbit/s in order to match the primary levels to the TAT-8 system interface of 140 Mbit/s. Access at 6·3 Mbit/s can also be arranged by bilateral agreement. This hierarchy has now been accepted by the CCITT†.

The TWP arranged for a request for quotation (RFQ) to be sent out to submarine cable system manufacturers in the Atlantic region in December 1982. Tenders were returned in May 1983.

The RFQ comprised a general specification which defined the user requirements for TAT-8. Compiling this document was particularly challenging because there were no previous systems to give a background for the specification, few CCITT criteria etc. Therefore, the owners had to develop their own standards for performance of the system. The RFQ also comprised route detail, proposed contractual conditions and other contractual details such as payment conditions, price schedules and programme details. Also, a section was written about the arrangements for final testing of the link

† CCITT—International Telegraph and Telephone Consultative Committee

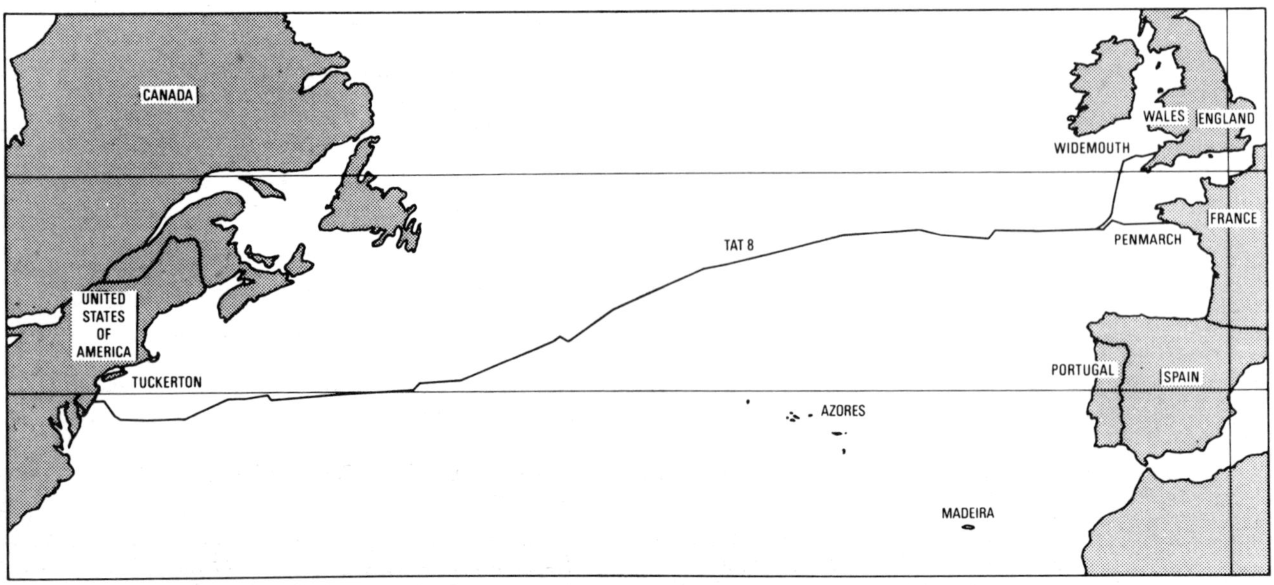

Fig. 3—TAT-8 configuration

leading to acceptance by the owners and the liquidated damages to be applied if the system did not comply with performance or delivery requirements. Warranty requirements were also detailed, the owners requesting prices for warranty periods of 2, 5 and 10 years.

Tenders were received from AT&T-Communications (AT&T-C), Submarcom (a French company combining the resources of Les Cables de Lyon and CIT-Alcatel), and STC plc of the UK. The tenders were backed up by a presentation of their facilities and resources to representatives of the co-owners in June/July 1983.

After a detailed assessment of the tenders and discussion of various adjudication methods, recognition of the effect of varying exchange rates, the closeness of the bids, and national interests, it was decided to proceed with a system design integrating the systems of the three tenderers.

This alternative solution was for each of the tenderers to supply part of the system. The individual system designs had to be modified to accommodate this arrangement and AT&T were charged to be the lead supplier, responsible for co-ordinating integration activities with the other two suppliers.

The resulting system arrangement is shown in Fig. 4.

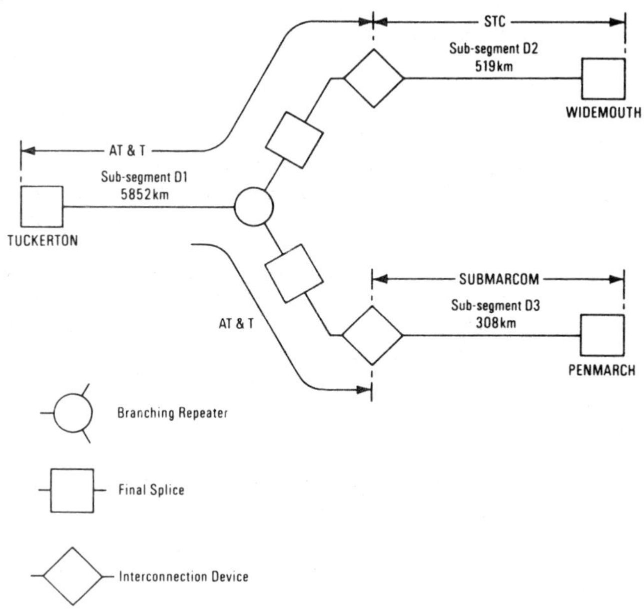

Fig. 4—TAT-8 system route

CONTRACT PLACING

The planning phase of the project came to an end in January 1984, when a construction and maintenance agreement was initialled by the potential owners. The decisions forming the basis for this agreement are summarised below:

(*a*) an optical system was feasible for service in 1988;
(*b*) the use of a branching unit was feasible, but for TAT-8 only a simple branch should be considered;
(*c*) the configuration was agreed;
(*d*) a solution to the digital hierarchy incompatibility was noted, at the time, in principle;
(*e*) financial parameters for ownership and investment were agreed; and
(*f*) an engineering solution based on the three tendered systems was agreed.

The next few months were taken up with discussions between suppliers to assess the means of arranging for actual compatibility of their product and deriving the specification for the interface conditions. Also, there was activity in finalisation of the contracts between the suppliers and the owners. These contracts were eventually signed on 12 July 1984. The contracts were actually placed between each supplier and the Co-signatories of TAT-8 on behalf of all the co-owners. The Co-signatories were AT&T-C, BT, and the French Administration des Postes, Télécommunications et Telediffusion (FPTT).

IMPLEMENTATION PHASE

To monitor and control the project during the implementation phase, the co-owners organised themselves on similar lines to previous transatlantic projects (see Fig. 5).

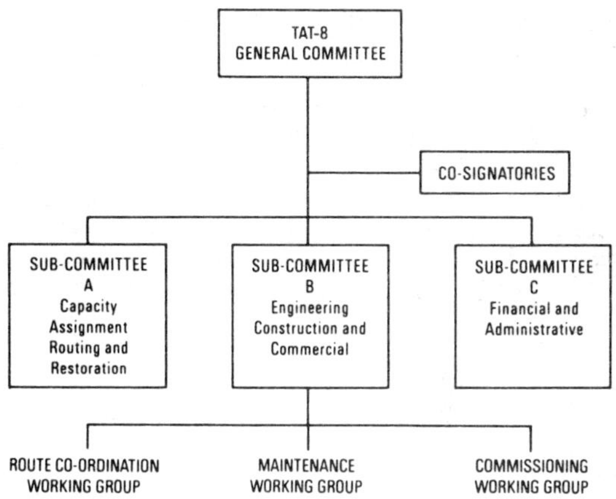

Fig. 5—Project control arrangements

The General Committee comprises senior managers from all the co-owners and meets when necessary, normally about once a year, to give overall direction to the project and sub-committees and to plan utilisation strategy.

The Co-signatories have a special role for interpretation of contractual details and, additionally, are directly responsible to the General Committee for integration matters.

Three sub-committees were set up:

Sub-Committee A to deal with the operational aspects of connecting a digital cable between the two continents,
Sub-Committee B to deal with the project management of the three contracts and the associated co-ordination, and
Sub-Committee C to monitor the cost of the project.

An important feature of the engineering sub-committee B is the identification of the methods that need to be developed to maintain the system. These include the means of fault localisation on the system as well as the equipment necessary to do this, and any special equipment needed to joint the optical-fibre cables, train the operatives etc. For this reason, a *maintenance working group* (MWG) has been established. Another sub-group of sub-committee B studies the survey requirements and difficulties of choosing a suitable place for laying the branching repeater. It has also been noted that topics relating to the commissioning and acceptance of the system will need to be studied by a *commissioning working group* (CWG).

The contracts were organised so that there were two major milestone reviews (31 March 1985 and 31 December 1985) at which the owners had the opportunity to review with the contractors the development progress and assess the build-up of reliability information.

ROUTE SELECTION AND MARINE CONSIDERATIONS

As mentioned in the previous paragraph, one of the key decision areas relates to the choice of a secure route, especially the location of the branching point.

With regard to the shallow water areas of the route, an

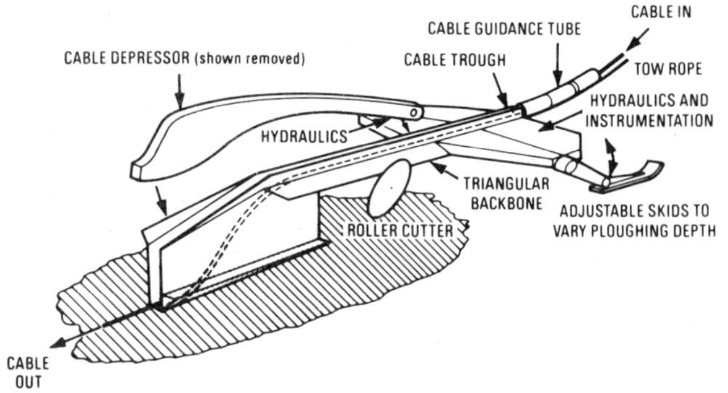

Fig. 6—British Telecom submarine cable plough

early decision was taken to protect the cable by burying it in the sea bed. BTI has, through British Telecom Research Laboratories (BTRL), developed a cable plough (Fig. 6) which has a number of potential advantages over the American and French equivalents. Various trials and early operational work have been undertaken to prove the new plough. In order to extend this method of protection, it is being further developed to bury cable at depths of 950 m. The BTI plough will operate from the *CS Alert* and will bury most of the STC-made part of the system from Widemouth, in Devon, to a point 950 m deep some 500 km from the shore.

The route from the UK to the branching unit will be surveyed by BTI using the latest navigational aids and sea-bed survey methods. Typically, side-scan sonar techniques and the use of the detrenching grapnel over the route enable sea-bed conditions to be ascertained. From this data, together with information concerning fishing and other sea-bed activities, the best practicable route for the cable is selected.

The originally passive branching unit, which is to be supplied by AT&T-C, now incorporates regenerators and is termed a *branching repeater*. As well as handling power-feed switching, it will contain some fibre flexibility points.

The original location for this branching repeater was chosen from a study of charts of the likely areas. A suitable location was found on a small plateau at about 2000 m depth around which the Continental Shelf quickly slopes down to the Abyssal Plain (Fig. 7). A special survey, however, found that suitable access into shallow water from the plateau was narrower than first thought. The French and UK spurs would therefore have to be closer together with the consequent potentially increased risk of a single hazard causing damage to both. The owners are presently studying whether adequate protection can be given or whether an alternative route will have to be found.

SYSTEM DESCRIPTION

The TAT-8 system comprises two pairs of fibres: one pair is routed from Tuckerton, USA, to Penmarch, France, and the second from Tuckerton to Widemouth, UK. The two pairs separate at a branching repeater off the European Continental Shelf and, to keep symmetry within the cable and facilitate restoration, a fibre pair is provided between the UK and France. The branching repeater provides for switching of the fibres between the different locations for operational security reasons. The branching repeater also ensures that power to the submerged plant can be fed from either the UK or France by means of a switch activated by the initiation of power. It is necessary to feed power from

Fig. 7—Three-dimensional route map showing location of branching repeater

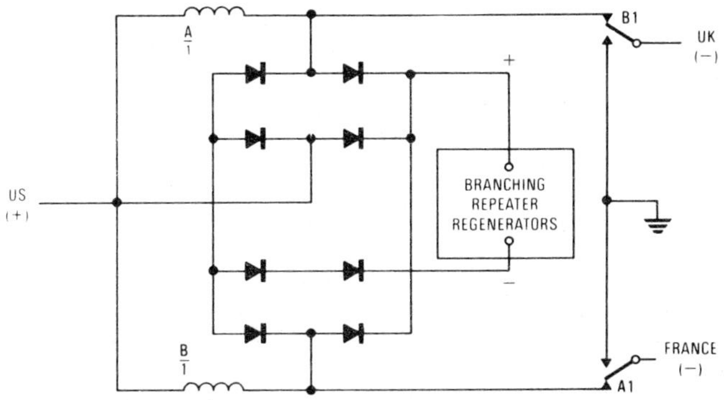

Fig. 8—Power-feed arrangements at branching repeater

both ends of the system for the transatlantic link. The spur not feeding to the USA feeds to an earth at the branching repeater (Fig. 8).

Each fibre pair carries two bothway 139·264 Mbit/s digital line sections. These two nominal 140 Mbit/s line sections are combined together into a 280 Mbit/s signal which, together with engineering facilities, is transmitted to line at a baud rate of 295·6 Mbaud.

The 140 Mbit/s line sections are made up by multiplexing 3 × 45 Mbit/s tributaries, each obtained from 21 × 2 Mbit/s blocks, which provides a total system capacity of 3780 × 64 kbit/s digital bearers per fibre pair. Digital circuit multiplication systems[1] can be used to provide the equivalent of 40 000 speech circuits. Although the multiplex equipment below 140 Mbit/s is not included in TAT-8, it can be seen from this paragraph how the problem of digital compatibility has been solved by a compromise between standards from Europe and North America.

The interface between the inland networks and the TAT-8 terminals is at 140 Mbit/s, and the performance of each 140 Mbit/s line system is specified according to CCITT Recommendation G.703.

A common line current and line code were the fundamental agreements forming the basis of the integrated system. The line current is 1·6 A and the line code is 24B1P.

As can be seen from Fig. 4, AT&T is to provide the system from Tuckerton up to and including the branching repeater. The Submarcom and STC systems connect from a position just east of the branching repeater to France and the UK, respectively.

The systems of each manufacturer differ in many respects; basically, however, the systems provide optical-fibre communications systems over single-mode fibre at a wavelength of about 1·3 μm.

Cable design is again different for the three suppliers. Each design provides a home for the fibres along the neutral axis of the cable. Features such as tensile strength, protection from pressure, hydrogen and water ingress are provided by other concentric layers of the cable. Hydrogen, an ever present hazard in submarine cables, was a major worry during the design of the optical cable when the effect of hydrogen on optical fibre became known[2].

A layer of insulation is provided to protect the high-voltage power-feeding current on the central package. In shallow water, the cable is further protected by a layer(s) of armour wires of varying size according to the type of protection required.

The three manufacturers have a different philosophy in their approach to achieving the specified system reliability. The owners' requirement was for the whole system to have a design life of 25 years and that only three ship repairs due to equipment failure were allowed.

To support their claim to meet this requirement, each of the manufacturers has set up a comprehensive life-testing programme of all the components used in the system. Also, the suppliers have used redundancy arrangements to a greater or lesser extent. The AT&T-C design, for example, has a separate pair of fibres completely equipped with regenerators which can be switched in at each repeater in order to by-pass any breakdown. All suppliers have provided cold stand-by laser transmitters which can be switched on and into circuit in the event of the failure of a working laser. Submarcom have used a measure of redundancy in the receiver and regenerator circuits and STC have duplicated the laser transmitters only.

The supervisory arrangements allow the system operators to monitor the performance of the undersea plant and take corrective action to restore service or locate the fault for maintenance action; this is described elsewhere in this issue of the *Journal*[3].

Typical facilities provided are an indication of the error rate at each regenerator, incoming light level and condition of the laser signal to line. In addition, the send and receive fibres can be looped at each repeater to enable absolute performance checks to take place from the terminal.

Operation of the terminal facilities to access repeater performance is controlled by computers at each terminal. Each digital line system terminal is duplicated with change-over points between set 1 and set 2 to provide minimum protection to service and meet the CCITT requirements for outage.

CURRENT STATUS

The current status of the project is that the main effort on the part of the suppliers has been to complete the development leading to a series of test programmes to demonstrate the compatibility of the three suppliers' designs. It is gratifying to be able to report that the first six phases of this work have been successfully completed. This means that repeater prototype models from AT&T, STC and Submarcom have successfully worked together and the terminal optical and multiplex equipment has also been shown to be compatible. The three manufacturers have also delivered all equipment for final testing in the USA.

Presently, the programme is on schedule for each sub-segment to be laid and accepted by 15 May 1988 in order that overall testing can be completed by 30 June 1988 for the system to be taken into service.

References

1 COTTERILL, D. A. User Requirements and Provision of Digital Circuit Multiplication Systems for TAT-8. *Br. Telecommun. Eng.*, July 1986 (this issue).

2 RUSH, J. D., *et al.* Hydrogen related degradation in optical fibres—system implications and practical solutions. *Br. Telecom Technol. J.*, Sept. 1984, **2**(4), p. 84.

3 DAWSON, P. A., and KITCHEN, J. A. TAT-8 Supervisory Subsystem. *Br. Telecommun. Eng.*, July 1986 (this issue).

Biographies

Roger Smith joined the Post Office, as it then was, as a Youth-in-Training in the Cambridge Telephone Area in 1958. He moved to Main Lines Development Department in 1965 and is currently Projects Manager for Submarine Systems and International Repeater Stations in BTI's Satellite and Lines Executive.

Dick Whittington, who is a Master Mariner joined the Post Office cable ships in 1971 as Third Officer, having spent the previous 12 years in the British Merchant Navy. He has served in all BTI cable ships in various ranks and has been involved in all aspects of planning, installing and maintaining submarine plant. He is currently Marine Planning Manager for submarine systems in BTI's Marine Services.

313 km TRANSMISSION EXPERIMENT AT 1 Gbit/s USING OPTICAL AMPLIFIERS AND A LOW CHIRP LASER

Indexing terms: Optical communications, Optical amplifiers, Optical transmission

The use of four in-line semiconductor laser amplifiers is reported in an optical fibre transmission experiment. Using direct detection at 1 Gbit/s, a transmission distance of 313 km of non-dispersion-shifted fibre was achieved. This is 142 km longer than previous experiments at similar data rates, and is the first time optical amplifiers have been used to improve transmission distance for fibre-optic communication systems.

Introduction: Semiconductor laser amplifiers can be used as linear repeaters in long-haul lightwave communication systems.[1] Two in-line laser amplifiers have recently been used in an intensity-modulated system at 1·5 μm wavelength[2] to overcome the system losses. However, the longest transmission experiments demonstrated so far have all been performed without optical amplifiers.

The low-loss region around 1·5 μm wavelength in conventional single-mode silica fibres is attractive for long-distance fibre-optic communication systems. The nonzero dispersion at these wavelengths requires single-mode lasers for the system to have a reasonably high capacity. However, lasers that are directly modulated show a dynamic linewidth broadening (chirp) that will limit the bit/s × km product at high bit rates.[3] Therefore, transmission systems operating at high bit rates and over very long fibre distances require lasers with very low chirp. To expand the transmission distance, a dispersion-limited optical signal requires regenerative repeating, while a loss-limited signal might use the much simpler technique of direct optical amplification.

In this letter we report the use of a directly modulated single-mode laser with very low chirp to transmit a signal through 313 km of conventional single-mode fibre at a rate of 1 Gbit/s. Four in-line laser amplifiers, having a total fibre-to-fibre gain of 40·9 dB, significantly increase the maximum transmission length. This result is to be compared with the longest directly modulated transmission experiment at 1·2 Gbit/s with 171 km of fibre,[4] using no optical amplifiers.

Experiment: The experimental configuration of the system is given in Fig. 1. The signal light source consists of an SCBR laser[5] emitting at 1·54 μm wavelength. The laser is made from a conventional channel substrate buried heterostructure[6] Fabry–Perot laser in contact with an external Bragg reflector on a silicon chip. The threshold current is 23 mA and the bias current used was 54 mA. Direct modulation through the threshold was applied by an NRZ $2^{15} - 1$ pseudorandom binary sequence signal at 1 Gbit/s. The modulation current was 75 mA peak-to-peak and the chirp was less than 80 MHz/mA. The fibre mean input power was +0·2 dBm.

A total of 313 km of conventional non-dispersion-shifted fibre was inserted between the laser and the receiver, with an accumulated loss (including splicing losses) of −69·9 dB, and a total dispersion of approximately 5 ns/nm. After various fibre lengths, as shown in Fig. 1, the signal level was amplified by optical amplifiers. Four InP/GaInAsP CSBH[6] semiconductor lasers with optical cavities about 500 μm long and antireflective facet coatings were used as the optical amplifiers. The first three amplifiers have facet reflectivities on the order of 0·5%, and their gain peaks are approximately 6 nm on the shorter-wavelength side of the signal, while the last amplifier has a facet reflectivity of about 3%, and the gain maximum coincides with the signal wavelength. The input signals to the amplifiers were TE-polarised by manual fibre polarisation controllers, placed just before each amplifier. For a TM-polarised input the gain in an amplifier decreased by approximately 7 dB. The fibre input power and the fibre-to-fibre gain for all amplifiers are given in Table 1. Microscope objectives in combination with optical isolators or lensed fibre ends were used to couple the light into and out of the amplifiers, as shown in Fig. 1. The total fibre-to-fibre coupling loss for each amplifier was estimated to be −10 dB, irrespective of the coupling method. Optical isolators were inserted between all amplifiers and a grating, acting as a wavelength filter, was inserted at the output of amplifier 3. The FWHM of the filter was 15 Å and all spontaneous emission, except the Fabry–Perot modes closest to the signal wavelength, was filtered out.

Table 1 AMPLIFIER PARAMETERS

	Amp. 1	Amp. 2	Amp. 3	Amp. 4
Signal input power, dBm	−18·4	−23·3	−23·7	−19·6
Fibre-to-fibre gain, dB	7·9	11·1	14·2	7·7

A *pin* GaInAsP diode with an integrated GaAs transimpedance circuit was used as the receiver. Fig. 2 shows the measured bit-error-rate characteristics of the receiver at 1 Gbit/s direct detection. Without any fibre or amplifiers, the receiver sensitivity ($P_e = 10^{-9}$) is −30·5 dBm. With 313 km of fibre and four laser amplifiers the receiver sensitivity is degraded by only 1·4 dB to −29·1 dBm, without any sign of error floor. The degradation is caused by the combined effects of signal-spontaneous beat noise, pattern effects due to gain saturation in the amplifiers, and dispersion. The penalty from spontaneous–spontaneous beat noise is negligible.

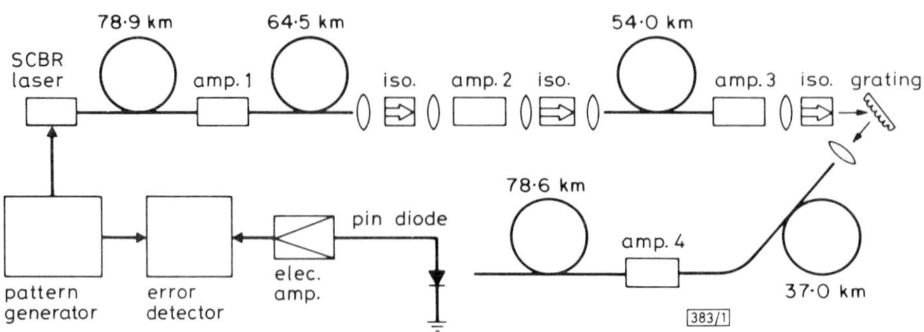

Fig. 1 *Experimental set-up*
Amp = laser amplifier, iso = optical isolator

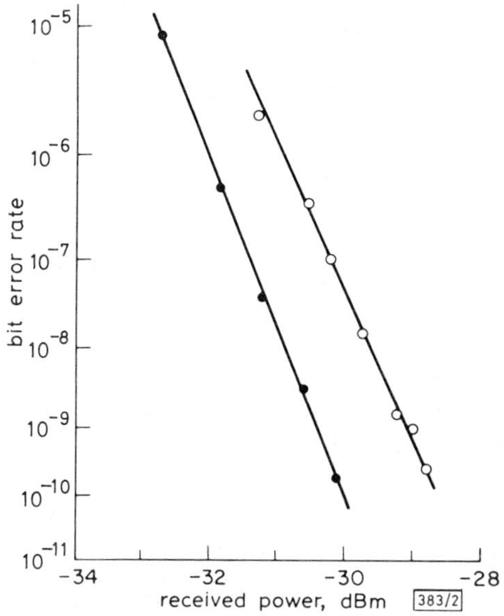

Fig. 2 *Measured bit error rate against received signal power for an NRZ* 2^{15-1} *PRBS signal at 1 Gbit/s*

○ no fibre or amplifiers ● 313 km fibre and four amplifiers

Conclusion: In this letter we have demonstrated the use of four laser amplifiers to transmit a 1 Gbit/s directly modulated signal through segments of conventional non-dispersion-shifted fibre over a total distance of 313 km. This is the longest nongenerative fibre transmission experiment using direct detection demonstrated so far. The low chirp of the SCBR laser source makes the nongenerative optical amplification useful, and the total transmission distance was expanded by 176 km by the linear optical amplification. The system penalty from amplifier noise and dispersion was only 1·4 dB.

Acknowledgment: We thank A. H. Gnauck and R. Giles for the loan of the low-loss fibre.

M. G. OBERG*
N. A. OLSSON
L. A. KOSZI
G. J. PRZYBYLEK

AT&T Bell Laboratories
Murray Hill, NJ 07974, USA

10th November 1987

* Permanent address: Institute of Microwave Technology, Box 1084, S-16421 Kista, Sweden

References

1 MUKAI, T., YAMAMOTO, Y., and KIMURA, T.: 'Optical direct amplification for fiber transmission', *Rev. Electr. Commun. Lab.*, 1983, **31**, pp. 340–348

2 MARSHALL, I. W., O'MAHONY, M. J., and CONSTANTINE, P. D.: 'Optical system with two packaged 1·5 μm semiconductor laser amplifier repeaters', *Electron. Lett.*, 1986, **22**, pp. 253–255

3 CHIDGEY, P. J., WHITE, B. R., BRAIN, M. C., HOOPER, R. C., SMITH, D. R., SMYTH, P. P., FIDDYMENT, P. J., NELSON, A. W., and WESTBROOK, L. D.: '1·2 Gbit/s optical fibre transmission experiment over 113·7 km using a 1·528 μm distributed-feedback ridge-waveguide laser', *ibid.*, 1984, **20**, pp. 707–709

4 SHIKADA, M., FUJITA, S., TAKANO, I., HENMI, N., MITO, I., TAGUCHI, K., and MINEMURA, K.: '1·5 μm high bit rate long span transmission experiments employing a high power DFB-DC-PBH laser diode'. IOOC/ECOC '85, Venice, Italy, 1985, pp. 49–52

5 OLSSON, N. A., HENRY, C. H., KAZARINOV, R. F., LEE, H. J., and ORLOWSKY, K. J.: 'Extremely low chirp narrow linewidth external silicon chip Bragg reflector (SCBR) laser'. OFC/IOOC '87, Reno, Nevada, 1987, PDP8, pp. 32–35

6 WILT, D. P., LONG, J., DAUTREMONT-SMITH, W. C., FOCHT, M. W., SHEN, T. M., and HARTMAN, R. L.: 'Channelled-substrate buried-heterostructure InGaAsP/InP laser with semi-insulating OMVPE base structure and LPE regrowth', *Electron. Lett.*, 1986, **22**, pp. 869–870

5 Gbit/s TRANSMISSION SYSTEM EXPERIMENT OVER 111 km OF OPTICAL FIBRE

Indexing terms: Optical communications, Optical fibres, Optical transmission

A 5 Gbit/s optical transmission system experiment is reported over 111 km of single-mode optical fibre, using a 1·5 μm directly modulated DFB Laser and a ternary APD. Full clock recovery as well as frequency division equipment is provided for demultiplexing purposes.

Introduction: Several high-speed, long-span optical transmission experiments have been reported recently using time-division-multiplexing techniques.[1–5]

With regard to the bit rate-distance product, the most powerful systems reported to our knowledge yield 546·4 Gbit km/s using external modulation[1] and 479·6 Gbit km/s using direct injection current modulation of the laser diode.[2]

In this letter we report on a transmission system experiment including clock recovery and using a directly modulated distributed feedback (DFB) laser, yielding a bit rate-distance product of 554 Gbit km/s.

Transmission system: The experimental set-up is displayed in Fig. 1. A microwave synthesiser (HP8341A) provides the

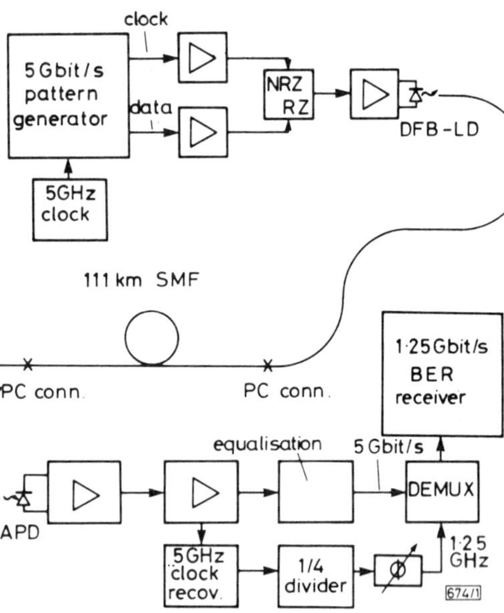

Fig. 1 *Block diagram of transmission system experiment*

5 GHz clock signal for the pattern generator (Anritsu MP1601A) that delivers a 5 Gbit/s NRZ electrical data stream. After amplification, the data are converted to RZ format using discrete GaAs circuitry. A GaAs MMIC amplifier is used to drive the 1·5 μm DFB laser diode ($\lambda = 1536$ nm). The optical signal is transmitted via a 110·8 km link consisting of 98 km of dispersion-shifted fibre ($\lambda_0 = 1548$ nm) and 12·8 km of standard monomode fibre. The whole link includes 19 splices and exhibits an attenuation of 26·9 dB. The receiver transimpedance front-end consists of a ternary APD in conjunction with a low-noise microwave HEMT. The main amplifier exhibits a bandwidth of 8 GHz. After filtering and equalisation the 5 Gbit/s data are provided for the 1 : 4 demultiplexer (Anritsu MP1603A). The clock regeneration circuitry is of the passive type using a 5 GHz resonator ($Q \simeq 250$). The clock division by four is done with a silicon integrated circuit[6] to provide the 1·25 GHz clock signal for the demultiplexer. The four 1·25 Gbit/s channels are selected by phase-shifting the clock signal. The bit error rate is measured using a BER receiver (Anritsu ME522 + MH577A).

Experimental results: In the experiment an electrical RZ data stream of 100 ps pulses is used to modulate the DFB laser diode. The optical output signal from the laser was measured using a fast PIN photodiode (rise time ≤30 ps) and a sampling oscilloscope. As a result, a pulse width of about 200 ps and an extinction ration of ~6 : 1 are observed. The mean optical output power in the fibre is +2·2 dBm. The received optical power at the output of the 111 km fibre link is −24·7 dBm. At the output of the receiver, the 5 Gbit/s eye diagram of Fig. 2 is measured.

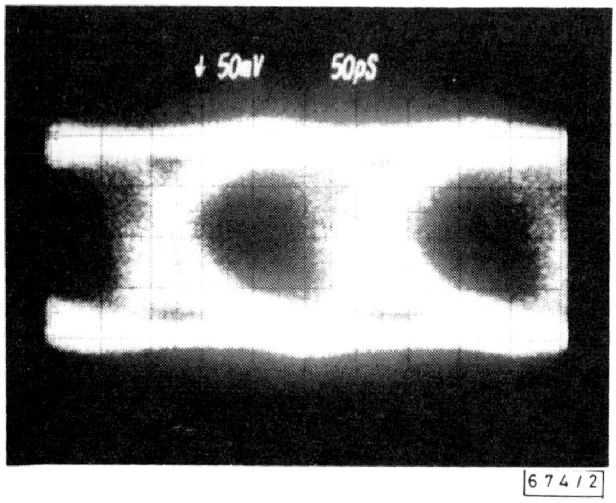

Fig. 2 *5 Gbit/s eye diagram at output of receiver after transmission via 111 km of optical fibre*

$2^{15} - 1$ PRBS data; vertical: 0·5 V/div; horizontal: 50 ps/div

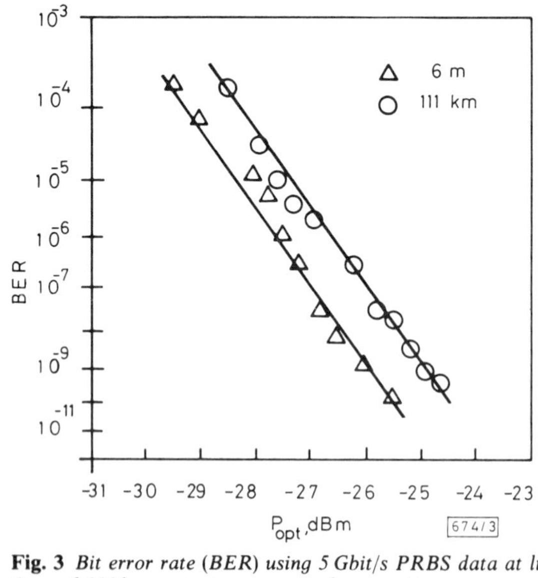

Fig. 3 *Bit error rate (BER) using 5 Gbit/s PRBS data at link lengths of 6 m and 111 km against mean optical power P_{opt}*

Fig. 3 displays BER measurements using 5 Gbit/s PRBS data ($2^7 - 1$) yielding a receiver sensitivity of −25·9 dBm at 10^{-9} error rate and 6 m of optical fibre.

Using the 111 km link, a degradation of 1 dB is observed. This penalty is due to the fibre dispersion of 130 ps/nm and the wavelength chirp of the current-modulated laser.

Conclusions: The experiments reported here demonstrate the feasibility of 5 Gbit/s transmission via 110·8 km of optical fibre. The transmission system includes a clock recovery circuit and uses a current-modulated laser. The resulting bit-rate × distance product of 554 Gbit km/s is, to our knowledge, the best value reported so far.

R. HEIDEMANN
U. SCHOLZ
B. WEDDING

Standard Elektrik Lorenz AG
Research Centre
D-7000 Stuttgart 31, W. Germany

6th July 1987

References

1 GNAUCK, A. H., KOROTKY, S. K., KASPER, B. L., CAMPBELL, J. C., TALMAN, J. R., VESELKA, J. J., and MCCORMICK, A. R.: 'Information-bandwidth-limited transmission at 8 Gbit/s over 68·3 km of single-mode optical fiber'. Digest OFC 86, Atlanta, GA, Feb. 1986, Postdeadline paper PD9-1

2 FUJITA, S., TAKANO, I., HENMI, N., SHIKADA, S., MITO, I., TAGUCHI, K., and MINEMURA, K.: '4 Gb/s long span transmission experiments employing high speed DFB-LDs and InGaAs–APDs'. ECOC '86, 1986, pp. 507–510

3 KOROTKY, S. K., EISENSTEIN, G., GNAUCK, A. H., KASPER, B. L., VESELKA, J. J., ALFERNESS, R. C., BUHL, L. L., BURRUS, C. A., HUO, T. C. D., STULZ, L. W., CIEMIECKI NELSON, K., COHEN, L. G., DAWSON, R. W., and CAMPBELL, J. C.: '4-Gb/s transmission experiment over 117 km of optical fiber using a Ti : $LiNbO_3$ external modulator', *J. Lightwave Technol.*, 1985, **LT-3**, pp. 1027–1031

4 GNAUCK, A. H., KASPER, B. L., LINKE, R. A., DAWSON, R. W., KOCH, T. L., BRIDGES, T. J., BURKHARDT, E. G., YEN, R. T., WILT, D. P., CAMPBELL, J. C., CIEMIECKI NELSON, K., and COHEN, L. G.: '4-Gb/s transmission over 103 km of optical fiber using a novel electronic multiplexer/demultiplexer', *ibid.*, 1985, **LT-3**, pp. 1032–1035

5 GNAUCK, A. H., BOWERS, J. E., and CAMPBELL, J. C.: '8 Gbit/s transmission over 30 km of optical fibre', *Electron. Lett.*, 1986, **22**, pp. 600–602

6 DERKSEN, R. H., REIN, H.-M., and WÖRNER, K.: 'Monolithic integration of a 5·3 GHz regenerative frequency divider using a standard bipolar technology', *ibid.*, 1985, **21**, pp. 1037–1039

TRANSMISSION AT 1 Gbit/s OVER 7·4 km OF SINGLE-MODE FIBRE USING AN EDGE-EMITTING LED AND PINFET RECEIVER

Indexing terms: Optical communications, Optical transmission

Transmission at 1 Gbit/s over 7·4 km of 1300 nm optimised single-mode fibre has been demonstrated using a standard, 1300 nm edge-emitting LED with 150 MHz 3 dB bandwidth and a PINFET receiver. Computer simulation was used to accurately assess each of the system degradations: LED rise/fall, extinction ratio, dispersion penalty and receiver noise.

Introduction: The combination of LEDs, single-mode fibres and PINFET receivers is attractive for local exchange and subscriber loop networks. Single-mode fibre offers low loss, large bandwidth, future upgrade capability and is less expensive than graded-index multimode fibre. Edge-emitting (EE-) LEDs have been demonstrated that couple sufficient power into single-mode fibre to allow transmission at data rates up to 560 Mbit/s over distances sufficient for most loop applications.[1,2] At the receiver, PINFET front ends are preferred for their lower cost, better reliability and simpler operation, in spite of their lower sensitivity compared to APDs. There has been considerable interest and progess in LED/single-mode technology,[3] and several experiments have been reported at distances and data rates ringing up to 100 km and 560 Mbit/s, respectively. This letter reports on 1 Gbit/s transmission over 7·4 km of single-mode fibre, using a simple transmitter circuit, a commercial edge-emitting LED and a PINFET receiver.

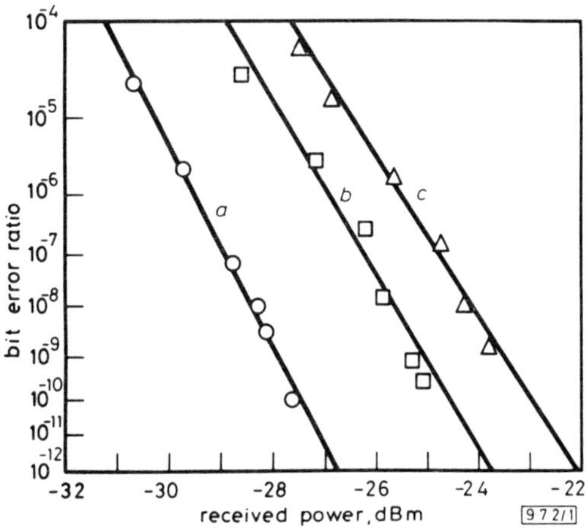

Fig. 1 *Transmission at 1 Gbit/s over 7·4 km SM fibre using 1300 nm EE-LED and PINFET receiver*

a No fibre
b 7·4 km fibre and 1250 nm highpass filter
c 7·4 km fibre without filter

Experiment: The source used in the experiment was a 1300 nm 'truncated stripe' geometry, GaInAsP edge-emitting LED mounted in a 12-pin dual in-line package (DIP), coupled to a single-mode fibre pigtail.* The average optical power output of the LED was −19·5 dBm at 100 mA drive current. The LED centre wavelength was 1294 nm and the FWHM spectral width was 70 nm. A 100 mA DC bias current and approximately ±50 mA average AC modulation was provided to the LED through the use of a simple bias-T circuit. The 3 dB optical roll-off of the LED frequency response was 150 MHz, which was inadequate for operation at 1 Gbit/s. A parallel *RC* equalising circuit in the AC path of the bias-T was employed to extend the 3 dB roll-off frequency to 680 MHz. This resulted in a large extinction ratio penalty (i.e. nonzero output power at the zero state), because of the need to avoid reverse-biasing the diode on the negative overshoot during turn-off.

* The edge-emitting LED and PINFET front-end receiver were provided by Lasertron Inc.

Table 1 POWER BUDGET FOR 1 Gbit/s, 7·4 km LED/SINGLE-MODE FIBRE EXPERIMENT

	With 1250 nm optical filter	Without filter
Power launched into SM fibre, dBm	−19·5	−19·5
7·4 km single-mode fibre, dB	3·6	3·6
Optical attenuator, dB	0·5	0·5
Highpass optical filter, dB	1·4	—
Dispersion penalty, dB	2·9	4·4
LED-induced ISI penalty, dB	0·7	0·7
Extinction ratio penalty, dB	5·0	5·0
System margin, dB	0·1	0·0
Receiver sensitivity (BER = 10^{-9}), dBm	−33·7	−33·7

The 1 Gbit/s NRZ modulated signal was transmitted over 7·4 km of standard, depressed-cladding, 1300 nm optimised single-mode fibre. A high-pass optical interference filter with cutoff wavelength at 1250 nm was employed to narrow the LED spectrum and to reduce the dispersion penalty.[4] Power budgets that compare the transmission with and without the filter are shown in Table 1. All the components were combined by biconic single-mode connectors, and the insertion loss of each component includes one connector. A variable optical attenuator was included in the circuit to facilitate bit-error-rate measurements.

A high-impedance PINFET receiver in a standard DIP, with a single-mode fibre pigtail was employed. The high-impedance front end was followed by an emitter-follower amplifying stage and by a wide-bandwidth postamplifier. The 3 dB roll-off of the receiver was 700 MHz and its sensitivity (10^{-9}) at 1 Gbit/s was measured to be −33·7 dBm with a high-speed laser source that did not introduce intersymbol interference (ISI). Bit-error-rate (BER) curves for the transmission experiment are shown in Fig. 1. The receiver sensitivity with the LED and no fibre (curve *a*) was measured to be −27·9 dBm, showing a 5·7 dB ISI and extinction ratio penalty. Curves *b* and *c* show the bit error rates after transmission over 7·4 km SM fibre with and without the 1250 nm highpass filter, respectively. The combined power penalty due to intersymbol interference and the extinction ratio penalty was 5·7 dB. All of these results were verified by computer simulation.[5] The receiver baseband filtering was simulated by an equivalent Butterworth filter and the total, measured, receiver thermal noise was used to normalise the measured BER. The simulation showed that 0·7 dB was due to the LED-induced intersymbol interference, and the other 5·0 dB to the extinction ratio penalty, corresponding to an extinction ratio of 2 : 1. The dispersion penalties with and without the filter are 2·9 and 4·4 dB, respectively. The dispersion penalties were simulated within a few tenths of a decibel of the measured

Fig. 2 *Eye diagrams for transmission at 1 Gbit/s using EE-LED and SM fibre*

a Transmitted eye diagram
b Received eye diagram after 7·4 km fibre and highpass filter

results. The 1·4 dB filter insertion loss is offset almost exactly by the 1·5 dB reduction in dispersion penalty due to spectral narrowing. This confirms previous simulation studies, indicating that spectral narrowing with a 1250 nm-cutoff highpass optical filter is optimum in balancing improved dispersion penalty against increased insertion loss for LEDs in the range 1270–1310 nm.[4] The transmitted and received eye diagrams with the highpass filter are shown in Figs. 2*a* and *b*, respectively.

Conclusion: 1 Gbit/s transmission has been demonstrated over 7·4 km of 1300 nm optimised single-mode fibre, using a commercially available edge-emitting LED and PINFET receiver. The experimental results were completely and accurately analysed by computer simulation. The LED required speed-up equalisation and the results showed that higher-speed LEDs could improve the system performance by up to 6 dB.

Acknowledgments: We would like to thank P. Kaiser, N. K. Cheung and M. B. Romeiser for their encouragement and support, and H. P. Leblanc for making the high-speed LED available. We would also like to thank L. Curtis and R. E. Spicer for providing the optical attenuator and precision single-mode connectors.

M. STERN
D. J. MILLICKER
A. E. ELREFAIE
K. RUNGE

Bell Communications Research
331 Newman Springs Road
Red Bank, NJ 07701, USA

20th February 1987

References

1 UJIE, T., SHIHADA, M., FUJITA, S., HAYASLUI, J., and ISODA, Y.: '565 Mbit/s-5 km and 140 Mbit/s-25 km single-mode fiber transmission using 1·3 μm mesa-structure surface-emitting LED'. ECOC/IOOC, Venice, Italy, Oct. 1985

2 GIMLETT, J. L., STERN, M., VODHANEL, R. S., CHEUNG, N. K., CHANG, G. K., LEBLANC, H. P., SHUMATE, P. W., and SUZUKI, A.: 'Transmission experiments at 560 Mbit/s and 140 Mbit/s using single-mode fibre and 1300 nm LEDs', *Electron. Lett.*, 1985, **21**, pp. 1198–1200

3 STERN, M., and GIMLETT, J. L.: 'Advances in LED/single-mode technology', *Proc. SPIE*, 1985, **715**

4 STERN, M., GIMLETT, J. L., and CHEUNG, N. K.: 'Dispersion-limited transmission and spectral filtering in LED/single-mode fiber optics communications systems', *Opt. Lett.*, 1985, **11**, pp. 584–586

5 ELREFAIE, A. E., and ROMEISER, M. B.: 'Computer simulation of single-mode fiber systems'. OFC '86, Atlanta, Jan. 1986

EFFECTS OF DYNAMIC SPECTRAL BEHAVIOUR AND MODE-PARTITIONING OF 1550 nm DISTRIBUTED FEEDBACK LASERS ON Gbit/s TRANSMISSION SYSTEMS

Indexing terms: Optical communications, Lasers and laser applications, Optical fibres

The performance of Gbit/s systems using 1550 nm distributed feedback lasers and conventional single-mode fibres depends critically on the dynamic behaviour of competing adjacent modes and the stopband characteristics, in addition to the usual criterion of the time-averaged side-mode suppression ratio.

Significant advances have been made in the application of directly modulated, 1550 nm distributed feedback (DFB) lasers in long-span, multi-Gbit/s transmission experiments using conventional single-mode fibre with minimum dispersion at 1300 nm.[1,2] These experiments in general employ state-of-the-art DFB lasers operating at carefully adjusted bias conditions with substantial extinction ratio penalty for bit rates in excess of 2 Gbit/s. The performance of commonly available DFB lasers, however, could be dispersion-limited at much shorter distances: bit-error-rate floors were recently reported for distances of only 4 km at 2·4 Gbit/s[3] and 30 km at 1·8 Gbit/s,[4] the latter having a high side-mode suppression ratio (SMSR) of 35 dB. We discovered that in addition to the usual criterion of time-averaged SMSR, the system performance of 1550 nm DFB lasers depends critically on the overall dynamic spectral behaviour. This includes the stopband characteristics of the diffraction grating, which are a measure of the coupling coefficient (κL), and the suppression of the residual DFB and Fabry–Perot (FP) modes, which give rise to mode-partition noise. We present measurement results of two DFB lasers having similar structures but different spectral characteristics. These differences led to drastically different system performance at 2·4 Gbit/s. Our results show that careful screening of DFB lasers is essential in the design of practical multi-Gbit/s systems. Their applications could otherwise be restricted to short fibre spans such as loop-feeders in broadband ISDNs, or the use of 1550 nm dispersion-shifted fibres would be required.

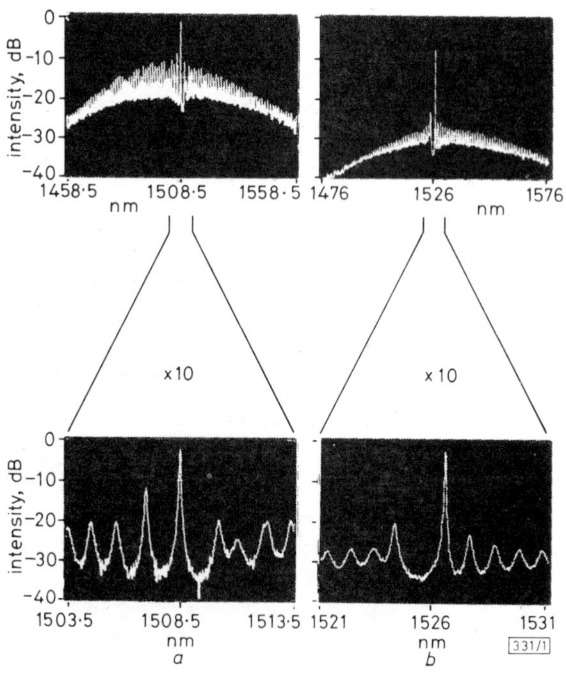

Fig. 1 *CW spectra of DFB lasers (a) A and (b) B biased at threshold*

The two lasers (A and B) used in the experiment are buried-heterostructure DFB lasers[5] having first-order corrugations with $I_{th} = 21$ and 13 mA and operating wavelengths of 1509 and 1527 nm, respectively. The front facets are AR-coated and the rear facets are as-cleaved. To isolate the effects of reflections,[6] a 30 dB optical isolator was used in all measurements. The CW spectra at threshold (Fig. 1) show that the side modes of laser A are at least 5 dB more prominent than those of laser B. The stopband of laser A is not clearly identified, possibly due to a low κL of below 1. The stopband of laser B is clearly visible, and a value of $\kappa L = 2$ was obtained by fitting the spectrum to the DFB laser theory.[7] Fig. 2 shows the dynamic spectra of two lasers with off-state bias current $I_{b0} = I_{th}$, and modulation current of 20 mA peak-to-peak at 2·4 Gbit/s NRZ format ($2^{15} - 1$ word pattern). Both the main mode and the dominant side mode of laser A exhibit considerable chirping, with a time-averaged SMSR of 30 dB. Laser B has a narrower main mode and insignificant side modes, resulting in an SMSR of 35 dB. Note that both lasers have a relatively high SMSR of above 30 dB, despite the significant differences of their spectral characteristics as shown in Fig. 1.

The system performance of the lasers at 2·4 Gbit/s is shown in Fig. 3. The bit-error-rate (BER) curves for laser A indicate a

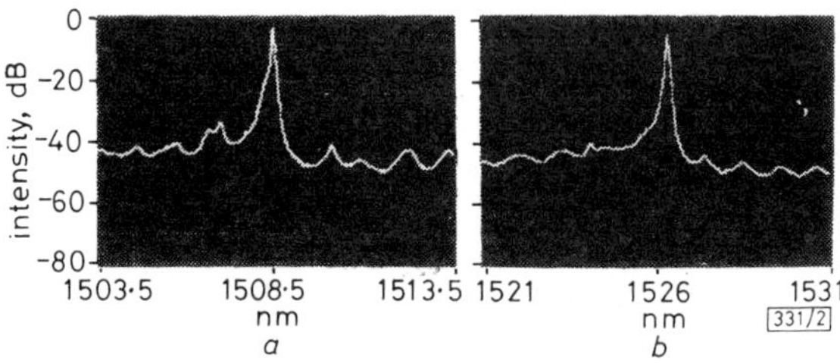

Fig. 2 *Dynamic spectra at 2·4 Gbit/s NRZ format ($2^{15} - 1$ pseudo-random word pattern)*

$I_{b0} = I_{th}$, modulation current = 20 $mA_{p\text{-}p}$
a Laser A *b* Laser B

floor of 10^{-6} with only 7·4 km fibre when $I_{b0} = I_{th}$. The floor is lowered to less than 10^{-9} when I_{b0} is increased to $1{\cdot}2 I_{th}$. These BER floors suggest the presence of a source-dependent noise, possibly due to the mode-partitioning between the main mode and the larger side modes. No BER floor was observed for laser B with 29·3 km of fibre, even though $I_{b0} = I_{th}$. Since the bit-error-rate curves for laser B are parallel, the power penalty of 0·85 dB is primarily due to laser chirp and chromatic dispersion.

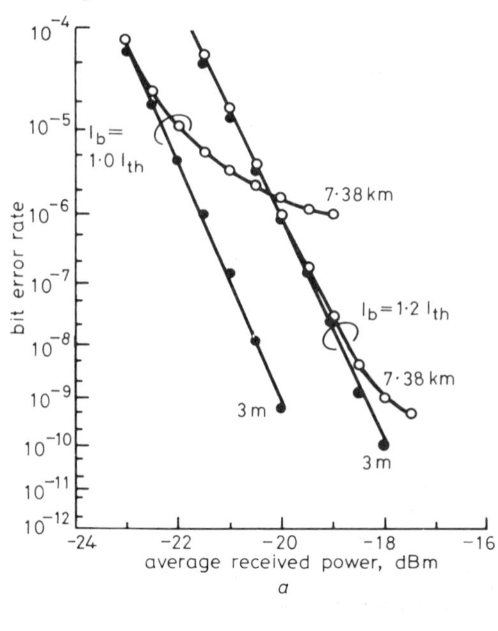

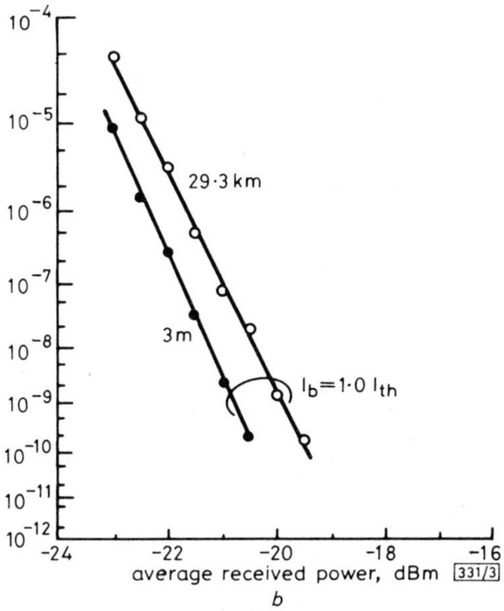

Fig. 3 *Bit-error-rate curves at 2·4 Gbit/s*

a Laser A *b* Laser B

The presence of significant mode-partitioning in laser A was confirmed by measuring its time-resolved photon statistics[8] of both the main mode (1508·5 nm) and the dominant side mode (1507·0 nm). Fig. 4*a* shows the measurement results at 300 ps after turn-on, with the off-state biased at $I_{b0} = I_{th}$. The results indicated that the intensity of the dominant side mode exceeded the main mode intensity with a probability of 10^{-6}. This, together with the 1·5 nm mode spacing of the two modes, and the chromatic dispersion of 15 ps/nm km, predicts remarkably well the observed BER floor of 10^{-6} after 7·4 km of fibre. However, the measured photon statistics of the dominant side mode do not follow a simple exponential function, as observed in a similar measurement at 500 Mbit/s.[9] It was also observed that when I_{b0} was increased to $1{\cdot}2I_{th}$, the above probability decreased to a value below 10^{-8} (limit of our apparatus), consistent with the observed BER floor of below 10^{-9} after 7·4 km in Fig. 3*a*. Similar photon statistics measurements for laser B are shown in Fig. 4*b*. The apparent absence of mode-partitioning explains the fact that no BER floor was observed for laser B as shown in Fig. 3*b*.

In conclusion, the performance of multi-Gbit/s systems could be severely limited by mode-partitioning between the main mode and the side modes for certain DFB lasers. It is inadequate to characterise the performance of DFB lasers by their time-averaged SMSR alone. Additional criteria such as stopband characteristics and the dynamics of the residual DFB and FP modes, as well as a realistic set of biasing conditions, must be specified to guarantee satisfactory performance of practical Gbit/s systems.

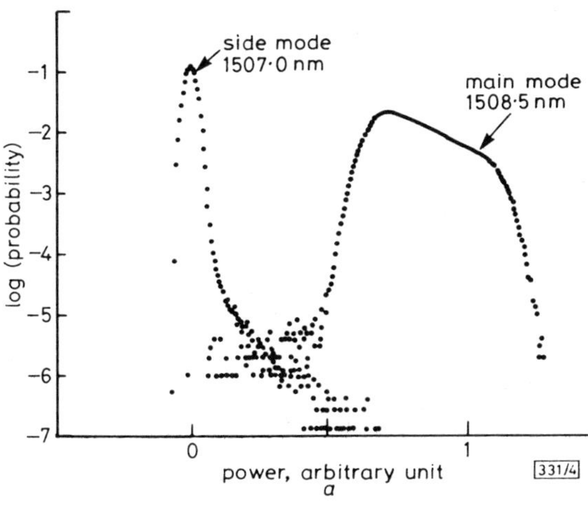

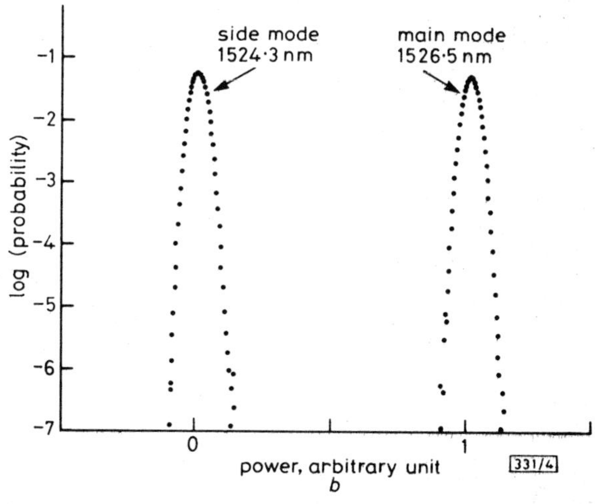

Fig. 4 *Time-resolved photon statistics at 300 ps after turn-on for both main mode and dominant side mode*

a Laser A *b* Laser B

Acknowledgment: We are indebted to Dr. P. Kaiser of Bell Communications Research, and Dr. M. Maeda and Dr. M. Nakamura of the Central Research Laboratory, Hitachi Ltd., for their encouragement. We also wish to thank Dr. Y. Koga of Takasaki Works, Hitachi Ltd, for the DFB lasers, and L. Curtis and R. Spicer of Bellcore for connectorising the fibre components.

S. SASAKI

Central Research Laboratory
Hitachi Ltd.
Kokubunji, Tokyo 185, Japan

M. M. CHOY
N. K. CHEUNG

Bell Communications Research
Red Bank, NJ 07701, USA

2nd November 1987

References

1 GNAUCK, A. H., *et al.*: '4 Gbit/s transmission over 103 km of optical fiber using a novel electronic multiplexer/demultiplexer', *J. Lightwave Technol.*, 1985, **LT-3**, pp. 1032–1035

2 MINEMURA, K., *et al.*: 'Optical devices and systems aspects in Gb/s range optical fiber transmission research'. Proc. ICC '87, Seattle, 1987, pp. 166–171

3 YAMAMOTO, S., WAKABAYASHI, H., and IWAMOTO, Y.: 'Studies of 1·2 Gb/s/2·4 Gb/s optical submarine cable systems'. Japan IEICE national convention, Tokyo, 1987, Paper 2455, p. 10-333

4 HAGIMOTO, K., KONDOU, N., OHKAWA, N., and OGISO, H.: 'Design and performance of 1·5 micron F-1·6G repeater'. Ibid., Tokyo, 1987 Paper 2456, p 10-334

5 TSUJI, S., *et al.*: 'Low threshold operation of 1·5 micron DFB laser diodes', *J. Lightwave Technol.*, 1987, **LT-5**, pp. 822–826

6 SASAKI, S., *et al.*: 'Bit-error-rate characteristics with optical feedback in a 1·5 μm DFB semiconductor laser'. Proc. ECOC '86, Barcelona, 1986, pp. 483–486

7 YARIV, A.: 'Quantum electronics' (J. Wiley & Sons, 1975), Chap. 19

8 CHOY, M. M., *et al.*: 'Measurement of dynamic photon fluctuations in a directly modulated 1·5 μm InGaAsP distributed feedback laser', *Appl. Phys. Lett.*, 1985, **47**, pp. 448–450

9 HENMI, N., *et al.*: 'The influence of directly modulated DFB LD sub mode oscillation on long span transmission systems'. ECOC '87, Helsinki, 1987

Chapter 9
Network Technology

The explosive growth of the fiber optics industry between 1982 and 1987 was fueled primarily by the long–haul market, as many nations both increased the capacity and improved the quality of their intercity communications. Some industry observers now feel that the next major growth opportunity is in the area of fiber optics networks. A network is a multiterminal system, as distinguished from the point-to-point transmission systems which have dominated the fiber optics markets until recently. Two categories of network application can be distinguished: The *subscriber loop* (SL) and the *local area network* (LAN). The term subscriber loop is used to denote a bidirectional network with switching capability, of which our present telephone system is by far the most important example. Present technology has the capability to construct a video-bandwidth version of our present telephone system by using fiber optics, but the component cost is still much too high for such wideband SLs to be economically feasible. In the near future, we can expect more commercial interest in LAN systems, which link many users with varying data transmission needs within a building or office complex.

Network technology is viewed from different perspectives in the first four papers of this chapter: **"High-speed and Broad-Band Communication Systems in Japan,"** by **S. Harashima and H. Kimura; "The Optical Fibre Overlay Network of the Deutsche Bundespost—An Approach from BIGFON Towards Broadband ISDN,"** by **K.E. Kneisel, W. Gerfen, and K. Hofig; "Lightguide Applications in the Loop,"** by **T.D. Nantz and W.J. Shenk**; and **"Optical LAN Activities in Europe,"** by **D. Rosenberger and H.H. Witte.** The important issue of network standardization is treated in the final paper, **"Standardized Fiber Optic Transmission Systems—Synchronous Optical Network View,"** by **R.J. Boehm, Y.C. Ching, C.G. Griffith, and F.A. Saal.**

High-speed and Broad-Band Communication Systems in Japan

SUSUMU HARASHIMA AND HIDETOSHI KIMURA, MEMBER, IEEE

Abstract —This paper presents the current state of each high-speed and broad-band system and the development of the high-speed and broad-band network (HBN) in Japan. System configuration of a video conference system, a video transmission system, a high-speed digital circuit system, and so on, are introduced. Fundamental technologies to realize HBN, 1) switching technologies to handle broad bit-rate class transmission, 2) high-capacity transmission and high-speed digital terminal technology, 3) economical fiber optic subscriber loop technology, and 4) high-performance video codec technology, are discussed. Based on the above discussion, configuration of HBN is proposed.

I. Introduction [1]

Visual communication services such as a video conference and a video program transmission are gradually utilized to improve business productivity. High-speed facsimile and data transfer services requiring a data transmission rate of more than several hundred kbits/s are also drawing attention.

The following developments during the last decade have renewed high-speed and broad-band communication systems.

- Large-capacity digital transmission systems are rapidly being introduced.
- Fiber optic subscriber loops have been introduced.
- The use of video codecs to reduce the video signal transmission bit rate has become a reality.

Up to now, each of the high-speed and broad-band communication systems has tended to be designed separately for some particular communication service. If communication facilities are shared by various communication systems, system cost will be reduced and more economical communication services can be provided. Furthermore, customer equipment can be used for multiple purposes.

These facts imply the necessity of a multipurpose high-speed and broad-band communication network (HBN).

This paper presents the current state of each high-speed and broad-band system and the development of the high-speed and broad-band network in Japan. The fundamental technologies to be utilized in the HBN are discussed. Next, the configuration and the expansion progress of the HBN are described. Finally, an experimental system is introduced.

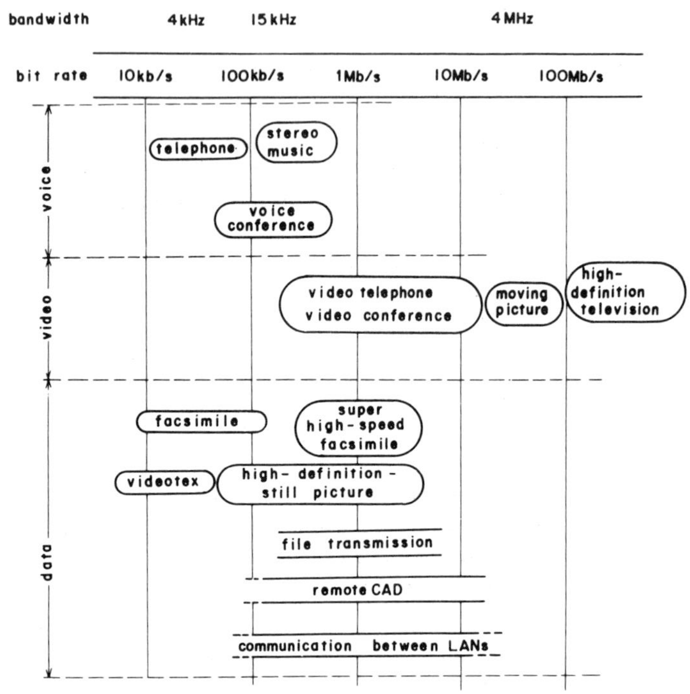

Fig. 1. Relation between telecommunication services and bit rate.

II. High-Speed and Broad-Band Communication Services

Current commercial or developing telecommunication services are shown in Fig. 1.

This paper defines the telecommunication services of more than several hundred kbits/s as "high-speed and broad-band communication services." They are classified into three services considering the kind of information:

1) video communication service,
2) still-picture communication service, and
3) high-speed data communication service.

Some examples of these services are shown in Table I.

The video communication service includes the end-to-end communication such as a video telephone system and a video conference system, and the center-to-end communication such as video response system (VRS) and both-way

TABLE I
KINDS OF HIGH-SPEED AND BROAD-BAND COMMUNICATION SERVICES

kinds of services		outline	bit rate (band width)
video communication service	video telephone	conversation by video and voice (one to one)	1.5-100Mb/s (4MHz)
	video conference	conference by video and voice (many to many)	
	television lecture	lecture by video and voice (one to many)	
	video signal transmission	remote supervising and rebroadcasting, etc by video and voice	32-100Mb/s (4MHz)
	video response system (VRS)	presentation of moving and still picture by request	
	both way CATV	distribution and presentation of moving picture signals	
	high-definition video signal transmission	high-definition video and voice signals transmission (high-definition television)	100-400Mb/s (20-30MHz)
still picture communication service	super high-speed facsimile	high-speed transmission of high-quality multi-level tone manuscript	several 100kb/s -32Mb/s
	high-definition still picture	high-speed transmission of high-definition color still picture	
high-speed data communication service	super high-speed data transmission	high-speed transmission of large amount of coded information as in a picture data file multi-channel multiplexed transmission between PBXs or LANs, etc.	several 100kb/s -32Mb/s

CATV. These services generally use a transmission bandwidth of 4–6 MHz as specified by modulations such as NTSC, PAL, SECAM, etc. However, the high-definition television system needs a bandwidth of 20–30 MHz.

Still-picture communication services include super high-speed facsimile and high-definition still-picture transmission.

High-speed data communication services transmit a large amount of data such as a file transmission and connecting PBX's or LAN's.

All of these services require transmission rates from several hundred kbits/s to 32 Mbits/s.

III. CURRENT STATE OF EACH SYSTEM

At present in Japan, high-speed and broad-band communication services are provided by each individual system. The outline of these systems is described below.

A. Video Conference System [2]

NTT initiated a new video conference service in 1983. This system consists of the following five components.

1) Video Conference Equipment (VCE): VCE is man–machine interface equipment consisting of cameras, displays, microphones, speakers, etc.

2) Video Conference Service Unit (VSU): VSU terminates local transmission lines and controls video signals according to the display mode.

3) Subscriber Transmission Lines: Subscriber transmission lines carry analog 4 MHz video and 7 kHz audio signals with the frequency-division multiplex.

In addition to the existing metallic pair-cable system (ITV-4M system), optical fiber cable systems (FV-4M-P system, FV-4M-A system), and a 26 GHz band radio system (26SS-A1 system) are now available.

4) Trunk Transmission Lines: Digitized video and audio signals are multiplexed by time division and transmitted at a bit stream rate of 6.3 Mbits/s. Interframe coding techniques are used for video digitization.

5) Network Control System Based on Reservations: This system automatically connects and disconnects circuits according to stored reservation data.

The video conference system configuration is shown Fig. 2.

B. Video Transmission Systems [3]

Video transmission services for ITV (industrial television) have been available in Japan since 1970. Such services use the ITV-4M system which utilizes twisted pair cable. NTT has developed two kinds of video transmission systems using optical fiber cable in 1982.

1) FV-4M-P System: In this system, the analog video signal is converted to a PCM signal, and is transmitted by an optical fiber system with laser diode.

The FV-4M-P system is employed in a medium-haul circuit such as a road observation because of its long repeater spacing and good signal-to-noise ratio characteristics. It is also employed in a medium-haul subscriber line.

2) FV-4M-A System: In this system, an LED optical source is directly modulated by the analog video signal.

The FV-4M-A system is simple and more economical, but its signal-to-noise ratio characteristics rapidly decrease in proportion to optical loss and the number of repeaters used. For this reason, it is used for a short-haul circuit.

A relation of the transmission costs between the ITV-4M, FV-4M-P, and FV-4M-A systems is shown in Fig. 3.

C. High-Speed Digital Circuit System [4], [5]

Recently, demands for high-speed digital circuits have arisen. In response to these demands, NTT implemented the high-speed digital circuit system in 1984. This system provides six speed classes: 64, 192, 384, 768, 1536, and 6144 kbits/s.

The system configuration is shown in Fig. 4.

The subscriber line sections consist of Digital Service Units (DSU) installed on the user's premises, and subscriber line terminating equipment (SLT) installed in a central telephone office or a repeater station. Three types of transmission systems using metallic cable, optical fiber cable, and digital subscriber radio have been developed for subscriber lines.

The metallic cable transmission system (ping-pong transmission system) is applied to 64 kbits/s and 192 kbits/s digital services, and the optical fiber cable transmission systems are used for digital services from 192 to 6144 kbits/s and digital subscriber radio systems are used for digital services from 64 to 6144 kbits/s. In actual application, the most suitable transmission system is

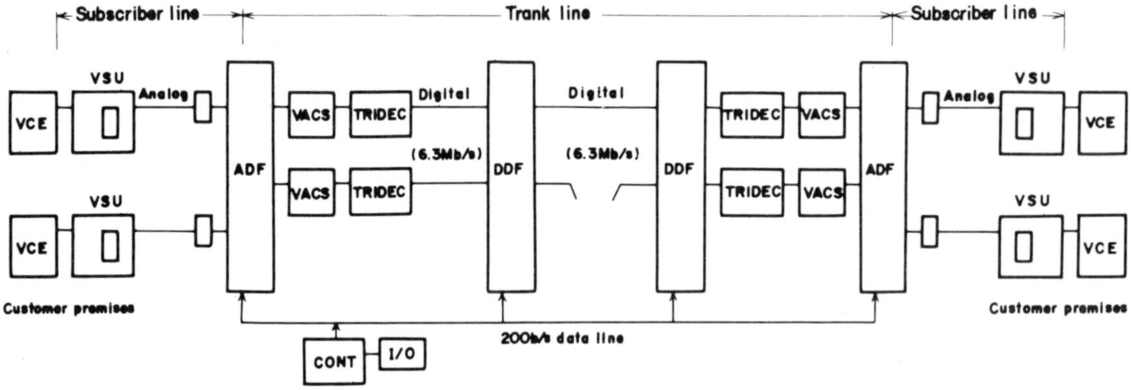

VCE : Video Conference Equipment
VSU : Video Conference Service Unit
VACS : Video and Audio signal Combiner/Separator
TRIDEC : Tri-parameter Interframe Coder/Decoder
CONT : Control Equipment
ADF : Automatic Analog Signal Distributing Frame } Network control system
DDF : Automatic Digital Signal Distributing Frame } based on reservation

Fig. 2. Configuration of video conference system.

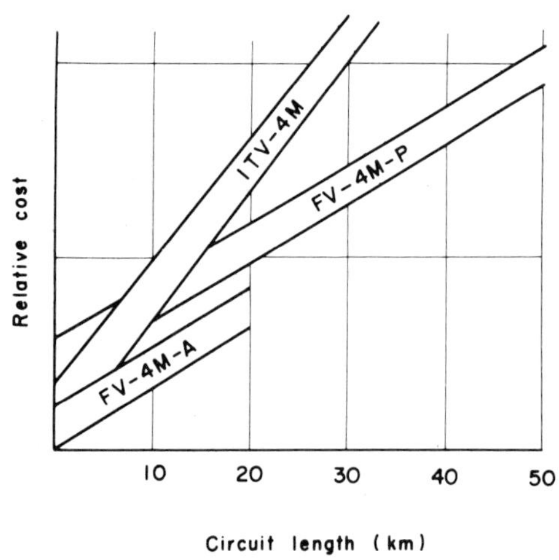

Fig. 3. Transmission costs.

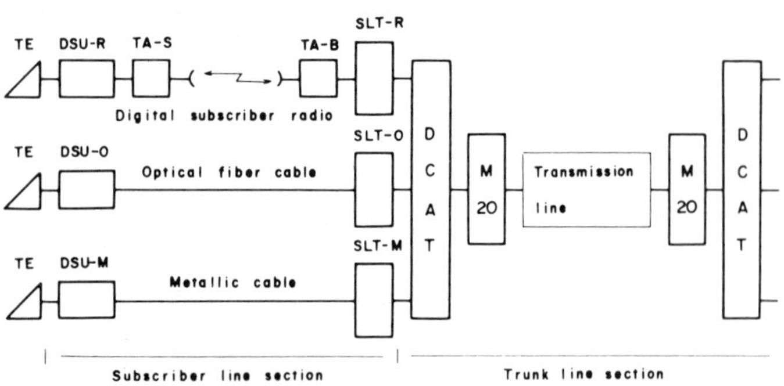

TE : Terminal Equipment
DSU-O : Digital Service Unit (Optical)
DSU-R : Digital Service Unit (Radio)
DSU-M : Digital Service Unit (Metallic)
SLT-O : Subscriber Line Terminating equipment (Optical)
SLT-R : Subscriber Line Terminating equipment (Radio)
SLT-M : Subscriber Line Terminating equipment (Metallic)
TA-B : Transmitter-receiver (Base station)
TA-S : Transmitter-receiver (Subscriber)
DCAT : Digital Circuit Access and Testing equipment
M20 : Secondary stage synchronous multiplexer

Fig. 4. Configuration of a high-speed digital circuit system.

TABLE II
CHANNEL CONTENTS

Channel	Contents
Channel H	· Information transfer channel at 64, 192, 384, 768, 1,536 and 6,144 kb/s bit rates
Channel D	· Not in use at present
Channel S	· Channel indicating circuit failure, unassigned circuit and circuit in-use or out-of-use status · The channel employs the following four signals ① DNR : Failure in network to terminal (Digital Terminal Equipment Not Ready) ② UNR : Unassigned satellite circuit (Uncontrolled Not Ready) ③ S : Circuit status indication (Status) ④ SEND : Alarm indication to terminal
F bit	· Bit for frame synchronization and CRV (Coding Rule Violation) in this system

selected according to the transmission condition of the subscriber line section.

The structure of the user–network interface consists of four components, the *H* channel (information channel), *D* channel (signal channel), *S* channel (service channel) and *F* bit (flame alignment signal). Specifications for the components at the user–network interface are shown in Table II.

The trunk line sections consist of digital circuit access and testing equipment (DCAT), and digital transmission systems. Secondary-stage synchronous multiplexers (M20) accommodate high-speed digital circuits, public switch telephone circuits, etc., to efficiently use the transmission lines.

D. Satellite Digital Communication System [6], [7]

NTT developed a satellite digital communication system (SDCS) at the end of 1984. This system communicates:

1) point to point,
2) point to multipoint, and
3) half duplex.

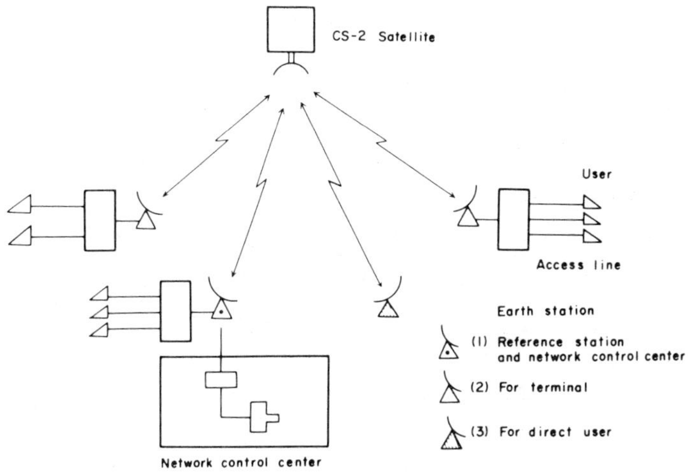

Fig. 5. System configuration of SDCS.

This system consists of the CS-2 satellite and some small 30/20 GHz Earth stations. The configuration of SDCS is shown in Fig. 5.

1) Earth Station: Three categories of Earth stations are used: the reference station and network control centers, Earth stations for terminals, and for direct users. The Earth stations will be initially located in Tokyo, Osaka, Tsukuba, Nagoya, Sapporo, and Fukuoka. NTT will make new Earth stations according to demands.

2) Access Line: The access line shares a transmission line with the high-speed digital circuit service to enhance compatibility. A digital service unit (DSU) is set on the user's premises. The access line bit rates are the same as that of the high-speed digital circuit service.

Service can easily be expanded because user–network interfaces are compatible with high-speed digital circuit services. NTT intends to begin demand-assignment TDMA service which is a more useful application for the near future.

IV. Main Technologies for Realizing HBN

In order to realize an HBN that can economically provide various services, the following technologies should be developed:

1) switching technologies to handle broad bit-rate class digital signals,

2) high-capacity transmission technology and high-speed digital terminal technology,

3) economical fiber optics subscriber loop technology, and

4) high-performance video codec technology.

A. Switching Technologies [8]

A high-speed and broad-band exchange is introduced to provide communication links between any terminals and to utilize transmission lines efficiently by concentrating communication traffic.

New types of connections such as reservation-based, unidirectional, and asymmetric connections, in addition to traditional no-delay and bidirectional connections, will be introduced to realize efficient transmission circuit operation. Therefore, mutually independent control mechanisms between upstream and downstream paths class are useful.

High-speed exchange ought to handle broad bit-rate class digital signals. A multiple bit-rate switching mechanism is suitable to increase efficiency of a circuit group partition on all bearer rates. For the switching of broadband signals such as full-motion video signals, space-division techniques are necessary for a provisional solution. However, in the development phase for five or ten years to come, high-speed time-division switching techniques are preferable in keeping with the trend toward digitalizing trunk and subscriber transmission systems. To realize a time-division switch which can handle information signals at bit rates of over 30 Mbits/s, very high speed time-division switching technologies with a 1 GHz operating frequency must be established. The key technologies are the applications of ultrahigh-speed LSI devices such as Si–bipolar and GaAs devices and assembling and packaging for the time switch.

Subscriber signaling systems should be unified in an integrated communication network. To establish international Recommendations, it is desirable to adopt the common channel-type subscriber signaling system studied by the CCITT.

Demands are arising for packet-switching-oriented communication with high-speed and low-density traffic in application areas such as inter-LAN connections, center-access-type CAD's newspaper editing, etc. In addition to present techniques based on a 48 kbit/s circuit, the following high-speed packet-switching techniques are necessary to establish how to use high-speed circuits with packet multiplexing techniques:

1) high-speed communication control LSI's with operating speeds of over several 10 Mbits/s,

2) multimedia protocol for efficient transmission, mixing various types of information such as video, audio, data, etc, and

3) parallel-processing architecture for increasing packet-switching throughput.

B. Trunk Line Transmission Technology [9], [10]

In an HBN, due to bandwidth considerations, the trunk line link cost is a governing factor in communication cost. The cost of the present transmission system, such as the F-400M, should be reduced. Moreover, it is necessary to develop advanced high-capacity optical transmission systems. Therefore, it is important to develop the following technologies:

1) a high-power coherent optical source (LD),

2) a high-sensitivity optical detector,

3) a low-loss optical fiber and connector,

4) coherent modulation,

5) wavelength multiplexing with narrow channel spacing and low loss.

By adopting the above-mentioned technologies, a high-capacity transmission system with a bit rate of 1–2 Gbits/s, which can transmit over 100 km without repeating, should be realized in ten years.

A 16-QAM long-haul digital radio system was developed for a microwave transmission system and is now in commercial use. This system has a capacity of 200 Mbits/s on 4/5 GHz bands. A 256 QAM 400 Mbit/s system is now being investigated to improve spectrum utilization efficiency and increase transmission capacity. The key technologies to develop the system are:

1) high-level modulation–demodulation (256 QAM modem),

2) a multipath fading correction technique,

3) a low sidelobe antenna with high cross-polarization discrimination. A high-speed digital terminal should be realized to efficiently accommodate a broad-band circuit of various bit rates to the trunk line. This terminal has the following functions:

1) multispeed circuit multiplexing,

2) asymmetric cross-connect,

3) intelligent multiplexing.

The following technologies are being investigated to realize such a high-speed digital terminal.

1) Synchronization Technology: A video coding method for color TV signals in the trunk line has a bit rate of 32 Mbits/s which this is the third bit rate of the digital hierarchy. Therefore, it is important to develop network synchronization technology for bit rates of more than 32 Mbits/s. The key technologies are the phase stabilization of the clock source and the wide-phase-lock-loop characteristic of the terminal.

2) High-Speed IC's: Logic IC's, elastic memories, and time-slot interchanges, which are components of a high-speed terminal, can operate at a speed of more than 8 Mbits/s. Therefore, it is important to develop high-speed components.

C. Subscriber Line Transmission Technologies [11]

Two systems have been developed for application to high-speed and broad-band subscriber line transmission systems:

1) an optical fiber cable-based high-speed digital transmission system,

2) a subscriber radio-based high-speed digital and analog video transmission system.

The subscriber optical fiber cable transmission system will be applied to most high-speed and broad-band services. On the other hand, the subscriber radio transmission system has an advantage, as it can rapidly respond to new or special requirements.

In the optical fiber system, it is desirable to transmit both-way video signals, high-speed digital signals, and control signals in an optical fiber. Therefore, it is important to apply digital-transmission, wavelength-multiplexing, and time-multiplexing technologies.

The subscriber radio transmission system is classified into: 1) an analog system suitable for video signal transmission, and 2) a digital system for high-speed data transmission.

In the analog system the base station and subscribers communicate in a point-to-point mode. In the digital system, the base station and subscribers communicate in point-to-multipoint mode in a time-division multiple-access (TDMA) format.

To provide easier access to the subscriber radio system and effectively use radio frequencies, steps should be taken to:

1) upgrade the current preassign time-division multiple-access system (PA-TDMA) into a demand-assign TDMA system (DA-TDMA). This is done by a preassignment of each subscriber as available, and an assignment of bursts at the request of subscribers;

2) reduce the cost of transmitters and receivers by incorporating monolithic microwave integrated circuits (MMIC) and installing TDMA equipment with LSI circuits.

D. Video Signal Coding [12]

The video signal coding is classified into three major categories.

1) Linear PCM Coding: Analog-to-digital converter,

2) Intraframe Coding: Redundancy reduction utilizing special correlation in a single video frame,

3) Interframe Coding: Redundancy reduction utilizing temporal correlation between successive video frames.

In conventional visual communication services, such as the industrial television (ITV) service, the optimum codec for each application can be applied. However, in the HBN, it is necessary to consider the following conditions for video signal transmission.

1) Standardization of Coding Algorithm: In order to insure an interconnection between any two terminals, it is necessary for all the codecs arranged in the network to work based on the same or compatible coding algorithms.

2) Tandem Connection Between Different Types of Codecs: It is likely that linear PCM, intraframe or interframe codecs will be selectively arranged according to the transmission length, and that these codes are tandem-connected via exchange switches. In this case, digital interconnection between different types of codecs is preferable to analog transmission to minimize picture-quality degradation. To satisfy this requirement, it is necessary to unify interface conditions and coding parameters, for all types of codecs. The sampling frequency is example of a coding parameter.

3) Economization: In the future, it is likely that the HBN will be entirely digitalized. Video codecs would then be installed at each customer side. In this case, it is strongly required to reduce the codec's cost by introducing LSI and/or mass production.

In addition to the above-mentioned subjects, it is hereafter necessary to develop coding algorithms that will attain a lower transmission bit rate, as well as ones that can be applied to ultrawide-band video signals such as high-definition television signals (HDTV).

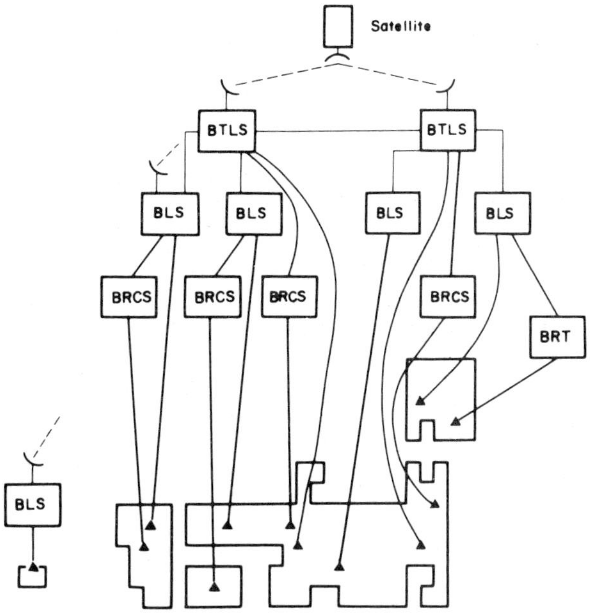

Fig. 6. Example of a high-speed and broad-band network configuration.

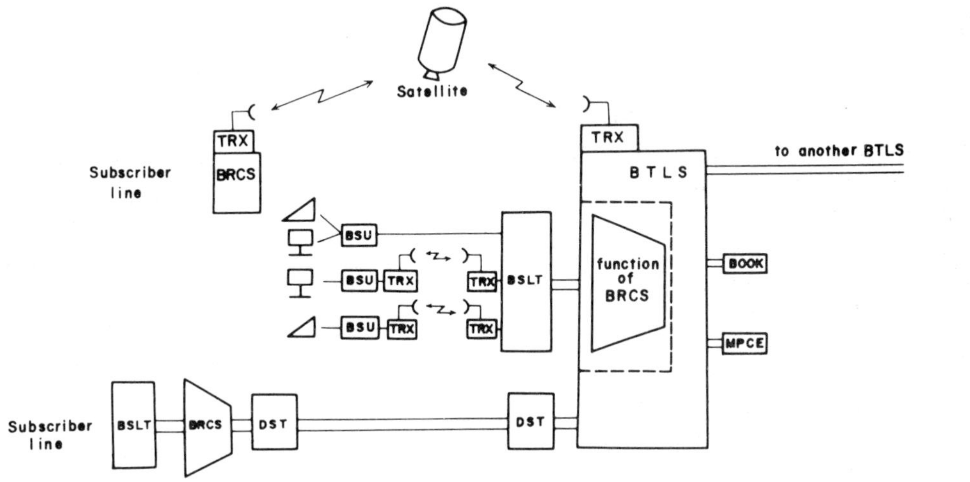

BTLS: Broadband Toll and Local Switch
BRCS: Broadband Remote Concentrate Switch
BSU : Broadband Service Unit
TRX : Transceiver
DST : Digital Synchronous Transmission Terminal
MPCE: Multi-Point Video Conferencing Connecting Control Equipment
BOOK: Call Booking and Reserving Equipment

Fig. 7. Configuration of high-speed and broad-band communication network version 1 (example).

V. HBN Configuration and Extent of HBN Expansion

NTT has a plan to construct a nationwide HBN. The HBN configuration is shown in Fig. 6. The local switching system (BLS) has the switch control function. A remote-control switching system (BRCS) and a remote terminal will also be introduced. They will offer considerable economic and operational benefits in areas where the number of subscribers requiring broad-band services is small. Moreover, it is effective to apply a satellite communication system for remote subscribers. The high-speed and broad-band network version 1 (HBN1) that will be constructed in the near future is shown in Fig. 7. The optical digital transmission system and subscriber radio transmission system will be applied to subscriber loops. The switch-

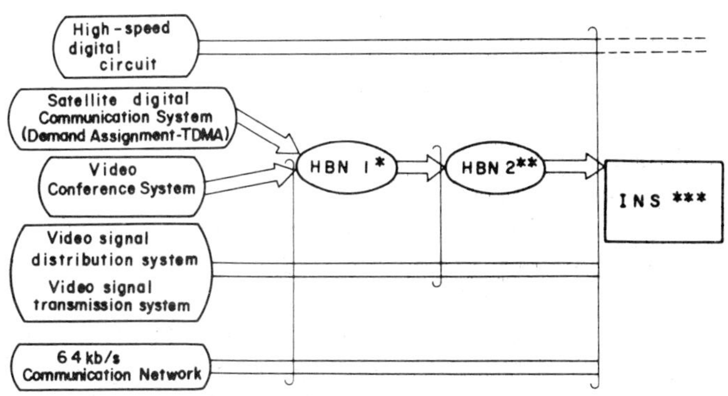

* High Speed and Broadband Communication Network Version 1
** Second Version of HBN 1
*** Information Network System

Fig. 8. Example of HBN expansion.

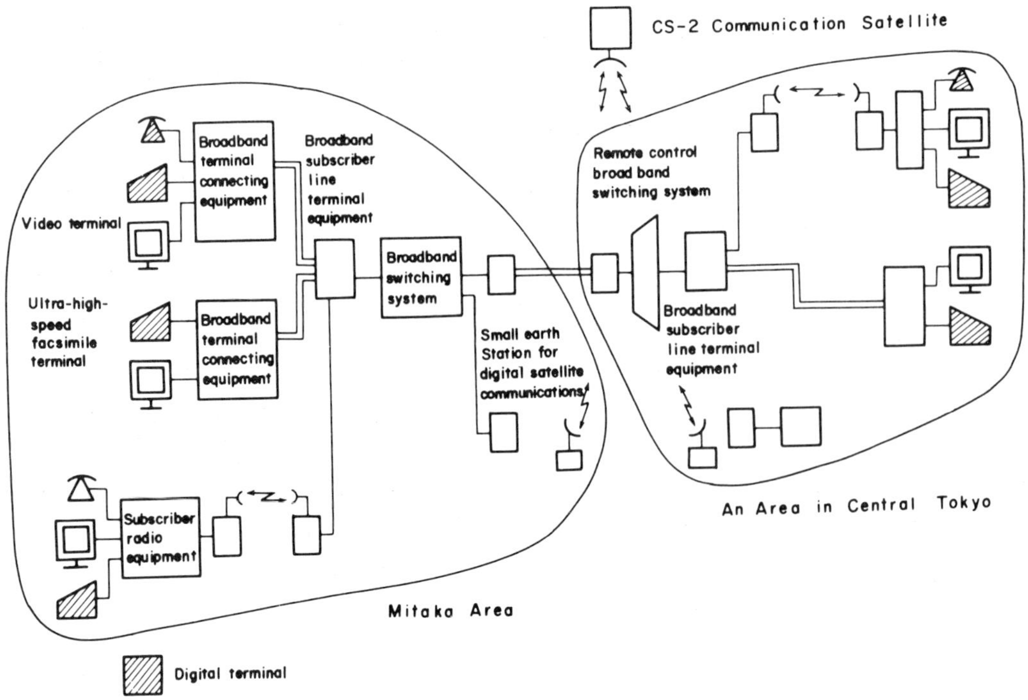

Fig. 9. Model system configuration.

ing system will be composed of BLS and BRCS. The BRCS is controlled by the host-office (BLS) controller. Intra-BRCS calls are connected within BRCS under the control of the host-office controller in order to reduce transmission cost. A reservation-based call request is transmitted to call booking equipment (BOOKS) via the telephone network. Connection control information is transmitted from the BOOKS to the central processor in the BLS. Reservation-based calls are connected in the BLS.

Anticipated expansion progress of HBN is shown in Fig. 8.

In this expansion progress, it is necessary to adequately examine the following:

1) CCITT is studying the integration of 64 kbit/s service and a higher bit-rate service.

2) Both the $H1$ 1 channel rate (1.536 Mbits/s) and 23 circuits of the B-channel rate (64 kbits/s) are the same structure of primary interface.

Network architecture for realizing an information network system (INS) is investigated based on the above-mentioned.

VI. Experimental System of the HBN [13], [14]

NTT has constructed an INS model system. An experimental broad-band switching system that offers such services as video conference, video transmission, and high-speed and high-definition facsimile, was put into service as a part of this model system. The broad-band switching system configuration of the model system was shown in Fig. 9. A broad-band toll and local switching system (BTLS) installed in the Mitaka office in Tokyo serves some 300 subscribers in the Mitaka area. It will be connected with a broad-band remote-control switching system (BRCS) in central Tokyo by an optical fiber transmission system (F-400M).

The broad-band communication service provided in the INS model system can be classified into two categories. One includes services using 4 MHz analog video channels, such as video conference and other moving picture services. The other includes services using 0.77 Mbit/s or 1.5 Mbit/s channels, such as high-speed and high-definition facsimile and color facsimile.

The transmission media between the broad-band switching system, and subscriber terminals will mainly be optical fiber cables. By applying wavelength-multiplexing technology, four different wavelengths not only for digital but also for analog information can be multiplexed onto one optical fiber. A subscriber radio and a satellite communication system is applied.

Information and signals transmitted within this system are composed of: 4 MHz analog video with high-quality voice information to and from the video terminal; 1.5 Mbit/s digital information to and from the digital terminal; 64 kbit/s control signals for end-to-end use; and 8 kbit/s signals for controling switching equipment.

VII. Conclusion [15]

This paper has presented the current state and trends of technological development and expansion of the high-speed and broad-band network. Moreover, an experimental system (Model System) for the HBN has been presented.

The following items are necessary to realize high-speed and broad-band communication services:

1) common use of communication network facilities,
2) standardization of user–network interfaces,
3) development of an efficient coding algorithm,

4) development of more inexpensive subscriber transmission.

For the time being, it is better to construct a high-speed and broad-band communication network as an individual network.

In the future, it will be important for such a network to be integrated with the 64 kbit/s network to provide economical and efficient services.

REFERENCES

[1] Y. Kitahara, "Information network system: Foundation of tomorrow's telecommunications world," *Telephony*, Nov. 22, 1982.

[2] T. Tanaka, "Video conference system for private use," *JTR*, vol. 25, no. 3, 1983.

[3] M. Yamashita, "Optical fiber video transmission systems—FV-4M-A and FV-4M-P systems," *JTR*, vol. 24, no. 1, 1983.

[4] S. Harashima, "Subscriber line transmission technologies toward information network system," *JTR*, vol. 25, no. 3, 1984.

[5] T. Yoshida, "Digital subscriber radio system in 26 GHz band," *JTR*, vol. 25, no. 3, 1984.

[6] T. Murakami, "Domestic satellite communication in NTT," *JTR*, vol. 24, no. 1, 1983.

[7] Y. Morihiro, "Satellite digital communication system for new business use," *JTR*, vol. 26, no. 4, 1984.

[8] S. Nakajima and K. Hiraide, "A broadband switching system for field trials," presented at ISS '84, Florence, Italy, May 1984.

[9] R. Kishimoto *et al.*, "The development of broadband transmission technologies in NTT," presented at the 3rd Int. Conf. Telecommun. Transmission, London, England, March 1985.

[10] S. Kato *et al.*, "Design and result of demand assignment-TDMA system," *Rev. ECL*, vol. 30, no. 10.

[11] J. Yamagata *et al.*, "A study on fiber optic multi-media access system," presented at ICC'85, Chicago, IL, June 1985.

[12] H. Kuroda *et al.*, "A 1.5 Mb/s interframe codec for video teleconferencing," to be published.

[13] K. Habara *et al.*, "The configuration of the INS model system," *ECL Tech. J.*, vol. 32, no. 9, 1983.

[14] M. Kuwabara, "Model system—Toward realizing the information network system," *JTR*, vol. 24, no. 1, 1983.

[15] NTT Int. Symp. '86, Tokyo, Japan, May 1986.

Susumu Harashima received the B.S. degree in electrical engineering from Tohoku University in 1960.

Since joining NTT in 1960, he has been engaged in development and design of analog coaxial cable system, and later became Director of the Plant Engineering Division at Kyusyu Telecommunication Bureau, and Director of the Visual Communication Division at Engineering Bureau, and then Director General of the Osaka Carrier Transmission Division. He is currently the Deputy Executive Manager of the Engineering Department, responsible for the development of line and radio transmission systems involving satellite communication and cables, including optical fibers.

Mr. Harashima a member of the Institute of Electronics and Communication Engineering of Japan.

Hidetoshi Kimura (S'62–M'66) received the B.S., M.S., and Dr. Eng. degrees in electrical engineering from Tohoku University in 1961, 1963, and 1966, respectively.

Since joining NTT in 1966, he has been engaged in developmental research on digital transmission systems and research on communication networks. He is currently Deputy Director of the Telecommunications Networks Laboratories engaged in INS (Information Network System) network planning.

Dr. Kimura a member of the Institute of Electronics and Communication Engineers of Japan.

The optical fibre overlay network of the Deutsche Bundespost-an approach from BIGFON towards broadband ISDN

K.-E. Kneisel, W. Gerfen, K. Höfig

FTZ, Projektorganisation für Diensteintegration
Am Kavalleriesand 3, D-6100 Darmstadt, Germany

Abstract

In its overall responsibility as the major network provider in Germany, the Deutsche Bundespost launches all highly sophisticated technologies in order to serve the telecommunication market with those promising services which the national economy require.
In this contribution some background information about the objectives can be found. The basic assumptions for the implementation of the future highly efficient, totally integrated universal network achievable by a step-by-step strategy are mentioned and the existing and intended activities pointed out in more detail.
As a result, the reader of the article may recognize that there exists a well defined German approach for handling the challenges of the future development in the telecommunication scenario, though not all the trends of technological development can be assessed in detail.

Introduction

The Deutsche Bundespost is of the opinion that micro-electronics and optical fibre technology are key elements allowing the implementation of a universal, interactive and fully digital Integrated Broadband Communication Network (IBCN) between subscribers, thereby fulfilling the prerequisites for the expansion of the 64-kbit/s ISDN into the "Broadband ISDN".
According to the plans of the Deutsche Bundespost, the IBCN should be able to provide not only a wide range of broadband dialoque services (moving picture dialoque services, moving picture retrieval services and high-speed data communication services), but also broadband distribution services (distribution of public and private radio and television programmes, possibly also HDTV broadcasts).
The realization of an IBCN, i.e. a network in which all forms of communication (broadband dialogue and broadband distribution services) are offered on the basis of a common infrastructure, is dependent not only on technology and the development of the demand and the decisions taken in the Federal Republic of Germany on media policies. The strategic concept of the Deutsche Bundespost /1/ for the transition from today's independent, service-related networks to a fully digital universal network is illustrated in Fig.1.

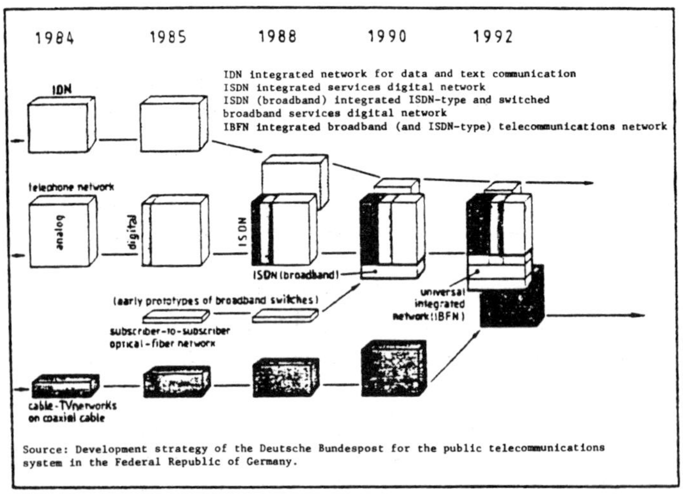

Fig.1: Development strategy for the transition from servicerelated telecommunication networks to a single universal digital network

Previous fibre optic activities

Optical fibre transmission tests

Since 1977 the Deutsche Bundespost has systematically undertaken optical fibre tests in its network in cooperation with a number of interested communication manufacturers in the Federal Republic of Germany. The positive results of these early point-to-point transmission tests at the bit rate of 34 Mbit/s were immediately adopted in a series of optical fibre cable projects involving live traffic (these projects replaced planned conventional copper cable projects).

The year 1980 marked a new era in the use of optical fibres. The first tests with optical fibre cables in the subscriber line network were carried out to assess the suitability of this new technology for the direct connection of subscribers to the exchange.

Two other transmission tests were carried out in the local network in Berlin. In the first, which took place at the end of 1982, 140 Mbit/s signals were transmitted over a minimum distance of 18 km by means of graded-index fibres. None of the 8 participating firms encountered any technical problems; some of the firms even bridged a distance of 36 km using commercial fibres. The Deutsche Bundespost has also been successful in its desire to induce German communication manufacturers to develop 140-Mbit/s transmission systems suitable for large-scale implementation as standard technology as early as possible. A 160-km optical fibre cable comprising 60 graded-index fibres and equipped with 8 regenerators was put into operation in autumn 1983.

In the second field trial, which started in Berlin at the beginning of 1985 and has not yet been completed, 565-Mbit/s transmission tests are being carried out over monomode fibres. The test specifications gave manufacturers a choice of several technical alternatives: the transmission of one 140 Mbit/s signal over at least 36 km, or of four 140-Mbit/s signals over at least 18 km using wavelength division multiplexing, or the transmission of 565-Mbit/s signals over also at least 18 km using time division multiplexing. The best results have been achieved by SIEMENS with a standard monomode fibre (0.16 dB/km (1500 nm), 3 ps/km.nm) available on the world market. The system uses wavelength division multiplexing for transmitting 2 x 565 Mbit/s at 1500 nm and 2 x 140 Mbit/s at 1300 nm over nearly 40 km with an adequate system reserve (6 dB). The system even allows 2 x 565 Mbit/s to be transmitted per transmission window. The results are so convincing and the practical problems of cable laying, jointing and plug attachment have been resolved with such success that the Deutsche Bundespost has not only decided to install the first operational 565-Mbit/s optical fibre cable for long-distance applications in the course of 1985, but is also considering the possible use of monomode fibres in the local network. From 1987 onwards, monomode systems only will be installed in the long-distance network.

BIGFON

Perhaps the internationally most well-known field trial is the BIGFON (broadband integrated optical fibre local telecommunication network) system trial.
The trial is based on 10 separate projects carried out in 7 large German cities in cooperation with 6 firms or groups of firms from the German communications industry. The BIGFON trial set into operation at the end of 1983 / beginning of 1984 will continue until the end of 1986, although its extension until the end of 1987 is currently under discussion. A total of the 320 subscribers are taking part in the system trial. For economic reasons, only 68 out of the 320 subscribers are equipped with a videophone terminal. The purpose of the trial is to enable a hardware concept for the future to be developed on the basis of technical means which are either already available today or are to be supplied within the period specified in order to establish the technical feasibility of an integrated broadband subscriber line system. The BIGFON system trial is not an acceptance test for determining the acceptance of broadband individual communication services. Since, with very few exceptions, international technical standards were not yet available at the time of planning of the project, the participating firms were given a free hand in developing individual concepts and prototype technology for achieving the set goal.

The basic specifications of the project are set out in Fig. 2.

service-specific signals		transmission		number of service-specific channels to subscriber	number of broadcast. progr. in centre	additional requirements
1		2		3	4	5
telephony (introduced services)	1	digital in ISDN channels		2 ≤ n ≤ 30		
teletex (introduced services)	2					
data services (introduced services)	3					
textcommunication services (Tx, Ttx, Btx) (introduced services)	4					
remote-control services	5					option
sound - progr. distribution	6	analogue or digital, uni-direct.	simultan. en bloc	24	24	
			distributive	≤ 4		remotely controllable backward channel (ISDN)
TV progr. distribution	7		switching	2 ≤ m ≤ 4	12	
			simultan. en bloc (WDM)	12		
picturephone	8	bidirect.		1		joint use TVchannel, if m≥2, PAL G. 5 MHz
videoconferencing	9	bidirect.				option

Fig.2: Basic requirements imposed on an integrated subscriber line system

The formulation of these requirements was based on the concept that in a future universal network the signals from all conceivable narrowband and broadband services - dialoque and distribution services - are to be transmitted over a single subscriber line consisting of preferably only one optical fibre.

Experimental videophone network

As mentioned earlier, a total of 68 subscribers connected to the 10 BIGFON systems are equipped with a videophone terminal. A videophone trial restricted to the limited number of videophone subscribers connected to a single BIGFON system (6 to 10 videophone terminals per system) would be unlikely to yield useful technical data since it is practically impossible to find a group of subscribers with a large volume of purely local traffic. It has therefore always been the Deutsche Bundespost's aim to interconnect the videophone subscribers within the BIGFON system trial by means of a nationwide experimental videophone network as shown in Fig.3.

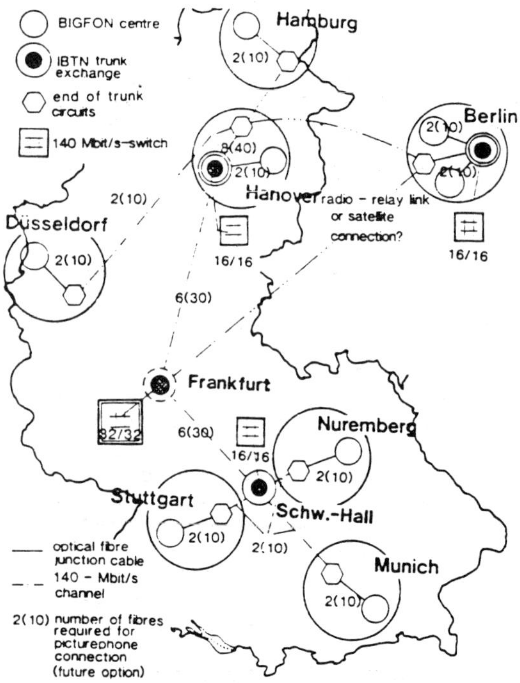

Fig.3: Configuration of the experimental videophone network

The experimental videophone network is a switched network and as such has a single-level hierarchical structure. It is digitally operated at a bit rate of 140 Mbit/s.

Since the BIGFON system trial is based on prototype technology developed by separate firms, the mode of signal processing in the individual BIGFON projects differs considerably. For the interconnection of the various BIGFON projects, however, a uniform transmission standard is required to allow videophone subscribers to communicate. This problem is overcome in the experimental videophone network by converting the video signals at the long-distance network port of each BIGFON exchange to the format of the composite signal is the only possible common denominator for the definition of an interface. The video signal is sampled at the rate of 13.5 MHz (sampling rate defined by CCIR for studio signal processing) to obtain a composite signal with a bit rate of 135 Mbit/s and is then transmitted over the network at a bit rate of 139.264 Mbit/s, which corresponds to the fourth level of PCM hierarchy.
The speech signal in the videophone service is transmitted over the existing telephone service. The video information is only transmitted if both videophone subscribers agree to the use of the videophone terminal.

The technical implementation of the experimental videophone network depends on the following components:
- a signal converter for the conversion of the video signals at the long-distance network port of the BIGFON exchanges,
- a network with a 140-Mbit/s transmission capability,
- broadband transit exchanges,
- a signalling system and the appropriate hardware for the signalling between the BIGFON exchanges and the broadband transit exchanges,
- measuring and test equipment for a phased entry into service of the network.

All these components have meanwhile been developed and procured.

The experimental videophone network serves purely experimental purposes, and videophone subscribers can therefore use the network free of charge.

The installation of the entire network will be completed in autumn 1986.

Encoding

The so-called videophone codec is used for the coding of the video signals. The codec consists of signal converters and a multiplexing unit. The functions of the codec include analogue/digital conversion, signal coding and the multiplexing of the signals of various information types (video, speech/sound, text/data) into a signal packet adapted to a specific service. The codec also provides a transparent 2-Mbit/s channel. Since the device converts composite colour video signals by direct sampling at a rate of 13.5 MHz (CCIR Rec. 601) into 135-Mbit/s digital signals, the structure of the codec allows it to be used for different purposes: for the transmission of moving picture dialogue service signals (videophone and videoconference services) and for sound/television signal distribution. The codec is based on a frame structure of 64 kbit/s, which is also suitable for the processing of component video signals.

Switching

Broadband exchanges can perform the functions of both local and transit exchanges.

They can be expanded gradually and are designed to switch 16 x 16 lines during the first phase of development and 128 x 128 lines upon completion. The switching networks use 3-stage space division multiplexing based on ECL technology, implemented either as a monolithic integrated crosspoint matrix or as a gate array. The switching networks are capable of switching bit rates between 2 and 140 Mbit/s, merely the line connection units differ accordingly.

The switching facilities can be synchronized by an external clock for operation in a synchronous network.

Signalling

The signalling for the experimental videophone network within the separate BIGFON projects conforms to the concepts designed by the individual companies. The broadband transit exchanges are controlled by a specially developed technique based on the HDLC procedure. Calls are currently set up by means of remote data units over 2.4-kbit/s point-to-point data circuits in the public data network via X.21 interfaces. Call set-up takes approximately 1 second.

The experimental videoconference network

The Deutsche Bundespost regards videoconference as to be the business application and therefore the first step of a general videophone service.
Apart from the experimental videophone network, which has been installed purely for experimental purposes, an independent experimental videoconference network is being established to allow videoconferences between groups of persons at different places. Although still at an experimental stage, the videoconference network is already to be used commercially. After the demonstration phase, customers will have to pay charges for the use of the network.
The implementation of the videoconference network follows along the same lines as in other European countries. The network basically uses the same prototype system components (videophone codec, broadband transit exchanges, optical fibre transmission systems) as the experimental videophone network. However, there are 2 major differences between the 2 networks:

- at present, videoconference connections cannot be set up on demand but only by reservation,
- since insufficient broadband transmission capacity is available in the network, the 135-Mbit/s video signals must currently be reduced to 2 Mbit/s on many of the trunk connections by means of redundancy reduction methods, which inevitably leads to a certain impairment of the picture quality. Redundancy reduction is achieved by codecs conforming to CCITT Recs. H.120/H.130.

The subscriber line of a videoconference terminal is generally routed to the broadband exchange via a 140-Mbit/s optical fibre transmission system. For its application in the videoconference network, the videophone codec is equipped with a sound signal interface unit and two 64-kbit/s-type X.21 interfaces. It therefore provides for the same interfaces as the videoconference codec 2Mbit/s conforming to CCITT Recs. H.120/H.130 which is used in international videoconference networks. The broadband exchange is remotely controlled by a central reservation system via a data circuit (see Fig.4).
The central reservation system stores all relevant datas about the network availability, call reservations, call charges etc. A special signalling system was developed for the control procedure. Since the experimental videoconference network in its initial phase of implementation is designed to cater for only a limited number of users and the configuration of the network, the broadband exchanges perform the functions of both local and transit exchanges. In view of the current lack of broadband capacity in the network, redundancy reduction is generally applied in the long-distance network by means of a so-called network codec which comprises both the above-described videophone codec and a codec conforming to CCITT Recs. H.120/H.130. At present, the network codec only permits the bit rate to be converted from 140 Mbit/s to 2 Mbit/s and vice-versa via an analoque interface for sound and video signals.

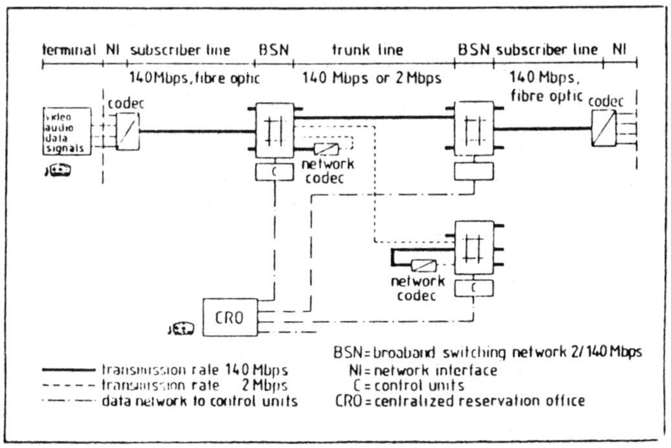

Fig.4: Technical concept of the experimental videoconference network

This implies that the switching system must be capable of processing 2-Mbit/s and 140-Mbit/s signals simultaneously. Signal transmission for national videoconference connections takes place primarily over terrestrial circuits supported to a minor extend by Satellite-links over TELECOM 1, where as international connections are realised primarily by using satellites (ECS and TELECOM 1 for Europe, INTELSAT V for intercontinental connections), see Fig.5

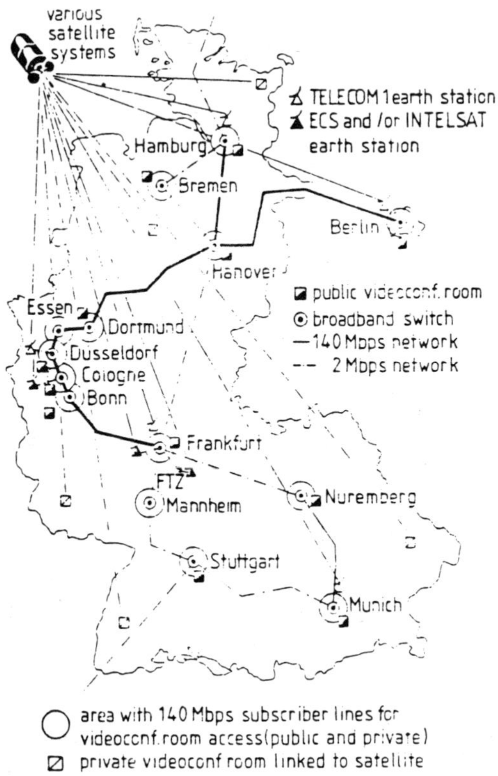

Fig. 5: Experimental videoconference network

Since the Deutsche Bundespost regards the videoconference service as the business application of the videophone service, the experimental videoconference network has from the outset been designed to remain in operation until broadband services can be offered in the so-called Broadband ISDN which will evolve from the 64-kbit/s ISDN. Today, mid-1986, 12 public videoconference studios are made available by the Deutsche Bundespost in various major German cities as a means of informing interested customers about videoconference applications and about 15 private videoconference terminals are even connected to the network, being operated for national as well as for international videoconferencing. In the final phase of implementation of the network in the early 1990s (date at which Broadband ISDN technology should be available), not many more than 300 videoconference subscribers should be connected to the network.

The optical fibre overlay network

As already mentioned the Deutsche Bundespost regards optical fibres as the transmission medium of the future. The Deutsche Bundespost therefore strives to implement this new medium on as wide a scale as possible, especially in the subscriber line network, because it assumes that there will be a high demand for broadband services, in particular broadband dialoque services, and in view of the fact that an optical fibre infrastructure is the prerequisite for the intoduction of widely used broadband applications. For this reason, a feasibility study was undertaken as early as 2 years ago to evaluate the use of optical fibres in the local network with a view to the introduction of a videophone service.

Within the scope of this study, the dynamics of the estimated development of demand for a videophone service in terms of the number of videophone subscribers were examined in relation to the dynamics of a mass-production-induced reduction in the price of the most important technical component of the system (optical fibre cables, transmission systems, coding facilities, etc.) The study was to determine the technical and economic conditions under which the risks involved in the network remained acceptable, and to ascertain the break-even point on the basis of an estimated increase in the revenues derived from the videophone service. In view of the results of the study, the decision was taken in spring 1984 to install roughly 300,000 to 400,000 fibre kilometers in specific large local networks in the 4-year period from 1986 to 1989. This decision is to be re-examined at the end of the 4-year period. Overall investments will amount to some DM 1.5 to 2 billions. Contrary to conventional methods in which the provision of telecommunication facilities was based strictly on customer requirements, specially set-up organizational units in 18 cities with large local networks are now actively enlisting subscribers for broadband applications - at present mainly videoconference users in the business sector - and have started planning the optical fibre cable facilities required. The cables are to be laid in 1986. Videophone and data connections, possibly on non-business user permises as well, are to be installed in addition to videoconference connections in mid-1987. The possibility of introducing optical fibre cables in 80 other large local networks is currently under discussion.
The overall number of subscribers who can be connected to the videoconference network, i.e. a maximum of about 300, is rather modest compared to the immense investments required for the basic infrastructure. However, it is assumed that there is a latent demand for additional connections in trade and business centres. In the light of this factor, a switched network will be established, using the prototype technology developed for the experimental videophone and videoconference networks by expanding the maximum capacity of the 16 broadband exchanges used for the experimental videoconference networks to 128 x 128 circuits. Only the videoconference service will continue to be offered on the basis of a reservation system. The switched network is intended to meet initial demand and will therefore be designed to cater for about 1000 subscribers. The unconventional approach adopted by the Deutsche Bundespost in undertaking substantial investments without an accurate picture of the demand to be anticipated may at first seem rather unusual. However, from the Deutsche Bundespost's point of view, it is the only way out of the chicken and egg dilemma: demand can also develop after investments have been made in a network since potential customers can very soon expect other customers to use the network, and secondly, because customer interest can be stimulated by demonstrating the facilities of the new service. To wait for subscribers first and then to provide a network meeting demand seems a

less appropriate course to take. The above approach will result in a filigree network, superimposed on the conventional network as an overlay network, which will initially provide services to business users. However, the network is designed to allow it to be expanded into a nationwide network for large-scale utilization at a later stage.

The introduction of optical fibres in the subscriber line network does not, however, fulfill all the requirements for communication. Thus transmission capacity is also being provided in the long-distance network and in other parts of the trunk network. Fig.6 illustrates the configuration of an overlay network as it is planned for 1987. At that time, the network will be capable of handling the long-distance broadband traffic expected in the initial phase.

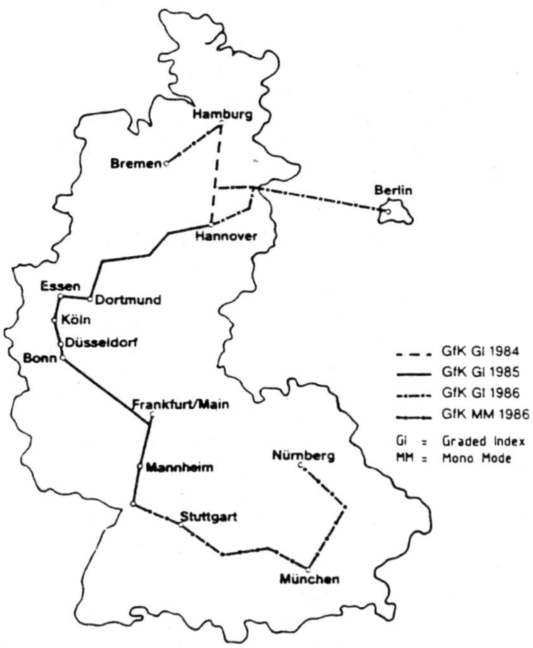

Fig.6: Planned expansion of the long-distance network

Outlook

Although the overlay network mentioned above is intended to be used not only for the videoconference service but also for other broadband services such as the videophone service and high-speed data transmission the general goal of introducing broadband services is to offer such services within the framework of an ISDN with broadband capacity from the early 1990s onwards. The work on the basic concept for such a network has already made considerable progress. The various broadband services, which still have to be standardized internationally, are to use transparent digital channels and, if possible, a universal user/network interface which, for reasons of compatibility, must also be capable of supporting 64-kbit/s terminal equipment. From today's point of view, an interface capable of handling a H4 + n x H1 + mB + D64 channel structure is not only technically and economically the most suitable solution but also the most future-oriented solution.

Such an interface could handle future picture signals of more advanced coding methods as well as radio and television signals and would thus contribute to the most significant step towards integration in a universal network as indicated in Fig.1, by paving the way for an Integrated Broadband Communication Network.

References

/1/ Development strategy of the Deutsche Bundespost for the public telecommunications systems in the Federal Republic of Germany (Brochure published by the Federal Minister of Posts and Telecommunications of the Federal Republic of Germany)

/2/ K.E. Kneisel: Goals and Strategies for the Introduction of Broadband Optical Communication Systems into the Deutsche Bundespost Telecommunication Networks, Conference Record (Vol 2) of ICC, Philadelphia, June 1982

/3/ K.E. Kneisel: BIGFON - THE FIRST STEPPING STONE TO AN INTEGRATED BROADBAND TELECOMMUNICATION NETWORK?
2nd Internat. Conf. on new Systems and Services in Telecommunication, Liege, Belgium, Nov. 1983

/4/ K.E. Kneisel: Broadband Communication Systems in Germany

The authors

Dipl.-Ing. K.E. Kneisel, aged 47 studied at the Technical University in Darmstadt from which he graduated with a degree in communication engineering. He joined the Faculty of Theoretical Electrical Engineering as Research Assistant for two years during which he was concerned with waveguide problems. In 1968, he joined the Fernmeldetechnisches Zentralamt of the Deutsche Bundespost where he has been working as the Head of various Sections dealing with cable and radio-relay systems and network planning. After having worked as Head of the Section "Coordination of the Introduction of Optical Communication Systems", he is now concerned with the coordination of system concept developments and introduction strategies of broadband ISDN.

Dipl.-Ing. K. Höfig, aged 34 studied at the Technical University in Aachen from which he graded with a degree in electrical engeneering. He joined a development department of AEG for 2 1/2 years. The main subjects he dealt with were power supplies and different types of industrial control units. In 1982, he joined the Deutsche Bundespost. He is now concerned with the coordination of system concept developments and introduction strategies of Integrated Broadband Communication Networks.

Dipl.-Ing. W. Gerfen, aged 42 joined the Deutsche Bundespost in 1960. Since 1971 he has been member of the Fernmeldetechnisches Zentralamt (FTZ). He worked in fields of local network planning, strategies and economical studies on digital trunk networks and implementation of fibre optic systems. He is now head of section for videoconferencing.

LIGHTGUIDE APPLICATIONS IN THE LOOP

AT&T TECHNICAL JOURNAL

Timothy D. Nantz and Walter J. Shenk

***Timothy D. Nantz** is a supervisor in the Loop Terminal Systems Engineering Department at AT&T Bell Laboratories in Whippany, New Jersey.* ***Walter J. Shenk** is a supervisor in the Product Development Department at AT&T Bell Laboratories in Holmdel, New Jersey. Mr. Nantz is responsible for systems support on digital loop carrier products. He has a BSEE from the University of Pennsylvania and an MSEE from Northeastern University. Mr. Shenk works on product development and design transfer methodologies and tools and is responsible for forward looking design and interconnection techniques for gigabit rate transmission systems. He has an AAS in design technology from the Academy of Aeronautics, a BSCE from Newark College of Engineering and an M. Eng. from Stevens Institute of Technology.*

Traditionally, the telephone network has been dominated by voice traffic. With the rapid growth in computers, integrated workstations, automated banking machines, digital PBXs (private branch exchanges), etc., today's voice network must evolve to a wideband network that can support voice, data, and video. This changing environment puts new demands on the telephone network, especially the local access portion. Accurate forecasts for new services and the mix of services are difficult to obtain. Therefore, a network must be developed that can accommodate a wide range of services and, at the same time, change as rapidly as new services evolve. Lightwave systems and digital loop carrier offer many of the characteristics necessary for this flexible network. This article describes AT&T's DDM-1000 lightwave multiplexer and SLC® Series 5 carrier system for loop applications.

Perspective

Since the early 1980s, the Bell Operating Companies (BOCs) have been actively deploying lightwave technology in the loop. Initially, systems used light emitting diodes (LEDs) transmitting on multimode fiber cables.[1] With the rapid advances in lightwave technology, it is now possible to use lasers or edge-emitting LEDs transmitting on single-mode fiber. Today, almost all systems in the loop use lasers or edge-emitting LEDs and single-mode fibers.

With this capability, the BOCs can deploy single-mode fiber media with almost limitless capacity and simply upgrade the end-point electronics to meet demand. Many BOCs are planning to deploy fiber media directly to the customer's location, which will give them the bandwidth necessary to provide a wide range of services that range from voice through video.[2] However, with today's technology, it is not economically possible to take fiber directly to most customer locations.

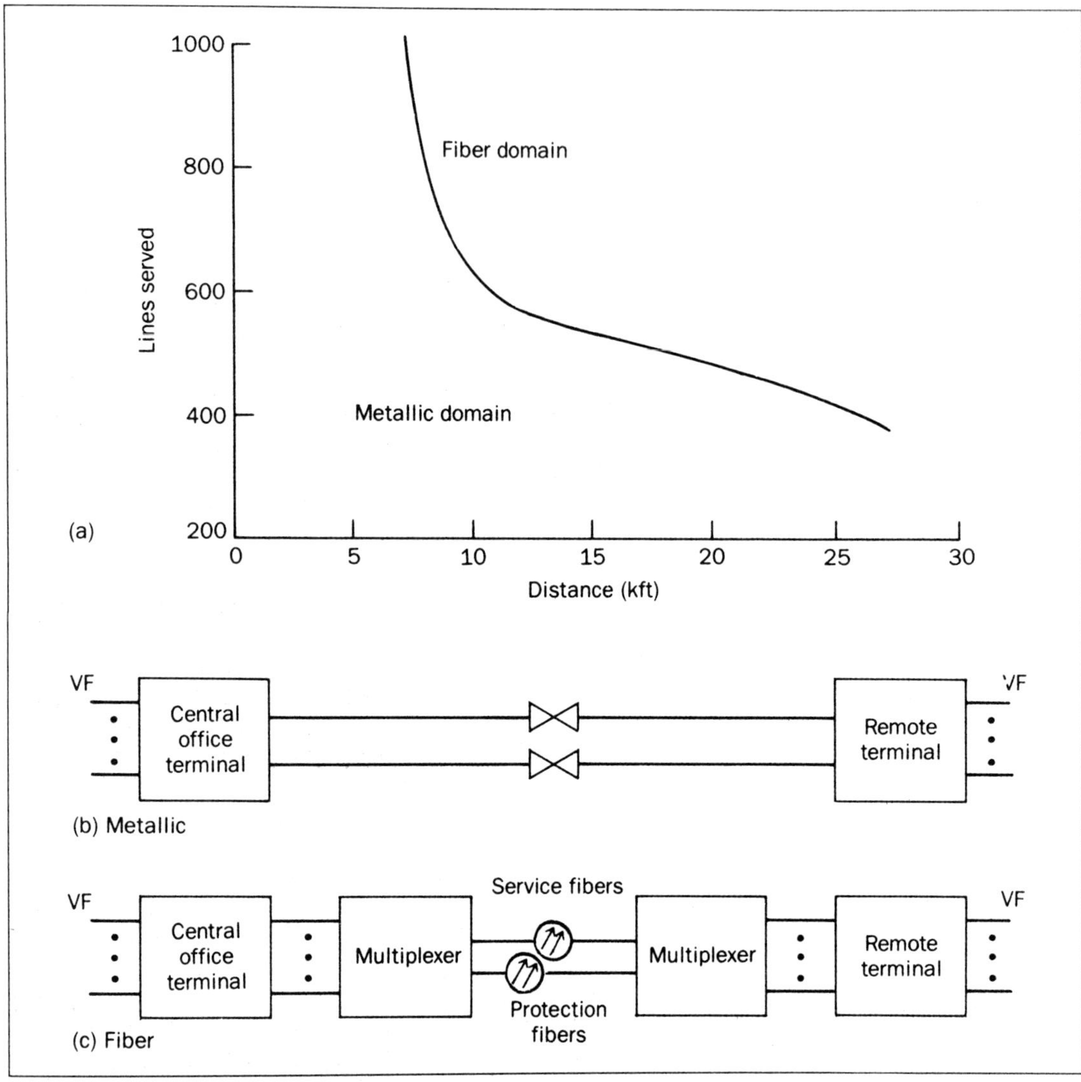

Figure 1. Economics of fiber as feeder cable. The curve shows crossover distance versus lines served for fiber loops. (COT = central office terminal; MUX = multiplexer; RT = remote terminal; VF = voice frequency signal.)

Therefore, an evolutionary strategy is being followed.

Most BOCs are deploying fiber in the feeder portion of the loop. They are taking economic advantage of the technology available today, while building a backbone network to support future digital services.

Figure 1 shows the economics of deploying fiber in the feeder portion of the loop. It compares the relative cost of installing new metallic T1 cables or fiber optic cable and multiplexers to provide the facilities for a digital loop carrier system, such as AT&T's SLC® Series 5 carrier system. For example, Figure 1 shows that, at 15 kft from the central office, it is more economical to use fiber if more than 550 lines are to be served.

The model used to generate Figure 1 did not include:

- Additional revenues that might be obtained from the fiber's bandwidth
- Maintenance benefits that occur from stabilizing the outside plant facilities
- The cost of rebuilding manholes to accomodate T1 repeaters
- Allocated cost of conduit.

Inclusion of these factors will move Figure 1's curve down and to the left, increasing the region of application for fiber.

This economic picture explains why, in Figure 2,

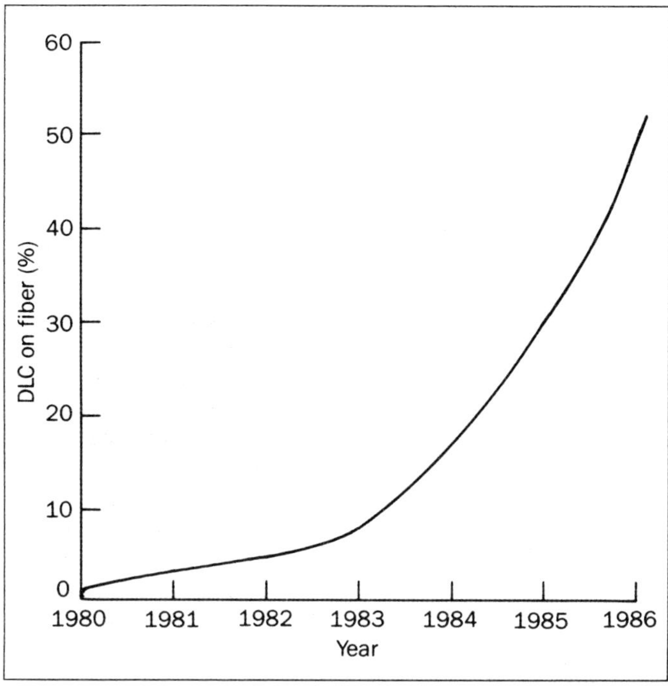

Figure 2. Deployment of digital loop carrier (DLC) on fiber since 1980.

we see a rapid increase in the percent of digital loop carrier systems being deployed on fiber between 1980 and 1986.

The key technical requirements for digital loop carrier and lightwave systems to be used in the local loop are:

- *Size*—Land costs generally are high, and the land for remote terminal enclosures is difficult to obtain in the short intervals often required to provide service to new residential areas or industrial parks. Component systems permit the use of cabinets on the existing right of way, reducing service intervals and lowering overall costs.
- *Flexibility*—Forecasts for the quantity and mix of services are difficult to obtain. Flexible systems that can support a wide range of services and data rates reduce the risk associated with inaccurate forecasts.
- *Environmental robustness*—Because many loop applications are in uncontrolled environments, the equipment must be environmentally robust. Operation from –40°C to +65°C at up to 100-percent humidity (noncondensing) must be assured. This requires special component selection and specification, as well as unique circuit and physical design. In addition, extensive testing must be done throughout the development process to validate thermal models.
- *Maintenance capabilities*—Both system and circuit maintenance capabilities are extremely important in loop applications. Alarms must be available to identify both service-affecting and potentially service-affecting conditions. The system must transmit alarms from the remote terminal to the central office location without the need for external telemetry systems. In addition, the system must provide circuit maintenance capabilities to the appropriate testing centers for both POTS (plain old telephone service) and special services.

This article focuses on AT&T's DDM-1000 lightwave multiplexer and SLC Series 5 carrier system. It tells how they meet the requirements for loop applications, and can be used as part of a fiber-evolution strategy.

DDM-1000 Multiplexer

The DDM-1000 is a compact, highly flexible and cost-effective multiplexer that contains two complete DS3 muldems in an 8.5-inch high shelf. (A DS3 signal is 44.736 Mb/s with 672 channels; a muldem is a multiplexer-demultiplexer.) Each DS3 muldem can accept up to 28 DS1 (1.544 Mb/s, 24 channels), 14 DS1C (3.152 Mb/s, 48 channels), or a mix of them. These two muldems can be connected to transmission systems through a DSX-3 cross-connect, or connected internally to a 90-Mb/s lightwave transmitter/receiver. Figure 3 shows a fully equipped DDM-1000 assembly.

The DDM-1000 was designed to meet the requirements of loop applications. Its extended temperature and humidity capabilities and rich feature set that combines central office and remote terminal functions provide flexibility in planning and implementing digital loop carrier installations.

Thermal Design. Early in the development, model and empirical analyses generated benchmarks for thermal design of the DDM-1000. Component selection, fan unit, and shelf design requirements were established, all focused

Table I. Line Capacities of DDM-1000 with SLC Carrier Equipment

Outside plant structure	DDM-1000 with SLC-96 carrier	DDM-1000 with SLC Series 5 carrier
80C/80D	Up to 288	384-576
Concrete hut	1300-1400	2600-2800
16-ft CEV	1500-1700	3200-3400
24-ft CEV	2600-2800	5200-5600
Maxi-Hut	3400-3500	6800-7000

NOTE: CEV = controlled environment vault.

on operation in a wide range of ambient temperatures, from –40°C to +65°C. Prototype and early production models were subjected to an extensive series of performance tests at the temperature and humidity extremes.

The major emphasis on thermal design required going a step beyond traditional development activities. To develop an economically attractive package, a balance had to be struck between robust thermal performance and selection of cost-competitive components. Thermal management influenced all aspects of the development.

An example of this systems approach is the development of the DDM-1000 shelf and fan unit designs. Existing hardware systems were inadequate from the standpoints of cost and percent of open area for convection and forced air cooling. A custom shelf design allowed flexibility in circuit pack placement, improved the cooling airflow, and resulted in lower cost.

Early models of the custom shelf, equipped with circuit packs, were tested on a flow bench to determine the air flow resistance of the assembled unit. Based on these results, we selected forced air cooling for uncontrolled environments.

The overall compact size of the DDM-1000 permits installing eight shelves in a standard network bay frame in a controlled environment. Three 3-inch high baffles are installed in this configuration, one between every two shelves, to keep the circuit-pack temperature within established limits. In the free-convection mode of operation, cooling air is drawn in from the front of the equipment and exhausted to the rear.

Thermal management also included specific design of many components. This was especially true in the design of the optical line interface units (OLIUs) that operate in uncontrolled environment applications. A custom gate array required a special heat sink that clips on the ceramic substrate and keeps junction temperatures within limits.

Table II. Optical Line Interface Unit Requirements

OLIU	Environment (°C)	System gain (dB)	Span length (miles)
16E	0 to +50	24.0	0 to 18
16F	−40 to +65	28.5	0 to 28
16G	−40 to +65	34.5	28 to 37

The laser transmitter package (Astrotec® laser transmitter system) has both heating and cooling capabilities for the laser. During manufacture, a single setpoint control temperature is established within the Astrotec package. During operation, temperatures above the setpoint start thermoelectric cooling, while temperatures below the setpoint initiate heating. With this arrangement, the laser temperature is not subjected to the wide temperature variations that other components experience.

Operation. Simplicity of operation was a key ingredient in the DDM-1000 design. This extends to installation, turn-up, documentation, and maintenance.

A user's manual that covers functional and maintenance descriptions, applications engineering and ordering information, and installation procedures is shipped with every DDM-1000. Appendix A presents a functional description of DDM-1000 operation, taken from the user's manual.

Three automated turn-up procedures reduce DDM-1000 installation time to a minimum. These tests can be started using an ASCII terminal or push buttons mounted on the front of the equipment. (ASCII is the American Standard Code for Information Interchange.)

The first procedure, the local test, verifies operation of the assembly as an isolated entity. The second procedure checks the wiring to the cross-connects, while the third procedure—the system test—verifies end-to-end operation of the facility.

The DDM-1000 provides an ASCII terminal interface that can be used to obtain detailed maintenance information and control the advanced features of the assembly. Many portable terminals and most display terminals can be used directly. Installations with more than one DDM-1000 per bay can share a link between shelves that allows the ASCII terminal to communicate with other shelves.

When the DDM-100 is installed in a central office/

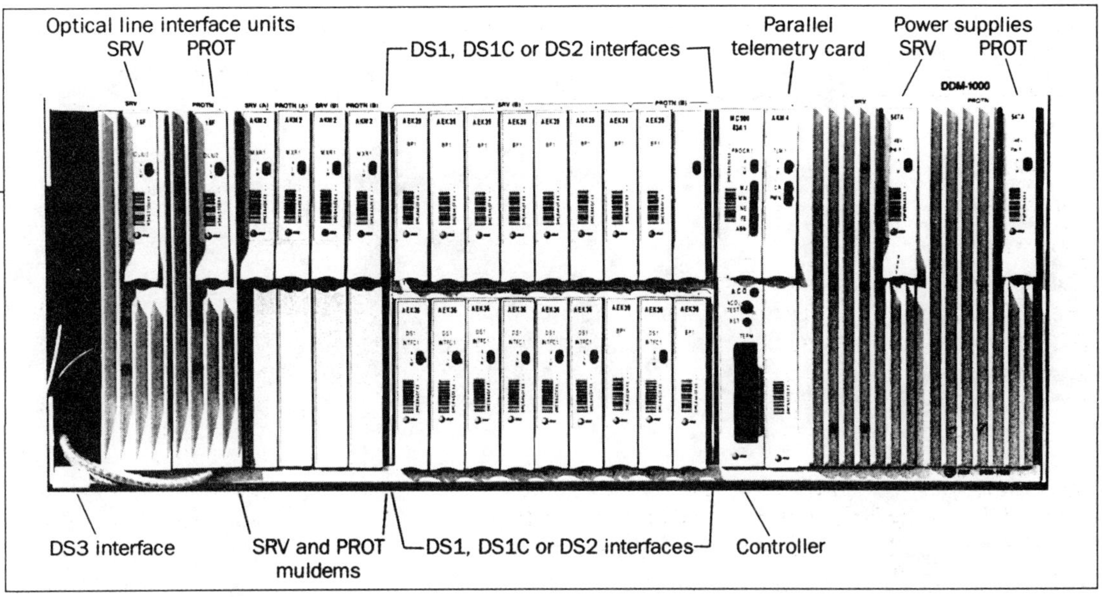

Figure 3. Fully equipped DDM-1000 assembly. SRV = service; PROT = protection.

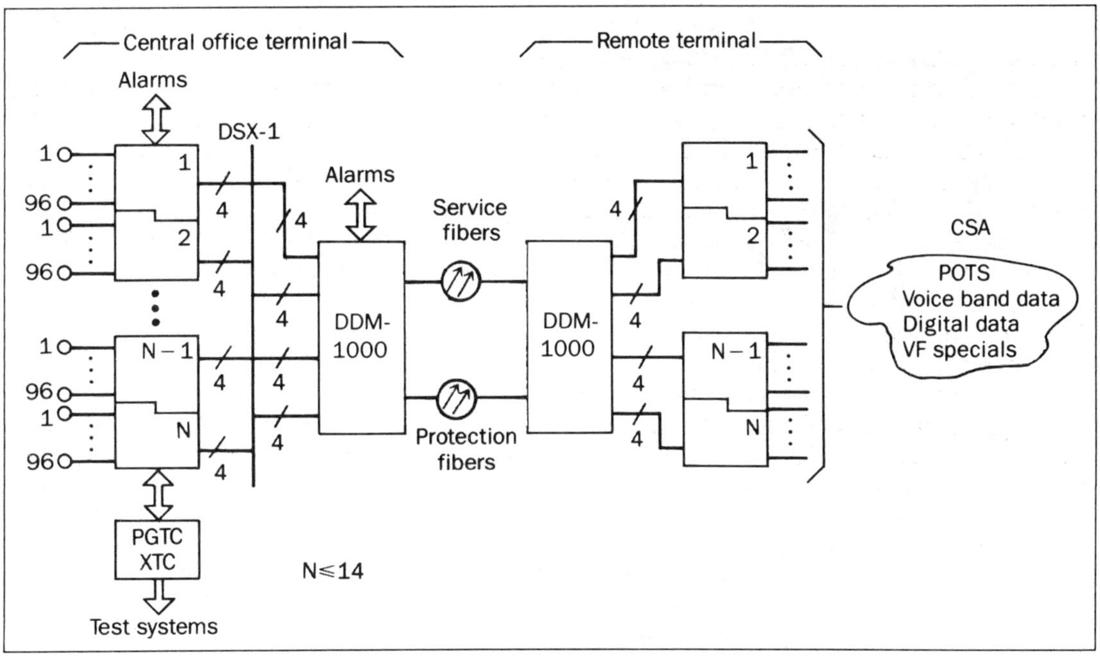

Figure 4. Typical SLC Series 5 carrier system application with DDM-1000 (for N ≤ 14). The pair gain test controller (PGTC) and extended test controller (XTC) provide test capabilities for all services. The carrier serving area (CSA) encompasses POTS (plain old telephone service), voiceband special services, and digital data services.

remote terminal application, an optional telemetry pack provides additional maintenance interfaces. This package includes additional alarms; an extensive, parallel telemetry set; and a bidirectional, overhead telemetry capability.

Various outside plant structures are available to house the DDM-1000 and SLC carrier equipment. Table I lists the approximate line capacities of these structures, using DDM-1000 together with SLC-96 carrier or the SLC Series 5 carrier system.

Applications of DDM-1000 in its optical mode are based on the operating environment of each assembly and the approximate span length between assemblies. Table II specifies the operating environment, system gain, and span length for the various OLIUs.

Figure 5. An 80D cabinet that contains six SLC Series 5 carrier systems and a DDM-1000 multiplexer serves up to 576 derived pairs.

SLC Series 5 Carrier System

The SLC Series 5 carrier system is a microprocessor-based, digital loop carrier system that supports a full range of voice and digital services in one-fourth to one-half the space that the SLC-96 system requires. (Hereafter, we will refer to the SLC Series 5 carrier system as Series 5.) The Series 5 architecture permits upgrading service capability with simple plug-in changes. Figure 4 shows a typical arrangement of Series 5 with DDM-1000. The interface between the two systems is at the DS1 rate (1.544 Mb/s).

Both Series 5 and DDM-1000 support traditional bipolar, alternate mark inversion (AMI) line coding or bipolar with eight zero substitution (B8ZS) for clear channel capability. (In the bipolar AMI format, positive and negative pulses—the 1s—that always alternate represent one state, and the absence of a pulse—the 0s—represents the other state. In the B8ZS format, 1s alternate in polarity but a special code word replaces eight 0s in a row in the original signal.) Line coding is selected with dip switches on the respective line interface units.

Appendix B discusses the internal architecture of the Series 5 channel bank.

Maintenance and Testing. System maintenance is accomplished by the internal system intelligence and the overhead message-oriented data link between the Series 5 remote terminal and the corresponding central office equipment (central office terminal or digital switch). Use of this data link enables the terminals to perform automated maintenance actions that sectionalize trouble to either the central office terminal, the remote terminal, or the facility. Alarms are available at the central office equipment for transmission via standard telemetry systems to the appropriate operations centers.

Circuit maintenance is accomplished via the bank microprocessor and a test controller that is located at the serving central office. Test capabilities for POTS-type lines, as well as locally switched special service lines, are supported by the pair gain test controller (PGTC) that was developed for the SLC-96 carrier system. Enhanced testing for all services is available through the extended test controller (XTC) that was developed specifically for the Series 5 system.

With the XTC and Series 5, all services may be tested. This enhanced testing capability is possible because of the microprocessor technology in the XTC and Series 5, the test access relays on the Series 5 channel units, and the VLSI (very-large-scale integration) chips in the channel bank that permit full splitting bit-stream access. System programming in Series 5 and the XTC initiates sequences and interprets results.

Thermal Design. As with the DDM 1000 development, thermal management was considered at all stages of Series 5 development. Extensive modeling and testing were done to assure operation in an uncontrolled environment. Heat baffles were developed to aid convection cooling at the central office terminal, and a fan unit and associated controls were designed for forced convection in remote terminal applications. Thermal setpoints for fan operation were established to maximize system reliability.

80D Cabinet Arrangement

The size and service capabilities of the DDM-1000 and the Series 5 provide increased opportunity for the BOCs to deploy fiber. In the arrangement shown in Figure 5, six Series 5 systems and a DDM-1000 multiplexer are placed in a single 80D cabinet, providing capability for 576 derived pairs (equivalent cable pairs).

The 80D cabinet, which is factory assembled and prewired for the Series 5 systems and the DDM-1000, is 66 inches high, 84 inches wide, and 27 inches deep. The DDM-1000 is powered from the Series 5 power shelf, and Series 5 fan units supply forced air cooling for both electronic assemblies. Three battery strings on each side of the cabinet provide eight hours of operation if ac power fails.

The tie block above the DDM-1000 is used to cross-connect the DS1 lines between the two systems. The telecommunications company may use the tie block to rearrange the DS1 lines should that be necessary.

Fiber cables enter the cabinet on the left and are connected to the DDM-1000 via lightguide interconnect hardware. The derived pairs are spliced to the distribution cables inside the cabinet, and protector blocks in the middle of the cabinet provide electrical protection for the pairs.

A second cabinet can be placed nearby with eight Series 5 systems, permitting full use of all 56 DS1 ports of the DDM-1000.

To provide flexibility for DS1 services or permit T1 extensions to remote terminal sites beyond the 80D cabinet, office repeater shelves can be provided in the space above the tie block (Figure 6).

Summary

The size, flexibility, environmental robustness, and maintenance capabilities of the DDM-1000 and the Series 5 systems have been designed to meet the requirements of the local loop. When used in structures or the 80D cabinet, they give the Bell Operating Companies a way to implement an economical, evolutionary strategy for fiber deployment. The microprocessor-based architecture of the systems allows upgrades of service and maintenance capabilities as needs are identified.

References

1. P. P. Bohn et al., "The Fiber SLC® Carrier System," *AT&T Bell Laboratories Technical Journal*, Vol. 63, No. 10, Part 2, December 1984, pp. 2389-2416.
2. R. K. Snelling and J. Chernak, "The Revolution in the Loop Network: The Evolving Broadband Technology," *Proceedings of the International Symposium on Subscriber Loops and Services*, Tokyo, Japan, September 29 to October 3, 1986.

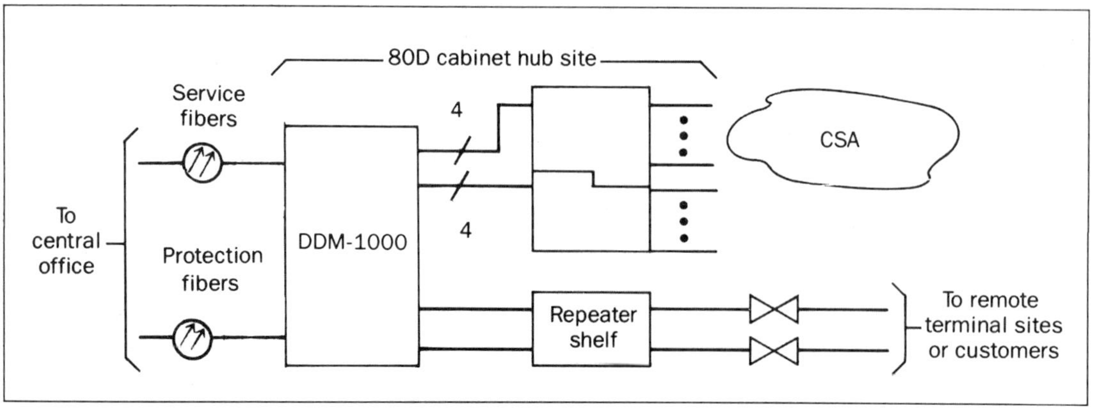

Figure 6. Office repeater shelves placed in the space above the tie block can provide DS1 services or permit extending T1 to remote terminal sites.

Appendix A. DDM-1000 Functional Description

As Figure A-1 shows, two muldems (multiplexers/demultiplexers) share high-speed interfaces, power, and a controller.

If the DDM-1000 is configured for DS3 (44.376 Mb/s) electrical terminations, each muldem is independent. If it is configured for lightwave terminations, Muldem B is synchronized to Muldem A to provide an economical 90-Mb/s lightwave system.

The controller provides all monitoring, protection switching, and display within the assembly. In addition, it contains interfaces for an ASCII terminal; office alarms; and serial, parallel, and overhead telemetry.

Separate optical line interface units (OLIUs) terminate each service or protection 90-Mb/s lightwave line in both the transmit and receive direction. Three codes are available:

- The OLIU1 (16E) can only be used in a controlled environment for short to moderate length spans.
- The OLIU2 (16F) can be used in an uncontrolled environment for short to moderate length spans.
- The OLIU3 (16G) can be used in an uncontrolled environment for moderate to long spans.

Transmission. In the lightwave transmitting direction, an 89.472-MHz clock is received from the function unit—service or protection, as appropriate—in Muldem A and Muldem B (Figure A-2) One of these clocks is selected for the OLIU pack's use as a clock source. This clock is divided by two, and the resulting 44.736-MHz clock is sent to the function units for their use to return synchronous data to the OLIU.

The OLIU also receives a unipolar, scrambled NRZ (nonreturn on zero) data and clock stream from each function unit. These synchronized DS3 signals are combined to form the 88.742-MHz NRZ data signal. However, the DS3 signal from Muldem B is marked to allow the separator at the far end of the system to identify Muldem B's signal.

A single-mode, 1.3-μm laser transmitter converts the electrical signal into information pulses of light that are coupled onto a single-mode fiber pigtail for connection to the transmitting single mode or multimode fiber.

Reception. In the lightwave receiving direction, the receiving single mode or multimode fiber is connected to a multimode fiber pigtail. A PIN (positive-intrinsic-negative) diode based lightwave receiver converts the pulses of light into electrical pulses. The clock is recovered from this 89.472-MHz unipolar, NRZ data stream using a surface acoustic wave (SAW) filter. The signal is separated into two unipolar, scrambled NRZ data and 44.736-MHz clock

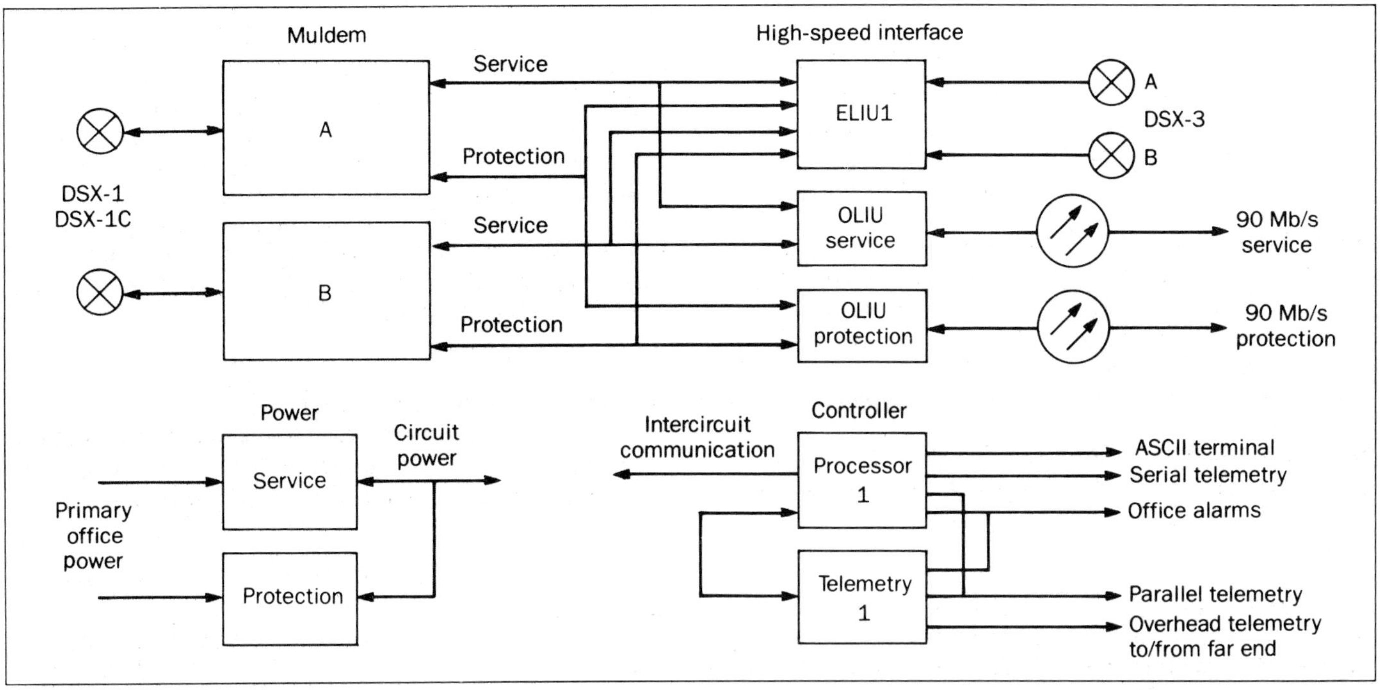

Figure A-1. DDM-1000 block diagram.

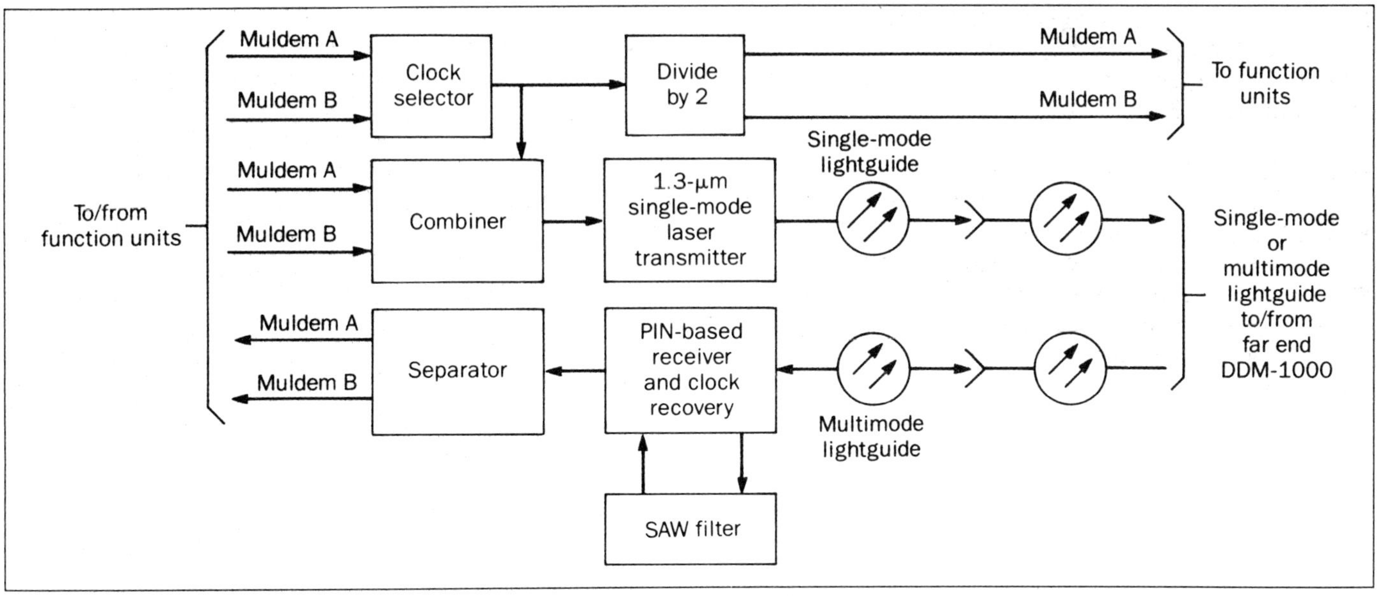

Figure A-2. OLIU block diagram.

streams that are then sent to the appropriate function units.

Processor. In the controller (Figure A-1), the processor provides control for all monitors, protection switching, alarms, and display within an assembly. In addition, it contains interfaces for an ASCII terminal, office alarms, serial telemetry, and a limited parallel telemetry set.

The heart of the processor is a sophisticated microcomputer. It consists of a 16-bit microprocessor with read only memory (ROM) for program storage, random access memory (RAM) for variable storage and state information, interrupt controllers, interval timers, an intelligent sanity timer, and an operating system designed for real-time applications.

Application-specific hardware links this microcomputer with the transmission hardware. A high-speed communication link that gives the processor complete status and control of the assembly is provided for each muldem. This link allows access to and control of the far end DDM-1000 terminal for many different features.

Dedicated monitors for parity error and loss of clock/frame ensure quick line switching for the line terminating option. LED (light emitting diode) drivers can turn on any LED to provide easy identification of failures. Primary-pack switch drivers control the protection switching on all packs. Pack-equipped sensors identify new packs as they are added to the assembly.

Figure B-1. Block diagram of a central office terminal for the SLC Series 5 carrier system. (ADU = alarm display unit; BCU = bank control unit; CU = control unit; LIU = line interface unit; LSU = line switch unit; TRU = transmit-receive unit.)

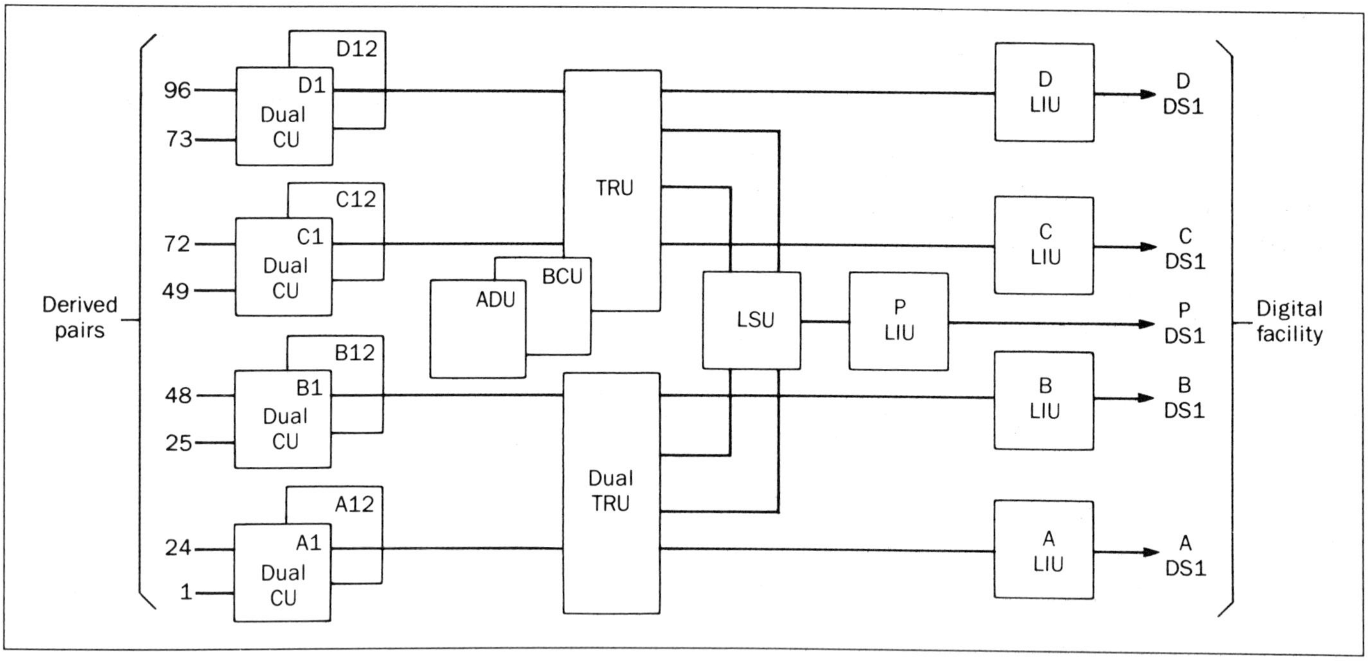

Appendix B. Series 5 Functional Description

As the block diagram (Figure B-1) shows, the channel bank forms the transmission and signaling interface between the derived cable pairs and the DS1 digital facility. It multiplexes 96 derived pairs (channels) onto four active DS1 lines. The alarm display unit (ADU) and bank control unit (BCU) circuit packs contain the system processor, system programming (firmware), memory, and intrasystem interfaces. Together, they form the system controller and control performance monitoring, protection switching, alarming, and testing. New service and maintenance capabilities can generally be obtained through changes in the controller.

The controller is an 8-bit microprocessor with electronically programmable read-only memory (EPROM) for program storage, random access memory (RAM) for variable storage, interrupt controllers, timers, and a serial interface to channel units and common circuit packs. The operating system is designed for real-time applications.

In the transmit direction (derived pairs to digital facility), voice frequency (VF) signals enter the channel units (CU), are sampled at an 8-kHz rate, and converted into pulse code modulation (PCM). The transmit-receive unit (TRU) acts as the traffic cop by telling each channel unit when to put its PCM data on the bus. The TRU also inserts the data link (used for terminal to terminal communications) and forwards the bit stream to the line interface unit (LIU). The LIU converts the internal system format into the standard bipolar (1.544 Mb/s) DS1 format and interfaces with the transmission facility. The reverse process is carried out in the receive direction (digital facility to derived pairs).

The two-way, terminal to terminal data link is used for transmitting status and alarm conditions between the central office terminal and the remote terminal. The data link is message oriented using the X.25 protocol and is transmitted on the A digroup.

The bank controller uses the data link to communicate with the other terminal's bank controller; each controller keeps its own common units informed of alarms and status conditions.

The LIU automatically uses internal bank loopbacks to sectionalize failures so that the bank controller can provide detailed information about failures to craftspeople.

The line switch unit (LSU), an optional plug-in, provides the Series 5 system's 4:1 protection switching feature. By setting the appropriate switches on the LSU faceplate, craftspeople can manually force a line to protection or deny line access to protection.

Optical LAN Activities in Europe

D. ROSENBERGER AND H.-H. WITTE

(*Invited Paper*)

Abstract—This paper provides an overview of European activities in the field of optical local-area networks (LAN's). The fully developed systems are currently at the process control and automation stage. Comparative estimates of the cost effectiveness of optical and electrical LAN's in industrial applications favor the optical systems. To date, deterministic protocols have predominately been used. Statistical access protocols are favored by Ethernet and are gaining in importance. Recent developments aim at mixed protocols which allow simultaneous transmission of both speech and data. The most important characteristics and results obtained with different optical LAN's are compared in a table.

Introduction

OPTICAL waveguide cables are distinguished by their great bandwidth, low attenuation, insensitivity to electromagnetic interference, freedom from ground-looping problems, low weight, and high flexibility. These properties make optical fibers an attractive transmission medium for the most varied fields of application. Depending on application, however, attention focuses on different features of optical fibers. In telecommunications, great bandwidth and low attenuation are paramount, whereas in industrial and military applications and in office automation the high degree of intercept security is the prime consideration.

This paper reports only on optical transmission networks extending a few kilometers (optical local-area networks) and is restricted to activities in Europe.

Local-area networks (LAN's) can basically be classified by network topology and access procedure, the choice of which is determined by the type of information to be transmitted.

The basic optical LAN topologies are the bus, the ring, and the star. In the bus, data circulates bidirectionally. Subscribers feed in and remove data to and from the bus via four-port couplers located at the nodes. Each coupling point reduces power. In the ring configuration, the information generally circulates unidirectionally. Data are looped through the nodes at each coupling point, and are thus repeatedly regenerated in phase and amplitude. Line loss is substantially greater in the optical bus than in the ring, but the bus is less prone to failure.

The optical star may be operated actively or passively. In passive operation, the central star coupler splits up the data in terms of power, so that it is available to all subscribers. Line loss is primarily determined by the degree of splitting (i.e., by the number of nodes). In active operation, virtually the only cause of line loss is fiber attenuation because the data is split electrically in the active star coupler. In terms of the power balance, the active star network behaves like the ring. Via the passive star configuration, at least one order of magnitude more of subscribers can be connected than via the optical bus because line loss increases only logarithmically in the star, but linearly with the number of subscribers in the bus. The star system continues to function should one subscriber fail.

The type of information largely determines the procedure for handling communications via the LAN. This is known as the *access method* and may be deterministic or statistical. The information may flow continuously or in bursts. Examples of continuous traffic are transmission of voice and moving images, whereas in the office sector, computer communication, or process control large quantities of data must be transmitted within fractions of a second.

In the continuous transmission category, it is important to keep delays caused by the LAN below a maximum limit, but transmission need not be 100 percent free from errors. When data are transmitted in bursts, on the other hand, e.g., between database processors or process-control units, error-free transmission is the prime consideration, but reaction times may generally fluctuate within wide limits. If LAN's are to be suitable for subscribers and equipment with different traffic properties, e.g., if real-time and non-real-time transmissions are to be possible, combined access procedures must be applied.

Deterministic access procedures can be divided into those with static and those with dynamic assignment. In static assignment, such as the time-division multiple access procedure (TDMA), every subscriber is assigned a fixed time slot regardless of whether he uses it. Better utilization of channels is obtained if a channel is not assigned until required. A switching center can thus scan subscriber requests, as in the polling technique, for example, or the nodes can themselves decide and seize any idle channel directly (from a central point), as in the empty-slot or token procedure. In the empty-slot procedure an empty slot of constant width passing a station wishing to transmit is filled by the latter with information and denoted as loaded. In the token procedure, a station can only load the channel if it possesses the authorization token. Both procedures assure orderly traffic in that no data collisions can occur. Finer differences in the protocol cause the average reaction time of the token procedure to be longer than that of the empty-slot procedure, but prevent a maximum network delay from being exceeded only in the token protocol, so that real-time compatibility is assured.

In statistical access procedures, collisions occur as a rule. The best-known nondeterministic procedure is the carrier-sense multiple access procedure with collision detection (CSMA/CD). When the network is detected as idle, subscribers wishing to transmit begin to do so. More than one subscriber can send data simultaneously, so that collisions may occur. LAN's that operate with such access procedures are thus unsuitable for transmission of information for which a certain reaction time must not be exceeded. If traffic density is low and data-packet length great in relation to the network round-trip time, data throughput in the system is higher in the CSMA/CD procedure than in the token procedure.

Classification by Access Procedures

One of the first local networks to be implemented was the electrically based Cambridge ring, now expanded into an optical Cambridge ring [1], [2]. The access protocol is the empty-slot procedure. The capacity of the ring is 1.3 Mbit/s. British Telecom were asked by Oxford University Computing Service for advice on how to add an optical-fiber section to a working ring over a distance of some 600 m through a hostile electrical environment. Various options for fiber transmission have been evaluated including the trivial substitution of two fibers for the two twisted pairs, which makes inefficient use of the available bandwidth and suffers from differential delay problems over the distances in question. An improvement on this involves recoding the data onto a single fiber, and a further possibility is to employ full-duplex working on a single fiber. The first two options are working at Oxford and, in one experiment, a ring was demonstrated with a number of 245 terminals. The maximum distance between adjacent nodes is 2 km. A graded-index fiber 50/125 μm and an LED were used in this experiment.

Some refinements are being carried out at present and it is hoped that full-duplex operation can be demonstrated in the near future.

The Fraunhofer-Institut für Informations- und Datenverarbeitung in Karlsruhe, West Germany, has developed a really distributed computer control (RDC) system comprising distributed microprocessor stations linked by a second redundant optical waveguide ring [3]. Active connection of subscribers means splitting the optical waveguide link at each microprocessor station with intermediate optic-electronic and serial-parallel conversion of the data transmitted. Each subscriber can send his own telegram as long as a telegram which is to be forwarded (i.e., is not intended for him) is not being received. If a telegram to be forwarded is received while the subscriber is sending his own telegram, it is buffered and sent when the subscriber has completed transmission (buffer insertion principle). If the network has n subscribers, up to n telegrams may be in circulation on the ring simultaneously. If the transmission breaks between telegrams, pause symbols are sent for the sake of error tolerance; from these every subscriber detects a ring interruption on the receive side. By means of switching commands and the use of a second ring (with transmission in the opposite direction), half-duplex working is obtained automatically, i.e., the direction of transmission alternates in the fixed time-slot pattern of 20 ms. A complete ring interruption is thus tolerated. In the event of multiple interruption, subscribers linked by coherent chains remain capable of communication with one another.

Since early 1979, a system with 30 subscribers has been working under the tough conditions of permanent industrial service to control pit furnaces. Six systems serving some 80 subscribers altogether have been supplied. Since the autumn of 1983 a successor, known as the communication microcomputer (KMR) system, with substantially improved performance has been in service in the automobile industry. Error tolerance to allow for failure of subscribers, transmitters, receivers, and line interruptions is assured by:

a) a redundant ring line,
b) decentralized control of transmission,
c) a decentralized diagnostic system,
d) a decentralized reconfiguration system for services of ISO layer 2, and
e) active subscriber coupling.

The main area of application of the RDC system is digital decentralized process automation, with a wide variety of devices connected via the optical LAN. If the fibers were replaced by electrical cables, the transmission length would become much longer, since an electromagnetically contaminated environment, for instance, a manufacturing shop, does not generally permit direct device-to-device connection. This is one of the reasons why economic aspects have led to the adoption of the optical LAN.

At one of Germany's leading large-scale chemical plants, two bus systems connecting different parts of the plant within a process-control system were installed in February 1984. The bus system is part of the Siemens Teleperm M system. Teleperm is a digital process-control system with a decentralized structure. It contains components to handle simple and complex process-automation tasks and equipment for process-control stations. The decentralized structure of process-related functions permits a clear-cut breakdown into technologically and functionally related areas known as automation islands.

Two bus systems with optical fibers were installed. One bus system connects three automation islands. The greatest distance between islands is 2.5 km. The second bus system links seven automation islands in one building. The maximum distance here is 500 m. The two bus systems are interconnected via a bus coupler in the computer center.

The TOP NET optical waveguide ring developed by the

Institute of Electronics of the Austrian Research Center in Seibersdorf operates on the static TDM procedure. TOP stands for time-segmented optical-pulse transmission and is used to couple locally distributed process-control computers and peripherals. An internal standard bus is provided for the interfaces with the interface modules. The network has a maximum capacity of 512 data channels capable of full-duplex working at a rate of 9.6 kbit/s each. TOP NET has a defined transmission time. The transmission time for a 10-bit word is about 1 ms, regardless of network utilization, which corresponds to the round-trip time of one frame. In the data ring several different protocols can be implemented simultaneously without influencing one another. With a receiver sensitivity of −30 dBm, up to 128 terminals can couple to the ring. An LED transduces the electrical 5-Mbit/s nonreturn-to-zero (NRZ) data. The main range of application is the control and automation of processes in an industrial environment.

Giozza and Noguez [4] have proposed a LAN in the form of an optical ring known as Fipnet for a speed of 10 Mbit/s. This is a TDM system in that time-synchronous frames are transmitted from central node to the ring, which is divided into three field types: synchronous channel, asynchronous channel, and token channel. The synchronous and asynchronous channels correspond to a point-to-point link and thus set up a connection between two subscribers which is maintained throughout communication and assigned only to them. In this manner traffic can be handled both in bursts and continuously.

In a joint project by two French subsidiaries of Philips, Laboratoires d'Electronique et de Physique Appliqué and Centre Technique Industriel de Philips Data Systems S.A., a closed optical bus has been developed [5]. The loss of the *T* couplers used is 2.5–3 dB. Eight to ten nodes can be coupled to the bus with a data rate of 340 000 16-bit words per second and per station. The bus length is several kilometers. The network can also be configured as a star. It can then supply 16 nodes at the above speeds with 85 000 16-bit words per second each. The transmission principle selected is TDM with 24-bit words per channel.

Beratungsgesellschaft für Informationsverarbeitung in Stuttgart, West Germany, has developed a ring structure known as light bus system (LIBSY) with access by the token procedure [6], [7]. This integrated information system offers a computer network which is based on microcomputers and optical waveguide links and has been specially developed to handle process-control data. LIBSY's central element is an optical-fiber ring line to which up to 255 computers or peripherals can be coupled per ring via intelligent interfaces known as Likoms. The interface spacing may be up to 2.5 km, resulting in a maximum ring length of 640 km. Data are transmitted to speeds from 64 kbit/s to 1 Mbit/s, depending on capacity stage. The problem of system availability is approached as follows: a second optical-fiber line is added to the first. If a fault occurs, whether caused by mechanical damage to the ring or failure of a Likom, the interfaces adjacent to the fault short the two parallel data lines. The closed ring structure is restored. Periodic self-tests on the Likoms initiate the procedure automatically. The system can be extended without interrupting operation.

The relatively low capacity of 100 kbit/s is sufficient for the exchange of data between the intelligent microcomputers. There is little communication between the self-contained, commercially available computers. Also, the low data rates permit the use of multimode step-index fibers with a core diameter of 200 μm. Cost analyses have shown that for this industrial application the fiber-optic version can be implemented at lower cost than the electrical LAN version.

With support from British Telecom [8], Plessey has developed a fiber-optic link suitable for use in network systems, particularly in token rings. The link consists of a single IC chip transmitter and an LED as well as a single IC receiver with a silicon photodiode detector. The link accepts data up to 50 Mbit/s. It can be used for direct data transmission bypassing the coder/decoder, in which case NRZ data up to 100 Mbit/s can be transmitted. The link's operating wavelength is 900 nm and the optical power is transmitted by a 133/250-μm step-index fiber. Other fiber types can be used. The link margin is 12 dB maximum. The receiver sensitivity is −27 dBm at 40 Mbit/s and −24 dBm at 50 Mbit/s. The components are packaged in a 16-p-i-n DIL package with 0.3-in lead spacing. The emergent fiber-optic pigtail is protected by a cable sheath and can be terminated with the customer's choice of connector.

The AEG-Telefunken company in Backnang, West Germany, has conceived its digital communication system Dikos as an information distribution system with multiple access for voice and data [9]. With a transmission rate of 10.24 Mbit/s, it has an effective bit rate of 8.192 Mbit/s. Information is transmitted in the network using the TDM procedure. The TDM frame contains 256 (128 or 64) time slots, of which 254 (126 or 62) are provided for transmission of useful information. Each time slot can carry a digitized information signal of 32 kbit/s (64 or 128 kbit/s). The decentralized open-loop station is not assigned to a fixed time slot in the multiplex frame, but can be switched to one or more channels as desired (variable TDM). Setting up a connection between two subscribers is simple, and the link is maintained during communication: the transmission route connecting all subscribers consists of a transmit line and a receive line. When the initiating subscriber station begins to set up a connection, it looks for an idle time slot on the receive line. When a slot has been found, it transmits a call to the desired station in the same time slot on the transmit line. In this handshake procedure it takes an average of 15 ms to set up a connection. Addressing while a connection is set up is done in the time-slot number. Overheads are low because the recipient's address does not have to be transmitted. A message of any length can be transmitted in the time slot marked. This orderly procedure is particularly suitable for continuous traffic, such as in telephony. This time-slot procedure with synchronized frames is a first step towards fully decentralized switching.

Siemens in Munich, West Germany, is developing optical LAN's which are compatible with Ethernet. One project involves an optical star with a passive star coupler. The number of subscribers can go up to 128. The distances are on the scale of 1 km and the data rates in the region of 10 Mbit/s, which, because of the Manchester coding used, corresponds to a line data rate of 20 Mbit/s [10].

The access procedure used is based on the CSMA/CD technique. Collisions thus occur in the optical LAN. The optical part of the network comprises the star coupler, the fibers, and the optical transceiver assigned to each subscriber, which acts as the interface between the fibers and the electrical controller. As the system is passive, collisions cannot be detected in the optical transceivers. In an earlier version, collision of data packets was detected by a listening method [11]. This procedure has two disadvantages:

1) collisions of data packets cannot be detected with 100-percent reliability; and
2) the tolerable dynamic range of the system is restricted to about 10 dB.

In the improved network, data-packet collisions are detected entirely in the transceiver by comparing the bits of the transmitting and receiving source addresses. Subscribers who thus detect a data-packet collision generate an optical JAM signal to notify all other subscribers of the collision. To permit a wide dynamic range (approximately 30 dB) in the system, the JAM signal is not detected with a level evaluation method, but by the difference between the frequency content of the JAM signal and the frequency spectrum of the data in Manchester code. With an error probability of 10^{-9}, the receiver sensitivity of the three-stage transimpedance amplifier is measured to $-44{,}5$ dBm. The fiber is a graded-index fiber with 100/140 μm, which has a bandwidth $\times$ length product of 100 MHz $\times$ km. LED's, p-i-n diodes or APD's are used, depending on system capacity stage.

In another project, Siemens is developing the optical connection of two electrical Ethernet bus segments. The controllers of the two Ethernet segments are each replaced by an optical repeater interfacing the fibers and the corresponding transceiver. In addition to equipment for optical-electrical conversion and regeneration by a PLL timing-recovery circuit, the optical repeater basically contains a circuit which prevents the "quasi-reflections" inevitable in the Ethernet bus from giving the appearance of a collision. These "quasi-reflections" occur when the transceiver is coupled to the Ethernet coaxial cable. Data transmitted by the transceiver to the coaxial cable automatically returns to the transceiver and thus to the optical repeater as "quasi-reflections." If this "reflected" data is not suppressed in the optical repeater, it produces a collision on the Ethernet segment from which it originates. The control circuit in the optical repeater also makes sure that a collision is recognized by all subscribers from the packet fragment size if a data-packet collision has taken place between two or more subscribers not belonging to the same segment. The features of the optical devices, which are very similar to those used in the passive optical star network, are listed in Table I. It is worth noting that the restriction in distance to about 1 km is dictated by the Ethernet protocol rather than by the characteristics of the optical components.

At the ECOC 1984 in Stuttgart, West Germany, L. A. Bergman and S. T. Eng of the University of Gothenburg in Sweden presented a decentralized ring LAN with synchronous working which is also suitable for mixed traffic communication. The frame is gain divided into time slots via which both real-time and packet switching can be handled. It should be stressed here that the fiber used is monomode, with a wavelength of more than 1.3 μm. The transmission rate is 3 Gbit/s, so that each of the 30 TDM channels is operated at 100 MHz.

Prospects

The major activities and development trends in optical LAN's have been outlined. Most of the optical LAN's in operation today still employ a ring topology and a deterministic access protocol. Since more and more findings indicate that for a fairly moderate traffic incidence the CSMA/CD protocols generally involve lower mean network delays than the TDM and token procedures, increasing use is being made of statistical access methods as well. This trend is also fostered by the broad introduction of the electrical Ethernet bus, for which controller modules are already available. Since the high path attenuation—there is no optical equivalent for the high-impedance tap—and the heavy dynamics in the optical bus restrict the number of terminals to about ten, the optical equivalent for the electrical bus is the star configuration. This is why more and more investigations are being carried out on optical star networks, with particular emphasis on the passive star configurations because these are less fault prone than the active ones. A further development step towards a LAN system that will accommodate all types of information consists in the implementation of protocols which support simultaneous speech and data transmission, and permit decentralized switching.

There is no doubt that the role which fiber optics play in LAN's is primarily determined by the insusceptibility of optical fibers to external electromagnetic influences. In process-control systems, the need to combat the electromagnetic contamination prevailing in the process environment makes the optical fiber first choice from an economic point of view.

In the years to come, multimode technology will certainly play the key part. The first indications of expansion of monomode systems can be seen. The greater bandwidth they yield and the reduced loss make it possible to cover greater distances and handle heavier network loads than

TABLE I
SUMMARY OF OPTICAL LAN CHARACTERISTICS

name	network	access protocol	distance between adjacent nodes	transm. speed	capacity	number of nodes	type of fiber	source	detector	receiver sensitivity
Optical Cambridge ring U.K.	ring	empty slot	2 km	10 Mbit/s	1.3 Mbit/s	245	G.I. 50/125 μm	LED		
RDC FHG; FRG	ring	modified empty slot	2.5 km	1.25 Mbit/s	0.8 Mbit/s	256	G.I. 100/140 μm	LED	PIN	-36 dBm
Teleperm Siemens; FRG	passive star	modified polling	1.5 km	250 kbit/s	160 kbit/s	16	St. I. 100/140 μm	LED	PIN	-36 dBm
TOP NET Austria	ring	static TDM	2 km	6.1 Mbit/s	5 Mbit/s	128	G.I. 100/140 μm	LED	PIN	-30 dBm
Fipnet France	ring	TDM		9.8 Mbit/s	6.5 Mbit/s		St. I. 100/140 μm	LED	PIN	-35 dBm
French subsidiary of Philips	closed bus / passive star	TDM	3 to 5 km		5.4 Mbit/s / 1.4 Mbit/s	10				
LIBSY FRG	ring	Token	2.5 km	128 kbit/s	100 kbit/s	64	St. I. 200 μm	LED	PIN	-30 dBm
Lanlink 50 Plessey and Brit.Telecom U.K.	ring	Token		100 Mbit/s	50 Mbit/s		St. I. 133/200 μm N.A. = 0.3	LED	PIN	-24 dBm
Dikos AEG; FRG	open loop	TDM/CSMA/CD	6 km	10,24 Mbit/s	8,2 Mbit/s	65	G.I. 50 μm St. I. 200 μm	LED	PIN	-40 dBm
Passive star (Optical Ethernet) Siemens; FRG	passive star	CSMA/CD	1 km	20 Mbit/s	10 Mbit/s	128	G.I. 100/125 μm 100 MHz x km	LD	APD	-44.5 dBm
Optical extention of Ethernet Siemens; FRG	bus	CSMA/CD	according to Ethernet specif.	20 Mbit/s	10 Mbit/s	accord. to Ethernet	St. I. 100/140 μm	LED	PIN	-44.5 dBm
University of Gothenburg Sweden	ring	TDM		3 Gbit/s			monomode fiber	LD		

ever. In future systems, even LAN's installed at different sites in a city will thus be able to link up. In such metropolitan area networks (MAN's), distances of several tens of kilometers are quite common.

Optical LAN's have not yet made a breakthrough on a wide front in the sense of completely replacing electrical systems by optical ones. This is explained not only by psychological inhibitions about new technologies, but also by the lack of optical components with the same degree of sound engineering and proven long-term reliability under tough environmental conditions as their electrical counterparts. The scope of application and pace of introduction of optical communication systems will hinge on the long-term economy of optical implementations in comparison with that of electrical ones.

REFERENCES

[1] G. W. Litchfield, P. Hensel, and D. J. Hunkin, "Application of optical fibres to the Cambridge ring system," in *Proc. 6th Int. Conf. Computer Communication*, Sept. 1982, pp. 513–517.
[2] D. Rowarth and M. Cole, "Fibre optic developments for the Cambridge ring," in *Proc. Localnet '83* (Europe), Mar. 1983, pp. 21–23.
[3] R. Bähre, D. Heger, P. Peschke, and F. Saenger, "The RDC ring, a fault-tolerant, decentrally controlled and event-driven fiber optic communication network," in *Proc. ECA LAN Seminar, RTS 83/1* (West Berlin, Germany), May 17–18, 1983.
[4] W. Giozza and G. Noguez, "Fibnet: A 10 Mbit/s Fibre Optic Local Network," *Computer Commun. Rev.*, vol. 12, no. 2, pp. 6–20, 1982.
[5] K. Dreyfack, "Fast fiber bus points to fast local nets," *Electronics*, pp. 64–66, Dec. 18, 1980.
[6] T. Wagner and H. Rebel, "LIBSY-ein Informations-Verbundsystem mit Lichtleiterbus," *Elektronic*, vol. 19, pp. 104–108, Sept. 1982.
[7] T. Wagner, "Licht-Bus-System in der Qualitätskontrolle," *Elektro-Anzeiger*, vol. 36, no. 7, pp. 58–60, 1983.
[8] J. R. Stern, "Optical wideband subscriber loops and local area networks in the UK," in *ICC '84, Dig. Tech. Papers* (Amsterdam, The Netherlands), May 1984, pp. 884–887.
[9] A. Reim and K. D. Schenkel, "Digitales Kommunikationssystem DIKOS," *Nachrichtentech. Z.*, vol. 34, no. 10, pp. 658–663, 1981.
[10] S. Moustakas, H.-H. Witte, V. Bodlaj, and V. Kulich, "Passive optical star bus with collision detection for CSMA/CD-based local area networks," *J. Lightwave Technol.*, vol. LT-3, no. 1, pp. 93–100, Feb. 1985.
[11] S. Moustakas and H.-H. Witte, "Novel collision detection technique for CSMA/CD-based fibre-optic local area networks," *Electron. Lett.*, vol. 19, no. 15, pp. 592–593, 1983.

Dieter Rosenberger was born in Grattstadt, Germany, on June 24, 1934. He received the Diplom and Ph.D. degrees in physics from Munich Technical University, Munich, Germany, in 1958 and 1961, respectively.

In 1961 he joined the Siemens Research Laboratories, Munich, where he has worked in the fields of laser research and laser application. He is currently Head of the Optics Depatment, Siemens Research Laboratories, in which capacity he is concerned with optical communication and optical data processing.

Dr. Rosenberger is a member of the German Telecommunication Society. He is Editor of the book *Technical Application of the Laser*.

Hans-H. Witte was born in 1938, in Dortmund, West Germany. He received the Dipl.-Phys. degree in 1963 and the Dr.rer.nat., in physics, in 1968 from Freiburg University.

Since 1969, he has been with the Central Research and Development Laboratories of Siemens AG in Munich. He has published papers in several fields including gas lasers, absorption measurements in bulk glasses, optical branching elements and fiber-optic LAN's. He is currently on the personal staff of the Director of the Siemens Central Research Laboratories.

Standardized Fiber Optic Transmission Systems—A Synchronous Optical Network View

RODNEY J. BOEHM, YAU-CHAU CHING, MEMBER, IEEE, C. GEORGE GRIFFITH, SENIOR MEMBER, IEEE, AND FREDERICK A. SAAL, MEMBER, IEEE

Abstract—**The deployment of fiber optic systems has drastically altered the complexion of the digital network, and standardization of optical interface parameters has assumed paramount importance. This paper reports on the activities in various standards organizations, with emphasis on a synchronous network proposal which is currently being discussed in the T1 Committee.**

STANDARDIZATION work on various fiber optical components has been active in the EIA and the CCITT standards arenas for several years. EIA has generated numerous FOTP's (fiber optic test procedures) which deal with standardizing definitions and test procedures for fiber optical components, while CCITT Recommendations G.651 and G.652 address similar topics. Standardization of fiber optical transmission systems is, however, still in its infant stages. There are three main reasons for this late start of system standardization process. First, systems are more complex than their components and therefore require more time to mature. Second, there are a number of manufacturers, both domestic and foreign, producting fiber optic transmission systems, but none seems to be dominant enough to establish a de facto standard. Third, the seven regional companies, the numerous independent exchange and interexchange carriers have diverse interests and preferences and have not been able to dictate a unified standard. With the rapid advances of fiber optic technology, new systems are being introduced to the marketplace steadily, and the consequences of these nonstandardized systems are a fragmented network and vast confusion to users.

For this rapidly changing technology and fragmented network, one might ask,'' Do we really need a standard for fiber optic systems?'' From the perspective of an operating telephone company, his goals are to generate new revenues and to reduce costs. New revenues can be generated either by introducing new services or by providing more cost-effective services to a wider customer base. Cost reduction can be achieved either through savings on operational expenses or through purchase of more competitively priced equipment. From a manufacturer's viewpoint, his goals are also to generate new revenues and to reduce costs. New revenues can be generated by introducing new products and by expanding market shares. Costs can be reduced by using advanced technology, ingenious system design, and cost-effective manufacturing processes. Cost can also be reduced by expanding the market and thereby achieving the economics of scale. Standards are beneficial to both groups in their pursuance of revenue enhancement and cost reduction.

Standards are valuable to operating telephone companies because they lead to the possibility of procuring compatible equipment from multiple vendors (which tends to imply more competitive pricing). Standardization also reduces operational costs by unifying operation and maintenance procedures, simplifying training, and lowering sparing requirements (because the equipment has a common maintenance interface and can be interconnected without special interfacing assemblies). For manufacturers, standards lead to economics of scale in production and engineering because multiple customers for the manufacturer's product have the same or nearly the same requirements.

The introduction of new services can benefit both the manufacturers and the operating companies alike by providing more product opportunities to manufacturers and more revenue to operating companies, and standardization is essential in introducing new services. A telecommunications network by its very nature requires coordination for any services. A national standard is therefore necessary for an orderly introduction of new services. Many of the new services introduced into the network in past years can be attributed to the vertically integrated Bell System. After the divestiture of AT&T, the need for such telecommunication standards was well recognized and Committee T1 was established by ECSA to formally set these standards so that a network can evolve and progress gracefully. The structure and missions of Committee T1 have been documented in [1] and will not be repeated here. However, much of this paper concentrates on the activities of T1, particularly that of the T1X1.2 working group (carrier-to-carrier interfaces) and T1X1.4 working group (digital hierarchical rates and formats).

I. WHAT NEEDS STANDARDIZATION?

There are two separate categories in a fiber optic system that need standardization: optical parameters and signal format. To start, let us consider the simple point-to-point transmission systems that use DS3 bundles either as multiplexer inputs or outputs. For optical transmission around the DS3 rate, a system usually starts with an M13 multiplexer. Additional overhead capacity is then added so that the transmission system can be maintained. For a higher speed system, several DS3 bit streams are fed into a multiplexer, maintenance overhead is added, and the resultant signal is then converted to an optical form. Due to the variations in the overhead channels and the stuffing mechanisms, most of these systems have unique rates and formats. Additionally, technological advances in fibers and transducers have resulted in vast variations in optical parameters such as optical wavelengths, spectral widths, attenuations, waveshapes, and dispersion characteristics. These different rates, formats, and variations in optical parameters lead to incompatibilities in fiber optic transmission systems and confusion to users.

II. Optical Parameters

Standardization of optical parameters is currently being addressed in three areas: standardization of optical parameter definitions and measurement techniques, standardization of fiber optic transmission system regeneration section design methodology, and standardization of the values of optical parameters for transmission systems.

The definitions and measurement techniques are least controversial in that there is consensus that such standards need be established. Difficulties arise from the rapid technological advances and the fact that many properties are not well understood. However, great progress has been made in this direction in IEEE, EIA, and CCITT [2]–[4]. Although these activities do not address the questions of fiber systems per se, they are nonetheless an integral part of the system standardization work.

The regenerator section design methodology is somewhat more controversial in that there is no agreement as to how a standard should be used. Regenerator section design methodology also encounters difficulties from rapid technological advances and the fact that many parameters are still not well understood. To date, the designs are nominally loss limited, with dispersion and bandwidth limitations serving only as checking mechanisms. Dispersion penalties are devised for regions where dispersion does not drastically affect the performance of the system. As system speed increases, dispersion becomes the limiting factor, and still there is no accurate and verifiable measure of dispersion limitation of a fiber optic system that a span designer can use as a guide [5].

There is also disagreement on whether a worst case approach or a statistical approach should be preferred. With the demise of the Bell System, each operating company has begun engineering its own fiber routes; formerly Western Electric usually served as the system integrator. Operating companies also frequently purchase fiber and terminal equipment separately from numerous suppliers. It is therefore imperative that the specifications of the products be consistent and complete to allow the telco personnel to design regenerator sections. A standard design methodology and the associated set of parameter specifications would allow such regenerator section design. In assuming the design responsibilities, however, the telco personnel also assume the responsibilities of guaranteeing the overall end-to-end transmission performance. A conservative approach to this is to apply a worst case design, using the worst case parameters. Since the worst case parameter values are generally guaranteed by the suppliers, the discrepancy between the design and the actual performance can be easily traced to individual components. The worst case design, suffers from overdesigning. The final architecture often contains too many regenerators and excessive loss and dispersion margins. A statistical method alleviates some of the overdesign problems, but introduces uncertainty in component performance as well as a small percentage of statistical failures.

Regenerator section design methodology is currently being addressed in the EIA FO2.1 working group. A multimode fiber system design methodology has been included in a document entitled "Optical fiber digital transmission systems—Consideration for users and suppliers" [6] and a methodology for single-mode systems is under development.

Standardization of the actual values of optical parameters is the most controversial. Since these parameters are changing drastically with technological advances, unless standards are implemented carefully, they might include unnecessary constraints which might limit innovation on the part of the manufacturer. Therefore, it is important to standardize only what is absolutely necessary to allow various manufacturers' products to interface and to interoperate properly. T1X1 has initiated a project in this direction, and a fiber optical subworking group in T1X1.2 has been established to address the questions. The original purpose of this group is to establish a standard so that a multiowner fiber system (also known as midfiber meet) can be installed without lengthy contract negotiations [7]. This standard, when established, may also have more far-reaching implications in that it would allow optical cross-connect capabilities in an operating company's own network, thus greatly simplifying the design restrictions and partially realizing the flexibilities of a fiber network.

Despite the rapid changes in technology, there are indications that optical parameters standards are beneficial at some low speeds (i.e., <150 Mbits/s) in a short-loop environment. For these applications, regenerators can be avoided. In fact, improvements of fiber and transducer manufacturing processes and quality control will make it possible to design a regenerator section not by detailed computation of the parameters, but by a simple table look-up procedure, in much the same manner as voice circuits are designed today. Because of the low-loss fibers, the high-power transmitters, and the high-sensitivity receivers, there will be enough flexibility in fiber optic transmission systems to allow the introduction of optical cross-connect systems without unduly limiting the applications of these systems. To ensure compatibility, only the signal rate and format and a minimum set of optical parameters need to be standardized.

The question of a standardized signal rate and format is being addressed by another subworking group in T1X1—subworking group on rates and formats for fiber optic network interfaces of the T1X1.4 working group [8]. Rates and formats also depend on how the fiber optic systems are deployed: as a point-to-point transmission system or as part of a fiber optical network. Presently, transmission systems are designed and deployed as point-to-point systems, without much consideration of the services they can carry or the network to which they belong. Transmission parameters and the signal format are largely determined by the characteristics of the transmission media. The introduction of fiber optic systems and the Integrated Service Digital Network (ISDN) may have changed the nature of the network, and the standardization process should be changed accordingly.

III. Point-to-Point Transmission Systems

Existing digital networks are delineated by digital cross-connect systems. A digital hierarchy is defined at these digital cross-connect points. In North America, there are DS1, DS2, and DS3 signals. Typically, the cross-connect serves as a maintenance span boundary. Span performance specifications need to be satisfied between the two cross-connect points. The means by which these performance specifications are satisfied is usually the function of the line-terminating equipment, which often also serves as a multiplexer. When the T1 carrier was first introduced, the maintenance features were few and simple. There were no error checks or operations channels. The order wire was provided via a separate copper wire pair. Fault location was performed off line by injecting a bipolar test signal into the line, and looping the signal back at the appropriate regenerator through a second separate pair. As technology advanced, i.e., as higher capacity systems were deployed, more automated features for maintenance were included. Error checking and fault location became part of the line-terminating equipment and the operations channel became part of signal overhead. Often, the order wire was also included in the overhead channels. Furthermore, the operations channels were expanded to include administration and provisioning. Since none of these functions has any standard features, each manufacturer was forced to choose its own interpretation and implementation. Often, the choice of overhead functions is dictated by the coding technique and the bit rates.

For a fiber optic system, standardization of these overhead functions need not be influenced by technological limitations. There is enough bandwidth in almost all fibers to allow as many overhead functions as necessary. The difficulty lies in the fact that none of these overheads has been standardized, and they will not be standardized in the near future. In the T1X1.4 subworking group, an approach was adopted to standardize only the channel capacity and the locations of overhead channels. The details of overhead have been left for future work, either in T1X1 or in T1M1 (internetwork operations, administrations, maintenance (OAM), and provisioning). The tentative list of the overheads as proposed by the subworking group is listed in Appendix A [9]. Note that sufficient overhead channels are being reserved for future expansion or for manufacturer and user proprietary uses. Note also that a channel called Span Channel Identification (SCID) is included in the list. The function of such a channel will become apparent in the section of synchronous multiplexing (Section VII).

Unlike other transmission media, the specific rate of a fiber optic system also does not depend on technology. A system at 140 Mbits/s and a system at 150 Mbits/s can both provide three DS3 signals as their information payload, with slightly different overhead capabilities. These two systems probably perform in the same manner and cost the same amount. Thus, the choice between these two options is almost arbitrary. Performance differences become apparent only when the rates are high (above 500 Mbits/s) and far apart. However, there are still great difficulties in arriving at a standard digital rate for the fiber optic systems, mainly due to political and business considerations. Any standard rate and format which do not agree with an existing product will introduce development and revision costs for respective manufacturers.

While the present arrangement for point-to-point systems is workable without a standard, the continued proliferation of incompatible systems has drawbacks. All high-speed signals must be demultiplexed down to the standard DS3 level for interconnection. All overhead signals are terminated at the line-terminating equipment and routed via some low-speed lines to their appropriate destinations. There are also possibilities of accepting the existing or future digital hierarchical signal rates and formats and concentrating only on arriving at a set of standard optical parameters. For example, the current DS3 format and the proposed SYNTRAN and DS4E formats can all be used for fiber optic systems [10], [11]. SYNTRAN and DS4E are particularly promising since the current activities in T1X1.4 also stress a set of standard operations channels for these signals. This approach is similar to work by CCITT on fiber optic systems [12]. However, this approach has its own limitations. First, DS3 signals do not have enough overhead capacity, even with the SYNTRAN modifications, to make them full line signals, so their application may be limited. Second, there are numerous asynchronous DS2 signals and some DS4E signals in the current network. Even if new standards can be adopted at these two rates, it is only the beginning. It would be difficult to carry these existing signals with the new standard systems. Third, newer systems at higher rates are being introduced rapidly and there is no standardization work at these high rates.

In addition, there is some apprehension about this laissez-faire approach. As more fiber systems are introduced into the network, there will be more incompatible systems that can be interconnected only via DS3 cross connects. Current requirements call for a distance limitation of 450 ft between the cross connect and the line-termination equipment and lead to congested Central Offices. At the DS4E rate, the distance limitation is even more restrictive (270 ft). There will also be more incompatible operations systems associated with these fiber systems. The overhead signals will still be carried around the network via outdated low-speed copper wires, not utilizing the enormous flexibility that a fiber can provide. Most of all, this approach is inconsistent with the philosophy of ISDN. Proliferation of incompatible fiber optic systems, which do not integrate ISDN services, will lead to a fragmented network and definitely stifle the development of ISDN. For these reasons, we will also examine standardization from a network approach. That is, instead of always interfacing at the standard electrical cross connect of a DS3 signal, we will examine the prospect of interfacing optically at some standard optical cross-connect systems. Because the maintenance and other signal associated overheads are embedded in the signal when it traverses the cross connect, the signal standards will be considered not only for the integrity of a transmission span, but also for the integrity of the network, the fiber network.

IV. A Network Approach

What is a fiber network? The deployment of fiber systems has drastically altered the complexion of the digital network. The two most outstanding features of these transmission systems are low loss and large bandwidth. The low loss allows us to build fiber transmission systems with few or no regenerators. The large bandwidth allows us to build very high-speed links. In addition, low loss and large bandwidth are obtained without special treatment to the fiber media, so that a low-speed link can be easily upgraded to a higher speed system without modifications to the media. Only the electronics need to be changed. To date, most fiber transmission systems are being deployed in trunks and long-haul links. However, there is a trend towards the penetration of the loop distribution plants. A most interesting and fruitful development of fiber systems will occur in the distribution plant. With fiber, wide-band services can be introduced into distribution plant economically. The bundle size of these wide-band services can be varied drastically and requires a flexible network. If these wide-band services are fed to hubbing points where they can be switched either manually or electronically, the hubbing configuration allows flexible capacity. The hubbing configuration may also be economical because it minimizes the construction cost, although using more fiber.

The cost of providing electrical to optical conversion is still relatively high, and will be for many years to come. Thus, it may be economical to provide an optical cross-connect system at the hub such that hubbing can be performed in the optical domain without incurring the back-to-back conversion cost. Because of the low loss of the fiber and therefore the long regenerator span length, an optical cross-connect system can cover a wide geographical area. Such wide coverage may lead to the deployment of optical cross-connect systems in areas with only medium densities of customers.

Whether or not the hubbing is performed at an optical cross-connect or with an electronic digital cross-connect system, the hub should provide flexible capacity. This capacity should be easily rearranged or reassigned. Standardization is essential in providing this flexible capacity. Using the existing fiber optic systems, all signals would have to be decomposed down to the DS3 level in the electrical form for interconnection. These DS3 signals, however, lack sufficient overhead functions to allow them to become line signals. Thus, a new standard signal may be necessary. To allow flexible rearrangement of the capacity, the high-speed signals should have the same format or they should have the same channel rearrangement capability, which calls for a standard overhead function. To have flexible capacity, it is advisable to have a number of bundle sizes available. However, under the existing hierarchical structure, these bundle sizes may be overly restrictive. Thus, a new hierarchical structure may also be beneficial. To simplify the administration procedures, it may also be advisable to devise a new basic bundle size and to provide services and to transport signals in multiples of this basic bundle. To simplify reassigning channel capacity, it is also proposed to use a synchronous multiplexing technique, which will be described in a later section. At this moment, the signal in a basic bundle size will be referred to as synchronous transport signal, level 1 (STS-1).

The STS-1 is a basic signal which transports 64 kbit/s voice services and connects to the existing voice-oriented network. Two requirements should be considered for this feature. First, the existing network is heavily DS1 and DS3 oriented. These DS1 and DS3 signals could be either synchronous or asynchronous. Regardless of the origins of its component signals, STS-1 should be easily decomposed into DS1's and DS3's. Second, savings can be realized through direct fiber termination to a voice switching system. STS-1 should be easily decomposed into DS0's at the switches. As mentioned earlier, this STS-1 should also incorporate features of an ISDN for easy and economical access for a variety of new services.

All of these arguments seem to call for a new fiber optic system rate and format standard. The T1X1.4 subworking group on rates and formats for fiber optic network interfaces has been working on such a project since April 1985. The basic concept came from a contribution by Bellcore in November 1984, which proposed a network called Synchronous Optical Network (SONET) [13], [14]. In the next section, we will briefly describe SONET. For identification purposes, we have retained the name of SONET for this discussion. It should be acknowledged that the present status of the project is the result of the joint ongoing efforts of the members of the T1X1.4 subworking group and the features described may be significantly different from both the original contribution and the final form.

V. Synchronous Optical Network (SONET) Components

SONET is a fiber network that consists of four standardized products families that could be easily interconnected. Of course, standardization only refers to the interface. The actual implementation should be left to the manufacturers to maximize the benefit of technological advances and designer's ingenuity.

To start, SONET offers a standard family of fiber optic transmission system product possibilities. These transmission systems perform the same functions as the existing transmission systems, with the exception that their interface specifications are standardized. It is recognized that the purpose of the standardization is to allow flexible network interconnectibility. Therefore, the rate and format are devised to allow such flexibility. The inputs to these transmission systems are DS3 signals, both synchronous and asynchronous. The output is a family of optical signals called optical carrier, level M (OC-M) supporting M signals at the DS3 level. Corresponding to each OC-M, there is an electrical signal called synchronous transport signal, level M(STS-M), which has the same rate, format, and functionalities as that of OC-M. The basic electrical signal, STS-1, is approximately 50 Mbits/s and can accept one DS3 signal. An STS-1 also has an abundance of overhead capacity for network administration and operations support. Therefore, the optical signal OC-1 can be transmitted as a line signal. A high-speed signal, OC-M, is generated by, first, bit-interleaved multiplexing of M STS-1's to form an STS-M and, then converting STS-M to an optical signal. With the capabilities of the overhead channels and the simple bit-interleaved multiplexing, these transmission systems can be easily interconnected.

The second family of SONET products are the OC-1/OC-M converters, the optical add–drop multiplexers, the optical cross-connect systems, and the electronic digital cross-connect systems. These are the devices by which the transmission systems are interconnected. OC-1/OC-M converters and optical add–drop multiplexers are currently perceived as electrical equipments that accept optical signals OC-1's and multiplex them to OC-M's. In the future, they may use optical processing as well as electrical processing. With these four types of equipment, an OC-M signal can be easily rearranged at the hub and an STS-1 signal can be easily reassigned to different routes. It should be noted that the equipment to function properly, the optical network should be synchronized; thus, we have the acronym SONET. Synchronization is necessary because a simple synchronous bit-interleaved multiplexing technique is employed to all multiplexing equipments.

A third family of products are those that convert services at sub-DS3 rates to OC-M transport directly without first converting to DS3's. They include the fiber optic transmission systems that accept DS1 signals (both line-terminating multiplexers and add–drop multiplexers) and DS0 signals. Should ISDN services be transmitted as a primary service ($23B + D$ channels at 1.536 Mbits/s), they will be carried by the fiber network as DS1-rate signals. For ISDN services at rates between that of DS1 and DS3, they will probably be multiplexed up to a DS3 signal first.

The fourth family of products are those that provide broad-band services that cannot be carried by one STS-1. A transparent DS4E signal or an enhanced quality video signal may require the capacity of an STS-3. A high-definition video signal may require the capacity of an STS-12. There are debates concerning how these services can be efficiently provided in SONET. Some suggest that the basic bundle size should be increased to around 140 Mbits/s so that both DS4E and enhanced quality video signal can be carried by STS-1. They also argue that equipment at approximately 140 Mbits/s is both technologically feasible and economically attractive. Defining STS-1 at this high speed may also simplify network administration procedure if these high-speed services are as widespread as someone has anticipated.

Another suggestion for these high-speed services is to devise a new signal called single service STS-3 (SSTS-3) and maintain STS-1 at 50 Mbits/s. An SSTS-3 retains the simplicity of an STS-3 in that it appears as three bit-interleaved multiplexed STS-1 signals as far as span maintenance is concerned. However, it retains the integrity of a single signal by not allowing the bit stream to be broken in the process of transmission. This last restriction is imposed so that elaborate and expensive timing recovery circuits can be avoided. The advantage of this approach is that both voiceband services and broad-band services can be provided with the same family of signals. Should a new wide-band service be introduced into the network at another rate, say STS-6, this approach can be easily extended to accept such a new service. The relative merits of choosing STS-1 at 140 Mbits/s or at 50 Mbits/s with SSTS-3 modifications is still under discussion in the T1X1.4 working group at the time of this writing.

Because of the network approach, the products necessary to form SONET include not only the transmission systems, but also distribution and switching systems. Although the SONET transmission systems per se may not have many economical advantages over that of existing fiber optic transmission systems when used point-to-point, the capability of easily interconnecting between them and the deployment of an associated interconnect and access system will lead to a simple, flexible, and economical network. In addition, there is also a consensus in the T1X1.4 subworking group that this is a golden opportunity to redirect the network development for future growth, and the opportunity may be lost forever if we do not act promptly.

With so many new products to be introduced, an easy deployment scheme which does not ignore the existing network is crucial. Initially, only point-to-point fiber optic transmission systems are deployed. These equipments can operate with an internal crystal-controlled clock. As soon as it becomes desirable to deploy some interconnecting systems such as OC-1/OC *M* converters, optical add–drop multiplexers, optical cross-connect systems, and electronic digital cross-connect systems, all SONET equipments would have to be synchronized. This second step probably starts with the deployment of OC-1/OC-*M* converters and optical add–drop multiplexers, coupled with some simple optical splicing and terminating devices. Only when the network is sufficiently full would an optical cross-connect system or an electronic digital cross-connect system be deployed. This completes the core work of SONET. However, not many new services have been introduced at this stage. Both voice and data signals are probably converted to DS3 for optical transmission. SYNTRAN terminals can be used efficiently to convert DS0 and DS1 signals to DS3.

The third step involves deploying new equipment that converts services directly to OC-*M*'s to take full advantage of SONET capabilities. Switching system interfaces to an optical signal could be produced cheaply if the rate and format are chosen carefully. Additionally, the standard can provide an optical ISDN interface to CPE which may be requested by customers in the future. Broad-band ISDN services can also be provided with optical lines. It is difficult to enumerate all new services that could be provided by the optical network. However, it should be emphasized that, once the interface to the customers is defined and the mechanisms of transporting the signal through the network is established, new services will proliferate rapidly.

VI. The SONET Rates and Formats

The goals which were established to guide the selection of STS-1 rate and format are

- the ability to transport DS3 signals,
- the ability to transport DS1 signals,
- the rate to be not much higher than 50 Mbits/s to take advantage of inexpensive VLSI processing for framing and related operations,
- the frame repetition rate of 125 μs to facilitate easy definition of 64 kbit/s channels, and
- a regular frame structure to separate overhead channels from information payload.

It was also desirable to select a frame format such that switch termination and broad-band ISDN access can be implemented simply and economically. However, it was realized that with the diverse requirements of switch termination and uncertainties in broad-band services, there is probably no perfect rate and format to satisfy all of these conditions. The signal currently being favored by the T1X1.4 subworking group is at 49.920 Mbits/s which can be separated into 26 overhead channels and 754 information channels. This STS-1 signal can accept one DS3 signal or 29 asynchronous DS1 signals. For signals other than the existing digital hierarchy, the information payload can be flexibly rechannelized.

An important feature of the overhead channels is the inclusion of a span channel identification marker (SCID). SCID is a bit sequence assigned to each STS-1 just prior to multiplexing and is used to quickly identify constituent STS-1's of a high rate signal. Each time an STS-1 is multiplexed, this SCID is rewritten. The inclusion of SCID allows an arbitrary numer of STS-1's to be multiplexed with a simple bit-interleaved multiplexing operation. The resultant high-speed signal does not require any additional overhead channels. Hence, the entire family of STS-*M* signals is defined once a standard rate and format for the STS-1 signal are adopted.

VII. SONET Optical Parameters

The development of optical parameter standards and the introduction of fiber optic equipments are intimately related. Since SONET products are deployed in three steps and in many different applications, the requirements of a standard on optical parameters also evolves in stages. In any case, because of the rapidly advancing technology, these standards should be minimal, flexible, and upgradable. Appendix B shows a minimum set of optical parameters which were discussed in the T1X1.2 subworking group. For a point-to-point fiber optic transmission system, each manufacturer could use parameters of his choice to fully exploit technology. However, the set of parameters in Appendix B could be used as a guideline since they may eventually be interconnected to other systems. This set is especially important for lower speed transmission systems since the technology has matured and stabilized at these speeds.

When an optical cross-connect system is deployed, it is also necessary to define the power level at the cross-connect points. Assuming that the optical cross-connect system is at the midpoint of a transmission system, we might arbitrarily set the power at, for example, −20 dBm. Since most transmitters have a power of higher than −5 dBm and most receivers have a sensitivity of better than −35 dBm, the optical cross-connect system could still find wide usage. Assuming further an average fiber loss of 0.5 dB/km, the optical cross-connect system could cover an area of up to 30 km in radius without using regenerators.

For high-speed systems, standardization still involves definitions, test procedures, and span design methodology. As new fibers and transducers are introduced, new definitions, test procedures, and span design methodology would have to be established. Fortunately for SONET, these high-speed transmission systems can still be used to interconnect two SONET hubs as long as they conform to the SONET format standards.

VIII. Conclusion

In this paper, we have reported various standards issues of fiber optic systems. In particular, we have addressed these issues from a synchronous optical network's viewpoint. The flexibility of such a network can only be realized if a standard exists, both in terms of signal format and optical parameters. However, such a standard also leads to a number of new products and new services. We have acknowledged the difficulties in adopting any standards, but we have also stressed the opportunities of such a standard.

Appendix A

Recommendation: Melding contributions T1X1.4/85-054 and T1X1.4/86-012 into the original T1X1.4/85-053 contribution resulted in the following recommended standard optical interface overhead structure (as of December 1985).

1) Framing: Two framing channels ($A1$, $A2$) are recommended:

Subframe	Pattern
$A1$	1111 0110
$A2$	0010 1000

The reframe is accomplished by examining the 8 bit blocks until one of the two framing patterns is identified. Then the other pattern is tracked. This framing pattern supports the DS3 and SYNTRAN requirements.

2) Channel Error Monitoring: This function consists of two 8 bit channels ($B1$ and $B2$), a cyclic redundancy check code-8 (CRC-8) for span error check, and a CRC-8 code for network error check.

3) Channel Identification: Two channels are recommended for user channel identification: one channel ($C1$) as a span information payload identification number—a unique number assigned just prior to multiplexing that stays with that payload until demultiplexing—and a network signal label channel ($C2$) that defines how the information payload is organized in the frame and how it is constructed.

This latter channel will be used to identify DS3 asynchronous, SYNTRAN, digital video, etc., payload—to include the need to group multiple payloads if such a need exists.

4) User Facility Maintenance and Control: Three channels for facility maintenance requirements are recommended. Two channels ($D1$ and $D2$) are allocated for span maintenance activities, e.g., parity, automatic protection switching, fault locating, terminal and switching office surveillance, and other span-to-span measurements. The T1M1 Working Group has a task (see attached T1M1.3 Task Proposal, dated June 12, 1985) to identify the specific channel definition and usage of these two channels, but the T1M1 representative in the Ad Hoc Group recommended that two channels be allocated to these functions, after the intricacies of protection switching were reviewed. A third channel ($D3$) is assigned for the monitoring, control, and analysis of the multispan networks. This latter channel is made available for packet switching and centralized maintenance applications.

5) Order Wire: Two order wire channels were allocated: local ($E1$) and express ($E2$). These are reserved for communication between central offices and, where required, repeater, hubs, and remote terminal locations. This reservation is in line with current industry practices.

6) User Proprietary Channels: Again, three channels have been allocated to this requirement: one channel ($F1$) is allocated to the EC/IEC user for his input of span information such as data communication for use in maintenance activities and remoting the alarms external to the span equipment, and two channels ($F2$ and $F3$) for the EC/IEC use as end-to-end VF communications channels for themselves and/or their end users.

7) Manufacturer's Proprietary Channel: Two channels ($G1$ and $G2$) are recommended to be reserved for manufacturer's use for unique enhancements he might introduce.

8) Frame Alignment Control: Three channels are recommended to allow for asynchronous impairments in an otherwise synchronous network. The first two ($H1$ and $H2$) are redundant (reducing error possibility) channels that are used for indicating the action to be performed, i.e., clocks in instantaneous synchronization (no action taken), add a sync bit, delete a bit, etc. The third channel ($H3$) is the bit to be acted upon. (Note: This channel will contain 7, 8, or 9 bits—depending on $H1/H2$. Seven bits of this channel are available for other assignments.)

9) Unreserved Channels: Seven channels ($J1$, $J2$, $J3$, $J4$, $J5$, $J6$, $J7$) allocated to future growth needs for the industry. Additional growth channels may become available as some of the above requirements are refined. The final format was designed with this latter factor in mind.

The final format of the recommended overhead is

$A1$	$D3$	$H1$	$H2$	$H3$	$J1$	$D1$	$D2$	$B1$	$B2$	$C1$	$F1$	J
1	2	3	4	5	6	7	8	9	10	11	12	13
$A2$	$F2$	$F3$	$J3$	$C2$	$E1$	$E2$	$J4$	$J5$	$J6$	$J7$	$G1$	$G2$
14	15	16	17	18	19	20	21	22	23	24	25	26

Appendix B
Transmitter Optical Parameters (Minimum Set)

	Ranges Under Consideration in T1X1.2		
Wavelength	Nominal	1310	nm
	Minimum	1280	nm
	Maximum	1340	nm
Spectral Width	Maximum	9	nm
Extinction Ratio	Minimum	10:1	
Coding	Nonreturn-to-Zero (NRZ)		
Pulse Shape	To be determined		

References

[1] I. Lifchus, "Committee T1 now ready to address interconnection standards," *Telephony*, 1984.

[2] "IEEE standard definitions of terms relating to fiber optics," IEEE STD 812-1984.

[3] J. R. Neigh, "Voluntary standards for fiber optics from EIA 7 IEC," FOC/LAN 84.

[4] G. Bonaventura and U. Rossi, "Standardization within CCITT of optical fibres for telecommunication systems," presented at GLOBECOM, 1984.

[5] K. Ogawa, "Considerations for single-mode fiber systems," *Bell Syst. Tech. J.*, vol. 61, pp. 1919–1931, Oct. 1982.
[6] EIA-522, Electron. Industry Association.
[7] T1X1.2 contributions, T1X1.2/84-017.
[8] T1X1.4 contribution, "Rates and formats for optical network interfaces," T1X1.4/85-017.
[9] T1X1.4 contribution, "Recommendations of ad hoc group on overhead structure," T1X1.4/85-23.
[10] T1X1.4 contribution, "Interface specification for synchronous DS3 format," T1X1.4/85-020.
[11] T1X1.4 contribution, "Fourth hierarchy level at 139264 kbits/s," T1X1.4/85-021.
[12] CCITT Recommendations G.951 and G.952.
[13] T1X1.4 contribution, "Synchronous optical network (SONET)," T1X1.4/85-005.
[14] Bell Commun. Res., Inc., "Synchronous optical network (SONET)," TA-TSY-000253, Apr. 1985.

Rodney J. Boehm received the B.S. and M.E. degrees from Texas A&M University, College Station, both in electrical engineering.

He worked at Bell Telephone Laboratories designing high-speed multiplexing equipment for digital microwave radio. In 1983 he joined Bell Communications Research in the fiber optic systems area where he helped develop the Synchronous Optical Network (SONET) concept. During this time, he helped introduce SONET to the T1 standards body to achieve a standardized optical interface. He is now a Product Line Administrator with the Collins Transmission Systems Division of Rockwell International, Dallas, TX.

Yau-Chau Ching (S'64–M'69) received the B.E.E.E. degree from City College of New York, New York, NY, in 1966 and the Ph.D. degree from New York University, NY, in 1969 under a National Science Foundation Fellowship.

From 1969 to 1984 he was with Bell Laboratories where he did a series of exploratory development work in data compression, such as interframe video coding, digital speech interpolation, embedded ADPCM coding, and fast packet network. Since the divestiture of the Bell System, he has been with Bell Communications Research where he is District Manager, Fiber Optic Systems Engineering, responsible for technical requirements of fiber optical systems and the SONET project.

Dr. Ching is a member of Tau Beta Pi and Eta Kappa Nu.

C. George Griffith (M'59–SM'67) graduated from the University of Missouri, Columbia, and received the M.S.E.E. degree from San Jose State University, San Jose, CA.

He joined GTE Lenkurt, San Carlos, CA, after service as a Fire Control Officer in the U.S. Navy, was assigned in 1976 as Vice President and Product Manager of GTE Communications Products, Stamford, CT, and in 1984, joined Bell Communications Research. He is Assistant Vice President in charge of interoffice technology. He has spent 30 years in research and development, manufacturing, and marketing in the field of telecommunications.

Mr. Griffith has organized International Conferences on Communications sponsored by the IEEE Communications Society.

Frederick A. Saal (M'84) received the B.S.E.E. and B.S.E.P. degrees from Lehigh University, Easton, PA, in 1954.

From 1954 to 1984 he was with Bell Laboratories where he was first associated with the development of the first commercial TASI system. From 1961 to 1984 he was responsible for development of various digital transmission equipments and transmission peripherals for digital switches. Since the divestiture of the Bell System, he has been with Bell Communications Research where he is Division Manager, Interoffice Transport Requirements, responsible for technical requirements of interoffice transmission systems.

Mr. Saal is a member of Tau Beta Pi, Eta Kappa Nu, and Phi Beta Kappa.

The Editor

Henry F. Taylor was born in Ft. Worth, Texas, on September 27, 1940. He received the B.A., M.A., and Ph.D. degrees in Physics from Rice University in 1962, 1965, and 1967 respectively. He was employed as a Research Physicist at the Naval Ocean Systems Center (formerly the Naval Electronics Laboratory Center) in San Diego, California, from 1967 to 1978. From 1978 to 1980 he was employed by Rockwell International in Thousand Oaks, California, where he was Principal Scientist of the Optoelectronics Department of the Microelectronics Research and Development Center. From 1980 to 1985 he was Head of the Optical Techniques Branch of the Naval Research Laboratory in Washington, DC. He joined the Electrical Engineering faculty at Texas A&M University as a Professor of Electrical Engineering and Director of the Institute for Solid State Electronics in November 1985.

Since 1970 his principal research interests have been in the fields of fiber optics, integrated optics, and diode laser applications. He has authored approximately 150 journal articles and conference presentations, and holds 21 U.S. patents. He was awarded a Civil Service Commission/Navy fellowship to study System Analysis at the Massachusetts Institute of Technology during 1971–72. He also received the Naval Electronics Laboratory Center Annual Science Achievement Award in 1974 and the American Society of Naval Engineers' Solberg Award for Applied Research in 1975.

He served as Conference Chairman for the Institute of Electrical and Electronics Engineers (IEEE)/Optical Society of America (OSA) Topical Meeting on Integrated and Guided Wave Optics for 1986, and was Program Chairman for that Conference in 1984. He is currently a member of the steering committee for the Optical Fiber Communication and Integrated Optics and Optical Communication Conferences. He was a member of the Program Committee for the International Microwave Symposium in 1986 and organized a workshop on "Fiber Optics for and interactive data bases of all sorts. This could eventually become the most important application of fiber optics, but the cost of the needed optical and elec-Microwave Applications" for that symposium. He was guest editor for a special issue of the *IEEE Transactions on Circuits and Systems* for December 1979 and for a special issue of the *Journal of Lightwave Technology* for March 1987.

Dr. Taylor is a fellow of the IEEE and of OSA, a life member of the American Society of Naval Engineers, and a member of the American Physical Society.